$= 3 \times 10^8$ m/s
10^{-34} J/Hz
$.6 \times 10^{-19}$ C
Rest mass of the electron, $m_e = 9.1 \times 10^{-31}$ kg
Rest mass of the proton, $m_p = 1836\, m_e$
Rest mass of the neutron, $m_n = 1839\, m_e$
Gravitational constant, $G = 6.7 \times 10^{-11}$ N·m²/kg²
Electric force constant, $K_e = 9 \times 10^9$ N·m²/C²
Boltzmann's constant, $k = 1.4 \times 10^{-23}$ J/K
Avogadro's number, $N_o = 6 \times 10^{23}$ units/mol
Acceleration due to gravity near the surface of the earth, $g = 9.8$ m/s²

COMMON SI PREFIXES

Prefix	Abbreviation	Factor
kilo	k	one thousand, 10^3
centi	c	one hundredth, 10^{-2}
milli	m	one thousandth, 10^{-3}

500.2
At52e

ESSENTIALS OF PHYSICAL SCIENCE

ESSENTIALS OF PHYSICAL SCIENCE

KENNETH R. ATKINS
UNIVERSITY OF PENNSYLVANIA

JOHN R. HOLUM
AUGSBURG COLLEGE

ARTHUR N. STRAHLER
SANTA BARBARA, CALIFORNIA

JOHN WILEY & SONS
NEW YORK · SANTA BARBARA
LONDON · SYDNEY · TORONTO

This book was printed by Murray Printing and bound by Book Press.
It was set in Baskerville by York Graphic Services.
The drawings were designed and executed by John Balbalis with the assistance of the Wiley Illustration Department.
Christine Pines supervised production.
Picture research was done by Kathy Bendo.

Copyright © 1978, by John Wiley & Sons, Inc.

All rights reserved. Published simultaneously in Canada.

No part of this book may be reproduced by any means, nor transmitted, nor translated into a machine language without the written permission of the publisher.

Library of Congress Cataloging in Publication Data:

Atkins, Kenneth Robert, 1920–
  Essentials of physical science.

  Includes index.
   1. Science.   I. Holum, John R., joint author.
II. Strahler, Arthur Newell, 1918–   joint author.
III. Title.
Q161.2.A84   500.2   77-12507
ISBN 0-471-03617-X

Printed in the United States of America

10 9 8 7 6 5 4 3 2 1

PREFACE

The physical science course is most often taught emphasizing the phenomena of science, rather than its structural unity. One of the reasons for this approach is that the unity of science is intuitively difficult to grasp without a thorough knowledge of the substance of all its parts. In this book our approach has been to present in sequence the fundamentals of the three major areas of physical science. A unique feature of this book flows from this approach: the physics section was written by a physicist, the chemistry section by a chemist, and the earth sciences section by an earth scientist, all of whom are, or have been at different institutions. The three parts of the book therefore have different flavors. Physicists, chemists, and earth scientists deal with different aspects of nature and, in general, have through their training developed different attitudes toward science. For a student whose contact with the physical sciences might be confined to this one single course, it seems desirable to present these different points of view.

We strive to follow a straight course from the physicist's world of the small and very small, through the small and medium-sized world of the chemist, to the large and very large worlds of the geologist and the astronomer. Giving attention to the *scale* of physical phenomena may well be easier for students to grasp than coping with a forced concept of the homogeneity of a single world of science. Our dimension-scale approach has a certain logic that may be welcome to teachers accustomed to presenting the course in terms of language that needs a simple theme to tie the topics together.

The physics section is mainly concerned with the basic concepts of both classical and modern physics. A determined attempt was made to include only an amount of material that might reasonably be absorbed by a student in this type of course. Many interesting topics had to be rejected. The material included emphasizes the fundamental laws of physics instead of their detailed applications to particular situations. However, some of the detailed applications are introduced in the chemistry and earth science sections.

Mathematics has not been entirely left out, but has been presented at the simplest possible level. Whenever an important equation is stated in algebraic symbols, it is also stated in words. A student unable to cope adequately with algebra would lose very little by ignoring the algebraic symbols completely. However, some students can cope with simple mathematics and find that it helps their understanding. Such students should not be ignored. For students with a strong aversion to mathematics, however, an equation expressed in words is introduced merely as a brief summary of an important physical principle.

The second area, chemistry, is about the science of substances and the changes they undergo. Since during many of these changes energy is absorbed or released and unbalanced forces generated, much of what physics tells us about our world applies also to chemistry. Therefore chemistry is presented as a broad application of physics. Chemical changes also produce new substances, and the contributions to science that are most uniquely those of chemistry are the laws of chemical combination and the recognition of the fundamental kinds of substances and kinds of chemical change.

To bridge the areas of physics and chemistry in a way that emphasizes both their uniqueness and their interrelations, we use a study of the most fundamental substances, the chemical elements, and the physical properties that make them particularly useful in human affairs. After dealing with such elementary substances as macroscopic materials and after describing some of their periodic properties, we plunge back into the world of the submicroscopic. We extend the fundamental concepts of atomic structure begun in the chapters on physics to a survey of the electronic configurations of the elements. That study is essential to a study of chemical bonds and molecular structure, and one of the cardinal principles of chemistry is that structure determines properties.

Our study of physics deals with the forces that hold nuclei and atoms together. In the chapters on chemistry we see how these forces are redirected to hold together compound substances—molecules, ions and their macroscopic aggregations, molecular and ionic compounds.

With an eye toward the study of the earth sciences that will follow, we survey several types of substances and their important properties. The earth's substances are both inorganic and organic, and no study of physical science in general could be complete without looking at both these in particular. We return to the very simple, the important metals and nonmetals, and discuss how we extract the useful elements from their chemical combinations in the earth. We pause to see how several of them are involved in the great geochemical cycles of nature and how others plague mankind in the form of various kinds of pollutants. Then we move to the complex, the organic materials both in our living cells and in the world of synthetics. Thus the chapters on chemistry emphasize familiar materials around us, their structures, their properties, and the forces that induce them to change, and we see that underneath the enormous complexity of the world of substances lies a harmony and order revealed grandly by both physics and chemistry.

The part of the book that deals with earth sciences covers a particularly broad domain. We have tried to show the unity of the earth's natural processes and their respective environments in terms of their connecting concepts. As with the section on chemistry special emphasis is given both to the composition and the processes of natural materials. Equal attention is given the solid earth and its constituents and processes, and to the fluid envelopes: the oceans and our atmosphere.

The book ends with a review of astronomy, beginning with the most familiar domain of space nearest us, the solar system. We then extend our range of vision to galactic space and its outer limits, concluding with a review of the leading cosmological models.

Our individual experiences as teachers and expositors of physical science have taught us that good textbooks can reach students in much the same way as a good teacher does. Our sincere hope in this effort has been that we have made the physical science teachers' task an easier one.

Kenneth R. Atkins
John R. Holum
Arthur N. Strahler

Acknowledgements:

Many people have contributed to this book. We thank all those who have reviewed the manuscript during its development. Many useful suggestions have come from them. In particular we thank Professors Gene Craven (Oregon State), David Maloney (Creighton University), Whit Marks (Central State University), David Newton (Salem State College), Faril Simpson (University of Central Arkansas), and Edward Zganjar (Louisiana State University).

September, 1977

Kenneth R. Atkins
John R. Holum
Arthur N. Strahler

CONTENTS

PART ONE
PHYSICS

1.	PARTICLES IN MOTION	3
2.	THE LAWS OF MOTION	16
3.	MOMENTUM, ANGULAR MOMENTUM, AND ENERGY	28
4.	ATOMS AND HEAT	43
5.	ELECTRICITY	56
6.	MAGNETISM	69
7.	WAVES	83
8.	RELATIVITY	100
9.	THE QUANTUM THEORY	115
10.	NUCLEI, AND FUNDAMENTAL PARTICLES	132
	GLOSSARY	150

PART TWO
CHEMISTRY

11.	THE CHEMICAL ELEMENTS	159
12.	CHEMICAL COMBINATION	173
13.	THE PERIODIC SYSTEM AND ELECTRONIC CONFIGURATIONS	183
14.	COMPOUNDS, CHEMICAL BONDS, AND CHEMICAL ENERGY	202
15.	THE EARTH'S SUBSTANCES—INORGANIC MATERIALS	220
16.	NONMETALLIC INORGANIC SUBSTANCES	239
17.	METAL REFINING	256
18.	THE EARTH'S SUBSTANCES—ORGANIC COMPOUNDS	271
19.	ORGANIC COMPOUNDS OF THE BIOSPHERE	289
20.	SYNTHETICS AND DRUGS	306
	GLOSSARY	325

PART THREE
EARTH SCIENCE

21.	ENERGY FLOW IN ATMOSPHERE AND OCEANS	341
22.	DYNAMICS OF THE WEATHER—PRECIPITATION AND STORMS	362
23.	EARTH MATERIALS—MINERALS AND ROCKS	379

CONTENTS

24.	THE DYNAMIC EARTH—PLATE TECTONICS	399
25.	EARTH HISTORY IN REVIEW	421
26.	GEOMORPHIC PROCESSES AND LANDSCAPE EVOLUTION	438
27.	THE SOLAR SYSTEM	460
28.	ASTROLOGY—THE GEOLOGY OF OUTER SPACE	479
29.	STARS, GALAXIES, AND THE UNIVERSE	496
GLOSSARY		514
INDEX		526

PART ONE
PHYSICS
by Kenneth R. Atkins

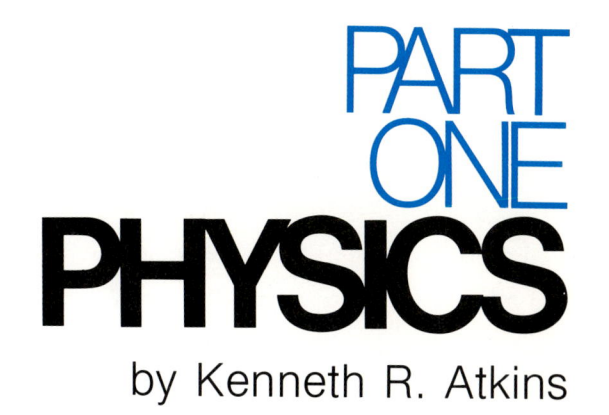

CHAPTER 1
PARTICLES IN MOTION

1.1 THE NATURE OF PHYSICS

Physics is an attempt to describe the nature and behavior of the world around us in as fundamental and penetrating a way as possible. This sounds like an abstract and profitless philosophical statement of the type that authors must use to get their books smoothly under way. Nevertheless, accept the challenge, lift your eyes from the page, and look at the world around. As I do this myself, I am first conscious of a multitude of shapes and colors. There is a subtle interplay of patches of color on the wallpaper. There is an expanse of bright red carpet. Through the window, the intricate shape of a green tree stands out against a flat expanse of blue sky. On the wall is a reproduction of Picasso's "Boy with a Horse." The most conspicuous item on the bookshelf is a bulky "Complete Works of Shakespeare" bound into one volume. From the record player I hear the sounds of Schubert's "Octet in F Major." Whatever the circumstances surrounding you, the reader, you will probably not have to look far to find things equally complicated and marvelous. Certainly, the mind you use to consider these matters is an example of one of the most complicated and marvelous structures that can be contemplated.

Obviously, a complete description of the world around us must include the tree, the human mind, the art of Picasso, the plays of Shakespeare, and the music of Schubert. Even more obviously, these things are not the primary concern of physics. However, if I lift my Picasso off its hook and then let it go, it falls to the ground. The same thing happens with the "Complete Works of Shakespeare" and the Schubert record. Moreover, if all three objects are held at the same height and released at the same time, they all strike the ground at the same time. This is the sort of thing that is the concern of physics.

Whatever is relevant to the aesthetic appeal of the Picasso is clearly irrelevant to the rate at which it falls to the ground. We are led to make a very general, fundamental statement that: "All unsupported objects fall toward the center of the earth at the same rate." Of course, we shall have to be more precise about the meaning of the phrase "at the same rate." We might also be well advised to specify that the objects are to fall through an evacuated space in order to eliminate the effect of the air, which, in an extreme case, makes a balloon rise.

Then, following in the steps of the seventeenth-century physicist Sir Isaac Newton, we must refine the idea to the point where the statement becomes: "Any two objects attract one another with a gravitational force that is proportional to the product of their masses and inversely proportional to the square of the distance between them." We have then arrived at one of the most important and basic laws of physics, Newton's law of universal gravitation; it is explained in more detail in Chapter 2.

The patches of color that are an essential part of our immediate experience can be analyzed by means of an instrument known as a spectroscope, which is essentially a prism of glass. If white light, such as the sunlight now streaming in through my windows, falls upon this glass

4   1 PARTICLES IN MOTION

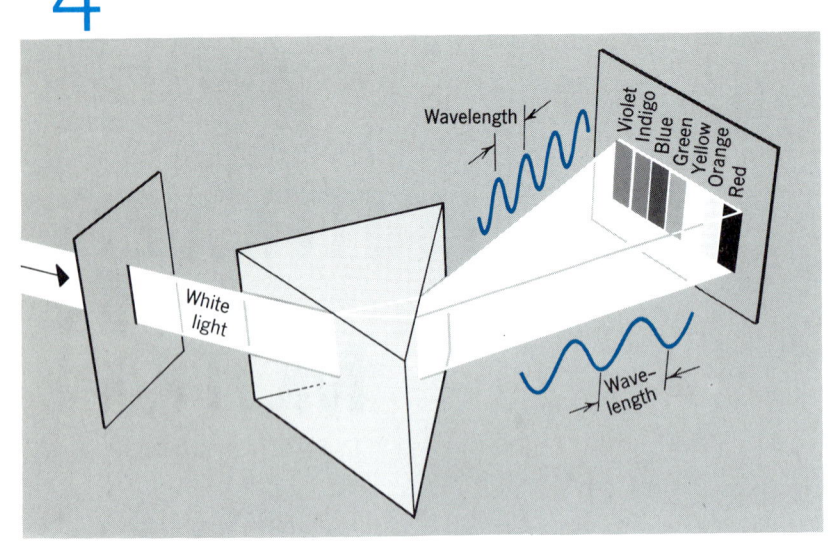

FIGURE 1-1
A glass prism revolves light into a spectrum of basic colors. The different colors correspond to different wavelengths.

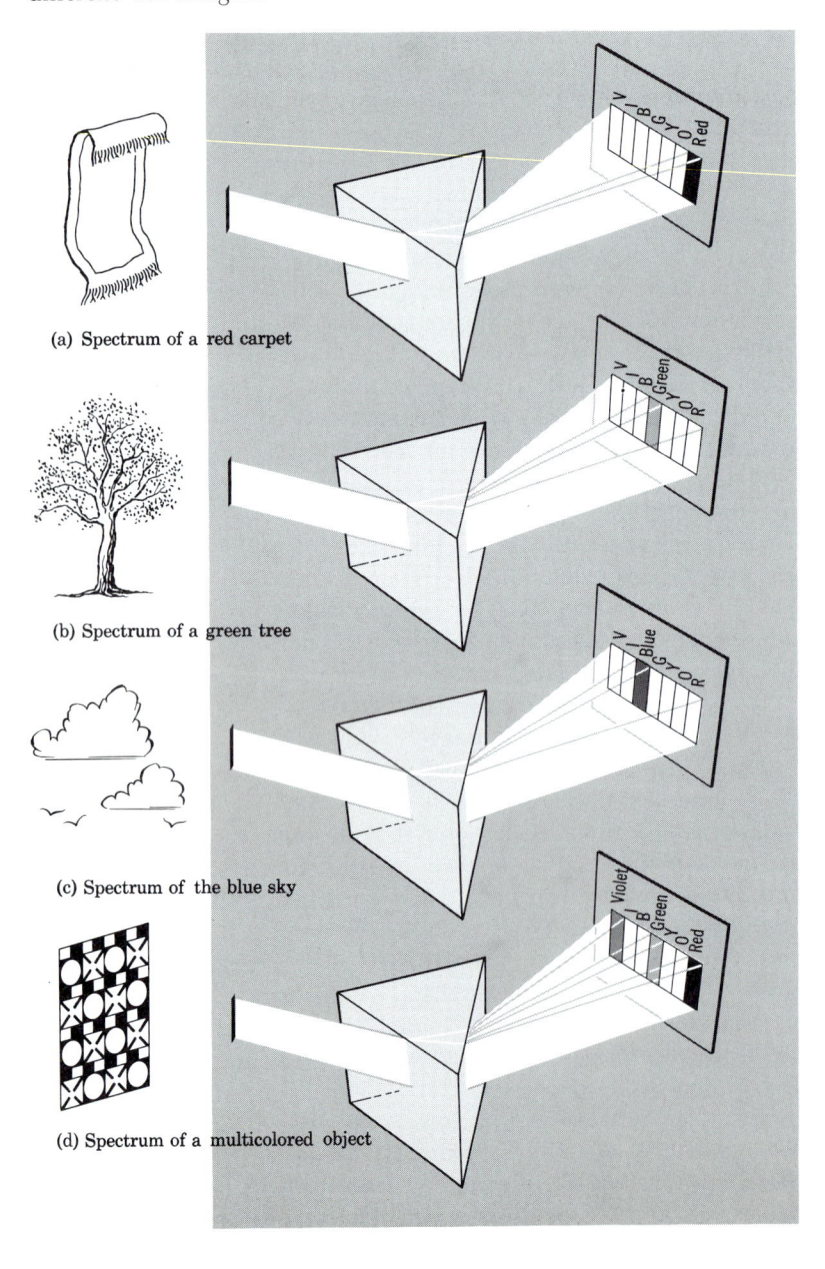

(a) Spectrum of a red carpet

(b) Spectrum of a green tree

(c) Spectrum of the blue sky

(d) Spectrum of a multicolored object

prism, it is spread out into a spectrum of the colors of the rainbow, as shown in Figure 1-1. All the rays of light are bent by the prism, but the red light is bent least, the violet light most, and the other colors in between, as shown in the figure. The colors of this spectrum are the basic components of all visible light.

If the spectroscope were arranged to accept only light from my red carpet, then only the red region of the spectrum would be illuminated. If the light came from the green tree, the spectrum would be illuminated mainly in the green region. If the light came from the blue sky, it would be the blue region that would be illuminated (see Figure 1-2). However, if the spectroscope were made to accept light from my multicolored wallpaper, then several regions of the spectrum would be illuminated, as shown in Figure 1-2d. In this way, any sample of visible light can be analyzed into some or all of the basic colors of the spectrum mixed in appropriate proportions.

We are again making the sort of analysis that is the concern of physics, but physics is not satisfied to stop at this stage. It must penetrate more deeply into the heart of the matter and ask, "What is the nature of light, and what is the difference between various basic colors?" This particular question led to several centuries of speculation, culminating in the early nineteenth century with a burst of ingenious experimental activity that provided the following answer: "Light is a wave, similar to a wave on the surface of the ocean." The quantity that characterizes the color is the wavelength, or the distance between adjacent crests of the wave. The distance for red light is about twice as great as for violet light (see Figure 1-1).

There is much more to be said about the nature of light. An ocean wave requires the presence of the ocean, but a light wave can travel through empty space, as it does between the sun and the earth. The explanation of this, provided later in the nineteenth century, was that "Light is a wave in an electromagnetic field." Now the analysis has become so profound that it can no longer be expressed in familiar terms. The beginning reader cannot be expected to understand even vaguely what is meant.

The explanation is in fact so lengthy that it will not be attempted until later in this book. Even so, it is a nineteenth-century explanation. Today a physicist would rather say that "Light consists of particles (or quanta or photons) whose behavior is governed by a wave in an

FIGURE 1-2
The glass prism analyzes samples of light into their basic color components.

electromagnetic field." We shall attempt to explain all of this later on.

Having thus overreached ourselves to illustrate how physics is always penetrating more and more deeply into the mystery of the behavior of things, let us also emphasize again the other aspect of the matter. Whatever is true in these statements about light is equally true of the light from the red carpet and the light from the painting by Picasso. The statements are an attempt to find something common to all visual experiences.

1.2 THE UNIVERSE AS A COLLECTION OF PARTICLES

Now let us settle down to our discussions in earnest. We shall start with the physics of objects in motion as developed by Galileo and Newton and their successors from the seventeenth century onward.

Scientists of the seventeenth century were particularly interested in astronomy. It is therefore easy to understand why some of them pictured the universe as a collection of isolated objects separated by vast regions of empty space. The appearance of the sky on a clear night illustrates this well (Figure 1–3). The stars are points of light whose diameters are clearly very much smaller than the distances between them. Most of the space in the sky appears to be empty. The actual situation is even more impressive than it appears to be. The distance from the sun to the nearest star, Alpha Centauri, is about thirty million times the diameter of the sun! If the sun were represented by a small black dot the size of a printed letter *o* on this page, then Alpha Centauri would have to be represented by a similar dot 20 miles away!

Let us consider how the solar system would appear to an observer viewing it from outside. Shrink the scale until the sun is about the size of a football placed in the center of a football field. The planets can then be represented roughly as follows. Mercury is a grain of sand at a distance of about 10 yards from the sun. Venus is an apple seed at a distance of about 20 yards from the sun. Earth is another apple seed at a distance of about 25 yards from the sun. Mars is a poppy seed at a distance of about 40 yards from the sun, just inside the field. Jupiter is a golf ball somewhere in the stands. Saturn is a table tennis ball in the parking lot outside. Uranus is a pea three blocks away. Neptune is a pea about half a mile away. The outermost planet, Pluto, is a grain of sand about three quarters of a mile away. The moon would be represented by a grain of sand about $2\frac{1}{2}$ inches from the apple seed that represents the earth.

FIGURE 1–3
A portion of the night sky. "Isolated objects separated by vast regions of empty space" (Mount Wilson and Palomar Observatories).

For some of the purposes of astronomy, then, the earth may be treated as a small particle in space. But we all know that it is much more complicated than that. What about the air, the rocks, the oceans, and the myriad forms of complicated living organisms that teem over the surface of the earth? In particular, what about the human mind? Physics tells us that any object, however complex, can be visualized as a structure built of atoms. The atom itself is a structure composed of the three fundamental particles, *electrons*, *protons*, and *neutrons*. The protons and neutrons are packed tightly together into the *nucleus*, which occupies a very small space at the center of the atom. The electrons are even smaller than the nucleus. They move around the nucleus inside a region of space which is about 10,000 times wider than the nucleus. If the nucleus were represented by a football placed in the center of a city, the electrons would be tennis balls bouncing around in the suburbs! The atom, like the solar system, is mainly empty space.

Consider the following suggestion for giving a complete description of the whole universe. Assume that the universe is composed entirely of small particles such as electrons, protons, and

A very accurate atomic clock in which cesium atoms are made to vibrate in a certain way. At the time of this writing, the second is defined as the time interval during which the cesium atom makes 9,192,631,770 of these vibrations. (National Bureau of Standards.)

neutrons. At a given instant of time, specify exactly the position of each particle in the universe. This gives a complete description of the instantaneous appearance of any structure in the universe. For example, given the positions of all the particles inside a human brain, we have a complete description of the state of that brain.

Now repeat this procedure at all instants of time. Then, according to this viewpoint, everything that can be known about the universe is known. For example, by describing in this way the changing pattern of particles in the human brain, we might hope to have a complete description of all the thoughts passing through that brain! Do you believe that?

1.3
UNITS

Locating an object in space involves the measurement of distances. The measurement of a distance presupposes a *unit* of length. Suppose I say that a drugstore is 3 miles north of my home. The number 3 is meaningless unless I know how long a mile is. Here the unit of length is the mile. If I say that it will take me 40 minutes to walk to the drugstore, I assume that I know how long a minute lasts. Here the unit of time is the minute.

Scientists rarely use miles or minutes. They confine themselves almost exclusively to the *metric system of units*. In this system the unit of length is the *meter*, which is slightly longer than one yard. The unit of time is the *second*. In addition, it is convenient to define a unit of mass, which is the *kilogram*. One kilogram is slightly larger than the mass of a quart of milk.

In order to achieve high precision in scientific measurements, these units are very carefully defined. We shall not concern ourselves with the complicated details, and in any case the definitions are continually being changed as methods of measurement improve. Next year's definition of the meter might be significantly different from this year's as far as the basic ideas are concerned. But the change would be made in such a way that the actual length of the meter would change by only a very small amount.

The metric system is a decimal system, and prefixes are used to indicate multiplication or division by factors of 10. For example, the prefix kilo means one thousand, so 1 kilometer is 1000 meters. Similarly, the prefix milli means one thousandth, so 1 millimeter is 0.001 meter. The prefix centi means one hundredth, so there are 100 centimeters in a meter. A length of 240 centimeters is the same as 2.4 meters.

A kilogram is 1000 grams. Unlike the words for the units of length and time, the basic word for the unit of mass has a prefix. This situation

At the center of this elaborate protective system of bell jars is the International Kilogram, chosen in 1889 as the standard for the unit of mass. It is kept at the Bureau International des Poids et Mesures (International Bureau of Weights and Measures) located at Sèvres near Paris.

TABLE 1.1
PREFIXES FOR MULTIPLES AND SUBMULTIPLES OF UNITS

Multiples

Prefix	Abbreviation	Multiplication Factor			
deka	da	ten	=	10	$= 10^1$
hecto	h	one hundred	=	100	$= 10^2$
kilo	k	one thousand	=	1,000	$= 10^3$
mega	M	one million	=	1,000,000	$= 10^6$
giga	G	one billion	=	1,000,000,000	$= 10^9$
tera	T	one trillion	=	1,000,000,000,000	$= 10^{12}$
peta	P	one million billion	=	1,000,000,000,000,000	$= 10^{15}$
exa	E	one billion billion	=	1,000,000,000,000,000,000	$= 10^{18}$

Submultiples

Prefix	Abbreviation	Multiplication Factor			
deci	d	one tenth	=	$\frac{1}{10}$	$= 10^{-1}$
centi	c	one hundredth	=	$\frac{1}{100}$	$= 10^{-2}$
milli	m	one thousandth	=	$\frac{1}{1,000}$	$= 10^{-3}$
micro	μ	one millionth	=	$\frac{1}{1,000,000}$	$= 10^{-6}$
nano	n	one billionth	=	$\frac{1}{1,000,000,000}$	$= 10^{-9}$
pico	p	one trillionth	=	$\frac{1}{1,000,000,000,000}$	$= 10^{-12}$
femto	f	one millionth of one billionth	=	$\frac{1}{1,000,000,000,000,000}$	$= 10^{-15}$
atto	a	one billionth of one billionth	=	$\frac{1}{1,000,000,000,000,000,000}$	$= 10^{-18}$

arose because it was customary in the past to define 1 gram as the mass of 1 cubic centimeter of pure water at a prescribed temperature and pressure. (One cubic centimeter is the volume of a cube with a side 1 centimeter long.) The procedure now is to define 1 kilogram as the mass of a certain carefully made cylinder of a special metallic alloy. This standard of mass is kept under very careful conditions at a laboratory near Paris. It has approximately the same mass as 1000 cubic centimeters of water. So the mass of 1 cubic centimeter of water is still approximately one thousandth of a kilogram, which is 1 gram.

The prefixes used to indicate various multiples and submultiples (fractional parts) are listed in Table 1.1. You need not memorize the whole table. The important prefixes, such as kilo and centi, will be encountered so often that you will soon become familiar with them. If you encounter an uncommon prefix, look it up in the table. For your convenience, this table is reproduced among the appendixes.

The table includes conventional abbreviations for the prefixes. For example, centi is abbreviated to c and kilo is abbreviated to k. The names of the units themselves may be abbreviated in a similar way. The abbreviation for meter is m. A length of twelve meters may be written 12 m. There are 100 centimeters in a meter, and this length can be expressed either as 12 m or as 1200 cm. The abbreviation for second is s. One minute is 60 s. The abbreviation for milli is m, so a time interval of five thousandths of a second is 5 milliseconds or 5 ms.

The abbreviation for gram is g. The abbreviation for kilogram is therefore kg. As we explained earlier, this is the strange anomaly that has grown out of past practices. The unit of mass, the kilogram, is expressed as a multiple of one of its submultiples, the gram.

1.4 POWERS OF TEN

The column on the extreme right of Table 1.1 gives the multiplication factor in the form of *powers of ten*. In physics we often deal with very large numbers or very small numbers. Powers of ten are then found to be very convenient. They are most easily understood by considering a few examples.

One thousand can be written 1000. Three zeros follow the number one. This could also be written 10^3, which means "multiply together three tens."

$$1000 = 10 \times 10 \times 10 = 10^3 \quad (1\text{-}1)$$

Here are some more examples.

$$10^9 = 1{,}000{,}000{,}000 = \text{one billion} \quad (1\text{-}2)$$

$$2.57 \times 10^6 = 2.57 \times 1{,}000{,}000 = 2{,}570{,}000 \quad (1\text{-}3)$$

When the power of 10 is negative, the implication is that we are dividing by 10 several times, not multiplying by 10.

$$10^{-3} = \frac{1}{10} \times \frac{1}{10} \times \frac{1}{10} = \frac{1}{1000}$$
$$= \text{one thousandth} \quad (1\text{-}4)$$

Notice that

$$10^{-3} = \frac{1}{1000} = \frac{1}{10^3} \quad (1\text{-}5)$$

Are the following now obvious to you?

$$200 \times 10^{-2} = 2 \quad (1\text{-}6)$$

$$\frac{1}{4 \times 10^3} = \frac{1}{4} \times 10^{-3} \quad (1\text{-}7)$$

$$= 0.25 \times 10^{-3} \quad (1\text{-}8)$$

$$= 2.5 \times 10^{-4} \quad (1\text{-}9)$$

We obtain Equation 1-9 from Equation 1-8 in the following way. Multiply 0.25 by 10 to obtain 2.5. Compensate for this by dividing by 10. Dividing by this one extra factor of 10 turns 10^{-3} into 10^{-4}.

1.5
SI UNITS

Even after we agree to use the metric system of units, there are still various ways in which we might proceed. It used to be very popular to express all lengths in centimeters and all masses in grams. This system of units was called the Centimeter-Gram-Second system or *CGS system of units*. The modern procedure is to express all lengths in meters and all masses in kilograms. This is the Meter-Kilogram-Second or *MKS system of units*. The second is the unit of time in both systems.

For example, in the CGS system a certain steel rod might be said to have a length of 37 centimeters and a mass of 823 grams. In the MKS system we would have to say that its length is 0.37 meter and its mass is 0.823 kilogram.

To appreciate the point of all this, consider the density of water. The *density* of a substance is defined as the mass of unit volume of that substance. In the CGS system, the unit of volume is 1 cubic centimeter. The mass of 1 cubic centimeter of water is almost exactly 1 gram. In the CGS system, the density of water is therefore 1 gram per cubic centimeter. This can be written 1 g/cc, or 1 g/cm^3.

In the MKS system, the unit of volume is the cubic meter. One cubic meter is the volume of a cube with a side of length 1 meter. The length of the side of this cube is 100 centimeters. Its volume is therefore $100 \times 100 \times 100 = 1{,}000{,}000$ cubic centimeters. Each cubic centimeter of water has a mass of 1 gram. A cubic meter of water therefore has a mass of 1,000,000 grams. The unit of mass in the MKS system is 1 kilogram, which is equal to 1000 grams. The mass of a cubic meter of water is therefore 1,000,000/1,000 or 1,000 kilograms. In the MKS system of units, the density of water is 1000 kilograms per cubic meter. This can be written 1000 kg/m^3 or, if you are very sophisticated, 10^3 kg/m^3.

It matters quite a bit whether you use the CGS or the MKS system. The density of water is either 1 g/cm^3 or 1000 kg/m^3.

An international agreement has decided in favor of the meter, the kilogram, and the second. The MKS system forms the basis of the system of units that is now almost universally accepted under the name of *SI units*. SI is an abbreviation of "Le Système International d'Unités," which is French for "International System of Units." In SI all lengths must be in meters, all masses in kilograms, and all time intervals in seconds. This convention leads to special units for other important physical quantities such as energy and pressure. We shall consider these special units as they arise.

The SI units of density are kilograms per cubic meter. As you will discover later in this book, scientists still have a fondness for expressing densities in grams per cubic centimeter. There are two reasons for this. The numbers are then not too different from one for most substances. Moreover, the density of water is very nearly 1 g/cm^3. The density of any other substance in grams per cubic centimeter therefore tells you immediately how many times more dense than water the substance is.

The rules can be broken in other small ways as long as you proceed with care and know what you are doing. Chemists like to use the *liter* as a unit of volume. One liter is exactly 1000 cubic centimeters. Convince yourself that it is 10^{-3} cubic meters.

A *metric ton* is exactly 1000 kilograms. It is

about 10 percent bigger than one U.S. ton, which is 2000 pounds. A small car has a mass of about 1 metric ton.

The kilometer is a convenient unit of length in geology. One kilometer (1000 meters, of course) is about three-fifths of a mile. To obtain a very rough estimate in miles, divide the distance in kilometers by two. Of course, a decade or two from now Americans may think more easily in kilometers than in miles. Most Europeans already do.

1.6 SPEED

Where are we? We soared to the heights in Section 1.2 when we considered the question of whether your brain could be described in terms of the behavior of electrons, protons, and neutrons. We then came down to earth again in Sections 1.3, 1.4, and 1.5, and we worried about such practical matters as whether the density of water should be stated as 1 g/cm^3 or 1000 kg/m^3. Physics is like that. We may have the noble ambition of understanding the hidden mysteries of matter. We may aspire to salvage the future of humanity by capturing the energy of the sun's rays. But, if we are to do it efficiently, we have to worry about such matters as how to measure a length accurately and how to make a clear, precise statement of the result of the measurement.

Nevertheless, the principal aim of this chapter is to describe objects in motion, and we shall return to it. We must first decide exactly what we mean by such concepts as speed and acceleration.

With the help of Figure 1–4, imagine a racing car speeding along a straight racetrack marked off in meters. Suppose that you first see it as it crosses the starting line, labeled 0 meters. You then look away for 10 seconds. You look back at the end of these 10 seconds and see the car crossing the 60-meter mark. You immediately conclude that the car has traveled 60 meters in 10 seconds and therefore has a speed of $\frac{60}{10} = 6$ meters per second. This can be written as 6 m/s.

However, this is only the *average speed* during the 10 seconds. In fact, the car may have sputtered its way to the 50-meter mark, stopped there for a couple of seconds, and then taken off with a sudden spurt. The average speed of 6 m/s then has no obvious relationship to the speed with which it crossed the starting line or the speed it had acquired by the time it reached the 60-meter mark.

The thing to do is to time the car over the next 10 meters. Suppose that it takes 0.5 second to go from the 60-meter mark to the 70-meter mark.

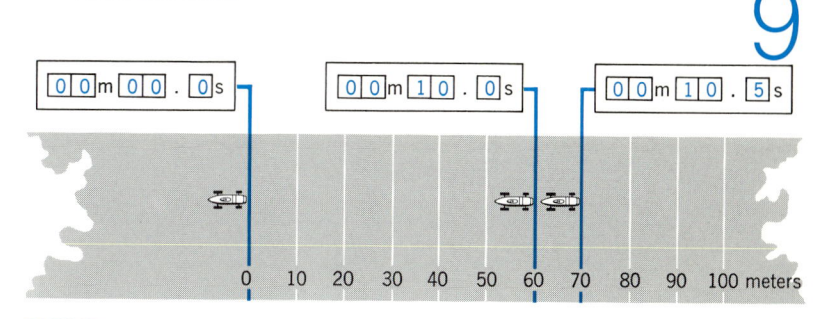

FIGURE 1–4
Illustrating the definition of speed.

mark. Its average speed over this short distance is $\frac{10}{0.5} = 20$ m/s. This is a much better estimate of the speed at the actual instant of crossing the 60-meter mark. We are still assuming, of course, that the speed was not varying rapidly even during this 0.5 second. Perhaps the driver was frantically moving his or her foot up and down on the accelerator pedal. The only way to be quite certain would be to measure the very short distance traveled by the car during a very short time interval immediately after crossing the 60-meter mark. During this very short time, say, one-thousandth of a second, the car would not have the opportunity to change its speed appreciably.

When a body is moving with a variable speed, its *instantaneous speed* at a particular instant is measured in the following way. The very short distance it travels during a very short time interval immediately following this instant is measured. The interval of time should be as short as we can possibly make it and yet still measure it with sufficient accuracy. Then,

$$\text{Instantaneous speed} = \frac{\text{very short distance traveled}}{\text{very short time taken}} \quad (1\text{-}10)$$

The frame surrounding this statement is there to emphasize its fundamental importance to the subject.

1.7 ACCELERATION

The *acceleration* of an object moving along a straight line is the rate at which its speed is changing. Suppose that we determine that the instantaneous speed of the car as it crossed the 60-meter mark is 20 m/s. Ten seconds later, we determine its instantaneous speed again and find that it has increased to 27 m/s (Figure 1–5). The speed has increased by 7 m/s in 10 s. It increased at the rate of 0.7 meters per second every second or 0.7 meters per second per second. This is the average acceleration during the 10 seconds. It can be written 0.7 m/s^2.

1 PARTICLES IN MOTION

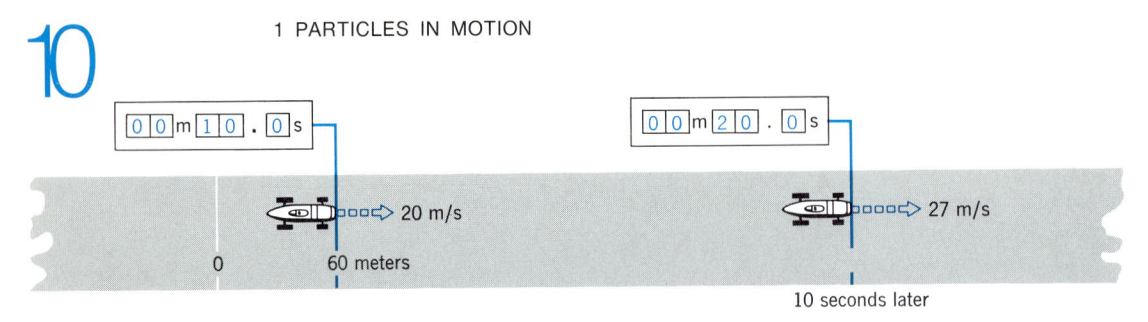

FIGURE 1-5
Illustrating the definition of acceleration.

The acceleration itself may not be constant. If the driver slowly moves his or her foot up and down on the accelerator pedal, a varying acceleration is produced. We should therefore use the same caution that we did when we defined speed. To determine the acceleration at a particular instant, we should measure the very small change in speed occurring during a very small interval of time.

$$\text{Instantaneous acceleration of an object moving along a straight line} = \frac{\text{very small change in speed}}{\text{very short interval of time}} \quad (1\text{-}11)$$

1.8 THE ACCELERATION DUE TO GRAVITY

The simplest case of accelerated motion in a straight line occurs when the acceleration is constant. The most important example of such a motion is an object falling freely under gravity. Suppose that the conditions are such that the resistance of the air can be neglected. Assume that the total height through which the object falls is small compared to the earth's radius. Then, as the object falls, its acceleration is always vertically downward and has the same value at all instants. Moreover, at a particular place on the earth's surface, this acceleration is the same for all falling objects, independently of their size, shape, or mass. It is called the *acceleration due to gravity*. It is usually represented by the symbol g and its value is approximately 9.8 meters per second per second or 9.8 m/s^2.

Figure 1-6 is a photograph of a ball falling vertically from rest. It is a superposition of a large number of snapshots taken at successive intervals of $\frac{1}{30}$ of a second. It brings out quite vividly the way in which the ball picks up speed.

If we are asked to calculate the speed after the ball has been falling for a certain time, the procedure is as follows. Each second the ball gains an additional speed of 9.8 m/s. So, multiply the

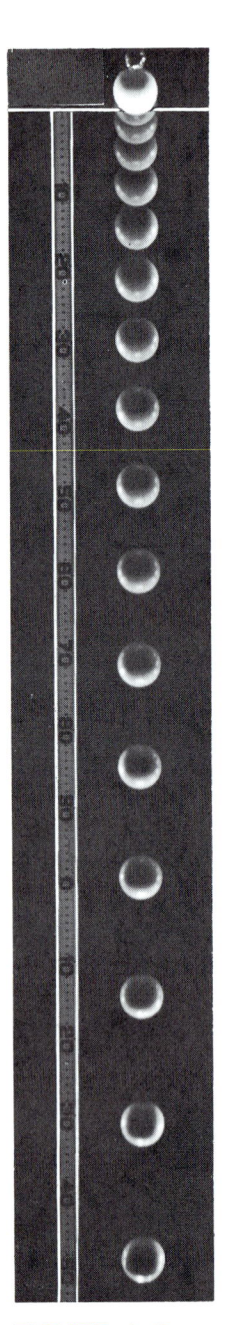

FIGURE 1-6
A ball falling from rest photographed at successive intervals of $\frac{1}{30}$th of a second (from P.S.S.C. *Physics*, D. C. Heath and Co., Boston, 1965).

acceleration g (= 9.8 m/s²) by the number of seconds t that the body has been falling. The product is the speed v at the end of this time. Algebraically, this would be written

$$v = gt \qquad (1\text{-}12)$$

As an example, one half a second after the ball has been released from rest, its speed v is $9.8 \times \frac{1}{2} = 4.9$ m/s

1.9 AN OBJECT THROWN HORIZONTALLY

In the previous section we discussed the straight-line motion of an object falling vertically downward. We shall now extend the discussion by introducing a horizontal component of the velocity so that the path is curved.

Consider, for example, a ball thrown over the edge of a cliff with an initial velocity in the horizontal direction, as in Figure 1–7. The acceleration produced by gravity g is entirely in the vertical direction and cannot change the horizontal component of the velocity. It follows that a fielder starting from the foot of the cliff at the instant the ball is thrown and running outward with a constant horizontal velocity the same as the initial velocity of the ball always remains directly underneath the ball.

Meanwhile, the ball continues to fall in the vertical direction in exactly the same way that it would if it had no horizontal velocity. Imagine a second ball dropped from rest at the top of the cliff at exactly the same instant that the first ball is thrown horizontally outward. The two balls fall vertically in exactly the same way and are always at the same height. A second fielder standing stationary at the foot of the cliff catches the second ball at exactly the same instant that the running fielder catches the first ball.

We discussed this situation by considering the horizontal and vertical motions separately. What is the total velocity at any time? In addition to its horizontal velocity, the ball builds up a downward vertical velocity that steadily increases. The horizontal and vertical velocities combine in some way to produce a velocity tilted downward from the horizontal. The ball starts moving horizontally, but as it falls its direction of motion tilts down more and more. The result is the curved path shown in Figure 1–7. The velocity is not adequately specified merely by giving the speed in meters per second; the direction of motion is equally important. A quantity with both direction and magnitude is called a *vector*.

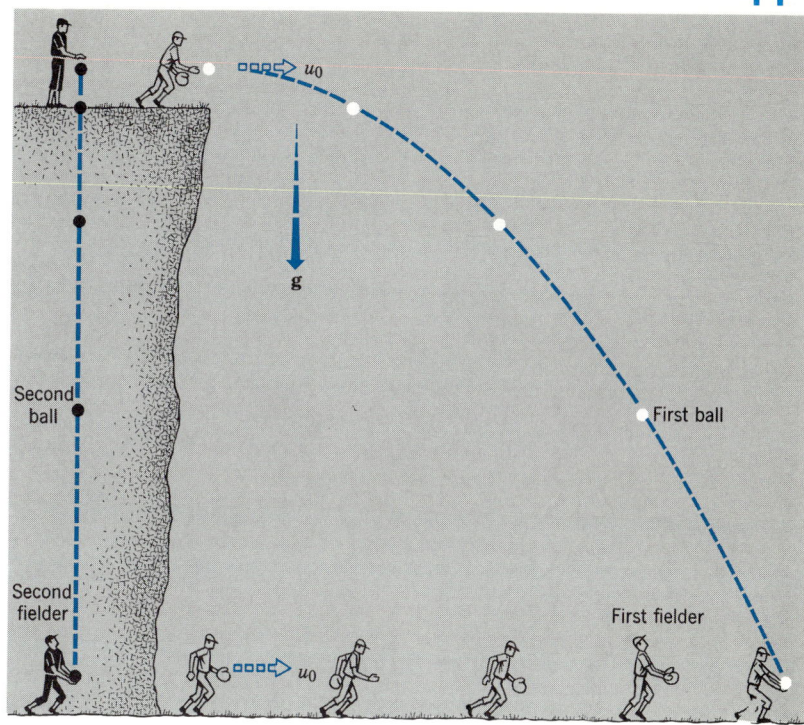

FIGURE 1–7
A ball thrown horizontally over the edge of a cliff.

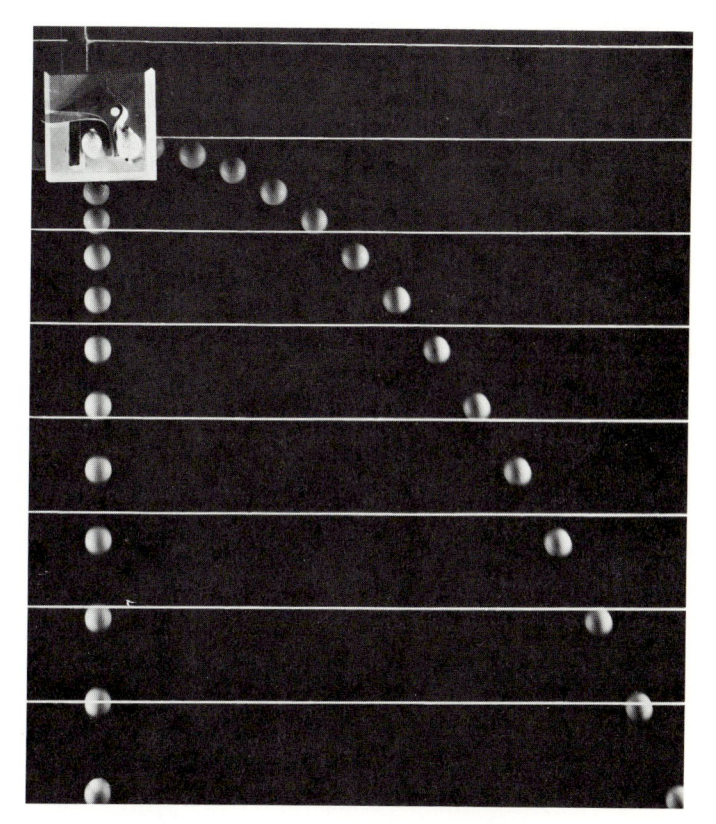

Flash photograph of two balls released simultaneously from the mechanism shown. One ball was released from rest and the other was projected horizontally. The light flashes were repeated at intervals of 1/30 second. (Reproduced by permission of D. C. Heath & Company, P.S.S.C. *Physics*, 1976.)

1.10
VECTORS

Many physical quantities have direction as well as magnitude; these quantities are called *vectors*. One of the simplest examples of a vector is the *displacement* of one point from another. The distance between Philadelphia and New York is 90 miles, but you cannot reach New York by traveling 90 miles in a southerly direction from Philadelphia. You must travel 90 miles in exactly the right direction, which is approximately northeast. The displacement of New York from Philadelphia is a vector, and it has both magnitude and direction. Its magnitude is 90 miles and its direction is northeast.

A second example of a vector is *velocity*. From a practical point of view, traveling due west at 60 miles per hour (mph) is very different from traveling due east at 60 mph. To describe completely the velocity of a body, one must give the direction in which it is moving in addition to the distance it travels in a given time. Strictly speaking, the word *velocity* should always indicate a vector. The word *speed* should be used to indicate the magnitude of the velocity, ignoring its direction. In this book we follow this convention whenever the distinction is important, but in common usage the word *velocity* is often used instead of *speed*.

A third example of a vector is a *force*. Until we have defined force precisely, you may rely on your intuitive notion of force as a push or pull. Clearly, if you are trying to push an automobile up a hill, the direction in which you push is as important as the strength exerted. It is not advisable that the person at the front should push downhill while the person at the rear pushes uphill.

A quantity that has magnitude but no direction is called a *scalar*. Examples of scalars are time, mass, speed, temperature, electric charge, volume, density, and energy.

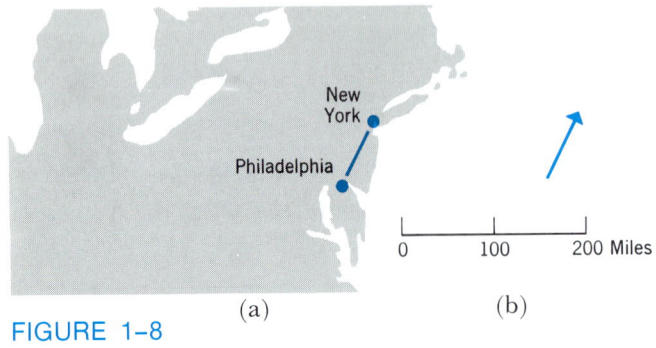

FIGURE 1–8
Illustrating the vectorial nature of a displacement. (a) The locations of Philadelphia and New York on a map. (b) The vectorial representation of the displacement of New York from Philadelphia.

1.11
REPRESENTATION OF A VECTOR

In print, the symbol for a vector is set in boldface type, whereas the symbol for a scalar is set in italics. The vector representing a velocity would therefore appear as **v**, whereas the corresponding speed would be *v*. In handwriting, this distinction is not possible, and it is convenient to indicate a vector quantity by placing an arrow above it—for example, $\vec{v}$. The vector displacement from point P to point Q might be written $\vec{PQ}$, with an additional implication that the displacement is in the direction from P to Q, rather than from Q to P. The magnitude of the displacement, a scalar quantity, is written PQ.

A vector is represented pictorially by a line drawn in the same direction as the vector with a length proportional to the magnitude of the vector on some agreed scale. An arrowhead indicates one of the two possible directions along the line. Figure 1–8a is a map of the northeastern United States drawn on a scale of 1 inch to 200 miles. Figure 1–8b shows how the displacement of New York from Philadelphia can be represented by an arrow. This arrow must be drawn in the right direction. It must have a length of 0.45 inches; according to the scale of the map, this represents the 90 miles between the two cities.

A velocity of 60 mph due north could be represented by a line with an arrowhead pointing due north and a length of 6 inches, if it had been agreed that 1 inch represented 10 mph. A velocity of 40 mph due west would then be represented by a line 4 inches long pointing due west and therefore at right angles to the first line.

1.12
ADDITION OF VECTORS

To add vectors together, we proceed as follows. Figure 1–9 is another map of part of the northeastern United States. The displacement from Philadelphia to New York is represented by the vector $\vec{PQ}$. The displacement from New York to Albany is represented by the vector $\vec{QR}$. From the point of view of the eventual location, a trip from Philadelphia to New York followed by a trip from New York to Albany has exactly the same end result as a trip directly from Philadelphia to Albany. The vector $\vec{PR}$ is therefore exactly equivalent to the sum of the vectors $\vec{PQ}$ and $\vec{QR}$.

This procedure works for the addition of any two vectors, whether they are displacements, velocities, forces, or any other type of vector. Join the two vectors together head to tail to form two sides of a *triangle of vectors*. The third side of

1.13 MOTION IN A CIRCLE

the triangle is then the sum of the two and is called their *resultant*. The procedure also works in reverse. The resultant vector $\vec{PR}$ may be replaced by the combined effect of the two vectors $\vec{PQ}$ and $\vec{QR}$, which are called its *components*. Replacing a vector by its components can be particularly helpful when the two components are at right angles.

Suppose, for example, that you are attempting to drag a heavy trunk along the floor by means of a rope over your shoulder as in Figure 1–10. The force **F** exerted on the trunk by the rope is at an angle to the horizontal. By constructing a suitable triangle of vectors as shown, **F** can be replaced by its horizontal component **H** plus its vertical component **V**. These two components have different independent effects. The lifting effect of the vertical component **V** reduces the friction between the trunk and the floor. The horizontal component **H** achieves the main objective of moving the trunk horizontally.

Later in this book we shall sometimes be interested in the effect of a vector in a direction at an angle to the vector. We shall use language such as "**H** is the component of **F** in a horizontal direction." This implies that **F** has been replaced by two components **H** and **V** at right angles.

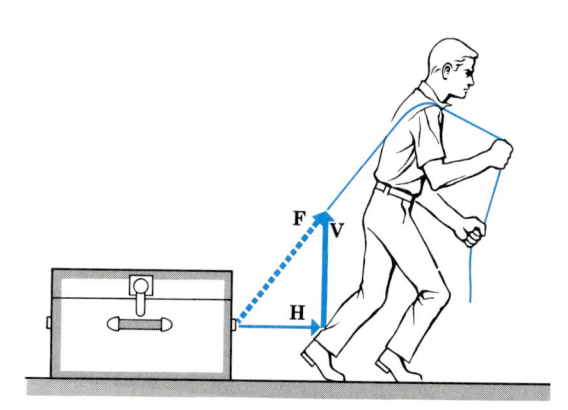

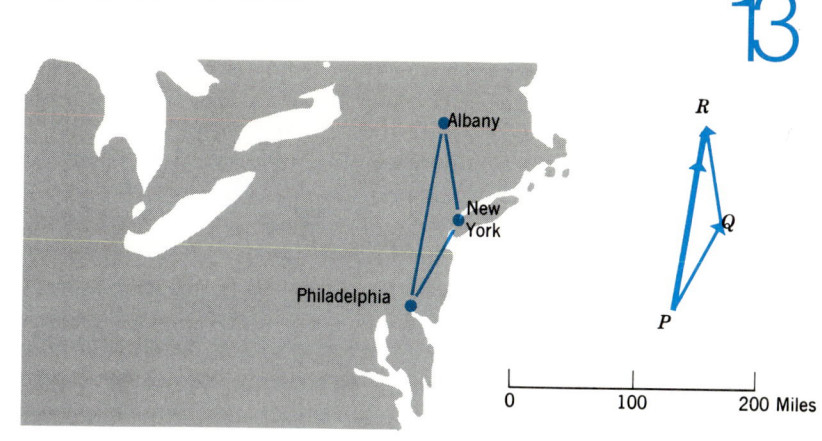

FIGURE 1–9
Using a map to illustrate the addition of vectors.

FIGURE 1–10
Replacing a vector by two components at right angles to each other.

1.13 MOTION IN A CIRCLE

An object moving in a circle with constant speed provides a good illustration of the vector nature of velocity and acceleration. The speed is easily obtained by measuring the distance the object moves along a short arc of the circle and dividing by the time taken to cover this arc. The numerical value of the speed, v, determines the scalar magnitude of the vector velocity **v** and hence the length of the arrow representing this velocity. The direction of the arrow is tangential to the circle and therefore, of course, perpendicular to the radius of the circle. If, for example, a ball is whirled around in a horizontal circle on the end of a string and the string suddenly breaks, then the ball flies off along a straight line tangential to the circle (Figure 1–11).

We must not assume that because the speed is constant there is no acceleration. As the object moves around the circle, the vector velocity **v** continually changes its direction, even though its scalar magnitude, the speed v, remains constant. A change in the direction of a velocity is just as real an acceleration as a change in magnitude.

In Figure 1–12 the arrow labeled $\mathbf{v}_P$ represents the velocity at the point P. The arrow labeled $\mathbf{v}_Q$ represents the velocity at a slightly later time when the object has moved to a neighboring point Q. The triangle of vectors is shown separately on the right side of the figure. In order to change the velocity $\mathbf{v}_P$ at P into the velocity $\mathbf{v}_Q$ at Q, it is necessary to add to $\mathbf{v}_P$ the vector labeled **V**. This extra velocity **V** is produced by the force that constrains the object to move in the circle, such as the pull of the string in Figure 1–11. The force acts on the object in such a way that it changes its velocity by **V** while it is

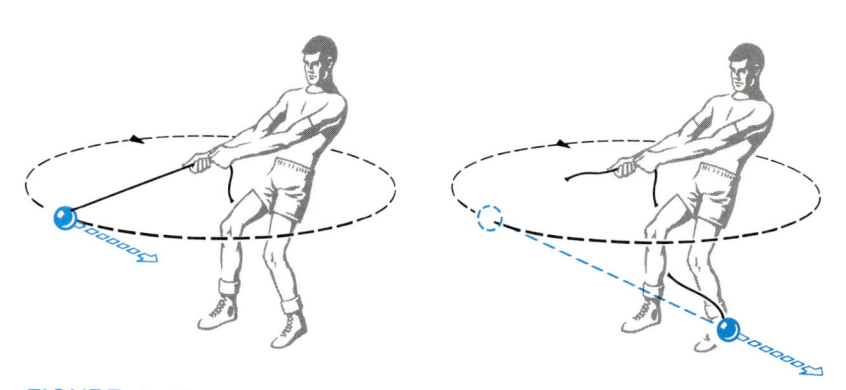

FIGURE 1–11
The instantaneous velocity is tangential to the circle.

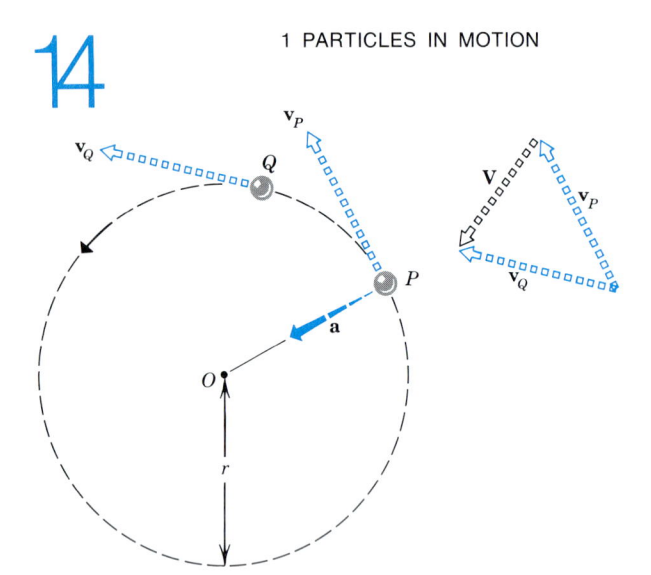

1 PARTICLES IN MOTION

FIGURE 1-12
The acceleration of an object moving in a circle with constant speed.

moving from P to Q. If we divide the magnitude of **V** by the time taken to go from P to Q, we obtain the magnitude of the rate of change of the velocity with time, which is the magnitude of the acceleration. But acceleration, like velocity, is a vector. Its direction is the direction of the change **V** in the vector velocity.

Of course, in order to obtain the instantaneous vector acceleration **a** at the point P, the time interval should be very short and the points P and Q should be very close together. As Q is moved nearer and nearer to P, the angle between the vectors $\mathbf{v}_P$ and $\mathbf{v}_Q$ becomes smaller and smaller. Perhaps you can see from the triangle of vectors that **V** then becomes more nearly perpendicular to $\mathbf{v}_P$. This means that **V** eventually points along the radius toward the center of the circle. It follows that the instantaneous acceleration points inward along the radius toward the center of the circle, as illustrated by the vector labeled **a** in Figure 1-12.

A mathematical treatment of the above arguments reveals that the magnitude of the acceleration is equal to the square of the speed divided by the radius r of the circle.

Acceleration of an object moving with constant speed in a circle

$$= \frac{\text{square of the speed}}{\text{radius of the circle}} \quad (1\text{-}13)$$

or

$$a = \frac{v^2}{r} \quad (1\text{-}14)$$

The acceleration points inward along the radius toward the center of the circle.

For example, if an object moves in a circle of radius 3 m with a constant speed of 6 m/s, the square of the speed is $6 \times 6 = 36$. Dividing this by the radius, the magnitude of the acceleration is seen to be $36/3 = 12$ m/s^2.

SUMMARY

A universe of particles. A possible description of the universe is that it is a complicated structure built of particles such as electrons, protons, and neutrons. A complete physical description of the universe might be obtained by specifying the positions of all the particles at every instant of time.

Units. Before we can make measurements we must define units of length, mass, and time. SI units are the ones now almost universally used. The unit of length is the meter (m). The unit of mass is the kilogram (kg). The unit of time is the second (s). A prefix before the name of the unit indicates multiplication or division by factors of ten (Table 1.1)

Speed. The instantaneous speed of an object is the distance it travels divided by the time taken when the time interval is made as short as possible. The units are meters per second or m/s.

Acceleration is the rate at which the velocity is changing. The instantaneous acceleration of an object moving along a straight line is the small change in its speed divided by the very short time interval during which this change takes place. The units are meters per second per second or m/s^2.

Acceleration due to gravity. At a particular place on the earth's surface all objects fall with the same acceleration. This downward acceleration is always represented by the symbol g. Its value is approximately 9.8 m/s^2. If an object starts from rest and falls for t seconds, it acquires a downward speed $v = gt$.

Vectors. A vector has both magnitude and direction. It can be represented by an arrow pointing in the right direction with a length proportional to the magnitude of the vector according to some agreed scale. Vectors can be added by drawing a triangle of vectors (Figure 1-9). The procedure can be reversed to replace a single vector by two components, usually at right angles to one another. Velocity is a vector; its scalar magnitude is the speed.

Motion in a circle. The instantaneous velocity is in a direction tangential to the circle and therefore perpendicular to the radius. Acceleration is a vector and is equal to the rate of change of the vector velocity. Even if the object moves around the circle with constant speed, the direction of its velocity is continually changing and the acceleration is not zero. The acceleration points from the object toward the center of the circle. Its magnitude is the square of the speed divided by the radius of the circle.

TERMS AND CONCEPTS

electron
proton
neutron
nucleus
units
metric system of units
meter
second
kilogram
powers of ten
SI units
MKS system of units
CGS system of units
density
liter
metric ton
average speed
instantaneous speed
acceleration
acceleration due to gravity
vector
displacement
velocity as a vector
scalar
triangle of vectors
resultant vector
components of a vector
acceleration as a vector
acceleration of a body moving in a circle

QUESTIONS

1. Name the three fundamental particles that are the building blocks for all atoms. Are these particles packed tightly together like the bricks of a house?

2. Name the system of units that has been almost universally accepted by international agreement. What are the units of length, mass, and time in this system?

3. How many grams are there in one kilogram? How many centimeters are there in one meter?

4. Define density. What is the density of water in (a) the CGS system of units? (b) the MKS system of units?

5. Define instantaneous speed. What are its SI units?

6. What is meant by acceleration? What are its SI units?

7. What is meant by the acceleration due to gravity?

8. Describe the motion of a ball thrown horizontally.

9. What is a vector? How are vectors added?

10. Explain how a vector may be replaced by two vectors at right angles to one another.

11. What is the direction of the velocity of an object moving in a circle?

12. What are the magnitude and direction of the acceleration of an object moving in a circle with constant speed?

PROBLEMS

1. Which of the following distances is comparable with the length of your index finger? (a) 10^6 cm, (b) 10 m, (c) 0.1 m, (d) 10^{-3} cm, (e) 10^{-6} m.

2. How many centimeters are in 1 kilometer?

3. How many micrograms are in 1 milligram?

4. The density of liquid mercury is 13.6 g/cm³. What is the mass in kilograms of 1 liter of mercury?

5. Simplify $\dfrac{(2 \times 10^9) \times (9 \times 10^{-3})}{6 \times 10^2}$.

6. A car moves along a straight road with a constant speed of 25 m/s. How far does it travel in one minute?

7. A ball is released from rest and falls vertically. (a) What is its speed after it has been falling for 3 seconds? (b) For how long must it fall before its speed builds up to 100 m/s?

8. A ball is thrown from the top of a vertical cliff with a horizontal velocity of 15 m/s. It strikes the horizontal plain at the foot of the cliff 2.5 s later. (a) What is its horizontal distance from the foot of the cliff at the instant it strikes the ground? (b) What is the vertical component of its velocity at this instant?

9. An object moves with a constant speed of 15 m/s around a circle of radius 2.5 m. What is the magnitude of its acceleration?

CHAPTER 2
THE LAWS OF MOTION

2.1 INTERACTIONS AND MOTION

In our plan of visualizing the universe as a collection of particles in motion, we have already considered concepts such as velocity and acceleration needed to describe this motion. We now turn to the basic question of why each particle moves along its particular path and how we might predict the details of the path. The underlying idea is that the details of the motion are determined by the interactions between the particle and the other particles in the universe. The other particles exert forces on the particle that push and pull it along the twists and turns of its path.

Before pursuing this point, a still more basic question must be answered. How would the particles behave if they did not interact with one another, if they were entirely oblivious of one another's existence? Would they all remain at rest, move in straight lines or circles, or oscillate backward and forward about fixed centers? The answer is not obvious and can only be found by carefully observing the universe and discovering which assumption best fits the facts. The answer is contained in *Newton's first law of motion*.

> Newton's first law of motion:
> In the absence of any interaction with the rest of the universe, an object would either remain at rest or move continually in the same straight line with a constant velocity. Its velocity, regarded as a vector, would remain constant. Rest would be the special case of zero velocity. The acceleration would be zero.

If they did not interact with one another, the particles in the universe would move in straight lines with constant velocities as in Figure 2–1a. The interactions change the velocities and produce accelerations and hence complicated paths, as in Figure 2–1b. A large part of physics consists of formulating the laws needed to calculate the acceleration produced in one particle by a second particle.

2.2 DETERMINISM

Suppose that, at a fixed instant of time, we know the positions and velocities of all the particles. Concentrate on the behavior of a particular particle such as the one in the center of Figure 2–2. Consider its interaction with any other particle in the universe. We must know the distance between the two particles, their velocities, and such properties of the particles as their masses and electric charges. The fundamental laws of physics then enable us to calculate the acceleration induced in the central particle by its interaction with the other particle. Similarly, we can calculate the acceleration induced in the central particle by each of the other particles in the universe in turn. The vector sum of all these accelerations then gives us the total acceleration of the central particle at the fixed instant of time we are considering.

The acceleration of each of the other particles can be calculated in a similar way. It can be shown that this procedure, when combined with some advanced mathematics, enables us to describe the path of each particle in complete detail. We can then calculate the positions, velocities, and accelerations of all the particles at all future instants of time and all past instants of time.

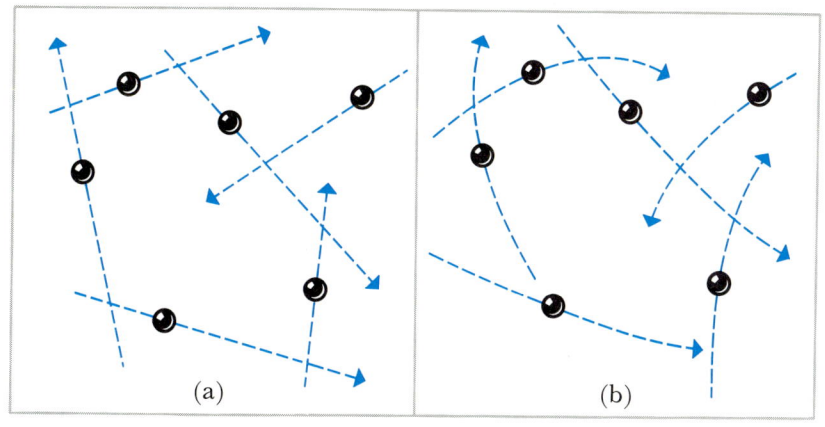

FIGURE 2-1

(a) If the particles did not interact they would move in straight lines with constant velocities. (b) The particles do in fact interact. They accelerate one another and distort one another's paths.

The idea that what will happen in the future is determined completely by what has happened in the past is called *determinism*.

> Determinism in physics:
> If the positions and velocities of all the particles are known at one instant of time, then the fundamental laws describing the interactions of the particles with one another can be used to calculate in precise detail the behavior of the particles at all past, present, and future times.

This statement has profound philosophical significance. It would seem to imply that the future of the universe is completely determined by its past and that there is no such thing as free will or human intervention. Suppose that I take the extreme point of view that my thoughts are merely a consequence of the positions and velocities of the electrons, protons, and neutrons in my brain. Suppose that I know their positions and velocities at 12:30 P.M. Suppose that I also know the positions and velocities of certain other particles in my immediate vicinity. With sufficient effort and the use of sufficiently complicated computers, I can presumably predict that at 12:55 P.M. the positions of the particles in my brain will be such as to cause me to make a decision to go out to lunch. Any feeling I have that this decision was made voluntarily is pure delusion. If, in the process of crossing the road to the lunch counter, I am struck by an automobile and killed, this could presumably have been predicted from the previous positions and velocities of the particles in the universe. My death was therefore quite inescapable.

These conclusions are very difficult to avoid if one accepts this basic approach to physics without question. Fortunately, the modern quantum theory provides a loophole, as we shall discover in a later chapter. Unfortunately, though, it is not clear that it provides us with an alternative that is any more comforting.

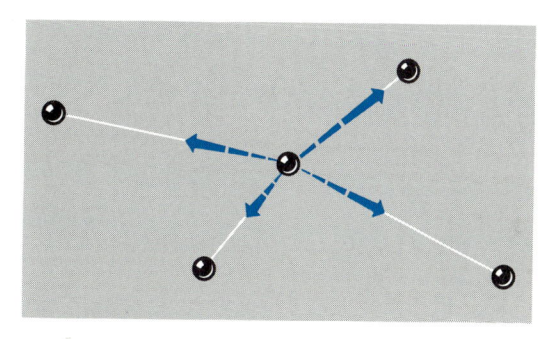

FIGURE 2-2

The acceleration of the particle in the center is the vector sum of contributions resulting from interactions with each of the other particles in the universe.

2.3 FORCE

What determines the acceleration of an object? How can the interactions between two objects best be described?

Imagine an athlete throwing a weight or, to

2 THE LAWS OF MOTION

An athlete putting a shot must accelerate the shot to as high a velocity as possible by exerting a force on it. (Walter Iooss, Jr., for *Sports Illustrated* © 1976 Time, Inc.)

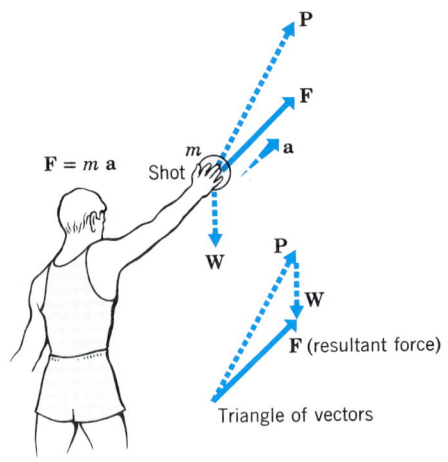

FIGURE 2-3
An athlete putting a shot.

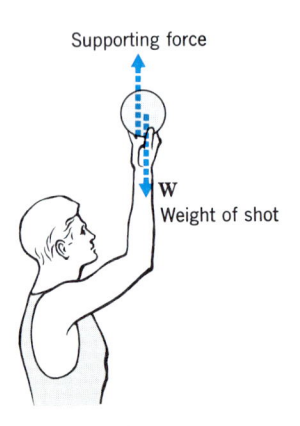

FIGURE 2-4
An athlete holding a shot stationary.

use a more exact phrase, putting a shot (Figure 2-3). He accelerates the shot by exerting a *force* **P** on it. The strain he feels in his muscles is a crude indicator of the strength of this force. The acceleration of the shot is produced by the force, and the velocity with which it is finally thrown therefore depends on the force.

We must not ignore one complication. **P** is not the only force on the shot. The earth exerts on the shot a gravitational force **W** that pulls it vertically downward toward the center of the earth. We normally call this the *weight* of the shot.

If the athlete were merely holding the shot stationary, as in Figure 2-4, he would have to exert an upward force of exactly the right size to counteract the weight. He would still feel a noticeable strain in his muscles.

When he is throwing the shot he must do a little better. He must apply a force that is large enough to overcome the weight with something left over to accelerate the shot in the desired direction. The acceleration is then determined by the combined effect of the force **P** he exerts and the gravitational force **W** exerted by the earth.

Force is a vector, and the two forces on the shot must be added together using the triangle of vectors, as shown on the right of Figure 2-3. The force **F** that finally determines the acceleration of the shot is the vector resultant of the two separate forces **P** and **W**.

The direction of the acceleration **a** of the shot is the same as the direction of the resultant force **F** on it. The magnitude a of the acceleration is proportional to the magnitude F of the resultant force. However, the magnitude of the acceleration also depends on the *mass* m of the shot. The larger the shot, the more difficult it is to set it in motion.

The difficulty in accelerating a more massive shot is not entirely due to its extra weight. By making a greater effort, the athlete could compensate for the extra weight and arrange to apply the same resultant force on the larger shot. In spite of this, the acceleration of the larger, more massive shot would be smaller. With a constant applied resultant force **F**, the magnitude of the acceleration is inversely proportional to the magnitude of the mass. If the mass is doubled, the acceleration is halved.

2.4 NEWTON'S SECOND LAW OF MOTION

In general, consider any body of mass m. Suppose that all the forces exerted on it from outside add together vectorially to produce a resultant force **F**. Then the numerical value of its acceleration is given by the equation:

$$\text{Acceleration} = \frac{\text{resultant force}}{\text{mass}} \quad (2\text{-}1)$$

or

$$a = \frac{F}{m} \quad (2\text{-}2)$$

We can multiply both sides of this equation by the mass and obtain an equation for the force.

$$\text{Resultant force} = \text{mass} \times \text{acceleration} \quad (2\text{-}3)$$

or

$$F = ma \quad (2\text{-}4)$$

This equation is often called *Newton's second law of motion*. If we multiply the mass of a body by its acceleration, the product is equal to the resultant force on the body. This force is produced by one or more surrounding bodies. The force is determined by the nature of these surrounding bodies and how far away they are. Exactly how that comes about involves a knowledge of much more physics. It will gradually emerge as this book proceeds.

Newton's second law of motion:

$$\text{Acceleration} = \frac{\text{resultant force}}{\text{mass}} \quad (2\text{-}5)$$

Resultant force =
$$\text{mass} \times \text{acceleration} \quad (2\text{-}6)$$

Force and acceleration are both vectors. The acceleration is in the same direction as the resultant force.

2.5 THE UNIT OF FORCE

Force is a very important quantity in physics. The unit of force therefore has a special name. If the mass m is measured in kilograms and the acceleration a in meters per second per second, then the force F is measured in newtons. The SI unit of force is the *newton*. Its abbreviation is N.

For example, if a mass of 3 kilograms has an acceleration of 4 meters per second per second, the force on it is 3×4 or 12 newtons.

If a force of 35 N is applied to a mass of 5 kg, its acceleration is 35/5 or 7 m/s².

2.6 GRAVITATIONAL FORCES

So far, what has been said amounts to multiplying the mass by the acceleration and giving the product a new name, the force. There is more to it than that. If the origin of the force is known, it can be calculated in a second way, using the physical properties of the surrounding objects that cause it. The two calculations must agree; therefore, the acceleration can be directly related to the properties of the surrounding objects. Unfortunately, the details are sometimes very complicated.

There are several kinds of force. The simplest kind are *gravitational forces*. All objects in the universe attract one another with gravitational forces. Figure 2-5 represents any pair of objects. They have been labeled 1 and 2. They might be two stars moving around one another, object 1 might be the sun and object 2 a planet, or object 1 might be the earth and object 2 might be falling freely toward the surface of the earth. Either object exerts on the other object an attractive gravitational force that pulls the other object directly toward it. A gravitational force always points straight toward the center of the object producing it.

2.7 NEWTON'S THIRD LAW OF MOTION

Although the gravitational forces on the two objects point in opposite directions, they have

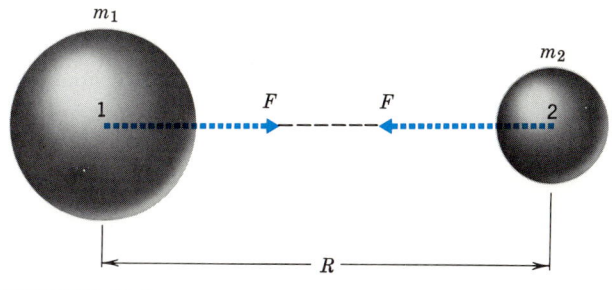

FIGURE 2-5
The gravitational forces between two objects.

2 THE LAWS OF MOTION

This group of stars is held together by the gravitational forces attracting the outer stars toward the stars nearer the center. (Lick Observatory.)

exactly the same magnitude. The two arrows representing the forces have exactly the same length. This is an example of a more general law proposed by Newton.

> **Newton's third law of motion:**
> The forces that two objects exert on each other are equal in magnitude but exactly opposite in direction.

Newton believed that this law applied to all kinds of force. It does apply to many forces, but occasionally breaks down. We will discover in Section 6.2 that it sometimes breaks down for magnetic forces.

2.8
NEWTON'S LAW OF UNIVERSAL GRAVITATION

The magnitude of the gravitational force is proportional to the mass of either object. It is therefore proportional to the product of the two masses. If the mass of the first object is doubled, the force becomes twice as big. If the mass of the second object is tripled, the force becomes three times as big. If the mass of the first object is doubled and the mass of the second object is simultaneously tripled, the force is multiplied by 2×3, which is 6.

The gravitational force is inversely proportional to the square of the distance between the two objects. If the distance is doubled, the force decreases, not to $\frac{1}{2}$ of its previous value, but to $\frac{1}{2^2}$ or $\frac{1}{4}$ of its previous value. If the distance is multiplied by 10, the force decreases to $\frac{1}{10^2}$ or $\frac{1}{100}$ of its previous value.

This *inverse square law* implies a very rapid falling off in the size of the gravitational force as the two objects are moved apart. Gravitational forces are strong for objects close together, but weak for objects far apart. The gravitational force on a falling object is mainly due to the nearby earth. Although the sun is much more massive than the earth, it is also much farther away. The sun's contribution to the gravitational force on a falling object is therefore less than 0.1 percent. A star such as Sirius, the brightest star in the northern skies, has a mass not very different from that of the sun. But it is so far away that its contribution to the force on the falling object near the earth is less than one millionth of a billionth (10^{-15}).

Because the force is directly proportional to the product of the two masses and inversely proportional to the square of the distance, it can be represented by the following algebraic equation, first suggested by Newton.

> **Newton's law of universal gravitation:**
> The gravitational force between any two objects with masses m_1 and m_2 at a distance R apart is
>
> $$F = \frac{G m_1 m_2}{R^2} \qquad (2\text{-}7)$$
>
> The force on either object points directly toward the other object.

This is our first complicated algebraic equation; let us consider its meaning carefully. The subscripts 1 and 2 on the masses m_1 and m_2 have no profound mathematical significance. They are there as a kind of shorthand to remind us that m_1 is the mass of object 1 and m_2 is the mass of object 2. The superscript 2 in the denominator is a different matter. Its function is to remind us that we must square the distance R between the two objects and insert the resulting number in the denominator.

The quantity G in the numerator is a very special number called the *gravitational constant*. It always has exactly the same numerical value for any two objects anywhere. If the masses m_1 and m_2 are measured in kilograms, the distance R in meters, and the force F in newtons, then the invariable numerical value of G is

$$G = 6.67 \times 10^{-11} \quad \text{SI units} \qquad (2\text{-}8)$$

Remember that 10^{-11} means $1/10^{11}$ or $1/100{,}000{,}000{,}000$ with 11 zeros.

The value of G can be looked upon as a measure of how strong gravitational forces are in general. It should not be confused with the acceleration due to gravity, g, which is not a universal constant but has a value that depends on the accidental circumstance of being located at a particular spot near the earth's surface. As we shall discover in the next section, the acceleration due to gravity, g, varies from place to place over the surface of the earth. The gravitational constant G never varies. G is often called a fundamental constant of physics, whereas g is a property of the environment.

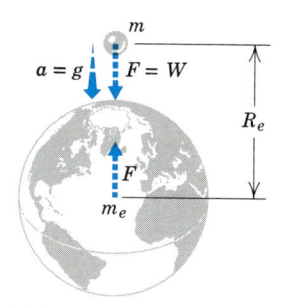

FIGURE 2–6
An object falling vertically near the surface of the earth.

2.9
THE EARTH'S GRAVITY

Suppose that one of our two objects is the earth and the other is an object falling freely near the surface of the earth, as in Figure 2–6. Instead of m_1 we can write m_e. The subscript e then reminds us that we are talking about the mass of the earth. Instead of m_2 we write m, which stands for the mass of any falling object. Since the falling object is near the surface of the earth, the distance between its center and the center of the earth is what we would normally call the "radius of the earth." To remind ourselves of this, we will represent it by R_e.

The gravitational force pulling the falling body downward toward the center of the earth is exactly what we mean by the *weight* of the falling body. To remind ourselves of this we will use the symbol W for this force. It is the same kind of force as the gravitational force labeled F in Figure 2–5.

Using these new symbols, Equation 2-7 for the gravitational force takes the form

$$W = \frac{Gm_e m}{R_e^2} \quad (2\text{-}9)$$

The weight of an object is proportional to its own mass m, but it is also proportional to the mass m_e of the whole earth. It is inversely proportional to the square of the radius of the earth.

2.10
WEIGHT AND MASS

Notice the distinction between weight and mass. The mass of an object is a scalar quantity measured in kilograms. It is an invariant property of that object and always has the same numerical value wherever the object is.

The weight of an object is a gravitational force measured in newtons. Strictly speaking, it is a vector. The direction in which the gravitational force points is as important as its magnitude. The weight depends on where the object is. If it were taken to the moon, its mass would remain exactly the same. On the other hand, its weight would be different, because the gravitational force attracting it to the moon would be different. To obtain the weight on the moon from an equation similar to Equation 2-9, we would have to replace the mass of the earth m_e by the mass of the moon and the radius of the earth R_e by the radius of the moon.

If the mass m is unsupported, it falls freely with an acceleration g—the *acceleration due to gravity*, which has a value of about 9.8 m/s² near the surface of the earth. The force F on the mass is the same as its weight W. The acceleration a of the mass is the acceleration due to gravity, g. Newton's second law states that the force is equal to the mass multiplied by the acceleration. It now takes the special form that the weight is equal to the mass multiplied by the acceleration due to gravity.

Weight:
$$\text{Weight} = (\text{mass}) \times (\text{acceleration due to gravity}) \quad (2\text{-}10)$$
$$W = mg \quad (2\text{-}11)$$

For example, if g is exactly 9.8 m/s², the weight of 50 kg is 50 × 9.8 or 490 newtons.

2.11
THE ACCELERATION DUE TO GRAVITY

We now have two separate equations for W, Equations 2-9 and 2-11. Both expressions represent the same quantity and so they must be equal.

$$mg = \frac{Gm_e m}{R_e^2} \quad (2\text{-}12)$$

Since m appears on both sides, it can be canceled out. What remains is an equation that shows how the value of the acceleration due to gravity depends on the mass of the earth and the radius of the earth.

$$g = \frac{Gm_e}{R_e^2} \qquad (2\text{-}13)$$

A very important aspect of this equation is that the acceleration of a freely falling object does not depend on any property of the object. At a particular place on the earth's surface, all objects fall with the same acceleration (if the effect of the air on the motion is eliminated or allowed for). This important fact was first realized by Galileo toward the end of the sixteenth century. It is immortalized in the famous story of how he dropped two very different objects from the top of the Leaning Tower of Pisa and observed that they reached the ground at the same time. Since then the point has been tested by experiments of increasing complexity and sophistication. We now know that g for various objects of different size and chemical composition does not differ by more than one part in a trillion parts. More generally, when an object is made to accelerate by the gravitational attraction of other objects, the acceleration does not depend on any property of the object that is being accelerated. This peculiar aspect of gravitational forces is not true of any other type of force. It is one of the fundamental principles underlying Einstein's general theory of relativity.

The earth is not a perfect sphere and its density is not uniform throughout, so Equation 2-13 is only approximately true. In fact, g varies from place to place on the earth's surface. At the top of a high mountain g is less than at sea level. The top of the mountain is farther from the center of the earth, and the inverse square law comes into play. There is a tendency for g to increase from the equator to the poles. There are also local variations in g caused by variations in the nature of the earth's crust. Characteristic local variations in g have been used to help locate oil deposits. None of these effects is large and g varies by only about 0.5 percent over the whole of the earth's surface.

2.12
A FREELY FALLING BOOK

We shall now attempt to obtain a better understanding of Newton's laws by applying them to some particular situations. Suppose that I lift up a book and let it fall to the floor (Figure 2–7). While it is falling the only force acting on it is its weight W, which is the downward force of gravitational attraction exerted on the book by the earth. According to Newton's third law, the book

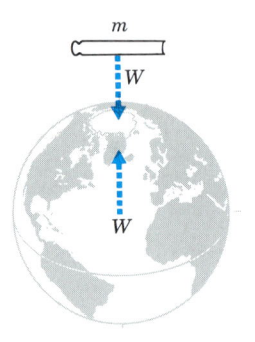

FIGURE 2–7
A freely falling book.

must exert on the earth an upward force that has the same magnitude W. The book pulls the earth upward just as strongly as the earth pulls the book downward.

However, the mass of the earth is very much greater than the mass of the book. The upward acceleration given to the earth is therefore very much smaller than the downward acceleration given to the book. Nevertheless, it is a fact that the earth does fall upward a little to meet the book.

2.13
A BOOK RESTING ON A DESK

Now suppose that I place the book on top of a desk, as in Figure 2–8. It is then stationary and has no acceleration. Newton's second law tells us that the resultant force on it is equal to the product of its mass and its acceleration. Since the acceleration is zero, the resultant force must be zero. The downward pull W of the earth must be compensated by an upward push R exerted on the book by the top of the desk. The magnitude of R must be exactly equal to the magnitude of W and they must be in exactly opposite directions. How does that come about?

R is clearly not a gravitational force. Gravitational forces are always attractive, whereas the

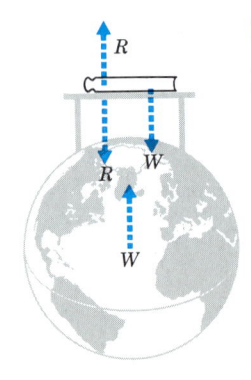

FIGURE 2–8
A book resting on a desk. The forces all act along the same vertical line but they have been displaced sideways for the sake of clarity.

2.15 THE SEPARATE ROLES OF THE SECOND AND THIRD LAWS

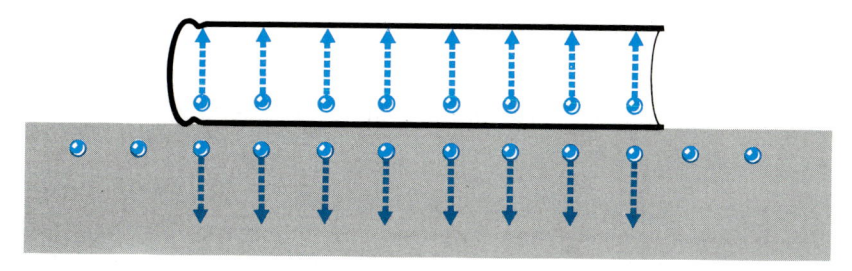

FIGURE 2-9
Interatomic forces.

desk repels the book. The forces involved are forces between the atoms on the top of the desk and the atoms on the bottom of the book (Figure 2-9). They are called *interatomic forces* and they are essentially electrical forces.

2.14
INTERATOMIC FORCES

Interatomic forces are produced in the following way. An atom consists of negatively charged electrons moving around a nucleus containing positively charged protons. As is explained more fully in Chapter 5, a negative electron repels another negative electron. A positive proton repels another positive proton. A negative electron and a positive proton attract each other. An atom contains equal numbers of negative electrons and positive protons. The electrical forces therefore tend to cancel out, but not quite.

When two atoms approach closely, the electrons on the outside of one atom come very close to the electrons on the outside of the other atom. The repulsive electrical forces between these electrons are then very strong and dominate the situation. Atoms close together repel each other.

When the atoms are farther apart, it turns out that the net electrical forces are attractive. The origin of this attraction cannot be understood without the quantum theory. It is very important, though, because it is the reason why atoms hold together in liquids and solids.

Therefore, when the book is first placed on the desk, its weight causes it to sink steadily down. The atoms on the bottom of the book approach closer and closer to the atoms on the top of the desk. The repulsive forces between these atoms become stronger and stronger. Eventually the total upward repulsive force on all the atoms on the bottom of the book is just sufficient to counteract the weight of the book. The resultant force on the book is then zero and its acceleration is zero. It settles down into a stationary position.

Interatomic forces obey Newton's third law. If a particular atom on the top of the desk exerts a certain upward force on its neighboring atom on the bottom of the book, then the atom in the book exerts a downward force of equal magnitude on the atom in the desk (Figure 2-9). The total downward force on all the atoms on the top of the desk has the same magnitude as the total upward force on all the atoms at the bottom of the book. Since the desk exerts a total upward force of magnitude R on the book, the book exerts a total downward force on the desk with the same magnitude R (Figure 2-10).

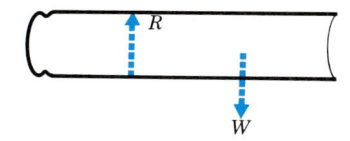

FIGURE 2-10
The forces exerted on the book by the earth and the desk. The two forces really act along the same vertical line but they have been displaced sideways for the sake of clarity.

2.15
THE SEPARATE ROLES OF THE SECOND AND THIRD LAWS

Notice the different ways in which Newton's second and third laws of motion enter into this discussion. Newton's third law tells us that the downward gravitational pull of the earth on the book has the same magnitude W as the upward gravitational pull of the book on the earth (Figure 2-7). It also tells us that the upward force exerted on the book by the desk has the same magnitude R as the downward force exerted on the desk by the book (Figure 2-8).

The forces R have their origin in interatomic forces (Figure 2-9) and are essentially electrical forces. In the third law, the two forces involved always act on different objects. Moreover, the two forces are always of the same kind. Both are gravitational or both are electrical.

It would be wrong to appeal to the third law to deduce that R has the same magnitude as W. The second law is involved. We concentrate our attention on the book only (Figure 2–10). Its acceleration is zero, so the second law tells us that the resultant force on it is zero. The two forces R and W exerted on it by its surroundings must therefore be equal in magnitude and act in opposite directions in order to cancel each other. In this argument, the two forces act on the same object. In this particular example, W is a gravitational force and R is an electrical interatomic force.

Moreover, the equality of R and W depends on the special circumstance that the acceleration is zero. They would no longer be equal if the book and the desk were inside a rocket being launched with an upward acceleration a (Figure 2–11). There would then have to be a net upward force $(R - W)$ on the book to produce its upward acceleration. Newton's second law would then tell us that

$$R - W = ma \qquad (2\text{-}14)$$

2.16
THE MATHEMATICS OF A SATELLITE IN ORBIT

An artificial satellite continues to orbit about the earth for many revolutions even though it is not powered by any fuel. On the other hand, an airplane must use up fuel if it is to maintain its altitude. There are some important differences between the satellite and the airplane. The presence of the air is essential to the flight of an airplane, but its motion is opposed by the frictional resistance of the air. It uses up its fuel to overcome this resistance. A satellite, however, must be placed in orbit at a very high altitude where the atmosphere is so rarefied that it exerts very little frictional drag on the satellite. (Even so, this small frictional drag eventually brings down the satellite after it has completed many revolutions around the earth.) Once it has reached this rarefied part of the atmosphere, the satellite moves in a stable orbit only if its velocity has exactly the right magnitude and direction for the orbit in question. The required speed is greater than can be achieved by an airplane.

To avoid complicated mathematics and yet still illustrate the physical principles involved, consider the particularly simple case of a circular orbit of radius R (Figure 2–12). Let m_e be the mass of the earth, m the mass of the satellite, and v the speed that the satellite must have if it is to remain in this orbit. The situation will be described from the point of view of an observer in space watching the satellite revolve around the earth.

The gravitational force pulling the satellite toward the center of the earth is

$$F = \frac{Gm_e m}{R^2} \qquad (2\text{-}15)$$

According to Newton's second law, this force gives the satellite an acceleration a pointing toward the center of the earth, where

$$F = ma \qquad (2\text{-}16)$$

However, when the satellite is moving in a circle of radius R with speed v, it has an acceleration toward the center of the circle given by an equation similar to Equation 1-13 of Section 1.13.

$$a = \frac{v^2}{R} \qquad (2\text{-}17)$$

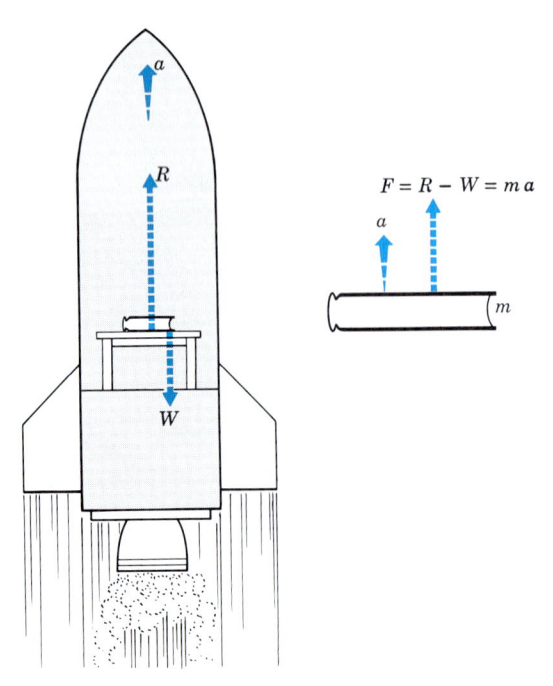

FIGURE 2–11
The book is on top of a desk inside a rocket with an upward acceleration a.

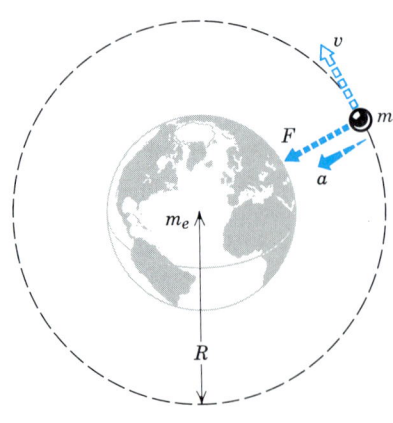

FIGURE 2–12
A satellite in a circular orbit.

The circle is a possible orbit only if the speed v has exactly the right value. The acceleration given by Equation 2-17 must be multiplied by the mass as in Equation 2-16 to give a force F with exactly the right value to satisfy Equation 2-15. The algebraic form of this last statement is

$$F = m\left(\frac{v^2}{R}\right) = \frac{Gm_e m}{R^2} \qquad (2\text{-}18)$$

A little algebraic manipulation reduces this to an equation for the speed v.

$$v = \sqrt{\frac{Gm_e}{R}} \qquad (2\text{-}19)$$

Remember that G is the gravitational constant and m_e is the mass of the earth. Equation 2-19 gives the unique value that the speed must have if the satellite is to remain in a circular orbit of radius R. The required speed is inversely proportional to the square root of the radius.

2.17
A QUALITATIVE DISCUSSION OF THE SATELLITE'S MOTION

In case the sophistication of this algebra has obscured the issue for you, let us revert to the often-asked question: "Why does the satellite not fall down to the earth?" The only requirement that the earth's gravitational pull imposes on the satellite is that it shall always have an acceleration directed toward the center of the earth. A body released from rest at a height achieves this acceleration by falling directly downward. A satellite can achieve it equally well by moving continually in a circle with just the right speed. Motion in a circle always involves an acceleration directed toward the center of the circle.

There is a sense in which the satellite is falling toward the earth. In the absence of the earth's gravitational attraction it would continue to move in a straight line tangential to the circle, in accordance with Newton's first law of motion. It would then occupy the successive positions P, Q, R shown in Figure 2-13. Actually, when it moves in a circular orbit it occupies the successive positions P, S, T. This can be imagined to be the result of having fallen through the distances QS and RT.

2.18
WEIGHTLESSNESS

We are now in a position to understand *weightlessness* inside a spaceship in orbit. Suppose that an astronaut holds up an object and then re-

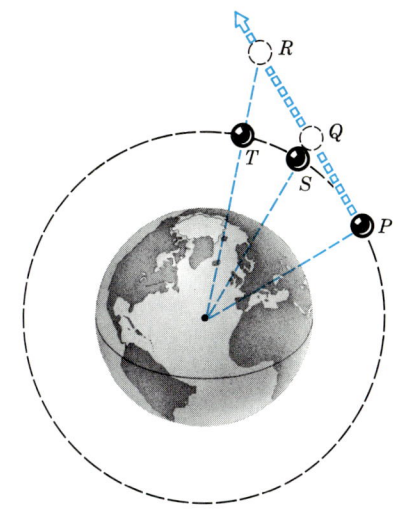

FIGURE 2-13
In a certain sense, the satellite is falling toward the earth.

leases it. Just before and immediately after being released, the object has the same velocity v as the spaceship. However, this is exactly the right velocity to keep it moving in the same orbit as before. Notice that Equation 2-19 does not contain the mass m of the orbiting object and therefore works just as well for the released object as for the spaceship.

The object and the ship therefore continue to move as before with the same velocity in the same orbit. They stay in the same positions relative to one another. Inside the spaceship the object appears to remain suspended in midair.

Finally, let us realize that the arguments applied above to the motion of an artificial satellite around the earth are equally applicable to the motion of the moon around the earth, the motion of any moon around any planet, or the motion of any planet around the sun. The theory of the motions of the planets and their moons in the entire solar system is a straightforward extension of the ideas just presented.

SUMMARY

Newton's first law of motion. In the absence of any interaction with the rest of the universe, an object either remains at rest or moves continually in the same straight line with a constant velocity.

Newton's second law of motion. The resultant force on an object is equal to the product of its mass and its acceleration. The SI unit of force is the newton (N).

Newton's third law of motion. The forces that two objects exert on one another are equal in magnitude but exactly opposite in direction.

Determinism in physics. If the positions and velocities of all the particles are known at one instant of

2 THE LAWS OF MOTION

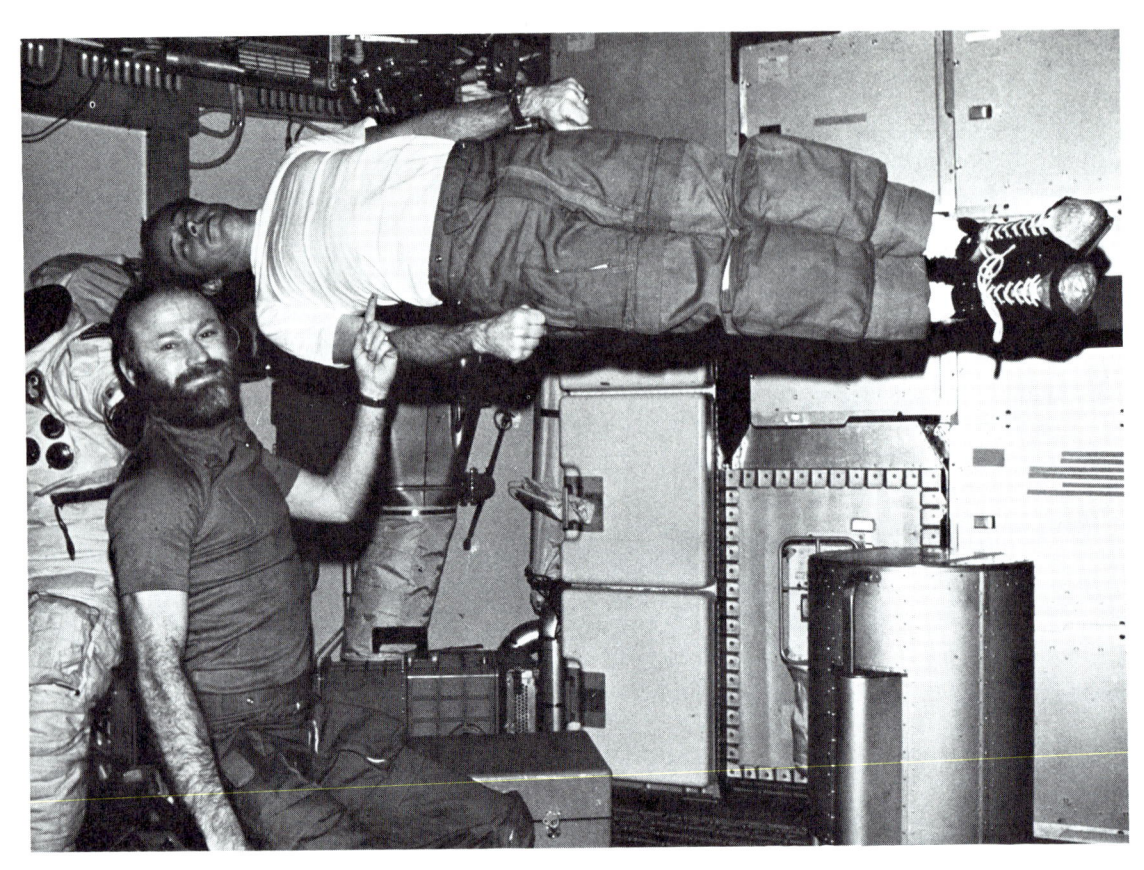

Weightlessness in Skylab 4. Astronaut Gerald P. Carr points to, but does not support, Edward G. Gibson, who is floating in midair. (NASA.)

time, the laws of physics enable their positions to be calculated at all past and future times.

Newton's law of universal gravitation. The gravitational force of attraction exerted on an object by any other object is proportional to the product of their masses and inversely proportional to the square of the distance between them. The force on either object points directly toward the other object.

$$F = \frac{Gm_1 m_2}{R^2}$$

The quantity G is called the gravitational constant. It always has the same numerical value.

Weight of an object is the gravitational force acting on it. The weight W is equal to the product of the mass m and the acceleration due to gravity g.

$$W = mg$$

Whereas the mass of an object is the same everywhere, its weight depends on the local value of the acceleration due to gravity.

Earth's gravity. The acceleration due to gravity near the surface of the earth depends on the mass of the earth and the radius of the earth. At a particular place it is the same for all objects. It varies from place to place over the surface of the earth.

Interatomic forces. When two objects are in close contact, the atoms of one exert repulsive forces on neighboring atoms in the other. These forces are electrical forces between the negatively charged electrons on the outside of the atoms. The interatomic forces obey Newton's third law. The net result is that the two objects in contact repel each other with equal and opposite forces.

Satellites in a circular orbit have acceleration toward the center of the circle by virtue of their curved motion. The orbit is stable if the gravitational force acting on the satellite is exactly equal to the product of its mass and the above acceleration.

TERMS AND CONCEPTS

Newton's first law of motion
determinism
force
weight
mass
Newton's second law of motion
newton (unit of force)
gravitational force
Newton's third law of motion
Newton's law of universal gravitation
inverse square law of gravitational force

gravitational constant, G
earth's gravity
acceleration due to gravity, g
interatomic forces
weightlessness

QUESTIONS

1. Explain carefully what determines the shape of the path of a moving object.

2. Which of Newton's laws of motion is equivalent to a definition of force? Give this definition.

3. If Newton's third law of motion applies to the earth and a falling object, why are we not aware that the earth falls upward toward the object?

4. Write down from memory the algebraic equation expressing Newton's law of universal gravitation. Now put this equation into words instead of symbols.

5. Why is the gravitational constant G called a fundamental constant of physics whereas the acceleration due to gravity g is not?

6. Define weight. What is the relationship between the weight of an object and its mass?

7. Describe the nature of interatomic forces.

8. Two satellites are in stable circular orbits around the earth. Does the satellite in the orbit of larger radius have a greater or a lesser speed than the satellite in the smaller orbit?

9. If an object is instantaneously at rest, is the force on it necessarily zero? Give a simple example.

10. A satellite is moving around the earth once every three hours in a circular orbit that passes over both the north and south poles. A hydrogen bomb hangs from a string below the satellite. When the satellite is over the north pole, the pilot cuts the string. The bomb is timed to explode 90 minutes later. Where will the explosion take place and who will suffer most from it?

11. A trapeze artist is hanging from a swinging trapeze at the end of a rope held between her clenched teeth. According to Newton's third law, which force is equal but opposite to her weight?

PROBLEMS

1. What is the magnitude of the force that produces an acceleration of 3.2 m/s^2 in a mass of 5 kg?

2. When a force of 25 N acts on an object it produces an acceleration of 150 m/s^2. What is the mass of the object?

3. A force of 200 N acts on an object of mass 15 kg. If the object is initially at rest, what is its speed 12 s later?

4. Two objects attract each other with a gravitational force of 6×10^{-3} N. What would be the magnitude of the gravitational force if the distance between the objects were doubled?

5. What is the magnitude of the gravitational force exerted by a mass of 25 kg on a mass of 3 kg at a distance of 5 m from it?

6. What is the approximate weight of one standard kilogram at a point on the earth's surface?

7. A mass of 2.5 kg has a weight of 6×10^3 N on a distant planet. What is the acceleration due to gravity on the surface of this planet?

8. What is the acceleration due to gravity at a distance from the center of the earth equal to twice the radius of the earth?

9. If a satellite is launched into a circular orbit just above the dense part of the earth's atmosphere, the radius of the circle is approximately 7×10^6 m. Calculate the speed of the satellite. How long does it take to complete one revolution?

10. A man with a mass of 100 kg is in an elevator accelerating upward with an acceleration of 4 m/s^2. What is the total force exerted on the soles of his shoes by the floor of the elevator?

CHAPTER 3
MOMENTUM, ANGULAR MOMENTUM, AND ENERGY

3.1 MOMENTUM

It is more painful to be hit on the head with a baseball than with a table tennis ball. The mass is not the only thing that matters, though. The velocity is also important. To be hit on the head by a baseball dropped from a height of one inch is much less painful than if the baseball were dropped from the roof of the house at a height of 20 feet and thereby given time to build up a greater velocity. The direction of the velocity is equally important. The real danger arises from a fast ball coming straight at you, whereas a glancing blow from the same ball can be comparatively harmless.

What matters is the product of the mass m and the vector velocit **v**. This is called the *momentum* and is represented by the symbol **p**.

$$\text{Momentum} = \text{mass} \times \text{velocity} \quad (3\text{-}1)$$
$$\mathbf{p} = m\mathbf{v} \quad (3\text{-}2)$$

Because the velocity **v** is a vector, the momentum **p** is also a vector and is in the same direction as the velocity. The plural of *momentum* is *momenta*.

Newton's second law of motion can be expressed in a different form using the concept of momentum. The resultant force on an object is equal to the rate of change of its momentum with time. Momentum is mass times velocity, but the mass does not change. The change in momentum per second is therefore equal to the mass multiplied by the change in velocity per second. The change in velocity per second is the acceleration. The change in momentum per second is therefore equal to the mass multiplied by the acceleration. But the original form of Newton's law tells us that the mass multiplied by the acceleration is equal to the resultant force. Finally, then,

> Newton's second law of motion in a new form:
>
> The resultant force on an object is equal to the rate of change of its momentum with time.

In the theory of relativity (Chapter 8), the mass is no longer constant; it varies with the velocity. The definition of force as mass multiplied by acceleration is then found to be unsatisfactory—the definition as rate of change of momentum with time is much to be preferred. Moreover, as our exposition of physical theory gradually unfolds, it will be seen that the concept of momentum becomes more and more useful and develops more and more physical significance.

3.2 RECOIL

One reason why the concept of momentum is so useful is that, under the right circumstances, the total momentum of a group of interacting objects remains constant as the motion proceeds.

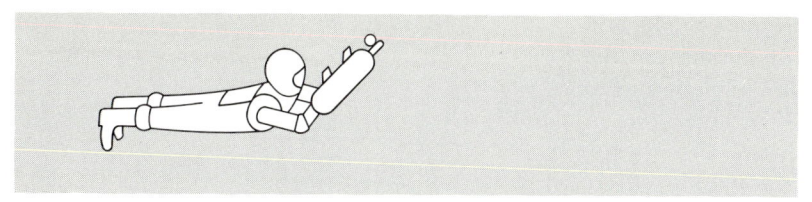

(a) Initially both astronaut and tank are at rest.

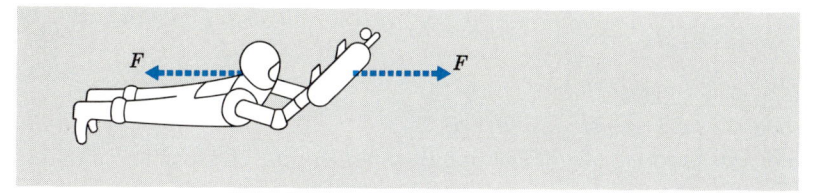

(b) The forces are equal and opposite.

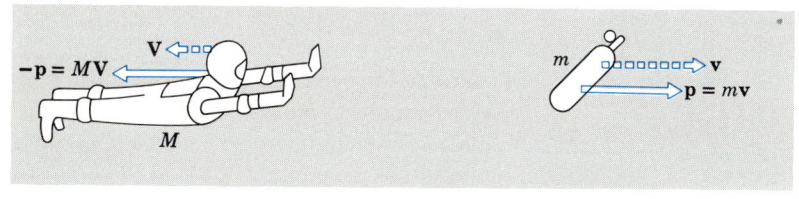

(c) The recoil momentum of the astronaut is equal and opposite to the momentum of the tank.

FIGURE 3-1
Recoil of an astronaut throwing a tank of oxygen in gravity-free space.

We say that the total momentum is conserved and refer to this as the *law of conservation of momentum*.

Imagine an astronaut floating in gravity-free space holding a tank of oxygen, as in Figure 3-1a. Initially she has no velocity. She then decides to throw away the tank. To accelerate the tank, she must apply to it the force F shown pointing toward the right in Figure 3-1b. We now know that this force is the sum of a large number of interatomic repulsive forces on those atoms of the tank that are in contact with atoms on the surface of the astronaut's gloves. Newton's third law of motion applies. The tank reacts back on the astronaut with a force of equal magnitude F in the opposite direction, pointing toward the left.

Force is equal to rate of change of momentum. The two forces have equal magnitudes but exactly opposite directions. The rate at which the tank gains momentum toward the right is therefore equal to the rate at which the astronaut gains momentum toward the left. After the tank has been released, the momentum it has acquired toward the right has the same magnitude as the momentum with which the astronaut recoils toward the left (Figure 3-1c). The arrows representing the two momenta have equal lengths but point in exactly opposite directions. They have been labeled $\mathbf{p}$ and $-\mathbf{p}$, the minus sign indicating a reversal of direction.

Consider the group of two objects consisting of the astronaut and the tank. When we calculate the total momentum of the group we must remember that momentum is a vector. We are not adding together the positive scalar magnitudes of the momenta, but are adding vector quantities using the triangle of vectors. In the present case, the final total momentum is zero because the two arrows of equal length in opposite directions cancel each other. The initial total momentum was also zero because the two bodies were at rest. Momentum has been conserved.

If the mass of the tank is m and it acquires a velocity $\mathbf{v}$, its momentum is $m\mathbf{v}$. If the mass of the astronaut is M and she acquires a recoil velocity $\mathbf{V}$, her momentum is $M\mathbf{V}$. The law of conservation of momentum states that

$$m\mathbf{v} + M\mathbf{V} = 0 \qquad (3\text{-}3)$$

This is not an ordinary algebraic equation. It is an equation for vectors. The addition sign implies that the quantities $m\mathbf{v}$ and $M\mathbf{V}$ are added together as vectors, with due regard to their directions. In terms of the speeds v and V, which are intrinsically positive scalar quantities,

$$mv = MV \qquad (3\text{-}4)$$

The two vectors must be of equal length in order to cancel each other.

Suppose that the tank has a mass of 5 kg and is given a velocity of 7 m/s. If the mass of the astronaut is 100 kg, Equation 3-4 enables us to calculate her recoil velocity V because

$$5 \times 7 = 100\, V \qquad (3\text{-}5)$$

so

$$V = \frac{5 \times 7}{100} \qquad (3\text{-}6)$$

or

$$V = 0.35 \text{ m/s} \qquad (3\text{-}7)$$

Notice that the more massive body always recoils with the smaller speed because the product of mass and speed must be the same for both bodies.

3.3 THE LAW OF CONSERVATION OF MOMENTUM

In general, consider a group of objects that is well defined in the sense that we clearly understand which objects belong to the group and which do not (Figure 3–2). Let A and B be any two of these objects. Then, according to Newton's third law of motion, the force exerted on A by B is equal and opposite to the force exerted on B by A. Any change in the momentum of A produced by B is canceled by the change in momentum of B produced by A. The forces between the two objects cannot change the vector sum of their two momenta.

The same argument applies to any pair of objects in the group. The forces that the objects in the group exert on one another all occur in equal and opposite pairs and cannot possibly change the momentum of the group as a whole.

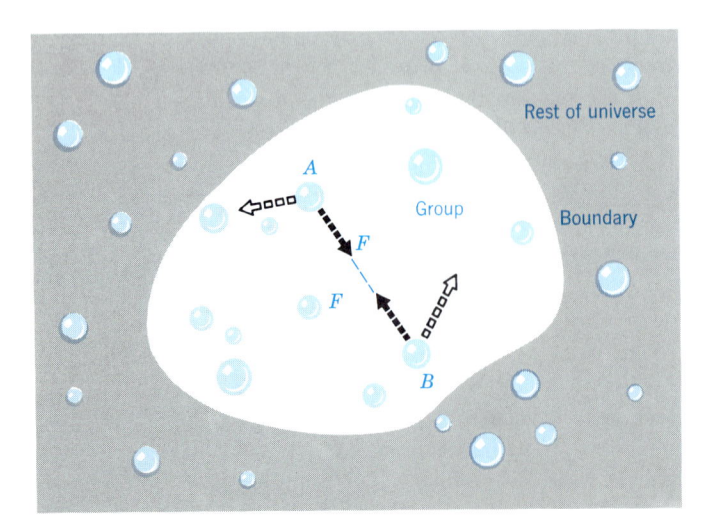

FIGURE 3–2
A group of objects isolated from the rest of the universe.

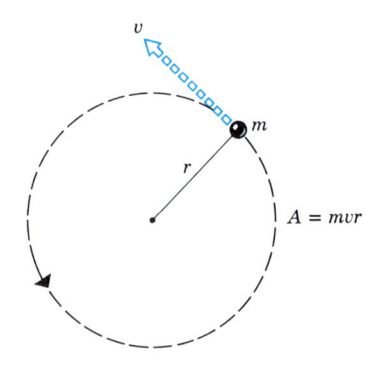

FIGURE 3–3
The angular momentum of a particle moving in a circle is mvr.

Of course, a force exerted on an object inside the group by an object outside the group is a different matter. In fact, the rate of change of the total momentum of the group is equal to the resultant of all the forces exerted on it from outside.

Frequently, though, the forces from outside either do not exist or are so small that they can be ignored. This was the case in the previous example of the astronaut and the oxygen tank isolated in gravity-free space. Under such circumstances the total momentum of the group has a zero rate of change. It remains constant—it is conserved. Remember that the total momentum means the vector sum of the individual momenta.

> Law of conservation of momentum:
>
> If the total resultant force exerted on a group of objects by the rest of the universe is zero, then the vector sum of the momenta of all the objects in the group is a constant vector that does not change with time.

Notice that the law of conservation of momentum is a consequence of Newton's third law of motion. But the third law sometimes breaks down, particularly for magnetic forces. The law of conservation of momentum nevertheless remains valid. Conservation of momentum appears to be more fundamental than the third law.

3.4 ANGULAR MOMENTUM

The momentum mv that we have just discussed is sometimes called linear momentum because it is associated with the motion of an object along a line. A similar quantity, called *angular momentum*,

3.5 TORQUE

is associated with rotation about an axis. In the simple case of a particle of mass m rotating with a speed v in a circle of radius r, the angular momentum A is defined as mvr (see Figure 3-3).

> Angular momentum
> $= $ mass $\times$ speed $\times$ radius (3-8)
> $A = mvr$ (3-9)

Now consider a rotating solid object, such as the earth rotating about an axis through its north and south poles, or a spinning top (Figure 3-4). Any very small part P of the spinning top moves around the axis of rotation in a circle, and its angular momentum is mvr. But v and r are different for different parts of the top. Nevertheless, the total angular momentum of the top may be defined simply by adding together the contributions mvr from all the small parts such as P that make up the whole top. Notice that m is not the mass of the whole top, but of a small part P of it. Moreover, v and r refer to the small part P and have no meaning for the top as a whole.

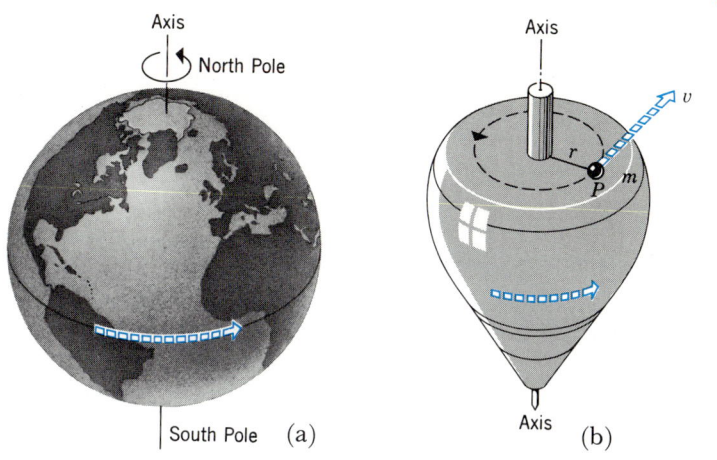

3.5 TORQUE

To change the angular momentum by changing the rotational velocities v, it is necessary to apply forces that have a "twisting" effect. To see how this can be done, grasp the bottom right-hand corner of this book in your right hand. Grasp the bottom left-hand corner in your left hand. With the right hand push the book directly away from you. At the same time pull it directly toward you with the left hand, using a force equal in magnitude but exactly opposite in direction to the force of the right hand (Figure 3-5). The two forces exactly compensate each other, so the resultant force on the book is zero and its acceleration is therefore zero. This procedure cannot start the book moving bodily away from you.

What happens, of course, is that the book rotates in a counterclockwise direction. The reason is that, although the forces are equal but opposite, they are displaced sideways from one another and therefore produce a turning effect.

A pair of equal but opposite forces displaced sideways from each other is called a *couple*. The turning effect of the couple depends on its *torque*, defined as the magnitude of either force multiplied by the perpendicular distance between the forces.

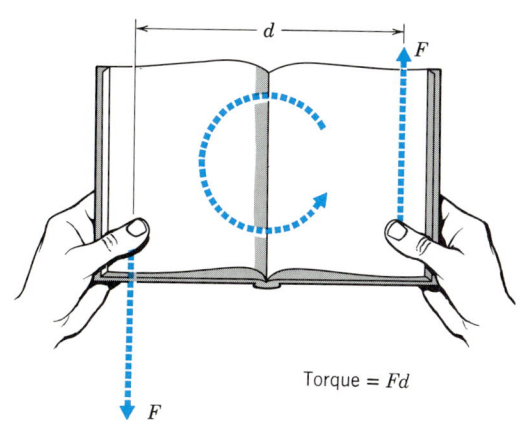

FIGURE 3-4
Rotating extended objects. (a) The earth rotates about an axis through its north and south poles. (b) A spinning top.

FIGURE 3-5
A couple rotating a book.

Torque $= Fd$

The rotating Ferris wheel has angular momentum about a horizontal axis through its center perpendicular to the plane of the photograph. (Kenneth Karp.)

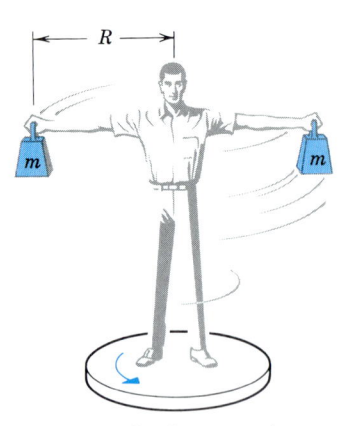

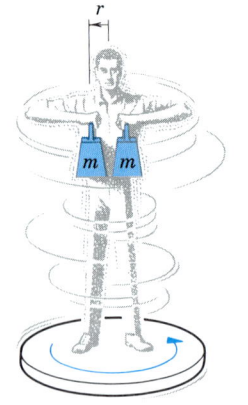

FIGURE 3-6
A demonstration to illustrate the law of conservation of angular momentum.

(a) Large R, slow rotation. (b) Small r, faster rotation.

The torque plays the same role for rotational motion that force plays for motion along a straight line. The resultant external force acting on a group of objects is equal to the rate of change of the total linear momentum. Similarly, the resultant external torque on a group of objects, or a solid body composed of many particles, is equal to the rate of change of the angular momentum.

3.6 CONSERVATION OF ANGULAR MOMENTUM

If there is no net external torque, the total angular momentum remains constant. This is the *law of conservation of angular momentum*. It can be demonstrated in the following way.

A man stands on a slowly rotating turntable mounted on ball bearings that introduce very little friction (Figure 3-6a). Initially his arms are outstretched and each hand holds a mass m at a distance R from the axis of rotation. The two masses make a contribution $2mvR$ to the angular momentum of the rotating system. He then bends his arms and pulls in the two masses until their distance from the axis has the much smaller value r (Figure 3-6b). Their contribution to the angular momentum is thereby reduced.

The torque exerted on the system by the frictional forces due to the ball bearings is negligibly small. Therefore, the total angular momentum of the system cannot change. The only way in which this can be achieved is by a sudden increase in the velocities of all parts of the rotating system. The result of the maneuver is that the man suddenly rotates much faster. If he extends his arms again, the rotation immediately slows down.

The principle is used by a figure skater executing a spin (Figure 3-7). If she starts her spin slowly with outstretched arms, she can increase her rate of rotation suddenly by folding her arms across her chest. She can slow down by extending her arms again.

3.7 ENERGY

Energy is another quantity that is conserved for a group of objects subject to no external influence. It is more complicated because there are several different kinds of energy. As the motion of a group of objects proceeds, one kind of energy may change into another kind. Thus, the amount of one kind of energy does not necessarily remain constant. It is the sum of all kinds of energy that is constant. We shall start by discussing *kinetic energy* and *gravitational potential energy*. Other kinds of energy will be introduced later when we deal with the relevant physical principles.

3.8 KINETIC ENERGY

Kinetic energy is energy of motion (Figure 3-8a). Its numerical value for a single object is equal to one half of the mass multiplied by the square of the speed.

$$\text{Kinetic energy} = \frac{1}{2} \times (\text{mass}) \times (\text{speed squared}) \quad (3\text{-}10)$$
$$= \frac{1}{2}mv^2 \quad (3\text{-}11)$$

FIGURE 3-7
A figure skater increasing her rate of rotation.

Energy is always a scalar quantity. No direction is associated with it. Kinetic energy is always positive.

Energy is important enough to merit a special name for the unit of energy. In SI, when the mass is measured in kilograms and the speed in meters per second, the kinetic energy is measured in joules. The SI unit of energy is the *joule*. Its abbreviation is J.

Thus, a mass of 4 kg moving with a speed of 3 m/s has a kinetic energy of

$$\frac{1}{2}mv^2 = \frac{1}{2} \times 4 \times (3)^2 \qquad (3\text{-}12)$$

$$= \frac{1}{2} \times 4 \times 9 \qquad (3\text{-}13)$$

$$= 18 \text{ joules or } 18 \text{ J} \qquad (3\text{-}14)$$

3.9 GRAVITATIONAL POTENTIAL ENERGY

Gravitational potential energy is a mutual property of a pair of objects resulting from the fact that they exert gravitational forces on each other (Figure 3–8b). Suppose that they have masses of m_1 kilograms and m_2 kilograms and the distance between them is R meters. Then their mutual gravitational potential energy in joules is given by the following equation.

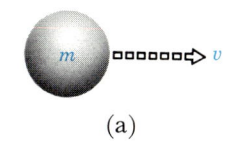

$$\text{Gravitational potential energy} = -\frac{Gm_1m_2}{R} \qquad (3\text{-}15)$$

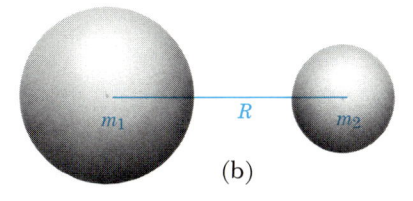

G is the gravitational constant that appears in Newton's law of universal gravitation (Section 2.8). The equation for the gravitational potential energy is very similar to Equation 2-7 for the gravitational force, but there are some important differences. Energy is a scalar quantity, whereas force is a vector. In the equation for the potential energy, the distance R in the denominator is *not* squared. Finally, and this is very important, the gravitational potential energy is negative.

When the two objects are a long way apart, the gravitational potential energy is very small. The R in the denominator is very large and so the fraction is very small. If the two objects were infinitely far apart, the potential energy would be zero. As the two objects approach closer together, R becomes smaller and the fraction becomes larger. That means that the gravitational potential energy becomes more and more negative.

Kinetic energy = $\frac{1}{2}mv^2$

(a)

Gravitational potential energy = $-\frac{Gm_1m_2}{R}$

(b)

FIGURE 3–8
Kinetic energy and potential energy. (a) Kinetic energy is energy of motion. (b) Gravitational potential energy is a property of a pair of objects exerting gravitational forces on each other.

3.10 CONSERVATION OF ENERGY

Imagine a group of objects exerting gravitational forces on one another, but free of interference from the rest of the universe. It might, for example, be an isolated cluster of stars, or it might be the sun, planets, and moons of our solar system. Each object is accelerated by the forces exerted on it by the other objects. It continually changes its velocity and may move along a complicated curved path. In principle, the laws of motion discussed in the previous chapter enable the motion to be calculated in precise detail. In practice, the mathematical calculations can be horrendous.

Nevertheless, in spite of the messy details, it is always possible to proceed in the following way. For each object write down an equation saying that its mass multiplied by its acceleration is equal to the vector sum of all the forces on it. The forces should be expressed in the form prescribed by Newton's law of universal gravitation (Equation 2-7).

The equations obtained in this way can always be manipulated mathematically and combined together to give a single equation. This equation says that the sum of several algebraic terms is equal to a constant number that does not change as the motion progresses. For each object, a term of the form $\frac{1}{2}mv^2$ represents its

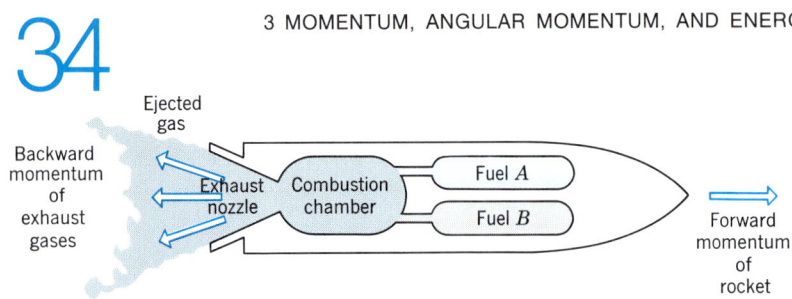

FIGURE 3-9
Rocket propulsion.

kinetic energy. For each pair of objects, a term of the form $-Gm_1m_2/R$ represents their mutual gravitational potential energy.

That is the reason why kinetic energy and potential energy are defined in the way they are. In the equation we are discussing, they turn up as terms similar in form to Equations 3-11 and 3-15. This equation is the mathematical expression of the *law of conservation of energy*.

We can therefore proceed as follows. At a particular instant of time we calculate the kinetic energy for each object in the group and add together all the resulting positive numbers to obtain the kinetic energy of the whole group. We then calculate the gravitational potential energy of each pair of objects in the group and add together all the resulting negative numbers to obtain the negative gravitational potential energy of the whole group. The sum of the kinetic energy and the potential energy is the *total energy* of the group at this instant of time. (Of course, adding the negative potential energy amounts to subtracting a positive number).

The law of conservation of energy tells us that we would always obtain the same numerical value of the total energy if we repeated this procedure at any other past or future instant of time. The individual kinetic energies would be different. The distribution of energy between kinetic energy and potential energy would be different. But the total energy would remain constant.

In the final analysis, the law of conservation of energy always takes the form of such an equation. In more complicated situations other kinds of terms may be present in the equation. They might represent heat energy, chemical energy, electrical energy, and so on. These other types of energy will be explained in due course.

3.11
CONSERVATION OF MOMENTUM FOR A ROCKET

Launching a rocket provides a good example of the application of the laws of conservation of momentum and energy. Figure 3-9 illustrates the basic principles underlying the operation of a rocket. The two fuels are mixed in the combustion chamber, where they undergo a violent chemical reaction. The products of this reaction are hot, high-pressure gases that escape through the exhaust nozzle with a high velocity and thereby acquire a large backward momentum. The total momentum of the system must remain constant. The backward momentum acquired by the ejected gases must be compensated by an equal and opposite forward momentum given to the rocket. The forward velocity of the rocket therefore continually increases as long as the fuel is being burned, but not, of course, after the fuel has been used up.

Consider a group of objects consisting of the rocket, the fuel, and the earth. The avantage of choosing this particular group is that the gravitational forces between the rocket and the earth are internal to the group. The external gravitational forces exerted by the other heavenly bodies are too weak to matter. It is therefore permissible to apply the laws of conservation of momentum and energy to this group.

Assume, as is usually the case, that the fuel is all spent before the rocket leaves the earth's

A rocket launch. The upward momentum given to the rocket is compensated by the downward momentum of the ejected gas. (NASA.)

atmosphere. During the firing the rocket acquires an upward momentum and the ejected gases a downward momentum. However, the ejected gases mingle with the earth's atmosphere and eventually transfer their momentum to the earth. Therefore, when the firing is complete and the rocket has escaped from the earth's atmosphere, the earth has recoiled in the opposite direction to the motion of the rocket. The earth then has a momentum exactly equal and opposite to the momentum of the rocket. The total momentum of the group was zero before the launching and remains zero after the launching.

During the subsequent flight of the rocket, the total vector momentum remains zero. The earth always has a momentum equal and opposite to that of the rocket. The motion of the earth is a miniature version in reverse of the trajectory of the rocket. However, the mass of the earth is so very much greater than the mass of the rocket that the velocity of the earth is very much smaller than the velocity of the rocket. If a rocket weighing a hundred tons were launched to a distance of one million miles from the earth, the earth would be made to move through a distance of only one billionth of an inch! That is less than the size of an atom.

In particular, the recoil velocity of the earth is so very small that is can be shown that the kinetic energy imparted to the earth is negligibly small compared with the kinetic energy of the rocket. In the following discussion of the application of the law of conservation of energy, the kinetic energy of the earth will be ignored.

3.12
CONSERVATION OF ENERGY FOR A ROCKET (MATHEMATICAL)

Consider the situation at the end of the launching, just after the fuel is spent (Figure 3–10a). A reasonable value for the mass m of the rocket is 40 thousand kilograms (4×10^4 kg or about 44 tons). A possible value of the speed v_1 of the rocket is ten thousand meters per second (10^4 m/s or about 22,000 mph).

Initial kinetic energy of the rocket

$$= \frac{1}{2} m v_1^2 \qquad (3\text{-}16)$$

$$= \frac{1}{2} \times (4 \times 10^4) \times (10^4)^2 \qquad (3\text{-}17)$$

$$= 2 \times 10^4 \times 10^8 \qquad (3\text{-}18)$$

$$= 2 \times 10^{12} \text{ joules} \qquad (3\text{-}19)$$

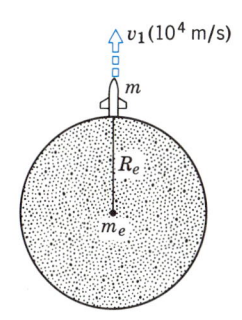

Kinetic energy	= +	2.0	trillion	joules
Potential energy	= −	2.5	trillion	joules
Total energy	= −	0.5	trillion	joules

(a) The initial launch, just after the fuel is spent.

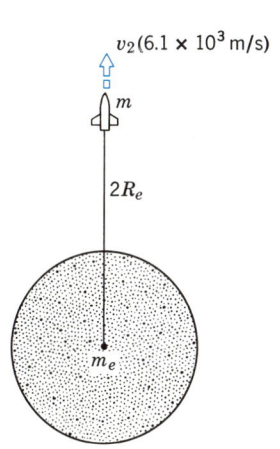

Kinetic energy	= +	0.75	trillion	joules
Potential energy	= −	1.25	trillion	joules
Total energy	= −	0.50	trillion	joules

(b) The rocket has doubled its distance from the center of the earth.

FIGURE 3–10
Conservation of energy for a rocket.

If you think about the meaning of powers of ten, that means two trillion joules. The kinetic energy of the earth can be ignored, and so we have just calculated the kinetic energy of the whole group containing the earth and the rocket.

Initial kinetic energy
$$= 2 \text{ trillion joules} \quad (3\text{-}20)$$

One of the tricks in the art of rocketry is to burn all the fuel as quickly as possible before the rocket has traveled very far. Therefore, at the instant the fuel is spent, the distance of the rocket from the center of the earth is not appreciably greater than the radius of the earth. The radius of the earth is represented by the symbol R_e, and its numerical value is 6.37×10^6 meters. To calculate the mutual gravitational potential energy of the rocket and the earth, we use Equation 3-15 of Section 3.9 and insert this value for R. For m_1 we insert the mass of the earth, $m_e = 5.98 \times 10^{24}$ kg. For m_2 we insert the mass of the rocket, $m = 4 \times 10^4$ kg. The value of the gravitational constant G is 6.67×10^{-11} SI units.

Initial potential energy
$$= -\frac{G m_e m}{R_e} \quad (3\text{-}21)$$

$$= -\frac{(6.67 \times 10^{-11}) \times (5.98 \times 10^{24}) \times (4 \times 10^4)}{6.37 \times 10^6} \quad (3\text{-}22)$$

$$= -2.5 \times 10^{12} \text{ joules} \quad (3\text{-}23)$$
$$= -2.5 \text{ trillion joules} \quad (3\text{-}24)$$

The total energy just after launching is the sum of the kinetic and potential energies.

Initial total energy
= initial kinetic energy
$\quad\quad$ + initial potential energy $\quad (3\text{-}25)$
$= (+2.0 - 2.5)$ trillion joules $\quad (3\text{-}26)$
$= -0.5$ trillion joules $\quad (3\text{-}27)$

Notice that the total energy is negative. The negative potential energy dominates this particular situation, but that is not always the case.

Assume that the rocket is fired vertically upward and travels radially outward away from the earth. In Figure 3-10b it has reached a distance from the center of the earth equal to twice the radius of the earth. The denominator R in the equation for the potential energy was initially R_e and has now been doubled to $2R_e$. The numerical value of the potential energy has therefore been halved.

New potential energy
$$= -\frac{1}{2} \times 2.5 \text{ trillion joules} \quad (3\text{-}28)$$
$$= -1.25 \text{ trillion joules} \quad (3\text{-}29)$$

The law of conservation of energy requires that the total energy still has the same value.

$$\text{Total energy} = -0.5 \text{ trillion joules always} \quad (3\text{-}30)$$

The energy accounting therefore tells us that

The new kinetic energy
$$= +0.75 \text{ trillion joules} \quad (3\text{-}31)$$

The sum of the kinetic and potential energies is then $+0.75 - 1.25 = -0.5$ trillion joules, as it should be.

If we are interested in the new speed v_2, we can place

$$\frac{1}{2} m v_2^2 = 0.75 \times 10^{12} \text{ J} \quad (3\text{-}32)$$

A little arithmetic then yields

$$v_2 = 6.1 \times 10^3 \text{ m/s} \quad (3\text{-}33)$$

3.13
CONSERVATION OF ENERGY FOR A ROCKET (DESCRIPTIVE)

Now that we have coped with the mathematics, let us put all of this in words in order to try to understand what is happening. As the rocket moves away from the earth, the gravitational attraction of the earth tries to pull it back again and steadily slows it down. As it increases its distance from the earth, its speed decreases. How can this fact be interpreted in terms of energy?

In our particular example, when the rocket doubles its distance from the center of the earth, its gravitational potential energy changes from -2.5 to -1.25 trillion joules. That is equivalent to an increase in potential energy of $+1.25$ trillion joules. We must add $+1.25$ to -2.5 to change it into -1.25.

However, the total energy must remain constant. The gain in potential energy must be offset by an equal loss of kinetic energy. The kinetic energy therefore decreases by 1.25 trillion joules and changes from $+2.0$ to $+0.75$ trillion joules. A decrease in kinetic energy implies a decrease in speed.

In general, as the rocket moves away from the earth, it gains potential energy at the expense of its kinetic energy and its speed therefore decreases.

3.14
THE ESCAPE VELOCITY OF A ROCKET

Something interesting happens when the rocket reaches a distance $5R_e$ from the center of the earth. The potential energy is then $\frac{1}{5}$ of its initial value or $-2.5/5 = -0.5$ trillion joules. But that is the value of the total energy. It follows that the kinetic energy is zero and the speed is zero.

The gain in potential energy has used up all of the kinetic energy and brought the rocket to rest. In different words, the backward pull of the earth has slowed the rocket to a stop. The rocket has reached its farthest distance from the earth and is about to fall downward again—back toward the earth.

How fast must we launch the rocket so that it escapes completely from the earth's gravitational pull and never turns back? Its distance R from the center of the earth would then increase to very large values, and the gravitational potential energy would become almost zero. In climbing from the surface of the earth to these very great distances, the rocket would gain 0.5 trillion joules of potential energy. This can be obtained only at the expense of the kinetic energy. The initial kinetic energy must therefore be at least $+0.5$ trillion joules. The speed of launching must give the rocket an initial kinetic energy large enough to overcome the initial negative potential energy. Otherwise the rocket would never reach infinity with any spare energy left over to keep going.

For a rocket of mass m, this condition can be stated mathematically in the form that $\frac{1}{2}mv^2$ must be at least as big as Gm_em/R_e. The smallest permissible velocity satisfies the equation

$$\frac{1}{2}mv^2 = \frac{Gm_em}{R_e} \qquad (3\text{-}34)$$

The mass m can be divided out of both sides and the equation rearranged to give

$$v = \sqrt{\frac{2Gm_e}{R_e}} \qquad (3\text{-}35)$$

This is called "the *escape velocity* from the earth," and its value is 1.11×10^4 m/s or 24,800 mph. It does not depend on the mass of the rocket, but it does depend on the mass and radius of the earth. The escape velocity from the moon or another planet has a different value.

3.15
CONSERVATION OF ENERGY DURING THE FIRING OF THE ROCKET

We can apply the law of conservation of momentum to the firing while the fuel is being burned, but we must be careful about applying the law of conservation of energy. Just before ignition, there is no kinetic energy—only potential energy. Immediately after the fuel is spent, there is a large kinetic energy, but the rocket has not moved very far and so its potential energy is essentially unchanged. The sum of kinetic energy and gravitational potential energy is clearly not constant.

Where does the kinetic energy come from? It comes from the *chemical energy* stored in the fuel and released by the burning. We are dealing with a situation involving a form of energy other than kinetic energy or gravitational potential energy.

Later in this book we shall arrive at a better understanding of chemical energy—energy residing in atoms and molecules. Analyzed still more deeply, it is mainly the kinetic energy of electrons and nuclei plus a new form of potential energy we have not yet encountered. This is the electrical potential energy of the electrically charged electrons and protons in the atoms and molecules. We shall discuss electrical potential energy in Chapter 5.

3.16
WORK

The law of conservation of energy applies to a group of objects free from outside interference. We frequently deal with a group that is interacting strongly with its surroundings. The total energy of the group does not then remain constant. Nevertheless, the concept of energy is still useful. It can be proved that the amount by which the energy changes is equal to the *work* done on the group by the external forces. An external force is a force exerted on an object in the group by an object outside the group.

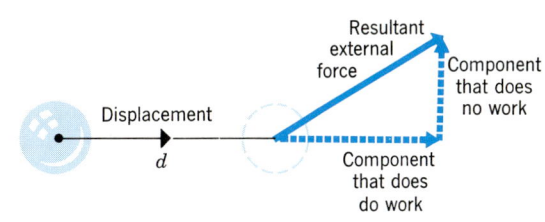

(a) A force doing work on an object.

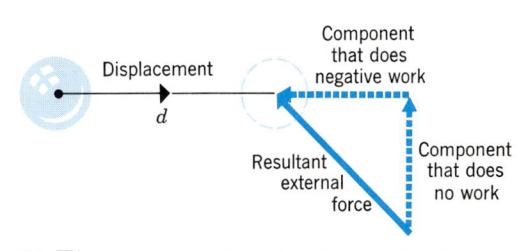

(b) The component of the force is in the opposite direction to the displacement. The work done by the force is negative.

FIGURE 3-11
Work.

The internal forces exerted by members of the group on one another are to be left out of the reckoning.

Figure 3–11a shows one of the members of the group and the resultant external force on it. In the course of its motion it has moved through the distance labeled d. The force should be replaced by two components at right angles as shown. (Refer back to Section 1.12 if you do not remember exactly what is meant by the component of a vector.) Only the component in a direction parallel to the displacement does work. The component at right angles to the displacement does no work.

The work done by the force during the displacement is equal to the product of its component parallel to the displacement and the distance d through which the object has moved.

> Work = (component of force parallel to displacement) × (distance moved)   (3-36)

Work, like energy, is a scalar quantity. Work produces energy and has the same units as energy. If the force is in newtons and the distance in meters, the work is in joules. If a force of 15 N moves an object through a distance of 6 m in a direction parallel to the force, the work done is 15 × 6 or 90 J.

Sometimes the component of the force is in the opposite direction to the displacement, as in Figure 3–11b. The work done on the object by the external force is then negative. Alternatively, we can say that the object has done positive work on its surroundings.

If we add together the work done on all the members of the group by all the external forces, the sum is the total work done on the group. It is equal to the increase in the total energy of the group. In the kind of situation we have been discussing so far, the total energy of the group is the sum of all the kinetic energies and all the potential energies. Other forms of energy may be involved in more complicated situations.

> The work done on all the members of a group by all the external forces is equal to the increase in the total energy of the group.

If the total work should turn out to be negative, that merely means that the total energy of the group decreases.

3.17
POWER

Power is the rate of doing work in joules per second. The SI unit of power is the *watt*. Its abbreviation is W.

$$1 \text{ watt} = 1 \text{ joule per second} \quad (3\text{-}37)$$
or
$$1 \text{ W} = 1 \text{ J/s} \quad (3\text{-}38)$$

> Power in watts
> $$= \frac{\text{work done in joules}}{\text{time taken in seconds}} \quad (3\text{-}39)$$

3.18
LIFTING A WEIGHT

In Figure 3–12a, a weight lifter is about to lift an object with a mass m and a weight W. We are going to consider a group containing the object and the earth, but excluding the weight lifter. The weight of the object is a downward gravitational pull exerted on it by the earth. It is an internal force and should not be included when we calculate the work done on the earth and the weight.

The relevant external forces are exerted by the weight lifter who is not a member of the group. To lift the weight, he must apply to it an upward force greater than W. If he is in no hurry and lifts the weight very slowly, the upward applied force exceeds W by a negligible amount. So we will put it equal to W.

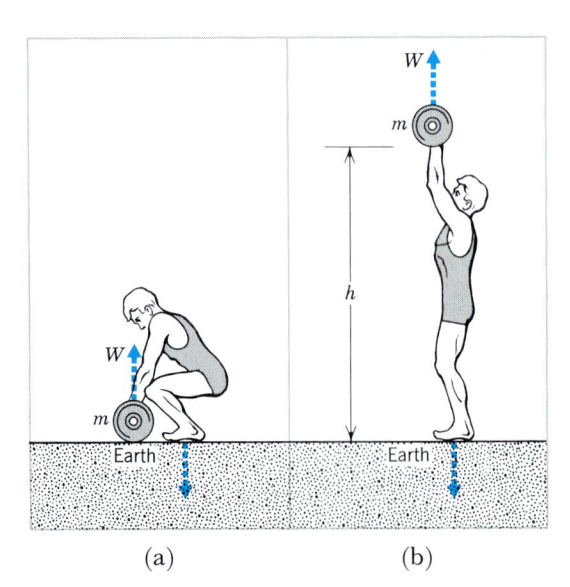

FIGURE 3–12

Lifting a weight. (a) A weight lifter about to lift a weight. Energy is stored in his muscles. (b) The gravitational potential energy has been increased by $Wh = mgh$.

As he braces himself to lift the weight, his legs exert a downward force on the ground. Clearly, he is not going to succeed in moving the very massive earth more than a very minute distance. The work done on the earth by this downward force may therefore be ignored.

The only external force that does work is the upward force W he exerts on the weight. If the weight is lifted through a height h, as in Figure 3-12b, the displacement is in the same direction as the force. The work done is the force multiplied by the distance, or Wh.

The total energy of the group containing the earth and the weight increases by an amount Wh. While the weight lifter holds the weight steady at a height h, it has no kinetic energy. The only part of the energy that has changed is the mutual gravitational potential energy of the weight and the earth. We have therefore shown that when a weight W is raised through a height h, its gravitational potential energy increases by Wh.

In Section 2.10 we pointed out that a mass m has a weight $W = mg$. The increase in gravitational potential energy may also be expressed in the form mgh.

If a mass m with a weight W is raised through a height h near the surface of the earth, its gravitational potential energy increases by Wh, which is the same as mgh.

Increase in gravitational potential energy of a raised weight

$$= \text{weight} \times \text{height} \qquad (3\text{-}40)$$
$$= (\text{mass}) \times (\text{acceleration due to gravity})$$
$$\times (\text{height}) \qquad (3\text{-}41)$$

A mass m at the surface of the earth, at a distance R_e from the center of the earth, has a gravitational potential energy of $-Gm_e m/R_e$. How much does this change if we raise the mass through a height h and increase its distance from the center of the earth to $R_e + h$? The mathematics is complicated, but the answer is mgh, in agreement with the result just obtained using the concept of work.

What was the source of the energy mgh given to the lifted weight? It was originally chemical energy stored in the food eaten by the weight lifter. The chemical reactions taking place as he digested this food released the energy. A series of very complicated physiological and biochemical processes then stored the energy in his muscles. The final act of lifting transfered the energy from his muscles to the weight.

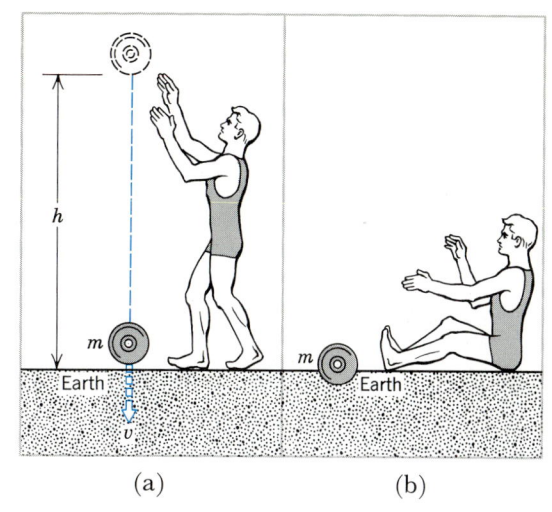

FIGURE 3-13

Dropping a weight. (a) The gravitational potential energy mgh has been converted into kinetic energy $\frac{1}{2}mv^2$. (b) The weight is embedded in the soil and the kinetic energy has been converted into heat.

3.19
DROPPING A WEIGHT

In Figure 3-13a, the weight lifter has released the weight and it is just about to strike the ground with a speed v. The weight lifter is no longer interfering, and so we can apply the law of conservation of energy to the group containing the weight and the earth. As the weight falls through the height h, it loses the extra potential energy mgh that the weight lifter had given to it. In its place it gains a kinetic energy of $\frac{1}{2}mv^2$. The gain must equal the loss and so

$$\frac{1}{2}mv^2 = mgh \qquad (3\text{-}42)$$

Divide both sides by m and multiply both sides by 2.

$$v^2 = 2gh \qquad (3\text{-}43)$$
or
$$v = \sqrt{2gh} \qquad (3\text{-}44)$$

This is a very simple way of obtaining the speed of a body after it has fallen from rest through a height h.

In Figure 3-13b, the weight has struck the ground and embedded itself in the soft soil. It has now lost both its potential energy and its kinetic energy. Where has the energy gone?

During the collision with the ground the kinetic energy of the weight is converted into *heat energy* which raises the temperature of the weight and the nearby soil. Here is yet another kind of energy. We discuss heat energy in the next chapter. Part of the heat energy is simply the kinetic energy of individual atoms. The atoms in the weight and the soil are made to jiggle around a little faster.

3 MOMENTUM, ANGULAR MOMENTUM, AND ENERGY

Four types of energy.

(a) The racing car has kinetic energy because it is in motion. (Gerry Cranham/Rapho-Photo Researches.)

(b) This car has gravitational potential energy because it is at a height above the ground. (Courtesy Volkswagen of America.)

(c) The gasoline in the can has chemical energy, some of which will eventually be converted into kinetic energy of motion of the car. (Everett C. Johnson/Leo de Wys.)

(d) Some of the chemical energy of the gasoline will reappear as heat in the engine, the radiator, and the tires. This is an infrared photograph (thermogram) of a car just after it was parked. The hot engine, radiator, and tires emit copious radiation and appear white on the thermogram. (Courtesy *Vanscan*, Continuous Mobile Thermography by Daedalus Enterprises, Inc.)

SUMMARY

Momentum of an object is the product of its mass and its vector velocity. Momentum is a vector. The resultant force on the object is equal to the rate of change of its momentum with time.

Law of conservation of momentum. If the total resultant force exerted on a group of objects by the rest of the universe is zero, then the vector sum of the momenta of all the objects in the group is a constant vector that does not change with time.

Angular momentum of a particle moving in a circle is equal to its mass multiplied by its speed multiplied by the radius of the circle. The angular momentum of an extended object rotating about an axis is the sum of the angular momenta of the small particles of which it is composed.

Torque. A couple consists of two parallel forces of equal magnitude in opposite directions but displaced sideways from each other. The torque of the couple is equal to the magnitude of either force multiplied by the perpendicular distance between the two forces. The resultant external couple on a group of objects, or an extended object composed of many particles, is equal to the rate of change of its angular momentum.

Law of conservation of angular momentum. If there is no net external torque on a group of objects or an extended object, its total angular momentum remains constant.

Energy is a scalar quantity that is conserved for a group of objects subject to no external interference. It is the sum of several different kinds of energy such as kinetic energy, gravitational potential energy, heat energy, chemical energy, and various other kinds to be discussed later in this book. The SI unit of energy is the joule, abbreviated J.

Law of conservation of energy. For a group of objects free from external interference, the total energy remains constant. The total energy is the sum of the various kinds of energy. One kind of energy may be converted into an equal amount of another kind, but the total sum of all kinds must remain constant.

Kinetic energy is energy associated with the motion of an object. It is equal to one half of the mass multiplied by the square of the speed ($\frac{1}{2}mv^2$).

Gravitational potential energy is a property of a pair of objects exerting gravitational forces on one another. It is negative and is given by the formula $-Gm_1m_2/R$.

Work is done when an external force acts on an object displaced through a certain distance. The work done is equal to the component of the force in the direction of the displacement multiplied by the distance. If the component of the force is in the opposite direction to the displacement, the work done by the force on the object is negative. The work done on all the members of a group of objects by all the external forces is equal to the increase in the total energy of the group. Work, like energy, is measured in units of joules.

Gravitational potential energy near the surface of the earth. If a mass m with a weight W is raised through a height h near the surface of the earth, its gravitational potential energy increases by $Wh = mgh$.

TERMS AND CONCEPTS

momentum
law of conservation of momentum
angular momentum
couple
torque
law of conservation of angular momentum
energy
kinetic energy
joule
gravitational potential energy
law of conservation of energy
total energy
escape velocity
chemical energy
work
power
watt
heat energy

QUESTIONS

1. Explain each of the following: (a) the law of conservation of momentum, (b) the law of conservation of angular momentum, (c) the law of conservation of energy.

2. Which of the following is a vector and which is a scalar? (a) resultant force, (b) momentum, (c) kinetic energy, (d) work, (e) chemical energy, (f) power.

3. How does the momentum of a satellite in a circular orbit vary with time?

4. A boy sleds down a slope and ends up in a snowbank. Where did his kinetic energy come from and where does it go to?

5. How does the law of conservation of momentum apply to a car accelerating along a straight horizontal highway?

6. If the car is traveling at a steady speed along the straight horizontal highway, what happens to the energy released from the gasoline used up?

7. The total energy of a group of objects was said to be constant if there were no external interference. Can you now be more precise about exactly what "no external interference" means? Hint: If the total energy of the group did change, what would the numerical value of the change be equal to?

8. Why is it difficult to design a helicopter with only one propeller?

PROBLEMS

1. A mass of 3 kg has a speed of 4 m/s. What is the magnitude of (a) its momentum? (b) its kinetic energy?

2. If the momentum of a moving object is doubled by changing its speed, by what factor is its kinetic energy multiplied?

3. If a mass of 7 kg has a kinetic energy of 56 J, what is its speed?

4. If a mass of 7 kg has a momentum of 56 SI units of momentum (kg m/s), what is its speed?

5. What is the angular momentum of a mass of 7×10^{-2} kg moving in a circle of radius 5 m with a speed of 2×10^2 m/s?

6. Two lumps of clay of equal mass are traveling in exactly opposite directions with the same speed. The mass of either is 0.2 kg. The speed of either is 10 m/s. They collide and stick together. (a) What is the speed of the single lump after the collision? (b) How much heat is developed by the collision?

7. The mass of the earth is 6×10^{24} kg. The radius of the earth is 6.4×10^6 m. What is the gravitational potential energy of 1 kg at the surface of the earth?

8. The mass of the earth is 81.5 times the mass of the moon. The radius of the earth is 3.7 times the radius of the moon. Is the escape velocity from the moon greater or less than the escape velocity from the earth?

9. A force of 20 N moves a body of mass 2×10^5 kg through a distance of 0.5 m in a direction parallel to the force. How much work is done by the force?

10. A satellite with a mass of 200 kg moves around the earth in a circular orbit of radius 7×10^6 m. How much work is done during each revolution by the gravitational force attracting the satellite toward the center of the earth?

11. A 0.2-kg book is carried from the street to a second floor room 4.5 m higher. What is the change in its gravitational potential energy?

12. A conveyer belt lifts bricks through a height of 16 m. Each brick weighs 5 N and the belt can handle 75 bricks per minute. What is the useful power developed by the machine driving the belt?

13. From what height must an object be dropped in order to build up a speed of 55 mph as it strikes the ground (55 mph = 24.6 m/s)? About how many stories would there be in a building this high?

CHAPTER 4
ATOMS AND HEAT

4.1 ATOMS

The laws of motion were discovered in the seventeenth century. The eighteenth century was devoted mainly to developing the mathematical techniques that enabled these laws to be applied to various problems. This was the age of the great mathematical physicists d'Alembert, Bernoulli, Euler, Lagrange, Laplace, and many others. They were interested in such problems as the detailed behavior of the solar system, the motion of a spinning top, and the flow of liquids. The mathematics was very difficult but the success of their investigations produced overwhelming confidence in the validity of Newton's laws.

However, all these successes were concerned with the behavior of macroscopic objects, and many important questions were left unanswered. The laws of motion predicted that a lead ball would fall to the earth at the same rate as an iron ball, but the difference between lead and iron was not understood. It was not known why one cubic centimeter of lead had a mass of 11.37 grams, whereas one cubic centimeter of iron had a mass of 7.87 grams. The first major step toward providing answers to questions of this kind was the emergence of the *atomic theory of matter* at the beginning of the nineteenth century.

The idea that matter was composed of atoms was proposed by the Greek philosophers Empedocles and Democritus between 500 and 400 B.C. But establishing a scientific theory involves more than a bright idea. It is necessary to show that this idea is useful in explaining known phenomena and in predicting new phenomena.

In the case of atoms, the initial evidence came mainly from chemistry. This evidence was first presented by Dalton at the beginning of the nineteenth century. The situation was finally clarified by Cannizzaro in the middle of the century. Once the chemical evidence had convinced scientists of the existence of atoms, atomic theory was found to provide a very successful explanation of the properties of gases and later of solids and liquids.

Chemical evidence will be developed in Chapter 12. Here, we show how the concept of atoms is very helpful in explaining the physical properties of gases.

4.2 LOOKING AT ATOMS

Atoms are very small. The diameter of an atom is only about 10^{-10} meter. If atoms were placed side by side along a line, it would take about ten billion of them to form a line only one meter long.

An atom is in fact too small to be seen directly, even with a microscope. A fundamental rule of microscopy is that two objects cannot be distinguished from each other if their distance apart is much less than the wavelength λ of the light illuminating them. (The wavelength was defined in Section 1.1; see also Figure 1–1.) An object smaller than λ is seen as a blur of light with a diameter of about λ. This is a consequence of the diffraction phenomenon, which will be discussed in Chapter 7. The smallest wavelength of visible light is several thousand times greater than the diameter of an atom. We

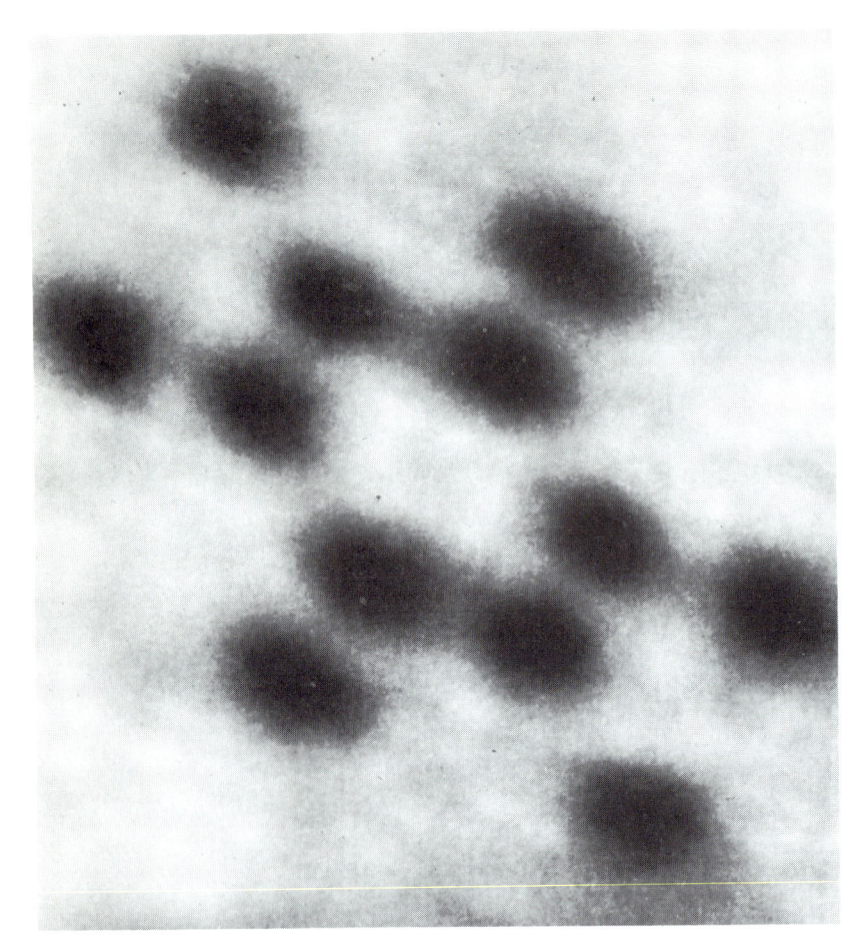

FIGURE 4-1

A "photograph" of a molecule containing 12 carbon atoms and 18 hydrogen atoms. The dark blobs are the 12 carbon atoms. The hydrogen atoms do not show up clearly. This is not a photograph of an individual molecule. It was obtained by gathering x-rays from a large number of molecules and converting the x-ray signal into a visible signal. (From Jay Orear, *Fundamental Physics*, John Wiley and Sons, Inc. The photograph was obtained by Dr. M. L. Huggins of the Kodak Research Laboratory.)

obviously cannot hope to use visible light to look at an atom and examine its structure.

Nevertheless, special techniques come very close to providing pictures of individual atoms. X-rays are similar to light, but they have much shorter wavelengths that can be much smaller than the diameter of an atom. It is therefore possible to design an x-ray "microscope" that will reveal the structure of the atom. It is still not possible to see an individual atom. The x-rays must be scattered by an orderly array of atoms in a solid. It is then possible to deduce from the detailed nature of the scattering the spatial arrangement of the atoms in the solid. Figure 4-1 is a photograph of a complicated organic molecule reconstructed from such an x-ray experiment.

One of the most impressive ways of "seeing" atoms is the *field ion microscope* invented by E. W. Müller. The explanation of how it works is complicated, but Figure 4-2 is easily appreciated. It is a "photograph" of the tip of a tungsten needle, and each bright dot is caused by an atom, or a group of two or three atoms, on the tip.

To illustrate the vast number of atoms in even a small particle of matter, let us suppose that a microscope had been invented that could distinguish the individual atoms. Suppose that you attempted to use this microscope to count the atoms in a raindrop. The psychological limit on the rate of counting could hardly be in excess of five per second. At this rate, it would take about 300 billion years to count all the atoms! Many astronomers believe that the universe is only about 20 billion years old. Therefore, if you had started to count these atoms the very instant the universe came into being, you would still be far from having completed the task!

4.3
THE IDEAL MONATOMIC GAS

A gas can be visualized as a collection of small, hard particles in rapid motion. For many gases the particle is a compound structure, or molecule, made of two or more atoms. If, instead, each particle is a single atom, traveling all alone, the gas is said to be *monatomic*. Examples of monatomic gases are helium, neon, argon, and the vapors of mercury, sodium, and potassium. We shall consider only monatomic gases, thereby

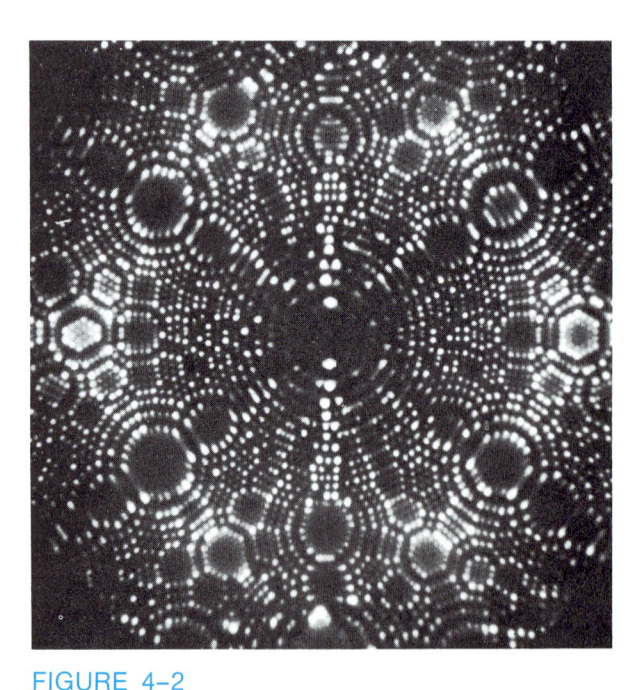

FIGURE 4-2

Atoms on the tip of a tungsten needle in a field ion microscope (courtesy of Dr. Erwin Müller).

avoiding some irrelevant complications that occur when the molecule has a more elaborate structure.

A gas is said to be *ideal* if there are no forces between its particles. Actually, all atoms and molecules exert forces on one another, but these forces diminish very rapidly as the distance between the particles increases. If the gas is expanded into a very large volume, its density becomes very small. The particles are then very far apart. The forces between the particles are negligibly small. An *ideal gas* is therefore the extreme gas of an actual gas with a very large volume and a very small density.

An *ideal monatomic gas* should therefore be visualized as a collection of small, hard particles, each of which is an individual atom. These atoms are separated by distances much larger than their diameters. They exert negligibly small forces on one another.

The atoms of an ideal monatomic gas are in rapid motion. They travel in straight lines until they rebound from the wall of the container—or very occasionally collide with one another (Figure 4–3). The distribution of the atoms in space is perfectly random. Their velocities point in all possible directions.

The atoms do not all have the same speed. They have all possible speeds from zero upward. However, there is an average speed, which we shall represent by the symbol v_a. Most of the atoms have speeds that are not very different from this average speed. Very few atoms have much smaller or much larger speeds. The average speed is related to the temperature of the gas. As the temperature is raised, the average speed increases.

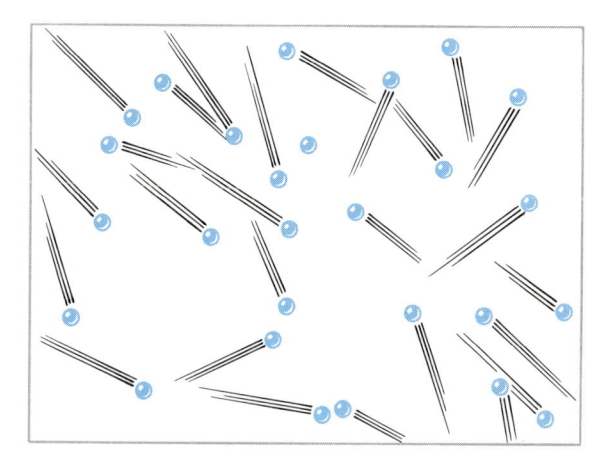

4.4 THE INTERNAL ENERGY OF AN IDEAL MONATOMIC GAS

In Section 3.9, we saw that the mutual gravitational potential energy of two objects is very small when they are very far apart. Interatomic forces are not gravitational. Nevertheless, it is still true that the mutual potential energy of two atoms is very small when they are far apart. In an ideal gas the atoms do stay far from one another most of the time. The potential energy associated with the interatomic forces is then negligibly small compared with the other forms of energy present.

If the ideal gas is monatomic, the only other important form of energy is the kinetic energy of the random motion of the individual atoms. We call this the *internal energy* of the ideal monatomic gas. It is a form of energy present within the gas that is not immediately obvious until we realize that the gas consists of atoms in random motion.

We shall now use a slightly different example to illustrate that internal energy can be very large. Take an object such as a baseball with a mass M and throw it with a speed V. We are normally content to say that its kinetic energy is $\frac{1}{2}MV^2$. However, if the baseball is visualized as a collection of atoms in rapid motion in all directions, the total kinetic energy of all these atoms is greatly in excess of $\frac{1}{2}MV^2$. In fact, the speed V is unlikely to be much greater than 50 meters per second. The speeds of the individual atoms are about 5,000 meters per second (10,000 miles per hour!), which is about 100 times larger than V. Because the kinetic energy depends on the square of the speed, the "hidden" internal kinetic energy is about 10,000 times as great as the "obvious" kinetic energy $\frac{1}{2}MV^2$.

There are some complications involved in the internal energy of a solid such as a baseball, so let us return to the ideal monatomic gas. Let m be the mass of each atom. The average speed of an atom is v_a. The average kinetic energy of an atom is $\frac{1}{2}mv_a^2$. The total internal energy is the average kinetic energy per atom multiplied by the number of atoms. We shall use the symbol N for the number of atoms.

FIGURE 4–3
An ideal monatomic gas. The tails are intended to create an impression of the motion of the atoms and do not imply that an atom leaves any sort of wake behind it.

Internal energy of an ideal monatomic gas
$\quad$ = (number of atoms) × (average kinetic energy per atom) $\quad$ (4-1)
$\quad$ = $N \dfrac{1}{2} m v_a^2$ $\quad\quad\quad\quad\quad\quad\quad\quad$ (4-2)

4.5 TEMPERATURE

The average kinetic energy of an atom of an ideal monatomic gas is a direct measure of the temperature of the gas. In order to relate that idea to methods of measuring temperature that have developed somewhat haphazardly over the centuries, it is necessary to introduce a new fundamental constant of physics called *Boltzmann's constant* and represented by the symbol k.

Boltzmann's constant, k
$$= 1.38 \times 10^{-23} \text{ SI units}$$
$$\text{(joule/kelvin degree or J/K)} \quad (4\text{-}3)$$

The *absolute temperature* T is then defined as the average kinetic energy per atom divided by $\frac{3}{2}$ times this constant.

Absolute temperature
$$= \frac{\text{average kinetic energy per atom}}{\frac{3}{2} \times \text{Boltzmann's constant}} \quad (4\text{-}4)$$

Average kinetic energy per atom
$$= \frac{3}{2} \times (\text{Boltzmann's constant})$$
$$\times (\text{absolute temperature}) \quad (4\text{-}5)$$

$$\frac{1}{2}mv_a^2 = \frac{3}{2}kT \quad (4\text{-}6)$$

This is not the temperature measured on the *Fahrenheit scale* used in the United States. Nor is it measured on the *Celsius scale*, sometimes called the Centigrade scale, used throughout the rest of the world. The temperature we have defined is measured on the *absolute scale of temperature* and is written T K. The SI unit of temperature (the "degree" of the absolute temperature scale) is the *kelvin*, with the abbreviation K. (The notation °K has been used in the past, but the superscript °, the degree sign, is now usually omitted.)

The absolute scale of temperature is much more commonly used in physics than either the Celsius or Fahrenheit scales. One of its advantages is that the zero of the absolute scale is really the lowest possible temperature that can ever be attained. When the absolute temperature is zero, the kinetic energy of a gas atom is zero and its speed is zero. The *absolute zero of temperature*, 0 K, is therefore the temperature at which the ideal monatomic gas has lost all its internal energy and its atoms have come to rest. Clearly, we cannot do any better than that.

The Celsius degree has the same size as the absolute degree, the kelvin. The temperature T K on the absolute scale is obtained by adding 273.15 to the temperature t° C on the Celsius scale.

$$T \text{ K} = t° \text{ C} + 273.15 \quad (4\text{-}7)$$

The Fahrenheit degree is $\frac{5}{9}$ the size of the absolute or Celsius degree. The relationship between the absolute temperature and the temperature t° F on the Fahrenheit scale is

$$T \text{ K} = \frac{5}{9}(t° \text{ F} + 459.67) \quad (4\text{-}8)$$

The relationship between the Celsius and Fahrenheit scales is

$$t° \text{ C} = \frac{5}{9}(t° \text{ F} - 32) \quad (4\text{-}9)$$

The absolute zero of temperature is

$$0 \text{ K} = -273.15° \text{ C} = -459.67° \text{ F} \quad (4\text{-}10)$$

The temperature at which ice melts under a pressure of one standard atmosphere (the *ice point*) is

$$273.15 \text{ K} = 0° \text{ C} = 32° \text{ F} \quad (4\text{-}11)$$

The temperature at which water boils under a pressure of one standard atmosphere (the *steam point*) is

$$373.15 \text{ K} = 100° \text{ C} = 212° \text{ F} \quad (4\text{-}12)$$

4.6 PRESSURE

The atoms of a gas are in rapid motion and are continually striking the walls of the container. This steady bombardment of atoms is equivalent to a force pushing the wall outward. It is the origin of the *pressure* of the gas. More precisely, the pressure is the force exerted on unit area of a plane wall. However complicated the shape of the container (Figure 4–4), consider any area of its wall that is small enough to be essentially plane. Atoms bombarding it exert an outward force that is perpendicular to the small portion of the wall. The magnitude of the force divided by the area is the same on all parts of the wall and is equal to the pressure.

Pressure = force per unit area   (4-13)

The SI units of pressure are clearly newtons per square meter or N/m². Pressure is one of those physical quantities that is considered important enough to have a special name for its unit. The SI unit of pressure is the *pascal*, abbreviated Pa.

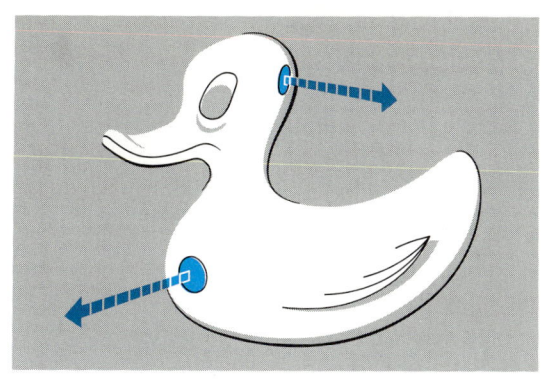

4.7 THE PRESSURE OF AN IDEAL MONATOMIC GAS

FIGURE 4-4
The pressure inside a rubber balloon with a complicated shape. The outward force per unit area is the same on all parts of the rubber.

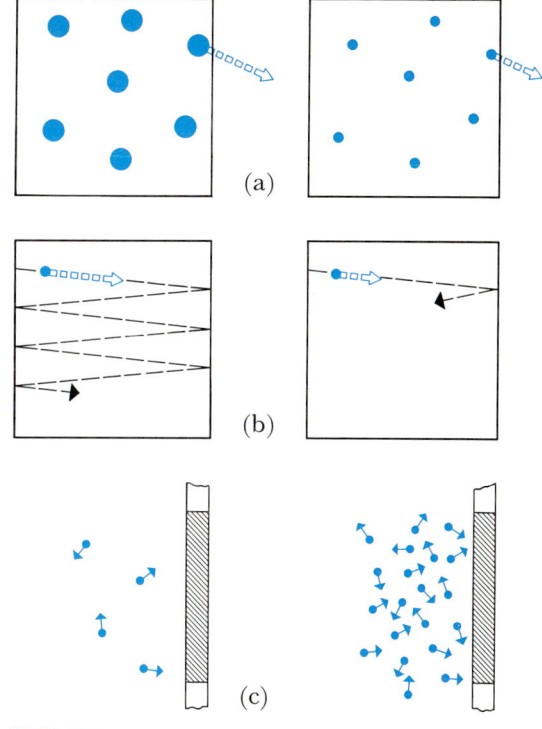

$$1 \text{ pascal} = 1 \text{ newton per square meter} \quad (4\text{-}14)$$

or

$$1 \text{ Pa} = 1 \text{ N/m}^2 \quad (4\text{-}15)$$

In this book, we shall often use newtons per square meter instead of pascals to emphasize that pressure is a force per unit area.

Two other units of pressure are commonly used. The *standard atmosphere,* abbreviated atm, is now defined as

$$1 \text{ atm} = 101{,}325 \text{ Pa} \quad (4\text{-}16)$$

The *bar,* no abbreviation, is now defined as

$$1 \text{ bar} = 100{,}000 \text{ Pa} \quad (4\text{-}17)$$

These last two units are not very different. The pressure of the earth's atmosphere near sea level is approximately 1 atm or 1 bar.

FIGURE 4-5
The three factors that determine the pressure of a gas. (a) The average momentum of an atom mv_a. The faster, more massive atom strikes the wall a harder blow. (b) The average speed v_a. The faster atom moves from side to side in a shorter time and therefore strikes the wall more often. (c) The number of atoms per unit volume, N/V. The more atoms there are per unit volume, the more collisions there are on unit area of the wall.

4.7 THE PRESSURE OF AN IDEAL MONATOMIC GAS

Figure 4-5 may help you to appreciate the three factors that determine the magnitude of the pressure of an ideal monatomic gas. The first is the average momentum of an atom, mv_a (Figure 4-5a). A more massive atom strikes the wall a harder blow; a faster atom strikes the wall a harder blow.

The second factor is the speed v_a, which again enters, but in a different way (Figure 4-5b). A faster atom moves from wall to wall in a shorter time and therefore strikes the walls more often.

Putting these first two factors together, the pressure is proportional to the momentum multiplied by the speed. That is $(mv_a) \times v_a$ or mv_a^2, which is just twice the average kinetic energy of an atom. The first two factors combined tell us that the pressure is proportional to the average kinetic energy of an atom.

The third factor is the number of atoms per unit volume. With the help of Figure 4-5c, visualize a small area of the wall and the nearby atoms that are about to strike it. The number of atoms striking the wall in a fixed interval of time clearly depends on how closely packed they are. If there are N atoms in a total volume V, the number of atoms per unit volume is N/V.

The detailed mathematical calculation puts all these three factors together and the pressure turns out to be given by the following equation.

Pressure =

$\frac{2}{3} \times$ (number of atoms per unit volume)

$\times$ (average kinetic energy per atom) $\quad (4\text{-}18)$

or

$$P = \frac{2}{3}\left(\frac{N}{V}\right)\left(\frac{1}{2}mv_a^2\right) \quad (4\text{-}19)$$

4 ATOMS AND HEAT

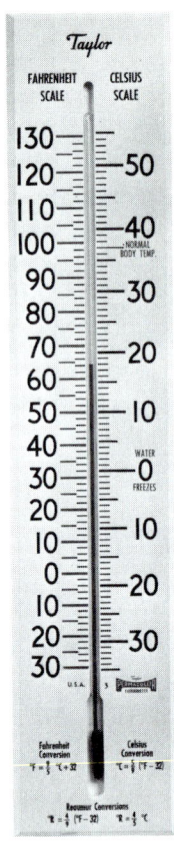

A liquid-in-glass thermometer. As the temperature rises, the liquid expands and its meniscus rises in the stem. The scale is calibrated by arbitrarily assigning numbers to special temperatures such as the freezing point of water. Notice the comparison between the Fahrenheit and Celsius scales. (Courtesy Taylor Instruments.)

The capital letter P is used for the pressure to avoid confusion with the small letter p reserved for momentum.

4.8
THE IDEAL GAS EQUATION

The average kinetic energy of an atom is related to the absolute temperature by Equation 4-6. If we substitute this into Equation 4-19, the factor $\frac{3}{2}$ in Equation 4-6 cancels the factor $\frac{2}{3}$ in Equation 4-19, and we obtain

$$P = \left(\frac{N}{V}\right)(kT) \quad (4\text{-}20)$$

If we multiply both sides of this equation by the volume V, the result is a very famous equation known as the *ideal gas equation*.

> Ideal gas equation:
> Pressure × volume
>   = (number of atoms)
>   × (Boltzmann's constant)
>   × (absolute temperature) (4-21)
> $$PV = NkT \quad (4\text{-}22)$$

Now let us pause and consider what we have achieved. We started with a model of an ideal monatomic gas as a collection of atoms in random motion. The pressure of the gas was then calculated by considering the bombardment of gas atoms on the walls of the container. The pressure was found to depend on the number of atoms per unit volume and the average kinetic energy of the atoms. The concept of temperature was introduced as a measure of the average kinetic energy. Finally, the ideal gas equation was obtained in a form involving measurable quantities such as the pressure, the volume, and the absolute temperature.

Actually, we have reversed the historical order of events. Originally, crude temperature scales were based on the fact that most substances expand as the temperature is raised, as in the familiar liquid-in-glass thermometer. On the basis of such crude temperature scales, the form of the ideal gas equation was discovered experimentally. It was then realized that a more fundamental temperature scale could be based on the properties of gases at very low densities. Finally, the atomic theory provided an explanation of the ideal gas equation and also revealed the relationship between absolute temperature and internal energy for an ideal monatomic gas.

4.9
THE LAWS OF BOYLE, CHARLES, AND AVOGADRO

If a fixed mass of gas is kept at a fixed temperature, everything on the right-hand side of Equation 4-21 is constant, and so

$$\text{Pressure} \times \text{volume} = \text{a constant} \quad (4\text{-}23)$$

or

$$\text{Pressure} = \frac{\text{a constant}}{\text{volume}} \quad (4\text{-}24)$$

This is *Boyle's law*, discovered experimentally by the English physicist Robert Boyle as early as the seventeenth century, a few years before Newton discovered the laws of motion. It tells us that if we keep the temperature of a sample of gas constant and decrease its volume, the pressure increases in inverse proportion to the volume.

4.10 THE NATURE OF HEAT

To obtain a vivid picture of what is happening, imagine the atoms to be analogous to a swarm of angry mosquitoes, trapped inside a rubber balloon, continually flying from side to side and battering themselves against the rubber in an attempt to escape (Figure 4–6). When the balloon has a large volume (Figure 4–6a) the mosquitoes spend most of their time flying across the balloon and only occasionally hit the rubber. When the volume is small (Figure 4–6b) they hit the rubber much more frequently and exert a greater outward pressure on it.

If the volume of the gas is kept constant and its temperature is varied, Equation 4-20 is equivalent to

Pressure = (another constant)
 × (absolute temperature)   (4-25)

The pressure of the gas is directly proportional to the absolute temperature if the volume is kept constant. This is essentially *Charles's law,* although the French physicist Jacques Charles did not express it in quite this way, since he was not familiar with the concept of absolute temperature. He proposed this law at the end of the eighteenth century, more than a century after the discovery of Boyle's law and a few years before Dalton's atomic theory. Our pictorial interpretation is that raising the temperature is equivalent to making the mosquitoes angrier. They then fly faster, hit the rubber more often and with a greater impact and therefore exert a greater pressure on it.

Another rearrangement of Equation 4-21 puts it in the form

Number of atoms

$$= \frac{(\text{pressure}) \times (\text{volume})}{(\text{Boltzmann's constant}) \times (\text{absolute temperature})}$$

(4-26)

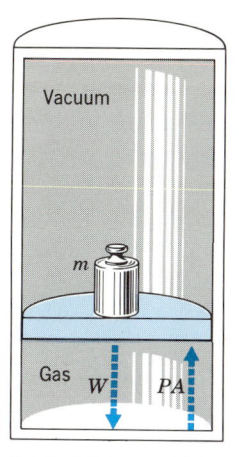

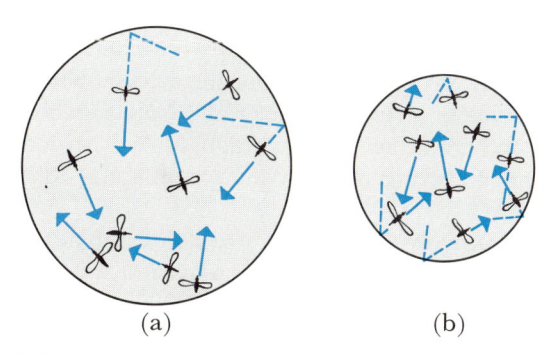

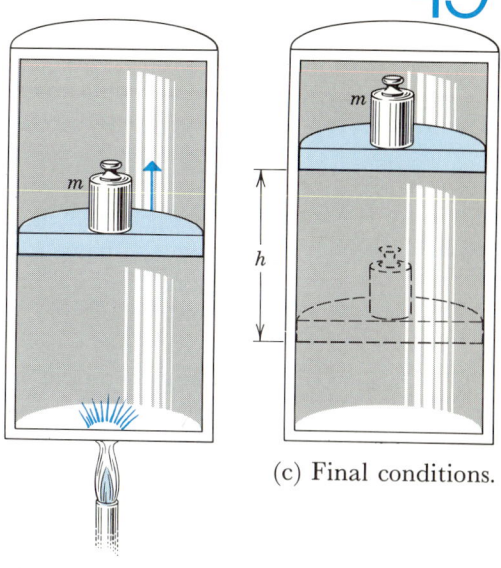

(a) Initial conditions.

(b) Heat is added.

(c) Final conditions.

FIGURE 4–7
Addition of heat to an ideal monatomic gas at constant pressure. The temperature and volume both increase.

Consider two different ideal monatomic gases corresponding to two different chemical elements. If they have the same pressure and temperature and occupy equal volumes, then everything on the right-hand side of Equation 4-26 is the same for the two gases. They must therefore contain the same number of atoms. This is *Avogadro's law,* named for Amadeo Avogadro, an Italian chemist and physicist (1776–1856). Equal volumes of two ideal monatomic gases at the same temperature and pressure contain the same number of atoms.

4.10
THE NATURE OF HEAT

Heat is a form of energy. When heat is added to an object, its internal energy increases. We must be careful, though, because some of the heat can be used up in other ways. We shall consider a simple example in which only part of the heat is converted into internal energy.

In Figure 4–7, the lower part of the cylinder contains an ideal monatomic gas and the upper part is a vacuum. The gas is separated from the vacuum by a piston, which is well lubricated so that it slides up and down without friction. The gas pushes upward on the piston with a force equal to the pressure P of the gas multiplied by the area A of the piston. If the piston is to remain stationary, this force must exactly balance the weight W of the piston. A unique value of the pressure $\left(P = \dfrac{W}{A}\right)$ makes this possible. The

(a)   (b)

FIGURE 4–6
An analogy between Boyle's law and the behavior of a swarm of mosquitoes inside a rubber balloon.

height of the piston changes until the volume of the gas has the right value to produce this pressure according to the ideal gas equation. The arrangement therefore ensures that the pressure always has the same value under all circumstances after the piston has come to rest.

Suppose that heat is added to the gas by applying a flame to the bottom of the cylinder. The temperature of the gas rises. The average kinetic energy per atom increases, and so the total internal energy of the gas increases. The flame must supply the energy to do this. But that is not all. As the temperature rises and the atoms move faster, the pressure of the gas starts to increase. The upward force on the piston then exceeds its weight and the piston moves upward. The resulting increase in the volume of the gas lowers the pressure again. Eventually, after the flame has been removed, the gas is at a higher temperature and the piston has settled down at a higher level. Because the pressure must return to its original value, the ideal gas equation implies that the volume must increase in proportion to the absolute temperature.

The gas has lifted the weight W of the piston through a height h and has therefore increased the gravitational potential energy of the piston by an amount Wh. Part of the heat energy fed into the gas from the flame is used to increase the gravitational potential energy of the piston. The law of conservation of energy must be stated carefully in the following way. The amount of heat energy added is equal to the sum of two parts. The first part is the increase in the internal energy of the gas. The second part is the increase in the gravitational potential energy of the piston.

4.11
THE FIRST LAW OF THERMODYNAMICS

Thermodynamics is the sophisticated name for the subject we are presently discussing. It is concerned with heat, temperature, and such properties of matter as pressure, volume, and density. The *first law of thermodynamics* is really a restatement of the law of conservation of energy for processes involving heat. It is a generalization of the considerations of the previous section.

If we wish, we may take the point of view that the heat added from the flame is first converted entirely into additional internal energy of the gas. The gas expands and raises the weight W through a height h. The gas does an amount of work on the piston equal to Wh. This work is done at the expense of the internal energy, which is reduced by an amount Wh. The net result is that the increase in internal energy is equal to the heat added minus the work done on the piston.

The heat added is equal to the increase in internal energy plus the work done by the gas on the piston. This is similar to the statement at the end of Section 4.10, except that the increase in gravitational potential of the piston is replaced by the work done on the piston by the gas. The two are known to be equal. However, in more complicated situations, the work done on the surroundings is not necessarily converted into gravitational potential energy. It may be converted into electrical energy, magnetic energy, or other forms of energy that we have not yet discussed in detail. The first law of thermodynamics is therefore stated in the following way.

> The first law of thermodynamics:
>
> Heat is a form of energy. The heat added to an object is equal to the increase in its internal energy plus the work done by the object on its surroundings.

Heat is energy, and the SI unit of heat is the joule. Following an old tradition, heat is sometimes measured in calories. One *calorie* is the heat required to raise the temperature of one gram of pure water from 14.5° C to 15.5° C. The conversion factor between calories and joules is

$$1 \text{ calorie} = 4.186 \text{ joules} \qquad (4\text{-}27)$$

The calorie referred to by nutrition experts and dieticians is not the same thing. They really mean the kilocalorie, sometimes called the "large" calorie and often written Calorie with a capital C. It is 1000 of the "small" calories defined above.

4.12
ADIABATIC PROCESSES

An *adiabatic process* is one in which no heat is allowed to enter or leave. Consider the arrangement of Figure 4–7. Remove the flame completely. Wrap the outside of the cylinder with a thermally insulating material to prevent heat from entering or leaving it (Figure 4–8). Then any change taking place in the gas must be adiabatic.

Normally the upward force PA exerted by the pressure of the gas on the piston is just sufficient to counterbalance its weight W. The piston remains at rest. In Figure 4–8a, the weight W_a of the piston is initially less than PA. The net upward force moves the piston upward. The volume of the gas increases and its pressure decreases. Eventually, the piston settles down again at a

4.13 A BRIEF COMMENT ON HEAT ENGINES

higher level when the smaller value P_a of the pressure again makes P_aA equal to W_a.

During this adiabatic expansion the piston is raised through a height h_a. The gas does work on the piston and increases its gravitational potential energy by an amount W_ah_a. Since no heat can enter the system, the only available source of this energy is the internal energy of the gas. The internal energy of the gas therefore decreases, its atoms move more slowly, and its temperature falls. When a gas expands adiabatically, it cools down.

In Figure 4-8b the weight W_b is greater than PA. There is a net downward force on the piston. The piston falls, the volume of the gas decreases, and its pressure increases. When the increased pressure P_b is large enough to support W_b, the piston settles down again at a depth h_b below its initial position.

During this adiabatic compression the piston loses gravitational potential energy by doing work on the gas. The work is done by the piston on the gas because it is the piston that is pushing the gas atoms downward. The internal energy of the gas increases, it atoms move faster, and it warms up. An adiabatic compression warms a gas.

4.13
A BRIEF COMMENT ON HEAT ENGINES

The device of Figure 4-7 can be seen as a simple heat engine. It uses heat energy to do work and raise a weight. Of course, practical heat engines are much more complicated. They are designed to work continuously, whereas our simple device is a one-shot affair. Nevertheless, it does illustrate some of the important principles and problems.

The original source of the energy is the chemical energy stored in the fuel that is burned in the flame. Only a part of this energy is used to raise the weight. The rest of it warms up the gas, which may be a less desirable objective. That turns out to be an unavoidable feature of even the very best continuously operating engines. There is a theoretical limit to the fraction of the energy in the fuel that can be used to do work.

Moreover, the process is a one-way street. There is no simple way of lowering the piston, cooling the gas, returning the heat to the flame, and remanufacturing the fuel. This is an example of a very important general principle that might be stated very crudely in the following way. Once a source of energy is used up, it cannot be replaced without using up even more energy than it yielded in the first place.

That sounds very reasonable, but it is not quite precise enough. It would take us too far

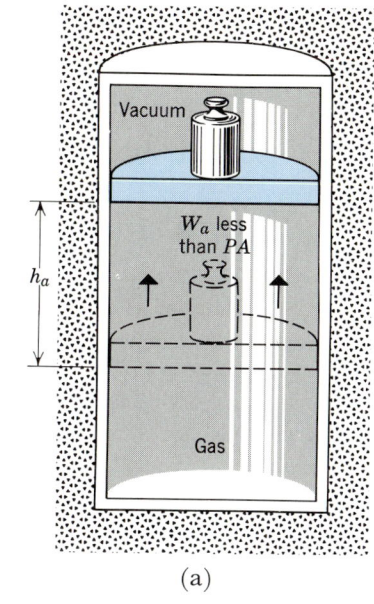

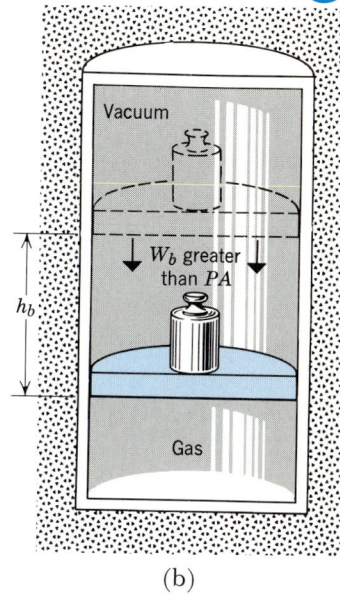

FIGURE 4-8

Adiabatic processes. No heat can enter or leave. (a) An adiabatic expansion. The gas does work and cools. (b) An adiabatic compression. Work is done on the gas and warms it.

afield to discuss in detail the application of thermodynamics to heat engines and the utilization of energy. In the remainder of this chapter we shall discuss the matter from a more fundamental point of view. We shall discover that when we use up fuel and other sources of energy, what we are wasting is orderliness. What we are creating is disorder.

Eventually, all order will be destroyed and there will be only chaos. Moreover, there seems to be nothing we can do about it! Philosophy is not the main purpose of this book, but the thoughtful reader may wish to ponder these matters.

4.14
ORDER AND DISORDER

Heat is disorganized energy. Consider again the case of a baseball of mass M thrown with a speed V. Its kinetic energy $\frac{1}{2}MV^2$ would not normally be called heat. However, suppose that it strikes the ground and comes to rest. The kinetic energy of its orderly motion is converted into the kinetic energy of the ramdom disorderly motion of its individual atoms. The latter is rightly called heat. This concept of randomness, or disorder, is very important in thermodynamics.

For an ideal monatomic gas at a temperature T, we have so far only specified that the average kinetic energy of an atom shall be determined by the temperature. This could be realized in many different ways. For example, the atoms might

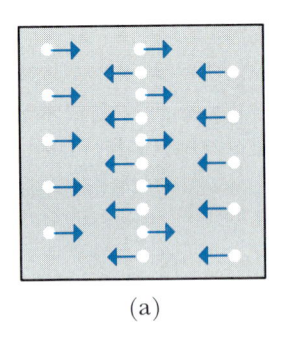

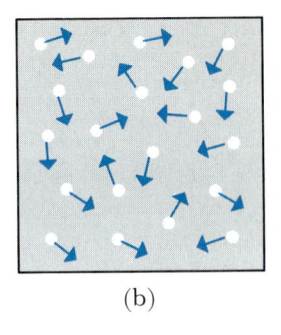

FIGURE 4-9

Two ways of achieving an average kinetic energy per atom of $\frac{3}{2}kT$ in an ideal monatomic gas. (a) Two opposing streams. An orderly motion with small entropy. (b) A random distribution of velocities. A disordered motion with high entropy.

divide into two equal streams moving in opposite directions as in Figure 4-9a. Each atom would then bounce backward and forward between the right-hand and left-hand walls of the container. They could all have the same speed v_a. The value of v_a would be determined by Equation 4-6.

It seems unlikely that a gas really looks like that. Why should there not be streams up and down, or back to front, or for that matter in any direction? Why must the atoms all have the same speed? It seems much more reasonable to assume that the atoms are moving in all possible directions with all possible speeds as in Figure 4-9b. The only restriction need be that the average kinetic energy is related to the temperature by Equation 4-6.

Suppose that the gas started with the orderly motion of Figure 4-9a. Every time two atoms made a glancing collision they would rebound at an angle. The atoms would soon be moving in all directions. Collisions would also lower the speeds of some atoms and increase the speeds of others. There would soon be a wide range of speeds. Within a fraction of a second the orderly motion would degenerate into a disorderly motion similar to that in Figure 4-9b.

As time went on, collisions would continually change one disordered motion into a different disordered motion. It is conceivable that the orderly motion of Figure 4-9a would eventually turn up again, but only after a very long time indeed. Moreover, the orderly motion would disappear again within a fraction of a second. If you followed the behavior of the atoms patiently, you would see that most of the time they would be in a random, disorderly motion. Only very, very occasionally would there be a brief period of orderliness. In fact, it can be shown that orderliness is such a rare event that you might as well forget about it altogether.

The situation is similar to taking a new pack of cards that is carefully ordered into suits and shuffling it. The initial order is very soon destroyed. But it would be a very long time indeed before the shuffling reproduced the initial order again. There are very many ways of arranging a pack in a random fashion. There is only one way of arranging it in the order in which it comes from the manufacturer.

What we are saying, then, is that the ordered arrangement of Figure 4-9a is quite consistent with the laws of physics, but it is very unlikely to occur in practice. On the other hand, the disordered arrangement of Figure 4-9b, or some similar random arrangement, is very likely to be found in practice. Moreover, a random arrangement is overwhelmingly more probable.

A student confronted with these ideas for the first time is likely to argue as follows. Although the ordered arrangement is improbable, it is possible. Then, why does it not sometimes occur? If you find this argument attractive, you should sit down with a pack of cards and try shuffling them until they are arranged into suits. Your patience with the argument will soon become exhausted.

4.15 ENTROPY AND THE SECOND LAW OF THERMODYNAMICS

It is possible to put these ideas on a firm mathematical basis and to define a quantity called *entropy*. Entropy is a measure of the probability that a particular type of motion will occur. Since a disordered situation is a more probable situation, entropy is also a measure of disorder. The situation in Figure 4-9a is an ordered situation but an improbable situation. It corresponds to a small value of the entropy. The situation in Figure 4-9b is disordered, but more probable. It corresponds to a larger value of the entropy.

As we have just seen, the ordered situation soon changes into the disordered situation with larger entropy. The chance of the disordered situation changing back into the ordered situation is so very small that it can be neglected. This is an example of a very general law, the *second law of thermodynamics*.

> The second law of thermodynamics:
>
> When an object containing a large number of atoms is left to itself, it assumes a state with maximum entropy. It becomes as disordered as possible.

The state with maximum entropy is the most probable state. The second law merely says that "what is most likely to occur will in fact occur." What is most likely to occur is overwhelmingly more probable than any other possibility. This other possibility can therefore be completely discounted.

4.16 THE THIRD LAW OF THERMODYNAMICS

As we increase the temperature of an object, its atoms move faster. Their motion becomes more chaotic and more disorderly. As the temperature increases the entropy increases.

Conversely, if we lower the temperature the entropy decreases. The lower the temperature, the more orderly the behavior of the atoms. There must be a point at which perfect order is achieved and the entropy is zero. The temperature then has its lowest possible value. It must be the absolute zero of temperature. At the absolute zero of temperature, perfect order is realized and there is no entropy.

> The third law of thermodynamics:
> An object at the absolute zero of temperature is in a state of perfect order and has zero entropy.

Perfection is difficult to achieve. The absolute zero of temperature can never be reached. The third law can be shown to have the following consequence. It is impossible to make an apparatus that will reach the absolute zero of temperature in a finite number of operations. Physicists who attempt to produce very low temperatures know from experience that this is true.

4.17 DEATH BY CHAOS?

Everyday experience contains many examples of processes that destroy order and create disorder. The reverse process to restore the order never occurs.

Is it likely that the smoke will ever recombine with the ash to form a new log? (K. T. Bendo.)

Is it likely that the passage of time will ever rebuild the ancient Inca city of Machu Picchu? (Tom Hollyman/Photo Researchers.)

Is it likely that the parts of this egg will ever come together again to form a whole egg capable of hatching into a chicken? (K. T. Bendo.)

Milk poured into coffee soon distributes itself uniformly. It is inconceivable that it should later separate out again into a stream of pure milk leaping up into the jug. A log readily burns to form ash and various gases, which mix with the air. It is inconceivable that these gases should return from the atmosphere to the flame and recombine with the ash to form a new log. Can you imagine any process that would reconvert an omelet into an egg capable of hatching into a chicken? Wind, rain, and frost can reduce an ancient city to rubble, but are unlikely ever to rebuild it again.

All these examples confirm the second law of thermodynamics. Order always changes into disorder. Entropy always increases. Everywhere throughout the universe order is continually being replaced by chaos. Eventually there will be no order, only chaos.

Life requires order. A living organism is a complex orderly arrangement of atoms set apart from its more disorderly surroundings. When there is no order left in the universe there will be no life. The second law of thermodynamics implies that life must ultimately disappear in the final reign of chaos.

Don't worry yet. It will probably take at least ten billion years. Moreover, there is a possible reprieve. We do not yet understand the universe as a whole. Perhaps the second law of thermodynamics does not apply to the whole universe. We are still too ignorant to be quite certain of our ultimate fate.

SUMMARY

Atoms are very small. It would take about ten billion atoms placed side by side to cover a distance of one meter.

An ideal monatomic gas consists of individual atoms moving randomly in all directions with all possible speeds. The atoms stay far apart, and the weak forces between them can be neglected.

The internal energy of an ideal monatomic gas is the sum of the kinetic energies of all its atoms. The average kinetic energy per atom is equal to $\frac{3}{2}$ times Boltzmann's constant k multiplied by the absolute temperature T.

The absolute temperature T is measured in degrees kelvin, abbreviated K. The absolute zero of temperature is the temperature at which the atoms of an ideal monatomic gas are all at rest. It is therefore the lowest attainable temperature. The absolute scale of temperature is related to the Celsius and Fahrenheit scales by Equations 4-7 through 4-12.

Pressure is force per unit area. The SI unit of pressure is the pascal, abbreviated Pa. It is equal to one newton per square meter.

The ideal gas equation says that pressure times volume is equal to the number of atoms multiplied by Boltzmann's constant multiplied by the absolute temperature. Boyle's law says that the pressure is inversely proportional to the volume at constant temperature. Charles's law says that the pressure is proportional to the absolute temperature at constant volume. Avogadro's law says that equal volumes of two ideal monatomic gases at the same temperature and pressure contain the same number of atoms.

Heat is a form of energy. It can be measured in joules but is often measured in calories. One calorie is the heat required to raise the temperature of one gram of pure water from 14.5° C to 15.5° C. One calorie is equal to 4.186 joules.

The first law of thermodynamics. The heat added to an object is equal to the increase in its internal energy plus the work done by the object on its surroundings. This is a form of the law of conservation of energy.

An adiabatic process is one in which no heat is allowed to enter or leave. An adiabatic expansion of a gas cools it. An adiabatic compression of a gas warms it.

Entropy is a measure of disorder and also of probability. States with high entropy are very disordered, but are more likely to occur.

The second law of thermodynamics. When an object containing a large number of atoms is left to itself, it assumes a state with maximum entropy. It becomes as disordered as possible.

The third law of thermodynamics. An object at the absolute zero of temperature is in a state of perfect order and has zero entropy. It is impossible to make an apparatus that will reach the absolute zero of temperature in a finite number of operations.

TERMS AND CONCEPTS

atomic theory of matter
field ion microscope
ideal monatomic gas
internal energy
Boltzmann's constant
absolute scale of temperature
Fahrenheit scale of temperature
Celsius scale of temperature
kelvin
absolute zero of temperature
ice point
steam point
pressure
pascal
standard atmosphere

bar
ideal gas equation
Boyle's law
Charles's law
Avogadro's law
thermodynamics
first law of thermodynamics
calorie
adiabatic process
entropy
second law of thermodynamics
third law of thermodynamics

QUESTIONS

1. What is an ideal monatomic gas? Under what circumstances is a real gas a good approximation to an ideal gas?

2. Discuss the concept of internal energy. What is the total internal energy of an ideal gas containing N atoms at a temperature T K?

3. Why do physicists prefer to use the absolute scale of temperature?

4. What is the temperature difference between the ice point and the steam point? Express your answer in (a) kelvins, (b) degrees Celsius, (c) degrees Fahrenheit.

5. Why does the pressure of a gas vary with its temperature?

6. Why does the pressure of a gas vary with its volume even if the temperature is kept constant?

7. An ideal monatomic gas is compressed in such a way that its temperature remains constant. How does its internal energy vary with its volume?

8. Comment on the following definitions of heat. (a) Another name for energy. (b) The change in the total kinetic energy of random motion of the atoms of an object. (c) The change in the total internal energy of an object.

9. If an ideal monatomic gas expands adiabatically, is its pressure inversely proportional to its volume?

10. Which has the greater entropy, (a) a pack of cards ordered into suits? (b) a shuffled pack?

PROBLEMS

1. (a) What is the approximate volume occupied by an atom? (b) In a solid the atoms almost touch one another. About how many atoms are there in one cubic meter of a solid?

2. The average speed of the atoms of an ideal monatomic gas is doubled. By what factor is the absolute temperature multiplied?

3. The weather forecast predicts a high temperature of 68° F. Convert this to (a) the Celsius scale, (b) the absolute scale.

4. A gas at a pressure of 5 N/m^2 is contained inside a cubic box. If the length of a side of the box is 0.2 m, what is the force exerted by the gas on one face of the box?

5. An ideal monatomic gas has a pressure of 20 N/m^2. With the temperature constant its volume changes from 2 m^3 to 0.25 m^3. What is its new pressure?

6. An ideal monatomic gas has a pressure of 8×10^3 N/m^2. If its volume is doubled and its absolute temperature is halved, what is its new pressure?

7. How many calories are equal to one joule?

8. Five hundred calories of heat are fed into a machine, and 5 percent of this heat is used to lift a mass of 4 kg. Through what height can the mass be lifted?

9. An ideal monatomic gas is initially at a temperature of 300 K and its total internal energy is 2 J. Its volume is kept constant and 1.2 J of heat are added to it. What is the final temperature of the gas?

10. An ideal monatomic gas has a pressure of 25 N/m^2, a volume of 15 m^3, and a temperature of 300 K. It expands adiabatically to twice its initial volume. How much heat is added during this expansion?

CHAPTER 5
ELECTRICITY

5.1
ELECTRIC CHARGE

When a glass rod is rubbed with a silk cloth, the rod acquires a charge of positive electricity and the cloth acquires a charge of negative electricity. They then exert attractive forces on each other (Figure 5–1). If another glass rod is rubbed with another silk cloth, the two positively charged rods repel each other. The two negatively charged cloths also repel each other. Like charges repel and unlike charges attract.

The modern explanation of this proceeds as follows. Atoms are composed of neutrons, protons, and electrons. Neutrons are electrically neutral, so we may ignore them at the moment. A proton is a unit of positive electricity and an electron is a unit of negative electricity. A proton and an electron attract each other, two protons repel, and two electrons repel (Figure 5–1 again).

A proton and an electron have exactly equal but opposite electrical effects. A negative electron placed on top of a positive proton cancels it electrically. An atom contains a certain number of protons in a central nucleus and an equal number of electrons orbiting around this nucleus. Except in the immediate vicinity of the atom, this turns out to be equivalent to putting the electrons on top of the nucleus. The atom as a whole is therefore electrically neutral.

When the silk cloth is rubbed against the glass rod, some of the electrons near the outside of the glass atoms are rubbed off and migrate into the silk. Electrons would rather be in silk than glass for reasons that are now understood but are too complicated to describe here. The silk cloth acquires an excess of electrons and a net negative charge. The glass rod has a deficiency of electrons or, if you prefer, an excess of protons. It has a net positive charge. The forces on the left of Figure 5–1 are therefore a consequence of the forces between electrons and protons shown on the right of the figure.

5.2
CONSERVATION OF CHARGE

An uncharged object contains equal numbers of electrons and protons. This is the case for both the silk and the glass before they are rubbed together. During the rubbing the number of electrons migrating to the silk is equal to the number of excess protons left behind in the glass.

Electricity. (Lester V. Bergman & Associates.)

The amount of negative electricity on an electron is equal to the amount of positive electricity on a proton. Therefore, when a measured quantity of negative electricity appears on the silk cloth, an equal amount of positive electricity appears on the glass rod.

If the measure of the charge on the glass rod is a certain positive number, the measure of the charge on the silk cloth is the same number with a minus sign in front of it. The sum of the two is zero. From this point of view, rubbing the rod with the cloth has not created any electric charge. It has merely transferred negatively charged electrons from one place to another.

The only way to create electric charge would be to create or destroy charged fundamental particles such as electrons and protons. This is now known to be possible. But the particles are always created or destroyed in pairs, one positive and one negative. The total amount of electric charge never changes. This is the *law of conservation of charge*. We might even go a little further than what we are quite certain of, and guess that the number of positive particles in the universe is exactly equal to the number of negative particles.

5.3 THE FORCE BETWEEN TWO STATIONARY CHARGES

Consider two stationary charges q_1 and q_2 a distance R apart (Figure 5–2). The forces they exert on each other have opposite directions but equal magnitudes, in accordance with Newton's third law of motion. For two positive charges, or two negative charges, the repulsive force on either charge points directly away from the other charge. For a positive and a negative charge, the attractive force on either points directly toward the other.

The law giving the magnitude of the force was discovered by the French physicist Charles Augustin Coulomb in the late eighteenth century. It is called *Coulomb's law* and is very similar to Newton's law of universal gravitation. The electric force is proportional to the product of the magnitudes of the two charges. It is inversely

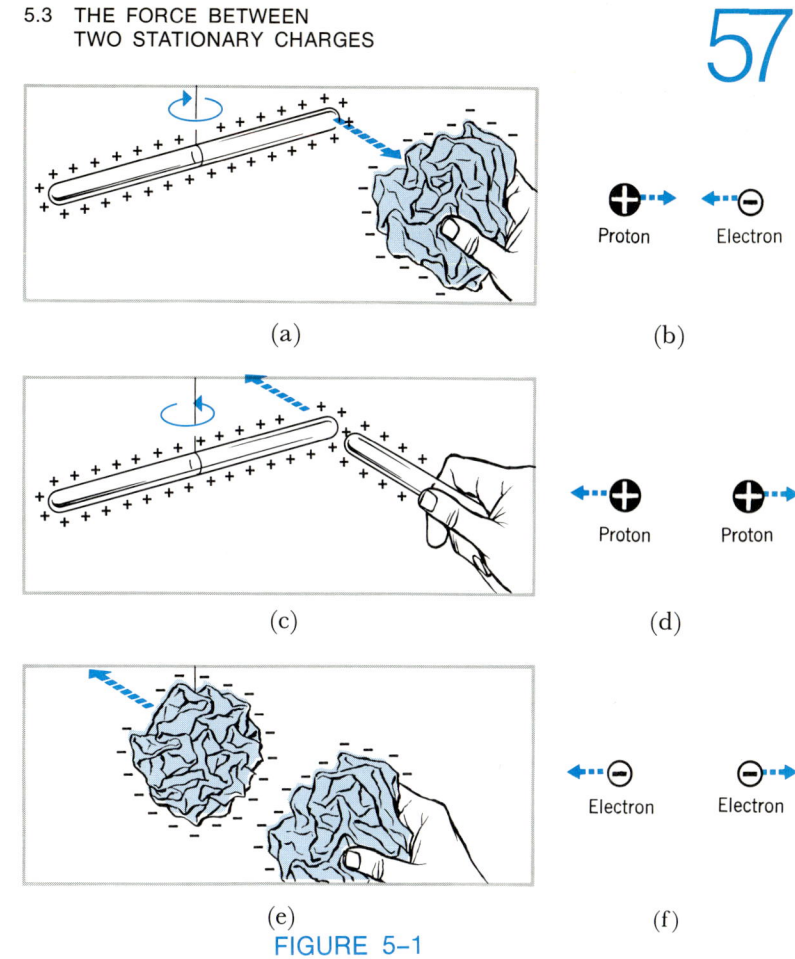

FIGURE 5–1

The basic facts of electricity. (a) The negatively charged silk cloth attracts the positively charged glass rod. (b) An electron attracts a proton. (c) Two positively charged glass rods repel each other. (d) Two protons repel each other. (e) Two negatively charged silk cloths repel each other. (f) Two electrons repel each other.

proportional to the square of the distance between the charges. The strength of the force falls off rapidly as the charges are moved apart.

The constant of proportionality is called *the electric force constant* and is represented by the symbol K_e. The subscript e reminds us that we are talking about electricity and also avoids confusion with the abbreviation K for the kelvin, the unit of absolute temperature. The electric force constant is a measure of how strong electric forces are. It is similar to the gravitational constant G, which appears in Equation 2-7.

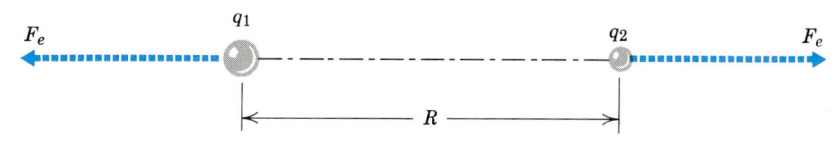

FIGURE 5–2

The electric forces between two stationary positive charges.

5 ELECTRICITY

> Coulomb's law:
>
> The electric force between two stationary charges is equal to the electric force constant multiplied by the magnitude of the first charge, then multiplied by the magnitude of the second charge, and finally divided by the square of the distance between them. (5-1)
>
> $$F_e = \frac{K_e q_1 q_2}{R^2} \quad (5\text{-}2)$$

The SI unit of electric charge is the *coulomb*, which is abbreviated to C. Its precise definition is too complicated to be attempted here. In Equation 5-2, the charges q_1 and q_2 should be measured in coulombs. The force F_e should be in newtons and the distance R in meters. The value of the electric force constant is then

$$K_e = 9.0 \times 10^9 \text{ SI units } (\text{Nm}^2/\text{C}^2) \quad (5\text{-}3)$$

The stipulation in Coulomb's law that the charges shall be stationary is very important. If the charges are in motion there are additional forces. As we shall explain in Chapter 6, these additional forces are responsible for magnetism.

The charge q on an object is equal to the number of its excess protons or electrons multiplied by the charge on a single one of these particles. The numerical value of the charge on either a proton or an electron is 1.60×10^{-19} coulomb. It is often called *the charge on the electron* and is represented by the italic letter e. It should not be confused with the upright Roman letter e which represents the electron itself in the same way that the letter H is used by chemists to represent a hydrogen atom.

We shall always look upon the number e as being intrinsically positive. The charge on the proton is then $+e$ and the charge on the electron is $-e$.

5.4
THE RELATIVE STRENGTHS OF ELECTRIC AND GRAVITATIONAL FORCES

Although electric forces are very similar to gravitational forces, they are much stronger. The electric force between a proton and an electron exceeds the gravitational force between them by the enormous factor of 2×10^{39}. That is 2 followed by 39 zeros. The electric force is two thousand trillion trillion trillion times stronger!

The only reason why gravitation ever gets a chance is that most matter is uncharged. Electric charge comes in positive and negative forms that cancel each other in ordinary uncharged matter. Mass is always positive and can never be canceled out in this way. Two uncharged bodies exert no electric forces on each other, but they always have mass and always exert gravitational forces. *electric forces may be either attractive or repulsive, whereas gravitational forces are only attractive.*

5.5
ELECTRIC POTENTIAL ENERGY

Two masses have a mutual gravitational potential energy given by Equation 3-15. Two electric charges at rest a distance R apart have a mutual *electric potential energy* given by a similar equation.

> The mutual electric potential energy of two charges is equal to the electric force constant multiplied by the magnitude of the first charge, then multiplied by the magnitude of the second charge, and finally divided by the distance between them, (5-4)
>
> $$= +\frac{K_e q_1 q_2}{R} \quad (5\text{-}5)$$

This is very similar to Equation 5-2 for the Coulomb force. However, in the equation for the force the distance is squared. Electric potential energy is a form of energy, and its SI units are joules.

Equation 3-15 for the gravitational potential energy has a minus sign. The equation for the electric potential energy has a plus sign. However, the charges q_1 and q_2 may be either positive or negative. If one is positive and the other is negative, the electric potential energy is negative. A positive charge attracts a negative charge. As in the case of gravitation, attractive forces give a negative potential energy. If both charges are positive, or both are negative, the electric potential energy is positive. Two positive charges, or two negative charges, repel each other. Repulsive forces give a positive potential energy. In all circumstances the potential energy becomes zero as the distance R becomes infinitely large.

5.6 THE CONCEPT OF A FIELD

So far we have been developing the idea that the universe is a collection of discrete particles moving through empty, featureless space. We have assumed that the behavior of the universe must be described in terms of the motion of these particles. We discovered that this motion can be explained and predicted if we know certain laws that tell us the nature of the forces between the particles. We have introduced the first two of these laws, Newton's law of universal gravitation and Coulomb's law for the force between two stationary charges.

The next step is to consider a third type of force that exists between two moving charges. It is the origin of magnetism. Unfortunately, the laws governing magnetic forces cannot be expressed as simply as the laws of gravitation and electricity. The next step in our program is therefore difficult.

In the early nineteenth century, when the laws of magnetism were being discovered, the discovery was considerably aided by a new attitude to the nature of the universe, an attitude involving the concept of electric and magnetic fields. Toward the end of the century, Clerk Maxwell was able to explain all the known phenomena of electricity and magnetism by a set of simple, elegant equations describing the behavior of the electric and magnetic fields. Simplicity and success are two powerful advocates for any scientific theory, so we must now turn our attention toward this new idea of a *field*.

Our previous attitude implied that two objects at a great distance from each other, with only empty, featureless space between them, can nevertheless influence each other's motion. This is called *action at a distance*. It worried many scientists and philosophers, who felt that two bodies could interact only if they were in "direct contact" with each other. To appreciate this point of view, imagine a railroad engine backing up to a line of railroad cars. The shock of impact is transmitted along the line as the first car hits the second, the second hits the third, and so on, until the disturbance eventually reaches the caboose (Figure 5-3). However, in a situation such as the gravitational attraction between the earth and the moon, there are no visible, touchable bodies in between to provide the connecting links (Figure 5-4).

To overcome this difficulty, it was suggested that the whole of space is filled with an invisible material called the *ether*. The ether was imagined to be similar to a jelly. If you press your finger into a jelly, the strain is transmitted to all parts of the jelly. A pressure is thereby exerted on a body embedded in the jelly some distance away (Figure 5-5). In the same way, the earth was imagined to produce a strain in the ether which extended to the moon and transmitted a force to the moon (Figure 5-6).

This idea of a material ether filling the whole of space no longer appears to be acceptable. Some of the most forceful arguments against it will be presented later in our discussion of the theory of relativity. Nevertheless, this line of thought is fruitful. It has taken our attention away from the material objects and discrete

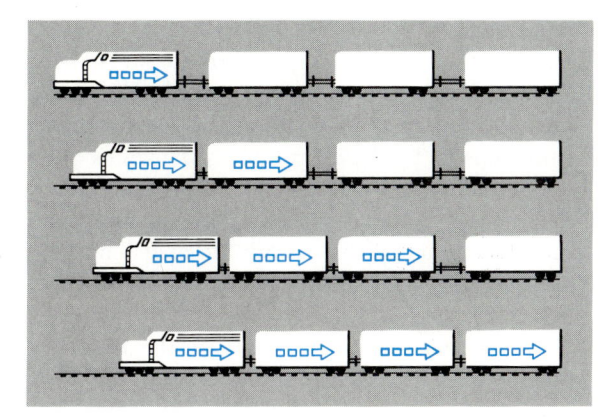

FIGURE 5-4
The moon and the earth exert forces on each other even though the space between them is apparently empty.

FIGURE 5-3
Transmission of an interaction along a line of railroad cars.

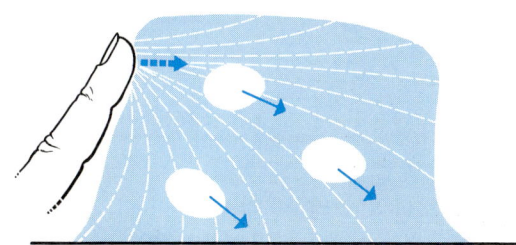

FIGURE 5-5
Strain is transmitted through a jelly to objects embedded in the jelly.

5 ELECTRICITY

FIGURE 5-6
Is there a jellylike ether between the earth and the moon? Is strain transmitted through this jelly to produce an attraction between the earth and the moon?

particles. It has redirected our attention toward the important question of whether space itself does not possess properties of interest to the physicist.

Our original prescription for describing the instantaneous state of the universe was to specify the position of each elementary particle. Now let us consider a possible alternative description in which physical properties are assigned to each point in space even if there is no particle located at that point. This is best appreciated by seeing how it is done in the case of the electric field. We introduce the magnetic field in Chapter 6.

5.7
THE ELECTRIC FIELD

A vector quantity called the *electric field* is associated with any point P in space. It is always represented by the symbol **E**. Its physical significance is as follows. Although there may not actually be a charge at the point P, imagine that a very small charge q is placed there. The electric forces exerted on it by the surrounding charges add together to produce a total resultant force **F**. The electric field at the point is defined as this resultant force divided by the charge on which it acts.

$$\mathbf{E} = \frac{\mathbf{F}}{q} \qquad (5\text{-}6)$$

Conversely, the resultant electric force on a charge placed at any point in space is the product of the charge and the electric field at that point.

Electric force on a charge at any point in space
 = (magnitude of charge)
  × (electric field at the point)
$$\qquad (5\text{-}7)$$
$$\mathbf{F} = q\mathbf{E} \qquad (5\text{-}8)$$

If the charge q is positive, the force **F** is in exactly the same direction as the electric field vector **E**. If the charge q is negative, the force **F** is in exactly the opposite direction to **E**.

5.8
ELECTRIC FIELD LINES

A very vivid pictorial representation of an electric field is obtained by visualizing *electric field lines* in space, as in Figure 5-7. The field lines show the direction of the electric field vector at all points. More precisely, the direction of the electric field at any point is tangential to the field line passing through that point. This is illustrated at the point P in Figure 5-7d.

Suppose that we explore the electric field with a small positive test charge q. If the test charge is placed near a positive charge, it experiences a strong repulsion in a direction pointing away from the positive charge. Electric field lines diverge from a positive charge. A positive charge is a source of field lines.

On the other hand, if the positive test charge is placed near a negative charge, it experiences a strong attraction in a direction pointing toward the negative charge. Electric field lines converge onto a negative charge. A negative charge is a sink for field lines. All this is obvious if you examine Figure 5-7 carefully.

Field lines also give an immediate impression of the strength of the field. When the lines are close together, as in the immediate vicinity of a charge, the field is strong. When the lines are far apart the field is weak.

5.9
THE ELECTRIC POTENTIAL AT A POINT

There is an alternative way of describing the electrical properties of space, using a scalar quantity called the *electric potential*. When the small test charge q is placed at the point P, it acquires a potential energy. This is the sum of its mutual electric potential energies with all of the surrounding charges. Dividing this electric potential energy by the magnitude of q gives us the electric potential at the point P. Conversely, the electric potential energy associated with the test charge is equal to the magnitude of its charge multiplied by the potential at the point where it is located.

Notice the distinction between the electric potential at a point and the mutual electric potential energy of two charges. The mutual potential energy is a property of two charges. It is an energy measured in joules. The electric potential is a property of a point in space. It is

5.9 THE ELECTRIC POTENTIAL AT A POINT

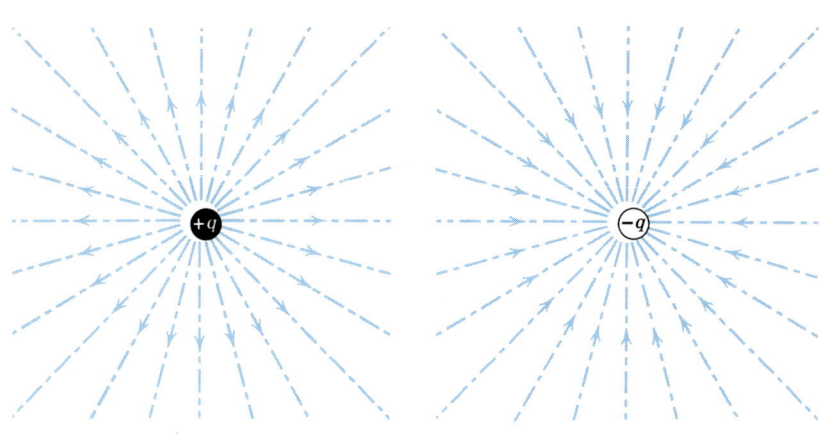

(a) A single point positive charge.

(b) A single point negative charge.

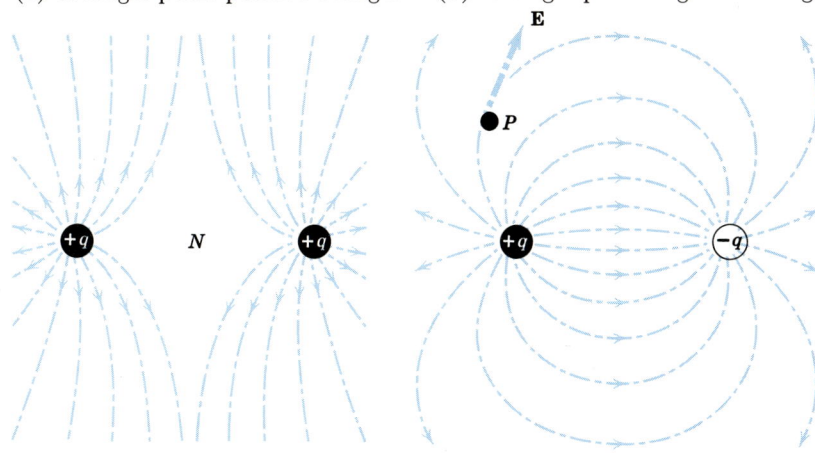

(c) Two equal positive charges.

(d) A positive charge and a negative charge of equal magnitude.

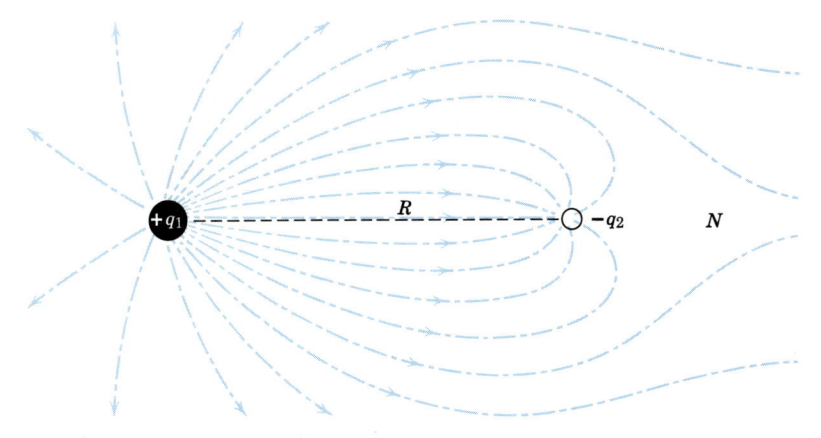

(e) A large positive charge and a small negative charge.

FIGURE 5-7
Some typical patterns of electric field lines.

an energy divided by a charge. Its units could be joules per coulomb, but it is important enough to have a special unit of its own. The SI unit of potential is the *volt*, abbreviated V.

5.10
POTENTIAL DIFFERENCE OR VOLTAGE

We shall often be interested in the difference between the potentials at two points in space. If the potentials are measured in volts, this potential difference is often referred to as the *voltage* between the two points. We shall represent it by the italic letter V, which should not be confused with the upright Roman letter V used as an abbreviation for the SI unit of potential, the volt.

The potential energy of a charge is equal to the magnitude of the charge q multiplied by the potential at the point where it is located. Suppose that the charge moves to another point where the potential is greater by an amount V. The increase in its potential energy is equal to its magnitude q multiplied by the potential difference, or voltage, V.

When an electric charge moves through a potential difference, the change in its potential energy in joules is equal to the magnitude of the charge q in coulombs multiplied by the potential difference in volts.

Change in potential energy
= (magnitude of charge)
$\quad\times$ (voltage) $\quad\quad\quad\quad\quad$ (5-9)
= qV $\quad\quad\quad\quad\quad\quad\quad\quad\quad$ (5-10)

5.11
ELECTRIC CURRENTS AND BATTERIES

The practical applications of electricity are numerous and well known, but they hardly ever make use of the electric forces between stationary charged bodies. They rely instead on the effects produced by an *electric current* flowing through a metal wire that has no net electric charge. When an electric appliance is plugged into a wall socket, electrons flow in through one pin of the plug, circulate through the wires inside the appliance, and leave via the other pin of the plug. As many electrons leave as enter, and

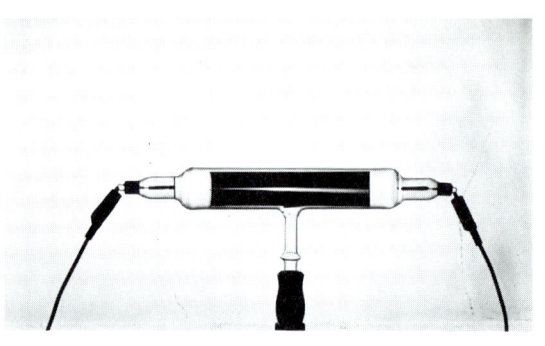

An electric current can flow through an evacuated glass tube as a beam of fast-moving electrons. When the electrons collide with the few remaining atoms in the vacuum, light is emitted and the beam glows. (Fundamental Photographs from The Granger Collection.)

there is never any net accumulation of charge inside the appliance. Actually, since the wall socket usually supplies 60-cycle *alternating current*, the electrons surge backward and forward 60 times a second. This is a nonessential complication. It will be simpler to discuss the *direct current* produced by a *storage battery* of the kind used to operate the ignition system of an automobile.

Figure 5-8 shows a storage battery with a copper wire connected between its terminals. The battery consists essentially of two metal plates immersed in dilute sulfuric acid. One of the plates is coated with sponge lead and the other is coated with lead dioxide. In Section 5.1 we stated that when a glass rod is rubbed with a silk cloth, the electrons would rather be in the silk than the glass and therefore migrate from the glass to the silk. Similarly, electrons prefer to be in lead than in lead dioxide. By a series of elaborate processes, electrons are transferred from the lead dioxide plate through the sulfuric acid to the lead plate. The lead dioxide plate then has a deficiency of electrons, an excess of protons, and a positive charge; it is the positive *terminal* of the battery. The lead plate acquires an excess of electrons and is negatively charged; it is the negative terminal of the battery.

The electric potential at the positive terminal is greater than the potential at the negative terminal. The battery operates in such a way that the potential difference between its terminals maintains an almost constant value of about 2 volts.

If a copper wire is connected between the terminals, electrons flow continuously from the negative terminal through the wire to the positive terminal. The electrons then return to the negative terminal through the battery. In copper, as in other metals, some of the electrons on the outside of the copper atom are very loosely

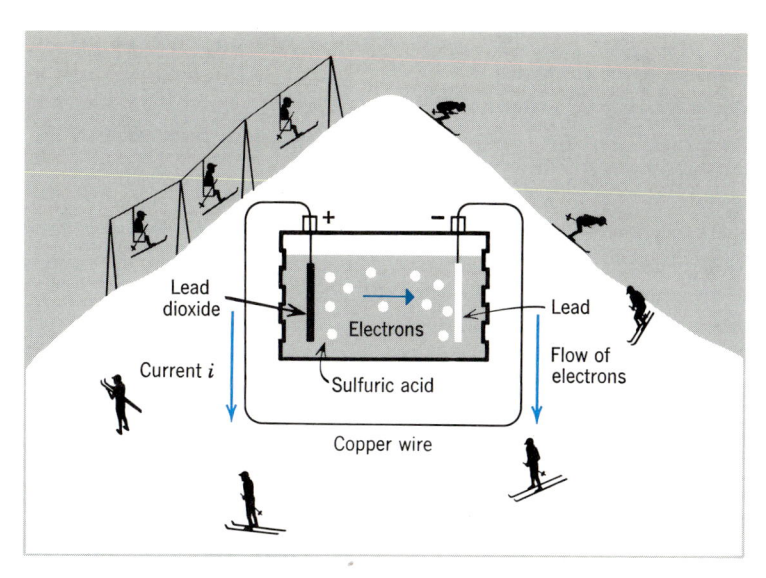

FIGURE 5-8
A storage battery with a copper wire connected between its terminals compared with a ski lift and a ski run.

held and are easily shaken free. These electrons do not remain attached to a particular copper atom, but wander freely throughout the whole of the copper wire. Because they are negatively charged, they are repelled by the negative terminal and attracted by the positive terminal. They are therefore continually accelerating toward the positive terminal. This does not mean that the electrons move faster and faster the nearer they get to the positive terminal because they frequently collide with the copper atoms. Although an electron acquires an extra velocity toward the positive terminal in between collisions, this extra velocity is destroyed at each collision. However, in between collisions the electron does get a chance to move toward the positive terminal. The net result is a slow drift of the electrons toward this terminal.

If a piece of wood were connected between the terminals, no current would flow. All the electrons in the wood are firmly held to their atoms and are not free to move from one terminal to the other. Wood is an *insulator* because it will not carry an electric current. Copper is a *conductor* of electricity. All metals are conductors.

When the electrons flowing through the copper wire arrive at the positive terminal, they must be returned to the negative terminal. Otherwise the charges on the two terminals would soon be neutralized, and the current would cease to flow. The transfer of electrons through the sulfuric acid to the negative terminal takes place against the repulsion of the negative charge on this terminal. This transfer is a consequence of certain complicated chemical processes that occur inside the battery.

In Figure 5-8 a storage battery is compared with a ski lift and a ski run. The chemical processes inside the battery that lift the electrons from the positive terminal to the negative terminal are analogous to the ski lift that lifts the skiers from the bottom to the top of the mountain. The electron being pulled through the copper wire by the electric attraction of the positive terminal is analogous to the skier gliding down the ski run by taking advantage of the gravitational attraction of the earth. The electron maintains a steady drift velocity because of its numerous collisions with the copper atoms. Similarly, the skier can maintain a steady velocity down the run by making skillful use of the frictional forces exerted on the skis by the snow.

5.12 THE AMPERE

The numerical measure of an electric current is the charge flowing past any point of the wire in one second. If a charge q flows past in t seconds, the current i is the charge divided by the time.

$$\text{Electric current} = \frac{\text{charge flowing}}{\text{time taken}} \quad (5\text{-}11)$$

$$i = \frac{q}{t} \quad (5\text{-}12)$$

Charge flowing past any point in a given time

$$= (\text{current}) \times (\text{time}) \quad (5\text{-}13)$$

$$q = it \quad (5\text{-}14)$$

The SI unit of current is the *ampere,* with the abbreviation A. If a current of one ampere flows in a wire, then one coulomb of electric charge flows past any point in the wire every second. A small flashlight bulb carries a current of about one ampere. About ten billion billion (10^{19}) electrons flow through the bulb every second.

An electric current is actually a flow of negatively charged electrons. It is equivalent to a flow of positive charge in the opposite direction to the flow of the negative electrons. It is traditional to pretend that positive charge is flowing and to take the direction of the current to be the direction of flow of this imaginary positive charge. The direction of the current is therefore always taken to be opposite to the direction of flow of the electrons.

The arrows in Figure 5-8 illustrate this point. Electrons flow through the copper wire from the negative terminal to the positive terminal. Nevertheless, we always pretend that the current is a flow of positive electricity from the positive terminal through the wire to the negative terminal.

5.13
OHM'S LAW

When we connect a wire between the terminals of a battery, what determines the size of the current flowing through the wire? The current is driven by the voltage between the two terminals. The size of the current i is in fact proportional to the voltage V.

On the other hand, the flow of electrons is hindered by their collisions with the atoms of the wire. The effectiveness of these collisions in limiting the size of the current is determined by a physical property of the wire called its *electric resistance*. We often omit the word *electric* and talk about "the resistance of the wire." The symbol R is commonly used to represent the numerical magnitude of this resistance.

The SI unit of electric resistance is the *ohm*. Its abbreviation is Ω, the Greek capital letter omega.

The current is inversely proportional to the resistance. The larger the resistance, the smaller the current. *Ohm's law* combines the two facts that the current i is proportional to the voltage V and inversely proportional to the resistance R.

Ohm's law:

$$\text{Current in amperes} = \frac{\text{voltage in volts}}{\text{resistance in ohms}} \quad (5\text{-}15)$$

$$i = \frac{V}{R} \quad (5\text{-}16)$$

The resistance of a wire depends on its dimensions. The resistance is proportional to the length of the wire. It seems very reasonable that a long wire should offer more resistance to the flow of the electrons than a short wire. You might also guess that a thin wire would offer more resistance than a fat wire. The resistance is in fact inversely proportional to the cross-sectional area of the wire.

The resistance is very dependent on the material of the wire. All metals are good conductors of electricity, but some are better than others. Copper is a better conductor than iron. A copper wire has a smaller resistance than an iron wire with the same dimensions. Nonmetals are very poor conductors under most circumstances. If a glass fiber is connected between the terminals of a battery, the current is so small that it can be considered to be zero for most practical purposes. The resistance of the glass fiber is enormously large. Glass is an insulator.

5.14
ELECTRIC POWER

Figure 5-9 shows an electrical device connected across the terminals of a storage battery. This device might be something simple like the copper wire we have previously discussed. Or it might be something complicated like a small electric motor. All that matters to us at the moment is that it draws a current i from the positive terminal and returns a current i to the negative terminal. The voltage applied to the device is V.

In one second a charge of i coulombs flows from the positive terminal through the device to the negative terminal. The potential at the neg-

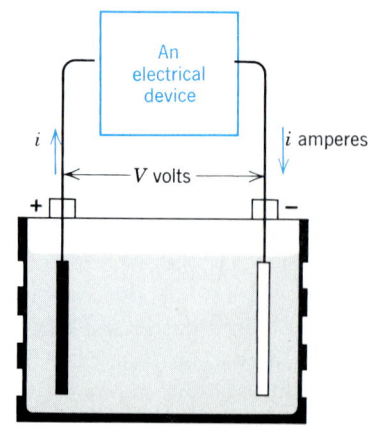

FIGURE 5-9
A voltage of V volts is applied to an electric device, and it draws a current of i amperes.

ative terminal is lower than the potential at the positive terminal by V volts. Therefore, according to Equation 5-10, this charge of i coulombs loses an amount of potential energy equal to iV joules.

The energy lost by the charge reappears inside the device and is used in some manner that depends on the nature of this device. Whatever the details, in one second an energy iV joules is delivered to the device. According to Equation 3-39, the energy in joules delivered per second is equal to the power in watts. The power delivered to the device is therefore iV watts.

> Electric power in watts
> $\quad$ = (current in amperes)
> $\quad\quad$ × (voltage in volts) $\quad$ (5-17)
> $\quad$ = iV $\quad\quad\quad\quad\quad\quad\quad\quad\quad$ (5-18)

5.15 ELECTRIC HEAT

Let us return to the simple case of a metal wire connected between the terminals of the battery. What happens to the power delivered by the battery to this wire?

Consider the behavior of the free electrons drifting through the wire. A negatively charged electron is repelled by the negative terminal and attracted by the positive terminal. It therefore accelerates toward the positive terminal. As it picks up speed its kinetic energy increases. However, it soon loses its extra kinetic energy when it collides with an atom of the metal wire.

An atom in a solid stays near a fixed point but vibrates rapidly around this point. The colliding electron makes the atom vibrate still more rapidly. The energy therefore finally turns up as the energy of random vibration of the atoms of the metal wire. That type of energy is exactly what we mean by heat. The power delivered by the battery appears as heat in the metal wire and raises its temperature.

Eventually, of course, the metal wire settles down at a steady high temperature. It may even glow red hot, as in the heater of an electric range. All the heat supplied by the electric current then leaks away to the surroundings.

Because all the power is used to develop heat, Equation 5-18 takes the form

$$\text{Heat developed per second} = iV \quad (5-19)$$

Equation 5-16 (Ohm's law) can be expressed in a slightly different form by multiplying both sides by R.

$$V = iR \quad (5-20)$$

If we insert this expression for V into Equation 5-19, we obtain the following:

$$\text{Heat developed per second} = i(iR) \quad (5-21)$$
$$= i^2 R \quad (5-22)$$

Heat developed per second
$\quad$ = (the square of the current)
$\quad\quad$ × (the resistance) $\quad$ (5-23)

5.16 ALTERNATING CURRENT

The current from a storage battery always has a steady value and always flows in the same direction. It is called a *direct current*, often abbreviated DC. Throughout most of the United States electric power lines carry 60-cycle *alternating current*, abbreviated AC.

To understand what this means, consider an electric appliance plugged into a wall socket. Label the two pins of the plug A and B. The potential difference between A and B does not have a steady value as in the case of a storage battery. It continually oscillates backward and forward; 60 times every second pin A has a higher potential than pin B, and 60 times every second pin B has a higher potential than pin A.

The current flowing through the appliance behaves in a similar way. It repeatedly reverses the direction of its flow; 60 times a second it flows in one direction, and 60 times a second it flows in the opposite direction.

Under these circumstances, what meaning can be given to the "voltage" supplied by the wall socket when in fact the potential difference between the pins is continually changing? What meaning can be given to the current drawn by the appliance if the current flows as often in one

Electric power lines leaving the Tennessee Valley Authority's Bull Run electric generating plant. In these metal lines, the electrons surge backward and forward 60 times per second.

direction as in the opposite direction? You might argue that a current flowing as often in one direction as in the opposite direction is equivalent to no current at all. A simple consideration shows that this cannot be so. As the current flows in one direction through a wire resistance it creates heat. If its direction is reversed, it does not take the heat away again. It continues to create more heat.

We can therefore define a useful *effective current* in the following way. It is the magnitude of the direct current that would create heat at the same rate. A useful *effective voltage* can be defined in a similar way. It is the magnitude of the steady voltage that would have to be applied across the resistance to create heat at the same rate as the alternating voltage. It can be shown that the effective current and the effective voltage defined in this way obey Ohm's law. The effective current is equal to the effective voltage divided by the resistance.

Therefore, as long as we use the effective current i and the effective voltage V, we can proceed in exactly the same way that we do for direct current. Equation 5-15 expressing Ohm's law is still true. Equation 5-17 still gives the power supplied to the electric appliance. Equation 5-23 still gives the rate at which heat is developed in a resistance. We usually take the above reasoning for granted and drop the word *effective*.

Consider a simple example. The electric power supplied to my house has an effective voltage of 110 volts. I have a small space heater rated at 110 volts and 1350 watts. It is important to me to know what effective current it will draw. If the current exceeds 15 amperes, a fuse will blow in my basement. In Equation 5-17, I put the power equal to 1350 watts and the voltage equal to 110 volts. Then

$$1350 = (\text{current in amperes}) \times 110 \quad (5\text{-}24)$$

$$\text{Current in amperes} = \frac{1350}{110} \quad (5\text{-}25)$$

$$= 12.27 \text{ A} \quad (5\text{-}26)$$

This is less than 15 A and the fuse will not blow.

I can now calculate the resistance of the heating element in my space heater. In Equation 5-15 (Ohm's law), I put the voltage equal to 110 volts and the current equal to 12.27 amperes. Then

$$12.27 = \frac{110}{(\text{resistance in ohms})} \quad (5\text{-}27)$$

$$\text{Resistance in ohms} = \frac{110}{12.27} \quad (5\text{-}28)$$

$$= 8.96 \, \Omega \quad (5\text{-}29)$$

SUMMARY

Electric charge. An object with an excess of electrons is negatively charged. An object with a deficiency of electrons, which is the same as an excess of protons, is positively charged. Like charges repel and unlike charges attract. The SI unit of charge is the coulomb, abbreviated C. The charge on a fundamental particle is represented by the symbol e and has a magnitude of 1.6×10^{-19} C.

Conservation of charge. Electricity cannot be created or destroyed. Charged fundamental particles are always created or destroyed in pairs, one positive and one negative.

Coulomb's law. The force between two stationary charges is equal to the electric force constant multiplied by the product of the two charges and divided by the square of the distance between them. Electric forces are much stronger than gravitational forces.

Electric potential energy is a mutual property of two charges. It is equal to the electric force constant multiplied by the product of the two charges and divided by the distance between them (not the square of the distance). It is positive for like charges and negative for unlike charges.

The concept of a field. Space is more than an empty featureless background through which particles move. It is possible to assign properties to each point in space. Forces are handed on from point to point through apparently empty space.

The electric field **E**. If a charge were placed at any point in space, the electric force on it would be equal to the magnitude of the charge multiplied by the electric field vector **E**. An electric field can be pictured as a pattern of electric field lines. At any point the electric field vector is tangential to the field line. When the field lines are closer together, the field is stronger. Field lines diverge outward from a positive charge and converge inward on a negative charge.

The electric potential at a point is obtained by placing a small positive charge at the point and dividing its total potential energy by the magnitude of its charge. Potential energy is an energy measured in joules. The potential at a point is energy per unit charge. The SI unit of potential is the volt, abbreviated V. The potential difference V between two points is frequently called the voltage between them. When a charge q is moved from one point to another point where the potential is greater by an amount V, the increase in its potential energy is equal to the product of the charge and the voltage, or qV.

Electric current is defined as the charge flowing past any point in one second. The SI unit of current is the ampere, abbreviated A. One ampere is one coulomb per second. All metals are conductors of electricity. Nonmetals are usually insulators. In a metal the current consists of free electrons drifting through the metal in the opposite direction to the conventional positive current.

Ohm's law. The current in amperes is equal to the voltage in volts divided by the resistance in ohms (Ω).

Electric power in watts delivered to any device is equal to the current through it in amperes multiplied by the voltage across it in volts.

Electric heat. The power delivered to a wire resistance is converted into heat. The heat energy created per second is equal to the square of the current multiplied by the resistance.

Alternating current (AC) periodically changes direction. The effective current is defined as the steady direct current (DC) that would create heat at the same rate. The effective voltage is defined as the steady one-way voltage that would create heat at the same rate. The equations applicable to direct current can be applied to alternating current if the effective current and effective voltage are used.

TERMS AND CONCEPTS

electric charge
conservation of charge
Coulomb's law
electric force constant
coulomb
charge on the electron, *e*
electric potential energy
field
action at a distance
ether
electric field
electric field line
electric potential
volt
voltage
electric current
storage battery
insulator
conductor
terminal
ampere
Ohm's law
electric resistance
ohm
electric power
electric heat
direct current (DC)
alternating current (AC)
effective current
effective voltage

QUESTIONS

1. If you were given a charged object, how would you decide whether its charge was positive or negative?

2. How can electric forces be related to the properties of fundamental particles?

3. Does the mass of an object change when it is given a charge? If so, does it increase or decrease?

4. What aspect of the behavior of fundamental particles explains the law of conservation of charge?

5. State Coulomb's law. Compare and contrast it with Newton's law of universal gravitation.

6. What is the basic idea underlying the concept of a field?

7. Define an electric field line. Why is it a useful concept?

8. Can two electric field lines cross at an angle?

9. Distinguish carefully between electric potential energy and electric potential.

10. Is a free electron attracted toward a point of greater or lesser potential? Is a positive charge attracted toward a point of greater or lesser potential?

11. You have a length of copper wire 1 meter long and 1 millimeter in diameter. You wish to obtain another length of wire with the same electric resistance. The only available copper wire has a diameter of 0.5 mm. Will you need a longer or shorter length of it?

12. In Figure 5–9, why is the current returning to the negative terminal the same as the current leaving the positive terminal? Does the electrical device "consume" electricity? If not, what does it consume?

13. Explain the difference between direct current and alternating current. Can alternating current deliver power?

PROBLEMS

1. A charge of +1.5 coulombs is 6 meters away from a charge of +4.0 coulombs. What is the magnitude of the electric force between them?

2. What is the electric potential energy of the two charges in the previous question?

3. When a charge of 0.1 C is placed at a point in space, the electric force on it is 3.5 N. What is the electric field at the point?

4. What is the electric force on a proton in an electric field of 3 N/C?

5. What is the electric potential at a distance of 5 m from a charge of 2.5 C if there are no other charges present?

6. A 2-volt storage battery promises to deliver 1 ampere for 100 hours before running down. If it does so, how many coulombs of charge will have passed between its terminals? How much energy will it have delivered?

7. A wire with a resistance of 10 ohms is connected between the terminals of a 2-volt battery. What is the current?

8. An electric light bulb has ratings of 110 volts and 60 watts. What current does it draw? What is its resistance?

9. The "unit" of electricity quoted on an electricity bill is one kilowatt hour, which is the energy consumed in one hour when the power delivered is one kilowatt. How many joules are there in one kilowatt hour?

10. If a current of 2 A flows through a resistance of 150 Ω, how much heat is created per second?

CHAPTER 6
MAGNETISM

6.1 FORCES BETWEEN MOVING CHARGES

Most people associate *magnetism* with *bar magnets*. A freely suspended magnet points north, as in a *magnetic compass*. A bar magnet attracts small iron or steel objects that stick to its ends. Two bar magnets attract or repel each other, depending on which way they are placed.

Physics has a more fundamental view of the matter. Magnetism is a consequence of the extra forces between moving charges. When the charges are stationary they exert electric forces on one another. When they are both moving, the electric forces are still present but there are also *magnetic forces*. The magnetic forces depend in a complicated way on the velocities of the charges. In the present chapter we first consider the nature of the magnetic forces between moving charges. We then explain how these forces are related to the behavior of bar magnets.

Figure 6-1 illustrates an experiment in which the moving charges are free electrons drifting through metal wires. Two adjacent vertical wires are freely hinged at the top. Their lower ends dip into mercury pools. This leaves the wires free to move sideways. Also, because mercury is a metal and conducts electricity, it enables electric currents to be fed into the wires from the two storage batteries shown. When the storage batteries are connected so that the two currents flow in the same direction, the wires move toward each other. When the currents flow in opposite directions, the wires move apart. Parallel currents moving in the same direction attract, and parallel currents moving in opposite directions repel. One way to memorize this is to notice that it is the other way around for electric forces. Like charges repel and unlike charges attract.

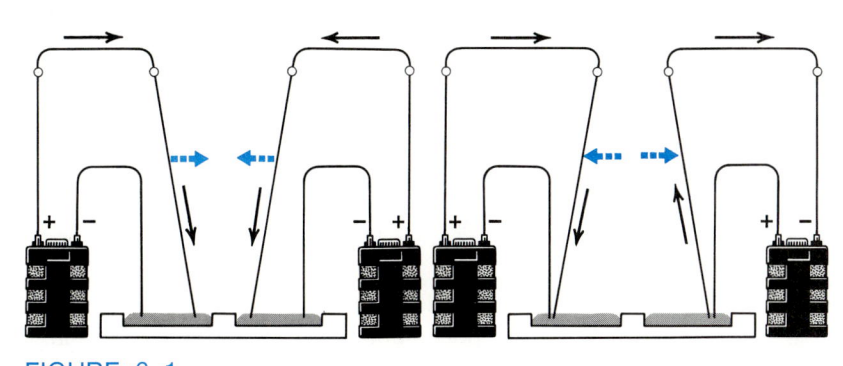

FIGURE 6-1
An experiment to demonstrate the magnetic forces between parallel electric currents.

A bar magnet picking up a chain of steel paper clips. (K. T. Bendo.)

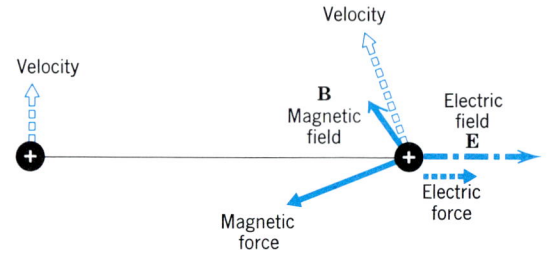

FIGURE 6-2
The magnetic field as an intermediate step in the calculation of a magnetic force.

In this experiment the wires always contain as many protons as electrons and do not accumulate any net charge. The forces between the wires are not the electric forces discussed in Section 5.1. They are a consequence of the fact that electrons are moving inside the wires. When two charged particles are both in motion, they exert on each other a new kind of force that depends on their velocities. This new force is zero if either velocity is zero. The forces in Figure 6-1 disappear if either current is switched off.

6.2
THE MAGNETIC FIELD

If we were to proceed as we have done previously, we would write down a law for this new magnetic force. It would be similar to Newton's law of universal gravitation or Coulomb's law for electric forces. It would relate the magnetic force to the charges, their velocities, and their distance apart. Unfortunately, it is not possible to formulate such a law in any simple form. Even a reasonable approximation to the law is very cumbersome. At this point, the concept of the *magnetic field* makes the discussion much easier.

Consider the particular case of Figure 6-2 in which two positive charges are both moving in the plane of the paper. Suppose we were interested in the electric force exerted on the right-hand charge by the left-hand charge. We could start by saying that the charge on the left produces an electric field **E** at the point where the charge on the right is located. The corresponding electric force is equal to the magnitude of the charge on the right multiplied by this electric field. Both the electric field and the electric force point directly away from the left-hand charge that produces them.

A similar procedure can be used to find the magnetic force acting on the charge on the right. The first step is to say that because the charge on the left is moving, it produces a magnetic field at the point where the charge on the right is located. A magnetic field is usually represented by the symbol **B**. A novel feature of this magnetic field is that its direction is not along the line joining the two charges. It is at right angles to this line. It is also at right angles to the direction of the velocity of the left-hand charge that produces it. In Figure 6-2 it goes perpendicularly into the plane of the paper.

The second step in the procedure is to say that because the charge on the right is moving, the magnetic field exerts a magnetic force on it. This force is not in the direction of the magnetic field. It is always at right angles to the magnetic field. It is also at right angles to the direction of the velocity of the right-hand charge on which it acts. In Figure 6-2, the magnetic force is in the plane of the paper, but tilts downward. Unlike the electric force, it does not necessarily point along the line joining the two charges.

The magnetic force on the left-hand charge can be obtained by a similar procedure. It is not equal in magnitude, but opposite in direction, to the magnetic force on the right-hand charge. It has an entirely different direction and a different magnitude. Magnetic forces do not obey Newton's third law of motion (Section 2.7). It is easy to see that the magnetic force on the left-hand charge is not opposite in direction to the magnetic force on the right-hand charge. The magnetic force on the left-hand charge must be at right angles to the velocity of this charge. In Figure 6-2 the magnetic force on the right-hand charge is clearly seen not to be at right angles to the velocity of the left-hand charge. A force in the opposite direction would also not be at right angles to the velocity of the left-hand charge.

Magnetic forces are obviously very complicated. In particular, the direction of a magnetic force cannot be obtained from any simple rule. In general, there are two steps in the calculation of a magnetic force, usually kept separate. First is the question of the magnetic field produced by an arrangement of moving charges or electric currents. Next is the question of the magnetic force that this magnetic field exerts on a moving charge or on an electric current.

We shall consider the production of the magnetic field in the next three sections. Then, in Section 6.6, we shall consider the magnetic force exerted by a magnetic field.

6.3 MAGNETIC FIELD OF A SINGLE CHARGE

The simplest way to appreciate the nature of a magnetic field is to consider the pattern of *magnetic field lines*. A magnetic field line is defined in a similar way to an electric field line. The direction of the magnetic field vector **B** at any point is tangential to the magnetic field line passing through that point. The direction of an electric force on a positive charge is the same as the direction of the electric field vector **E**. However, the magnetic force on a moving charge is not in the same direction as the magnetic field acting on it. The magnetic force is at right angles to the magnetic field. The magnetic field lines therefore give the direction of the magnetic field, but not the direction of the magnetic force.

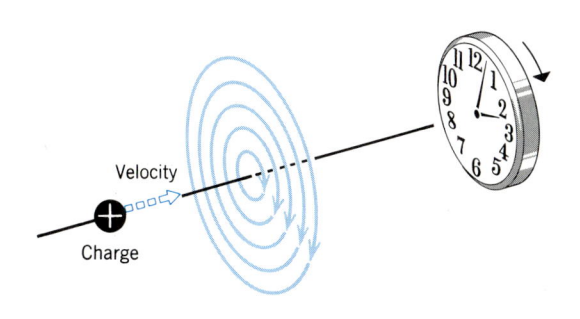

FIGURE 6-3
Magnetic field lines produced by a charge moving with constant speed along a straight line. Circular field lines like these surround each point on the straight-line path.

Figure 6-3 shows some of the magnetic field lines produced by a single positive charge moving along a straight-line path with constant speed. The magnetic field lines are circles centered on a particular point of the path. A similar set of concentric circles surrounds each point on the path. The field lines therefore occupy the whole of space.

To decide which way the arrowheads go around the circular field lines, look along the straight-line path in the direction in which the positive charge is moving. The magnetic field lines then go around in the same direction as the hands of a clock.

6.4 MAGNETIC FIELD OF A LONG, STRAIGHT CURRENT

When an electric current flows through a long, straight wire, the free electrons in the wire all move in a direction parallel to the length of the wire. The magnetic field lines in this case are

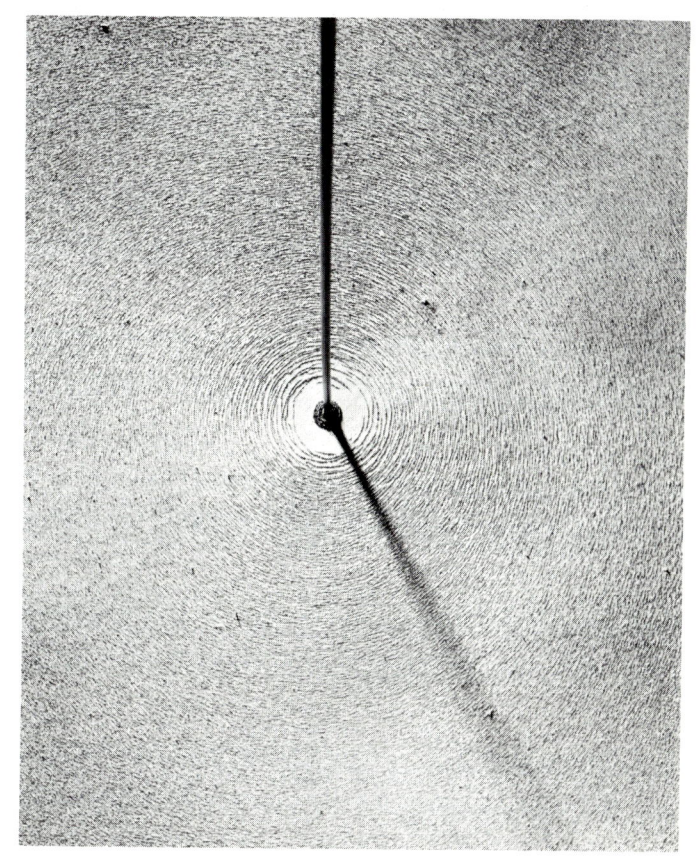

FIGURE 6-4
The circular magnetic field lines surrounding a straight wire carrying a current, as revealed by the iron filings technique (photograph by Berenice Abbott).

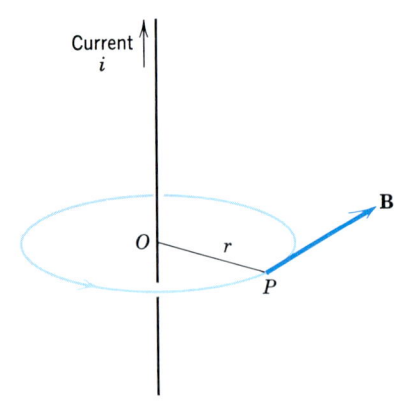

6 MAGNETISM

FIGURE 6-5
The magnetic field at a distance r from a long, straight current i.

therefore also circles centered on the wire. Figure 6-4 was obtained by sprinkling iron filings on a sheet of paper perpendicular to a straight wire carrying a current. Iron filings are long and thin. They align themselves in the direction of the magnetic field, in the same way that a compass needle points along the earth's magnetic field. The circles centered on the wire are easily seen in this photograph.

Figure 6-5 shows the magnetic field **B** at a point P which is at a distance r from the long, straight current. The magnetic field line through P is a circle of radius r with its center O on the wire. The magnetic field is tangential to this circle and therefore at right angles to the radius OP.

We quote the equation giving the magnitude of the magnetic field B in order to raise a very important point.

Magnetic field

$$= \frac{2 \times \text{(the electric force constant)} \times \text{(the current)}}{\text{(the square of a certain constant)} \times \text{(the radius)}} \quad (6\text{-}1)$$

$$B = \frac{2K_e i}{c^2 r} \quad (6\text{-}2)$$

The magnetic field is proportional to the current i and inversely proportional to the distance r from the wire. The electric force constant K_e is the same constant that appears in Coulomb's law.

The very important point promised above concerns the constant c that is squared in the denominator. Its numerical value is found to be exactly the same as the *speed of light!*

$$c = \text{speed of light} = 2.9979 \times 10^8 \text{ m/s} \quad (6\text{-}3)$$

There is obviously some connection between the nature of light and the laws of electricity and magnetism. The explanation that will emerge toward the end of the present chapter and in the following chapter is anticipated in the sentence "Light is an electromagnetic wave, obeying the laws of electricity and magnetism." Electricity and magnetism are intimately related to each other. The words *electromagnetism* and *electromagnetic* are used to include both electricity and magnetism.

SI units must be used in Equation 6-2. The current i is in amperes. The distance r is in meters. The electric force constant has the value given in Equation 5-3. The constant c has the value given above in Equation 6-3.

The SI unit of magnetic field is the *tesla*. The abbreviation for tesla is T. An old-fashioned unit of magnetic field, the *gauss*, became so well known that it is still frequently used.

$$1 \text{ tesla} = 10^4 \text{ gauss} \quad (6\text{-}4)$$

6.5 MAGNETIC FIELD OF A CIRCULAR CURRENT

Figure 6-6 shows the pattern of magnetic field lines in the important case of a current in a circular wire. The field lines are no longer perfect circles. Nevertheless, if we stand at any point on the wire and look in the direction of the current, the field lines go around the wire in a clockwise direction.

The reader may already have noticed one important difference between electric fields and magnetic fields. Electric field lines sometimes start on a positive charge and end on a negative

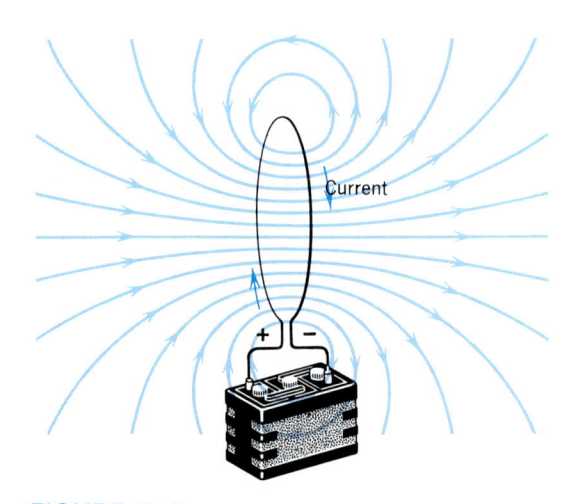

FIGURE 6-6
Magnetic field lines produced by a current in a circular wire.

6.7 BAR MAGNETS

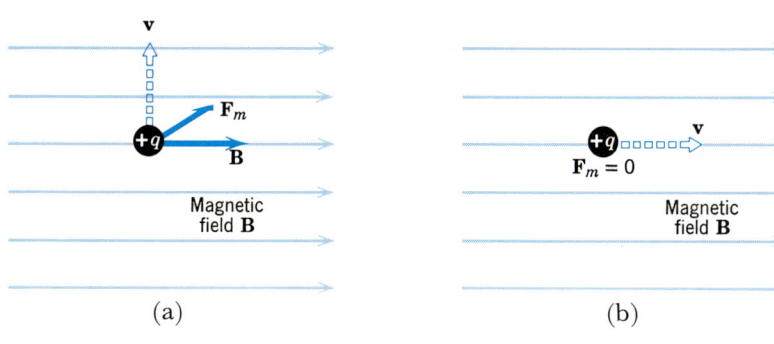

FIGURE 6-7
The component of the velocity of the charge in a direction at right angles to the magnetic field determines the magnetic force. (a) The charge moves at right angles to the magnetic field. There is a magnetic force. (b) The charge moves parallel to the magnetic field. Its velocity has no component at right angles to the field. There is no magnetic force.

charge. Magnetic field lines never start or end, but always form closed loops. Some of the field lines in Figure 6-6 appear to have a beginning and an end. They would also form closed loops if we could continue them beyond the edges of the figure.

6.6
THE MAGNETIC FORCE

The magnetic force on a moving charge is determined by the magnetic field B at the point where the charge is instantaneously located. It is not necessary to know anything about the origin of B as long as its magnitude and direction are known.

The force is proportional to the magnitude of the charge q and to the magnitude of the magnetic field B. As far as the velocity of the charge is concerned, the force depends only on the magnitude u of the component of this velocity in a direction at right angles to the magnetic field. Consequently, a charge moving at right angles to a magnetic field experiences a magnetic force. A charge moving parallel to a magnetic field experiences no magnetic force (see Figure 6-7). The equation for the magnitude of the magnetic force follows.

Magnetic force
= (charge) × (component of velocity at right angles to field)
× (magnetic field)   (6-5)

$$F_m = quB \qquad (6\text{-}6)$$

The direction of the magnetic force is at right angles to the direction of the velocity of the charge. It is also at right angles to the direction of the magnetic field. A discussion of the situation in Figure 6-8a may help you to understand this. The magnetic field is uniform, which means that it has the same magnitude and direction everywhere. The charge is moving at right angles to the magnetic field. The path of the charge can be shown to be a circle in a plane at right angles to the field. The magnetic force must always be at right angles to both the velocity and the field. If you look carefully at the figure, you will see that the force always points toward the center O of the circle. That is exactly the condition for a circular path (refer back to Section 1.13).

The situation is very similar to the case of a body whirled in a circle on the end of a string as in Figure 6-8b. The string keeps the body moving in a circle by always exerting a force pointing toward the center of the circle. The magnetic field performs the same function.

Figure 6-9 is a photograph of the curved tracks of charged particles moving in a magnetic field. The white tracks are trails of bubbles left behind when a charged particle passes through a liquid on the verge of boiling.

6.7
BAR MAGNETS

The magnetic properties of a bar magnet are a consequence of the fact that its electrons are *spinning* like tops. The electron may be visualized as a small spherical cloud of negative charge, as in Figure 6-10. It spins about the axis labeled

6 MAGNETISM

(a)

(b)

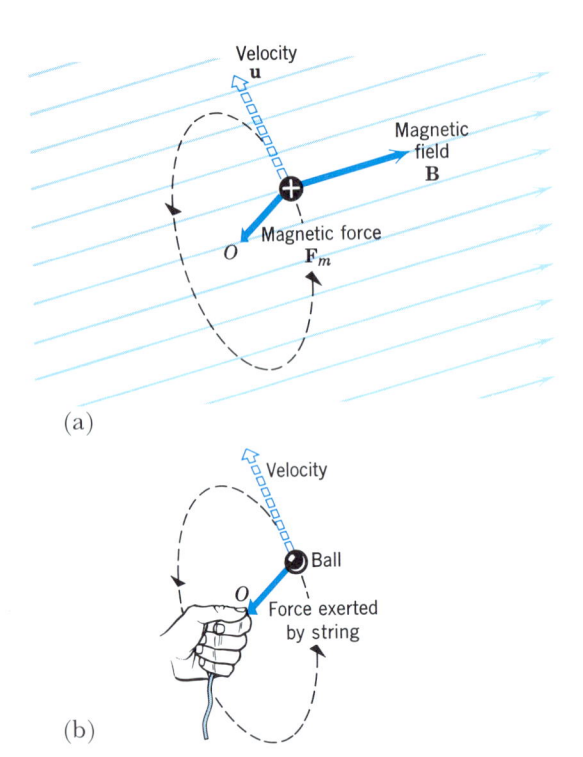

FIGURE 6-8
A charged particle moving in a circle in a uniform magnetic field compared with a ball whirled on the end of a string. (a) The circular path of a charged particle in a uniform magnetic field. (b) A ball whirled in a circle on the end of a string.

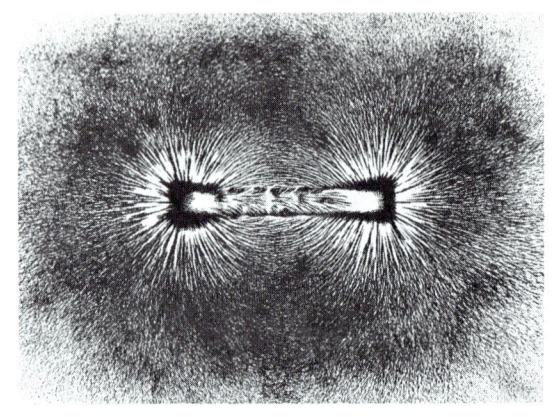

The magnetic field lines of a bar magnet as revealed by the iron-filings technique. (K. T. Bendo.)

SN. Any small portion of the negative charge describes a circular path about this axis. Charge moving in a circle is equivalent to the circular current of Figure 6-6 and produces a similar magnetic field. All the circulating portions of the charged electron combine together to produce the magnetic field shown in Figure 6-10. We should remember, though, that the electron is negatively charged and is equivalent to a positive current circulating the other way round.

Most materials show no net magnetic effects because the electrons in them cancel in pairs. The two members of a pair spin in exactly opposite directions and produce magnetic fields in

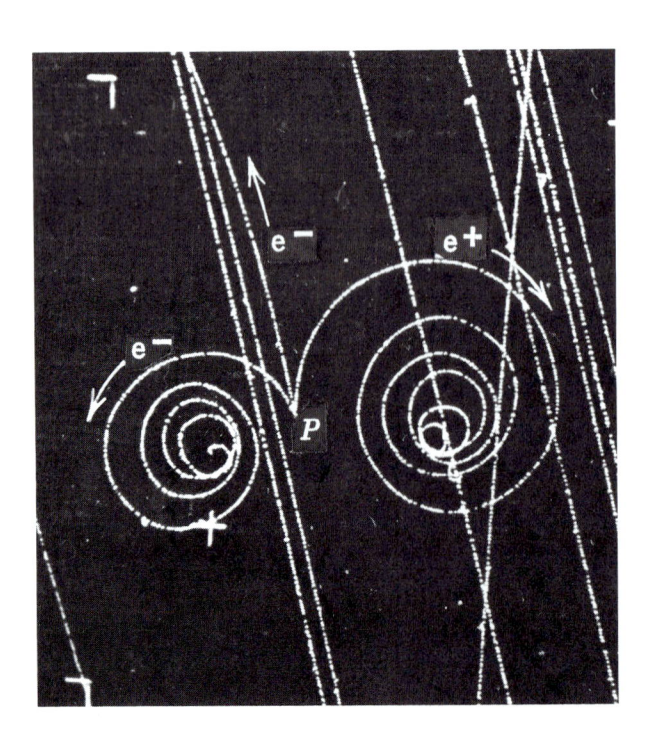

FIGURE 6-9
Curved tracks of electrons in a magnetic field perpendicular to the page. At *P* a certain process suddenly produces two negatively charged electrons labeled e^- and a positron labeled e^+. A positron is identical to an electron except that it is positively charged. A white track is a trail of bubbles left behind after the passage of a charged particle. The electron e^- that moves off to the left initially follows a circular path. It slows down as it collides with the atoms of the liquid through which it is passing. As its speed decreases, the radius of its circular path decreases. It spirals inward. The positron e^+ that moves off to the right behaves in a similar way. Because it is positively charged, its path curves in the opposite direction. The electron e^- in the center has a very high speed. The radius of its circular path is therefore very large. (Courtesy of the Lawrence Radiation Laboratory, University of California, Berkeley.)

6.8 THE MAGNETIC COMPASS

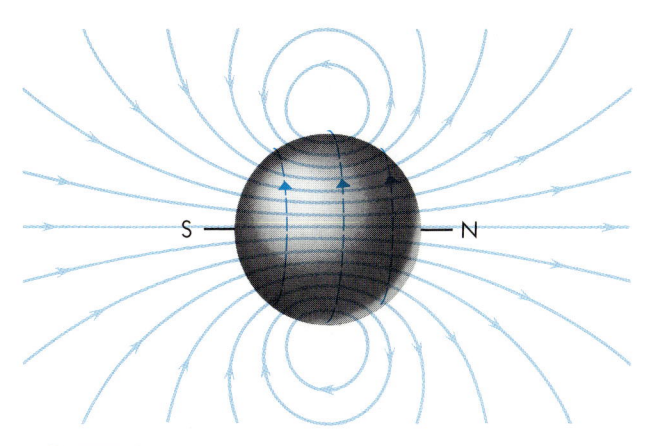

FIGURE 6–10
The electron is a spinning cloud of negative charge. It produces a magnetic field in the same way as the circular current shown on the right. Compare Figure 6–6.

opposite directions to cancel each other. The important exceptions are certain metals such as iron, cobalt, and nickel and some nonmetals. Among the nonmetals is the historically important material magnetite, which was used by the Chinese when they invented the magnetic compass. The exceptional feature of one of these magnetic materials is that some of its electrons decide to cooperate and all spin in the same direction. Each of these electrons produces a magnetic field of the kind shown in Figure 6–10. In a bar magnet made of the material, the small fields of the individual cooperating electrons add together to produce a large magnetic field, as in Figure 6–11. Notice that the magnetic field of a bar magnet is similar to the field of a circular current. Compare Figures 6–6 and 6–11.

A magnetic compass. (K. T. Bendo.)

6.8
THE MAGNETIC COMPASS

If a bar magnet is freely pivoted at its center, it becomes a magnetic compass. It turns until it points in an approximately north-south direction. The end of the bar pointing in a northerly direction is called the *north pole* of the magnet. The other end is called the *south pole*. If the magnet is suspended in any magnetic field, its north pole points in the direction of the field. The earth produces a magnetic field with field lines running approximately, but not exactly, from south to north. This magnetic field lines up a magnetic compass.

To explain all of this, consider the bar magnet of Figure 6–12 inclined at an angle to a magnetic field. The field shown here is an externally produced magnetic field. It should not be confused with the field produced by the bar magnet itself,

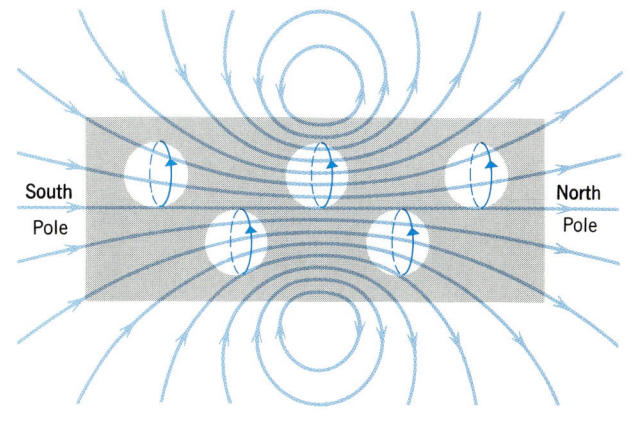

FIGURE 6–11
The magnetic field produced by a bar magnet. Five spinning electrons are shown. In a bar magnet of normal size, there would be about a trillion trillion (10^{24}) spinning electrons.

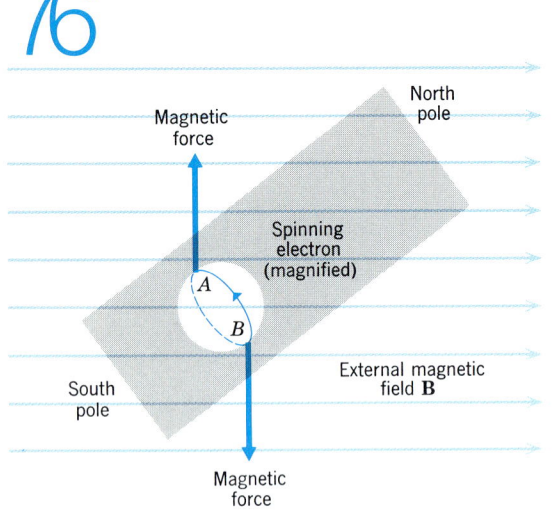

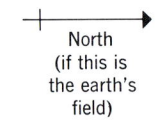

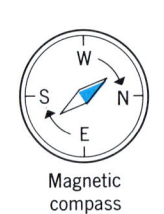

FIGURE 6-12
The origin of the couple that turns a bar magnet until it points in the direction of the external magnetic field. In this way the behavior of a magnetic compass is related to the behavior of the spinning electrons.

as in Figure 6-11. Concentrate on one of the spinning electrons. At point A a small portion of the negative electric charge moves perpendicularly into the plane of the paper. The magnetic force on it is at right angles to the direction of this motion and also at right angles to the magnetic field. In fact, the force is in the plane of the paper and points toward the top of the page.

At point B a small portion of the negative electric charge moves perpendicularly out of the plane of the paper. The velocity is in exactly the opposite direction to the velocity at A. The magnetic force at B is therefore in exactly the opposite direction to the force at A. The force at B points toward the bottom of the page.

The two forces together constitute a couple (Section 3.5). This couple turns the magnet until it points in the direction of the magnetic field, as in Figure 6-13.

Notice that the line joining the south pole to the north pole is the axis of rotation of the electrons. Imagine that you are looking along this line from the south pole toward the north pole. The electrons are seen to be spinning counterclockwise, in the opposite direction to the way the hands of a clock go around. Now compare Figures 6-11 and 6-13. Observe that the magnetic field lines produced by the bar magnet itself in Figure 6-11 enter the south pole and leave the north pole.

6.9 THE FORCES BETWEEN BAR MAGNETS

If the north pole of a magnet is brought near the south pole of another magnet, they attract each other. If two north poles or two south poles are brought together, they repel each other. This can be understood with the help of Figure 6-14.

In Figure 6-14a, the north pole of one magnet is adjacent to the south pole of the other magnet. The electrons in the two magnets are spinning in the same direction. The situation can be compared with two parallel circular currents flowing in the same direction. Refer back to the case of two parallel straight currents in Figure 6-1. When the currents flow in the same direction they attract each other. This is still true when the wires are bent into circles as in Figure 6-14a. Extending the argument to the electrons in the bar magnets, two electrons spinning in the same direction attract. The two magnets attract; a north pole attracts a south pole.

In Figure 6-14b, a north pole has been brought up to another north pole. The electrons in the two magnets are then spinning in opposite directions. The corresponding circular currents flow in opposite directions. The corresponding straight currents flow in opposite directions and repel each other. The spinning electrons therefore repel; the north pole repels the north pole.

A similar argument applies to adjacent south poles. A south pole repels another south pole.

6.10 ELECTROMAGNETIC INDUCTION

Consider the two circular wire loops of Figure 6-15. Loop I is connected to a storage battery that drives a current around it when switch S is

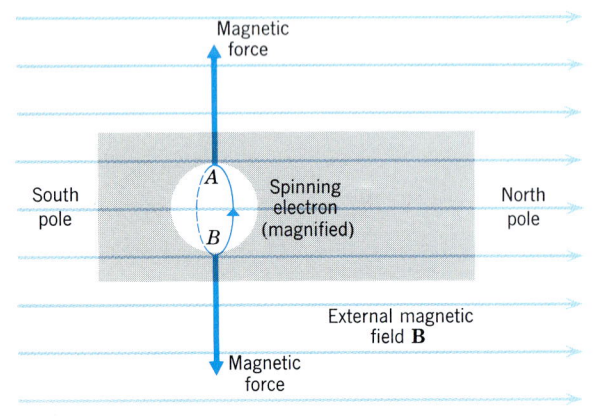

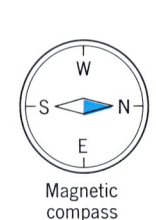

FIGURE 6-13
The bar magnet or magnetic compass at rest pointing in the direction of the external magnetic field. The line from the south pole toward the north pole points in the direction of the external field.

6.10 ELECTROMAGNETIC INDUCTION

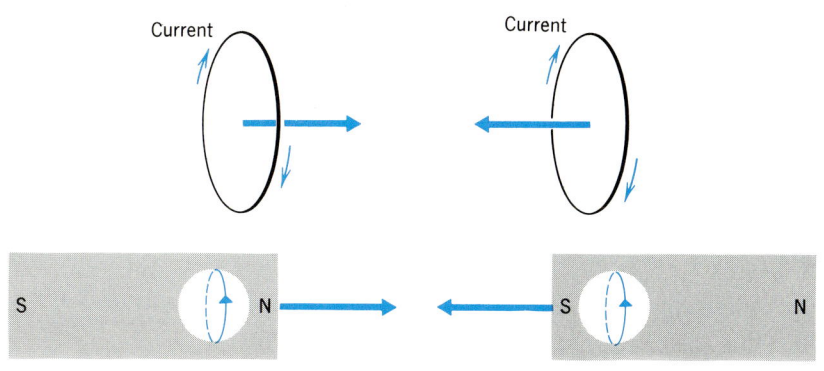

(a) A north pole attracting a south pole is analogous to the attraction between parallel circular currents in the same direction.

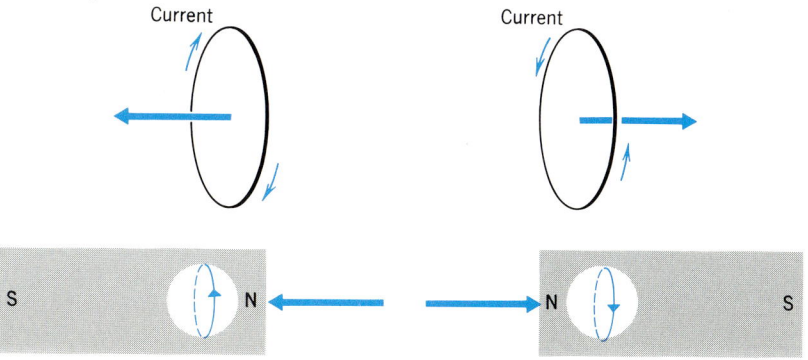

(b) A north pole repelling a north pole is analogous to the repulsion between parallel circular currents in opposite directions.

FIGURE 6-14
Forces between bar magnets related to the forces between spinning electrons.

closed. Loop II has no storage battery but is connected to a galvanometer. A galvanometer is an instrument for detecting the flow of a current. The pointer of this galvanometer is normally at the center of its scale. It deflects to the right if a current flows clockwise around loop II. It deflects to the left if a current flows counterclockwise around loop II.

When there is a steady current in loop I, it produces the magnetic field shown in the figure. Since there is no current in loop II, this magnetic field does not exert a force on loop II. However, if the current in loop I is switched off, an entirely new effect occurs. The galvanometer pointer deflects momentarily to the right, but soon returns to its midposition. While the magnetic field is dying down, a clockwise current flows temporarily around loop II.

Similarly, if the current in loop I is switched on again, the galvanometer is again observed to deflect momentarily but to the left this time. As before, it returns almost immediately to its midposition. While the magnetic field is building up, a counterclockwise current flows temporarily around loop II. As soon as the magnetic field

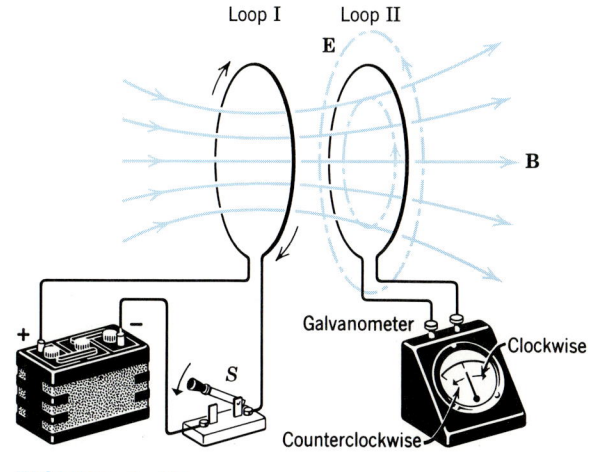

FIGURE 6-15
Electromagnetic induction. When the current in loop I is changing, a current is induced in loop II. When the current in loop I settles down to a steady value, there is no current in loop II.

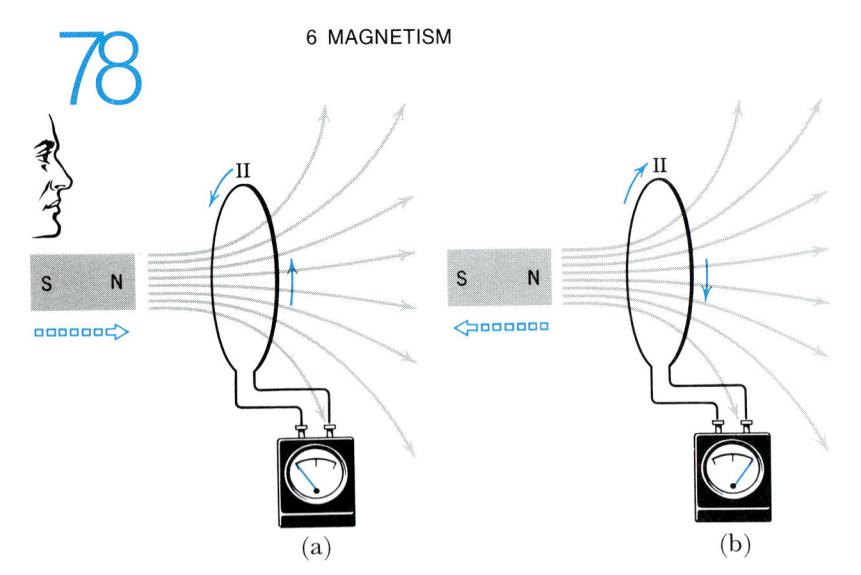

6 MAGNETISM

FIGURE 6-16
A current is induced in loop II by a moving bar magnet. (a) The north pole of the magnet approaches loop II. (b) The magnet is taken away again.

settles down to a steady value, there is no longer a current in loop II.

The current is obviously caused by changes in the magnetic field. What happens is that the changing magnetic field produces electric field lines that encircle the magnetic field lines. Two of these circular electric field lines are shown in Figure 6-15. The circular electric field lines lying inside the metal wire of loop II exert electric forces on the free electrons in the metal. The free electrons are driven around the wire, and the resulting current deflects the galvanometer needle. The electric field exists only while the magnetic field is changing. When the magnetic field is steady there is no electric field and no current.

> A changing magnetic field produces an electric field.

The production of a current by a changing magnetic field is called *electromagnetic induction*. It was discovered independently by Michael Faraday in England and Joseph Henry in the United States near the beginning of the nineteenth century.

There are other ways of demonstrating electromagnetic induction. Suppose that a bar magnet is brought up to loop II of Figure 6-16. As it approaches, the magnetic field it produces in the vicinity of loop II grows stronger. While the magnet is moving, the magnetic field through loop II is increasing. Circular electric field lines are created and they drive a current around the loop. As soon as the magnet stops, the magnetic field no longer increases and the current ceases to flow. If the magnet is taken away again, the magnetic field decreases and a current flows around the loop in the opposite direction.

6.11 ELECTROMAGNETIC INDUCTION AND RELATIVITY

In Figure 6-17a, loop I carries a steady current but moves toward loop II. The magnetic field produced by loop I in the vicinity of loop II increases as loop I approaches. As might be expected, a current is induced in loop II while loop I is in motion.

Now we can ask a very profound question—the kind of question asked by Albert Einstein that led him to the *theory of relativity*. What will happen if we leave loop I stationary and move loop II toward it?

The new situation is shown in Figure 6-17b. The magnetic field does not change at any point in space. Consequently, there are no circular electric field lines to drive a current around loop II. Does that mean that there is no induced current if we move loop II, whereas there is an induced current if we move loop I? Or is it only the relative motion of the two loops that matters? If so, there must be an induced current in both cases.

If we actually do the experiment we discover that there is an induced current in both cases. Only relative motion matters!

However, although there is in fact an induced current when loop II moves, it can no longer be explained in terms of circular electric field lines. It can be explained in terms of something else we already know. Refer again to Figure 6-17b. When loop II is moving, the free electrons in the metal wire of this loop are moving in the steady magnetic field of loop I. A charge moving in a magnetic field experiences a magnetic force. The magnetic force on a free electron in loop II is in the right direction to drive this free electron around the loop.

Moreover, the force turns out to have exactly the right strength to produce the same current that is produced by an electric force when loop I is moving, as in Figure 6-17a. As far as the end result is concerned, it does not matter which loop moves. The induced current depends only on the *relative velocity* of the two loops. Relative velocity means the rate at which the distance between the loops is changing, independently of which one is actually moving.

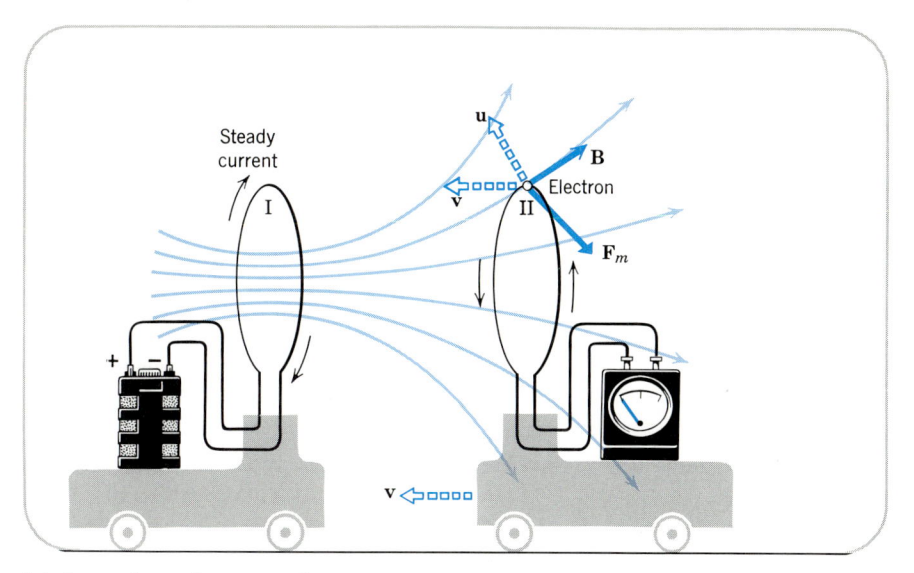

(a) Loop I carries a steady current and moves toward loop II.

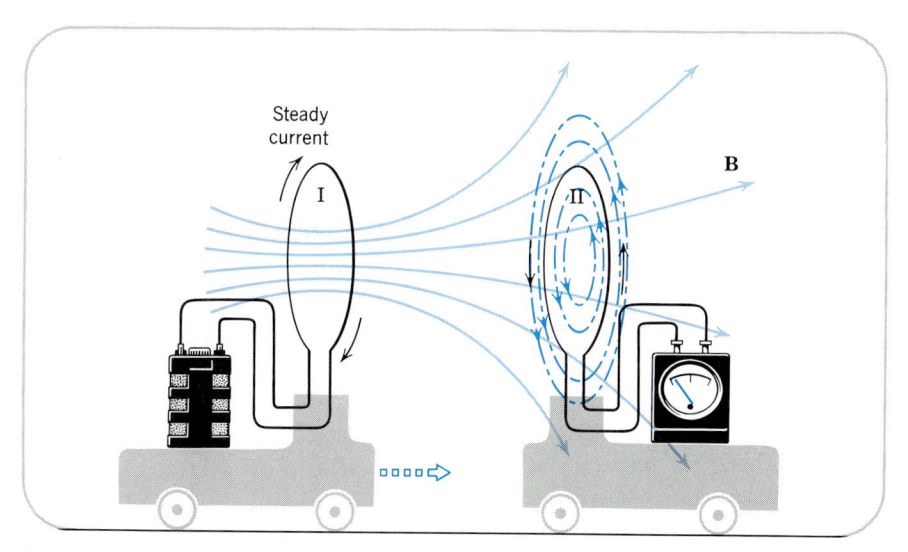

(b) Loop I carries a steady current and is stationary. Loop II moves toward it.

FIGURE 6–17
Two kinds of electromagnetic induction.

6.12
MAXWELL'S EQUATIONS

In the late nineteenth century, Clerk Maxwell was able to show that the complicated laws of electricity and magnetism can be summarized in a set of elegant equations. These equations describe the behavior of electric and magnetic fields in space. They do not relate the fields at a point to the behavior of charged objects at a distance from this point. Instead, the fields at one point are related to the behavior of the electric and magnetic fields at neighboring points in the immediate vicinity of the first point.

This is a complete swing away from our earlier action at a distance approach to the universe. Our previous attitude assumed interactions between discrete particles at a distance from one another. It did not worry about the space in between them. The present attitude might be described as the "handing on" of interactions from a point in space to a neighboring point.

Before he could formulate his equations, Maxwell had to guess a new law of nature. It is a kind of inverse of electromagnetic induction, but it cannot be so readily demonstrated by simple experiments. Electromagnetic induction is based on the idea that a changing magnetic field produces an electric field. Maxwell's new law is that a changing electric field produces a magnetic field.

> A changing electric field produces a magnetic field.

Without advanced mathematics, it is not possible to state *Maxwell's equations* in a precise form. We have to be satisfied with an oversimplified descriptive statement hinting at the meaning of each equation.

> Maxwell's equations:
>
> Equation 1—Electric field lines diverge from a positive charge and converge on a negative charge.
>
> Equation 2—Magnetic field lines always form closed loops.
>
> Equation 3—A changing magnetic field produces an electric field.
>
> Equation 4—A magnetic field is produced either by a moving charge or by a changing electric field.

6.13
THE TIME LAG

To appreciate the new attitude underlying Maxwell's equations, let us consider the magnetic field produced by a moving charge (Figure 6–18). Earlier in this chapter we assumed that a moving charge produces a magnetic field at a distant point P by action at a distance. We shall now point out that the magnetic field at P may be looked upon as a consequence of a changing electric field at P. As the charge moves past P, its distance from P and its direction relative to P continually change. The electric field it produces at P therefore continually changes in magnitude and direction. According to Maxwell's fourth equation, this changing electric field produces a magnetic field in the vicinity of P.

We have still not entered into the true spirit of Maxwell's equations. We have talked about the electric field at P as though it were produced by action at a distance. We ought really to describe the behavior of both the electric and magnetic fields at P only in terms of the behavior of the fields at points in the immediate vicinity of P.

We should proceed somewhat as follows. In the immediate vicinity of the charge q, electric field lines diverge from the charge as required by Maxwell's first equation. Also, since the charge is moving, Maxwell's fourth equation tells us that there are circular magnetic field lines in its immediate vicinity. As the charge moves by, the electric and magnetic fields in its immediate vicinity change. In accordance with Maxwell's third and fourth equations, these changing fields produce electric and magnetic fields at points farther away from the charge. These fields in their turn change and produce fields at points still farther away.

In this way the disturbance in the electromagnetic field travels outward from the charge until it reaches the distant point P. Anticipating some of the things to be discussed in the next chapter, the disturbance travels outward as an *electromagnetic wave*. It travels with the speed of light, c. There is a *time lag* before it reaches P.

This new concept has revolutionary importance to our thinking about the nature of the universe. Electromagnetic interactions are not instantaneous, but travel from one particle to another with the speed of light.

The resulting time lag is not important in the experiments on electricity and magnetism that have been described so far in the present chapter. An electromagnetic disturbance can travel right across a typical laboratory apparatus in about one billionth of a second (10^{-9} s). Such a short delay is not detectable in the experiments.

However, light itself is an electromagnetic wave traveling with the speed c. In the next chapter we shall turn our attention to the nature of waves and of electromagnetic waves in particular.

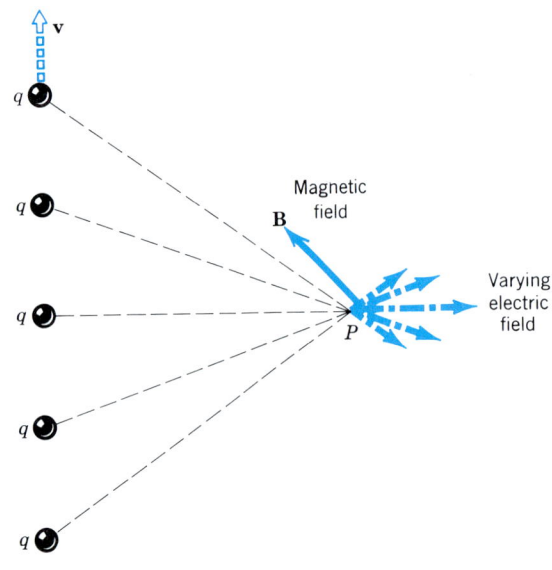

FIGURE 6–18

As the charge q moves past the point P, the electric field it produces at P varies. The changing electric field produces a magnetic field.

SUMMARY

Forces between moving charges. Magnetism is a consequence of the extra forces that exist between two charges when they are in motion. Parallel straight currents attract when they flow in the same direction and repel when they flow in opposite directions.

The magnetic field **B** is an intermediate step in the calculation of forces between moving charges. The SI unit of magnetic field is the tesla, abbreviated T.

Magnetic field lines. The magnetic field at any point is tangential to the magnetic field line through the point. A single charge moving with constant speed in a straight line produces magnetic field lines that are circles centered about points on the path of the charge. An electric current in a long straight wire produces magnetic field lines that are circles centered about points on the wire. If you look in the direction of motion of a single positive charge or the direction of a current, the magnetic field lines go around clockwise. The magnetic field lines of a circular current are shown in Figure 6–6.

A new constant c enters into the equations giving the magnetic field produced by moving charges or electric currents. The numerical value of this constant is exactly the same as the speed of light.

The magnetic force on a moving charge is at right angles to the velocity of the charge and also at right angles to the magnetic field acting on the charge. Its magnitude is equal to the charge multiplied by the component of the velocity in a direction at right angles to the field and also multiplied by the magnitude of the magnetic field. A charge moving with constant speed at right angles to a uniform magnetic field moves in a circle.

Bar magnets. An electron is a spinning cloud of negative electric charge. It produces a magnetic field similar to the field of a circular current. In a bar magnet some of the electrons all spin in the same direction, and their fields combine to produce a large magnetic field.

The magnetic compass. If a bar magnet is freely pivoted at its center in the earth's magnetic field, the end called the north pole points toward the north. The end called the south pole points toward the south. In any magnetic field the north pole of the magnet points in the direction of the field. This behavior can be explained in terms of the couple exerted on each spinning electron by the external magnetic field. The magnetic field lines produced by the bar magnet itself enter the south pole and leave the north pole.

Forces between bar magnets. If the north pole of one bar magnet is brought near the south pole of another bar magnet, an electron in one magnet spins in the same direction as an electron in the other magnet. The two electrons attract each other in the same way that parallel currents flowing in the same direction attract. A north pole therefore attracts a south pole; unlike poles attract.

TERMS AND CONCEPTS

If the north pole of one magnet is brought near the north pole of another magnet, electrons in the two magnets spin in opposite directions. They repel each other in the same way that parallel currents flowing in opposite directions repel. The same is true if a south pole approaches another south pole; like poles repel.

Electromagnetic induction. If the magnetic field through a circular loop of wire is changing, a current is induced in the wire. A changing magnetic field produces an electric field. This electric field drives the induced current around the loop.

Electromagnetic induction and relativity. The induced current in Figure 6–17 depends only on the relative motion of the two loops. The detailed explanation of what causes the induced current depends on which loop is moving.

Maxwell's equations are based on the idea that interactions can be handed on from one point to a neighboring point in empty space. Equation 1— Electric field lines diverge from a positive charge and converge on a negative charge. Equation 2— Magnetic field lines always form closed loops. Equation 3—A changing magnetic field produces an electric field. Equation 4—A magnetic field is produced either by a moving charge or by a changing electric field.

The time lag. When one charge exerts a force on a second charge, an electromagnetic disturbance travels outward from the first charge with the speed of light. There is a time lag before it influences the second charge.

TERMS AND CONCEPTS

magnetism
bar magnet
magnetic compass
magnetic force
magnetic field
magnetic field line
speed of light c
electromagnetism
tesla
gauss
spinning electron
north and south poles of a magnet
electromagnetic induction
theory of relativity
relative velocity
Maxwell's equations
time lag of electromagnetic interactions
electromagnetic wave

6 MAGNETISM

QUESTIONS

1. Refer to Figure 6–1 and decide whether the magnetic forces between two positive charges are attractive or repulsive when the charges move parallel to each other (a) in the same direction, (b) in opposite directions. What happens when one charge is positive and the other is negative?

2. A long horizontal wire carries a current flowing from the east toward the west. What is the direction of the magnetic field at a point vertically above the wire?

3. A long vertical wire carries a current flowig upward. What is the direction of the magnetic field at a point due east of the wire?

4. A current flows around a circular wire in a vertical plane. You view it from a direction in which it is seen to be flowing clockwise. Do the magnetic field lines produced by the current pass through the loop in a direction pointing toward you or away from you?

5. A long, horizontal, straight wire is vertically above a small magnetic compass. Initially the compass points in a direction parallel to the wire. What happens to the compass when a current is switched on in the wire?

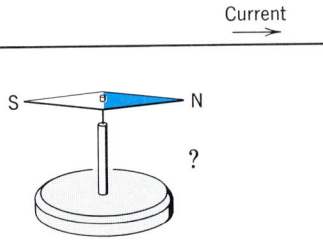

Question 6–5

6. Suppose that we try to explain the earth's magnetic field by assuming that the earth is a huge bar magnet. Is the geographic north pole the north pole or the south pole of this magnet?

7. Suppose that the earth's magnetic field is produced by a circular loop of current inside the earth in a plane passing through the equator. Looking from the north pole toward this circular current, does the current go around clockwise or counterclockwise?

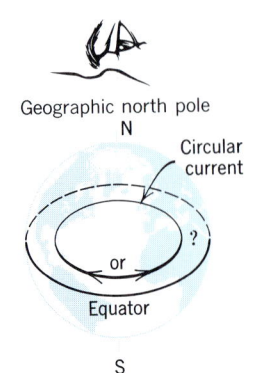

Question 6–7

8. Draw a figure similar to Figure 6–14b for the case of a south pole approaching another south pole.

9. Can you figure out why a bar magnet always attracts small iron or steel objects and never repels them?

10. Two circular loops are side by side as in Figure 6–15. An alternating current flows in loop I. What would you expect to happen in loop II? Explain your answer thoroughly.

PROBLEMS

1. The quantity $\frac{K_e}{c^2}$ appearing in Equation 6-2 could be replaced by a single constant. Calculate the numerical value of this new constant (take $c = 3.0 \times 10^8$ m/s).

2. What is the magnitude of the magnetic field at a distance of 4 m from a long, straight current of 3 A?

3. At a distance of 0.3 m from a long, straight current the magnitude of the magnetic field is 2×10^{-2} tesla. What is the magnitude of the magnetic field at a distance of 0.5 m from the current?

4. A charge of $+2$ C has a vertical velocity of 3 m/s. What is the magnitude of the magnetic force exerted on it by a horizontal field of 0.15 tesla?

5. A charge of $+0.3$ C has a velocity of 7 m/s vertically downward. What is the magnitude of the magnetic force exerted on it by a field of 3 teslas pointing vertically upward?

6. A charge of $+6$ C is 0.5 m due east of a long vertical wire carrying a current of 1.5 A flowing upward. If the charge has a horizontal velocity of 0.2 m/s toward the north, what is the magnitude of the magnetic force on it?

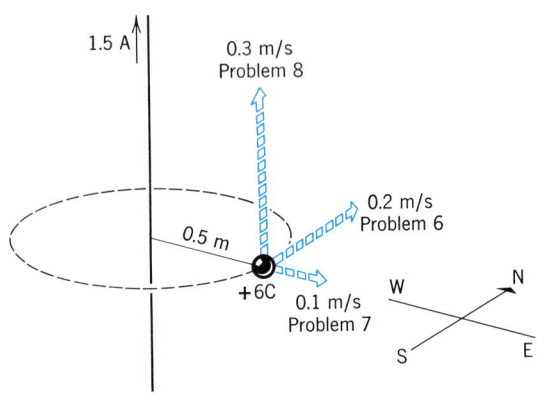

Problems 6–6, 6–7, and 6–8

7. Repeat question 6 when the charge has a velocity of 0.1 m/s pointing directly away from the wire.

8. Repeat question 6 when the charge has a velocity of 0.3 m/s vertically upward.

CHAPTER 7
WAVES

7.1 SIMPLE HARMONIC MOTION

In a wave something oscillates and it usually performs *simple harmonic motion*. Three typical examples of simple harmonic motion are illustrated in Figure 7-1. Figure 7-1a is the backward and forward swing of a pendulum; Figure 7-1b is the up and down motion of a weight hanging on a spring; Figure 7-1c is the twisting and untwisting motion of a horizontal bar suspended by a wire.

These motions are called periodic motions because they obviously repeat themselves periodically. The *period* T is defined as the time taken to complete one oscillation. More precisely, the period is the time to go from the extreme position A to the other extreme position B and back again to A.

The *frequency* is defined as the number of oscillations in one second. It is represented by ν (the Greek letter nu). Obviously, the frequency is the reciprocal of the period. If one oscillation takes $\frac{1}{10}$ second, then 10 complete oscillations are per-

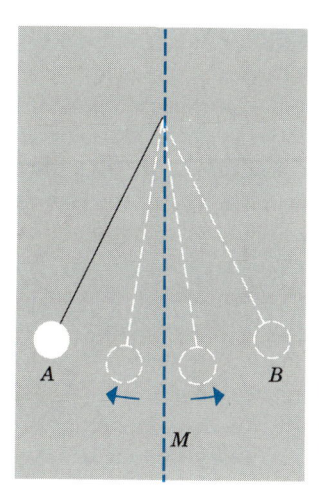

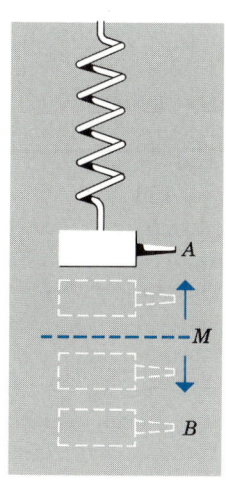

(a) A pendulum.   (b) A weight on a spring.   (c) A bar performing torsional oscillations.

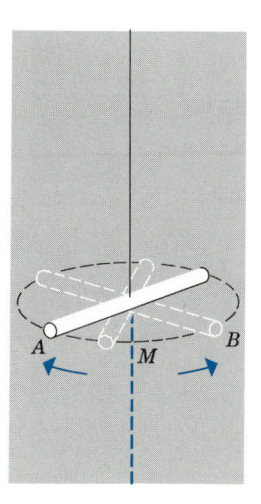

FIGURE 7-1
Three examples of simple harmonic motion.

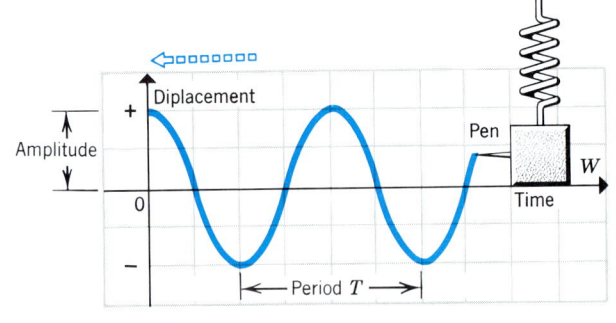

FIGURE 7-2

The variation of displacement with time for simple harmonic motion is a curve with a sinusoidal shape.

formed in 1 second. The SI unit of frequency is the *hertz*, abbreviated Hz. For example, the oscillation just mentioned has a frequency of 10 hertz or 10 Hz.

$$\text{Frequency} = \frac{1}{\text{period}} \quad (7\text{-}1)$$

$$\nu = \frac{1}{T} \quad (7\text{-}2)$$

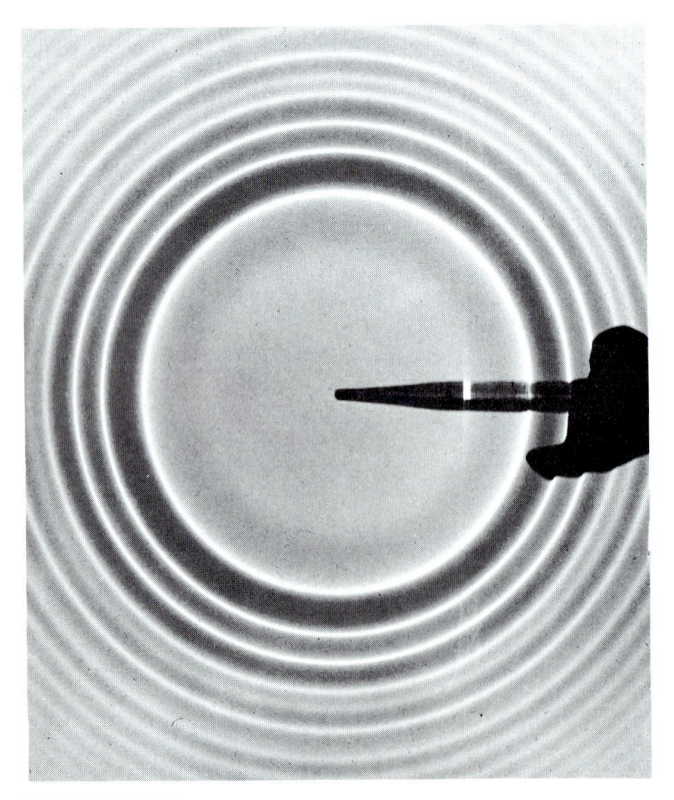

FIGURE 7-3

Circular ripples on a ripple tank. This photograph was obtained by allowing a single drop of water to fall into the tank. (Photograph by Berenice Abbott.)

The *amplitude* of an oscillation is the distance between the central position and an extreme position. In Figure 7-1b, the amplitude is the distance from the mid-position M to the highest position A or the lowest position B. The total distance covered, from A to B, is twice the amplitude.

Suppose that we attach a pen to a weight oscillating at the lower end of a spring, as in Figure 7-2. A piece of paper is pulled over the pen at a constant speed in a horizontal direction. The pen traces out a curve showing how the displacement of the weight from its midposition varies with time. The shape of this curve is said to be *sinusoidal*.

7.2 TRANSVERSE WAVES

The most familiar examples of waves in everyday life are ocean waves and ripples on water. The most striking characteristic of a water wave is that it appears to be moving outward from its source. Figure 7-3 is a photograph of the ripples produced as a single water droplet falls into a tank of water. Everyone has seen this happen and can imagine the pattern of ripples moving outward from the point of impact. However, only the shape of the surface of the water moves. A small drop of water at the surface merely bobs up and down as the wave passes over it.

To obtain a better appreciation of this last point, imagine a long horizontal spring with one end attached to the wall. The other end is held in the hand and pulled taut. By moving the hand up and down rapidly two or three times, a wave can be made to travel along the spring, as shown in Figure 7-4. One of the coils of the spring, P, is painted ▮ so that its motion can be easily observed. Although the wave travels along the spring in a horizontal direction, the single coil P does not move in a horizontal direction. It oscillates up and down in a vertical direction. The motion of P is, in fact, simple harmonic. When the oscillatory motion is at right angles to the direction in which the wave is traveling, the wave is called a *transverse wave*.

At a fixed instant of time, the spring has a sinusoidal shape similar to the curve of Figure 7-2. However, the difference between Figures 7-2 and 7-4 should be clearly understood. Figure 7-2 shows how the displacement of the weight varies with time. A similar curve would show how the vertical displacement of the coil P of the spring varies with time. Figure 7-4, on the other hand, shows the displacements of all the coils at a fixed instant of time. It is a kind of "snapshot" of the shape of the spring at a fixed instant of time.

7.3 THE VELOCITY OF A WAVE

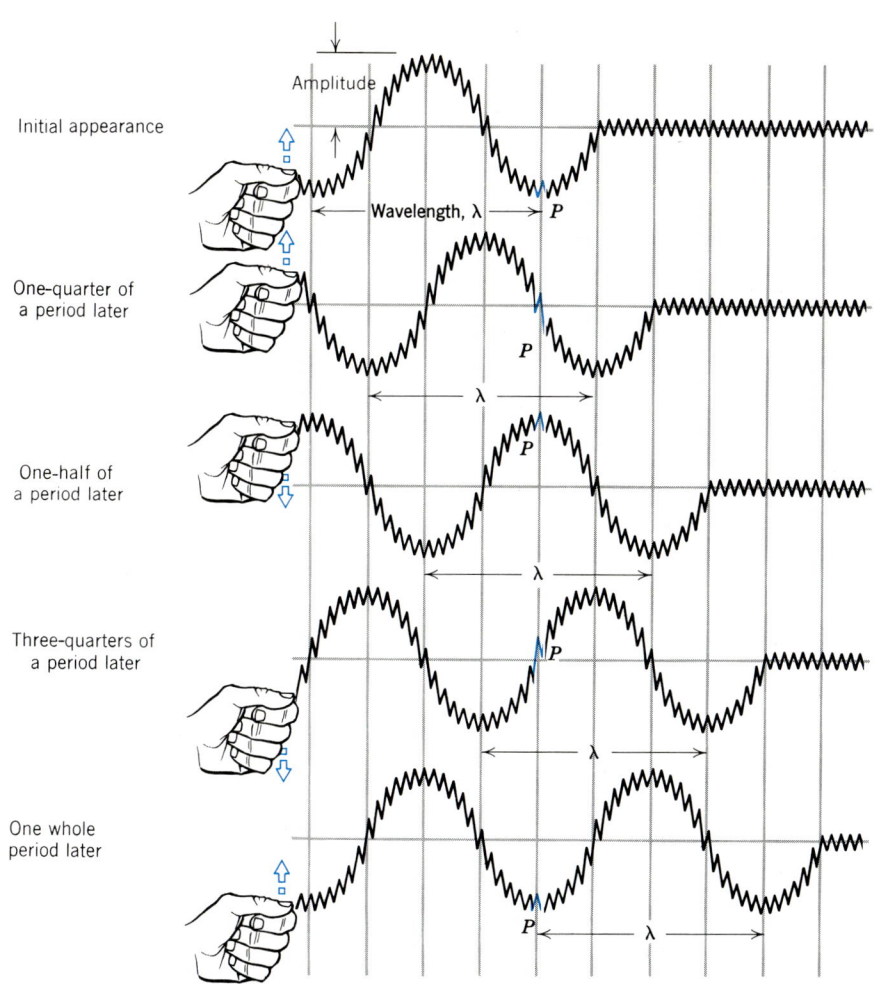

FIGURE 7-4
A transverse wave on a long spring.

In Figure 7-2, the distance between adjacent peaks or adjacent valleys of the curve represents the period T of the oscillation. In Figure 7-4, the horizontal distance between adjacent peaks or adjacent valleys is a real distance in space. It is called the *wavelength* of the wave and is usually represented by the Greek letter lambda, λ.

7.3 THE VELOCITY OF A WAVE

Study Figure 7-4 carefully and consider the motion of the coil P as the wave travels past it. While one wavelength is passing over P, the coil goes through one complete oscillation. During one complete oscillation, the wave moves forward through a distance equal to the wavelength. In one second, the number of complete oscillations is equal to the frequency. In one second, the wave travels a distance equal to the frequency multiplied by the wavelength. The distance the wave travels in one second is the *wave velocity* V. Therefore, the wave velocity is equal to the frequency multiplied by the wavelength.

$$\text{Wave velocity} = (\text{frequency}) \times (\text{wavelength}) \quad (7\text{-}3)$$
$$V = \nu\lambda \quad (7\text{-}4)$$

This important equation is true for all types of waves.

For all the waves discussed in this present chapter, the wave velocity V is a constant. It does not vary with the wavelength or the frequency. The product of frequency and wavelength is then constant. The wavelength is inversely proportional to the frequency. Long wavelengths correspond to low frequencies with few oscillations per second. Short wavelengths correspond to high frequencies with many oscillations per second.

7.4 LONGITUDINAL WAVES

Let us repeat the experiment with the long horizontal spring but oscillate its free end backward and forward in a horizontal direction (Figure 7–5). The spring remains straight, but a wave of alternating compressions and extensions travels along its length. The painted coil performs simple harmonic motion in a horizontal direction. When the oscillatory motion is in the same direction that the wave is traveling, the wave is called a *longitudinal wave*.

The wavelength is the distance from a point on the spring where the coils are closest together to the next nearest point where the coils are closest together. You can easily convince yourself that Equation 7-3 is valid for the longitudinal wave of Figure 7–5. You need only apply the same arguments that were used for the transverse wave of Figure 7–4.

7.5 SOUND

Sound is usually transmitted as a longitudinal wave in the air. Consider what happens when you speak to a friend. Air is expelled from your lungs past two elastic membranes, known as *vocal cords*, inside your windpipe (Figure 7–6). The flow of air past the vocal cords causes them to vibrate. The size of the opening between them varies periodically. The air is consequently expelled in "puffs" that follow one another in rapid succession. The number of puffs per second determines the frequency of the sound emerging from your mouth. The frequency of a sound is what we often call its *pitch*.

The puffs of air do not travel all the way to your friend's ear. As a puff emerges it increases the pressure of the air in the immediate vicinity of your mouth. In between puffs the molecules of air near your mouth move farther apart to re-

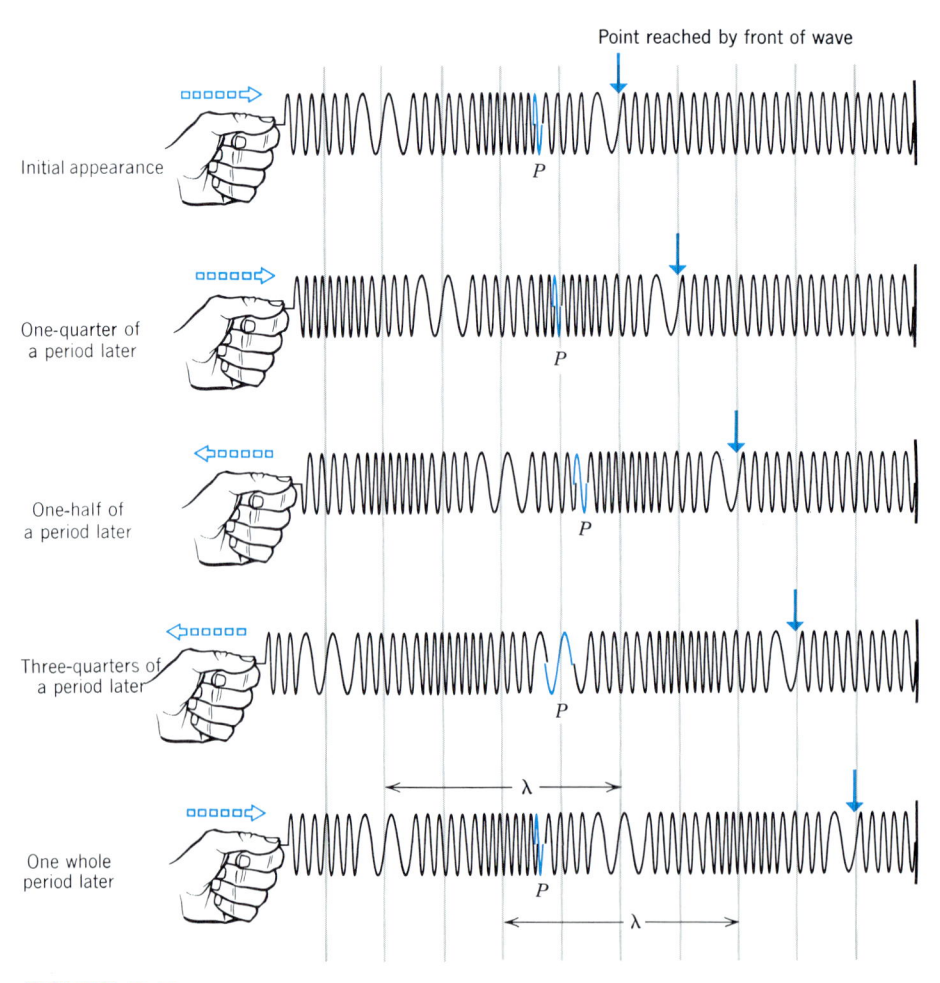

FIGURE 7–5
A longitudinal wave on a long spring.

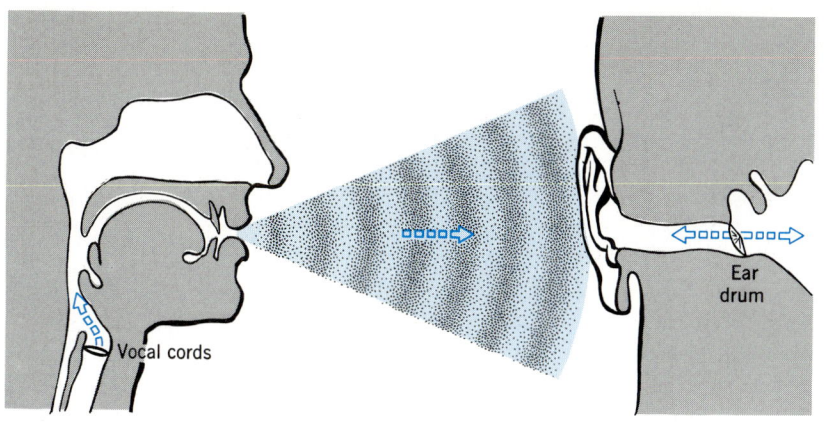

FIGURE 7-6

Speech, sound, and hearing. The anatomy is simplified in order to illustrate the basic principles.

lieve the excess pressure. As these molecules move outward they collide with more distant molecules and knock them outward. The more distant molecules crowd together and create a region of high pressure farther from your mouth. The process continues and the region of high pressure moves outward from your mouth with the speed of sound.

Each puff creates such a rgion of high pressure moving outward. The sound wave therefore consists of a train of high-pressure regions following one another in rapid succession. They are separated by regions of lower pressure (see Figure 7-6). The number of high-pressure regions moving past any point in one second is equal to the frequency of the sound.

The individual molecules do not travel with the high-pressure regions. They merely oscillate backward and forward parallel to the direction in which the sound is traveling. The behavior of the molecules is similar to the behavior of the coils of the spring in Figure 7-5. A region of high pressure is analogous to a part of the spring where the coils are close together. A region of low pressure is analogous to a part of the spring where the coils are far apart.

Eventually the high-pressure regions impinge on a membrane known as the *eardrum* in your friend's ear. They cause it to oscillate backward and forward with the same frequency as your vocal cords. The mechanism of your friend's ear converts this oscillation into a train of pulses of electric current traveling along one of her nerves to her brain. The frequency of arrival of the electric pulses is exactly the same as the frequency of vibration of your vocal cords. She is thereby made aware of what you are saying.

7.6 ELECTROMAGNETIC WAVES

In order of increasing frequency and decreasing wavelength, the important types of *electromagnetic waves* are radio waves, infrared radiation, visible light, ultraviolet light, x-rays, and γ-rays. Their production can always be ultimately related to the acceleration of electrons, protons, or other charged fundamental particles. When an electric charge accelerates, it radiates an electromagnetic wave.

To illustrate the production and character of electromagnetic waves, let us consider the transmission and reception of a radio wave (Figure 7-7). The transmitting antenna is a long metal rod or wire carrying an alternating electric current oscillating backward and forward. In a typical case the frequency might be a million hertz. Each free electron in the antenna performs a simple harmonic motion with a frequency of one million hertz.

During its simple harmonic motion, the electron accelerates. It therefore radiates an electromagnetic wave. The oscillating electron continually changes its position and velocity. The electric and magnetic fields it produces in its immediate vicinity continually change. In accordance with Maxwell's equations and the considerations of Section 6.13, these changing fields produce other fields in their vicinity. The disturbance travels outward with the speed of light.

When Maxwell's equations are applied to this situation and solved, the outgoing electromagnetic wave is found to have the form illustrated in Figure 7-7. This diagram shows the electric

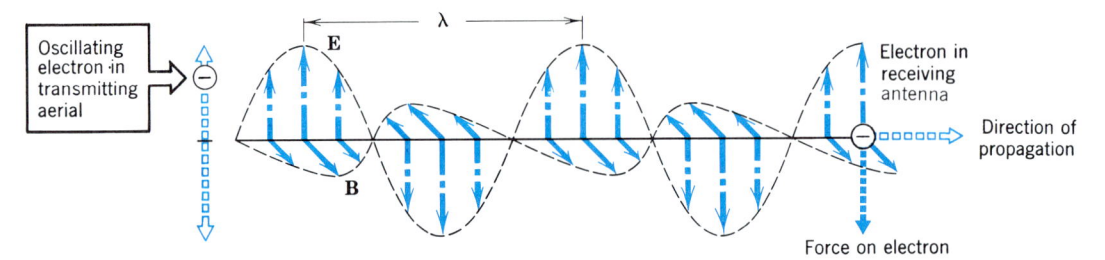

FIGURE 7-7

The electromagnetic wave produced by an oscillating electron. In the interests of clarity, the diagram ignores the fact that as the wave travels away from the electron, its energy spreads over a greater area. Consequently, the amplitudes of the oscillating fields decrease with distance.

and magnetic fields at a single instant of time at points along a line. For convenience, this line is taken to be in a direction at right angles to the motion of the electron. At each point, the electric and magnetic fields are represented by two vectors. The electric field **E** is seen to be parallel to the motion of the electron. Its direction and magnitude vary along the line in such a way that the tips of the vectors trace out a sinusoidal curve. The magnetic field **B** is at right angles to the electric field and to the direction of travel. It traces out a similar sinusoidal curve.

The diagram in Figure 7-7 represents the electric and magnetic fields at points along the line at a fixed instant of time. The whole pattern should now be imagined to move along the line away from the electron with the speed of light. Consider what happens as various parts of the pattern pass over a fixed point. The vector representing the electric field oscillates between a maximum value in an upward direction and an equal value in a downward direction. Similarly, the magnetic field oscillates in and out of the plane of the page.

This behavior is similar to the behavior of the spring in Figure 7-4. It is important to realize, however, that we are describing the electric and magnetic fields in empty space at points along a straight line. Nothing is moving in a direction at right angles to this line. The significance of the picture is the following. If an electron were present at any point in space, the oscillating electric field at that point would exert an oscillating force on it. The electron would be made to perform a simple harmonic motion similar to the motion of the electrons in the transmitting antenna. This actually happens to the free electrons in the receiving antenna. It is the mechanism whereby the behavior of the electrons in the transmitting antenna is reproduced by the electrons in the receiving antenna. That is how communication becomes possible through the intervening empty space.

An electromagnetic wave is always a transverse wave. Both the electric and magnetic vectors are always at right angles to the direction in which the wave is traveling. The electric and magnetic vectors oscillate in step with each other. They both assume a maximum value at the same time and are zero at the same time.

7.7 THE CONSTANT c

We first encountered the constant c in Equation 6-2 for the magnetic field produced by a long, straight current. Because it is present there, it is also present in Maxwell's equations. When these equations are manipulated mathematically to deduce the existence of electromagnetic waves, the wave velocity in a vacuum is found to be precisely this same constant c. In the first instance, therefore, the value of c can be determined by experiments that measure the electric and magnetic forces between charges. In the second instance, once it has been demonstrated that electromagnetic waves can be generated by oscillating electric currents, their velocity can be measured directly. This second step was first taken by Hertz in 1885, twelve years after Maxwell had published the final form of his equations. The resulting good agreement between the two entirely different methods of measuring c inspired confidence in those equations.

The really striking aspect of the situation, though, was that the velocity c of an electromagnetic wave was exactly the same as the velocity of light. The velocity of light had been deduced from astronomical observations by Roemer in 1675 and by Bradley in 1729. I had then been measured in a terrestrial experiment by Fizeau in 1849. At the beginning of the nineteenth century, experiments on interference and diffraction of light (to be described later in this chapter) had convinced scientists that light is a wave. One of Maxwell's outstanding achievements was the realization that light is an electromagnetic wave. Figure 7–7 also serves as a picture of a light wave. The difference between light and radio waves is that the frequency of a light wave is much greater than the frequency of a radio wave. The wavelength of a light wave is much shorter than the wavelength of a radio wave.

The most accurate modern method of determining the value of c is to measure both the wavelength and frequency of a light wave. As Equation 7-3 indicates, the product of the wavelength and the frequency is equal to the velocity of the light wave. At the time of this writing the best available value of c is 2.99792458×10^8 m/s. Throughout the rest of this book it will be good enough to use the approximate value of 3×10^8 m/s.

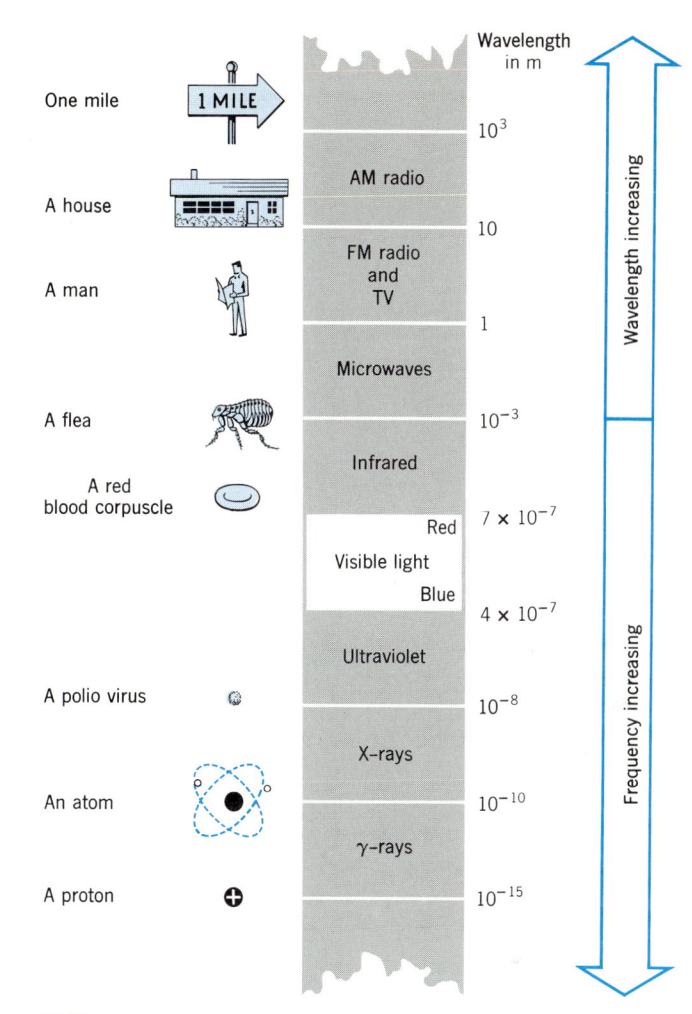

7.8 TYPES OF ELECTROMAGNETIC WAVES

Figure 7–8 tabulates the various types of electromagnetic waves and shows the range of wavelengths associated with each type. The sizes of some familiar objects are included for comparison. There is no particular scale associated with this diagram. In fact, it is necessary to vary the scale from one part to another. Except in the case of visible light, the limits are only approximate.

Electromagnetic radiation visible to the human eye covers only a small range of wavelengths in the vicinity of 5×10^{-7} m (one fiftieth of one thousandth of an inch). This is approximately the size of the smallest object that can be seen in an optical microscope. The longest wavelength of *visible light* is red. As the wavelength is decreased, the color changes and passes through the sequence red, orange, yellow, green, and blue to violet, which has the shortest wavelength.

Beyond the red end of the visible spectrum, where the wavelengths are becoming longer, there is first *infrared* radiation. It is sometimes called "heat radiation," because it is strongly emitted by a fire or central heating radiator. A radiator emits no visible light and cannot be seen in a dark room. But the heat radiation from it can easily be felt. The phrase "heat radiation"

FIGURE 7–8
Types of electromagnetic waves.

A radar installation used to observe severe storms. A shortwave radio beam is sent out, reflected from the storm, and detected when it returns. (National Center for Atmospheric Research.)

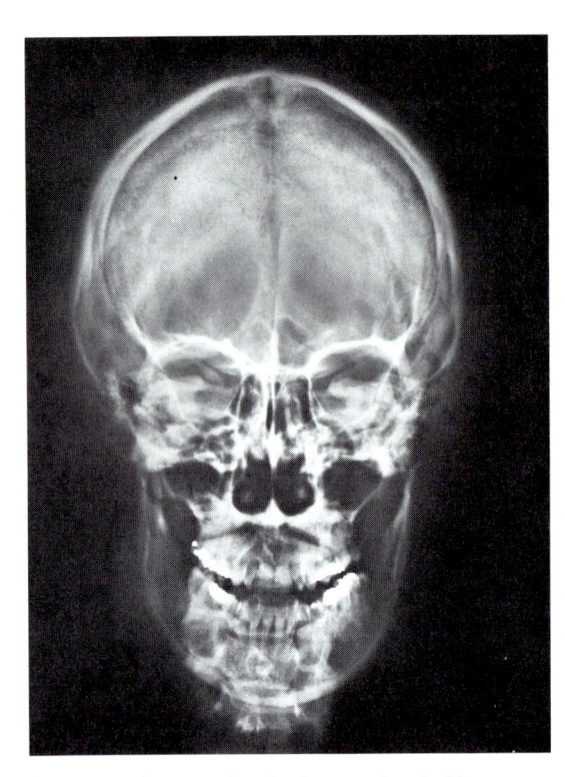

An x-ray photograph of a human head. X-rays passing through the head are more readily absorbed by the bones than by other tissues. (Eastman Kodak Company.)

can be misleading. As we shall explain in Section 9.2, hot bodies emit all types of electromagnetic radiation. Any type can be absorbed by the human body to create a sensation of warmth.

At still longer wavelengths there are the *microwaves* used in radar. *Radio waves* have the longest wavelengths, ranging from a yard to a mile.

Beyond the blue end of the visible spectrum, where the wavelengths are becoming shorter, there is first *ultraviolet* light. This is strongly emitted by the sun, but almost completely absorbed by the earth's atmosphere.

Then, at wavelengths comparable with the size of a molecule, there are *x-rays*, which can be generated by allowing high-speed electrons to bombard a metal target. The rapid deceleration of the electrons as they are stopped by the target atoms produces electromagnetic radiation. As we have already emphasized, an accelerating or decelerating charge always radiates an electromagnetic wave.

Finally, at wavelengths comparable with the size of a proton or even less, there are *gamma rays*. The greek letter gamma is γ. It is common practice to say gamma rays, but to write γ-*rays*. γ-rays are emitted by radioactive nuclei.

7.9
THE PRESSURE OF LIGHT

Electromagnetic radiation from the sun warms you. It follows that electromagnetic waves transport energy through empty space. When the sun's radiation falls on your body, the radiation is absorbed. Its energy is converted into heat energy, which creates a sensation of warmth.

An electromagnetic wave also transports momentum. When it falls on an object and is absorbed, this momentum is transferred to the object. Transferring momentum to an object at a steady rate is equivalent to exerting a steady force on it (Section 3.1). When light falls on an object it exerts a force on it. The force per unit area is called the *pressure of light*.

The pressure exerted on you by sunlight is negligibly small. To detect the pressure of light it is necessary to use a sensitive device such as the one shown in Figure 7–9. The two mirrors joined by a horizontal rod are suspended from a very thin fiber inside a very good vacuum. A beam of light falls on one of the mirrors. The force it exerts is seen to twist the suspended system. If the beam is switched to the other mirror, the system twists in the opposite direction.

7.10
ENERGY AND MOMENTUM OF AN ELECTROMAGNETIC WAVE

As an electromagnetic wave passes by, the electric field vector at a point oscillates up and down. The tip of the arrow representing the vector performs simple harmonic motion. The maximum length of the arrow is analogous to the amplitude of a simple harmonic motion, as illustrated in Figure 7–1. We call this maximum length the "maximum electric field" and represent it by the symbol E_o.

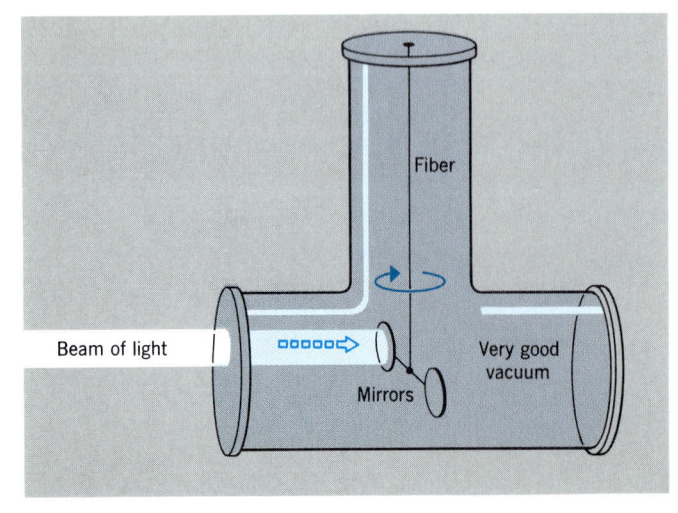

FIGURE 7–9

A demonstration of the pressure of light.

Imagine an electromagnetic wave traveling at right angles to a flat surface. The *intensity* of the wave is defined as the energy flowing in one second across unit area of the surface. Its SI units are clearly joules per second per square meter, which is the same as watts per square meter or W/m². In the case of visible light, the intensity is a measure of its brightness.

The intensity of an electromagnetic wave can be shown to be proportional to the square of the maximum electric field E_o. The constant of proportionality is $\frac{c}{8\pi K_e}$.

Intensity of an electromagnetic wave

= The energy flowing perpendicularly across unit area in one second (7-5)

= (a constant) × (the square of the maximum electric field) (7-6)

= $\frac{cE_0^2}{8\pi K_e}$ (7-7)

An electromagnetic wave transports momentum as well as energy. The momentum is equal to the energy divided by the speed of light, c.

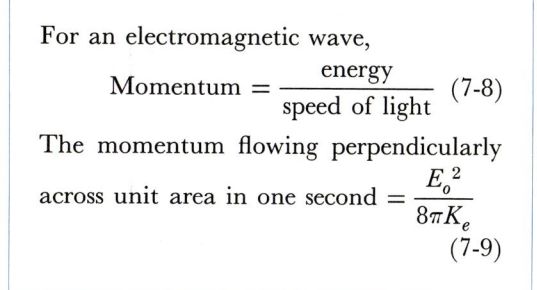

For an electromagnetic wave,

$$\text{Momentum} = \frac{\text{energy}}{\text{speed of light}} \quad (7\text{-}8)$$

The momentum flowing perpendicularly across unit area in one second = $\frac{E_o^2}{8\pi K_e}$

(7-9)

7.11
THE WAVE NATURE OF LIGHT

In the seventeenth and eighteenth centuries it was generally believed that light is a stream of particles. Newton tended to favor this point of view. When sunlight passes through a hole in a blind it forms a well defined sunbeam (Figure 7–10a). The patch of light on the floor has sharp edges and has about the same size as the hole. A very simple explanation might be that the sunlight consists of a stream of particles traveling in straight lines from the sun through the hole to the floor.

A wave passing through a narrow opening behaves in a very different way. In Figure 7–10b, S is the source of circular ripples on the surface of water. Each circle centered on S corresponds

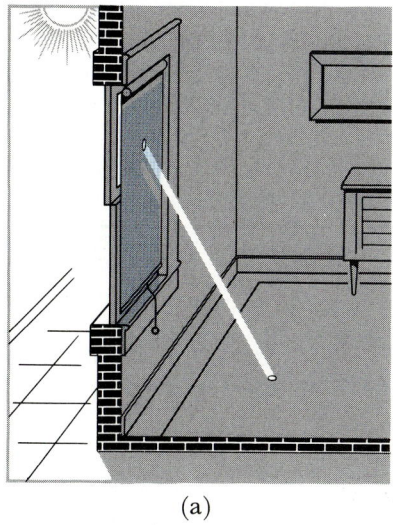

FIGURE 7–10
The apparent straight-line propagation of a sunbeam contrasted with the spreading of a ripple passing through a gap. (a) Sunlight streaming through a hole in a blind. (b) A circular ripple passing through a very narrow gap.

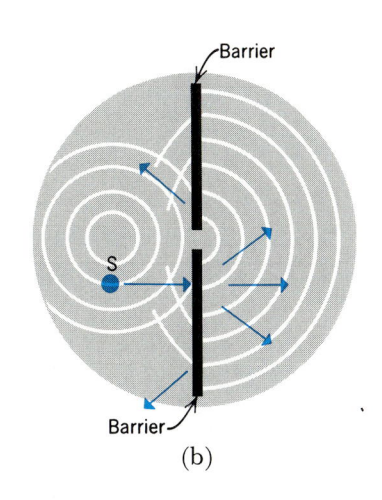

to a wave crest where the upward displacement of the water is greatest. These circles are sometimes called *wavefronts*. They move radially outward with the wave velocity. If a wavefront impinges on a very narrow gap in a barrier, one might expect a small portion of the wavefront to pass through the gap and to continue on its way undisturbed. What actually happens is quite different. On the far side of the gap, the wave spreads out in all directions. New circular wavefronts centered on the gap are formed. This property of a wave that makes it spread out after passing through a narrow opening is called *diffraction*. It is one of the most distinctive characteristics of the behavior of waves.

A good way to visualize what is happening is as follows. Imagine each very small portion of a wavefront to be a disturbance that becomes the center of an outgoing circular *wavelet*. The barriers of Figure 7–10b block out everything except the wavelet from the small portion of the wavefront entering the gap. Only this wavelet is seen on the other side of the gap.

In the absence of the barriers, however, each small portion of the wavefront sends out its own circular wavelet. The wavelets from all parts of the wavefront add together to produce a new wavefront centered on the original source (Figure 7–11). This is, in fact, the way in which the original wavefront is maintained on its outward motion.

If light is a wave passing through the hole in the blind, why does the sunbeam not spread out and illuminate the whole room? The explanation is that the wavelength of light is very small compared with the size of the hole in the blind. As we shall soon discover, the spread of the light wave is then too small to be noticeable. On the other hand, the wavelength of the ripple is longer than the width of the gap. The spread of the wave is then very pronounced.

Because the wavelength of light is so short, it is not easy to do an experiment to demonstrate the wave nature of light. Diffraction of light was actually discovered in the early seventeenth century by Francesco Maria Grimaldi. The concept of wavelets was introduced later in the same century by Christiaan Huygens, so they are often called Huygens' wavelets. Nevertheless, the scientific world remained unconvinced of the wave nature of light until the early nineteenth century. In 1801, Thomas Young did a famous experiment on the *interference* of light. Then Augustin Jean Fresnel did a series of brilliant experiments on the *diffraction* of light. Between them they established the wave theory of light on a firm footing.

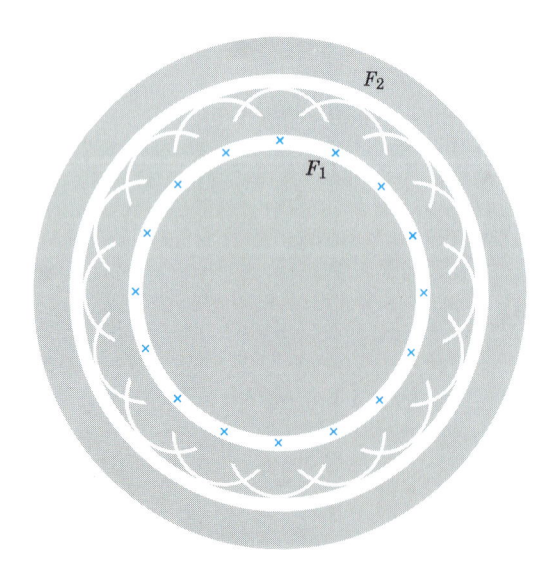

7.12
YOUNG'S INTERFERENCE EXPERIMENT

Young's experiment was concerned with the interference of two light beams emerging from two narrow slits side by side. To avoid nonessential complications, we shall not describe Young's original experiment, but instead give a simplified version of it (Figure 7–12). Light from a lamp is passed through a filter that transmits only one color, giving an electromagnetic wave of definite wavelength. The light then passes through a long, very narrow slit S_o. It emerges with circular wavefronts, which are viewed from above in the lower part of the figure. It then passes on to a second screen in which there are two very narrow parallel slits S_1 and S_2 very close together. Each slit acts as a source of its own circular wavefronts. The waves from the two sources interfere with each other. The pattern seen on the viewing screen consists of a set of equally spaced, vertical, bright strips separated by dark strips. These are called *interference fringes*. Figure 7–13 is a photograph of a set of actual fringes.

You can obtain a hint of how the interference of the waves can produce this effect in the following way. Hold this book horizontal and raise it to the level of your eye. View the right-hand side of the lower part of Figure 7–12 in light that skims over the surface of the page into your eye. You should see a bunch of alternately bright and dark beams radiating outward from the point M midway between the two slits.

FIGURE 7–11
Wavelets adding together to form a new wavefront. Points marked X on the original wavefront act as sources of wavelets. These wavelets add together to form the new wavefront, which has moved ahead of the original wavefront.

7.13 INTERFERENCE OF RIPPLES ON WATER

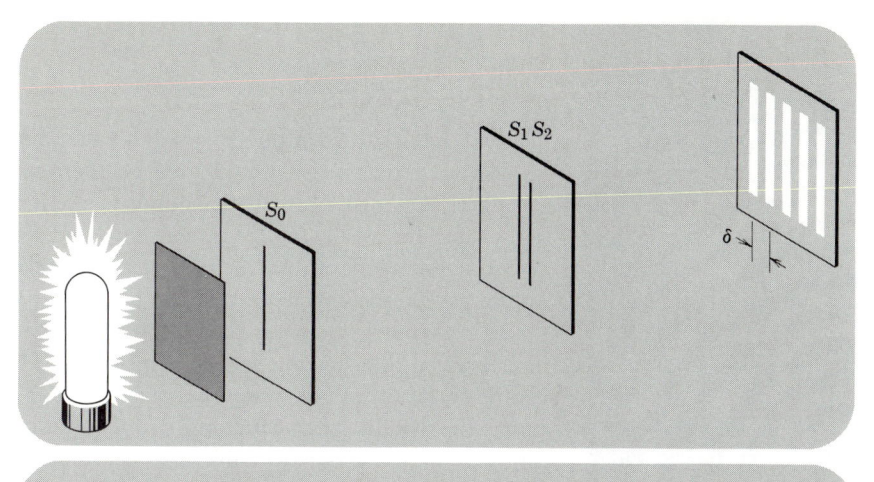

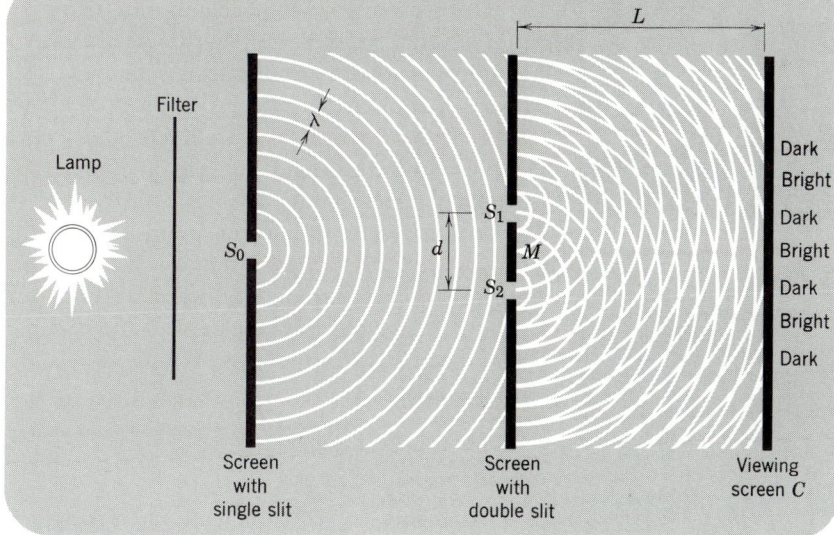

FIGURE 7-12
Young's experiment on the interference of light passing through a double slit.

7.13 INTERFERENCE OF RIPPLES ON WATER

The same kind of interference may be obtained with ripples on water. Dip two probes into the water side by side and oscillate them up and down in unison. Photographs of this effect in Figure 7-14 clearly show the bright and dark radial beams. (We shall return shortly to the difference between the two photographs.) The explanation in this case is as follows. Along the bright beam, the crests of the waves from one probe coincide with the crests of the waves from the other probe. The two waves augment each other to produce a crest twice as high. At the same time, the troughs coincide to produce a trough twice as deep. Along a dark beam, however, a crest from one probe coincides with a trough from the other probe. They cancel each other, with the result that the surface of the water is neither raised nor lowered.

The explanation of Young's double-slit experiment is similar. Light is an electromagnetic wave consisting of oscillating electric and magnetic fields. Along a bright beam, which produces a bright fringe, the oscillating electric fields from the two slits are in the same direction and augment each other. The magnetic fields behave in the same way. Along a dark beam,

FIGURE 7-13
Interference fringes produced by passing light through a double slit. (Photograph by Dr. Brain Thompson.)

94   7 WAVES

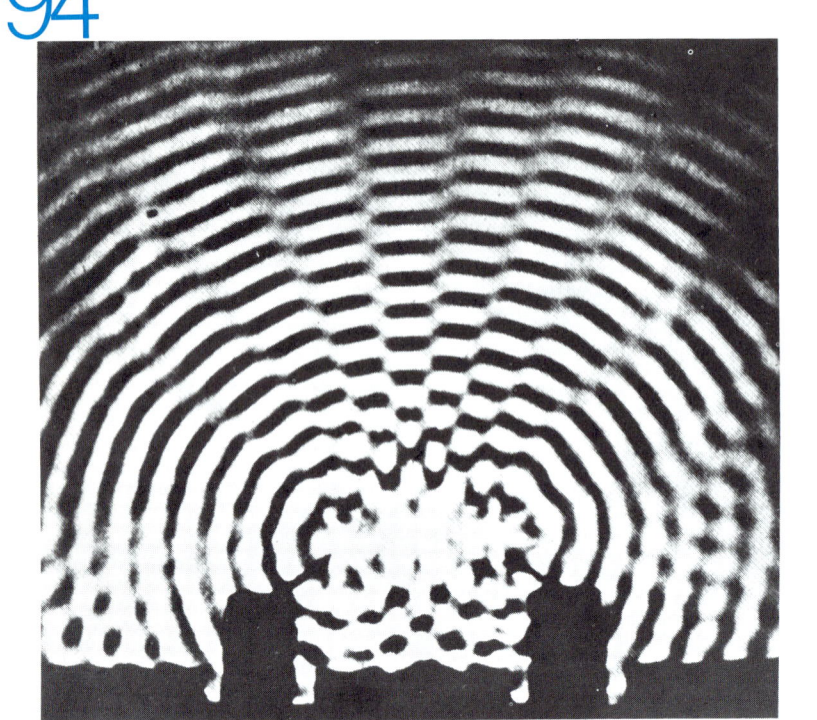

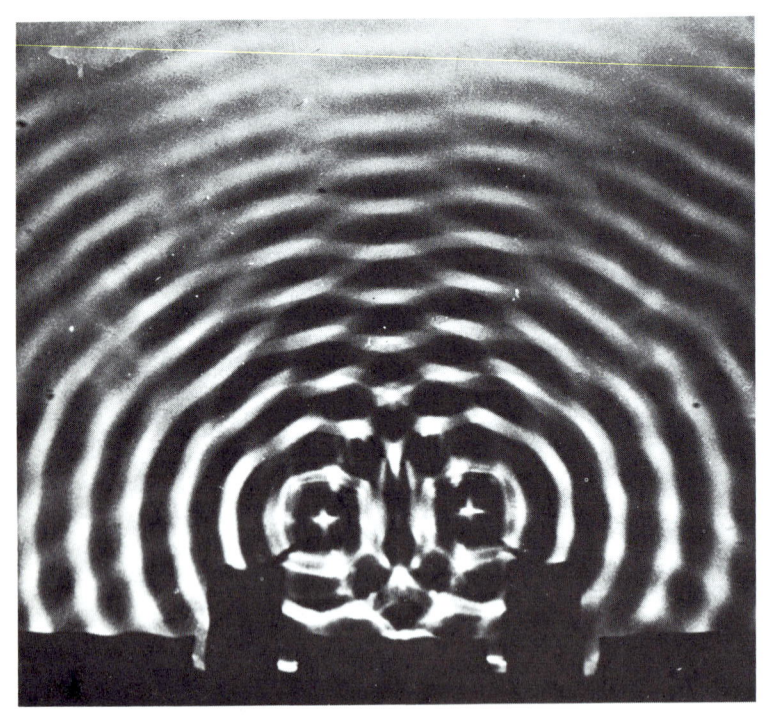

FIGURE 7-14
Interference of ripples from two adjacent sources. Notice that when the wavelength is longer, the beams spread farther apart. (Reproduced with permission from P.S.S.C. *Physics,* D.C. Heath and Co., Boston, 1965.)

7.14 THE THEORY OF INTERFERENCE FRINGES

To find the positions of the fringes, it is necessary to solve a moderately complicated geometrical problem. We must locate all the points where the crests of the waves from the two sources coincide. We shall not attempt this calculation, but merely quote the important aspects of the results.

The distance between neighboring fringes increases as the viewing screen is moved away from the two slits. The reason for this is obvious, since the bright beams diverge from a point midway between the two slits.

The distance between neighboring fringes is proportional to the wavelength. As the wavelength is made longer, the fringes move farther apart. A comparison of the two parts of Figure 7-14 illustrates this very well. For the longer wavelength, the angle between the diverging beams is greater.

A final, very important point is that the fringes are farther apart if the two slits are closer together. The distance between neighboring fringes is inversely proportional to the distance between the two slits.

Interference fringes:

If the wavelength is increased, the fringes move farther apart.

If the slits are placed closer together, the fringes move farther apart.

7.15 FRESNEL'S DIFFRACTION EXPERIMENT

In Figure 7-10b, the wave spreads out in all directions on the far side of the gap. This happens only if the width of the gap is much smaller than the wavelength. The portion of the wavefront incident upon the gap is then narrow enough to be considered the source of a single wavelet.

If the width of the gap is comparable with the wavelength, or larger, the diffraction effects are more complicated. The wide portion of the wavefront incident upon such a wide gap must be divided up into very narrow parts. It is then permissible to consider each very narrow part to be the source of a single wavelet. However, the various wavelets interfere with one another to produce a complicated diffraction pattern.

which produces the dark region between two bright fringes, the electric fields from the two sources are in opposite directions and cancel each other; the magnetic fields also cancel.

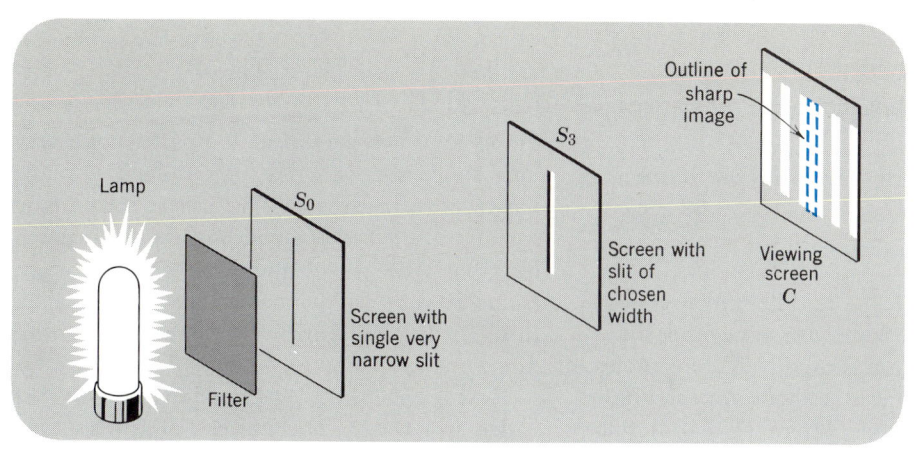

7.16 THE THEORY OF DIFFRACTION FRINGES

FIGURE 7-15
Fresnel's experiment on diffraction of light by a single slit.

Diffraction of light was extensively investigated by Fresnel in a series of experiments such as the one illustrated in Figure 7-15. It is similar to Young's experiment, except that the middle screen does not have two narrow slits but a single wider slit S_3.

Consider the theory that light is a stream of particles like bullets from a machine gun. These particles either would be stopped by the screen or would pass straight through the slit S_3. The ones passing through would travel on along straight-line paths. They would strike the viewing screen within a well-defined area and would form a sharp image of the slit. The expected outline of this sharp image is shown as a broken colored line in Figure 7-15.

When the experiment is performed, the pattern of illumination on the screen consists of a set of *diffraction fringes* spreading far beyond the sharp image. Figure 7-16 is a photograph of such a set of diffraction fringes. The graph above the figure indicates how the brightness of the screen varies with distance from the center of the pattern. The central fringe is broad and very bright. The side fringes are only half as wide. They become fainter and fainter as the distance from the center of the pattern increases.

7.16
THE THEORY OF DIFFRACTION FRINGES

A detailed theoretical treatment of diffraction by a slit yields the two features we have already encountered in the case of interference fringes. If the wavelength is made longer, the central fringe becomes wider and the rest of the pattern broadens out in proportion. The width of the central fringe is proportional to the wavelength.

If the slit is made wider, the central fringe becomes narrower and the rest of the pattern shrinks in proportion. The width of the central fringe is inversely proportional to the width of the slit. This is the opposite of what we would expect if the light traveled in straight lines to form a sharp image of the slit on the screen. The image would then become wider as the slit became wider.

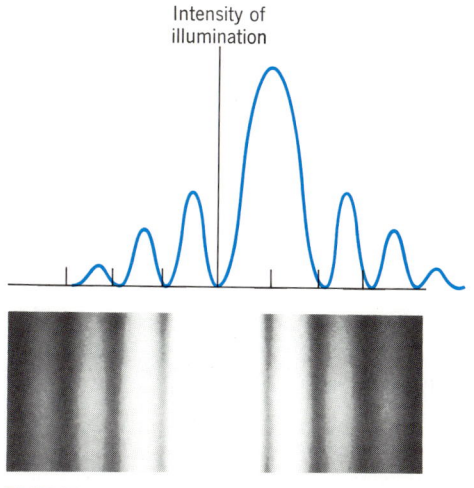

FIGURE 7-16
The diffraction pattern produced by a single slit. For the sake of clarity, the graph has been distorted. The ratio of the intensity of a side fringe to the intensity of the central fringe is actually even smaller than it appears to be on the graph. (Photograph by Dr. Brian Thompson.)

> Diffraction by a single slit:
>
> If the wavelength is increased, the pattern broadens out.
>
> If the slit is made narrower, the pattern broadens out.

If the opening in the middle screen is not a single slit but an aperture with some other shape, the diffraction pattern is usually more complex. It often has considerable aesthetic appeal. Figure 7–17 is the diffraction pattern of a triangular aperture. It might be described as a tapered slit that is wide at the bottom and narrow at the top. The figure illustrates admirably the fact that as the slit becomes narrower, the diffraction fringes move farther apart.

Figure 7–18 is the diffraction pattern of a rectangular slit, with the length and width not very different. It is similar to the pattern for a very long, very narrow slit, but shows fringes in both the horizontal and vertical directions. The shape of the rectangular slit is shown in the bottom right-hand corner. The fringes are farther apart in the direction parallel to the narrower side of the rectangular slit.

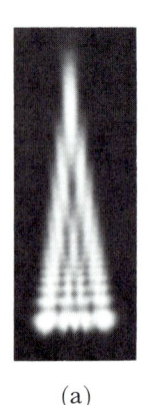

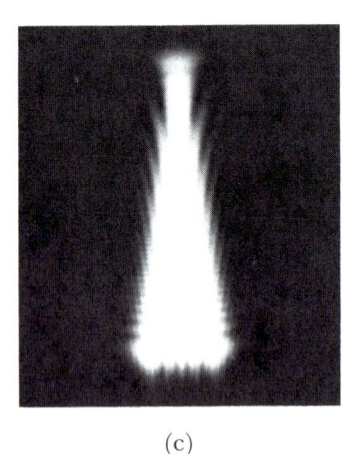

7.17 WHEN IS DIFFRACTION IMPORTANT?

Let us return to the diffraction pattern of a very long single slit. Imagine that the slit is made narrower and narrower. The pattern continually spreads out. When the width of the slit is much less than the wavelength, the central bright fringe is very wide indeed. The light passing through the very narrow slit spreads out in all directions as in Figure 7–10b.

Now reverse the procedure and make the slit wider and wider. The pattern of fringes then shrinks. However, when the width of the slit is many times the wavelength, the theory becomes more complicated. The width of the central fringe is then approximately the same as the width of the sharp image that would be formed if the light traveled along straight lines. The outer fringes are very close together just outside the edge of this sharp image. They are too faint to be seen except in a very narrow region near the edge. The fringes are in fact quite inconspicuous. The general appearance is close to what would be expected if light were a stream of particles traveling along straight lines.

We can now understand the sunbeam of Figure 7–10a. In everyday life, the light does appear to travel along straight lines, and diffraction effects are rarely encountered. The wavelength of light is very small, about 5×10^{-7} m. The objects we normally look at are about a million times larger. We therefore normally encounter the situation in which the diffraction fringes are very closely spaced and confined to a narrow region near the edge of the image. We rarely notice them. This circumstance explains why the particle theory of light was so popular until Young and Fresnel performed their careful experiments with very narrow slits.

An optical microscope cannot be used to obtain a clear image of an object whose size is much less than the wavelength of light. Illuminating such an object with light waves is similar to passing light through a slit of width less than the wavelength. The broad diffraction fringes formed spread out beyond the edges of the sharp image. The actual image appears blurred. Similarly, if we try to distinguish two small objects separated by a distance less than the wavelength, the two sets of diffraction fringes overlap. A single blur is seen. In general, if we wish to obtain a clear picture of an object, we must illuminate it with electromagnetic radiation of wavelength less than the size of the object.

(a) (b) (c)

FIGURE 7–17
Diffraction of light by an aperture with the shape of a long thin triangle. Part b is the image that would be formed if the light traveled strictly in straight lines. Part a, taken with a short exposure, shows the formation of crossed diffraction fringes inside the triangular image. Part c, taken with a longer exposure, shows the faint outer fringes. Considering the aperture to be a tapered slit, notice how the diffraction fringes move farther apart as the slit becomes narrower at the top. (Photograph by Dr. Brian Thompson.)

SUMMARY

Simple harmonic motion. The period T is the time to complete one oscillation. The frequency ν is the number of complete oscillations per second. The frequency is the reciprocal of the period. The SI unit of frequency is the hertz, abbreviated Hz. The amplitude is the distance from the midposition to an extreme position. A graph of the displacement plotted against time has a sinusoidal shape.

Transverse waves have their oscillatory motion in a direction at right angles to the direction in which the wave is traveling. A graph of the displacement plotted against distance at a fixed instant of time has a sinusoidal shape. The distance between adjacent peaks is the wavelength λ.

Longitudinal waves have their oscillatory motion in a direction parallel to the direction in which the wave is traveling.

The wave velocity V is equal to the product of the frequency ν and the wavelength λ. Short wavelengths correspond to high frequencies. Long wavelengths correspond to low frequencies.

Sound is a longitudinal wave traveling through the air. The pitch of a sound is the same as the frequency of the wave.

Electromagnetic waves. When an electric charge accelerates, it radiates an electromagnetic wave. An electromagnetic wave is a pattern of oscillating electric and magnetic fields. The fields are at right angles to the direction in which the wave is traveling and so it is a transverse wave. The wave velocity is equal to the constant c, which enters into the equations for magnetic forces. The constant c is also equal to the speed of light because light itself is an electromagnetic wave.

Types of electromagnetic waves. In order of increasing frequency and decreasing wavelength they are: radio waves, microwaves, infrared, visible light, ultraviolet, x-rays, and γ-rays. Their wavelengths are shown in Figure 7–8. Red light has a longer wavelength than blue light.

Energy and momentum of an electromagnetic wave. An electromagnetic wave transports energy and momentum. When an electromagnetic wave is incident on the surface of an object, it exerts a force on the object. The force per unit area is called the pressure of light. The intensity of an electromagnetic wave is defined as the energy flowing perpendicularly across unit area in one second. It is proportional to the square of the maximum value of the oscillating electric field. The momentum is equal to the energy divided by the speed of light c.

The wave nature of light is demonstrated by experiments on interference and diffraction.

Interference. If light passes through two narrow slits side by side, the waves from the two slits interfere.

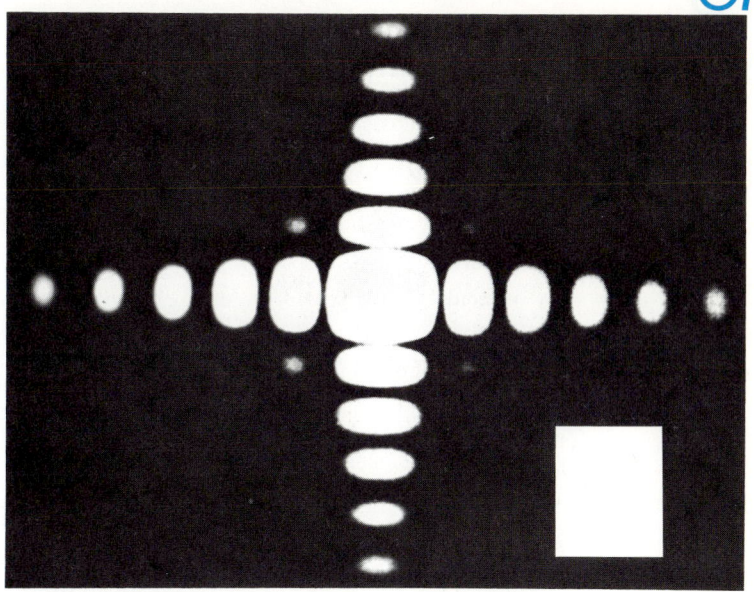

FIGURE 7–18

The diffraction pattern of a rectangular aperture 0.8 cm $\times$ 0.7 cm. A magnified drawing of the aperture is shown in the bottom right-hand corner of the photograph. Notice that the diffraction pattern is more spread out in a direction parallel to the narrower side of the aperture. (Photograph by Dr. Brian Thompson, from Wolf and Born, *Principles of Optics*, Pergamon Press, Ltd., 1959.)

Beams radiate out from the line midway between the slits and form interference fringes on a screen. The distance between neighboring bright fringes is proportional to the wavelength of the light and inversely proportional to the distance between the slits.

Diffraction. If light passes through a slit of width neither too large nor too small compared with the wavelength of the light, diffraction fringes are formed. The central fringe is twice as wide as the outer fringes. The central fringe is very bright and the fringes become fainter as the distance from the center increases. The width of the central fringe is proportional to the wavelength and inversely proportional to the width of the slit. Diffraction effects are important when the object being viewed has a size comparable with the wavelength. They are not conspicuous in everyday life because the wavelength of light is very small compared with the objects we normally observe.

7 WAVES

TERMS AND CONCEPTS

simple harmonic motion
period T
frequency ν
hertz
amplitude
sinusoidal
transverse wave
wavelength λ
wave velocity V
longitudinal wave
sound
pitch
vocal cords
eardrum
electromagnetic wave
speed of light c
radio waves
microwaves
infrared
visible light
ultraviolet
x-rays
gamma rays (γ-rays)
pressure of light
intensity of an electromagnetic wave
energy of an electromagnetic wave
momentum of an electromagnetic wave
wavefront
wavelet
interference
diffraction
interference fringes
diffraction fringes

QUESTIONS

1. Referring to Figure 7–1, can you think of any other examples of simple harmonic motion?

2. In the simple harmonic motions of Figure 7–1, in which of the positions A, M or B is the speed (a) greatest? (b) zero?

3. Compare and contrast sound waves and light waves.

4. State at least three different ways in which the constant c enters into physics.

5. Which of the following has the lowest frequency? (a) γ-rays, (b) blue light, (c) infrared light, (d) ultraviolet light.

6. Which of the following has the shortest wavelength? (a) radio waves, (b) x-rays, (c) red light, (d) ultraviolet light.

7. An electromagnetic wave has a wavelength comparable with the size of an apple. What type of electromagnetic radiation is it?

8. As the sun radiates away energy, does it also lose momentum and slow down?

9. In Young's double-slit experiment, why should the viewing screen be placed as far from the slits as possible?

10. In Fresnel's experiment on the diffraction of light by a single slit, why is it desirable to use as narrow a slit as possible?

11. In the interference and diffraction experiments of Young and Fresnel, why is it desirable to use light of a single color?

12. If white light were used in a Fresnel experiment on diffraction by a single slit, can you guess what the fringes would look like?

13. Imagine that an astronomer has invented a "camera" that does not use visible light but radio waves of wavelength 10 meters. Do you think that it would be suitable for photographing two astronauts shaking hands on the moon?

PROBLEMS

1. What is the frequency of a sound wave with a period of one thousandth of a second?

2. What is the period of 60-cycle alternating current?

3. The central note on a piano, middle C, has a frequency of 256 Hz. The speed of sound in air is 340 m/s (about 760 mph). What is the wavelength of middle C as it travels through the air?

4. Assuming that the lowest audible note has a frequency of 16 hertz, and the highest audible note has a frequency of 20 kilohertz, find the corresponding wavelengths in air. In each case think of a familiar object that has a size comparable with the wavelength. Take the speed of sound in air to be 340 m/s.

5. In an experiment to measure the speed of sound through a liquid, the frequency is 800 Hz and the wavelength is found to be 2.25 m. What is the speed of sound in the liquid?

6. What is the frequency of a radio wave with a wavelength of 2 meters?

7. What is the frequency of a light wave with a wavelength of 5×10^{-7} m?

8. Approximately how long does it take an electromagnetic wave to travel (a) from this book to your eyes? (b) one city block? (c) to your radio receiver from a radio station 10 km away? (d) from New York to Los Angeles? (e) from the moon to the earth? (f) from the sun to the earth? (Some parts of this question may require a little library research. Do not look up the answer directly. Look up the distance and then calculate the answer.)

PROBLEMS

9. The intensity of a light wave is 20 watts per square meter. What would the intensity be if the amplitude of the oscillating electric field were doubled?

10. In a Young's double-slit interference experiment, the wavelength of the light is 4×10^{-7} m. The distance between the slits is 0.5 mm. The distance between neighboring fringes is 2.4 mm. What would be the distance between neighboring fringes if a new color filter were inserted to pass light with a wavelength of 6×10^{-7} m?

11. In the experiment of question 10, the filter passing blue light is used. The distance between the slits is changed to 1.0 mm. What is the new distance between neighboring fringes?

CHAPTER 8
RELATIVITY

8.1
THE QUESTION

Is there a state of *absolute rest*? Is there, as Newton believed, a background space at absolute rest through which all objects move? If so, each object has a unique velocity measured relative to this space. An observer at absolute rest in this space obtains a very special view of the universe.

The alternative is *Einstein's theory of relativity*. This theory maintains that there is no *absolute space* and no absolute rest. The concept of the velocity of an object relative to a background space has no physical significance. Only the velocity of one object relative to another object has any significance. All that matters to a particular observer is the velocity of an object relative to the observer.

Even if there is an absolute space, it is most unlikely that an observer on the surface of the earth is at rest in this space. The rotation of the earth whirls us around with speeds up to a thousand miles per hour. The motion of the earth around the sun hurtles us through space with a speed of about sixty thousand miles per hour. The sun is a member of a group of stars, called the Milky Way galaxy, which is spinning like a wheel. This spin gives the sun a speed of about half a million miles per hour.

If we have such a high speed relative to absolute space, is there any physics experiment that will reveal it and enable us to measure it? Newton's laws of motion give us very little help in this respect. These laws are concerned with acceleration not velocity. Nothing in the behavior of a group of moving objects provides any clue to their absolute velocities relative to absolute space. The observable effects involve only the rates at which these velocities are changing.

The laws of electromagnetism appear to be more promising at first sight. Magnetic forces do depend on the velocities of the charges. On closer examination, magnetism provides no help either. The effects observed in actual experiments turn out to be independent of the absolute velocities. We obtained a small insight into this when we were discussing electromagnetic induction in Section 6.11. The induced current was found to depend only on the relative motion of the two loops.

Perhaps, however, some other electromagnetic phenomenon might help us to resolve the issue. In particular, light is an electromagnetic wave. We can restate our crucial question in the form, "Relative to what should the velocity of light be measured?"

In the case of sound, the answer to a similar question is well understood. Sound travels with a certain velocity relative to the material through which it is traveling. The velocity of sound through air is 750 mph. Suppose that I speak to you with a 50-mph gale blowing from behind me directly toward you (Figure 8–1). The sound emerging from my mouth travels at 750 mph relative to the air. It therefore travels from me to you with a speed of $750 + 50 = 800$ mph. On the other hand, if you speak to me against the direction of the wind, the sound travels from you to me with a speed of $750 - 50 = 700$ mph.

Perhaps the situation is the same for light. Perhaps, as was suggested in Section 5.6 where we were discussing the concept of a field, space is

8.2 THE SEARCH FOR THE ANSWER

FIGURE 8-1
The speed of sound is 750 mph relative to the air through which the sound is traveling.

filled with a mysterious, invisible, weightless, untouchable, jellylike substance called the *ether*. Perhaps light travels through the ether in the same way that sound travels through the air. The velocity of light c would then be 3×10^8 m/s relative to the ether. If that were so, absolute rest could mean "at rest relative to the ether." Only an observer at rest relative to the ether would always see light moving past with a speed c. To an observer moving relative to the ether, the velocity of light would be different in different directions, as in the case of sound discussed above.

8.2 THE SEARCH FOR THE ANSWER

At the end of the nineteenth century and the beginning of the twentieth century, the issues we have just raised became important in physics. Many experiments were performed to investigate whether any physical significance could be attached to the concept of an ether or a state of absolute rest.

The experiment that is most often quoted in connection with the theory of relativity was first performed by Michelson and Morley in 1887. The principle of this experiment is illustrated in Figure 8-2. Imagine that two flashes of light are sent out at right angles and are reflected back from two mirrors at equal distances from the source of the light. They might be expected to arrive back at the same time. However, if there were an ether and if the apparatus were moving through this ether, they would not arrive back simultaneously. Refer to Figure 8-2b and imagine that you are at rest relative to the ether watching the apparatus move past. One flash of light travels along the zigzag path *ABC*. The other flash travels along the path *ADC*. These paths are not equal in length. The two flashes would therefore be expected to arrive back at different times.

The apparatus of Michelson and Morley was capable of detecting a difference in the time of arrival of about one hundredth of one millionth of one billionth of a second (10^{-17} s)! Nevertheless, when the experiment was performed, no time difference was observed. The velocity of the apparatus relative to the ether could not have been greater than 10^4 m/s. But the velocity of the earth in its orbit around the sun is 3×10^4 m/s. By chance, the earth could have been almost at rest relative to the ether when the

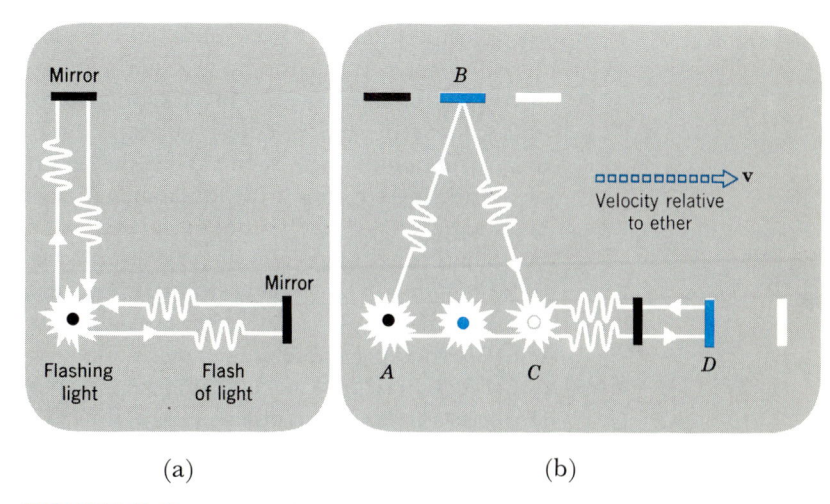

FIGURE 8-2
The principle of the Michelson-Morley experiment, assuming that the ether exists. (a) The apparatus is stationary relative to the ether. The two flashes travel the same distance and arrive back at the same time. (b) The apparatus is moving through the ether. The two flashes travel different distances and do not arrive back at the same time.

experiment was first performed. However, six months later the earth would have reversed its orbital velocity. It would then have had a velocity of about 6×10^4 m/s relative to the ether, large enough to produce an easily detectable time difference. When the experiment was repeated six months later, there was still no detectable effect. It began to look as though the idea of a velocity relative to the ether had no physical meaning.

Many experiments have been performed to detect motion relative to the ether. In all cases the anticipated effect has been completely absent. No experiment has yet provided evidence for the existence of an ether. No experiment has given meaning to the concept of absolute velocity through absolute space.

8.3
EINSTEIN'S ANSWER

Einstein's answer is that the concept of velocity relative to absolute space has no meaning. The ether does not exist. All motion is relative to the observer who measures it.

The observer we refer to is a physicist observing the universe and performing physical experiments on it. No experiment can possibly give a result that depends on the observer's velocity through absolute space, because such a velocity is a meaningless concept. There is therefore nothing the observer can do to measure this velocity.

Imagine several observers moving with various velocities relative to one another. Suppose that they all perform the same experiment to discover a fundamental law of physics. The results cannot depend on their velocities; they must all obtain the same result. When they complete their investigations, the fundamental laws of physics they discover must be exactly the same for all of them. In particular, there is no observer for whom the fundamental laws have a specially simple form because this observer is at absolute rest. The concept of absolute rest is meaningless.

Before proceeding, we must understand that the acceleration of an observer is a different matter. As we shall explain later in the present chapter, it is possible to measure the absolute acceleration of an observer. It is permissible to talk about observers with no acceleration. The *special theory of relativity* confines itself to unaccelerated observers. The *general theory of relativity* is needed to deal properly with accelerating observers. The special theory is our immediate concern.

> The basic principles of special relativity:
>
> The fundamental laws of physics are the same for all unaccelerated observers. It is impossible to perform an experiment that will measure the velocity of the observer relative to absolute space. Absolute rest is meaningless.

The fundamental laws of electromagnetism are contained in Maxwell's equations. They imply that an electromagnetic wave travels with a definite speed c. This conclusion must be valid for any unaccelerated observer. All observers must obtain exactly the same value when they measure the speed of any electromagnetic wave traveling through a vacuum. The speed of light should not be related to the ether or to absolute space. It is always 2.9979×10^8 m/s relative to the observer who measures it.

> The constancy of the speed of light:
>
> Any unaccelerated observer measuring the speed of any light wave obtains exactly the same numerical answer as any other unaccelerated observer, even though the two observers are moving relative to each other.

8.4
THE BOLDNESS OF EINSTEIN'S ANSWER

Consider the "common-sense" attitude to the following situation. A baseball player can throw a ball horizontally with a speed of 30 mph. Suppose that he stands up in a car that is traveling at 20 mph and throws the ball forward (Figure 8-3). The ball is given a speed of 30 mph relative to the car, but the car itself is moving in the same direction at 20 mph. An observer standing still and watching this event "obviously" sees the ball fly past at $30 + 20 = 50$ mph.

Now consider a similar situation involving light (Figure 8-4). Spaceship B is traveling past spaceship A at the very high speed of 2×10^8 m/s relative to A. B sends out flashes of light in the forward direction. According to Einstein's hypothesis, any observer must say that the flashes of light have a speed of 3×10^8 m/s. The pilot of B therefore says that the flashes are moving ahead at a speed of 3×10^8 m/s. The

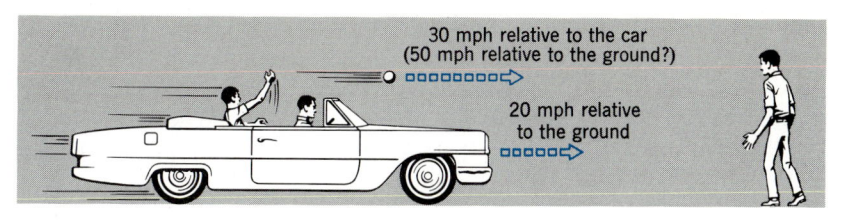

FIGURE 8–3
A problem in the addition of velocities.

pilot of B should be compared with the baseball player in the car. The flashes of light should be compared with the ball.

Can we apply to the pilot of the spaceship A the same argument that we applied to the observer standing by the side of the road? If so, the pilot of A should see the flashes of light travel past at a speed of $(3 \times 10^8) + (2 \times 10^8) = 5 \times 10^8$ m/s. However, that would be contrary to Einstein's hypothesis. All observers, measuring the speed of any light beam, must obtain a value of 3×10^8 m/s.

The natural reaction to this discussion is that it proves the absurdity of Einstein's theory. Nevertheless, physicists now believe, with a high degree of confidence, that the pilot of A would obtain a value of 3×10^8 m/s. It is common sense that is wrong. The outstanding achievement of Einstein's genius was the realization that the arguments applied to the car and the ball are based on certain subtle fallacies.

We now believe that the speed of the ball relative to the stationary observer would not be 50 mph but 49.999999999999933 mph. The correction introduced by the theory of relativity is so very small that it is clear why we are not aware of it in our everyday experience of cars and baseballs. The reason for this, as we shall see later, is that the speeds of the car and the ball are very small compared with the speed of light. In the example of the two spaceships and the beam of light, the speeds are not small compared with c.

What Einstein clearly realized was the following. We cannot start to understand a concept such as the velocity of an object without devising some procedure for determining time at distant places. We must know where the object is at what time. If we look at it, we rely on the light traveling from the object to our eyes. There is a time lag while the light travels to us from the object. We do not see the object where it is now, but where it was at an earlier time when the light left it. To correct for this time lag, it is necessary to know how fast the light traveled. This immediately raises all the questions about the velocity of light with which we have been preoccupied in this chapter.

Einstein's theory leads to a complete revolution in our ideas about space and time. When first encountered it is very hard to swallow. All we can do is work out the consequences of the theory and see if they are consistent with observations. Then we can work out the consequences of alternative theories and show that they are inconsistent with observations. So far Einstein's theory has always been victorious. Nevertheless, its consequences are very strange. We must be prepared to abandon some of our cherished prejudices.

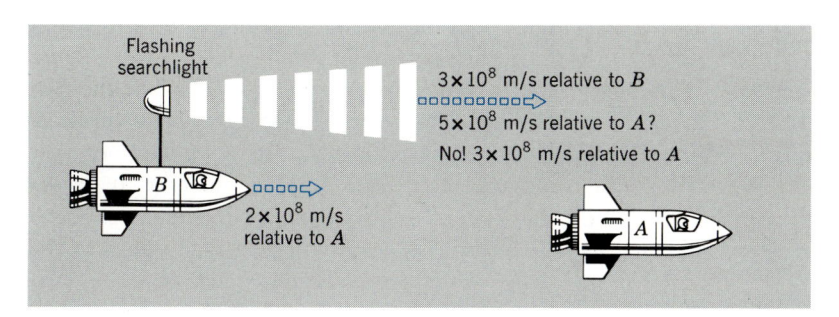

FIGURE 8–4
Even though B has a speed of 2×10^8 m/s relative to A, both A and B say that the light beam has a speed of 3×10^8 m/s.

8.5 MOVING CLOCKS RUN SLOW

One of the consequences of Einstein's theory is that two observers moving relative to each other cannot agree about physical measurements. In particular, each believes that the other's clock is slow, as illustrated in Figure 8–5. The two observers in spaceships A and B have a constant relative velocity **v**. Their clocks are identical copies. If A and B were at rest side by side, their clocks would agree exactly.

When they are in relative motion, observer A must observe B's clock from a distance through a telescope (Figure 8–5a) and must make allowance for the time taken for light to travel from B's clock to the telescope. A sees B's clock as it was at some earlier time just as the light left it. To calculate the necessary correction, A must assume that the light has traveled with a speed c, in accordance with Einstein's assumption.

It can be shown that after carefully taking these considerations into account, A finally concludes that B's clock is slow. While A's own clock registers a certain time interval, B's clock appears to register a time interval that is shorter by the factor $\sqrt{1 - \frac{v^2}{c^2}}$.

> Slowing down of a moving clock:
> Time interval registered by a moving clock
> $$= \sqrt{1 - \frac{v^2}{c^2}} \text{ (time interval registered by a stationary clock)} \qquad (8\text{-}1)$$

For example, if the relative velocity is $\frac{4}{5}$ of the velocity of light, $v = \frac{4}{5}c$ and $\sqrt{1 - \frac{v^2}{c^2}}$ is easily shown to be $\frac{3}{5}$. While A's own clock is registering a time interval of 5 minutes, it appears to A that B's clock registers only $\frac{3}{5} \times 5$ or 3 minutes.

It is not permissible to say that B's clock is slow in any absolute sense. That would introduce a basic difference between the two observers contrary to the principle of relativity. The correct statement is that according to A's observations, B's clock is slow. B forms exactly the opposite point of view. When B makes observations on A's clock, as in Figure 8–5b, B comes to the conclusion that A's clock is slow by the same factor $\sqrt{1 - \frac{v^2}{c^2}}$. From B's point of view, B's own clock is stationary and gives the correct time. A's clock is the moving clock and it therefore runs slow.

The disagreement between A and B is basic. If A and B were back on earth at rest, and A had a well-made wristwatch, whereas B had a poorly made one, they could easily come to an agreement. They could synchronize their watches so that they both read 12 noon when the sun was at its highest position in the sky. They might then discover that A's watch registered 1 P.M. when B's watch registered 12:55 P.M. A might then say that B's watch was slow, whereas B might say that A's watch was fast, which is logically consistent. However, they could come to an agreement in favor of A's watch by noticing that A's watch again registered 12 noon when the sun was at its highest position the following day. An adjustment in the rate of B's watch would make it agree with A's from then on.

When they are moving relative to one another, however, and they both have the best watch they can make, A always says that B's watch is slow, whereas B insists that A's watch is slow. No adjustment of their watches can make them agree. They must agree to disagree.

The slowing down of a moving clock has been observed. A very accurate atomic clock flown around the world in fast jet planes was found to lose a little.

Certain fundamental particles are unstable and rapidly break up into other particles. If one of these unstable particles has a speed near the speed of light, it is observed to break up much less rapidly. The "inner clocks" that regulate its decay slow down.

8.6 PAST, PRESENT, AND FUTURE

Observers moving relative to each other can also have interesting disagreements about the sequence of events in time. Imagine two events occurring in two different distant places. Before determining the time of an event, an observer must wait for light to travel from the event, and must then correct for the time delay. After doing this, one observer might conclude that the two events occurred at the same instant of time. However, it is possible for a second observer moving relative to the first observer to decide that event 1 occurred some time before event 2. A third observer moving relative to both of the first two might conclude the very opposite, that event 2 preceded event 1. Past, present, and future are all mixed up. There is no absolute time. Different observers make different estimates of the time and inevitably disagree with one another.

In view of these startling conclusions, it is relevant to ask if there is any limit to the pecu-

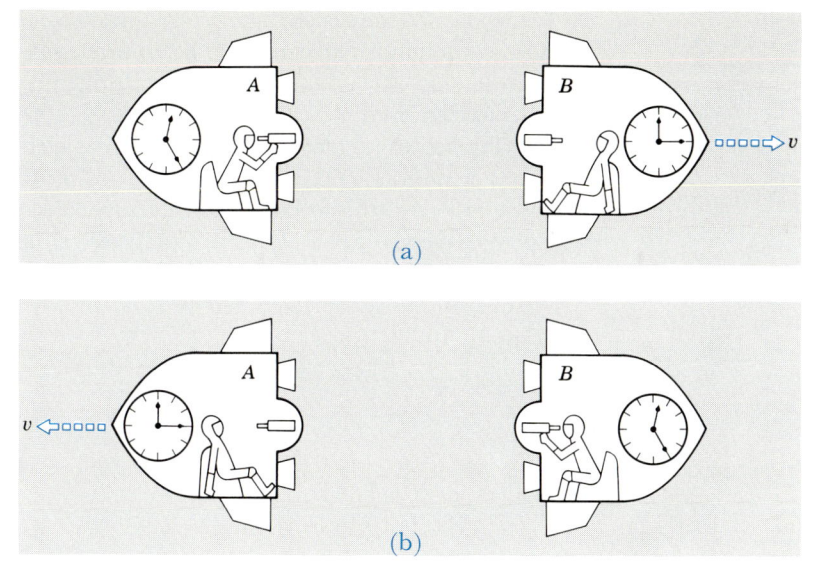

FIGURE 8–5

A moving clock runs slow. (The contraction effect of Figure 8–6 has been temporarily ignored.) (a) From *A*'s point of view, *B*'s clock is slow. (b) From *B*'s point of view, *A*'s clock is slow.

liar behavior of time. There are, in fact, severe restrictions on what is possible. Certain possibilities are so unreasonable that not even the theory of relativity will permit them to occur. The discrepancies about past, present, and future arise only when all observers agree that the events occurred at two different places. There is no problem for events that some observers say all occurred at the same place. If, for example, *B* talks to *A* over the radio, *B*'s vocal cords are always in the same position relative to *B*. *A* always hears the words in the correct sequence and does not hear *B* talking backward under any circumstances!

8.7 MOVING OBJECTS SHRINK

Two observers moving relative to each other disagree about space as well as time. They make different estimates of distances in space. They disagree about lengths of objects.

When they look into the matter, each decides that this is caused by the very peculiar behavior of the other observer's measuring rule (Figure 8–6). As long as a moving rule is perpendicular to the direction of its velocity, it has the right length. But when it is parallel to the direction of its velocity, it appears to shrink!

This effect is called the *Fitzgerald-Lorentz contraction*. When any object moves relative to the observer, its dimensions are unchanged in the direction perpendicular to its velocity. In the direction parallel to its velocity, all its dimensions shrink by the factor $\sqrt{1 - \frac{v^2}{c^2}}$. If its speed is $\frac{4}{5}c$, this factor is $\frac{3}{5}$. All lengths in the direction of the velocity then appear to be multiplied by a factor of $\frac{3}{5}$. A meter rule at rest has a length of 100 centimeters. If it moves with a speed of $\frac{4}{5}c$ in a direction parallel to itself, its apparent length becomes $\frac{3}{5} \times 100$ or 60 centimeters.

> Contraction of a moving object:
>
> When an object has a speed v relative to the observer its length in the direction of its velocity is multiplied by the factor
>
> $\sqrt{1 - \frac{v^2}{c^2}}.$

8.8 MASS, MOMENTUM, AND FORCE

In the light of these new ideas of time and space, Newton's laws of motion must be reconsidered. The significance of such concepts as mass, force, momentum, and energy must be carefully examined. New definitions must be provided for these quantities.

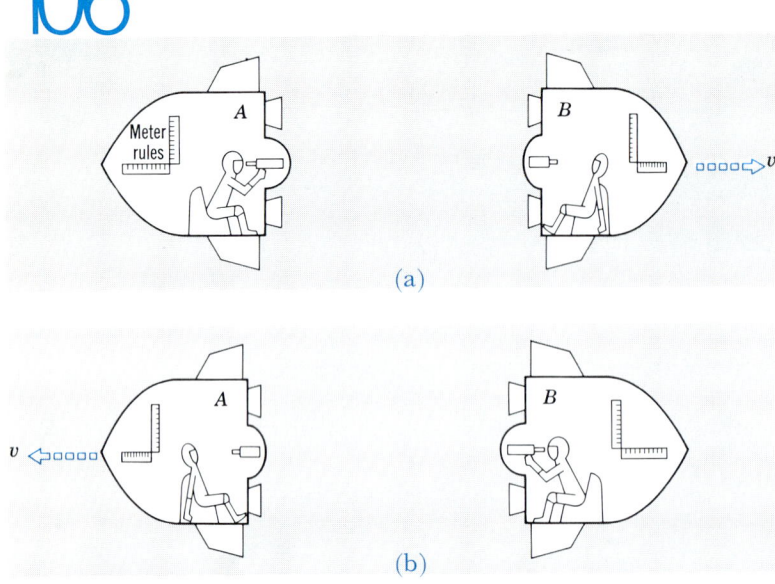

FIGURE 8-6
The contraction of a moving object. (a) A concludes that all objects in B's spaceship contract in the direction of the relative velocity. (b) B concludes that all objects in A's spaceship contract in the direction of the relative velocity.

It is necessary to abandon the concept of mass as an invariable property of a body that remains constant under all circumstances. The mass of a body increases as its speed relative to the observer increases. The mass of a body at rest is called its *rest mass* and is usually represented by the symbol m_o. When the body moves relative to the observer with a speed v, its mass m is equal to its rest mass divided by the factor $\sqrt{1 - \frac{v^2}{c^2}}$.

Variation of mass with speed:
$$m = \frac{m_o}{\sqrt{1 - \frac{v^2}{c^2}}} \quad (8\text{-}2)$$

Using this variable mass m, the momentum of a particle can still be defined as the product of mass and velocity.

$$\mathbf{p} = m\mathbf{v} \quad (8\text{-}3)$$

With this definition of momentum, the law of conservation of momentum is still true.

The force on an object must now be defined as the rate at which the momentum of the object changes with time. As the force changes the speed v, it also changes the mass m. Both factors influence the rate at which the momentum changes. It is no longer permissible to equate the force to a constant mass multiplied by the rate at which the velocity is changing. Newton's second law of motion is not valid in the form that says that the force is equal to the mass multiplied by the acceleration.

Charged fundamental particles, such as electrons or protons, can easily be accelerated up to speeds close to the speed of light. Many experiments have now been performed to measure the mass of such a particle at various speeds. The results are always in complete agreement with Equation 8-2.

8.9 THE MAXIMUM PERMISSIBLE SPEED

The factor $\sqrt{1 - \frac{v^2}{c^2}}$ appears persistently in all the relativistic effects we have discussed. When the speed v is very small compared with the speed of light c, the ratio v/c is very small compared with 1 and v^2/c^2 is even smaller. The factor $\sqrt{1 - \frac{v^2}{c^2}}$ is then only very slightly less than 1 and the relativistic effects are small.

The highest speed familiar to us in everyday life is the speed of a rocket. This might be as high as 10,000 meters per second or about 22,000 miles per hour. Nevertheless, the speed of light is about thirty thousand times greater. At the speed of a rocket, the fractional change in the rate of a clock, the length of a rule or the size of a mass is less than one part in a billion. A clock loses one second in about 50 years. A meter rule shrinks by about one-billionth of a meter. That is about the length of ten atoms laid side by side in a row. Relativistic effects are not conspicuous in everyday life.

On the other hand, when the speed v approaches the speed of light, v/c and v^2/c^2 each approach the value 1. The factor $\sqrt{1 - \frac{v^2}{c^2}}$ then approaches the value zero. Clocks become very slow and eventually stop when v is equal to c. At the same time meter rules shrink to zero length.

In Equation 8-2 for the variation of mass with speed, the factor $\sqrt{1 - \frac{v^2}{c^2}}$ appears in the denominator. As v approaches c, the denominator becomes smaller and smaller. The fraction therefore becomes larger and larger. As the speed of an object approaches the speed of light, its mass increases to infinitely large values.

It is now easy to see that a material object cannot have a speed greater than the speed of light. Suppose that we persist in trying to accelerate the object to as high a speed as possible. As its speed approaches the speed of light its mass becomes larger and larger. It becomes increasingly more difficult to accelerate it further. In fact, since the mass becomes infinite when its speed v is equal to c, we can never quite accelerate the object up to the speed of light c. We can approach closer and closer to c, by making greater and greater efforts, but we can never quite get there.

> The speed of an object can never exceed the speed of light.

8.10 MASS AND ENERGY

One of the major revelations of Einstein's theory is the relationship between mass and energy. The first hint of this comes from the properties of light discussed in Sections 7.9 and 7.10. When light is absorbed by an object, it warms up the object and exerts a pressure on it. The explanation is that light has both energy and momentum. Let us represent the energy by the symbol $\mathcal{E}$ and the momentum by the symbol p. Because momentum is usually the product of mass and velocity, it seems reasonable to consider the light to be equivalent to a mass m moving with a speed c.

Momentum of light
$$= (\text{mass of light}) \times (\text{speed of light}) \quad (8\text{-}4)$$
or
$$p = mc \quad (8\text{-}5)$$

Equation 7-8 tells us that the momentum of the light is also equal to its energy divided by the speed of light.

$$\text{Momentum of light} = \frac{\text{energy of light}}{\text{speed of light}} \quad (8\text{-}6)$$
or
$$p = \frac{\mathcal{E}}{c} \quad (8\text{-}7)$$

Comparing these two different expressions for the momentum, we see that

$$(\text{Mass of light}) \times (\text{speed of light})$$
$$= \frac{\text{energy of light}}{\text{speed of light}} \quad (8\text{-}8)$$
or
$$mc = \frac{\mathcal{E}}{c} \quad (8\text{-}9)$$

Divide both sides of this equation by c to obtain

$$m = \frac{\mathcal{E}}{c^2} \quad (8\text{-}10)$$

Mass of light
$$= \frac{(\text{energy of light})}{(\text{the square of the speed of light})} \quad (8\text{-}11)$$

The light can be considered to have a mass equal to its energy divided by the square of the speed of light.

In the theory of relativity, this equation is found to have a much deeper significance. Mass and energy can be considered to be different manifestations of the same physical quantity. A quantity of energy $\mathcal{E}$—whether it be kinetic energy, potential energy, chemical energy, nuclear energy, or any other form of energy—can always be considered to have a mass $\mathcal{E}/c^2$. Conversely, the mass m of a material object can be considered to be equivalent to an amount of energy equal to mc^2. Under some circumstances it can even be converted into this amount of energy. Even when an object is at rest, its rest mass m_o is equivalent to an energy $m_o c^2$, which is called its *rest mass energy*. Experiments on fundamental particles have discovered many processes in which a particle is destroyed and its rest mass is converted into an amount of energy $m_o c^2$.

> Interchangeability of mass and energy:
>
> Energy = (mass)
> $\times$ (the square of the speed of light)
> $$\quad (8\text{-}12)$$
> $$\mathcal{E} = mc^2 \quad (8\text{-}13)$$

This is the most famous equation of modern physics. Its connection with atomic and hydrogen bombs and nuclear power is well known. The full significance of the equation will become apparent as we apply it to various phenomena in subsequent discussions. We shall see, for example, that a ray of light is bent in the gravitational field of the sun as though the light were a massive body. We shall see that the energy of electromagnetic radiation can be converted into the mass of fundamental particles. Conversely, fundamental particles can disappear and their mass reappear as the energy of electromagnetic radiation. The mass of a nucleus will be shown to be due, in part, to the energy of the particles inside it.

In the nineteenth century it was taken for granted that mass could not be created or destroyed. There was a *law of conservation of mass* similar to the *law of conservation of energy* discussed in Chapter 3. Now that mass and energy have been found to be interchangeable, these two laws must be combined into a single *law of conservation of mass plus energy*. In the equation expressing this law, the mass must be multiplied by c^2 in order to enter on the same footing as the energy.

8.11 ACCELERATING OBSERVERS

So far we have been discussing the *special theory of relativity,* which is confined to observers moving with constant velocities. We now turn to the *general theory of relativity,* which extends the discussion to accelerating observers. We shall see that the general theory is primarily a theory of gravitation.

Although no experiment can distinguish between an observer at rest and an observer with a constant velocity, it is very easy for you to decide whether you are accelerating or not. If you are not accelerating, when you look at an isolated object, subject to no forces from other objects, it appears to move with a constant velocity in a straight line. That is Newton's first law of motion (Section 2.1). On the other hand, if you are accelerating, when you look at an isolated object, it appears to have an acceleration. Moreover, all isolated objects appear to have this same acceleration. The reason, of course, is that you have an acceleration in the opposite direction. This method of detecting acceleration is simple and reliable. You, the nonaccelerating observer, see all isolated objects moving along straight lines, whereas the accelerating observer sees most isolated objects moving along curved paths.

It is also very easy for you to decide whether you are rotating. If a vessel of water is rapidly rotated, the water is thrown outward and the surface becomes curved. The level near the wall is higher than the level in the center (Figure 8-7). You should therefore look at the surface of water in a vessel that is stationary relative to yourself. If the surface is flat, you are not rotating. If the surface is curved, you are rotating. This phenomenon is related to a very simple, commonsense attitude to the question of how you decide whether you are rotating. If you are rotating, you become dizzy! Rotation causes the fluid in the ducts of your inner ear to be thrown outward. The pressure of this fluid is responsible for the sensation of dizziness.

We have been discussing very real effects, and the observations cannot be denied. Special relativity is based on the idea that no meaning can be attached to the concept of "motion of an object relative to space," but only to the relative motion of two observers or the relative motion of two objects. Do the considerations of the present section imply that we can nevertheless attach physical significance to the concepts of "acceleration relative to space" or "rotation relative to space"?

8.12 MACH'S PRINCIPLE

After you have performed various physical experiments to establish that you are not rotating, you will discover a very important fact. The universe is not rotating either. The stars stay fixed in the same positions in your sky. On the other hand, if your experiments establish that you are rotating, the stars appear to rotate around you in the opposite direction. This is one of the reasons why we believe that the earth is rotating. To an observer on earth the sun and the stars appear to rotate around the sky once every day. "Rotation relative to space" seems to mean rotation relative to the matter in the universe.

Similarly, if you have performed various physics experiments to establish that you have a constant velocity, you then discover that this means a constant velocity relative to the matter in the whole universe. If the physics experiments establish that you are accelerating, you discover that you are accelerating relative to the universe. The stars on one side of you appear to be accelerating toward you. The stars on the other side appear to be accelerating away from you. Newton's "acceleration relative to absolute space" is the same as acceleration relative to the matter in the universe.

The Austrian physicist and philosopher Ernst Mach interpreted all this as meaning that we cannot detect acceleration and rotation relative to "absolute space," but only relative to the matter in the universe. He suggested that there is a hitherto unrecognized interaction between a moving object and all the other matter in the universe, the most distant matter being the most important. This interaction depends on the relative acceleration of the object and the distant

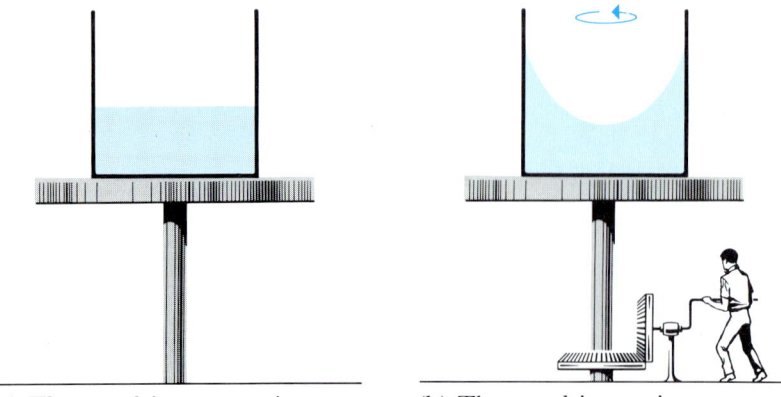

(a) The vessel is not rotating.   (b) The vessel is rotating.

FIGURE 8-7

When the vessel is rotating, the surface of the water inside it is curved.

matter. It is absent when the object moves with a constant velocity relative to the distant matter.

The isolated object of Newton's first law of motion is not really being left to itself. It experiences no strong force from nearby matter, but it is still subject to the influence of all the matter in the universe. The distant matter constrains it to move with a constant velocity relative to the universe.

8.13 A NEW INTERPRETATION OF NEWTON'S SECOND LAW OF MOTION

What happens if nearby matter exerts a force on the object? In Figure 8–8a, the force is exerted by a compressed spring. An observer who is not accelerating can apply Newton's second law of motion and conclude that

The force exerted by the spring
= (the mass of the object) × (the acceleration of the object relative to the universe) (8-14)

Can we adopt the underlying philosophy of relativity and describe the situation from the point of view of an observer sitting on the object and moving with it (Figure 8–8b)? From the point of view of this observer, the object has no acceleration. Therefore, Newton's second law suggests that there is no resultant force acting on the object. What cancels the force exerted by the compressed spring?

Mach's answer is that the distant matter in the universe appears to be accelerating relative to the observer moving with the object. The

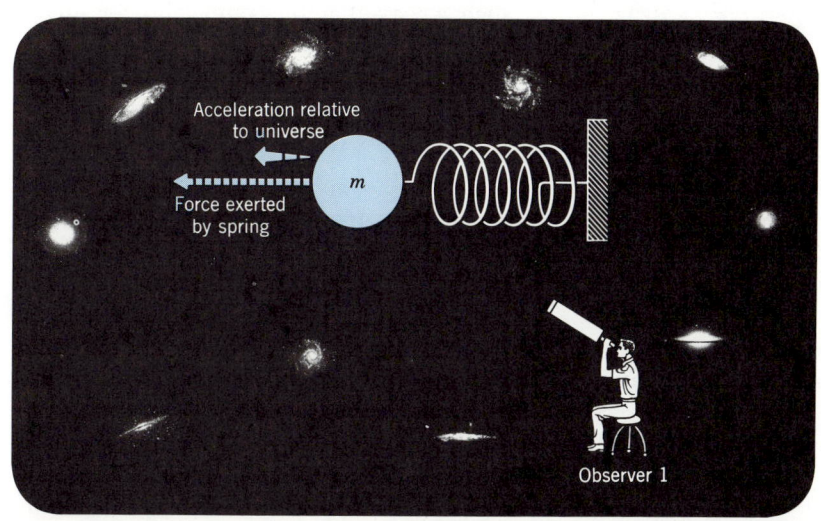

(a) Point of view of an observer at rest relative to the universe.

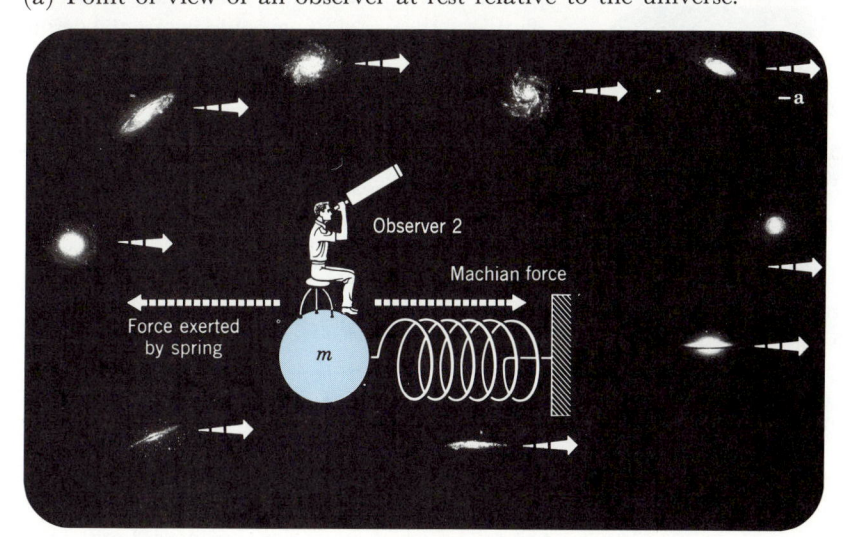

(b) Point of view of an observer moving with the object.

FIGURE 8–8
Mach's principle and Newton's second law of motion.

accelerating distant matter exerts a Machian force on the object. This force is proportional to the acceleration. It is also proportional to a certain number associated with the object that is a measure of how strongly the object interacts with the universe. This number is what we normally call the mass of the object.

The Machian force = (the mass of the object)
× (the acceleration of the universe relative to the object)

$$(8\text{-}15)$$

The observer moving with the object concludes that the body has no acceleration because the Machian force exactly cancels the force exerted by the spring. Equation 8-15 giving the magnitude of the Machian force therefore implies Equation 8-14 for the magnitude of the force exerted by the spring. Equation 8-14 is Newton's second law of motion, which therefore assumes a more profound significance in terms of the interaction of an accelerating object with distant matter in the universe.

Notice that if Mach's ideas are correct, we have the first real hint of the true physical significance of mass. It has something to do with the Machian force on an object when it accelerates relative to the matter in the universe.

The same ideas can be applied to rotation. Reconsider the curved surface of water in a rotating vessel from the point of view of Mach's principle (Figure 8–9). If you are rotating with the vessel the universe appears to be rotating around you. The rotating distant matter exerts Machian forces on the water. These forces pile the water up against the sides of the vessel.

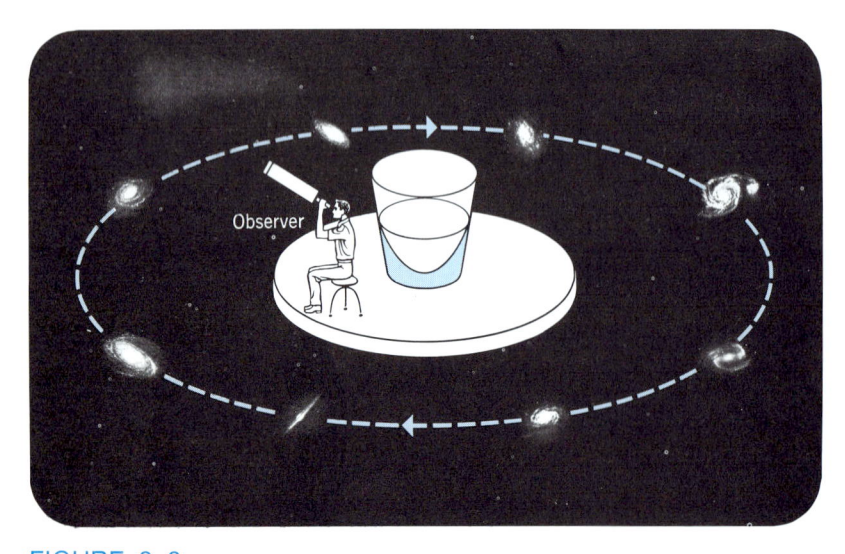

FIGURE 8–9
An observer rotating with the vessel sees the universe rotating around him. The rotating universe exerts Machian forces on the water and piles it up against the side of the vessel.

8.14
THE BASIC PRINCIPLE OF GENERAL RELATIVITY

Mach's principle shows how we can escape completely from the concept of absolute space and consider only the relative motion of different material objects. It encouraged Einstein to extend his principle of relativity to accelerating observers. One of the basic ideas underlying general relativity is that when the laws of physics are properly understood, they have the same form for accelerating observers as for nonaccelerating observers.

> The basic principle of general relativity:
> The laws of physics can be formulated in such a way that they are valid for any observer, however complicated this observer's motion.

The laws of physics might seem to be different for rotating and nonrotating observers. When a nonrotating observer looks at water in a vessel, its surface is flat. When a rotating observer looks at water in a vessel, its surface is curved. However, the laws of physics should include the fact that the behavior of the water is influenced by the motion of distant matter. The curved surface of the water is not a consequence of rotation relative to "absolute space." It is instead a consequence of the rotation of the whole universe relative to the vessel. If we could rotate the vessel of water and at the same time rotate all the matter in the universe with it, the surface would presumably remain flat! Actually, the concept of rotation of the whole universe relative to "absolute space" is meaningless, since no experiment could be performed to detect it. Only rotation relative to the matter in the universe is detectable.

8.15
LIGHT FALLS IN A GRAVITATIONAL FIELD

All massive objects fall with the same acceleration in a gravitational field. A beam of light has energy, which is equivalent to mass. Light falls in a gravitational field in the same way that material objects do.

A beam of light projected horizontally falls and moves along a curved path in the same way as the baseball thrown horizontally in Figure 1–7. We do not normally notice this because the speed of light is so large. The beam of light covers a very great horizontal distance in a very short time. During this very short time the distance through which it falls vertically is too small to measure.

Light has been observed to bend in the strong gravitational field of the sun. During an eclipse of the sun by the moon, it is possible to see stars very nearly in line with the edge of the sun. Light from these stars has passed very near the sun and has been bent in its gravitational field. The stars therefore appear to be displaced from their normal positions in a direction away from the sun (Figure 8–10). The observed magnitude of this displacement agrees with the value predicted by the general theory of relativity.

8.16 WARPED SPACE

We have just seen that light is bent in a gravitational field and no longer travels along a straight line. Let us pause and consider what is meant by a straight line. A good practical definition of a straight line, which might have been accepted without question before the general theory, is that it is the path followed by a ray of light in a vacuum. A farmer erecting a fence can check that it is straight by sighting along the posts to see that they are all "in line." Suppose, however, that the light is appreciably bent by the gravitational attraction of a nearby cow (Figure 8–11a). Clearly the method breaks down. When the cow walks away the fence no longer appears straight (Figure 8–11b).

All other possible definitions of a straight line lead to similar difficulties. The general theory of relativity copes with this situation by abandoning the *Euclidean geometry* that most of us learned in high school. Euclidean geometry is the geometry of flat space. In two dimensions it deals with the properties of diagrams that can be drawn on a flat piece of paper. If the diagrams were drawn on a bumpy piece of paper, the lengths of the lines would be different and the theorems would no longer be true. It is easy to draw a straight line on a flat piece of paper. On a bumpy piece of paper, there is no such thing as an ordinary straight line.

Einstein's idea is that a gravitational field warps space and makes it bumpy. A new *non-Euclidean geometry* must be used to describe this bumpy space. For example, the gravitational field of the sun warps space throughout the solar system. In Figure 8–12 the three-dimensional space around the sun is compared with an imaginary two-dimensional surface. In the absence of the sun this surface would be flat like a plain. The strong gravitational field near the sun warps the plain and produces a mountain with its peak at the location of the sun.

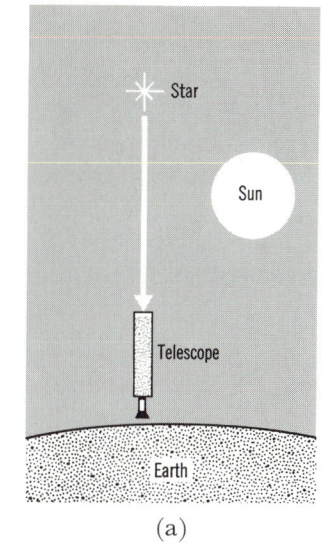

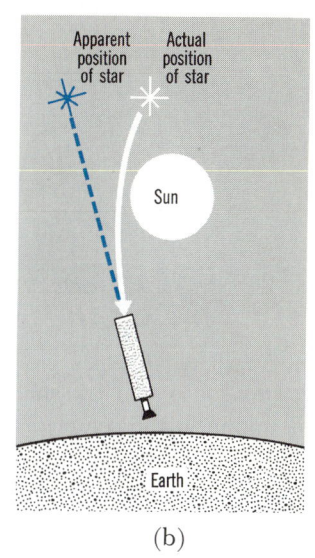

FIGURE 8–10
The bending of a ray of light traveling through the sun's gravitational field. (a) The ray of light from the star passes through a weak region of the sun's gravitational field and is not appreciably bent. (b) When the ray of light from the star passes near the sun, the bending is detectable.

FIGURE 8–11
Lining up the posts of a fence by sight. (In practice the bending would be negligibly small, but it has been exaggerated to demonstrate the effect.) (a) The ray of light is bent by the gravitational attraction of the cow. (b) The cow moves away and the posts no longer appear to be in line.

8 RELATIVITY

FIGURE 8-12
The warped space near the sun is represented by a mountain in the center of a plain. A radar pulse passing near the sun is delayed because it has to climb the mountain.

8.17
OBSERVATIONS IN WARPED SPACE

Figure 8-12 illustrates one observational test of these ideas about warped space. A radar pulse is sent out from the earth, is reflected off the planet Mercury, and is received back again on earth. When Mercury is in a suitable part of its orbit, the path of the radar pulse passes near the sun and is forced to climb the mountain. This slows it down and delays its arrival time on returning to the earth. The time delay is found to have a value in good agreement with the prediction of the general theory of relativity.

The effect we have just described should not be confused with the bending of a ray of light as it passes near the sun. The sideways bending of Figure 8-10 eventually determines the direction from which the light appears to be coming when it reaches the earth. The upward bending of Figure 8-11 merely delays the light and determines its time of arrival at the earth.

Another important effect caused by the warping of space near the sun is the *precession of the perihelion of Mercury*. The planet Mercury has an orbit that is not quite a circle. The distance of Mercury from the sun varies a little as it goes round its orbit. The point on the orbit nearest to the sun is called the *perihelion*. If Mercury were the sun's only planet, Newton's laws of motion would predict that the perihelion should be a fixed point in space. In fact the perihelion slowly revolves around the sun. This slow revolution is called the precession of the perihelion.

Most (99.2 percent) of the precession is caused by the gravitational forces exerted on Mercury by the other planets. The remaining 0.8 percent

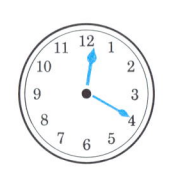

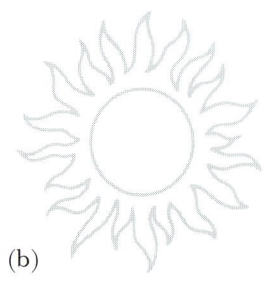

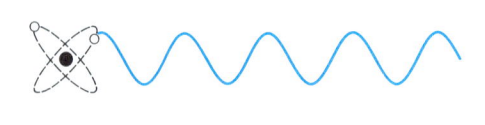

FIGURE 8-13
The gravitational red shift. (a) A stationary clock in a gravitational field runs slow. (b) An atom in a gravitational field emits light of lower frequency and longer wavelength.

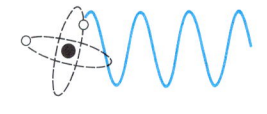

is a consequence of the warping of space by the sun's gravitational field. The general theory of relativity is able to predict the exact numerical value of this residual 0.8 percent.

8.18 THE GRAVITATIONAL RED SHIFT

A gravitational field distorts time as well as space. A stationary clock runs slow in a gravitational field. In Figure 8–13a, the clock close to the large object ticks more slowly than the clock farther away in a weaker gravitational field.

An oscillating electron in an atom is a kind of clock. In a gravitational field it oscillates more slowly and emits light of a lower frequency and longer wavelength. Compare the two atoms in Figure 8–13b. When the wavelength of visible light is lengthened, the color of the light moves toward the red end of the spectrum. The effect is therefore called the *gravitational red shift*. This red shift has been observed for light emitted by stars with strong surface gravitational fields.

The gravitational red shift has also been observed for the γ-rays emitted by radioactive nuclei in the earth's gravitational field. A subtle effect, known as the Mössbauer effect, makes it possible to compare γ-ray wavelengths with very great accuracy. The earth's gravitational field grows weaker as we move upward away from the earth. A nucleus at the surface of the earth is found to emit γ-rays of longer wavelength than a nucleus higher up.

SUMMARY

The Michelson-Morley experiment and all other experiments designed to measure the velocity of the apparatus relative to the ether have failed to detect this velocity.

Einstein's theory of relativity assumes that the concept of velocity relative to an ether, or relative to absolute space, is meaningless. The ether does not exist. All motion is relative to the observer who measures it.

The special theory of relativity is primarily concerned with observers who are not accelerating.

The general theory of relativity extends the ideas to accelerating observers. It is a theory of gravitation.

The basic principles of special relativity. The fundamental laws of physics are the same for all unaccelerated observers. It is impossible to perform an experiment that will measure the velocity of the observer relative to absolute space. Absolute rest is meaningless.

The constancy of the speed of light. Any unaccelerated observer measuring the speed of any light wave obtains exactly the same answer as any other unaccelerated observer, even though the two observers are moving relative to each other.

Moving clocks run slow. The time interval registered by a moving clock is shorter than the time interval registered by a stationary clock by a factor $\sqrt{1 - \frac{v^2}{c^2}}$. If two observers are moving relative to each other, each thinks that the other's clock is slow. Observers moving relative to each other have different ideas above the sequence of events in time. Past, present, and future can become mixed up.

Fitzgerald-Lorentz contraction. When an object has a speed v relative to the observer, its length in the direction of its velocity is multiplied by the factor $\sqrt{1 - \frac{v^2}{c^2}}$.

Variation of mass with speed. The mass of a moving object is its rest mass divided by the factor $\sqrt{1 - \frac{v^2}{c^2}}$. The mass increases as the speed increases and becomes infinite when the speed is equal to the speed of light. No object can have a speed greater than the speed of light.

Interchangeability of mass and energy. Energy has mass. Mass can be converted into energy. The energy is equal to the mass multiplied by the square of the speed of light. $\mathcal{E} = mc^2$. Mass and energy must be combined in a single law of conservation of mass plus energy.

Acceleration and rotation relative to the distant matter in the universe can be detected in physical experiments.

Mach's Principle. When an object accelerates relative to distant matter, the distant matter exerts a Machian force on it. This Machian force counterbalances the force exerted by nearby matter. Newton's second law of motion is explained if the Machian force is equal to the product of mass and acceleration.

The basic principle of general relativity. The laws of physics can be formulated in such a way that they are valid for any observer, however complicated this observer's motion.

Light falls in a gravitational field. A ray of light from a distant star passing near the sun is bent by the sun's gravitational field. During an eclipse of the sun by the moon, stars near the sun appear to be displaced away from the sun. In a gravitational field it becomes difficult to define a straight line. Space is warped and non-Euclidean geometry must be used. An electromagnetic wave (a radar pulse) is observed to slow down in the warped space near the sun. The warping of space by the sun's gravitational field produces a precession of the perihelion of Mercury.

The gravitational red shift. A clock slows down in a gravitational field. An atom in a gravitational field emits light of longer wavelength. A radioactive nucleus in a gravitational field emits γ-rays of longer wavelength.

TERMS AND CONCEPTS

absolute rest
Einstein's theory of relativity
ether
absolute space
Michelson-Morley experiment
Special theory of relativity
General theory of relativity
constancy of the speed of light
slowing down of a moving clock
Fitzgerald-Lorentz contraction
variation of mass with speed
rest mass, m_o
interchangeability of mass and energy
rest mass energy
law of conservation of mass plus energy
Mach's principle
Machian force
warped space
Euclidean geometry
non-Euclidean geometry
precession of the perihelion of Mercury
gravitational red shift

QUESTIONS

1. Why is the Michelson-Morley experiment so important in a discussion of the theory of relativity?

2. Why must the Michelson-Morley experiment be performed at different times throughout the year?

3. Describe two experiments that verify the prediction that a moving clock runs slow.

4. In addition to the two experiments of question 3, describe any other experiments you know that verify the unusual predictions of the special theory of relativity.

5. How does the theory of relativity change our ideas about the nature of time?

6. Why are the unusual consequences of the special theory of relativity not apparent in everyday life?

7. Mention two different ways in which light behaves like a massive object.

8. How does the theory of relativity change our ideas about the nature of space?

9. List and briefly describe all the experiments and observations that test the unusual predictions of the general theory of relativity.

10. Describe two different relativistic effects that make a clock run slow.

11. How can the gravitational red shift be investigated experimentally?

PROBLEMS

1. A spaceship is sent to a distant star with a constant speed of $\frac{4}{5}c$. According to the clocks on earth the journey takes 15 years. How long does the journey last according to the clocks in the spaceship?

2. A spaceship maintains a steady speed of $\frac{4}{5}c$ relative to the earth. The average life span of a baby born in the spaceship is 90 years according to the spaceship clocks. What is the average life span of a baby born in the spaceship according to the clocks on earth?

3. A meter rule moving in a direction parallel to its length appears to be only 80 cm long. What is its speed?

4. A meter rule has a speed of 1.0×10^8 m/s in the direction of its length. How long does it appear to be?

5. The rest mass of a proton is 1.67×10^{-27} kg. What is its mass when it has a speed of $\frac{4}{5}c$?

6. A meter rule is moving parallel to its length. What is its apparent length when its mass is twice its rest mass?

7. What is the speed of an object when its mass is three times its rest mass?

8. How many joules of energy are equivalent to a mass of 1 kg?

9. The cost of electrical energy supplied by a public utility company is one-half cent for a million joules. What is the cost of one kilogram of energy?

10. When it is stationary an object has a cubic shape with a side of length 1 meter. What is its apparent volume when it moves with a speed of $\frac{4}{5}c$ parallel to one side?

11. At rest the object in question 10 has a density of 9×10^3 kg/m^3. What is its apparent density when it moves with a speed of $\frac{4}{5}c$ parallel to one side?

CHAPTER 9
THE QUANTUM THEORY

9.1 A TWENTIETH-CENTURY REVOLUTION

So far we have formed the following picture of the universe. Matter consists of small particles moving along well-defined paths. The space between them is not featureless; it is occupied by various fields such as electric and magnetic fields. The electromagnetic interaction between two charged particles is not instantaneous; it is transmitted from one particle to the other by an electromagnetic wave traveling at the speed of light. Light is an electromagnetic wave. So are radio waves.

At the end of the nineteenth century this picture of the universe was well established. Then, in the early twentieth century, the whole beautiful scheme fell apart. Certain experiments were performed in which light behaved, not like a wave, but like a stream of particles! It began to look as though the universe could again be considered to be a collection of particles. However, a few years later, other experiments indicated that electrons could sometimes behave like waves! Both light and material particles sometimes exhibited the characteristics of waves and at other times exhibited the characteristics of particles. This was called the *dual nature of light* and the *dual nature of matter*. These developments had the virtue of putting material particles and electromagnetic radiation on the same footing, but they introduced a new dilemma. How was it possible for any entity to behave sometimes like a localized particle and at other times like a diffuse wave?

Let us compare and contrast the properties of particles and waves. Moving particles have energy and momentum. So do electromagnetic waves. Particles have mass. According to Einstein's principle of the interchangeability of mass and energy, light also has mass. The path of a moving particle is bent in a gravitational field. So is the path of a beam of light.

In one important respect, however, particles and waves are very different. A particle is a highly localized entity, which is situated at a particular point in space at a particular instant. A wave is a diffuse entity spread over a whole region of space. How can an electron be both located at a definite point and also spread out over space?

In the following section we shall consider these new developments in more detail. We shall show how they led to the *quantum theory* and to a complete revolution in our ideas of the nature of the physical universe. We shall present the decisive evidence that convinced physicists of the particle nature of light. We shall show how it was immediately followed by decisive evidence for the wave nature of matter. Then we shall try to resolve the dilemma presented by this dual nature of light and matter. In doing so we shall introduce probability and uncertainty into the basic laws of physics!

9.2 BLACK-BODY RADIATION

The first step in the direction of the quantum theory was taken by Max Planck in the year 1900. He was trying to explain how the energy of the electromagnetic radiation emitted by a hot body is distributed among the various possible frequencies.

The electromagnetic radiation emitted by a hot surface usually depends on its chemical composition. Nevertheless, it is possible to imagine an object called a *black body,* which is a perfect emitter of radiation at all frequencies. The name *black body* comes from the fact that it is also a perfect absorber of radiation. Imagine a cold black body in a dark room. Because it is cold, it emits no visible light of its own and cannot be seen. Moreover, even if it is illuminated by a beam of light, it still cannot be seen. Being a perfect absorber, it absorbs all the light falling on it. No light is reflected into the eye of the observer and the body appears to be perfectly black.

What we have just said is strictly true only if the black body is at the absolute zero of temperature. At any higher temperature a perfect black body emits *black-body radiation.* The nature of black-body radiation depends only on the temperature. The amount of energy radiated increases very rapidly as the temperature is raised. The *Stefan-Boltzmann law* states that the energy emitted by unit area of the surface of the black body in one second is proportional to the fourth power of the absolute temperature (T^4). Suppose that the black body is initially at 300 K, which is the temperature of a warm summer day. We double its absolute temperature to 600 K, which is about the temperature of molten lead. Since $2^4 = 16$, the black body radiates 16 times as much energy.

Figure 9-1 shows how the energy radiated by a black body is distributed over the various frequencies of the emitted electromagnetic waves. At any temperature all frequencies from zero to infinity are emitted. However, at a particular temperature T, most of the energy is radiated at frequencies not very different from a particular frequency ν_m. About three-quarters of the energy is contained within the frequency range between $\frac{1}{2}\nu_m$ and $2\nu_m$.

As the temperature is raised, the energy emitted at any frequency rapidly increases. At the same time the bulk of the radiated energy shifts toward higher frequencies. *Wien's law* states that the frequency ν_m, near which most of the energy is concentrated, is directly proportional to the absolute temperature T.

$$\nu_m = 5.9 \times 10^{10}\, T \text{ hertz} \quad (9-1)$$

We sometimes talk about the wavelength λ_m instead of the frequency ν_m. The wavelength of an electromagnetic wave is inversely proportional to its frequency (see Section 7.3). Wien's law therefore states that the wavelength λ_m is

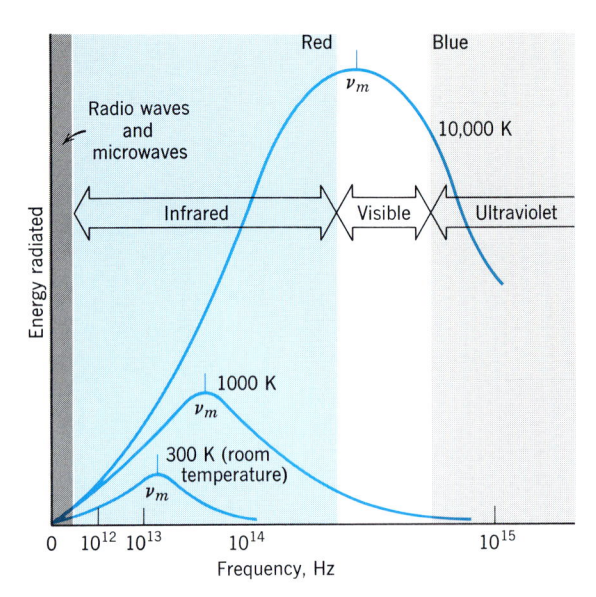

FIGURE 9-1
Each curve shows how the energy of black-body radiation is distributed over the various frequencies at a particular temperature. The three curves correspond to black-body temperatures of 300 K, 1000 K, and 10,000 K. In order to bring out the main features, some distortion has been introduced. In reality, the curve at 10,000 K is about 37,000 times higher than the curve at 300 K. The radio and microwave region occupies only $\frac{1}{300}$th of the horizontal axis.

proportional to the reciprocal of the absolute temperature $1/T$. As the temperature is raised, the bulk of the radiated energy shifts toward shorter wavelengths.

To appreciate the significance of Wien's law, imagine a blackened steel ball hanging in a dark room. A blackened steel ball is a good approximation to a perfect black body. As you well know, the ball is invisible if it is at the same temperature as the room. At room temperature, which is about 300 K, the frequency ν_m lies well inside the infrared region (see Figure 9-1). The steel ball emits mainly invisible infrared radiation and very little visible light.

Now imagine that the steel ball is gradually warmed up. As the temperature rises the bulk of the radiation shifts toward higher frequencies and shorter wavelengths. The frequency ν_m moves through the infrared toward the red end of the visible region. The first thing to happen is that red light is copiously emitted. The ball glows "red hot." At a still higher temperature, ν_m is in the middle of the visible region, and all colors of visible light are emitted. The ball glows "white hot." At a very high temperature ν_m is near the blue end of the spectrum. Blue light is emitted

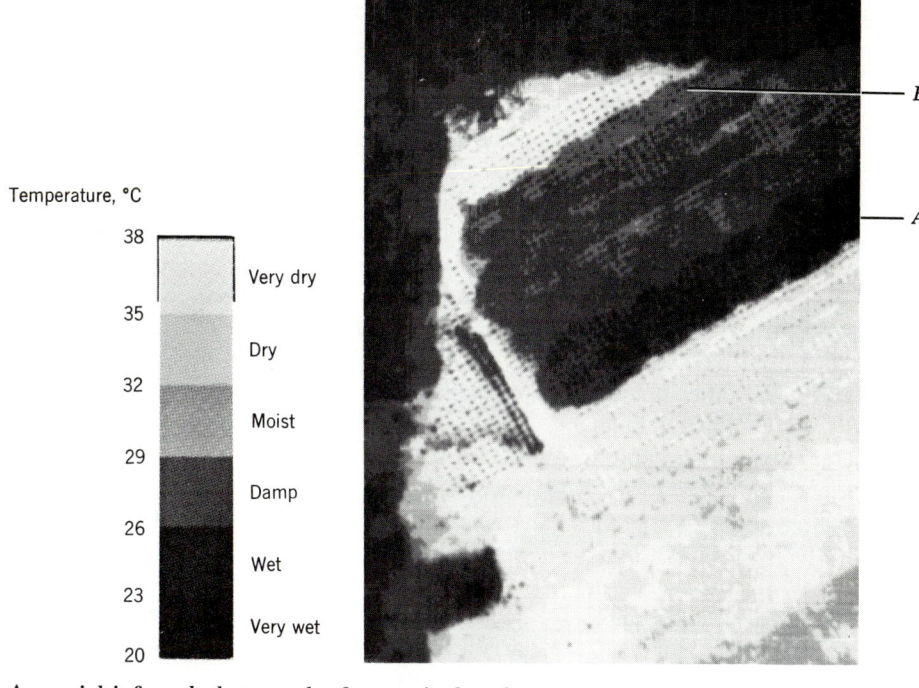

An aerial infrared photograph of an agricultural region near Saginaw, Michigan. Hot, dry areas emit more infrared radiation and show up as a lighter tone in the photograph. Cool, wet areas are darker. *A* is a row of crops that has just been irrigated and cooled. *B* is a row of crops that was irrigated earlier and has had time to dry out and warm up. (Courtesy Daedalus Enterprises, Inc.)

preferentially and the ball glows "blue hot." Molten steel pouring out of a steel furnace often has a bluish appearance.

Similar considerations apply to stars. The temperature of the surface of the sun, about 6000 K, is such that it radiates all colors of visible light. Sunlight is more or less white light. A red star has a lower surface temperature. A blue star has a higher surface temperature. A rough estimate of the surface temperature of a star can be made by observing its color.

9.3 PLANCK'S QUANTUM OF ENERGY

At the end of the nineteenth century, theoretical physicists were unable to explain these facts about black-body radiation. In particular, their theories seemed to predict that an infinite amount of energy would be radiated at the highest frequencies!

The bold step taken by Planck in the year 1900 was to try out a startling assumption. When a black body emits electromagnetic radiation, it does not emit it continuously but in bursts of well-defined energy. The quantity of energy in a burst is called a *quantum* of energy. We shall represent this amount of energy by the symbol ε (the Greek letter epsilon). It is equal to the frequency ν of the radiated wave multiplied by a certain fundamental constant of physics. This very important constant is called *Planck's constant* and is always represented by the symbol h. Its value is 6.63×10^{-34} SI units (joule/hertz).

Planck's quantum of energy
   $=$ (Planck's constant)
      $\times$ (frequency of wave)   (9-2)
$\varepsilon = h\nu$ (9-3)

Planck's idea was that the black body cannot emit as small an amount of energy as it wishes. It must either emit a quantum of amount $h\nu$ or nothing. It can then go on to emit subsequent quanta, each of amount $h\nu$. But it cannot, for example, emit a total amount $\tfrac{3}{2}h\nu$. On a fine scale the emission of radiation is discontinuous.

Notice, however, that the scale is very fine. For visible light of frequency about 5×10^{14} Hz, the energy in a quantum is 3.31×10^{-19} J. To acquire a kinetic energy of this magnitude, a grain of sand would have to fall through a height of only about 3×10^{-12} m, which is much less than the diameter of a single atom! In our everyday experience of large objects the energies involved are so large compared with Planck's quanta that we are not normally aware of the discontinuous nature of energy.

It is only when we deal with very small objects like atoms, or fundamental particles, that the quantum aspects of nature become apparent. If we ask how far a single proton must fall before it acquires enough kinetic energy to be able to create a quantum of visible light, the answer is interesting. A proton starting at rest a long way from the earth, at an effectively infinite distance, and falling freely all the way to the earth's surface, would still not have quite enough kinetic energy to create a quantum of visible light.

Planck's bold assumption worked. It gives a completely satisfactory explanation of the nature of black-body radiation. Wien's law takes on a new significance. The quanta emitted most copiously have an energy near $h\nu_m$. Planck's theory reveals that this energy is equal to $2.82\,kT$, where k is Boltzmann's constant. Refer back to Section 4.5 and notice the interesting similarity with the value $\frac{3}{2}kT$ for the average kinetic energy of an atom of an ideal monatomic gas.

$$h\nu_m = 2.82\,kT \tag{9-4}$$

If we divide both sides of this equation by h we obtain

$$\nu_m = \left(\frac{2.82\,k}{h}\right)T \tag{9-5}$$

This equation is similar to Equation 9-1, which was derived experimentally. The two equations can be made exactly the same by choosing the value of Planck's constant that we have already quoted.

9.4
THE PHOTOELECTRIC EFFECT

Planck was reluctant to go all the way and say that the black body emits particles of light with energy $h\nu$. We now accept this as a permissible description of the situation. The particles of light are called *photons*. In 1905, the same year he published his first paper on special relativity, Einstein pointed out that the concept of photons could explain certain puzzling aspects of a phenomenon known as the *photoelectric effect*. When electromagnetic radiation of a sufficiently high frequency falls on the surface of a metal, electrons are ejected from it (Figure 9-2). This is the photoelectric effect, and the elected electrons are called photoelectrons.

Two aspects of the photoelectric effect could not be explained by the electromagnetic wave theory of light. First, the light is able to eject photoelectrons only if its frequency exceeds a certain threshold frequency, ν_c. Second, ignoring some minor complications, the ejected photoelectrons have a definite kinetic energy that depends only on the frequency of the light.

Einstein's explanation is simple. When a photon strikes the metal, it is completely absorbed and gives up all its energy to the photoelectron. A part of this energy is needed to push the electron out of the metal. The remaining part appears as the kinetic energy of the escaped electron.

If the electron is to escape at all, the frequency of the incident light must be at least large enough so that the energy of the photon exceeds the energy needed to lift the electron out of the metal. Planck's constant multiplied by the threshold frequency $h\nu_c$ is therefore the energy of a photon that is just able to lift the electron out of the metal. This tells us that an amount of energy $h\nu_c$ is needed to lift the electron out of the metal.

At a higher frequency, the total amount of energy supplied by a photon is equal to $h\nu$—Planck's constant multiplied by the frequency. If we subtract from this the energy $h\nu_c$ used up in lifting the electron out of the metal, the remainder should be the kinetic energy of the electron as it emerges. This kinetic energy can be measured. Its value is always in excellent agreement with the above explanation.

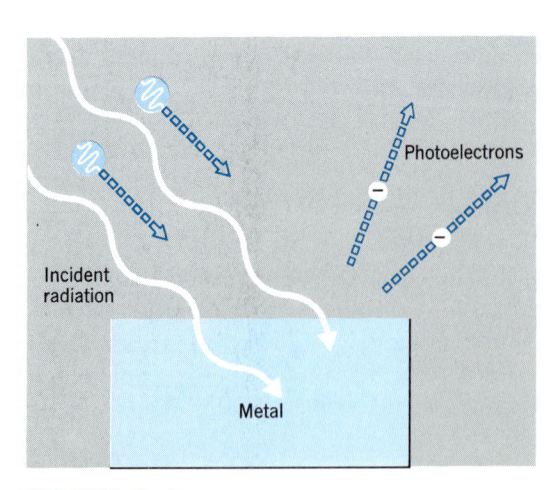

FIGURE 9-2
The photoelectric effect.

9.5 THE COMPTON EFFECT

The clinching argument for the existence of photons came from an experiment performed by Arthur Compton in 1922. When x-radiation passes through a solid—typically graphite—part of it is scattered in all directions. The scattered radiation has two components. One component has exactly the same frequency as the incident radiation. The other component, the one that interests us, has a frequency slightly smaller than that of the incident radiation (Figure 9–3a). Moreover, the decrease in frequency becomes greater as the angle of scattering increases.

This effect is easily understood if we assume that it is a consequence of a collision between a photon of the incident x-rays and an electron in the solid. This collision may be discussed in exactly the same way as the collision of two billiard balls. The collision is depicted in Figure 9–3b. Before the collision the electron is at rest. After the collision it has been knocked on with a large velocity and a large kinetic energy.

Because the collision must conserve energy, the kinetic energy of the electron can be gained only at the expense of the energy of the photon. The scattered photon has therefore lost some energy. But the energy of a photon is Planck's constant multiplied by the frequency of the wave. The scattered photon is therefore associated with a wave of lower frequency.

This approach leads to a completely satisfactory explanation of the dependence of the change in frequency on the angle of scattering. All that needs to be done is to apply the law of conservation of energy and also the law of conservation of momentum to the collision. The momentum of a photon must be taken to be equal to Planck's constant divided by the wavelength of the associated electromagnetic wave. We shall return to this in Section 9.7.

Notice an important difference between the photoelectric effect and the Compton effect. In the photoelectric effect, the photon completely disappears and all of its energy is given to the photoelectron. In the Compton effect, there is still a photon after the collision, but its fre-

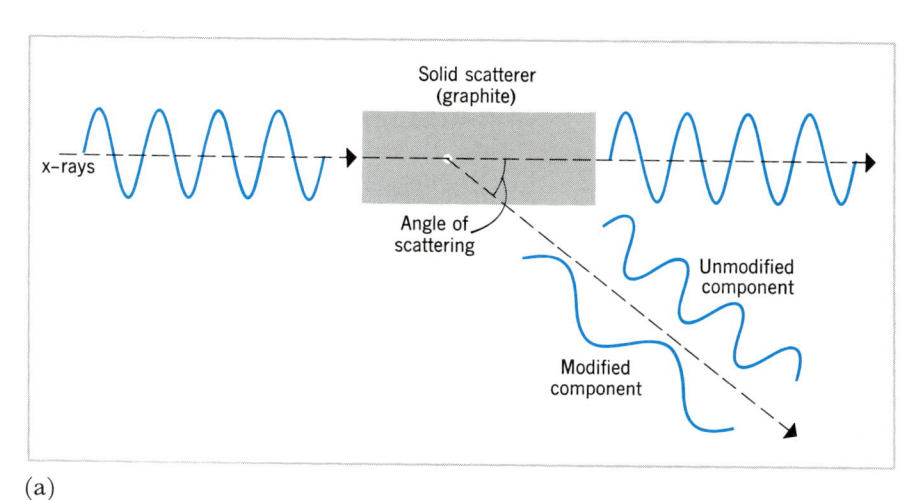

FIGURE 9–3

The Compton effect. (a) The wave point of view. (b) The particle point of view.

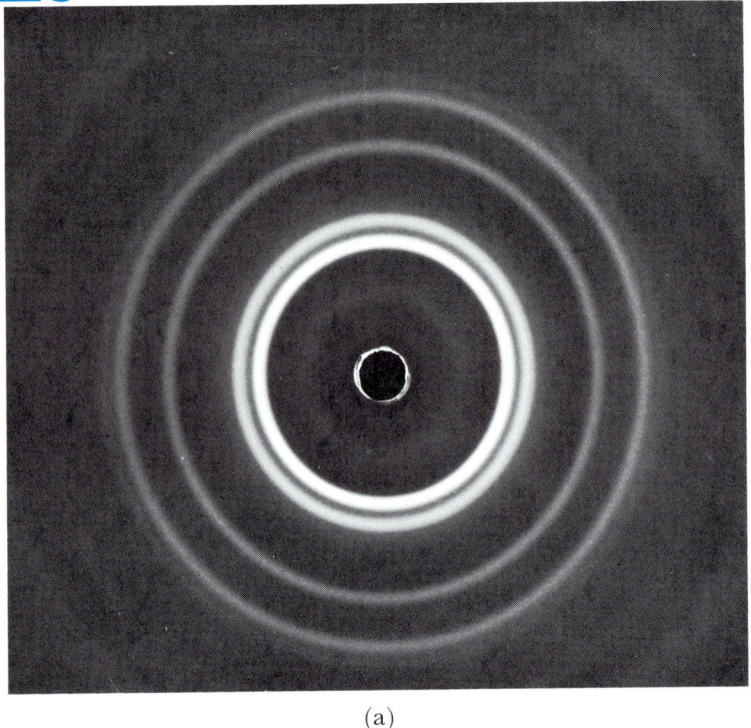

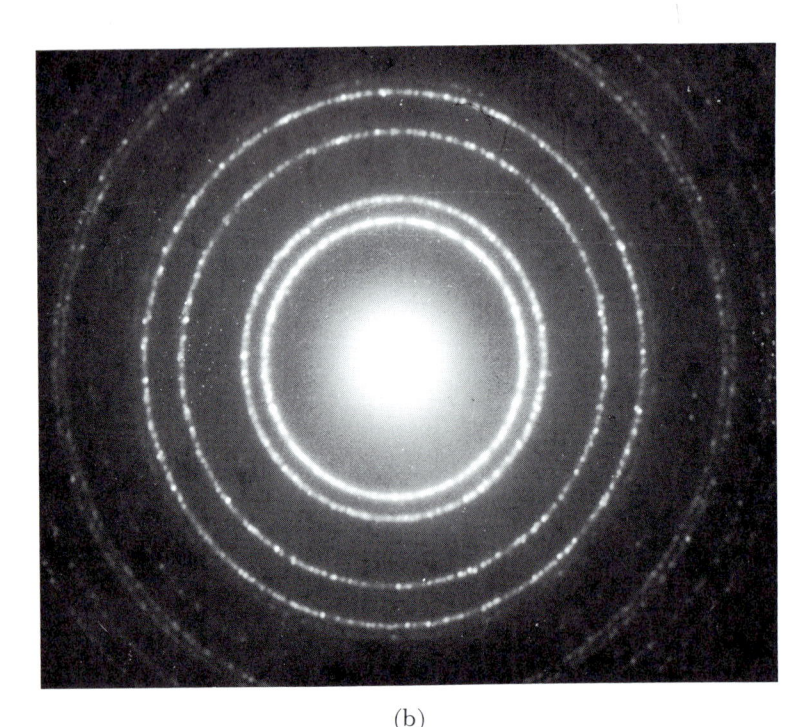

quency is less than that of the incident photon. Part of the energy of the incident photon has been given to the Compton electron.

The concept of photons with energy and momentum enables us to form a vivid picture of the pressure of light (Section 7.9). It can now be visualized as a hail of photons striking the surface on which the pressure is exerted. It is therefore very similar to the pressure exerted by the atoms of a gas on the walls of their container (Sections 4.6 and 4.7).

9.6 THE WAVE NATURE OF MATTER

In 1922 the Compton effect provided conclusive evidence that electromagnetic radiation sometimes exhibits the properties of particles. Yet interference and diffraction experiments clearly indicate that electromagnetic radiation also exhibits the properties of a wave. In some sense it must combine the properties of both particles and waves. In 1924 Louis de Broglie made the bold suggestion that electrons and protons, which had been shown in many experiments to behave like particles, might also behave like waves. By 1927 Davisson and Germer in the United States, and G. P. Thomson in Scotland, had demonstrated interference and diffraction effects with electrons.

In these experiments the diffraction was not produced by slits, as in the experiments of Young and Fresnel with light, but by crystalline solids. A crystal is an ordered array of atoms. When a wave falls on this array, each atom is the source of a scattered wavelet. The wavelets interfere constructively in some directions and destructively in other directions. The result is a complicated diffraction pattern, the details of which need not concern us here. Diffraction by crystals was first observed for x-rays. It has been successfully used to measure the wavelengths of the x-rays and also to study the spatial arrangement of the atoms in the crystals.

Figure 9–4a shows a set of circular diffraction fringes obtained by passing x-rays through an aluminum foil. Figure 9–4b shows a set of fringes produced by passing a beam of electrons through the same foil. The speed of the electrons was adjusted to ensure that the wavelength associated with the electron beam was the same as the wavelength of the x-rays. The similarity of the two sets of fringes is striking evidence that electrons can be diffracted in exactly the same way as electromagnetic waves.

Subsequent investigations have revealed the existence of diffraction effects with other fundamental particles, such as protons and neutrons,

FIGURE 9–4
A direct comparison of electron diffraction and x-ray diffraction. The wavelengths are the same in both cases. (From the P.S.S.C. film "Matter Waves".) (a) The diffraction pattern obtained by passing x-rays through an aluminum foil. (b) The diffraction pattern obtained by passing a beam of electrons through the same foil.

9.7 WAVE PROPERTIES AND PARTICLE PROPERTIES

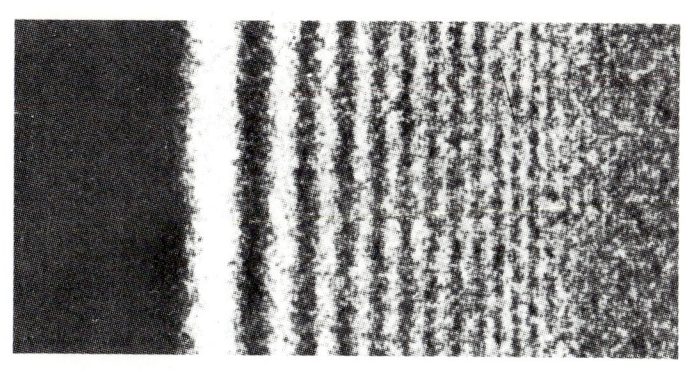

(a) Diffraction fringes produced when a beam of electrons passes near the straight edge of a small cubic crystal. (Reproduced by permission of D. C. Heath & Company, P.S.S.C. *Physics*, 1960.)

(b) The diffraction pattern produced when light passes near a straight edge. (Courtesy Brian Thompson.)

Comparison of the diffraction patterns produced by electrons and by light passing near a straight edge.

and even with atoms and molecules. All material particles sometimes behave like waves. The wave associated with a material particle is called a *de Broglie wave*. The quantity that oscillates in a wavelike fashion is called the *wave function*. It is analogous to the vertical displacement of the surface of the water in the case of an ocean wave, or to the oscillating electric vector **E** in the case of an electromagnetic wave. Its physical significance will be discussed in Section 9.8.

9.7 WAVE PROPERTIES AND PARTICLE PROPERTIES

The principal properties associated with a particle are its mass m, its velocity v, its momentum p, and its total energy ε. If the same particle behaves like a wave, the principal properties of this

A diffraction pattern obtained by passing neutrons through a single crystal of salt. (Courtesy E. O. Wollan.)

wave are its frequency ν and its wavelength λ. The quantities describing the particle are related to the quantities describing the wave by the following equations.

> Total energy of particle
> $\quad =$ mass
> $\quad\quad \times$ (square of the speed of light) (9-6)
> $\quad =$ (Planck's constant)
> $\quad\quad \times$ (frequency of wave) (9-7)
> $\varepsilon = mc^2 = h\nu$ (9-8)
> Momentum of particle
> $\quad\quad =$ mass $\times$ velocity (9-9)
> $\quad\quad = \dfrac{\text{Planck's constant}}{\text{wavelength}}$ (9-10)
> $p = mv = \dfrac{h}{\lambda}$ (9-11)

Equations 9-6 to 9-11 are valid for all particles, including photons. In the case of a photon, the significance of the mass m requires special consideration. For an ordinary "material" particle, like an electron or a proton, the mass m depends on the speed v.

$$m = \frac{m_o}{\sqrt{1 - \dfrac{v^2}{c^2}}} \quad (9\text{-}12)$$

If v were to become equal to c, the denominator would become zero and the mass m would become infinite. But the speed of a photon is equal to c. The only way the photon can avoid having an infinite mass m is by having a rest mass m_o equal to zero. The mass m is then $0/0$, which is indeterminate and may have any value whatsoever depending on the exact circumstances. Its value is actually $h\nu/c^2$ for an electromagnetic wave of frequency ν.

> A photon is able to move with the speed of light because it has no rest mass.

9.8
RECONCILING WAVES AND PARTICLES

What is the physical significance of the wave function, the quantity that oscillates in a de Broglie wave? In what sense is it possible for an entity like an electron to combine the properties of a localized particle and a diffuse wave?

In the case of light, the oscillating quantity is the electric field vector **E**. If we produce a diffraction pattern with light, the brightness of the screen at any point depends on the rate at which energy flows into a small area of the screen surrounding the point. According to Equation 7-7, this energy flow is proportional to the square of the oscillating electric field, E^2. From a particle point of view, since each photon has a definite energy, the rate of flow of energy is proportional to the number of photons striking the small area of the screen per second. It follows that the square of the oscillating electric field, E^2, is a measure of the number of photons striking the small area per second.

Similar reasoning can be applied to electrons or particles of any kind. Suppose that we allow a beam of electrons to pass through a single narrow slit and then fall onto a fluorescent screen, like the screen of a television tube (Figure 9–5). On the screen we shall see a diffraction pattern similar to the one shown in Figure 7–16. In both Figure 7–16 and Figure 9–5 the graph adjacent to the screen shows how the brightness of the pattern varies across the screen.

The brightness of any part of the screen clearly depends on the number of electrons striking a small area there per second. From the wave point of view, however, the brightness of any part of the screen depends on the square of the wave function. The square of the de Broglie wave function is seen to be a measure of the number of electrons striking a small area per second.

In front of the fluorescent screen set up a row of counters, as shown in Figure 9–5. A counter is a device that registers the passage of each electron and counts the total number of electrons arriving in any desired time interval. When a beam of electrons passes through the slit and produces a diffraction pattern, the number of electrons striking a particular counter is proportional to the square of the wave function in its vicinity. Suppose that 50 percent of the electrons strike the central counter labeled C. Meanwhile 1 percent strikes the side counter labeled S. This indicates that the square of the wave function is 50 times greater near the central counter C than near the side counter S.

9.9
THE ROLE OF PROBABILITY

We seem to have arrived at a very satisfactory explanation of the relationship between the wave and the particles. The square of the wave amplitude is a measure of the number of particles arriving per second. That explanation serves

9.9 THE ROLE OF PROBABILITY

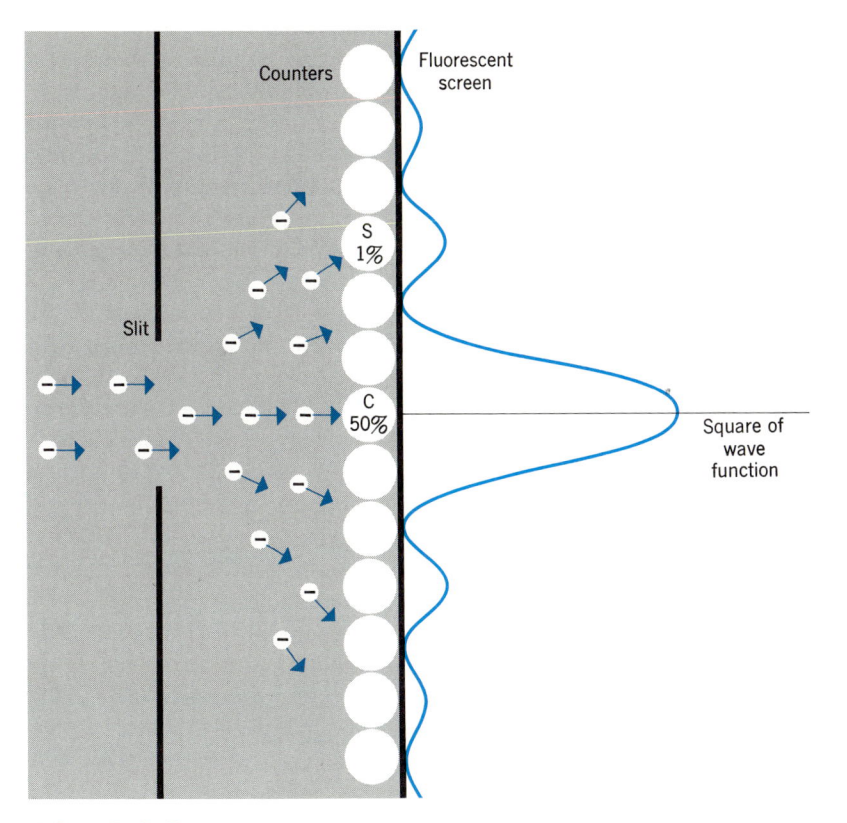

FIGURE 9-5
The physical significance of the de Broglie wave function.

us well until we ask a very nasty question. What would happen if we were to send a single electron through the slit on its own?

Would the single electron spread itself over the whole screen and trigger all the counters? Would 50 percent of it strike the central counter C and 1 percent of it strike the side counter S? A counter either responds to an arrival or does not respond at all. It cannot register a "partial" arrival. Perhaps the electron would strike a definite point on the screen and trigger only one counter. If so, which counter would it trigger? Would it settle for the popular center counter?

The answer, which is supported by an impressive amount of experimental evidence, is as follows. Only one counter will respond, as might be expected if the electron is a particle that can be in only one place at a time. However, it is quite impossible to predict in advance which counter this will be. The best that can be done is to assess the *probability* that the electron will strike a particular counter. This probability is proportional to the square of the wave function in the vicinity of the counter.

The significance of the probability in this instance is the following. If we perform a single experiment in which one electron is sent through the slit, it is completely impossible to predict which counter it will strike. However, suppose that the experiment is repeated many times. Many electrons are eventually sent through the slit one at a time. In 50 percent of the experiments the electron will strike the central counter C. There is a 50 percent probability that a single electron will strike the central counter. In 1 percent of the experiments the electron will strike the side counter S. There is a 1 percent probability that a single electron will strike the side counter. It is not possible to predict in advance which particular experiments will make up this 1 percent. The result of a single experiment is unpredictable. Only the net effect of a large number of experiments is predictable.

Our final conclusion is that the square of the de Broglie wave function at a point is a measure of the probability that the single particle will be found near that point. A de Broglie wave is a wave of probability.

Similar reasoning can be applied to photons of light. The oscillating electric field of an electromagnetic wave acquires a new significance. The square of the oscillating electric field E^2 at a point is a measure of the probability that a photon will be found near that point. In a certain sense, a light wave is also a wave of probability.

9.10 THE PHILOSOPHICAL IMPLICATIONS

According to the present interpretation of the quantum theory, it is quite impossible to predict in advance which counter a particular electron will strike. This is not believed to be a consequence of the presence of unknown factors that guide the electron along its path by means that we have not yet discovered. Instead, the idea is that the future behavior of an electron is not completely determined by its past history. Several possibilities are open to it. One of these is chosen purely by chance, for no reason that can ever be determined. This applies to all aspects of the behavior of an electron, not only to its diffraction by a slit. It applies to all moving bodies. However, its consequences are more important for bodies of small mass, such as fundamental particles, atoms, and molecules.

This indeterminism of quantum theory is in sharp contrast to the determinism discussed in Section 2.2. There it was suggested that the past, present, and future behavior of all the particles in the universe is completely determined. In principle, by applying the laws of motion to these particles, I could predict their future behavior and discover that I am predestined to be killed by a car as I cross the road in search of my lunch. Moreover, it would be beyond my power to do anything to prevent this.

The quantum theory suggests a very different state of affairs. Whatever the past behavior of the particles in the universe, several possibilities are available for their future behavior. The possibility that is actually realized is purely a matter of chance. The universe may not have predetermined my death in an automobile accident. But it may well be "playing roulette" to decide whether I shall be killed by the automobile or not.

The behavior of electrons is very remote from our everyday experience of human bodies and automobiles. Perhaps these considerations about probability and chance are important only to the physicist in calculations of the behavior of small particles. Perhaps they have no relevance to the larger objects of normal experience. The following experiment is designed to demonstrate that the indeterminism of the behavior of electrons may have serious consequences on a large scale.

The electron gun at the extreme left of Figure 9–6 ejects single electrons at a rate slow enough to ensure that only one electron is in the apparatus at any time. Consider the very first electron ejected after the gun is switched on. This electron passes through a very narrow slit that produces a broad diffraction pattern. If it should happen to be diffracted to the left, it strikes counter 1. The arrival of the electron at counter 1 results in a pulse of electric current, which travels to the component labeled "Trigger." This in its turn explodes a hydrogen bomb. If the electron should happen to be diffracted to the right, it strikes counter 2, which delivers its pulse of current to the component labeled "Deactivator." This component operates a mechanism inside the hydrogen bomb that renders it incapable of exploding at any future time.

You sit on the hydrogen bomb contemplating the philosophical implications of the quantum theory. The outcome of the experiment is a matter of great consequence to you. Yet, according to the contemporary point of view, there is no way of predicting whether the first electron will be diffracted to the right or to the left. The best team of scientists available, having at their disposal all that they might require in the way of technical or computational facilities, could still not predict the outcome. Moreover, rightly or wrongly, the current belief is that it will never be possible to predict the outcome of such an experiment, however much our understanding of the physical nature of the universe progresses. The element of chance is fundamental to the behavior of nature. All that can be said is that you have a 50 percent chance of surviving.

9.11 HEISENBERG'S UNCERTAINTY PRINCIPLE

In 1927 Heisenberg tried to resolve the problem of exactly what is meant by saying that the electron is a particle. He decided that a particle must have one essential characteristic. At a fixed instant of time, it must have a definite location at a definite point in space and must have a well-defined velocity. He then asked how it might be possible to determine experimentally

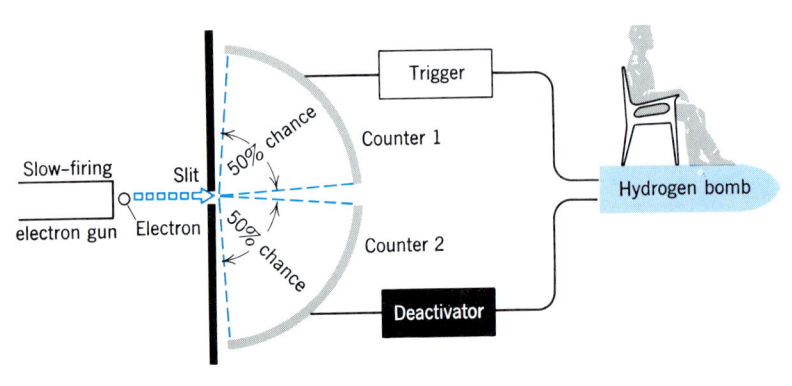

FIGURE 9–6
An experiment in philosophy.

this definite position and definite velocity. He came to the startling conclusion that in the case of a small particle such as an electron, it is not possible to measure precisely both its position and its velocity. In any conceivable experiment, there is always an uncertainty in the position, or the velocity, or both. This is not caused by imperfections in the design or construction of the apparatus used in the experiment; with the best possible apparatus that could ever be constructed, the uncertainty would still be present. It is an unavoidable consequence of the way in which nature behaves.

A precise theory reveals that the uncertainty in momentum is the important quantity, not the uncertainty in velocity. In general, whatever procedure we think up for simultaneously measuring the position and momentum of any particle, we find that we can measure either of these quantities as accurately as we wish, but only at the expense of a large uncertainty in the other quantity. The uncertainty in position multiplied by the uncertainty in momentum is always approximately equal to Planck's constant h. This means that if either uncertainty is made small, the other must be large in order to keep their product constant.

A similar difficulty applies to measurements of energy and time. It is not possible to measure precisely the energy of a physical system at an exact instant of time. There is always an uncertainty in the value of the energy and an uncertainty in the instant of time when the system has this energy. The uncertainty in energy multiplied by the uncertainty in time is approximately equal to Planck's constant h. Accuracy in the measurement of the energy is obtained at the expense of not knowing exactly when the system had this energy. Pinning down the time to some instant within a very short time interval is made possible only by sacrificing accuracy in the measurement of the energy.

Planck's constant h is very small. The uncertainty in momentum or energy is therefore also very small. What usually matters is the uncertainty in momentum or energy as a fraction of the total momentum or energy. For the objects we deal with in everyday life, the fractional uncertainty in momentum or energy is negligibly small. The uncertainty principle is not conspicuous in everyday life. It becomes important only when we deal with very small objects such as atoms, molecules, or fundamental particles. The total momentum and total energy of such a small particle are themselves very small. The Heisenberg uncertainty principle can then produce a very large fractional uncertainty in these quantities.

9.12 ATOMIC SPECTRA

If the light emitted by a hot monatomic gas is passed through a spectroscope, as in Figure 9–7a, the spectrum is observed to consist of several narrow bright lines separated by dark regions. The atoms do not emit all frequencies of electromagnetic radiation; they emit only a few distinct frequencies.

Quantum theory provides the following simple explanation. The electrons in an atom cannot move in any way they please. They must choose one of a limited number of special motions. This kind of special motion is called a *state* of the atom. Each state has a definite energy. It follows that the energy of the atom must have one of a limited number of special values.

Suppose that the atom is in a particular state with one of the permissible values for its energy. It then switches over to another state with a smaller permissible energy. A definite amount of energy is released and escapes from the atom as a photon with a very special frequency. This special frequency corresponds to one of the lines of the spectrum. The various lines of the spectrum correspond to the various possibilities for the initial and final states of the atom.

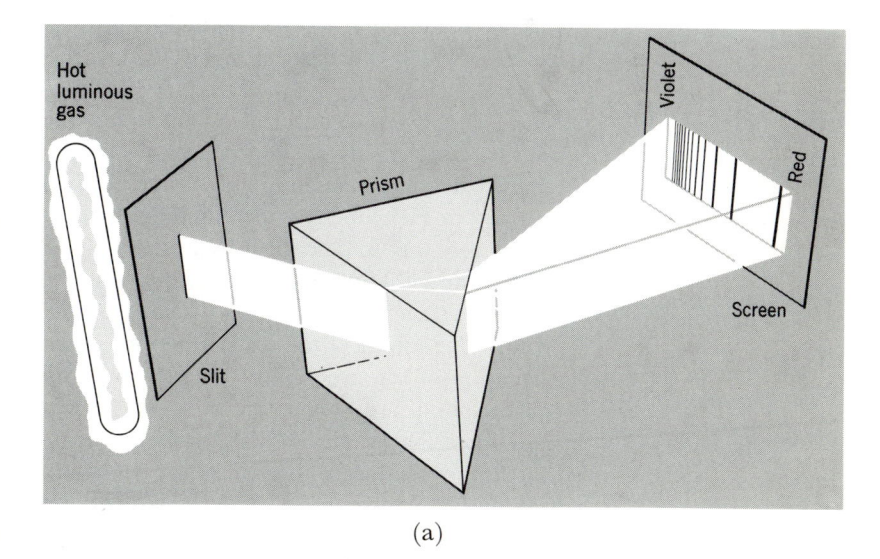

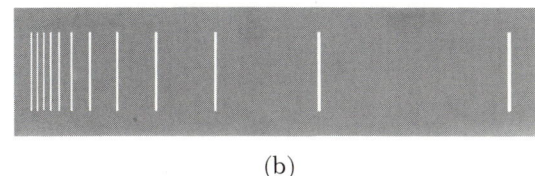

FIGURE 9–7
An atom can emit only certain special frequencies of light. (a) A simple method of displaying the line spectrum of a hot monatomic gas. (b) The spectrum of monatomic hydrogen in the visible region.

The next step, of course, is to explain why the atom is restricted to a limited number of special states of motion. The energy of each state should then be calculated. The energy $h\nu$ of a photon corresponding to any observed spectral line ought then to be equal to the difference between the energies of two of the permissible states. For the simpler atoms containing only a few electrons, the quantum theory is able to predict precise numerical values of the frequencies for all observed spectral lines. For atoms containing many electrons, there is no reason to doubt the underlying physical ideas, but the computations are very difficult.

9.13
THE NATURE OF AN ATOM

The simplest atom is hydrogen. Its nucleus is a single positively charged proton. This nucleus jiggles around a little, but it can be assumed to be almost stationary at the center of the atom. A single negatively charged electron moves around much more rapidly over an extensive region of space surrounding the nucleus. What we say about the behavior of this electron in hydrogen provides a good guide to the behavior of any electron in any atom.

The most familiar picture of a hydrogen atom is one in which the electron moves in an orbit around the proton. This picture is oversimplified. The concept of an orbit assumes that the position and velocity of the electron are known precisely at each precise instant of time. That is contrary to Heisenberg's uncertainty principle. The correct way to describe a hydrogen atom is to give the value of the wave function of the electron at each point in the vicinity of the proton. The square of the wave function is then a measure of the probability that the electron will be found near this point.

Suppose that we devise a method of locating exactly the position of the electron in the atom at a particular instant of time. This does not violate the uncertainty principle as long as we do not demand any knowledge of the velocity of the electron. It is not possible to predict where the electron will be found, although it is more likely to be found in those regions where the square of the wave function is large. Having found the electron, let us mark its position with a dot. If we repeat the experiment a second time, the electron will be found in an entirely different position, which we again mark with a dot. After repeating the experiment many times, we obtain a cloud of dots occupying a region of space all around the proton. The density of this cloud in the vicinity of a point is proportional to the square of the wave function at that point.

The atom may therefore be visualized as a "cloud of negative electricity" surrounding a small positive nucleus. We must remember, though, that different parts of the cloud do not coexist simultaneously. A slightly better description of this picture of an atom is that it is a *"cloud of probability."*

The various permissible states of the hydrogen atom correspond to clouds of various sizes and shapes. Some typical clouds are shown in Figure 9–8.

9.14
QUANTUM NUMBERS

A cloud of a particular size and shape can be specified by quoting the values of three integers, represented by the symbols n, l, and m_l. These integers are called the *quantum numbers* of the electron in the atom. The reader may find it easier to understand the following discussion of their significance by referring frequently to Figure 9–8.

The *principal quantum number* n is a positive integer such as 1, or 2, or 3, and so on. It determines the size of the cloud. In fact, the size of the cloud is proportional to the square of the principal quantum number, n^2. This means, for example, that a cloud with n equal to 2 is four times as big as a cloud with n equal to 1. The principal quantum number also determines the *shell* in which the electron is located. An electron with n equal to 1 is said to be in the K shell. The L shell corresponds to n equal to 2, the M shell to n equal to 3, the N shell to n equal to 4, and so on. Because the size of the cloud is proportional to n^2, the K, L, M, and N shells are progressively farther away from the nucleus in that order.

The principal quantum number n is the main factor determining the energy of the cloud. The energy is always negative, because it is dominated by the negative electric potential energy of the positively charged nucleus and negatively charged electron. The energy becomes less negative as n increases and the electron spends most of its time farther away from the nucleus. A photon is emitted when a cloud changes into a smaller cloud with a smaller value of n. The energy of the atom then becomes more negative. Positive energy is released in the form of the photon. The energy and hence the frequency of this photon both depend primarily on the values of n for the initial and final clouds.

For a fixed value of the principal quantum number n, there can be several clouds. They all have the same size but various shapes (examine

9.14 QUANTUM NUMBERS

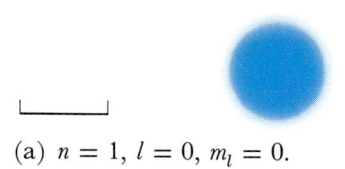

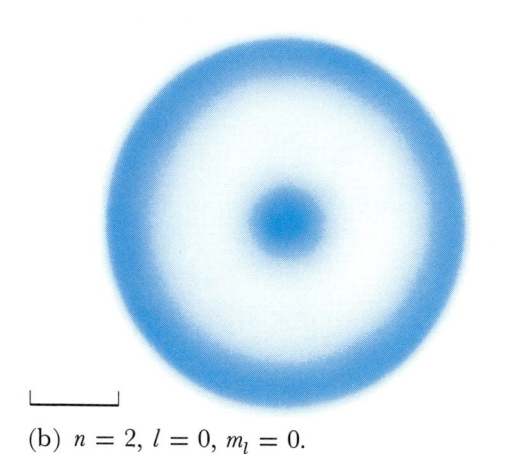

(a) $n = 1, l = 0, m_l = 0$.

(b) $n = 2, l = 0, m_l = 0$.

(c) $n = 2, l = 1, m_l = +1$.

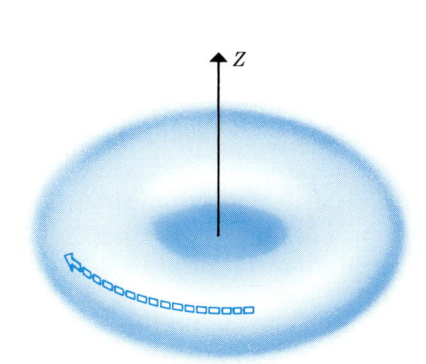

(d) $n = 2, l = 1, m_l = -1$.

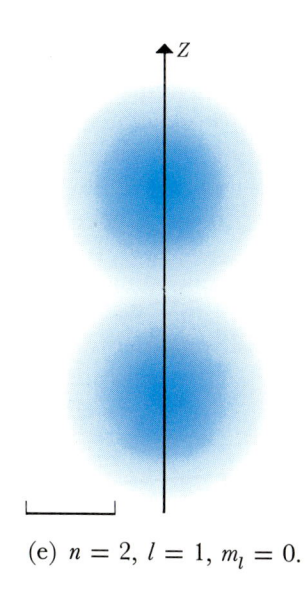

(e) $n = 2, l = 1, m_l = 0$.

FIGURE 9-8
Probability clouds for the first few states of the hydrogen atom. In part (a) the electron is in the K shell. In parts (b), (c), (d), and (e) the electron is in the L shell.

Figure 9-8 carefully). The different shapes correspond to different possible values of the *orbital quantum number*, l. These values are all positive integers ranging from 0 up to $n - 1$. For example, if n is 2, l can be 0 or 1. If n is 5, l can be 0, 1, 2, 3, or 4.

A very important function of the orbital quantum number l is to determine the *orbital*

9 THE QUANTUM THEORY

angular momentum of the cloud. Angular momentum was explained in Section 3.4. If the orbital angular momentum is not zero, the cloud is rotating about an axis through the nucleus of the atom. In the quantum theory, angular momentum, like energy, cannot have any value whatsoever. There is a unit of angular momentum equal to Planck's constant divided by 2π, or $h/2\pi$. The number of these units in the orbital angular momentum of a cloud is equal to the orbital quantum number. The orbital angular momentum of a cloud is $lh/2\pi$.

The *magnetic quantum number* m_l is concerned with the direction in which the cloud is rotating. If you compare Figures 9–8c and 9–8d, you will notice that the values of m_l are $+1$ and -1. The two doughnut-shaped clouds look similar, but they are rotating in opposite directions.

The above discussion assumes that we are considering a single isolated atom in empty space. The special direction labeled Z in Figure 9–8 might be the direction of an electric or magnetic field in the vicinity of the atom. If the atom is a constituent of a molecule, the situation is different. The magnetic quantum number m_l is no longer relevant. The doughnut-shaped clouds shown in Figures 9–8c and 9–8d do not exist. They must be replaced by dumbbell-shaped clouds similar to the one in Figure 9–8e. However, in part e the line joining the centers of the two halves of the cloud lies along the direction labeled Z. For either of the other two dumbbells, which replace the doughnuts, the line joining the centers of the two halves is at right angles to the Z direction. The three dumbbells are all mutually at right angles, like the three lines joining at the corner of a room.

The clouds we have just been discussing are intimately related to the *orbitals* that will be described in greater detail in Chapter 13.

9.15 SPIN

Our description of an atom is still not complete. A cloud for which the values of n, l, and m_l are all specified can still have either of two slightly different energies. The explanation depends on a property of electrons already introduced in Section 6.7 when we discussed bar magnets. An electron is spinning like a top. As shown in Figure 9–9, an electron in an atom is allowed to spin in two possible ways—clockwise or counterclockwise. The energy is slightly different in the two cases.

The orbital angular momentum of an electron is a consequence of its bodily motion around the nucleus. It is analogous to the angular momentum of the earth about an axis

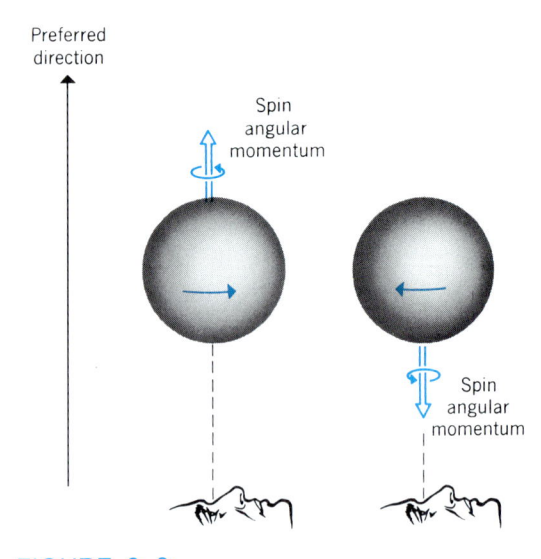

FIGURE 9–9
An electron in an atom can spin clockwise or counterclockwise. The energy is slightly different in the two cases.

through the sun due to the annual orbital motion of the earth around the sun. It is well known that the earth also performs a daily rotation about an axis through its north and south poles, which results in an angular momentum about this axis. An electron also rotates about an axis through itself and consequently has a *spin angular momentum*. This spin angular momentum is always present, even when the electron has no translational velocity. In terms of the fundamental unit of angular momentum $h/2\pi$, the spin angular momentum of an electron is surprisingly found to be one-half of a unit or $\frac{1}{2}h/2\pi$.

A proton, or a neutron, is exactly like an electron in having an intrinsic spin with an angular momentum of one-half a unit. A photon, on the other hand, has an intrinsic angular momentum of one whole unit, $h/2\pi$. We are not normally aware of the angular momentum of light because ordinary light contains equal numbers of photons spinning in both directions. They cancel in pairs and the total angular momentum is zero.

Spin:

The spin angular momentum of an electron, a proton, or a neutron is one-half a unit or $\dfrac{1}{2}\dfrac{h}{2\pi}$.

The spin angular momentum of a photon is one whole unit or $\dfrac{h}{2\pi}$.

We sometimes say, rather loosely, that an electron, proton, or neutron has a spin of $\frac{1}{2}$, whereas the photon has a spin of 1. There is an implicit assumption that the unit of angular momentum is $h/2\pi$.

9.16 THE PAULI EXCLUSION PRINCIPLE

In an atom containing several electrons, the electrons exert forces on one another and perturb one another's motion. Nevertheless, it is found to be a good approximation to treat the state of a single electron as though it were alone in a hydrogen atom. It can still be described by the quantum numbers n, l, and m_l and by the direction of its spin. However, from a consideration of atomic spectral lines and the set of energy values needed to explain them, Wolfgang Pauli discovered in 1925 that there was an important restriction on the way electrons are allowed to behave in atoms.

> Pauli's exclusion principle:
> No two electrons in the same atom may be in exactly the same state.

This means, for example, that the two electrons are allowed to be in clouds of different sizes or at least different shapes. If they are in clouds of the same size and shape, they have identical values of the three integers n, l, and m_l. Since the two electrons must behave differently in some respect, they are obliged to spin in opposite directions.

Although the Pauli exclusion principle may seem very erudite and far from everyday life, it is actually essential to the nature of everyday life. Suppose that we are building up a complicated atom by putting in the electrons one at a time. The first electron goes into the K shell with $n = 1$ to form a cloud of the smallest possible size. The Pauli exclusion principle forbids us to do exactly the same thing with the second electron. It can go into the K shell, but it must spin in the opposite direction. The third electron has no choice but to go into the L shell with $n = 2$ to form a larger cloud. Altogether, eight electrons can be put in the L shell in clouds of various shapes and with two possible directions of spin. The eleventh electron must then go into the M shell, and so on.

In Chapter 13, we shall see how these considerations lead to an explanation of the chemical properties of the atoms of the various elements. Without the exclusion principle, all the electrons would form small clouds with $n = 1$. The nature of atoms would then be entirely different. Chemistry would be entirely different, if it existed at all. The complicated processes taking place in living organisms would not be possible. In all probability, life would not exist. We would not exist. We owe it all to the Pauli exclusion principle.

SUMMARY

Black-body radiation. A black body is a perfect emitter and perfect absorber of electromagnetic radiation. The energy it radiates is proportional to the fourth power of the absolute temperature (Stefan-Boltzmann law). The bulk of the radiated energy is concentrated near a frequency that is proportional to the absolute temperature (Wien's law). As the temperature of a hot surface is raised, the electromagnetic waves emitted by it shift toward higher frequencies and shorter wavelengths.

Planck's quantum of energy. The hot surface emits energy in bursts or quanta. The amount of energy in a burst is equal to Planck's constant h multiplied by the frequency of the emitted electromagnetic wave ($\varepsilon = h\nu$).

The photoelectric effect was explained by Einstein on the assumption that a quantum of electromagnetic energy is actually a particle called a photon. When a photon is completely absorbed by a metal, part of its energy is used to lift an electron out of the metal. The rest of the energy appears as the kinetic energy of the ejected photoelectron.

The Compton effect. An x-ray photon collides with an electron and gives part of its energy to the electron. After the collision, the photon moves off at an angle to its original direction with reduced energy and frequency. Therefore, when x-rays are scattered by a substance such as graphite, part of the scattered radiation has a lower frequency than the incident radiation. This effect provides decisive evidence for the particle nature of electromagnetic waves.

The wave nature of matter. Particles such as electrons, protons, atoms, and molecules sometimes exhibit wave properties. A diffraction pattern is obtained when these particles are scattered by an orderly array of atoms in a crystal. The wave associated with a material particle is called a de Broglie wave. The quantity that oscillates in a wavelike fashion is called the wave function. The total energy of a particle is equal to Planck's constant multiplied by the frequency of the de Broglie wave. The momentum of a particle is equal to Planck's constant divided by the wavelength. The same equations apply to a photon. Because a photon moves with the speed of light, its rest mass must be assumed to be zero.

Probability and indeterminism. The square of the wave function at a point is a measure of the probability that the particle will be found near that point. The future behavior of a particle cannot be predicted in precise detail. The best that can be done is to estimate the probability of each of the possibilities open to it.

Heisenberg's uncertainty principle. It is not possible to measure both the exact position and the exact velocity of a particle. The uncertainty in the position multiplied by the uncertainty in the momentum is approximately equal to Planck's constant. Increased precision in the measurement of one of these quantities can be obtained only at the expense of a greater uncertainty in the value of the other quantity. The same restriction applies to measurements of energy and time. The uncertainty principle is important only for small particles such as electrons, protons, atoms, and molecules.

Atomic spectra. The spectrum of a hot monatomic gas consists of several narrow bright lines separated by dark regions. An atom is able to emit only a few special frequencies of electromagnetic radiation. An electron in an atom must choose one of a limited number of special motions called states. Each state has a definite energy. A photon with a definite energy and a definite frequency is emitted when the electron switches from one state to another state with lower energy.

The nature of an atom. The behavior of the single electron in a hydrogen atom provides a good guide to the behavior of any electron in any atom. Motion in a well-defined orbit violates Heisenberg's uncertainty principle. The electron is best visualized as a cloud of probability. The density of the cloud at any point is a measure of the probability that the electron will be found near that point.

Quantum numbers. The principal quantum number n is a positive integer such as 1, 2, or 3. It determines the size of the cloud and the energy associated with the state. The orbital quantum number l is an integer in the range from 0 to $n-1$. It determines the shape of the cloud. The orbital angular momentum is $l\frac{h}{2\pi}$. For an isolated atom in a magnetic or electric field, the magnetic quantum number m_l determines the direction of rotation of the cloud.

Spin. Electrons spin like tops. An electron in an atom can spin clockwise or counterclockwise. The energy is slightly different in the two cases. The spin angular momentum of an electron, a proton, or a neutron is one-half a unit, or $\frac{1}{2}\frac{h}{2\pi}$. The spin angular momentum of a photon is one whole unit, or $\frac{h}{2\pi}$.

The Pauli exclusion principle says that no two electrons in the same atom may be in exactly the same state. It is essential to the nature of an atom and the chemical properties of the corresponding chemical element.

TERMS AND CONCEPTS

quantum theory
dual nature of light
dual nature of matter
black body
black-body radiation
Stefan-Boltzmann law
Wien's law
quantum
Planck's constant h
photoelectric effect
photon
Compton effect
de Broglie wave
wave function
probability
Heisenberg's uncertainty principle
atomic spectra
state
cloud of probability
quantum number
principal quantum number n
shell
orbital quantum number l
orbital angular momentum
magnetic quantum number m_l
orbital
spin angular momentum
Pauli exclusion principle

QUESTIONS

1. Recently astronomers have discovered stars that are very strong emitters of x-radiation. Would you guess that these stars are very hot or very cold?

2. Explain how Max Planck made the first tentative step toward the quantum theory. What was he trying to explain? What assumptions did he make in order to explain it?

3. Why did Einstein first propose that light consists of particles?

4. Compare and contrast the photoelectric effect and the Compton effect.

5. List and briefly describe all the experiments you know that provide evidence for (a) the wave nature of light, (b) the particle nature of light, (c) the wave nature of electrons.

6. Which of the following types of radiation has a photon with (a) the least momentum? (b) the greatest momentum? (i) Radio waves, (ii) ultraviolet light, (iii) red light, (iv) violet light, (v) γ-rays.

7. Explain carefully why a photon can travel with the speed of light, whereas an electron cannot.

8. What is the physical significance of the wave function?

9. The classic example of determinism in Newtonian physics is the accuracy with which it is possible to predict the time of a future eclipse of the sun. Imagine that you have at your disposal all technical facilities that the human race might be expected to develop in the foreseeable future. Devise a method of perturbing the moon's orbit in order to change the times of future eclipses. Arrange for the exact result of this procedure to depend on an unpredictable factor, as in Figure 9–6.

10. Why is Heisenberg's uncertainty principle not conspicuous in everyday life?

11. What is meant by the "state" of an electron in an atom. What quantities must be known to specify a state completely? Discuss the physical significance of each quantity.

12. Why is the Pauli exclusion principle so important?

PROBLEMS

1. A black body radiates 176 W at 300 K. How many watts does it radiate at 150 K?

2. Show that the wavelength λ_m for black-body radiation is approximately $\dfrac{1}{200\,T}$ meters. At what temperature does the black body radiate mainly microwaves with a wavelength of 0.1 m?

3. What is the energy of a photon of an electromagnetic wave with a frequency of 10^{14} hertz?

4. A photon has an energy of 4×10^{-18} joule. What is the energy of a photon whose wavelength is twice as long?

5. If a particle has a de Broglie wavelength of 2×10^{-34} m, what is its momentum?

6. What is the de Broglie wavelength of a mass of 1 kg when its speed is 1 m/s?

7. To meet the requirements of the United States Golf Association, a golf ball must have a mass of 45.6 grams and acquire a speed of 76 m/s when tested on a certain machine. What is its de Broglie wavelength under these circumstances?

8. In Figure 9–5, what is the ratio of the de Broglie wave amplitude near C to the amplitude near S?

9. What is the numerical magnitude of the unit of angular momentum?

10. What is the numerical magnitude of (a) the spin angular momentum of a proton? (b) the spin angular momentum of a photon?

11. What is the numerical magnitude of the orbital angular momentum of an electron in a state with $l = 5$?

12. What is the numerical magnitude of the orbital angular momentum of an electron in the K shell?

CHAPTER 10
NUCLEI AND FUNDAMENTAL PARTICLES

10.1 NUCLEI

If an atom were magnified to the size of Mount Everest, its nucleus would still be only about as large as a football. Most of the mass of an atom resides in its nucleus, but the volume of the nucleus is less than one-trillionth of the total volume of the atom. It follows that the density of matter inside the nucleus is very high. It is, in fact, about two hundred trillion (2×10^{14}) times greater than the density of water.

The size of a nucleus is conveniently measured in femtometers, abbreviated fm. One femtometer is one millionth of one billionth of a meter. 1 fm = 10^{-15} m. The largest nuclei have diameters of about 15 fm. The diameter of an atom is about 100,000 fm.

The nucleus is believed to be composed of protons and neutrons. A proton has a positive electric charge that is equal in magnitude to the negative charge on the electron. The rest mass of the proton is 1836 times the rest mass of the electron. The neutron has no electric charge, and its rest mass is slightly larger than the rest mass of the proton. It is common practice to refer to the protons and neutrons in the nucleus collectively as *nucleons*. The justification for this is that the properties of the proton and the neutron are almost identical except for their electric charges.

An atom is usually electrically neutral, so the number of positively charged protons in its nucleus is equal to the number of negatively charged electrons moving around the nucleus. This number is always represented by the symbol Z and is called the *atomic number*. The total number of protons and neutrons inside the nucleus is represented by the symbol A and is called the *mass number*. It is an exact integer and is entirely different from the mass of the nucleus, which has a very small numerical value in units of kilograms. If the chemical symbol for an element is X, it is customary to represent its nucleus as $_Z X^A$. The subscript in front of the chemical symbol gives the number of protons. The superscript following the chemical symbol gives the total number of nucleons (protons and neutrons).

For a particular chemical element, the number of electrons in the atom, and hence the number of protons in the nucleus, is fixed. But the number of neutrons in the nucleus may vary. The resulting nuclei all have the same value of Z but different values of A and different atomic masses. They are called the *isotopes* of the element. For example, ordinary uranium is $_{92}U^{238}$. Its nucleus contains 92 protons and $238 - 92 = 146$ neutrons. The isotope "uranium 235" is $_{92}U^{235}$. Its nucleus still contains 92

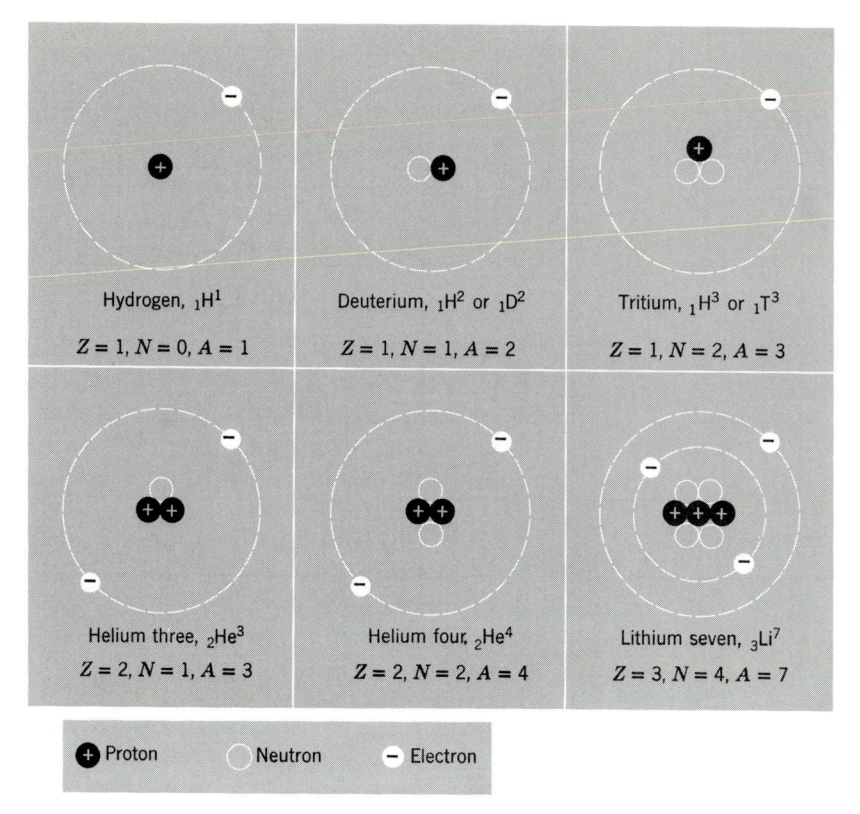

FIGURE 10-1
Composition of some of the lighter atoms. Z = number of protons, N = number of neutrons, $A = Z + N$ = total number of nucleons.

protons, but only $235 - 92 = 143$ neutrons.

A few of the smaller nuclei are illustrated in Figure 10-1. Notice in particular the three isotopes of hydrogen. They are called hydrogen, deuterium, and tritium.

10.2 NUCLEAR FORCES

The gravitational attraction between the nucleons is far too weak to hold the nucleus together.

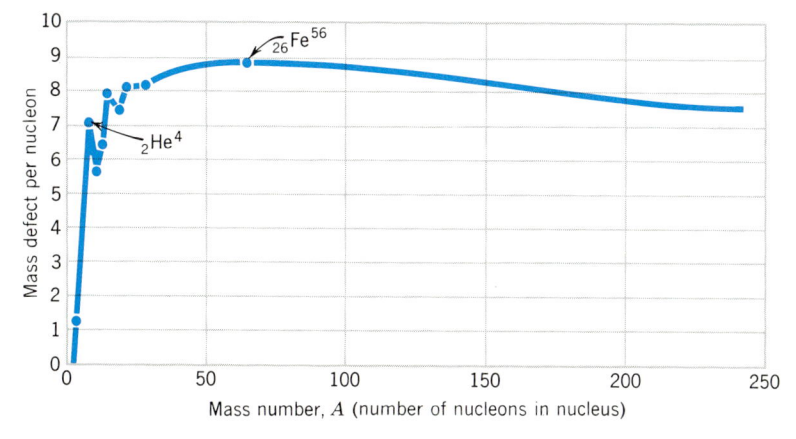

FIGURE 10-2
The mass defect per nucleon and its variation with mass number for naturally occurring nuclei. The special unit used here for the mass defect is approximately twice the rest mass of an electron.

10 NUCLEI AND FUNDAMENTAL PARTICLES

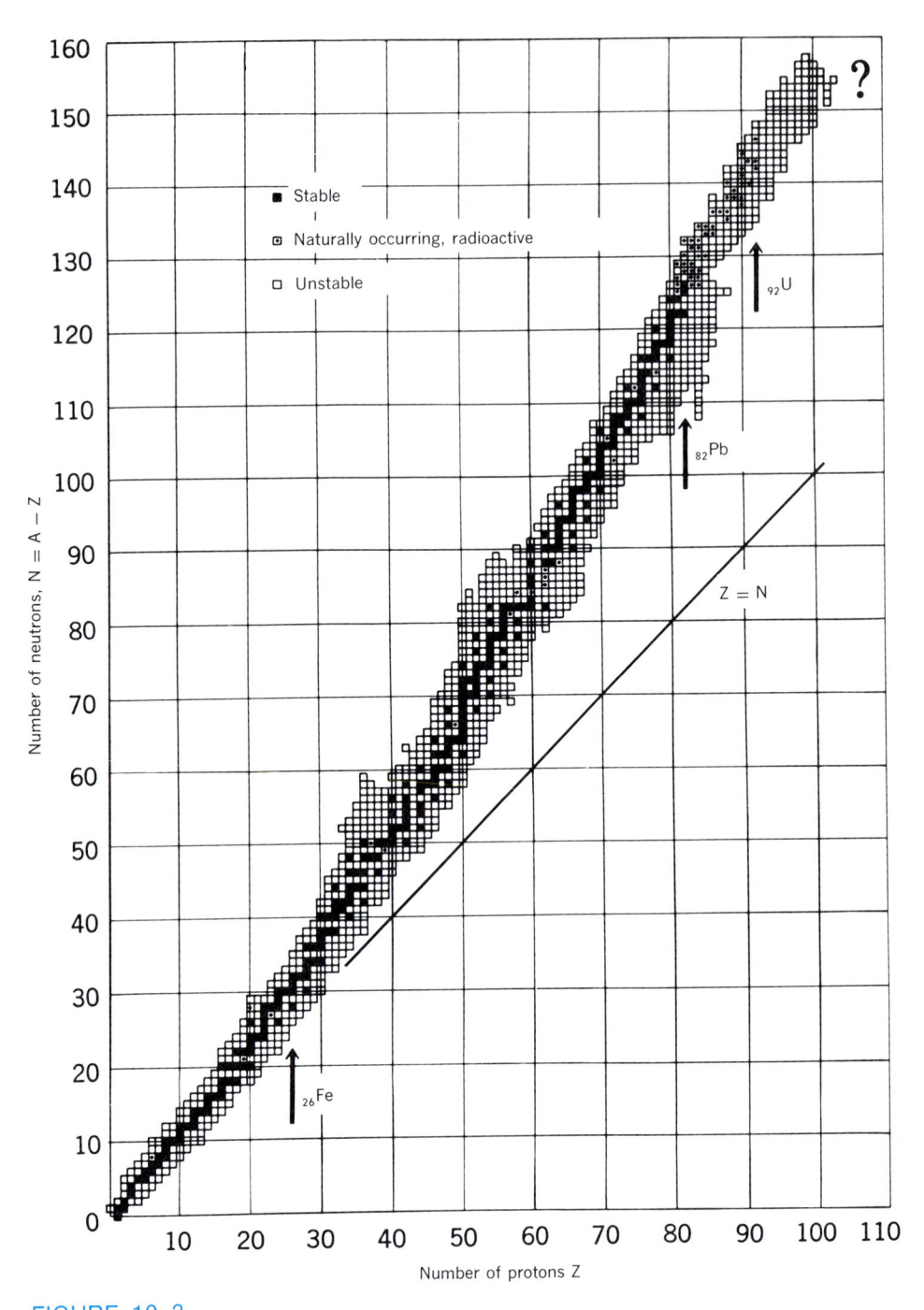

FIGURE 10-3
Chart of the known nuclei.

The electric repulsion between its protons would only blow it apart. The forces that hold the nucleons together are very different from any forces we have encountered so far. They are called *nuclear forces*.

The nuclear force between two nucleons seems to be very nearly the same for two protons, two neutrons, or a proton and a neutron. The way in which the force varies with the distance between the two nucleons is not yet fully understood. It definitely does not vary inversely as the square of this distance, as in the case of gravitational forces and electromagnetic forces. When the two nucleons are farther apart than 10 fm (femtometers), the nuclear force is negligibly weak. As they are brought closer together, to distances less than 10 fm, an attractive force builds up rapidly. Its rate of increase with decreasing distance is much faster than an inverse square law. However, when the two nucleons approach very closely and are less than about 0.2 fm apart, the nuclear force changes into a strong repulsion.

In most nuclei the distance between neighboring nucleons is about 2 fm. At this distance the attractive nuclear force is much stronger than the electric repulsion between two protons at the same distance apart.

10.3 MASS DEFECTS

The rest mass of a nucleus is always less than the sum of the rest masses of its constituent nucleons. The difference is called the *mass defect*. It is a measure of how strongly the nucleus is held together.

Remember that mass and energy are interchangeable. We can talk about them both at the same time. Apart from the rest mass energy of its neutrons and protons, a nucleus has two other sources of energy—the positive kinetic energy of the nucleons as they move around inside the nucleus and, more important, the negative potential energy associated with the nuclear forces. This potential energy is negative because the nuclear forces are attractive. Compare the case of gravitational forces, which are also attractive and also produce a negative potential energy (Section 3.9). The negative potential energy of the nucleons inside the nucleus makes a negative contribution to the mass of the nucleus that outweighs the positive contribution of the kinetic energy. This is the origin of the mass defect.

The mass defect per nucleon is the mass defect divided by the number of nucleons in the nucleus. It is a good indicator of how strongly each nucleon is bound inside the nucleus. The mass defect per nucleon is plotted against the mass number A in Figure 10–2. As the number of nucleons increases, the binding of the nucleus first becomes stronger. It reaches a peak for iron 56 ($_{26}Fe^{56}$), which is the most strongly bound nucleus. After that the strength of the binding steadily falls off as the nucleus becomes larger.

10.4 STABLE AND UNSTABLE NUCLEI

If the number of protons and the number of neutrons could be varied freely, the number of possible nuclei would be extremely large. Only a few of these possibilities have been found to exist. They are all shown in Figure 10–3. Each nucleus is represented by a square symbol—■, □, or ▣. The center of this square has a horizontal coordinate equal to the number of protons Z in the nucleus and a vertical coordinate equal to the number of neutrons N. All known nuclei lie within a narrow band, which should be compared with the straight line $Z = N$. For the lighter nuclei, the number of neutrons is never very different from the number of protons. For the heavier nuclei, there is a tendency for the number of neutrons to exceed the number of protons, but never by more than 60 percent.

The nuclei of Figure 10–3 have been divided into three categories. Those represented by the symbol ■ are *stable nuclei* found to occur in nature. They tend to lie near the center of the band.

The nuclei represented by the symbol ▣ are unstable and have to be produced artificially. They are very loosely bound and have an exces-

FIGURE 10-4
A modern atom-smashing machine—the tandem accelerator at the University of Pennsylvania.

$_2\text{He}^4 + {_7\text{N}}^{14}$ → $[_9\text{F}^{18}]$ → $_8\text{O}^{17} + {_1\text{H}}^1$

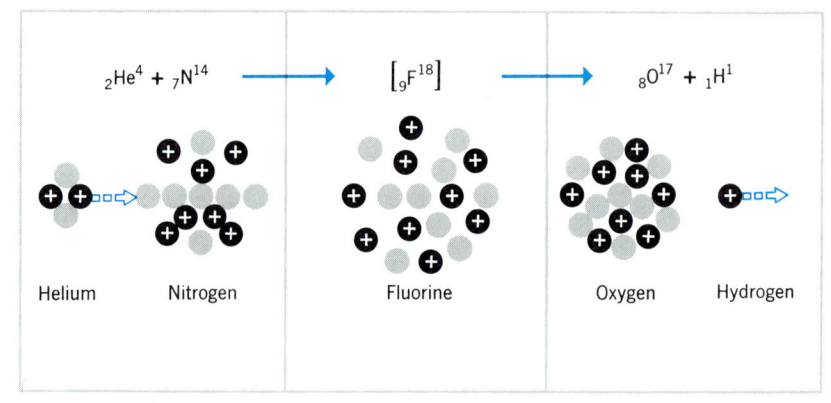

Helium   Nitrogen   Fluorine   Oxygen   Hydrogen

(a) The helium nucleus strikes the nitrogen nucleus.

(b) A compound nucleus is formed and has a brief existence.

(c) A proton is ejected. The nucleus left behind is an isotope of oxygen.

FIGURE 10-5
A nuclear reaction.

sive amount of energy. They break up to form stable nuclei, which have less energy. The excess energy is converted into the kinetic energy of the fragments, enabling them to fly apart. The various ways in which this disintegration can occur will be discussed later. If these *unstable nuclei* ever existed in nature, they would long ago have disintegrated to form stable nuclei.

Some unstable nuclei do occur naturally. They are the *naturally radioactive nuclei* represented by the symbol □. There are two reasons for their natural occurrence. Some of them disintegrate very slowly, taking several billion years to disappear. Since there is reason to believe that the earth and its elements were formed only a few billion years ago, these long-lived unstable nuclei have not yet had time to disappear completely. Others of the naturally occurring radioactive nuclei are short-lived but are produced during the disintegration of the long-lived radioactive nuclei. Thus, although they disappear soon after being formed, they are constantly being replenished by disintegration of the long-lived nuclei.

Notice that there are no stable nuclei beyond lead, which has $Z = 82$. There are no naturally occurring nuclei at all beyond uranium, which has $Z = 92$.

10.5
NUCLEAR REACTIONS

A chemical reaction is the reshuffling of the atoms of two reacting molecules to form different product molecules. A *nuclear reaction* is the reshuffling of the nucleons of two colliding nuclei to form different nuclei. In order to bring the two reacting nuclei close enough together, one of them must be accelerated to a high velocity. Otherwise, it will be turned back too soon by the strong electric repulsion between the two positively charged nuclei. A simple method of achieving a high impact velocity is to accelerate the particle through an electric potential difference of several million volts. An "atom-smashing" machine of this kind is shown in Figure 10-4.

A typical nuclear reaction is shown in Figure 10-5. A helium nucleus, $_2\text{He}^4$, strikes a nitrogen nucleus, $_7\text{N}^{14}$, and merges with it. The two protons and two neutrons of the helium nucleus mix with the seven protons and seven neutrons of the nitrogen nucleus. The resulting *compound nucleus* has nine protons and nine neutrons. It is therefore an isotope of fluorine, $_9\text{F}^{18}$. The fluorine nucleus is not formed in its normal state but in a state of much higher energy. It is therefore unstable and quickly gets rid of its excess energy by ejecting a fast proton. The eight protons and nine neutrons remaining form an isotope of oxygen, $_8\text{O}^{17}$. The equation representing this nuclear reaction is therefore

$$(\text{Helium 4}) + (\text{Nitrogen 14}) \rightarrow [\text{Fluorine 18}] \rightarrow$$
$$_2\text{He}^4 + {_7\text{N}}^{14} \rightarrow [_9\text{F}^{18}] \rightarrow$$
$$(\text{Oxygen 17}) + (\text{Hydrogen 1})$$
$$_8\text{O}^{17} + {_1\text{H}}^1$$
$$(10\text{-}1)$$

The total number of protons in the two colliding nuclei must be equal to the total number of protons in the two product nuclei. It follows that the sum of the subscripts on the left-hand side of the equation must be equal to the sum of the subscripts on the right-hand side of the equation. A similar argument applies to the total number of nucleons. The sum of the superscripts on the left-hand side of the equation must be equal to the sum of the superscripts on the right-hand side of the equation.

10.6 FISSION

Notice one feature of Figure 10–2: the mass defect per nucleon of a large nucleus containing about 200 nucleons is less than the mass defect per nucleon of a nucleus containing about 100 nucleons. If the large nucleus were to split up into two approximately equal fragments, the total rest mass would decrease. Some mass would be converted into energy. It would be used to give kinetic energy to the fragments, enabling them to fly apart. This type of disintegration, when a large nucleus splits up into two nuclei of intermediate size, is called *fission*.

Although fission is possible in naturally occurring large nuclei, such a nucleus usually waits for many billions or even trillions of years before disintegrating by fission. A large nucleus may be induced to undergo fission more rapidly by allowing an extra neutron to enter it. The neutron is uncharged and is not repelled by the nucleus, and it can therefore trickle into the nucleus at a low speed. It need not be accelerated to a high speed like a charged bombarding particle. Once inside the nucleus, the extra neutron upsets the balance between the number of neutrons and the number of protons. The enlarged nucleus is unstable and readily undergoes fission. Figure 10–6 shows the fission reaction that results when a neutron enters a nucleus of uranium 235. Notice that the symbol for a neutron is n.

Large nuclei contain a higher proportion of neutrons than nuclei of intermediate size (see Figure 10–3). It follows that the fission fragments initially have too many neutrons, and they rapidly jettison the excess ones. Each fission usually results in the liberation of two or three free neutrons. Some of these neutrons may enter other fissionable nuclei to produce further fissions and still more neutrons.

If, on the average, more than one neutron from each fission goes on to produce further fissions, the number of fissions taking place at each succeeding stage increases very rapidly (Figure 10–7). This is called a *chain reaction*. Within a small fraction of a second, the number of nuclei having undergone fission corresponds to many grams of material. The energy released, initially in the form of the kinetic energy of the fission fragments, is equivalent to the explosion of many thousands of tons of TNT. This is the principle underlying the atomic bomb.

If, on the average, exactly one neutron from each fission goes on to produce a further fission, the number of fissions occurring per second remains constant. The situation does not get out of hand (Figure 10–8). This is the principle of a *nuclear reactor*, which is a controllable source of useful nuclear energy. A careful balance is achieved in its design. Each fission produces more than one neutron that could produce a further fission. However, a sufficient number of these neutrons are absorbed in nonfission processes to ensure that exactly one does go on to produce a further fission.

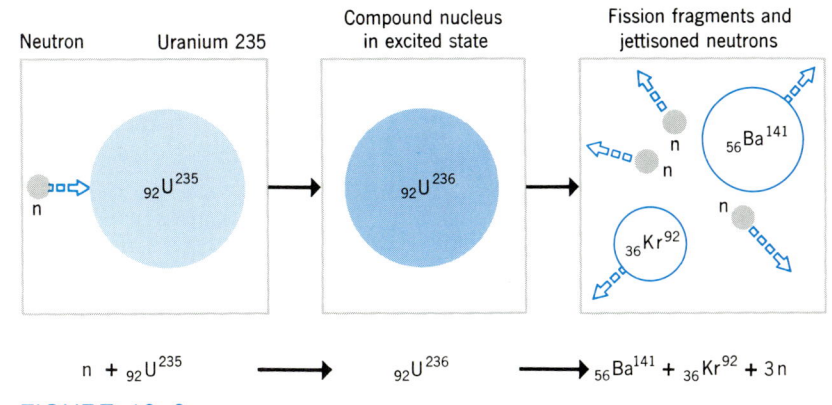

$$n + {}_{92}U^{235} \longrightarrow {}_{92}U^{236} \longrightarrow {}_{56}Ba^{141} + {}_{36}Kr^{92} + 3n$$

FIGURE 10–6

Fission induced in uranium 235 by the addition of a neutron.

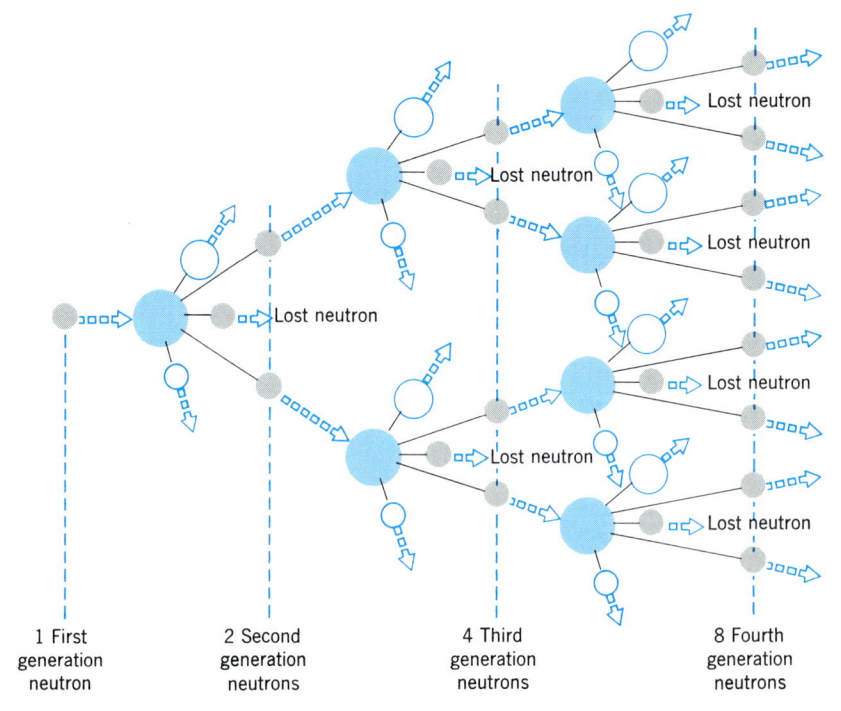

FIGURE 10-7
An uncontrolled fission chain reaction. The principle of the atomic bomb.

10.7
FUSION

Another important feature of Figure 10-2 is a strong tendency among the smaller nuclei for the mass defect per nucleon to increase as the nucleus becomes larger. If two small nuclei are fused together, the average mass per nucleon decreases. The lost mass is converted into energy. This is called *nuclear fusion*—the principle underlying the hydrogen bomb and *thermonuclear energy*.

Before fusion can take place, the nuclei must be brought together against the opposition of their mutual electric repulsion. At least one of the small nuclei must have a very high speed. A sufficiently high speed can be produced by an "atom-smashing" machine, but these machines deal with a small number of nuclei and yield small amounts of energy. The only way to achieve fusion in a large mass of material is to give the nuclei very high speeds by warming the material to a very high temperature. The required temperature is about ten million kelvin. Nuclear fusion produced by a very high temperature is called a *thermonuclear reaction*.

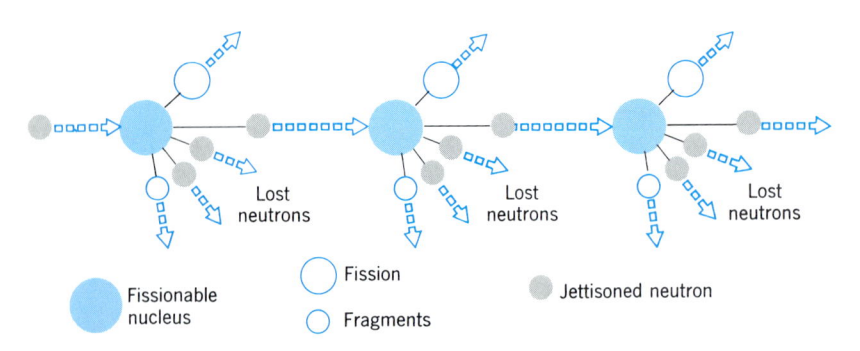

FIGURE 10-8
A controlled fission chain reaction. The principle of a nuclear reactor.

Temperatures of ten million kelvin are not easy to achieve on earth, but they occur naturally near the center of a star, such as the sun. Thermonuclear reactions therefore proceed readily near the center of the sun. They are the source of most of the energy poured out by the sun, or any similar star.

The explosion of an atomic bomb produces a temperature of about fifty million kelvin, which is high enough to make fusion possible. A hydrogen bomb consists of a mixture of light nuclei, such as deuterium, tritium, and lithium (see Figure 10–1). It uses an atomic bomb to initiate the fusion of these light elements. The size of an atomic bomb is limited by the following design problem. Before firing there must be a situation in which most neutrons are lost and less than one per fission produces a further fission. The firing must produce a sudden change to a situation in which fewer neutrons are lost and more than one per fission produces a further fission. The size of a hydrogen bomb is not limited by these considerations. Its destructive power depends only on the total quantity of light nuclei that can be incorporated into its design.

Vigorous attempts are being made to produce controlled thermonuclear reactions in the laboratory, with a view to using them as a possible future source of commercial thermonuclear power. At the time of this writing the technical problems have not yet been solved.

Fuel elements are loaded into the core of a nuclear reactor. (AEC.)

10.8 RADIOACTIVITY

A naturally occurring radioactive nucleus can disintegrate in three different ways. It can emit an *alpha particle* (α-particle), a *beta particle* (β-particle), or a *gamma ray* (γ-ray). An alpha particle is a helium nucleus, a beta particle is an electron, and a gamma ray is a high-energy photon. We discuss alpha emission and gamma emission briefly in the present section. In the following section we concentrate on beta emission because it will introduce us to some important new ideas.

Alpha emission occurs only in very large nuclei. It is the ejection by the nucleus of a small helium nucleus containing two protons and two neutrons. The large nucleus loses four nucleons, and its mass number A thus decreases by four. It loses two protons, and its atomic number Z decreases by two. It therefore becomes the nucleus of a different chemical element. Although the possibility of one chemical element changing into another is now common knowledge, its initial discovery had a dramatic impact on a scientific world that had come to regard the chemical elements as unchangeable.

Alpha emission resembles fission inasmuch as it does not always occur readily. For example, a nucleus of bismuth, $_{83}Bi^{209}$, waits on the average three hundred million billion years before emitting an alpha particle. This is about one hundred million times the age of the earth.

After one of the many kinds of nuclear transformations we have been discussing, the product nucleus is sometimes formed in a state with excess energy. This excess energy is usually jettisoned in the form of one or more gamma-ray photons as the nucleus changes to its normal state of lowest energy. Nuclear energies are many times larger than the energies of electrons in atoms. Gamma-ray photons emitted by nuclei therefore have much larger energies and consequently much higher frequencies than photons of visible light emitted by electrons in atoms.

The *half-life* of a radioactive material is the length of time that elapses before half of its nuclei have disintegrated. The individual nuclei decay at random at unpredictable times. One of the consequences of indeterminism in the quantum theory is that it is not possible to predict the exact instant at which a particular nucleus will disintegrate. When we are dealing with a very large number of nuclei, we can be sure that during the half-life time 50 percent of them will disintegrate and 50 percent will not. But we cannot know in advance which nuclei will disintegrate and which will survive.

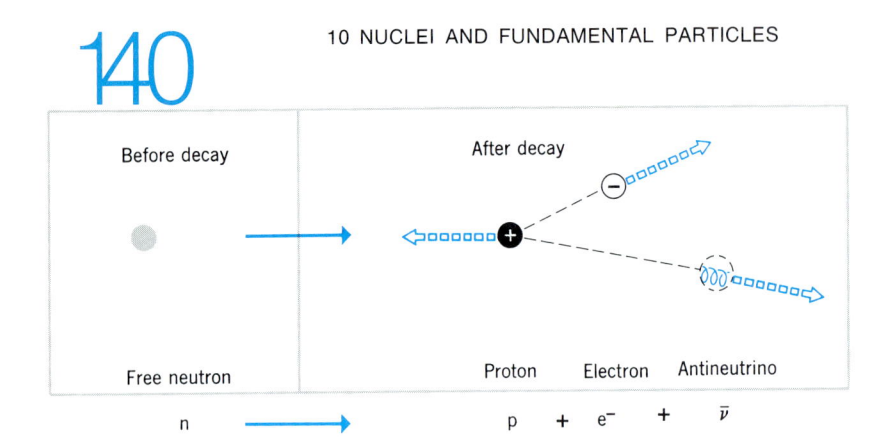

FIGURE 10-9
Beta decay of a free neutron.

10.9
BETA DECAY

When a nucleus *beta-decays*, it emits a β-particle, which is an electron. All other properties of nuclei firmly indicate that they are composed of protons and neutrons, but contain no electrons. Where, then, does the electron come from? The only satisfactory explanation is that electron emission occurs when a neutron inside the nucleus changes into a proton by emitting two small particles, an electron and an antineutrino. (The nature of neutrinos and antineutrinos will be explained in due course.) This process is essentially different from the processes we have discussed so far, which have merely involved the reshuffling of neutrons and protons. We are now introducing a new and very important type of process in which a fundamental particle changes into other fundamental particles. It is called a *fundamental process*.

Experiments on beams of free neutrons have shown that a neutron can indeed change into a proton with simultaneous emission of an electron. On the average a free neutron exists for about 10 minutes before disintegrating in this way. A free neutron is therefore an unstable particle that does not last very long by human standards.

Figure 10-9 illustrates the decay of a free neutron and gives the associated equation. It is conventional to represent the neutron by the symbol n, the proton by p, the electron by e^-, and the neutrino by ν. The superscipt associated with e indicates that the charge on the electron is negative. The proton might therefore be represented by p^+, but this is not normally done. The bar above the ν indicates that this particle is an antineutrino rather than a neutrino. The distinction will be explained later.

Why, then, do not all nuclei containing neutrons emit electrons? Why do not all neutrons decay into protons so that after a time not very much longer than 10 minutes, very few neutrons are left and the universe is composed almost entirely of protons and electrons?

The answer is that a neutron in a nucleus spontaneously beta decays into a proton if energy is released in the process, but not if additional energy must be supplied to the nucleus before the decay can take place. The rest mass of a free neutron is greater than the sum of the rest masses of its decay products. Some mass is therefore available to be converted into the kinetic energies of the products, enabling them to fly apart. Inside a nucleus the situation is more complicated. The replacement of a neutron by a proton sometimes results in a sizable increase in the energy of the nucleus. Under these circumstances, changing a neutron into a proton requires the addition of energy to the nucleus. It does not occur spontaneously. Electron emission is most likely to occur in a nucleus with too many neutrons. Such a nucleus lies above the black band of Figure 10-3.

10.10
THE NEUTRINO

When a nucleus emits an electron, the energy made available depends on the difference between the mass of the disintegrating nucleus and the slightly smaller mass of the product nucleus. The emitted electron usually receives only part of this available energy. To avoid a violation of the law of conservation of energy, Wolfgang Pauli suggested in 1930 that a second particle is emitted simultaneously with the electron and receives the rest of the energy. Enrico Fermi developed this idea and called the new particle a *neutrino*. (We now know that the particle accompanying electron emission is an *antineutrino*, but we shall ignore this distinction at the moment.)

Further study of electron emission reveals that the neutrino is also needed to salvage the law of conservation of momentum. Moreover, if the law of conservation of angular momentum is to be retained, the neutrino must be assumed to have a spin angular momentum of one-half a unit, $\frac{1}{2}\frac{h}{2\pi}$, like an electron or a proton.

The fraction of the available energy given to the neutrino varies from one disintegration to another. Sometimes, however, the electron takes all the available energy and the neutrino receives none. This must mean that the neutrino has no rest mass. Otherwise, even if it were emitted with zero speed, the neutrino would be obliged to claim a share of the energy at least equal to its rest mass energy.

The photon also has a rest mass equal to zero (Section 9.7). Like the photon, the neutrino must always move with the speed of light in order to

10.11 THE POSITRON

avoid having zero energy. The difference between a neutrino and a photon is that the neutrino has a spin of one-half a unit, whereas the photon has a spin of one whole unit.

Once it has escaped from the nucleus, the neutrino is a very elusive particle that does not make its presence felt in any obvious way. It rarely interacts with ordinary matter and is therefore very difficult to detect. The rarity of this interaction can be emphasized in the following way. A photon of appropriate energy passing through solid lead would travel on the average for about ten-billionths of a second and cover a distance of about 3 meters before interacting with an atom. A neutrino with the same energy would travel for about fifty years before interacting! In order to stop the neutrino, we would have to use a lead shield with a thickness more than ten times the distance between the sun and the nearest star.

The weakness of its interaction with matter proves that the neutrino carries no electric charge. Any charged particle moving rapidly through matter exerts strong electric forces on the atoms as it passes by and very soon pries loose an electron. The absence of charge on the neutrino may also be deduced from the law of conservation of charge. There is no known process that creates or destroys electric charge. Consider the decay of a neutron, as illustrated in Figure 10–9. A neutron has no electric charge. After the neutron decays, the positive charge of the proton is canceled by the negative charge of the electron. If the net charge is to remain zero, the neutrino must be uncharged.

The neutrino ν:

$$\text{Charge} = 0 \quad (10\text{-}2)$$
$$\text{Rest mass} = 0 \quad (10\text{-}3)$$
$$\text{Speed} = c \text{ always} \quad (10\text{-}4)$$
$$\text{Spin angular momentum} = \frac{1}{2}(h/2\pi) \quad (10\text{-}5)$$

10.11
THE POSITRON

A nucleus that lies below the black band of Figure 10–3 has an excess of protons. A proton in such a nucleus often changes into a neutron by emitting a *positron* and a neutrino. This neutrino really is a neutrino, not an antineutrino.

The positron is identical with the electron in all respects, except that it is positively charged. It has exactly the same rest mass as the electron. Its spin angular momentum is one-half a unit. But its charge is $+e$ as compared with $-e$ for the electron. It is represented by the symbol e^+, whereas the electron is represented by e^-. The equation for positron emission is

$$\begin{array}{ccccc} \text{proton} & \rightarrow & \text{neutron} + \text{positron} + \text{neutrino} \\ \text{p} & \rightarrow & \text{n} \quad + \quad e^+ \quad + \quad \nu \end{array}$$
$$(10\text{-}6)$$

In some sense, not yet fully understood, a positron is the exact opposite of an electron. It is called the *antiparticle* of the electron. If a positron encounters an electron, they cancel each other

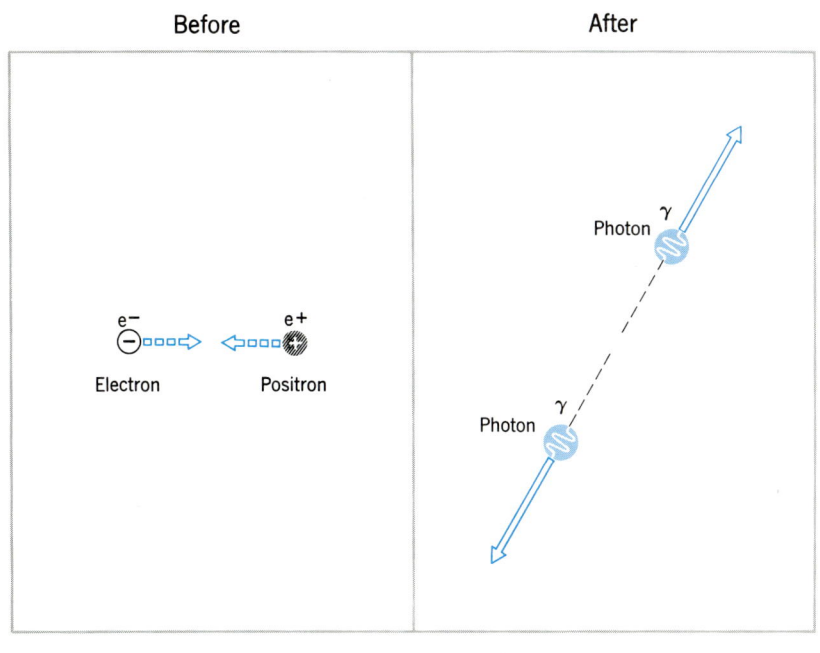

FIGURE 10–10
Annihilation of matter.

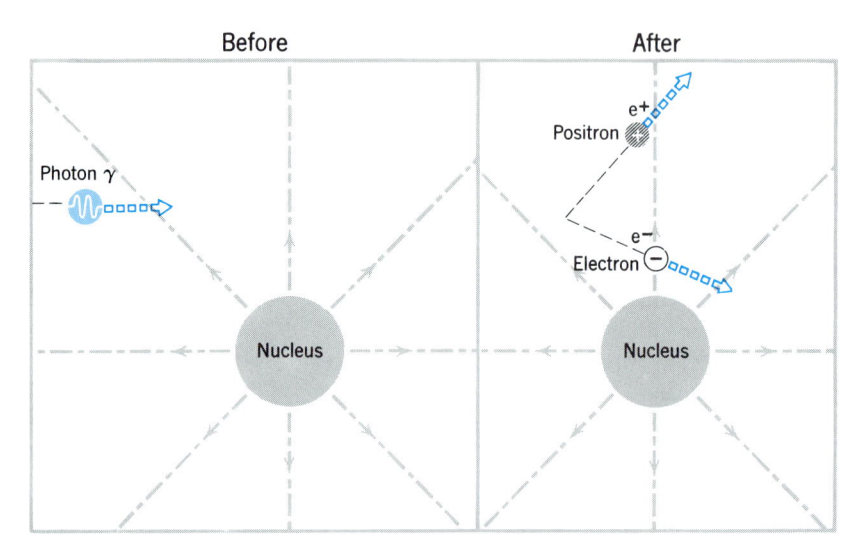

FIGURE 10–11

Materialization of light. Electron-positron pair production.

out and both disappear. Their energy, momentum, and angular momentum usually reappear in the form of two photons. This is called the *annihilation of matter* and is illustrated in Figure 10–10. Using the conventional symbol γ to represent a photon, the equation describing the process is

$$\text{positron} + \text{electron} \rightarrow 2 \text{ photons}$$
$$e^+ + e^- \rightarrow 2\gamma \quad (10\text{-}7)$$

This annihilation process is different from any other process that we have so far discussed. The rest mass is completely destroyed and completely converted into the energy of electromagnetic radiation. It therefore provides the final vindication of Einstein's principle of interchangeability of mass and energy.

The reverse process of conversion of energy into rest mass occurs in *pair production*, which is sometimes called the *materialization of light*. If a photon strikes a nucleus, the strong electric field in the vicinity of the nucleus induces the photon to change into an electron-positron pair (Figure 10–11).

$$\text{photon} \rightarrow \text{positron} + \text{electron}$$
$$\gamma \rightarrow e^+ + e^- \quad (10\text{-}8)$$

10.12
ANTIPARTICLES

Nearly all the fundamental particles have antiparticles. The most noteworthy exception is the photon. For some special reason, the antiphoton seems to be identical with the photon itself.

The antiparticle of the proton is the *antiproton*, represented by the symbol $\bar{p}$. A bar over the symbol for a particle always denotes it is an antiparticle. The antiproton is identical with the proton in all respects, except that it is negatively charged. However, the difference between a particle and its antiparticle is more profound than a mere reversal of the sign of the charge. The uncharged neutron has an antiparticle, the *antineutron*, $\bar{n}$. Although they cannot be distinguished by the sign of their electric charge, there is at least one way in which the neutron and the antineutron are clearly different. They are both spinning like tops and each has a spin angular momentum of one-half a unit. Also, they both behave like small bar magnets. However, if a neutron and an antineutron are both spinning in the same direction, their bar magnets point in opposite directions (Figure 10–12).

There is also a way to distinguish clearly between an uncharged neutrino and an antineutrino. If you could see a neutrino moving directly away from you, you would observe it to be

(a) Neutron.   (b) Antineutron.

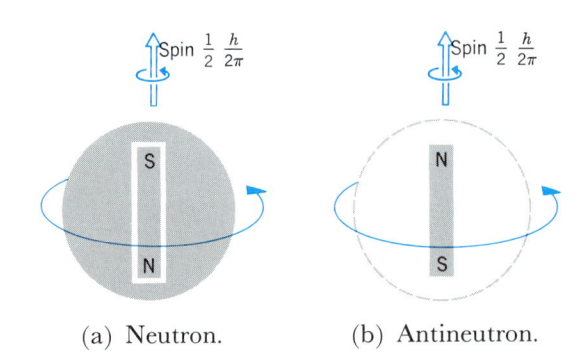

FIGURE 10–12

If a neutron (a) and an antineutron (b) spin in the same direction, their equivalent bar magnets point in opposite directions.

FIGURE 10–13
The "Bevatron" at the University of California Lawrence Radiation Laboratory, Berkeley. With this machine, the antiproton and numerous other particles have been discovered. Compare the size of the machine with the man in the bottom right-hand corner. (Courtesy of the Lawrence Radiation Laboratory, University of California, Berkeley.)

spinning in a counterclockwise direction. An antineutrino moving directly away from you would be spinning in a clockwise direction.

Protons, neutrons, and electrons are the principal constituents of our environment. Antiprotons, antineutrons, and positrons are very rare. A proton-antiproton pair or a neutron-antineutron pair can be created by a photon in a process similar to electron-positron pair production. Since the rest mass of a proton or a neutron is almost two thousand times larger than the rest mass of an electron, the photon must have a very high energy indeed. Nuclei do not emit gamma rays with sufficient energy. The machine that first created antiprotons and antineutrons is shown in Figure 10–13. It did not use photons to create the particle-antiparticle pairs. Instead, a proton was accelerated to a very high kinetic energy. It then collided with a stationary proton. Both protons emerged with comparatively low kinetic energies. The lost kinetic energy supplied the rest mass energy needed to create a proton-antiproton pair or a neutron-antineutron pair.

If a proton encounters an antiproton, each annihilates the other. Sometimes very high-energy photons are produced. Another possibility is the production of π-mesons or *pions*. A pion is a fundamental particle with a rest mass about 270 times the rest mass of the electron. It exists in a positively charged form π^+, a negatively charged form π^-, and an uncharged form π^0. As we shall see in Section 10.15, it plays a very important role in the theory of nuclear forces.

Figure 10–14 is a bubble chamber photograph of the annihilation corresponding to the equation

$$\begin{aligned} \text{proton} + \text{antiproton} &\rightarrow \\ p + \bar{p} &\rightarrow \\ 2 \text{ positive pions} &+ 2 \text{ negative pions} \\ \pi^+ + \pi^+ &+ \pi^- + \pi^- \end{aligned}$$
(10-9)

Notice how this equation conserves charge. Similar processes occur when a neutron and an antineutron annihilate each other.

10.13 ANTIMATTER

The atoms of ordinary matter have positively charged nuclei composed of protons and neutrons, surrounded by negatively charged electrons. *Antimatter* has atoms with negatively charged nuclei, composed of antiprotons and antineutrons, surrounded by positively charged positrons (Figure 10–15). Its properties are identical with those of ordinary matter in almost all respects. It would be difficult to distinguish between them. Of course, if antimatter came into contact with ordinary matter, each would annihilate the other. There would be an explosive release of large amounts of energy in the form of

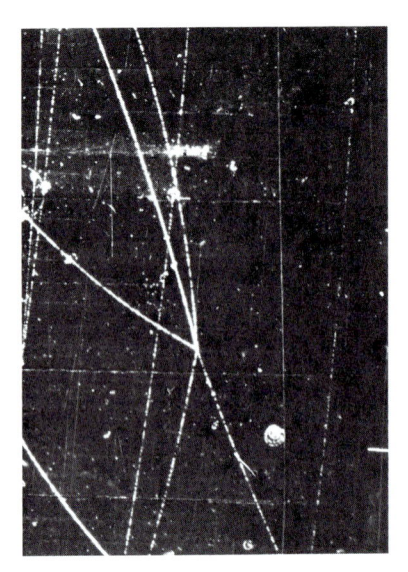

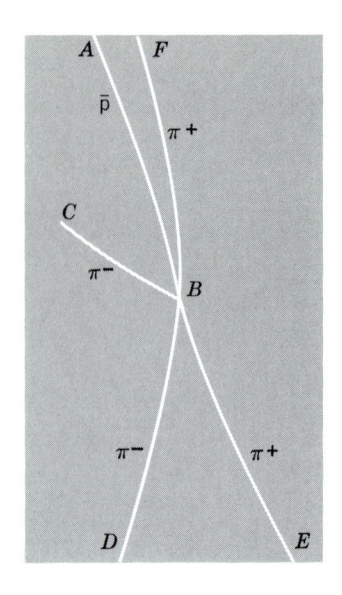

FIGURE 10-14
On the left is a bubble chamber photograph of a proton annihilating with an antiproton. A white track is a chain of bubbles left behind by a charged particle as it passes through a liquid on the verge of boiling. The drawing on the right interprets the photograph. The incoming antiproton enters at A and collides with a proton at B. Because the proton is at rest it does not produce a track of bubbles. The four pions resulting from the annihilation produce the tracks BC, BD, BE, and BF. (Courtesy of the Lawrence Radiation Laboratory, University of California, Berkeley.)

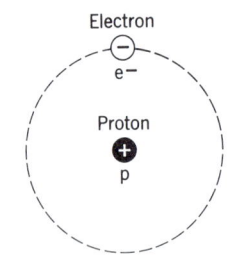

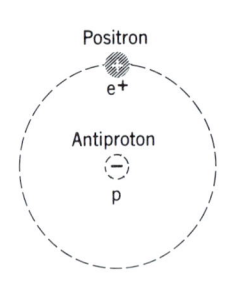

(a) A hydrogen atom.   (b) An antihydrogen atom.

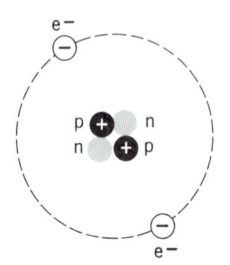

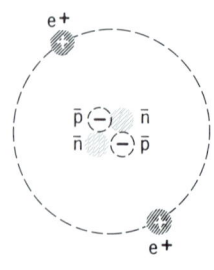

(c) A helium atom.   (d) An antihelium atom.

FIGURE 10-15
Matter and antimatter.

photons and pions. It is conceivable that some isolated regions of the universe are composed exclusively of antimatter. We may well be able to see some of these regions. Their visual appearance would be exactly the same as if they were composed of ordinary matter.

10.14
A PROFUSION OF PARTICLES

Throughout most of science the important fundamental particles are the photon, the electron, the proton, and the neutron. The photon, the electron, and the proton are *stable particles*. The only other stable particles are neutrinos. A stable particle left alone in empty space lives forever. A neutron lives on the average for about 10 minutes and then beta decays into a proton, an electron, and an antineutrino. The reason why neutrons are important in nature has already been given in Section 10.9. Neutrons inside certain nuclei can liver forever. They have been

10.14 A PROFUSION OF PARTICLES

deprived of the extra energy they need in order to β-decay.

More than one hundred *unstable particles* have now been discovered. Unstable particles live for a very short time and then break up into different particles. Many of them have half-lives of only a few trillionths of a trillionth of a second (about 10^{-23} s). The particles listed in Figure 10-16 are comparatively long-lived in the sense that their half-lives are much longer than 10^{-23} second. For example, a charged pion, π^+ or π^-, lives for about twenty-billionths of a second (2×10^{-8} s). Although this is an inconceivably short time, it is nevertheless two thousand trillion times longer than 10^{-23} second.

Figure 10-16 has the following characteristics. The rest mass increases from the top downward. Particles are on the left, and their corresponding antiparticles are placed opposite to them on the right. The three particles inside a circle have properties identical with those of their antiparticles. For them, particle and antiparticle are indistinguishable. All the particles in the figure have an electric charge of either $+e$, 0, or $-e$. If a particle is charged, its antiparticle has a charge of the opposite sign.

The fundamental particles can be divided into three broad classes. The photon is in a class of its own. The *leptons* are particles similar to the electron, the positron, and the neutrinos. The *hadrons* are particles similar to the proton, the neutron, and the pion. The hadrons can be further subdivided into *mesons* and *baryons*. Mesons, such as the pion, have no spin angular momentum. Baryons, such as the proton and the neutron, have half-integral spins of $\frac{1}{2}, \frac{3}{2}, \frac{5}{2}$, and so on.

Broad Classification		Name	Particles Charge +e	Particles Charge 0	Particles Charge -e	Antiparticles Charge +e	Antiparticles Charge 0	Antiparticles Charge -e	Rest Mass (Units Such That Rest Mass of Electron = 1)	Spin. Units of $\frac{h}{2\pi}$
Photon		photon		(γ)			(γ)		0	1
Leptons		neutrinos		ν			$\bar{\nu}$		0	$\frac{1}{2}$
Leptons		μ-neutrinos		ν_μ			$\bar{\nu}_\mu$		0	$\frac{1}{2}$
Leptons		electrons			e^-	e^+			1	$\frac{1}{2}$
Leptons		muons			μ^-	μ^+			207	$\frac{1}{2}$
Hadrons	Mesons	π-mesons (pions)	π^+	(π^0)			(π^0)	π^-	264 / 273	0
Hadrons	Mesons	K-mesons (Kaons)	K^+	K^0			$\bar{K}^0$	K^-	966 / 974	0
Hadrons	Mesons	η-meson (eta)		(η^0)			(η^0)		1074	0
Hadrons	Baryons	nucleons	p	n			$\bar{n}$	$\bar{p}$	1836.1 / 1838.6	$\frac{1}{2}$
Hadrons	Baryons	Lambda		Λ^0			$\bar{\Lambda}^0$		2183	$\frac{1}{2}$
Hadrons	Baryons	Sigma	Σ^+	Σ^0	Σ^-	$\bar{\Sigma}^+$	$\bar{\Sigma}^0$	$\bar{\Sigma}^-$	2328 / 2334 / 2343	$\frac{1}{2}$
Hadrons	Baryons	Xi		Ξ^0	Ξ^-	$\bar{\Xi}^+$	$\bar{\Xi}^0$		2573 / 2586	$\frac{1}{2}$
Hadrons	Baryons	Omega			Ω^-	$\bar{\Omega}^+$			3276	$\frac{3}{2}$

FIGURE 10-16

The fundamental particles that live much longer than a few trillionths of a trillionth of a second.

10.15
FORCES AND FUNDAMENTAL PROCESSES

There is an intimate relationship between forces and fundamental processes. For example, the electromagnetic force between two electrons may be described in the following way with the help of Figure 10–17. One of the electrons emits a photon. The corresponding fundamental process is represented by

(The first electron) →
$\quad$ e⁻ $\quad$ →
$\quad\quad$ (an electron with less energy) + (a photon)
$\quad\quad\quad$ e⁻ $\quad\quad\quad$ + $\quad$ γ
$\hfill$ (10-10)

The photon travels to the second electron, which absorbs it. The corresponding fundamental process is represented by a similar equation.

(The photon) + (the second electron) →
$\quad$ γ $\quad$ + $\quad$ e⁻ $\quad$ →
$\quad\quad$ (an electron with more energy)
$\quad\quad\quad$ e⁻
$\hfill$ (10-11)

Photons continually travel backward and forward between the two electrons. They transfer energy and momentum from one electron to the other. The rate at which either electron changes its momentum is equal to the electromagnetic force exerted on it by the other electron. The electromagnetic force between any two charged particles can be explained in terms of the interchange of photons between them.

The nuclear forces between nucleons can be described in a similar way. The particle thrown backward and forward between the two nucleons is a pion. For example, the following fundamental processes produce a force between a proton and a neutron. The proton emits a positive pion and changes into a new neutron.

(The proton) →
$\quad$ p $\quad$ →
$\quad\quad$ (a new neutron) + (a positive pion)
$\quad\quad\quad$ n $\quad\quad$ + $\quad\quad$ π⁺
$\hfill$ (10-12)

The pion then travels to the original neutron, which absorbs it and changes into a new proton.

(The positive pion) + (the original neutron) →
$\quad$ π⁺ $\quad$ + $\quad$ n $\quad$ →
$\quad\quad$ (a new proton)
$\quad\quad\quad$ p
$\hfill$ (10-13)

Positive pions are continually thrown backward and forward between the two nucleons. They transfer energy and momentum to produce forces between the nucleons. There are similar

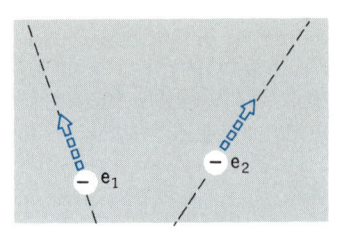

(a) Two electrons e_1 and e_2.

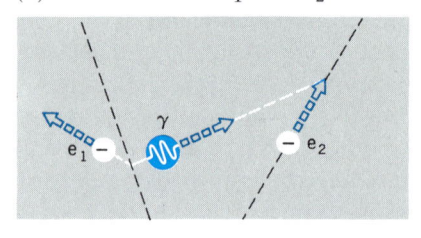

(b) e_1 emits a photon.

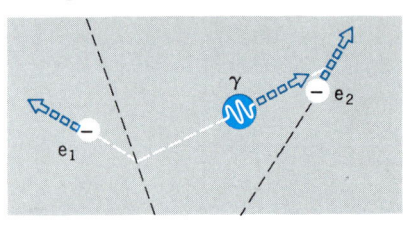

(c) The photon in flight.

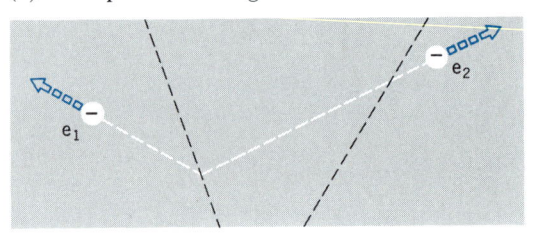

(d) The photon is absorbed by e_2.

FIGURE 10–17
The electromagnetic force between two electrons may be described in terms of the interchange of photons between them.

processes involving the transfer of a negative pion π^- or an uncharged pion π^0. Pions have been described as "the glue that holds the nucleus together."

SUMMARY

Nuclei are composed of protons and neutrons, which are collectively called nucleons. The atomic number Z of a nucleus is the number of protons it contains. The mass number A is the total number of protons and neutrons. If the chemical symbol for an element is X, its nucleus is represented as $_Z X^A$. The isotopes of an element have nuclei with the same number of protons but different numbers of neutrons.

TERMS AND CONCEPTS

Nuclear forces are negligibly small when the distance between the two nucleons is greater than 10 fm. The two nucleons attract each other when the distance between them is greater than about 0.2 fm. They repel when they are closer than about 0.2 fm.

The mass defect is obtained by subtracting the rest mass of the nucleus from the sum of the rest masses of its constituent protons and neutrons. It is a consequence of the mass equivalent of the negative potential energy associated with the attractive nuclear forces. The mass defect per nucleon is a convenient measure of how strongly the nucleus is held together.

Stable and unstable nuclei. Most nuclei contain approximately equal numbers of protons and neutrons. Apart from the stable nuclei that occur in nature, there are many unstable nuclei that disintegrate rapidly. The naturally radioactive nuclei are found in nature either because they disintegrate very slowly or because they are continually replenished by the decay of other nuclei.

Nuclear reactions. Two nuclei collide and reshuffle their nucleons to form different nuclei. At least one of the two colliding nuclei must have a high speed in order to overcome the electric repulsion between them.

Fission. A neutron enters a large nucleus, which then splits into two approximately equal fragments plus a few leftover neutrons. If, on the average, more than one of these neutrons goes on to produce a further fission, a chain reaction occurs as in an atomic bomb. A nuclear reactor is designed so that exactly one neutron goes on to produce a further fission, giving a controlled source of nuclear energy.

Fusion. When two very small nuclei fuse together to form a larger nucleus, energy is usually released. Only at temperatures of about ten million kelvin do the nuclei have high enough speeds to overcome their mutual electric repulsion and approach closely enough. Temperatures this high occur naturally at the center of the sun or a similar star. The sun's energy is thermonuclear energy. In a hydrogen bomb, the high temperature is produced by first exploding an atomic bomb.

Radioactivity. A naturally occurring radioactive nucleus emits either an α-particle, which is a helium nucleus, or a β-particle, which is an electron, or a γ-ray photon. The half-life is the length of time that elapses before half of the nuclei have disintegrated.

Beta decay. A neutron in the nucleus changes into a proton with the emission of an electron and an antineutrino. In this process a fundamental particle changes into other fundamental particles, and so it is called a fundamental process. A free neutron beta decays in this way after about ten minutes. A neutron inside a nucleus beta decays only if it has sufficient energy.

Neutrinos are needed to conserve energy, momentum, and angular momentum during beta decay. A neutrino has no charge, no rest mass, and one-half a unit of spin; it always moves with the speed of light. It rarely interacts with matter.

A positron is a positively charged electron. It is the antiparticle of the electron. An excess proton in a nucleus sometimes changes into a neutron with the emission of a positron and a neutrino.

Antiparticles. When a particle is charged, its antiparticle has a charge of the opposite sign. Even when the particle is uncharged, it is usually possible to find some way in which the antiparticle is the opposite of the particle. A particle and an antiparticle can annihilate each other. Their rest mass energy reappears in the form of photons or pions. In pair production, a photon changes into a particle plus its antiparticle. Antimatter is similar to ordinary matter except that the protons, neutrons, and electrons are replaced by antiprotons, antineutrons and positrons.

Fundamental particles. More than a hundred fundamental particles have now been discovered. The only stable ones are the photon, the electron, the proton, and the neutrinos. All the others disintegrate rapidly. They can be divided into three classes: The photon; leptons similar to the electron, the positron, and the neutrinos; and hadrons similar to the proton, the neutron, and the pions. The hadrons can be further subdivided into mesons, like the pion, or baryons, like the proton and the neutron.

Forces and fundamental processes. The electromagnetic force can be described in terms of the continual interchange of photons between two charged particles. The nuclear force can be described in terms of the continual interchange of pions between two nucleons.

TERMS AND CONCEPTS

nucleons

atomic number Z

mass number A

isotopes

nuclear forces

mass defect

stable nuclei

unstable nuclei

naturally radioactive nuclei

nuclear reaction

compound nucleus

fission

chain reaction

nuclear reactor

nuclear fusion

thermonuclear energy

thermonuclear reaction

alpha particle (α-particle)

beta particle (β-particle)

gamma ray (γ-ray)

half-life

beta decay

fundamental process

neutrino

antineutrino

positron

antiparticle

annihilation of matter

pair production

materialization of light

antiproton

antineutron

pion

antimatter

stable particle

unstable particle

lepton

hadron

meson

baryon

QUESTIONS

1. How many protons are there in $_{86}Rn^{219}$?

2. How many nucleons are there in $_{96}Cm^{240}$?

3. How many neutrons are there in $_{95}Am^{240}$?

4. Would you expect the electric potential energy of the protons to increase or decrease the mass defect of a nucleus?

5. According to Figure 10–3, which element has the most isotopes?

6. Give a reason why each of the following nuclei is unlikely to be found in nature: (a) $_{62}Sm^{57}$, (b) $_{73}Ta^{95}$, (c) $_{12}Mg^{56}$, (d) $_{123}Ap^{301}$.

7. Which of the following are not possible reactions?
   (a) $_1H^1 + {}_2He^3 \rightarrow {}_2He^4$.
   (b) $_1H^1 + {}_3Li^7 \rightarrow {}_2He^4 + {}_2He^4$.
   (c) $_{88}Ra^{224} \rightarrow {}_{86}Rn^{219} + {}_2He^4$.
   (d) $_{94}Pu^{238} + n \rightarrow {}_{54}Xe^{141} + {}_{40}Zr^{97} + 2n$.
   (e) $_5B^{11} + {}_1H^1 \rightarrow {}_4Be^8 + {}_2He^4$.
   (f) $_2He^4 + {}_{13}Al^{27} \rightarrow {}_{15}P^{30} + n$.
   (g) $_1H^2 + {}_{15}P^{31} \rightarrow {}_{13}Al^{29} + {}_2He^4$.

8. What is the difference between an atomic bomb and a nuclear reactor?

9. What is the difference between an atomic bomb and a hydrogen bomb?

10. Would you guess that the tritium nucleus is more likely to emit an electron or a positron?

11. Identify each of the following particles: (a) The antiparticle of the photon, (b) a lepton moving with the speed of light, (c) the antiparticle of the positron, (d) a particle moving with the speed of light and having a spin angular momentum of one whole unit, (e) the nucleon of largest mass.

12. List and briefly describe all the phenomena you have encountered in this book that provide evidence for interchangeability of mass and energy.

PROBLEMS

When needed, the atomic numbers and chemical symbols of the elements are listed in Table 13.4.

1. Insert the missing symbol in the following:
   (a) $_2He^4 + {}_6C^{12} \rightarrow {}_7N^{15} + ?$
   (b) $_{84}Po^{210} \rightarrow {}_{82}Pb^{206} + ?$
   (c) $? + {}_6C^{12} \rightarrow {}_6C^{13} + {}_1H^1$.
   (d) $_1H^1 + ? \rightarrow {}_2He^3 + n$.
   (e) $_{96}Cm^{240} + n \rightarrow ?$

2. What is wrong with the following statement? "The half-life for α-decay of deuterium is 10.2 s and so 25 percent of the original sample remains after 20.4 s."

3. "Just before falling under the death-rays of the Martian's guns, the scientist has time to scribble down part of his secret formula:

$$_{151}X^{362} \rightarrow {}_{150}Y^{362} + \ldots\text{''}$$

Can you guess what type of process he is describing and fill in the missing part?

4. When uranium 238 ($_{92}U^{238}$) alpha decays, what does it change into?

5. If a tritium nucleus were to emit an electron, what would it change into? See Figure 10–1.

6. When cobalt 60 ($_{27}Co^{60}$) beta decays, what does it change into?

7. If the half-life of a radioactive substance is 10.2 s, what fraction of the original sample remains after (a) 10.2 s? (b) 20.4 s? (c) 30.6 s?

8. At the end of two minutes, seven-eighths of a radioactive material has disintegrated. What is its half-life?

9. Calculate the frequency and wavelength of a photon that has just sufficient energy to create an electron-positron pair. What type of electromagnetic radiation is it?

10. You may have to do a little digging and a lot of thinking, but there is enough information in this chapter to enable you to answer the following question: Approximately, what is the mass equivalent of the nuclear potential energy in one kilogram of iron? You need not be very accurate. Is it about 100 grams? 10 grams? 1 gram? one millionth of a gram? or what?

11. If you were able to answer the previous question, you may not find this one too difficult: "Approximately what fraction of the mass is converted into energy in a fusion process?" Use Figure 10–2.

ANSWERS TO ODD NUMBERED PROBLEMS

CHAPTER 1
1. (c)
3. 1000
5. 3×10^4
7. (a) 29.4 m/s, (b) 10.2 s
9. 90 m/s^2

CHAPTER 2
1. 16 N
3. 160 m/s
5. 2×10^{-10} N
7. 2400 m/s^2
9. 8.28×10^3 m/s, 5.3×10^3 s or about $1\frac{1}{2}$ hours

CHAPTER 3
1. (a) 12 kg·m/s, (b) 24 J
3. 4 m/s
5. 70 kg·m^2/s
7. 6.3×10^7 J
9. 10 J
11. 8.8 J
13. 31 m. About 10 stories

CHAPTER 4
1. (a) 10^{-30} m^3, (b) 10^{30}
3. (a) 20° C, (b) 293 K
5. 160 N/m^2
7. 0.239
9. 480 K

CHAPTER 5
1. 1.5×10^9 N
3. 35 N/C
5. 4.5×10^9 V
7. 0.2 A
9. 3.6×10^6 J

CHAPTER 6
1. 10^{-7} N/A^2
3. 1.2×10^{-2} T
5. 0
7. 3.6×10^{-7} N

CHAPTER 7
1. 1000 Hz
3. 1.33 m
5. 1800 m/s
7. 6×10^{14} Hz
9. 80 W
11. 1.2 mm

CHAPTER 8
1. 9 years
3. $\frac{3}{5}c = 1.8 \times 10^8$ m/s
5. 2.78×10^{-27} kg
7. 0.943 c = 2.83×10^8 m/s
9. 450 million dollars
11. 2.5×10^4 kg/m^3

CHAPTER 9
1. 11 W
3. 6.6×10^{-20} J
5. 3.3 kg·m/s
7. 1.9×10^{-34} m
9. 1.05×10^{-34} kg·m^2/s
11. 5.3×10^{-34} kg·m^2/s

CHAPTER 10
1. (a) $_1$H^1, (b) $_2$He^4, (c) $_1$D^2, (d) $_1$T^3, (e) $_{96}$Cm^{241}
3. Positron emission
 $_{151}$X$^{362} \rightarrow {}_{150}Y^{362} + e^+ + \nu$
5. $_2$He^3
7. (a) $\frac{1}{2}$, (b) $\frac{1}{4}$, (c) $\frac{1}{8}$
9. 2.5×10^{20} Hz. 1.2×10^{-12} m. γ-ray.
11. 0.1%

GLOSSARY

Absolute zero of temperature, 0K, is the lowest attainable temperature. It is the zero of the absolute scale of temperature that has been adopted as part of SI units. At absolute zero, the atoms of an ideal monatomic gas come to rest.

Acceleration is the rate of change of velocity with time. Velocity is a vector, so a change in its direction is just as real an acceleration as a change in its magnitude.

Acceleration due to gravity (g) is the acceleration of an unsupported object falling freely in a vacuum. It usually refers to bodies falling near the surface of the earth, and its value is then approximately 9.8 meters per second per second.

Adiabatic process is one in which no heat is allowed to enter or leave the system.

Alpha particles (α-particles) are emitted by some radioactive nuclei. An alpha particle is the same as the nucleus of a helium atom. It is composed of two protons and two neutrons.

Alternating current (AC) is an electric current that surges backward and forward, periodically reversing its direction of flow.

Ampere (A) is the SI unit of electric current.

Angular momentum of a small object moving in a circle is equal to its mass multiplied by its speed and further multiplied by the radius of the circle. When an extended object rotates about an axis, it can be divided into many small parts, each of which moves in a circle. The total angular momentum is the sum of the contributions from all these small parts.

Annihilation of matter occurs when a fundamental particle encounters its antiparticle. They both disappear and their mass is changed into the energy of electromagnetic radiation (photons).

Antimatter. Every fundamental particle has an antiparticle, which is a kind of opposite of it. Whereas ordinary matter is composed of protons, neutrons, and electrons, antimatter is composed of antiprotons, antineutrons, and positrons (the antiparticle of the electron).

Atomic number (Z) of an element is equal to the number of protons in the nucleus of its atom. If the atom has no net electric charge, the atomic number is also the number of electrons moving around the nucleus.

Atomic spectra. An atom of a particular chemical element is able to emit only certain special frequencies of electromagnetic radiation. When light from a hot monatomic gas of this element is passed through a spectroscope, each special frequency corresponds to a narrow line with a definite color.

Avogadro's law. Equal volumes of two ideal gases at the same temperature and pressure contain the same number of molecules.

Bar. A unit of pressure approximately equal to atmospheric pressure (1 bar = 100,000 newtons per square meter = 100,000 pascal).

Baryon is a member of a group of fundamental particles. The best known members of the group are protons and neutrons.

Beta particle (β-particle) is an electron emitted by a radioactive nucleus.

Black body is a perfect emitter and absorber of electromagnetic radiation. It absorbs all the radiation falling upon it. The radiation emitted by a black body is distributed over all frequencies in a way that depends only on the absolute temperature of the black body.

Boltzmann's constant (k) is an important constant in the theory of heat. At an absolute temperature TK, the average kinetic energy of an atom of an ideal monatomic gas is $\frac{3}{2}kT$.

Boyles's law. When its temperature remains constant, the pressure of an ideal gas is inversely proportional to its volume.

Calorie. A unit of heat, originally defined as the heat that must be added to one gram of pure water to raise its temperature from 14.5°C to 15.5°C (1 calorie = 4.186 joule).

Charge of the electron (e) is approximately 1.6×10^{-19} coulomb. All fundamental particles carry an electric charge of $+e$, 0, or $-e$.

Charles's law. When its volume remains constant, the pressure of an ideal gas is proportional to the absolute temperature.

Compton effect. A photon collides with an electron. The electron is kicked away with increased energy. The photon is scattered sideways with reduced energy. The scattered photon has a lower frequency and longer wavelength than the original photon.

Conductors are materials that offer very little resistance to the flow of electricity. Most good conductors are metals.

Coulomb (C) is the SI unit of electric charge. When one ampere flows for one second, the charge passing by is one coulomb.

Coulomb's law. The force between two electric charges is proportional to the product of the magnitudes of the two charges and inversely proportional to the square of the distance between them. Like charges repel. Unlike charges attract.

Couple consists of two forces of equal magnitude acting in opposite directions, but displaced sideways from each other. The torque of the couple is equal to the magnitude of either force multiplied by the distance between their lines of action.

de Broglie wave. In certain experiments, electrons, protons, neutrons, and other small particles behave like waves, called de Broglie waves.

Density is mass divided by volume.

Diffraction. When a wave passes through an opening or round an obstacle, its behavior on the other side is very complicated. There is a sharp "shadow" only when the wavelength is very much smaller than the size of the opening or obstacle.

Direct current (DC) is an electric current that always flows in the same direction.

Dual nature of light and matter. In some experiments light behaves like a wave, but in other experiments it behaves like a stream of particles. Conversely, electrons, protons, and neutrons often behave like particles, but sometimes behave like waves.

Electric charge. When an object has an excess of protons, it is positively charged. When it has an excess of electrons, it is negatively charged. Like charges repel. Unlike charges attract. The SI unit of electric charge is the coulomb.

Electric current. A flow of electric charge. The SI unit of electric current is the ampere.

Electric field (E) is a property of a point in space. It is a vector. If an electric charge q is placed at the point, the electric force on it is equal to the product of the charge and the electric field. $\mathbf{F} = q\mathbf{E}$.

Electric potential is a property of a point in space. If an electric charge is placed at the point, its potential energy in joules is equal to its charge in coulombs multiplied by the potential at the point in volts.

Electric power is the electric energy consumed or delivered in one second. The power in watts consumed by an electric device is equal to the number of volts applied to it multiplied by the current in amperes flowing through it.

Electric resistance is a measure of the ability of an object to resist the flow of electric current through it. See *Ohm's law*.

Electromagnetic induction. When a magnetic field changes, it produces an electric field. When the magnetic field through a wire loop changes, an electric current is induced in the loop. When the loop moves through a steady magnetic field that changes from place to place, an electric current is induced in the loop.

Electromagnetic wave. When an electric charge oscillates or accelerates, it radiates an electromagnetic wave. This wave consists of oscillating electric and magnetic fields and can travel through empty space. In order of decreasing wavelength, the various types of electromagnetic wave are radio waves, microwaves, infrared, visible light, ultraviolet light, x-rays, and gamma-rays (γ-rays).

Electromagnetism includes both electricity and magnetism, which are intimately related.

Electron. A small negatively charged fundamental particle that is a basic constituent of all atoms.

GLOSSARY

Energy can take many different forms: kinetic energy, potential energy, heat energy, chemical energy, energy of electromagnetic radiation, nuclear energy, and rest mass energy. Each is discussed separately in this glossary. Although energy can change from one form into another, the total amount of energy of all forms must remain constant. The SI unit of energy is the joule.

Energy of electromagnetic radiation. An electromagnetic wave transports energy through empty space.

Entropy is an important concept in thermodynamics. It is a measure of disorder. Disordered states with high entropy are more likely to occur in nature.

Escape velocity is the minimum speed with which a rocket must be launched in order to escape completely from the gravitational field of an astronomical object without falling back again.

Ether. An invisible, untouchable, massless substance imagined to fill the whole of space. Its existence is denied by the theory of relativity.

Field. Empty space between particles is not featureless, but possesses physical properties. At each point in space values can be assigned to such physical quantities as the electric field and the magnetic field.

Field lines provide a picture of an electric field, a magnetic field, or any field represented by a vector. At any point the field vector is tangential to the field line.

Fission. A large nucleus breaks up into two approximately equal parts plus a few spare neutrons. This is the process underlying a nuclear reactor or an atomic bomb.

Fitzgerald-Lorentz contraction. When an object moves relative to the observer, the theory of relativity predicts that all dimensions parallel to the direction of its velocity will shrink.

Force. Objects interfere with one another's motion by exerting forces. The resultant force on an object is a vector equal to the product of its mass and its vector acceleration. (Newton's second law of motion). The SI unit of force is the newton.

Fundamental process. A fundamental particle breaks up into different fundamental particles.

Fusion. Two small nuclei join together to form a larger nucleus. This is the process underlying thermonuclear energy and the hydrogen bomb. It probably occurs at the center of stars like the sun and is the source of solar energy.

Gamma rays (γ-rays). Electromagnetic radiation of very high frequency and very short wavelength. Gamma rays are often emitted by radioactive nuclei.

Gauss is a popular unit of magnetic field not included in the SI system of units. The SI unit is the tesla, which is equal to 10,000 gauss.

Gravitational constant (G) is the constant of proportionality in Newton's law of universal gravitation.

Gravitational red shift is an effect explained by the general theory of relativity. In a gravitational field, stationary clocks run slow and atoms emit electromagnetic waves of longer wavelength.

Hadron is a member of a group of fundamental particles including both baryons and mesons.

Half-life. During the half-life time of a radioactive substance, 50 percent of the number of radioactive nuclei originally present disintegrate. The term may be applied to any process of disintegration, such as a fundamental process in which one fundamental particle changes into different fundamental particles.

Heat is a form of energy that cannot be defined precisely in any simple way. It is closely related to the "internal energy" of the individual atoms in a substance. It is often associated with the disorderly motion of these atoms.

Heisenberg's uncertainty principle. The position and momentum of a particle cannot be simultaneously measured with unlimited accuracy. The uncertainty in position multiplied by the uncertainty in momentum cannot be much smaller than Planck's constant h. Simultaneous measurements of energy and time are subject to the same restriction.

Hertz (Hz) is the SI unit of frequency. It is equal to one complete oscillation per second.

Ideal monatomic gas is composed of single atoms that do not exert forces on one another.

Infrared radiation consists of electromagnetic waves with wavelengths somewhat longer than the wavelength of red visible light.

Insulators are substances through which an electric current flows with great difficulty. Good insulators are often plastics, glasses, or ceramics.

Intensity of a wave is the energy transported by the wave perpendicularly across unit area in one second.

Interference is a consequence of the overlap of two waves from separate sources oscillating in unison.

Internal energy of an object is the energy of the atoms of which it is composed.

Isotopes of a chemical element have nuclei containing the same number of protons but different numbers of neutrons.

Joule (J) is the SI unit of energy.

Kelvin (K) is the SI unit of temperature, the "degree" of the absolute scale of temperature.

Kinetic energy is energy associated with the motion of a single object. It is equal to one-half of its mass multiplied by the square of its speed.

Lepton. A member of a group of fundamental particles including electrons, positrons, and neutrinos.

Liter. One thousand cubic centimeters or one thousandth of a cubic meter.

Mach's principle. The acceleration of an object is resisted by a new kind of long-range interaction with distant matter in the universe.

Magnetic field (B) is a useful concept that simplifies the calculation of magnetic forces. A charge in motion produces a magnetic field throughout space. The magnetic force on a second moving charge can be found if we know the magnetic field at the point where it is located. The SI unit of magnetic field strength is the tesla.

Magnetism is a consequence of the extra forces between two electric charges when both are in motion. The magnetism of a bar magnet can be ultimately traced to the fact that its electrons are spinning like tops and therefore consist of negative electric charge in motion.

Mass defect of a nucleus is obtained by subtracting its actual mass from the sum of the rest masses of its constituent protons and neutrons. It is a measure of how strongly the nucleus is held together.

Mass number (A) of a nucleus is equal to the total number of protons and nucleons it contains.

Maxwell's equations are the basic equations describing the electromagnetic field.

Mesons belong to a group of fundamental particles including pions. Mesons have no spin.

Metric ton is 1000 kilograms.

Microwaves are electromagnetic waves with wavelengths intermediate between radio waves and infrared. They are used in radar.

Momentum of an object is a vector equal to the product of its mass and its velocity.

Neutrino is an uncharged fundamental particle with zero rest mass. It always moves with the speed of light. It has half a unit of spin. When a radioactive nucleus emits an electron in beta decay, an antineutrino is emitted simultaneously.

Neutron is an uncharged fundamental particle similar to a proton. It is a constituent of the nuclei of all atoms except hydrogen.

Newton (N) is the SI unit of force.

Newton's law of universal gravitation. The attractive gravitational force between two objects is proportional to the product of their masses and inversely proportional to the square of the distance between them.

Newton's laws of motion. The first law states that an object left to itself continues to move in a straight line with constant velocity. The second law states that the resultant force on an object is equal to its mass multiplied by its acceleration. The third law states that the forces that two objects exert on each other are equal in magnitude but exactly opposite in direction.

Nuclear energy is released when nuclei react with one another and their neutrons and protons are reshuffled to form new nuclei.

Nuclear fission. See *fission*.

Nuclear fusion. See *fusion*.

Nuclear reactor. A device in which fission is kept under control to provide a steady source of useful nuclear energy.

Nucleons. A term used to include both protons and neutrons when they are constituents of nuclei.

Nucleus is a positively charged object located at the center of an atom. It is very much smaller than the complete atom but nevertheless has more than 99.9 percent of the total mass of the atom. The nucleus of a hydrogen atom is a single proton. All other nuclei contain protons and neutrons.

Ohm (Ω) is the SI unit of electric resistance.

Ohm's law states that the current in amperes is equal to the applied voltage divided by the resistance in ohms.

Pair production. A photon disappears and its energy is used to create a particle and its antiparticle.

Pascal (Pa) is the SI unit of pressure. It is equal to one newton per square meter.

Pauli exclusion principle states that no two electrons in an atom may have exactly the same set of quantum numbers.

Period of an oscillatory motion is the time taken to complete one oscillation and to return to the original position.

GLOSSARY

Photoelectric effect. A photon is absorbed by a metal. It disappears and its energy is used to eject an electron from the metal.

Photon is the particle of light or of any type of electromagnetic radiation.

Pion is a fundamental particle with no spin and a mass intermediate between the mass of the electron and the mass of the proton. Nuclear forces are a consequence of the interchange of pions between nucleons.

Planck's constant (h) is a fundamental constant required by the quantum theory. A photon associated with an electromagnetic wave of frequency ν has an energy $h\nu$. The basic unit of angular momentum is $h/2\pi$.

Positron is the antiparticle of the electron. It has the same mass as the electron and is similar to the electron in many other ways, but it is positively charged.

Potential energy is a mutual property of two objects arising from the forces they exert on each other. Gravitational potential energy, electric potential energy, and nuclear potential energy correspond to the various kinds of force. An object near the surface of the earth acquires extra gravitational potential energy when it is raised to a height.

Power is rate of doing work. The power in watts is equal to the work done in joules divided by the time taken in seconds.

Pressure is force per unit area. The SI unit of pressure is the pascal which is one newton per square meter.

Pressure of light. When electromagnetic radiation falls on a surface, it exerts a pressure on it.

Proton is a positively charged fundamental particle that is a basic constituent of all atoms. It is about 1836 times more massive than an electron.

Quantum is a small bundle of energy. Hot bodies do not emit electromagnetic radiation continuously, but in bursts. The amount of energy in a burst is called a quantum of energy. It is equal to Planck's constant h multiplied by the frequency of the electromagnetic radiation.

Quantum numbers of an electron in an atom. The *principal quantum number* (n) determines the energy and the shell in which the electron is located. The *orbital quantum number* (l) determines the orbital angular momentum of the electron. The *magnetic quantum number* (m_l) determines the direction of the axis about which the electron is orbiting. The *spin quantum number* (m_s) determines the direction in which the electron is spinning.

Radio waves are electromagnetic waves of long wavelength.

Relativity. The theory of relativity denies the existence of absolute space. All motion is relative to the observer. The laws of physics can be formulated in such a way that they are valid for any observer, whatever his motion.

Resistance. See *electric resistance*.

Rest mass (m_o) is the mass of an object at rest. As predicted by the theory of relativity, the mass increases with the speed of the object. The mass approaches infinity as the speed approaches the speed of light.

Rest mass energy. The rest mass m_o of a particle is equivalent to an energy $m_o c^2$. In an annihilation process, the particle disappears and an energy $m_o c^2$ appears in its place.

Spin. Many fundamental particles are spinning like tops. The quantum unit of angular momentum is Planck's constant divided by 2π. Electrons, protons, neutrons, and neutrinos have half a unit of spin. A photon has one whole unit of spin.

State. An electron in an atom must choose one of a limited number of special motions called states. Each state has a definite energy and a definite angular momentum.

Stefan-Boltzmann law. The energy emitted by unit area of the surface of a black body in one second is proportional to the fourth power of the absolute temperature.

Temperature scales. The *absolute scale of temperature* has its zero at the lowest attainable temperature. Its unit, or "degree," is the kelvin (K), which is the SI approved unit. The ice point is 273.15K and the steam point is 373.15K. The *Celsius scale* (in the past often called the Centigrade scale) has an ice point of 0°C and a steam point of 100°C. The *Fahrenheit scale* has an ice point of 32°F and a steam point of 212°F.

Tesla (T) is the SI unit of magnetic field strength.

Thermodynamics is the theory of phenomena connected with heat and temperature. The *first law of thermodynamics* states that the heat added to an object is equal to the increase in its internal energy plus the work done by the object on its surroundings. The *second law of thermodynamics* states that when an object containing a large number of atoms is left to itself, it assumes a state with maximum entropy and becomes as disordered as possible. The *third law of thermodynamics* states that an object at the absolute zero of temperature is in a state of perfect order and has zero entropy.

Thermonuclear. At very high temperatures, nuclei are moving fast enough to be able to approach close to one another against the opposition of their electric repulsion. Thermonuclear reactions can then take place, releasing thermonuclear energy.

Torque. See *couple*.

Ultraviolet light is electromagnetic radiation with wavelengths somewhat shorter than the wavelengths of visible light.

Vector has both magnitude and direction.

Velocity is a vector equal to the rate of change of displacement with time.

Volt (V) is the SI unit of electric potential.

Watt (W) is the SI unit of power. It is equal to one joule per second.

Wave is an oscillatory disturbance propagating through space. In a *transverse wave* the oscillation is at right angles to the direction of propagation. In a *longitudinal wave* the oscillation is parallel to the direction of propagation.

Wave function is the quantity that oscillates in a de Broglie wave associated with a particle of matter. The square of the wave function at any point is a measure of the probability of finding the particle near that point.

Wavelength is the distance between adjacent crests of the wave.

Weight of an object is the gravitational force acting on it.

Wien's law. Most of the electromagnetic energy radiated by a black body is concentrated near a particular frequency. This frequency is proportional to the absolute temperature.

Work is done when a force acts on an object that is displaced. The work is equal to the component of the force in the direction of the displacement multiplied by the distance through which the object is displaced. The SI unit of work is the joule.

X-rays are electromagnetic waves with very short wavelengths, but not quite as short as gamma-rays.

PART TWO
CHEMISTRY
John R. Holum

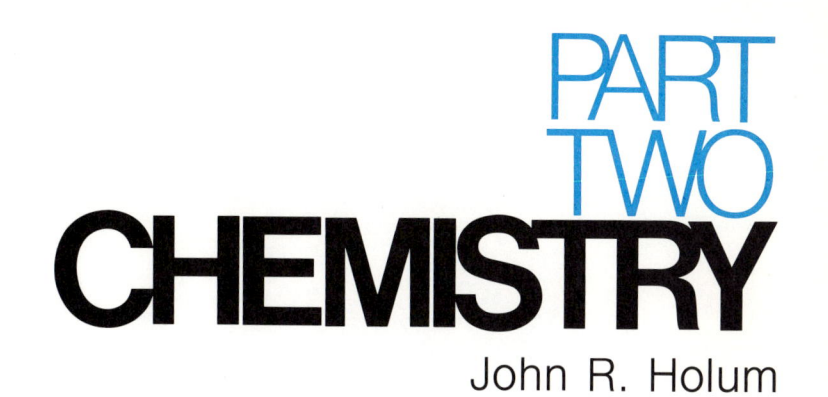

CHAPTER 11
THE CHEMICAL ELEMENTS

11.1 CHEMISTRY AND THE STUDY OF MATTER

Ten thousand years ago there were roughly as many lions as people, and neither seemed to have the upper hand. The ancients could not then realize that their ability to use several chemical elements would be a deciding factor in their survival. What they knew all too well was that lions could eat people and grow. Of course, from pinching hunger and desperate experience, they probably also knew that people could eat lions and grow. In more tranquil times individuals among the ancients may have wondered at the fact that people and lions must have something in common. Somewhere, at some deep level of existence, we can exchange parts with lions (and with lilies, too)! What are those parts? How can nature shuffle them and use them in so many different ways? How deep is our kinship with all of nature? With these questions, we move our study from physics to chemistry.

Chemistry is the study of matter, its nature, and the changes it undergoes. During our study of physics, we learned about energy and forces—concepts that are fundamental to all the sciences simply because they are basic to all of nature. Chemistry is also fundamental to all the sciences, partly because we can have neither energy nor forces apart from matter. We asked earlier about the nature of energy. We now ask about the nature of matter.

We shall study matter not only at the level of its atoms, but also in bulk—matter at its useful level. For all of us, daily contact with matter is with matter in bulk. We have learned to identify and distinguish different things in bulk by noticing differences in properties.

11.2 PROPERTIES

A property is a quality, a trait, or a characteristic that we use to recognize a thing (or a person or a place) and to tell it apart from all others. We need a list of several properties because one seldom is enough for identifying something. The properties most useful for identifying or using a substance fall into two classes. One includes those that can be studied without changing the substance into something else, classified as *physical properties*. A *physical change* (in contrast to a property) is any change we make in the substance in studying, measuring, or taking advantage of a physical property. To see if we can dent silver without shattering it, we hit it with a hammer. We thereby change its appearance, but we still have silver (Figure 11-1). To see if ice can absorb heat, we heat it, melt it, and then cool it back down again. We recover ice, the solid form of water. Because we do not recover the solid form of a different substance, the changes we put the ice through—melting and refreezing—are physical changes. Not even by changing the physical state of water—from solid to liquid to gas or back again—do we change the water into something that is not water.

Heat, of course, can sometimes cause major changes in things. It is one of the valued uses of heat. We change bread dough to bread, for example, by baking it (Figure 11-2). At a high enough temperature (above 2000°C), water "cracks." It literally breaks apart into the two elements, oxygen and hydrogen. We could not get water back again simply and solely by cooling this new mixture. We now have radically different substances. We have not simply changed the physical state of water; we have changed its very nature. Any change in which a substance is converted into another substance is called a *chemical change,* or a chemical reaction. The potential a substance has for undergoing a particular chemical change is a *chemical property,* the other kind of property a substance can have. It is a chemical property of water that at ultrahigh temperatures it breaks down into hydrogen and oxygen.

There are, then, two kinds of properties— physical and chemical—and two kinds of changes—physical and chemical. The property is itself not the change; the property is simply the potential for responding in a particular way to some change in the environment of the substance. How do we recognize that a change is chemical instead of physical? Occasionally this is difficult. Usually, however, most chemical changes are accompanied by large changes in physical appearances. Making wood split and making it burn are two changes you can cause in wood. When you split wood, you still have wood.

FIGURE 11-1
A physical change. The skilled work of this craftsman changes the appearance of the silver tray but not its chemical composition. (Robert Doisneau/Rapho-Photo Researchers.)

FIGURE 11-2
A chemical change. The loaf of bread bears no resemblance to the liquid or solid ingredients. In the oven the sugar, flour, and salt undergo numerous chemical changes. Chemical changes initiated by the yeast prior to baking release carbon dioxide, which makes bread rise. Nothing of a physical nature can be done to the bread to recover its ingredients. (Your mouth probably waters when viewing the bread, but not the ingredients, indicating your body's conditioned response to food as it prepares to work chemical changes of its own.) (K. T. Bendo.)

When you burn it, you have smoke, heat, new substances, new odors and finally ashes. Thus, in most chemical changes there are not only obvious changes in appearances, but also, at least usually in those chemical reactions that go spontaneously, some release of energy—heat, light, and sound—and some changes in color or odor. If any doubt about the nature of the change remains, careful chemical analyses must be done.

We must emphasize that not all spontaneous changes that release energy are chemical changes. When a gas changes to a liquid—called condensing—or when a liquid changes to its solid form—called fusing or freezing—heat is given off to the surroundings. Huge flows of heat energy occur in the atmosphere each day as water changes from one state to another, as we shall study in Chapter 22. We introduce the basic principles next.

11.3 LIQUIDS AND SOLIDS— A KINETIC VIEW

In Chapter 4 we learned how the kinetic theory of gases explained a number of properties concerning the ways that volume, pressure, and temperature of a gas affect each other. One assumption of this theory was that the particles of a gas are in a state of constant, chaotic, utterly random motion. We may extend this idea to other states. Individual entities of matter—atoms or molecules—are in motion in all three, but motions become more and more restricted as we consider first the gaseous, then the liquid, and finally the solid state.

In liquids molecules move about randomly and chaotically, but the distances between them are much shorter than in gases. Molecules in a liquid are always in contact but always shifting. In solids, molecules stay in fixed positions. They have constant neighbors, but they jiggle and vibrate constantly about fixed points.

11.4 MECHANICAL PROPERTIES OF LIQUIDS

Because there is very little free space within a liquid, a liquid is virtually incompressible. Unlike a gas a liquid does not shrink in volume when pressure is applied. (Solids are also virtually incompressible.) Because of its weight a liquid can, of course, be a source of pressure. At each point below the surface of a liquid there is a pressure, called the *hydrostatic pressure*. Like gas pressure, hydrostatic pressure is exerted equally in all directions. An object placed at some depth in a liquid will experience a pressure that is the sum of the hydrostatic pressure and the atmospheric pressure above the liquid. Because liquids are incompressible (or almost entirely so), the *density* of a liquid—its mass per unit volume—is independent of pressure, in great contrast to gases. Therefore, hydrostatic pressure is directly proportional to depth. Water has a density of 1 gram per cm^3. The pressure at a depth of 1 cm is therefore 1 g/cm^2; at a depth of 2 cm, 2 g/cm^2, and so on.

11.5 EFFECT OF TEMPERATURE ON DENSITY

As a general rule, the density of a liquid increases as it becomes colder until it freezes; the solid form, as a general rule, is more dense than the liquid form. A rare but important exception to these rules is water.

Water follows the pattern up to a point. As water is chilled its density increases to a maximum of 1.000 g/cm^3 at 4°C. Continued abstraction of heat from water does not yet change it to ice. The temperature continues to fall from 4°C to 0°C, and in this range its density decreases. The volume of a given amount of water actually expands as it is cooled from 4°C to 0°C. As it changes to ice at 0°C the density becomes about 0.9 g/cm^3. Ice, therefore, floats on water—a fact of incalculable importance to life on our planet. Were ice more dense than water, the oceans and lakes in the coldest climates would almost certainly be permanently frozen solid. As ice formed, it would sink to the bottom where no melting action could occur. Warmer water would stay at the surface. Ice formed in the winter would not melt in the summer and eventually all or nearly all would be frozen. The variation of density with temperature for salty water is different than that for pure water; this has implications for the oceans, as we shall discuss in Section 21.11.

What needs to be answered here is the question: what causes ice to float on water? How does it happen that a less dense object will float on a fluid of greater density?

11.6 ARCHIMEDES' PRINCIPLE

The fact that the pressure exerted by a fluid is exerted equally in all directions accounts for the buoyant force that the fluid generates. *Buoyancy* is the property of floating or tending to be supported by a liquid, and a *buoyant force* is the upward force responsible for it. The Greek mathe-

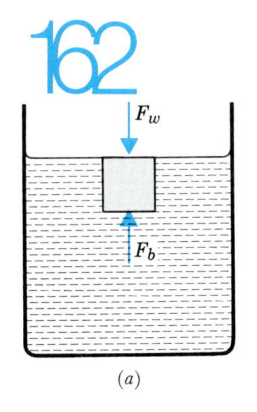

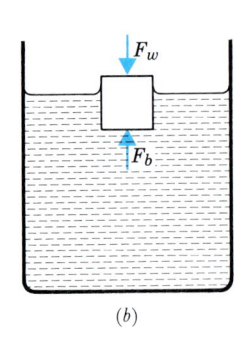

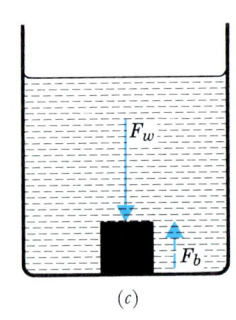

FIGURE 11-3

Archimedes' principle. The cube has the same volume in all parts but is of different materials and therefore different densities in *a*, *b*, and *c*. It weighs least in *b* because it has the lowest density there. In *c* it weighs most because it has the highest density in *c*. An intermediate stage is shown in *a*, where the weight of the cube creates a force, F_w, exactly equal to the weight of a corresponding volume of the fluid. The buoyant force created, F_b, is equal to F_w when the cube sinks to the level shown. In *b*, less liquid has to be displaced to create a buoyant force equal to the weight of this lighter cube, and the cube rests only partly submerged. In *c* the weight of the liquid displaced does not create a buoyant force large enough to support the heavy cube and it sinks to the bottom.

matician Archimedes (287–212 B.C.) is credited with the discovery, now called *Archimedes' principle,* that the buoyant force exerted by a fluid on a submerged object equals the weight of the fluid that the object has displaced (see Figure 11-3). We use a cube to simplify the discussion. A cube with a density exactly equal to that of the fluid will sink until its upper surface is even with the liquid's surface. The buoyant force now exactly equals the weight of the cube, and because the two forces are balanced the system is stable. A cube of the same volume but less density weighs less. This cube cannot sink as far as the first because a buoyant force equal to this lesser weight is reached at a lesser depth. A cube of material more dense than the fluid must initially exert a downward force greater than the fluid's upward force. Hence, it sinks until a depth is reached where the two forces are once again equal.

Submarines can assume any desired depth between the surface and the sea floor by changing their densities. They do this by filling or emptying large tanks with sea water. Ships made of steel, which is denser than water, float because the steel encloses a large volume of air making the overall density of the ships less than that of water. In northern climates during the fall, the water at the surface of a body of water becomes colder than water at a lower depth. Being colder (but not below 4°C) it is more dense and it sinks. By this action, called the fall overturn, the water of the lake becomes mixed. An object of greater density will always sink in a fluid of lesser density whether that fluid is a gas or a liquid. Sometimes the object is an air mass that is chilled and made more dense by being at a high elevation in a mountainous region. It can then slip down the slopes displacing warmer, less dense air in the valley.

11.7 THERMAL PROPERTIES OF LIQUIDS

A *thermal property* is one that concerns the physical response of an object to its being in contact with another object that is warmer or colder. Heat can cause chemical changes, but we assume in this section that they do not occur. Because heat can be transmitted even through a vacuum by electromagnetic radiation, the contact between the two objects need not be physical. We say that they are in *thermal contact* if heat can be exchanged between them by any means. The principal means are by radiation, convection, and conduction. *Conduction* occurs at the molecular or atomic level as particles of high kinetic energy bump into those of lower energy and transfer some of their energy. Heat can thereby be conducted; for instance, down a metal rod held so that one end is in a flame. Conduction is therefore the flow of heat by the transfer of kinetic energy from particle to particle. *Convection* is the flow of heat by the transfer of bulk quantities of heated material into a cooler zone by means of circulating currents. Only liquids and gases can have currents, so convection occurs only in matter in these states. An example of convection is the Gulf Stream, which carries heat from the Caribbean Sea to the coasts of the British Isles and Scandinavia.

When heat is transferred by any means to an object, one of two events will occur. Either its temperature will rise or it will undergo a change in state. We now have two questions: how much heat produces a given change in temperature and how much heat will cause a given amount of change in state?

11.8 HEAT CAPACITY

The amount of heat that is needed to change the temperature of one gram of a substance by one degree Celsius is called the *specific heat* of the substance. Sometimes the term *specific heat capacity* is used, and sometimes *heat capacity*. The units most commonly employed by scientists today are calories per gram per degree Celsius.

The heat capacities or specific heats of several common substances are given in Table 11.1, where the striking feature is water's relatively high specific heat. Because water can absorb more heat per gram than most other substances while undergoing only a 1°C change in temperature, it serves especially well in a number of ways important to life on earth. These range from the ability of large bodies of water to soak up solar energy without eventually boiling to aid in controlling the interior temperatures of plants and animals. Chemical reactions in the body produce heat. An adult weighing 75 kg (165 lb) generates each day about 2400 kilocalories even while at rest. If the heat capacity of the body were more like that of most substances, instead of being near that of water, this heat would be sufficient to raise the body temperature about 100–150°C. Actually, thanks to the high heat capacity of water, it could raise body temperature about 30°C. Of course, that would also be fatal, and the body takes advantage of other thermal properties of water to help get rid of its heat and stabilize its temperature.

11.9
LATENT HEAT

Sometimes the transfer of heat into or out of a solid or a liquid will not change its temperature. If ice at 0°C is placed in thermal contact with a warmer object, the temperature of the ice does not change; instead, the ice melts. The newly formed liquid in contact with the ice is also at 0°C. The heat that changed the ice to liquid at 0°C seemingly disappears. We call this heat *latent heat*. If we place water at 0°C in contact with something colder, the colder object will accept the liquid's latent heat. The liquid's temperature does not drop, however. Instead, the liquid undergoes a change in state from liquid to solid.

TABLE 11.1
SPECIFIC HEATS OF VARIOUS SUBSTANCES[a]

Substance	Specific Heat
Water, liquid	1.00
Ethyl alcohol	0.58
Olive oil	0.47
Granite	0.19
Cast iron	0.12
Gold	0.03

[a] The units are calories per gram per degree Celsius.

TABLE 11.2
LATENT HEATS OF FUSION OF VARIOUS SUBSTANCES IN CALORIES PER GRAM

Substance	Latent Heat of Fusion
Water	80
Iron	66
Benzene	30
Ethyl alcohol	25
Gold	15
Sulfur	11

The amount of heat per gram that will change a solid to a liquid at a constant temperature is called the *heat of fusion* (short for latent heat of fusion). The most common units are calories per gram. Some typical values are given in Table 11.2, where water is again seen to be exceptional. Water has a much higher heat of fusion than most substances. That is why, for example, ice packs are efficient in cooling an inflamed area of the body. The melting of a little ice pulls out a great amount of heat. If all the heat needed to melt one ordinary ice cube weighing 30 grams at 0°C were taken from 250 ml (about one cup) of water at a temperature initially of 25°C (room temperature), the water temperature would drop to about 15°C (59°F), a drop of 10°C or 18°F.

At the liquid-vapor transition, we have a situation similar to the solid-liquid transition. The latent *heat of vaporization* of a substance is the heat that will change a given quantity of liquid to the vapor state (at a pressure of one atmosphere) without any change in temperature. The units are usually calories per gram. Hence, when a liquid boils, or when it evaporates at some other temperature, it absorbs heat. This heat is latent in the vapor and is released when the vapor condenses back to the liquid form. As seen in the data of Table 11.3, water again is unusual. It has a very high heat of vaporization. The

TABLE 11.3
LATENT HEATS OF VAPORIZATION OF VARIOUS SUBSTANCES IN CALORIES PER GRAM AT BOILING POINTS[a]

Substance	Latent Heat of Vaporization at Boiling Points[a]
Water	540
Ethyl alcohol	204
Benzene	94
Gasoline	76–80
Chloroform	59

[a] At other temperatures values for latent heats of vaporization will be different.

evaporation of a little water, in other words, transfers a relatively large amount of heat to the vapor. Over 40 percent of all the water we ingest each day is evaporated at the surface of the skin and the inner surface of the lungs. In this way much of the heat we must remove from our body is carried away as latent heat in the water vapor. If water had a low heat of vaporization we should have to ingest much larger quantities of water to manage our heat budgets. Similarly, in the global management of heat energy, enormous quantities of water evaporate from tropical oceans as they absorb energy from the sun. Circulation in the atmosphere carries this water vapor to the cooler regions of the earth where the vapor condenses as rain or snow. As it condenses it releases all the latent heat of vaporization into the atmosphere. With this activity, heat received from the sun in low latitudes helps to warm the two polar regions of our planet. In Section 4.12 we learned that when an air mass rises it cools because the process occurs mostly adiabatically. Water vapor in the air mass may precipitate as a result of cooling. When it does it releases its latent heat of vaporization. As we shall explain in more detail in Section 22.13, this often happens as a moisture-laden air mass moves onto the slopes of a mountain range. One slope receives much of the precipitation. The other slope experiences strong, hot winds. As the air mass crosses the range, it can subside to a lower elevation. But now it carries the heat of vaporization as sensible heat. Moreover, as the air mass sinks it compresses adiabatically, and this is a further source of heat. That is why the air on the other side of the mountain is warm and dry (Section 22.14). Indeed, the famous dry chinook winds that sometimes occur in winter in certain parts of the Rocky Mountains obtain their heat largely from these sources.

11.10
VAPOR PRESSURE

One of the consequences of a high heat of vaporization for water is the fact that the *vapor pressure* of water can be changed by relatively large amounts for small changes in temperature. The vapor pressure of a liquid is simply a measure of its escaping tendency. The vapor pressure of a liquid at a given temperature is the pressure exerted by its vapor when an equilibrium exists between the escape of the liquid's molecules and their return to the liquid state. Figure 11–4 illustrates how vapor pressure is measured. Table 11.4 gives the vapor pressure of water at various temperatures. Up to 30°C, each 10°C rise in temperature roughly doubles the vapor pressure of water. The quantity of water vapor air can

TABLE 11.4
VAPOR PRESSURE OF WATER AT VARIOUS TEMPERATURES

Temperature (°C)	Vapor Pressure (mm Hg)
0	4.6
10	9.2
20	17.5
30	31.8
40	55.3
50	92.5
60	149
70	234
80	355
90	526
100	760

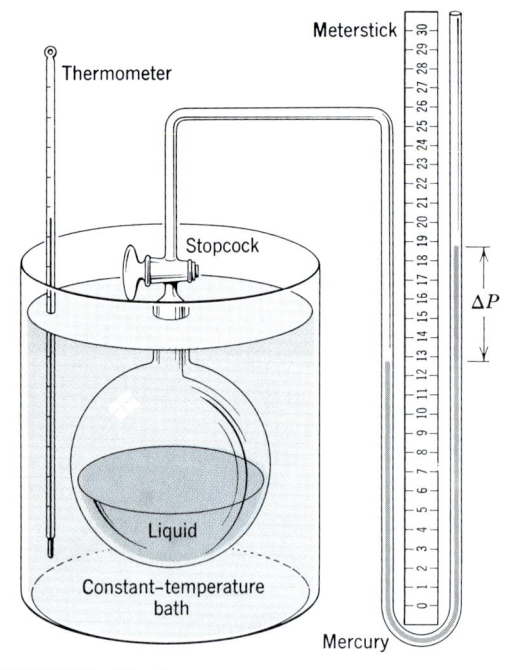

FIGURE 11–4
Measuring the vapor pressure of a liquid at various temperatures. The experimenter can make the bath temperature whatever is desired by regulating a heater (not shown). Allowing time for thermal equilibrium to be reached within and without the flask, the stopcock is then opened. The change in pressure caused by partial evaporation of the liquid eventually becomes a steady value. Equilibrium between the vapor and the liquid is reached. The vapor pressure of mercury is so low that, except for the most careful work or at greatly elevated temperatures, it may be ignored. (From J. R. Holum. *Principles of Physical, Organic, and Biological Chemistry*, 1968. John Wiley & Sons, Inc. New York.)

hold therefore varies greatly with small changes in temperature. That is why the temperature of the air need not drop very much to cause precipitation if the air is particularly humid. We shall continue our discussion of humidity and water vapor in the atmosphere in Chapter 22.

When the vapor pressure of a liquid exactly equals the pressure of the surrounding atmosphere the liquid boils. Liquid, in other words, changes to vapor not just at the surface but throughout the entire body of the liquid. Hence, vapor pockets or bubbles form within the liquid, and their appearance and surge to the surface produce the turbulence we see when *boiling* occurs.

Throughout several sections, we have been operating at the traditional borderline between physics and chemistry. Several of the previous topics, in fact, are considered the domain of the field of physical chemistry—a recognition that borderlines are not sacred, only useful. We shall now make a more definite move into chemistry because some of the remaining physical properties apply mostly to metals, and metals are chemical elements. We turn, therefore, to an introduction to the chemical elements, describe a few more physical properties, and increasingly shift our attention to chemical properties as we move into Chapter 12. No matter where we go, however, we never leave physics because the concepts of energy and force in many of their forms are indispensable to understanding chemical features of our universe.

11.11
ELEMENTS

An *element* is a substance whose atoms all have the same amount of positive charge on their nuclei. Any other kind of matter includes atoms from two or more elements. Water is not an element because it can be broken down into other substances—oxygen and hydrogen—that are elements. Water, instead, is a compound substance or, simply, a compound. We shall have much more to say about the nature of a chemical compound in Chapter 12. All we need now is the idea that in some way two or more elements are involved in forming a compound.

The number of elements is truly paltry when compared with the huge number of substances in the universe. Counting synthetic elements, we have only 105. Even so, elements are important not only as "parents" of all other substances, but also just as themselves. The names of two of the three ages of civilization—the Bronze Age and the Iron Age, which followed the Stone Age in that order—are tributes to elements or very close relatives.

FIGURE 11-5
From the most ancient of times, people have prized certain metallic elements. Seen here, left to right, are coins bearing the likenesses of Ptolemy II and Cleopatra VII on ancient Egyptian coins—silver, gold, and bronze (a copper-tin alloy). (Courtesy American Numismatic Society, New York City.)

11.12
METALLIC ELEMENTS

Most elements are *metals*. Some very familiar examples are gold, silver, copper, platinum, tin, iron, aluminum, lead, and nickel. All these are both metals and elements, and most were known to ancient people who valued them for several special, physical properties (Figure 11-5).

Malleability is one such property. Metals are malleable, which means they can, without shattering, be hammered into thinner and thinner pieces or to sharper and sharper edges. Gold, the most malleable of all substances, can be hammered into sheets so thin you can see through them. The gilt on roofs of some famous buildings or on interior decorations of cathedrals and palaces is made from gold leaf (Figure 11-6). *Ductility* is another physical property of metals. To greater or lesser degrees, metals can be pulled into wires after being softened by heat.

Metals generally can conduct heat and electricity well. Metals can be melted and cast into various forms ranging from statues to bullets. Metals have luster. They reflect light, sometimes in a dazzling way.

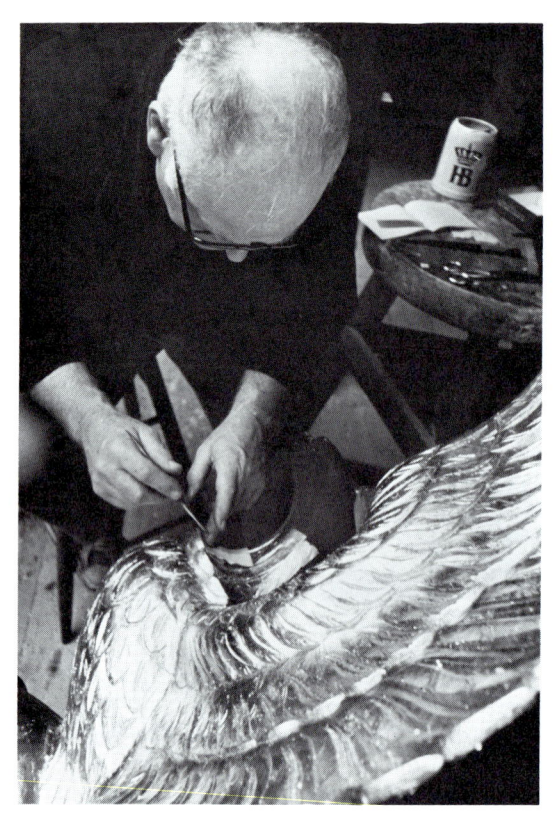

FIGURE 11-6
Gold is so malleable that ultrathin leafs, stored between leather "pages" bound as a book (on table), allow a little gold to cover a large surface. The artisan here is applying gold foil to the art object. (S. Sweezy/Stock, Boston.)

FIGURE 11-7
This huge specimen of native copper weighing three tons was unearthed in the early 1870s from an ancient mine pit at McCargo Cove on Isle Royale, Lake Superior. The dents from ancient stone hammers may be seen. Only a few metals occur naturally in uncombined form. Most copper, in fact, is mined as sulfide and oxide ores. (Burton Historical Collection, Detroit Public Library.)

11.13 ALLOYS

One of the most important discoveries in the long history of civilization occurred after two metals, very likely copper and tin, were heated and mixed together, probably accidentally. We now call a blend of two or more metals an *alloy*.

Copper, one of the few metals to occur as a free element in nature (Figure 11-7) as well as in ores, has been known since the late Stone Age. Finding "native" copper (free copper), members of some ancient tribe discovered that if they pounded it with a stone they could shape it into something they could use, perhaps a cutting tool or a weapon. Pounding copper makes it hard enough to make a fine, sharp edge. Without cutting tools superior to stone there could have been no advances in agriculture, and therefore no advances in civilization. Warfare, of course, became bloodier, and lions lost more of their advantages against people.

Uses for copper were not discovered all at once nor all over the earth simultaneously, but as these uses spread the beginning of the end of the Stone Age took place.

About 8000 years ago, probably through carelessness, someone dropped a copper tool into a campfire and discovered that heating melts copper and cooling brings its solid form back. Until then hammering and forging were the only means of working with metals, but now people realized molten metal could be cast into molds. This new process meant that small pieces of copper could be gathered together, melted, and cast into tools much larger than had been possible before. The next fortuitous step—the details are lost in history—was the mixing of two molten metals, copper and tin, to form what was probably the first known alloy, now called *bronze*. Those who were experienced in hammering, forging, and casting copper were no doubt surprised to discover that this copper-tin alloy melted more easily (at a lower temperature) than copper alone. Yet the alloy was harder and tougher than copper. Bronze can be sharpened

11.13 ALLOYS

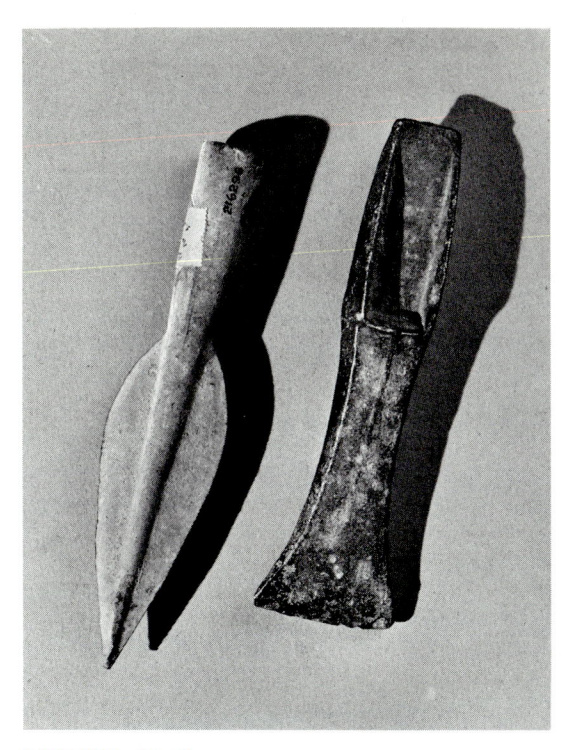

FIGURE 11-8
Bronze spearhead (left) and chisel (right) from the middle Bronze Age (about 1000 B.C.) found at Morsang-Saintery near Morel, France. (Field Museum of Natural History, Chicago.)

to a keener edge that holds longer than that of copper (Figure 11-8). So deeply did this development affect civilization that historians call the long period during which the alloy was dominant the Bronze Age.

There are many varieties of bronze depending on the ratio of copper to tin. Commonly, bronze has 5-10 percent tin. Still other bronzes are made by mixing a third metal with copper and tin. Coinage bronze, for example, is 95 percent copper, 4 percent tin, and 1 percent zinc.

Brass is an alloy similar to bronze, and in ancient times the two were sometimes confused. Brass is an alloy of copper and zinc, and several varieties are made, differing in the ratio of copper to zinc. The percent of copper, however, must be over 55 percent if the alloy is to be malleable and otherwise workable. Cartridge brass has about 65 percent copper. Brass containing 80 percent copper is closer to gold in sheen and color than any other metal or alloy.

Brass can be made stronger than bronze, and it is used when strong, corrosion-resistant items are needed, such as faucets, cartridge casings, bolts, pins, ornamental items, screws, tubes, and rods. Brass containing a trace of another metallic element, manganese, is exceptionally corrosion-resistant, and it is used to make ship's propellers and rudders.

When ancient peoples discovered iron and how to free it from red-colored minerals, they found a material stronger and better able to hold an edge than bronze or brass and more abundant than copper, tin, or zinc. The discovery created opportunities for tools, machines, and weapons of such value that it opened the third age of civilization, the Iron Age. Inevitably, given the earlier successes of bronze and brass, several alloys of iron were made and tried. We now have almost as many alloys of iron as we have other metallic elements, each alloy enjoying a particular use. One of the most important of all, *steel*, is an alloy of iron with a nonmetal, carbon. Often another element is added for special purposes, and many of the steel alloys are named after the second metal they include.

Nickel steel has the strength needed for long-span bridges (Figure 11-9) and armor plate. Bicycle chains are made of nickel steel.

Tungsten steel can withstand very high temperatures without losing an edge, and it is used in high-speed drill bits and other tools (Figure 11-10).

Manganese steel can take constant battering without breaking. The dipper teeth of power shovels used in mining are made of this alloy (Figure 11-11).

Vanadium steel can be cast into particularly large, hard castings.

Steel containing a trace of niobium (formerly called columbium) has great strength that holds over long periods at high temperatures, and this steel is used in atomic reactors and gas pipelines.

Cobalt steel makes excellent permanent magnets. It also stands up well under high temperatures, and high-speed cutting tools, jet engines, and gas turbines are made from it.

Silicon steel is used in electrical equipment.

Nickel-chromium steels constitute another family of alloys of iron. The original stainless steel was an alloy of iron and chromium containing more than 10 percent chromium. It is exceptionally resistant to both heat and corrosion. When nickel is added, the alloy can be made harder.

Molybdenum steel, similar to tungsten steel, can hold a cutting edge even when red hot. Molybdenum, when added to many kinds of steel, increases their hardeningability and their resistance to corrosion.

Many other kinds of iron and steel alloys are known. This brief survey merely shows how important are several metallic elements and their alloys, and it strongly hints at how much civili-

11 THE CHEMICAL ELEMENTS

FIGURE 11-9

To support long span bridges, alloy steel of exceptional strength is needed. Seen here is the Verrazano-Narrows Bridge at the entrance to New York harbor. Its main span (between towers) is 4260 feet, making it the world's longest suspension bridge. Over 80,000 tons of steel products went into its making. (Bethlehem Steel Corporation.)

zation now depends on metals, their ores,[1] and a world market in which ores and finished products may be bought and sold. By the year 2000, the United States will be importing 70 to 80 percent of its needs for metals, excluding iron.

11.14
NONMETALLIC ELEMENTS

Elements without the physical properties of metals are called *nonmetals*. A few elements, called semimetals or metalloids, have some properties of each group and form a borderline between them.

A gas is most obviously not metallic in any way, and all the gaseous elements are nonmetals. The air we breathe consists chiefly of two—nitrogen (79 percent) and oxygen (21 percent). Hydrogen is another nonmetallic element. Because it burns cleanly in air with a very hot flame, it

FIGURE 11-10

Clamped in the viselike carrier is a sharpened piece of tungsten steel, an alloy that can hold a cutting edge even as its temperature rises and it becomes very hot. (Bethlehem Steel Corporation.)

[1] An ore is any mineral from which a metal can be obtained at a profit.

11.14 NONMETALLIC ELEMENTS

FIGURE 11–11
An alloy of manganese and steel can take the punishment of hard rock mining without breaking and is often used in the teeth of the power shovel. (Reserve Mining Company.)

may someday be an important fuel. Using nuclear or solar energy, we could crack water to make hydrogen and oxygen, separate these two, and pipe the hydrogen to users somewhat as we pipe natural gas. This sounds easier than it actually is at the moment, but research is being done to solve the problems.

Chlorine, a greenish-yellow gaseous element, is widely used to kill bacteria in drinking water supplies and to bleach fabrics.

Fluorine, a colorless, gaseous element, is the most chemically reactive of all the nonmetals. Many things burst into flame when thrust into a fluorine atmosphere. Even asbestos is attacked by it. It has been used extensively in making various Freon gases, the fluorocarbon compounds used as propellants in aerosol spray cans (see Chapter 16).

The remaining gaseous elements—helium, neon, argon, krypton, xenon, and radon—are unusual because they have almost no chemical properties. Only a handful of compounds containing them are known, and none has yet been made from either helium or neon. Perhaps because, like royalty, they stand aloof from the humdrum turmoil of ordinary reacting chemicals, they are called the *noble gases*. Helium, less dense than air, is chosen for balloons because it is not flammable like hydrogen. When lightbulbs are made, the air inside the bulb is replaced entirely by argon (at low pressure). The glowing metallic filament cannot react with this gas. When an electric current is forced from one end to the other of a vacuum tube containing a trace of neon, the tube gives off a bright reddish-orange glow, a "neon light." Xenon is used in electronic flash bulbs.

Only two elements—bromine and mercury—are liquids at room temperature. Two others, gallium and cesium, have melting points of only 86°F (30°C). Partly because mercury, gallium, and cesium are silvery substances that conduct electricity, and partly because their chemical properties are like those of the rest of the metals, these three are classified as metals. Bromine, on the other hand, does not conduct electricity, and it is a reddish-brown liquid. It also evaporates readily, producing a reddish vapor with a terrible odor. (Its name is from the Greek, *bromos,* meaning stench.) It is a dangerous substance capable of causing very painful sores on the skin, and its vapor is very irritating both to the eyes and the throat. It is used in making dyes, drugs, lead scavengers in leaded gasoline, flameproofing compounds, and photographic chemicals.

Boron, carbon, phosphorus, sulfur, and iodine are five solid, nonmetallic elements. None is malleable, ductile, or silvery; none, except carbon, conducts electricity, although even carbon does not do this well. These elements furnish some of the most dramatic examples of an interesting aspect of nature, *allotropy*, which is the phenomenon of the existence of a substance in two or more physical forms, generally called allotropic forms. Carbon has three: the amorphous form, graphite, and diamonds, the most famous. (*Amorphous* here means that its particles have no definite characteristic shape. See Section 14.26.) All three are elemental carbon, but each has a different arrangement of carbon atoms. A diamond is one of the hardest substances known; graphite is one of the softest of all solids.

Boron exists in an amorphous form and three *crystalline* forms. Three common allotropic forms of phosphorus are white, red, and black phosphorus. White phosphorus has to be stored underwater; it bursts into flame in air. Neither red nor black phosphorus presents this problem. In the form of phosphates, which are oxygen-phosphorus compounds, phosphorus is essential to life (see Section 19.21, for example). It is a necessary ingredient in some fertilizers and is widely used in synthetic detergents. Sulfur probably has 20 allotropic forms, but at ordinary temperatures only one is common—orthorhombic sulfur, a familiar, bright yellow solid (Figure 11–12). It is used to make sulfuric acid, one of the world's most important industrial chemicals. A nation's per capita use of sulfuric acid is as accurate a measure of its industrial activity as its use of steel. Sulfur occurs in the free state in nature and it is produced as a by-product of refining crude oil. Two of its compounds with oxygen, sulfur dioxide and sulfur trioxide, are notorious air pollutants. They are made when sulfur-contain-

11 THE CHEMICAL ELEMENTS

FIGURE 11-12
Orthorhombic crystals of sulfur, the most common allotropic form of this nonmetallic element. (Joel E. Arem.)

ing oil and coal are burned to generate heat and electricity.

Iodine exists as gray-black, lustrous crystals that will evaporate if left exposed to the air. It passes from the solid state directly to the vapor state without first liquifying. Some other solids will do this, too; the phenomenon is called *sublimation*. We say that iodine sublimes. In a chemically combined form, iodine is essential to good health. Its absence from the diet can lead to a goiter—a swelling of the thyroid gland. Until iodized salt became widely available in the United States, a large number of people suffered from goiters. "Tincture of iodine" is a solution in alcohol of iodine plus one of its compounds, potassium iodide.

11.15
SEMIMETALS OR METALLOIDS

Creating systems of classification has been a human activity of immeasurable value in handling enormous amounts of data and facts. These systems, however, are seldom perfect. Borderlines between classes are often fuzzy, and some things stand with their feet in two camps. When we classify elements into the families of metals and nonmetals, we are left with a few elements that do not fit either. They form a third family—the *semimetals* or the *metalloids*. They often look like metals, but they have poor malleability, poor ductility, and chemical properties of both the other families. The following are semimetals.

Silicon (Figure 11-13) is the second most abundant element by weight in the earth's crust. (Oxygen is the most abundant.) Silicon combined with oxygen as silicon dioxide may take the form of sand, quartz, rock crystal, flint, and opal and other semiprecious stones. Pure silicon exists either as an amorphous powder or as a solid with every appearance of a metal. Superpure silicon metal is a key ingredient in transistors and other solid-state devices needed in miniature electronic instruments.

Arsenic is more nonmetallic than metallic. When pure it has the steel-gray appearance of a metal but it is very brittle and it sublimes. In chemically combined forms, its uses range from agricultural poisons to transistors to laser components.

Antimony is a brittle, blue-white, lustrous solid that does not conduct electricity very well. Its alloy with lead, which has greater mechanical strength and hardness than pure lead, is used in making battery casings and cable sheathing.

Selenium looks like a metal in one allotropic form, but it also exists as an amorphous powder. Because it has the unusual property of being able to capture light and change light energy into electricity, selenium is used in electric eyes, solar cells, television cameras, and exposure meters. Selenium is also used in xerography, the process of making photocopies used in Xerox machines.

FIGURE 11-13
Silicon, a semimetal or metalloid, has a lustrous, metallic sheen in the pure state, but is brittle. Seen here are pieces of silicon metal resting on white sand or silica. (General Electric.)

SUMMARY

Chemistry is principally the study of matter and the changes it can undergo, both physical and chemical.

Properties. Any characteristic of an object or substance that we use to recognize it is called a property. The two kinds of properties are physical and chemical. Physical properties are all those that can be measured or otherwise studied without causing any permanent change of the substance into something else. Physical properties include color, luster, physical state (solid, liquid, or gas), temperature, density, heat capacity, latent heat of fusion, latent heat of vaporization, melting point, boiling point, vapor pressure, thermal conductivity, electrical conductivity, malleability, and ductility. Chemical properties include all those capacities for undergoing changes into other substances. Every chemical reaction a substance can undergo is one of its chemical properties. Heat is one of the most important agents for bringing about chemical changes, but heat also is a factor in changes of state and other physical properties or changes.

Liquids. Substances in the liquid state consist of elementary entities (atoms or molecules) that are constantly in chaotic motion and nearly always in mechanical contact with some neighboring entities. Hence, liquids (like solids) are incompressible. Their densities—their mass per unit volume—do not vary with pressure but do vary with temperature. Usually the density of a liquid increases as it becomes colder. Water is an exception between 4–0°C, in which range its density decreases as the temperature is reduced. The hydrostatic pressure of a liquid is directly proportional to depth. This pressure is responsible for the buoyant force a liquid exerts on objects placed on or in it. Archimedes's principle says that the buoyant force exerted by a fluid equals the weight of the fluid displaced by an object placed in it.

Thermal properties. The ability of an object to conduct heat—its thermal conductivity—depends on the ability of its atoms or molecules to pass some of their kinetic energy along to neighbors. The specific heat of an object is the number of calories it can absorb per gram per degree rise in temperature (Celsius). The latent heat of fusion is the heat absorbed (or released) when one gram of a substance changes from solid to liquid (or liquid to solid in the case of heat released). At the transition between liquid and vapor, the latent heat of vaporization is the heat absorbed (in vaporization) or released (in condensation) by one gram of the substance. Liquids exert a vapor pressure which varies with temperature because molecules of the liquid at any temperature move back and forth from the vapor or gaseous state.

Substances. The three broad classes of substances are elements, compounds, and mixtures. An element is any substance all of whose atomic nuclei have identical nuclear charges. Compounds and mixtures are substances that include nuclei from at least two elements, usually more. (We shall have much more to say about these later.) Three kinds of elements are metals, nonmetals, and semimetals (metalloids). Several substances are mixtures of metals, called alloys, with physical properties that make them especially useful for different functions in commerce and industry.

TERMS AND CONCEPTS

property
physical property
chemical property
physical change
chemical reaction
hydrostatic pressure
density
buoyancy
buoyant force
Archimedes' principle
thermal property
thermal contact
conduction
convection
heat capacity
specific heat capacity
specific heat
latent heat
fusion
heat of fusion
heat of vaporization
vapor pressure
boiling
element
metal
malleability
ductility
alloy
bronze
brass
steel
nonmetal
noble gases
allotropy
amorphous
crystalline
sublimation
semimetal
metalloid

QUESTIONS AND EXERCISES

1. Describe two consequences for life on this planet of the fact that water has (a) a relatively high heat capacity, (b) a relatively high heat of fusion, (c) a relatively high heat of vaporization.

2. If a liquid is placed in thermal contact with a colder object and the temperature of the liquid does not change, what eventually happens to the liquid (apart from some evaporation, which we ignore)?

3. Which will be more effective in lowering the temperature of a cup of water at 25°C, adding 50 grams of cold, liquid water at barely above 0°C or 50 grams of solid ice at 0°C? Explain.

4. Give a reason why you will get a feeling of coldness faster if you sit on a chunk of ice at 0°C than if you sit on the same size chunk of granite at 0°C.

5. How much heat, in calories, is required to raise the temperature of 1 kilogram of water (about 1 quart) from 10°C to 100°C?

6. How much heat, in calories, is required to convert 1 kilogram of ice at 0°C to water, also at 0°C?

7. How much heat, in calories, does it take to convert 1 kilogram of water at 100°C to water vapor, also at 100°C?

8. How much heat is released to the surroundings (assumed initially to be much colder) by the change of 1 kilogram of water vapor at 100°C all the way down to 1 kilogram of ice at 0°C?

9. What is the major difference at the atomic level between the elements and all other substances?

10. List some of the most striking differences between metals and nonmetals.

11. What is the major difference in composition between bronze and brass?

12. Why was the discovery of brass such an important step in the development of civilization?

13. What were some advantages of iron over bronze and brass, as civilization developed?

14. What two elements are the most abundant, by weight, in the earth's crust?

15. Without the work of chemists and metallurgists in finding ways for making ultrapure elements, we could have had no "Age of Computors." Name an element essential in this respect.

16. In spite of being a liquid at room temperature, the element mercury is still classified as a metal. Explain.

17. In commerce the "ferrous metals" are iron, chromium, cobalt, niobium, manganese, nickel, tungsten, and vanadium. Why are the metals on this list besides iron important in an industrialized society?

CHAPTER 12
CHEMICAL COMBINATION

12.1 SOME FEATURES OF CHEMICAL CHANGES

In the last chapter we surveyed some physical properties and uses for several important elements. In this chapter we introduce certain basic features of chemical reactions.

When the elements of nature enter into chemical combination, what do we see and what can we measure as being consistently true about all chemical changes? We know that physical properties usually change as the starting materials—the *reactants*—"disappear" and the new substances—the *products*—emerge. In addition, when spontaneous chemical events occur, energy is usually given off as heat, sound, light, and sometimes as electricity (as in a battery running down). Changes in physical appearance and transfers of energy, therefore, are two phenomena we usually observe when substances react chemically. However, the most universal and fundamental characteristic of a chemical change is that when chemicals react, they consistently do so in very precise proportions, not in just random and varying proportions. In the realm of human well-being, this feature of our world is as important as gravity and sunshine.

12.2 LAWS OF CHEMICAL COMBINATION

Heat can cause both physical and chemical changes. Let us study one example that illustrates in a simple way several important laws of chemical combination.

A white, powdery substance is known to chemists as platinum(IV) oxide. When it is heated to 600°C, it beaks down. A metal of high density makes up the residue—the element platinum, one of the very precious metals. A gas has escaped as well. It is the same gas, oxygen, which is present in air and necessary for life. The decomposition of platinum(IV) oxide qualifies on nearly all counts as a chemical event. Physical properties change dramatically as the white powder gives way to a metallic residue. New substances have formed.

Because we get platinum and oxygen by heating platinum(IV) oxide, we say that platinum and oxygen are both simpler substances than the oxide. To determine if the two are in fact elements would require further experiments, and these have been done. Both platinum and oxygen are elements. It took a long time in the history of thought for curious minds to measure

12 CHEMICAL COMBINATION

TABLE 12.1
LAW OF DEFINITE PROPORTIONS

Weight of Sample of Platinum(IV) Oxide Used (g)	Weights of Recovered Substances Platinum (g)	Oxygen (g)	Sum of Weights of Products (g)	Ratio of Weights (g Pt:g O)
10.0	8.59	1.41	10.0	6.09:1
12.5	10.7	1.76	12.5	6.08:1
15.0	12.9	2.12	15.0	6.08:1
Average ratio				6.08:1

and appreciate the importance of the weight changes in a reaction. Table 12.1 contains data, reconstructed to serve as an illustration, of what you might observe about weight changes in the decomposition of platinum(IV) oxide. Weighed samples of the oxide are heated to decompose them, and both the platinum and the oxygen are collected and weighed. Two features stand out. First, the sum of the weights of the products equals the weight of the oxide taken; second, the elements are released in a definite ratio by weight.

12.3
LAW OF CONSERVATION OF MASS

Antoine Lavoisier (1742–1794), A French scientist and one of the great figures in the history of chemistry (Figure 12–1), was the first to measure carefully the weights of all substances involved in a chemical change. To Lavoisier belongs most of the credit for the evidence that compelled students of nature to accept the *law of conservation of mass*. According to this law of nature, mass is neither gained nor lost in a chemical change; mass is conserved. Only its form changes. In practical terms, we cannot create matter and we cannot destroy it by chemical means.[1] All we can do is transform it; for example, we may change something not directly useful such as an ore into something we need, such as a metal. The most efficient transformation of one kind of matter into other kinds is in every case most easily achieved only through the application of another law of chemical change.

[1] We may ignore the infinitesimal difference in masses—the mass defect—between the separated atoms and their combined form in a compound. The mass defect is too small to make any difference whatsoever in a study of weight changes in chemical reactions. Very strictly speaking, however, since Einstein's theory of relativity, the law of conservation of mass is actually the law of conservation of mass-energy.

12.4
LAW OF DEFINITE PROPORTION

No matter what the size of the sample of platinum(IV) oxide taken (Table 12.1), platinum and oxygen are produced in the same proportion by weight—6.08 weight units of platinum to 1 weight unit of oxygen. Evidently, in platinum(IV) oxide, the two elements are combined

FIGURE 12–1
Antoine Laurent Lavoisier (1742–1794), a French lawyer and chemist, who placed chemistry on a quantitative basis and laid the foundation for naming compounds, died at the hands of a mob at the guillotine. (The Bettmann Archive.)

TABLE 12.2
LAW OF MULTIPLE PROPORTIONS

Weight of Sample of Platinum(II) Oxide Used (g)	Weights of Recovered Substances Platinum (g)	Oxygen (g)	Sum of Weights of Products (g)	Ratio of Weights (g Pt:g O)
10.0	9.24	0.76	10.0	12.1:1
15.0	13.9	1.14	15.0	12.2:1
18.0	16.6	1.36	18.0	12.2:1
Average ratio				12.2:1

in a definite proportion by weight. Joseph Louis Proust (1754–1826), a Frenchman working at the Royal Laboratory of Madrid, Spain, is given most of the credit for establishing a second major law of chemical combination, the *law of definite proportions*. It states that in a given chemical compound, the elements are always combined in the same proportion by weight.

12.5 LAW OF MULTIPLE PROPORTIONS

Sometimes a pair of elements will form two different compounds. Platinum and oxygen are such a pair. We can have platinum(IV) oxide and a different powdery material called platinum(II) oxide. When the latter is broken back down to its elements, the weight data of Table 12.2 can be observed. We note that the law of conservation of mass holds, and we can also see that the law of definite proportions holds.

John Dalton (1766–1844), an English chemist and physicist (Figure 12–2), was the first student of nature to realize the importance of a third feature of the data of Tables 12.1 and 12.2 when taken together. The weight ratio of elements in platinum(IV) oxide is 6.08 to 1. In platinum(II) oxide the ratio is 12.1 to 1. Compare 12.1 to 6.08; these two numbers are in a ratio of 2 to 1—a ratio of simple, *whole* numbers in other words. Tin and oxygen are also a pair of elements that form two oxides. In one the weight ratio of tin to oxygen is 7.4 to 1. In the other oxide, the ratio is 3.7 to 1. Compare 7.4 to 3.7; these two are in a ratio of 2 to 1—two simple, *whole* numbers again. Results such as these (and there are many other examples) form the basis for a third law of chemical combination, the *law of multiple proportions*. According to this law of nature, whenever two elements form more than one compound, the different weights of one combining with the same weight of the second are in the ratio of small whole numbers. This fact about our earth is just that: a fact. It ex-

plains nothing. The more examples discovered, the more scientists of the time wondered at the absence of complicated ratios—ratios involving something besides simple whole numbers. It was a puzzle needing an explanation. *Dalton's atomic theory* was that explanation.

12.6 DALTON'S THEORY

Dalton asked: what must be true about elements if they behave in the ways described by the laws of chemical combination? The idea of an atom, as we have indicated, had been around for centuries. As early as 500 B.C., Greek philosophers, driven by both logic and everyday evidence, had

FIGURE 12–2
John Dalton (1766–1844), English chemist and physicist. (Brown Brothers.)

12 CHEMICAL COMBINATION

FIGURE 12-3
The slow imperceptible wearing away of stone steps was one observation used by the ancient Greeks to postulate that matter is made of invisible atoms. Seen here are steps leading into the New York Public Library. (K. T. Bendo)

speculated that tiny, invisible particles made up all matter, particles that could not be cut. (Hence, "atom," from *atomas* meaning, in Greek, "not cuttable.") They were trying to explain, for example, the very slow wearing away of stone steps (Figure 12–3), a process that surely happened but that could not be seen to happen from one day to the next.

Dalton was sure he saw in the weight data of chemical changes evidence that compelled belief in atoms. Moreover, these atoms must have certain properties.

A list of the supposed properties of atoms constitutes much of Dalton's theory. Before going into it, we must note that Dalton's idea of an atom embraces some other kinds of small particles we now call "molecules." To Dalton an atom was simply a fundamental particle. The following, restated in condensed form, are Dalton's main postulates:

1. Matter consists of atoms.
2. Atoms are indestructible. Chemical reactions are nothing more than atoms being rearranged.
3. All atoms of one element are identical in weight and other properties.
4. Atoms of different elements are different in weight and other properties.
5. When a compound forms from its elements, a definite but small number of atoms of each element join to make the more complex but fundamental particles of the compound.

Postulate 2 was made necessary by the observation of the conservation of mass in a chemical change. If postulates 3, 4, and 5 are correct, then the laws of definite and multiple proportion are explained. Thus, if a "*definite* but small number of atoms of each element" makes up each unit or fundamental particle of a compound, then a compound must contain its elementary atoms in a *definite* ratio. If atoms have unchanging weights, then *a definite ratio of atoms cannot help but result in a definite ratio by weight,* which is what we observe (law of definite proportions). If we cannot have a fraction of an atom—if atoms are indestructible—then the atom-to-atom ratio in which they are present in compounds can be a ratio expressible only in *whole* numbers. If one platinum atom can combine with, say, two oxygen atoms in one compound and with four oxygen atoms in another compound (but never with, say, 4.65 oxygen atoms), then the ratio of the oxygen atoms in the two compounds is 4 to 2 (or, simplified, 2:1)—small, whole numbers. Thus, constant weight ratios on one hand, plus the absence of fractional ratios on the other (in the law of multiple proportions), put the idea of atoms on a sound experimental basis.

It is interesting that Dalton's theory was wrong on two points and yet it worked to explain much about the laws of chemical combination. Atoms, of course, are not indestructible; they can be broken into subatomic particles (but not by chemical changes). All the atoms of an element are not identical in weight; isotopes exist, but they are essentially identical, chemically. Thus, the value of a theory is not entirely a function of its being correct in every way. It may be incorrect in some aspects and yet be a powerful tool to explain many observations and, most importantly, suggest new areas of study and research.

We have assumed all along that we have dealt with pure elements and pure compounds. Just what do we mean by "pure" in a chemical sense? We have reached the point where we must make the fundamental distinction between pure substances and mixtures.

12.7 COMPOUNDS VERSUS MIXTURES

Suppose we have a sample of iron in the form of small iron filings or even small brads, and suppose that someone has mixed pieces of sulfur with them. Just by looking at this, we should say that the iron is impure (Figure 12–4). (Of course, so is the sulfur.) We know, however, that iron is attracted to a magnet and sulfur is not. By using a magnet we can pull the iron away from the sulfur. The example is crude, but it illustrates one important difference between a *mixture* and a

12.7 COMPOUNDS VERSUS MIXTURES

FIGURE 12-4

Mixture versus compound. On the left, a mixture of sulfur powder (light color) and iron filings. Iron responds to a magnet and sulfur does not, furnishing a difference that makes the separation of this mixture by a magnet possible. The iron filings are attracted to the magnet. On the right is iron sulfide, a chemical compound of iron and sulfur that cannot be separated by a magnet or by any other purely physical means. (Mimi Forsyth/Monkmeyer.)

compound. To separate the parts of a mixture from each other, we find and use some difference in their *physical* properties. The difference is not often in response to a magnet, but certain iron compounds in iron ores are attracted to a magnet, which is used to separate iron-rich ore particles from dirt and gravel.

Differences in boiling points are often used to separate liquids. Because different components of crude oil boil at different temperatures, petroleum refiners can provide commerce and industry with a variety of petroleum "fractions," ranging from natural gas to gasoline to diesel oil to residual heating oil. Sometimes components of a mixture have different solubilities in some liquid; one component will dissolve, the other will not. You may have used lighter fluid to dissolve and remove a grease stain from a fabric. You separated the "mixture" of grease and fabric by taking advantage of the fact that the two have different solubilities in lighter fluid.

Unlike compounds, mixtures do not obey the law of definite proportions. We can make a mixture of iron filings and sulfur in any proportion we wish and still call it a mixture of iron and sulfur. Whatever the ratio of iron to sulfur, a magnet will pull the iron filings from the sulfur, or a special solvent called carbon disulfide will dissolve the sulfur (but not the iron). If we try to heat the mixture, however, we do not get a separation. Remember that heat often causes chemical changes, and it does here. When we heat a mixture of iron and sulfur, a change eventually begins that soon continues by itself with considerable evolution of heat. A black substance that can be broken and crushed into a powder forms. It does not dissolve in carbon disulfide and is not attracted to a magnet. We cannot separate it into iron and sulfur by any physical means whatsoever. This new substance is a compound called iron sulfide (Figure 12-4). It is a compound, not a mixture, because it obeys the law of definite proportions and because we cannot separate it (break it down) by merely physical changes.

Because iron sulfide obeys the law of definite proportions, we can get essentially pure iron sulfide only by heating a mixture of iron and sulfur prepared in that "definite proportion." It happens to be a ratio of 1.744 grams of iron to 1.000 gram of sulfur. If we take a greater relative amount of sulfur, the extra sulfur is left over and contaminates the compound. If we take a greater proportion of iron, its extra amount will be left over, too. We cannot prepare pure iron sulfide in just any ratio of iron to sulfur. Iron has a certain combining ability with sulfur, an ability expressed in terms of the weight ratio of 1.744 to 1.000.

We are moving to the strategy scientists use to solve the problem of mixing chemicals in exactly the right proportions to ensure that when they react, the desired product will be as pure as possible. The need for this strategy is important. The less pure the product, the more effort and energy needed to make it pure and the more waste of raw materials. That strategy, however, is most easily studied after we learn more about chemical symbols, the scientific shorthand we need for further study.

12.8 SYMBOLS OF ELEMENTS

Several of the more common elements are listed in Table 12.3 together with the abbreviations chemists use for them. The elements in the first column have the simplest symbols, merely the capitalized first letter of the name. In the second column are several elements whose symbols are the first two letters of the name. The first letter is always capitalized and the second is always lower case. The symbol for cobalt is Co, not CO, which is the formula of carbon monoxide, as we shall see. In the third column we find several elements where the first two letters cannot be used. Some elements have the identical first two letters. Therefore, for many we have to use the first letter and some letter other than the second. Some elements, illustrated in the last column, were named when Latin was the universal language of educated people. That is why the symbols are not related to the English names of the elements.

12.9 SYMBOLS FOR COMPOUNDS

The symbol of a chemical compound is called a *formula*. There are three kinds of formulas—the empirical, the molecular, and the structural. We shall learn here only about the first. "Empirical" in this context refers to factual data obtained by a chemical analysis. An empirical formula is one that uses symbols for elements and numbers called subscripts to show which atoms exist in a compound and in what ratio. The empirical formula—we shall hereafter simply call it the *formula*—of some of the compounds we have already mentioned are

platinum(IV) oxide PtO_2
platinum(II) oxide PtO
iron sulfide FeS.

(We shall not worry here about the meaning of the IV and the II.)

In iron sulfide the elements are combined in the ratio of one *atom* of iron to one *atom* of sulfur. We need not write Fe_1S_1. That is, we do not use subscripts for "1." In platinum(IV) oxide the subscript 2 tells us that for every atom of platinum (whose subscript 1 is understood) there are two atoms of oxygen.

12.10 SYMBOLS FOR REACTIONS

The symbol of a chemical reaction is called a *chemical equation*. The symbols of the reactants are written with plus signs between them, suggesting that we are adding them together. Then there is an arrow pointing to the products, also repre-

TABLE 12.3
SYMBOLS FOR COMMON ELEMENTS

C Carbon	Al Aluminum	Cl Chlorine	Ag Silver (*argentum*)
F Fluorine	Ba Barium	Mg Magnesium	Cu Copper (*cuprum*)
H Hydrogen	Br Bromine	Mn Manganese	Fe Iron (*ferrum*)
I Iodine	Ca Calcium	Pt Platinum	Pb Lead (*plumbum*)
N Nitrogen	Li Lithium	Zn Zinc	Hg Mercury (*hydrargyrum*)
O Oxygen	Ra Radium		K Potassium (*kalium*)
P Phosphorus			Na Sodium (*natrium*)
S Sulfur			

sented by symbols with plus signs, as needed, between them. The equation for the reaction of iron with sulfur is

$$Fe + S \rightarrow FeS$$

The decomposition of platinum(IV) oxide is symbolized as

$$PtO_2 \xrightarrow{heat} Pt + O_2$$

(Why oxygen must be written as O_2 instead of some other way will be studied later.)

12.11
BALANCED EQUATIONS

We call the symbol for a chemical change a *chemical equation,* borrowing the word *equation* from mathematics, because what is on one side of the arrow must in a special way equal what is on the other side. (In many books, an equals sign (=) is used instead of an arrow.) Because matter for all practical purposes is neither created nor destroyed in a chemical change, all atoms listed among the reactants must exist somewhere among the products. The two equations given above are obviously balanced in this sense. Suppose, however, we had mixed aluminum with sulfur and heated the mixture. A chemical change does take place, and a compound forms called aluminum sulfide with the formula, Al_2S_3. The subscripts 2 and 3 mean that aluminum atoms and sulfur atoms combine in a ratio of 2:3 in this compound. In words, we describe the reaction this way:

aluminum	combines with		
⋮	⋮		
Al	+		
⋮	⋮		⋮
S	→		Al_2S_3
⋮	⋮		⋮
sulfur	to form		aluminum sulfide

This is not a balanced equation. To balance an equation, let us be aware, first, of one thing we may never do. Once we have set down the correct formulas of the substances involved, we may not change any of their subscripts. We may not balance the above equation by erasing the 2 and the 3 in Al_2S_3. The reason is that AlS, the result of this erasing, does not form; it does not even exist. What we may do is insert numbers, called *coefficients,* in front of chemical symbols. If we place the coefficient 2 in front of Al and 3 before S we have

$$2Al + 3S \rightarrow Al_2S_3$$

The coefficient 1 is not written in front of Al_2S_3; it is understood, we take it for granted. We now have a balanced equation, and this relatively simple symbol says: aluminum combines with sulfur in a ratio of two atoms aluminum to three of sulfur to form aluminum sulfide of the formula Al_2S_3. The chemical equation says it much more neatly and clearly, and you can "read" it much faster.

One final point about formulas and equations. We use whole numbers, not fractional numbers, to express ratios, whether the numbers are used as subscripts within a formula or as coefficients in an equation. The formula for aluminum sulfide is written as Al_2S_3. The ratio 2:3 could also be expressed as 4:6 or as 1:1.3333. In terms of what is true about the ratio of aluminum to sulfur in aluminum sulfide, there is nothing fundamentally wrong in writing the formula as Al_4S_6 or even as $AlS_{1.333}$. However, the convention chemists use—and it is just that, a convention, not a law of nature—is to select the lowest *whole* numbers that correctly express the ratio. Where chemists make exceptions there are good reasons, as we shall see later. In equations, too, the coefficients used are the lowest whole numbers that express the proportions. We could have written the equation for the formation of aluminum sulfide as

$$4Al + 6S \rightarrow 2Al_2S_3$$

However, we can divide each coefficient by 2, and the equation is therefore written as earlier shown. Dividing by 2 does not change the fundamental proportions at all.

12.12
COMBINING WEIGHTS OF ELEMENTS

If we want to make a compound from elements and make it as pure as we can with nothing left over contaminating the product, we have to arrange a meeting of the elements in a very definite ratio. If we could see atoms and pick them up with tweezers, we could get any ratio of atoms we wished simply by counting them, much as we count out apples one by one and wooden sticks one by one when we want to make candy apples on a stick. To make apples on a stick requires one stick per each apple. We do not care what each apple or each stick weighs. But if we had to get this ratio of one to one without counting, then we should care about the weights. If we know that apples weigh, say, 100 grams each and sticks 10 grams each, we can count them out in "particle" ratios of 1:1 by weighing them. If we weigh apples and sticks in weight ratios of 100:10 g (or 1000:100 g or 10,000:1000 g—all represent the same ratio) we get one apple for each stick with nothing left

12 CHEMICAL COMBINATION

over. We do not have to do this with apples and sticks, but we could. Because atoms are so small, we have no choice. To count them in ratios by *atoms*, we are forced to use information about the relative weights of atoms—atomic weights.

12.13
ATOMIC WEIGHTS

Chemists speak of *atomic weights* instead of atomic masses for reasons that need not concern us now. On earth the two are identical by definition.

An atomic weight is a relative weight. An atomic weight is NOT the weight of a single atom; it is a number that tells us how *relatively* heavy an atom is, compared to some reference. Since 1961 that reference is the carbon-12 isotope, which is assigned an atomic weight of exactly 12.000 units. This number is chosen for the weight of carbon-12 because the lightest of all atoms, a hydrogen atom, is one-twelfth as heavy as an atom of carbon-12. It simplifies calculations if every element has an atomic weight of at least one whole unit. A hydrogen atom, one-twelfth as heavy as an atom of carbon-12, can be assigned an atomic weight of 1.0 (more accurately, 1.007825).

Oxygen atoms are heavier than those of carbon, $\frac{16}{12}$ths times as heavy. Hence, the relative weight or the atomic weight of oxygen is 16. In other words, an atomic weight is a numerator over 12 as a denominator, and this fraction tells us how much heavier an atom of some element is compared to a carbon-12 atom. One problem is that atomic weights are seldom whole numbers! We shall see why next.

Usually an element occurs in nature as a mixture of isotopes, a mixture whose *proportion* of isotopes is the same no matter where in the entire earth a sample of the element is obtained. Chlorine, for example, exists as a mixture of 4 atoms of chlorine-35 to 1 atom of chlorine-37. The average weight is 35.4 (it is actually 35.453, partly because the ratio is not exactly 4:1). Wherever on the planet a sample of chlorine is prepared, its atomic weight is 35.453. At least, that has so far been the experience of chemists.

The atomic weight of iron (rounded off) is 56; that of sulfur (also rounded off) is 32. Iron atoms evidently weigh more than sulfur atoms. We cannot take equal *weights* of the two and expect to get equal *numbers* of atoms of each, anymore than we could take equal weights of apples and sticks and get equal numbers of them. To get iron atoms and sulfur atoms in a ratio of 1:1 by atoms, we must weigh them in the ratio of 56 weight units of iron to 32 weight units of sulfur. Thus, in effect, we may obtain elements in any desired ratio by their atoms simply by weighing them in corresponding ratios by atomic weights.

12.14
GRAM-FORMULA WEIGHT

In Section 12.7 we learned that iron combines with sulfur in a weight ratio of 1.744 Fe to 1.000 S. That ratio is the same as 56 Fe to 32 S, the ratio of their atomic weights ($\frac{56}{32} = 1.744$). Atomic weights are always available in a table. Hence, we use those numbers expressed in grams to define one "reacting unit" of an element. We call that reacting unit the gram-atomic weight or, even better (as we shall see), the *gram-formula weight*.[2] It is easy to compute a gram-formula weight of an element: simply write the word *grams* after the atomic weight. The gram-formula weight of iron is 56 grams; of sulfur, 32 grams. When we use iron as a chemical, one "reacting unit" is 56 grams iron. Two reacting units of iron weigh 56 g $\times$ 2 = 112 g. Half a reacting unit of iron weighs one-half of 56 g or 28 g. With sulfur, one reacting unit weighs 32 g; two reacting units are 64 g; half a reacting unit weighs 16 g. Each element has its own weight taken in grams for its reacting unit.

12.15
MOLE

The name of the reacting unit is *mole*. One mole of iron weighs 56 g, two moles weigh 112 g, half a mole weighs 28 g; one mole of sulfur weighs 32 g.

How many atoms of iron are there in one mole or 56 g of iron? A great number, no doubt, and however large, it is the same as that of sulfur atoms in one mole of sulfur. When we take iron and sulfur in *weight* ratios according to their atomic weights, we get them in *atom* ratios of one to one. In general, *equal numbers of moles contain the same number of fundamental particles*—atoms in our example.

12.16
FORMUA WEIGHTS OF COMPOUNDS

The *formula weight* of a compound is the sum of the atomic weights of all the elements represented by the formula. The formula weight of iron sulfide, FeS, is simply

$$56 + 32 = 88$$
$$(Fe) \quad (S) \quad (FeS)$$

---

[2] This simply enlarges the idea of *formula* to include the elements as well as the compounds. The formula of iron is thus simply Fe.

The formula weight of aluminum sulfide, Al_2S_3, is

$$2 \times 27 + 3 \times 32 = 54 + 96 = 150$$
$$(Al) \quad\quad (S) \quad\quad\quad\quad\quad\quad Al_2S_3$$

Just as an element has a fundamental particle we call an atom, a compound can be considered to have a fundamental particle that for the moment we shall simply call its *formula unit*. The formula of the compound indicates the composition of the formula unit.[3] One formula unit of Al_2S_3 consists of (or is made from) 2 atoms of aluminum and 3 atoms of sulfur.

The gram-formula weight of a compound, like that of an element, is simply the formula weight written with "grams" after it. Hence, the gram-formula weight of aluminum sulfide is 150 grams. This amount of the compound makes up one reacting unit of aluminum sulfide or one mole. And 150 grams of aluminum sulfide, or one mole of it, contains the same number of formula units as 56 grams of iron, 27 grams of aluminum, and 32 grams of sulfur contain atoms. Each and every mole of all compounds and elements contains the same number of formula units. The identities and weights of these formula units vary a great deal, of course. We are referring here only to their relative *numbers*. Out of curiosity, we next ask, "how large is that number?" How many formula units are there in a mole?

12.17 AVOGADRO'S NUMBER

How many atoms are there in 56 grams of iron (1 mole Fe) or 32 grams of sulfur (1 mole S) or 1 mole of anything? By a number of indirect measurements, all of which agree, 1 mole of any substance has 6.0238×10^{23} formula units, a number too large to grasp. The total mass of the earth is about 10^{28} grams. The mass of all the water in the world's oceans is about 10^{23} g. (Do not confuse *grams* with *atoms*.) The mass of one *atom* of carbon-12 is only about 2×10^{-23} g, and a large number is needed to make 12 g of carbon.

We give names to many special numbers: 12 is a dozen; 144 is a gross. The number 6.0238×10^{23} has the name *Avogadro's number*, after Amedeo Avogadro, an Italian physicist who did pioneering research that led eventually to ways of measuring equal numbers of formula units. When we have Avogadro's number of formula units of any chemical, the sample will weigh the same as the formula weight expressed in grams (see Figure 12-5). We may, of course, have any size sample we wish, samples of fractions or multiples of moles. The relation between the size of a sample given in moles and that given in grams is expressed by this equation.

Number of moles present in a sample of a chemical

$$= \frac{\text{weight of the sample in grams}}{\text{gram-formula weight of the substance}}$$

or

$$\text{moles} = \frac{\text{weight (in grams)}}{\text{gram-formula weight}}$$

The whole idea of the mole concept in chemistry is to have a way to use *weights* of substances to get substances measured in whatever proportions *by formula units* that we want for a particular chemical reaction.

FIGURE 12-5
The number of water molecules in the ice cube equals the number of sugar molecules in the mound of sugar. Less water by weight is present because each water molecule weighs only about 5 percent as much as each sugar molecule (the formula weight of water is 18; of sugar, 342). (Mimi Forsyth/Monkmeyer.)

---

[3] Depending on the compounds, the formula unit might actually be a *molecule* or it might be a small cluster of *ions*. We make these distinctions in Chapter 14.

SUMMARY

Chemical changes are events in which new substances are produced. In most chemical changes, two readily observable results are changes in physical appearance such as a change in color, physical state, or temperature. Most but not all spontaneous chemical changes are accompanied by evolution of heat. A third important feature of a chemical change requiring careful measurement to detect is that very precise proportions of reactants and products are involved.

Laws of chemical combination. In chemical changes, mass is neither created nor destroyed; only the forms change (law of conservation of mass). In a given chemical compound, the elements are always combined in the same proportion by weight (law of definite proportions). When two elements form more than one compound, the different weights of one that combine with the same weight of the second are in the ratio of small whole numbers.

Dalton's theory. To explain the laws of chemical combination, John Dalton postulated that all matter consists of indestructible atoms differing from substance to substance principally in weight. In this theory, chemical changes became a matter of atoms rearranging themselves.

Formulas and equations. The shorthand symbol of a compound consists of the symbols of its constituent atoms combined with subscripts indicating their relative proportions by atoms. The shorthand symbol of a chemical reaction is a chemical equation consisting of the symbols of the reactants, separated by plus signs, the symbols of the products, likewise separated, and an arrow pointing from reactants to products. Coefficients—numbers standing before symbols of compounds—disclose the relative proportions of the chemicals involved. An equation is balanced only when all nuclei on one side of the arrow are found somewhere among the products on the other side.

Mixtures are the third kind of matter, after elements and compounds. These consist of two or more substances combined only in a physical way and not in definite proportions.

Gram-formula weight or mole. One mole of any substance is the number of grams corresponding numerically to its formula weight. Equal numbers of moles contain equal numbers of formula units, whatever the particles that make up the formula. In one mole of any substance, there are Avogadro's number (6.10×10^{23}) of formula units. The coefficients in a balanced equation give the proportions by moles involved in a reaction.

TERMS AND CONCEPTS

reactant
product
Law of conservation of mass
Law of definite proportions
Law of multiple proportions
Dalton's atomic theory
compound
mixture
formula, empirical
equation, chemical
coefficient
atomic weight
formula weight
gram-formula weight
mole
Avogadro's number

QUESTIONS AND EXERCISES

1. What are three important characteristics of chemical changes, and which one is consistently true about all chemical changes?

2. Chromium and oxygen make up two oxides. The composition of each compound is given by

Compound A	Compound B
61.90% chromium	76.47% chromium
38.10% oxygen	23.53% oxygen

These percents mean that in 100 g of compound A, you could obtain 61.9 grams chromium and 38.10 grams oxygen. Use a simple proportion to calculate how much oxygen goes with 1.00 g chromium. Do this calculation for compound B, too. Are the amounts of oxygen that can combine with the same amount of chromium (1 gram) related in a ratio of simple whole numbers? If so, what whole numbers?

3. What is the weight of a sample of iron consisting of 3.01×10^{23} atoms of iron?

4. What are the weights of the following samples (include the correct unit of weight): (a) 1 mole Li? (b) 10 moles Na? (c) 0.4 moles S? (d) 0.001 mole Al_2S_3.

5. Using a trial-and-error juggling of coefficients, balance each of these equations. (a) $Ca + C \rightarrow CaC_2$ (calcium carbide), (b) $Al + Cl_2 \rightarrow AlCl_3$ (aluminum chloride), (c) $C + O_2 \rightarrow CO$ (carbon monoxide).

6. How many atoms of all kinds are present in one formula unit of $CrC_4H_4O_6$ and $CoSnO_4$?

CHAPTER 13
THE PERIODIC SYSTEM AND ELECTRONIC CONFIGURATIONS

13.1 ON CLASSIFYING THINGS

We earlier noted that the number of elements is small, only 105 counting synthetic elements. (Element 106 has been claimed.) The number of compounds, however, is immense. Several million compounds include the element carbon. Compounds form by chemical reactions, and the number of possible chemical changes is likewise enormous. In the face of handling huge quantities of data and facts, what do we often do? Whether the facts are about plants, animals, minerals, insects, bacteria, compounds, or elements, we often begin by classifying them. We try to find patterns leading to explanations as we sort things or events into groups. We have already done some of that. We sorted elements into families of metals, nonmetals, and the rest (metalloids). While these families are helpful in some ways, they do not suggest theories and explanations and reasons. Because we seek reasons, we shall try another way of classifying elements in this chapter.

13.2 PERIODIC CHANGES IN PROPERTIES OF ELEMENTS

As scientists studied elements—both their physical and chemical properties—a startling pattern emerged. There seemed to be a cycle among the elements if they were arranged in a particular order. Dimitri Mendeleev (1837–1907) in Russia, Lothar Meyer (1830–1895) in Germany, and J. A. R. Newlands (1838–1898) in England were three of several investigators in this field. Working independently in the 1860s and 1870s, they studied a large number of physical and chemical properties of the elements known in their time. Each made a table of the elements in the order of their increasing atomic weights. It struck them as particularly interesting and probably very significant that every now and then, in a fairly regular way, some properties of lighter elements seemed to return among heavier elements. Mendeleev (Figure 13-1), given major credit for developments, put it this way, "the elements arranged according to the size of their atomic weights show a periodic change of properties."

The boiling points of the elements—that is, the temperatures at which they change from liquids to vapors—were one such property (see Figure 13-2). Boiling points do not simply increase and increase as the atomic weights become larger and larger. Instead, they rise and fall and rise again.

Any time something happens again and again we say it is periodic—it has *periodic properties*. The tides are a periodic event; they are a periodic function of time. Street intersections in a city occur again and again; they are a periodic function of distance. While the boiling points of the elements do not reoccur exactly, they do rise and fall with increasing atomic weights. They are a periodic function of atomic weight. Table 13.1 contains a list of several of the elements of low atomic weights as we now know them. Also tabulated are the corresponding boiling points, densities, and, to bring in chemical properties,

13 THE PERIODIC SYSTEM AND ELECTRONIC CONFIGURATIONS

FIGURE 13-1
Dimitri Ivanovich Mendeleev (1834–1907), Russian chemist who developed a periodic system. (Culver Pictures.)

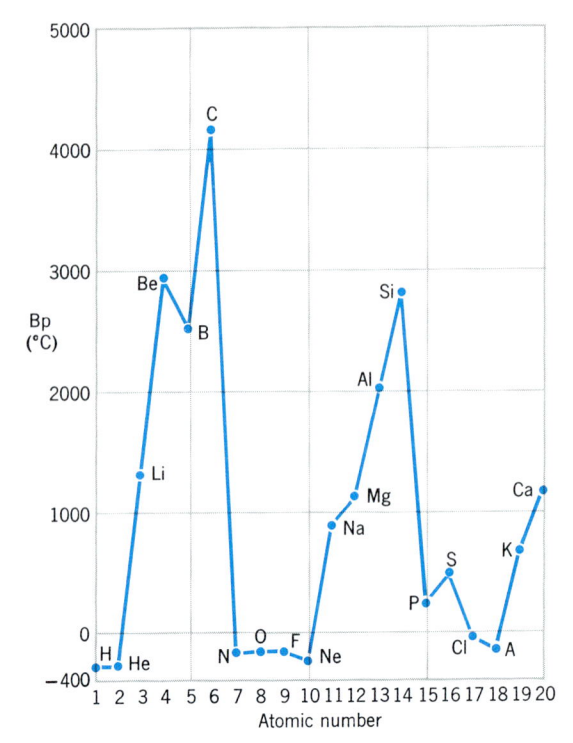

FIGURE 13-2
The boiling points of the elements (1 to 20) do not change smoothly, either up or down, with increasing atomic number. They fluctuate widely in a roughly periodic manner.

TABLE 13.1
PROPERTIES OF SOME LIGHTER ELEMENTS

Atomic Number	Atomic Weight	Element	Boiling Point (°C)	Density (g/cm³, 20°C)	Oxide	Hydride
3	7	Lithium	1317	0.534	Li_2O	LiH
4	9	Beryllium	2970	1.85	BeO	BeH_2
5	11	Boron	2550 (sub)	2.34	B_2O_3	BH_3 (as B_2H_6)
6	12	Carbon	4827	3.51	CO_2	CH_4
7	14	Nitrogen	−196	0.0013	N_2O_5	NH_3
8	16	Oxygen	−183	0.0014	O_3	H_2O
9	19	Fluorine	−188	0.0017	F_2O_2 (also F_2O)	HF
10	20	Neon	−246	0.0009	—	—
11	23	Sodium	892	0.97	Na_2O	NaH
12	24	Magnesium	1107	1.74	MgO	MgH_2
13	27	Aluminum	2467	2.702	Al_2O_3	AlH_3 (in $LiAlH_4$)
14	28	Silicon	2355	2.32	SiO_2	SiH_4
15	31	Phosphorous	280	2.34 (P_4)	P_2O_5	PH_3
16	32	Sulfur	445	2.07 (S_8)	SO_2 (also SO_3)	H_2S
17	35.5	Chlorine	−35	3.2 (0°)	ClO_2, Cl_2O_7, Cl_2O	HCl

13.5 CHEMICAL FAMILIES

	Group I	Group II	Group III	Group IV	Group V	Group VI	Group VII	
Element	Lithium	Beryllium	Boron	Carbon	Nitrogen	Oxygen	Fluorine	
Atomic Weight	7	9	11	12	14	16	19	
Formula of Hydride	LiH	BrH$_2$	BH$_3$ (B$_2$H$_6$)	CH$_4$	NH$_3$	H$_2$O	HF	Series 1
Formula of Oxide	Li$_2$O	BrO	B$_2$O$_3$	CO$_2$	N$_2$O$_5$	O$_3$	F$_2$O$_2$ (also F$_2$O)	
Density (g/cc)	0.534	1.85	2.34	3.51	0.0013	0.0014	0.0017	
Element	Sodium	Magnesium	Aluminum	Silicon	Phosphorus	Sulfur	Chlorine	
Atomic Weight	23	24	27	28	31	32	35.5	
Formula of Hydride	NaH	MgH$_2$	AlH$_3$	SiH$_4$	PH$_3$	H$_2$S	HCl	Series 2
Formula of Oxide	Na$_2$O	MgO	Al$_2$O$_3$	SiO$_2$	P$_2$O$_5$	SO$_2$	ClO$_2$, Cl$_2$O$_7$, Cl$_2$O	
Density (g/cc)	0.97	1.74	2.70	2.32	2.34	2.07	3.2 (O)	

FIGURE 13-3
This arrangement of the lighter elements helps to show how several properties "come back again." If they did not, we might expect the boiling points of elements to become higher and higher as we go to the heavier elements. We might also have expected the hydrides to contain ever-increasing proportions of hydrogen.

formulas of their compounds with oxygen (oxides) and with hydrogen (hydrides). To discuss one pattern we need a technical term, *valence*, which we shall only introduce here.

13.3 VALENCE

One striking fact about the chemical formulas in Table 13.1 is that individual elements have very definite combining abilities toward oxygen or hydrogen. Beryllium (Be) can bind twice as many hydrogens as lithium (Li), as we see by the formulas BeH$_2$ and LiH. Boron (B) evidently can combine with three times as many hydrogens as can lithium. The technical term for "combining ability" is *valence*, from the Latin *valere*, meaning "to be strong." In Chapter 14 we shall explain how elements have particular valences.

13.4 EARLY ATTEMPTS AT A PERIODIC CHART

Meyer and Mendeleev took special note of the apparent valences or combining abilities of the then-known elements toward hydrogen and oxygen. We may recapture their insight by starting with lithium in Table 13.1 and examining the formulas of the compounds that successive elements form with hydrogen. If we say that lithium has a relative combining ability or a valence of 1 and beryllium, 2, and so on, then these valences seem to occur in two identical series. From lithium to fluorine we have valences of

1 2 3 4 3 2 1

Then from sodium to chlorine we have the same

1 2 3 4 3 2 1

The formulas of the oxides, while sometimes a bit more complicated than those of the hydrides, also seem to be in two series.

13.5 CHEMICAL FAMILIES

Meyer and Mendeleev constructed a table of the elements by arranging the members of each series on horizontal rows making sure that elements having the most similar properties, especially chemical properties, came out beneath each other in vertical columns. Sodium, on the basis of similar chemical and physical properties, simply had to be associated with and placed

TABLE 13.2
PROPERTIES OF GERMANIUM COMPARED WITH MENDELEEV'S PREDICTIONS

Properties	Predicted	Found
Atomic weight	72	72.3
Density	5.5	5.47
Formula of oxide	MO_2	GeO_2
Density of oxide	4.7	4.703
Formula of chloride	MCl_4	$GeCl_4$
Boiling point of chloride, °C	less than 100	86

beneath lithium. From the way that elements sort into vertical columns came our concept of families of elements. It was an amazing development. By placing elements in horizontal rows in order of atomic weights, breaking rows, and starting new ones where some invisible "pendulum" of properties swung back again, elements fell into place into vertical columns whose members shared many very similar chemical and physical properties. Figure 13-3 shows such an arrangement for the elements in Table 13.1.

To get some elements into those vertical columns where they seemed to belong because of valences and other properties, Mendeleev had to take several important liberties with his original intention—to arrange the elements strictly in order of increasing atomic weights. In some cases he argued that certain atomic weights must be incorrect, and when the measurements were redone, he was right. At other places to get the vertical columns to "come out right" he left gaps and predicted that there were still-to-be-discovered elements for those gaps. At one point he even left two spaces, between zinc and arsenic. Mendeleev even went so far as to predict the properties of some of these undiscovered elements. In 1871 he predicted that standing just before arsenic and beneath silicon there must be an element of atomic weight 72. He called it eka-silicon. In 1886 the German scientist C. Winkler discovered it and named it germanium after his country. Table 13.2 contains a comparison of its properties with those Mendeleev had predicted. In similar manner, he predicted the existence and properties of elements later named scandium and gallium.

After Mendeleev had fixed the chart as much as he could by predicting new elements and by urging better research on atomic weights, the ordering of three pairs of "problem elements" remained a stubborn problem. Tellurium (Te, atomic weight 127.61) has a higher atomic weight than iodine (I, atomic weight 126.92). However, if for that reason you insist on placing iodine *before* tellurium in their series or period, iodine is beneath selenium and sulfur, whereas tellurium is beneath bromine and chlorine (refer to Figure 13-4, the modern form of the *periodic table*). The trouble is that iodine most resembles bromine and chlorine, chemically, and for that reason "belongs" beneath them (the respective hydrides are HCl, HBr, and HI). Tellurium similarly resembles selenium and sulfur chemically and belongs with them (the hydrides, for example, are H_2S, H_2Se and H_2Te). This did not stop Mendeleev. He cheerfully abandoned a rigid insistence on his original intent and boldly put tellurium and iodine "out of order" in terms of their atomic weights.

13.6
PERIODIC LAW— MODERN FORM

Mendeleev had said that the properties of the elements are a periodic function of their atomic *weights*. This is the old form of the periodic law. It was just about true. Telluium and iodine did not "fit". Neither did cobalt and nickel (58.93 versus 58.71 in atomic weights) nor argon and potassium (39.95 versus 39.10). Scientists had become so accustomed to order and harmony in nature that these particular exceptions rankled. A remedy was found in the early part of this century in the work of H. G. J. Moseley, a British scientist who studied the x-ray spectra of the

TABLE 13.3
PRINCIPAL ENERGY LEVELS

Principal quantum number (energy level number)	1	2	3	4	5	6	7
Letter designation	K	L	M	N	O	P	Q
Maximum number of electrons (found in nature)	2	8	18	32	32	10	2

Maximum Population of Sublevels

s	2	(1 orbital)
p	6	(3 orbitals)
d	10	(5 orbitals)
f	14	(7 orbitals)

elements. He discovered that the frequencies of certain lines in these spectra were proportional to the squares of whole numbers. These numbers increased by one unit from element to element, provided the elements were arranged in the order of atomic weights—or nearly so. These simple, whole numbers are the atomic numbers.

One of the surprises of Moseley's work occurred with the "problem pairs" of elements, those Mendeleev placed out of order when he arranged the elements in a general order of increasing atomic weight. Moseley found that these pairs were precisely in order if the elements were arranged according to increasing atomic number. At the heart of the periodic variation in the properties of the elements is the atomic number, not the atomic weight. To Moseley we owe credit for the modern version of the *periodic law:* the properties of the elements are a periodic function of their atomic numbers.

One eventual result of Moseley's work was that we now can be certain that between elements 1 and 105 there are no gaps. There are no new elements in that series. Any new elements will have to have higher atomic numbers.

The realization that the elements fall naturally into just a few families not only greatly simplified the study of the elements, but also was a tremendous aid in research. It made possible a much greater use of *reason* in planning experiments. Mendeleev reasoned that germanium and gallium and scandium existed. Had he not so reasoned, perhaps no one would actually have looked deliberately for them. These elements would have awaited purely accidental discovery. He predicted several properties of these new elements and their compounds. Countless chemists have similarly used information about known properties of some compounds in a family of elements to predict the properties of others yet to be made.

13.7
THE PERIODIC TABLE— ONE MODERN FORM

One of the most common forms of the periodic table is given in Figure 13–4. The horizontal rows are called *periods*. The vertical columns containing the families of elements are called *groups*. One striking feature is that the periods are not of the same lengths. The overriding concern is that groups consist of elements that really do have similar chemical properties. The only way this can happen is to make periods 4 and 5 longer than 2 and 3. Between elements 20 and 31 there are ten metallic elements in a zone of transition in the chart—*transition elements*. Period 6 works out only if you set 15 elements, those of numbers 57 through 71, outside the chart. Called the lanthanide series or the "rare earths," they are literally in a class by themselves. Similarly, elements 89 through 103, the actinide series or the transuranium elements, also form a class by themselves. Neither the lanthanide series nor the actinide series will fit into the regular periods in such a way that elements in vertical columns make up a chemical family.

The element hydrogen, the first one, does not fit too well anywhere. It does not belong to any family, which is why it is set aside near the top of the chart.

The chart is constructed without worrying about which elements are metals, nonmetals, or metalloids. Another interesting feature of this periodic arrangement is that the elements in these classes are grouped together, as seen in Figure 13–5. The metals are all on the left side and the nonmetals on the right, with the metalloids forming an actual borderline. Moreover, if you study the groups on the right you will see that elements become more and more metallic as you go down the chart. In Group IV we move down from carbon—a nonmetal—to tin and lead—definitely metals.

The periodic table, by itself, does not explain anything. The periodic law likewise explains nothing. However, the two are far more than just facts about the world. They are facts crying out for explanations. What is true deep down within the atoms of elements that results in a periodic law and a periodic table being possible? Modern knowledge of atomic structure and electronic configuration answers this question. It is not just a quirk of nature that properties of elements are a periodic function of atomic number, and we shall see how in the remainder of this chapter.

13.8
ATOMIC ORBITALS

In Section 9.13 we learned something about the nature of the hydrogen atom, the simplest of all atoms. This quantum mechanical picture came out of an intensive search for an equation or a set of equations that could be used to express the energy of an atom in each of its states. For each allowed energy state there was one equation or one set of equations. The equations included something quite unique in science—the several quantum numbers discussed in Section 9.14.

Each possible state was expressed by an equation with its unique set of four quantum numbers. If an electron were to be in a particular state, the equation for that state could be used to

13 THE PERIODIC SYSTEM AND ELECTRONIC CONFIGURATIONS

Periodic Chart of

Group → / Period ↓	I	II								
1										1 H 1.00797 (1)
2	2,1 · 3 Li 6.939	2,2 · 4 Be 9.0122								
3	2,8,1 · 11 Na 22.9898	2,8,2 · 12 Mg 24.312	TRANSITION ELEMENTS							
4	2,8,8,1 · 19 K 39.102	2,8,8,2 · 20 Ca 40.08	2,8,9,2 · 21 Sc 44.956	2,8,10,2 · 22 Ti 47.90	2,8,11,2 · 23 V 50.942	2,8,13,1 · 24 Cr 51.996	2,8,13,2 · 25 Mn 54.9380	2,8,14,2 · 26 Fe 55.847	2,8,15,2 · 27 Co 58.9332
5	2,8,18,8,1 · 37 Rb 85.47	2,8,18,8,2 · 38 Sr 87.62	2,8,18,9,2 · 39 Y 88.905	2,8,18,10,2 · 40 Zr 91.22	2,8,18,12,1 · 41 Nb 92.906	2,8,18,13,1 · 42 Mo 95.94	2,8,18,13,2 · 43 Tc (97)	2,8,18,15,1 · 44 Ru 101.07	2,8,18,16,1 · 45 Rh 102.905
6	2,8,18,18,8,1 · 55 Cs 132.905	2,8,18,18,8,2 · 56 Ba 137.34	57–71 *	2,8,18,32,10,2 · 72 Hf 178.49	2,8,18,32,11,2 · 73 Ta 180.948	2,8,18,32,12,2 · 74 W 183.85	2,8,18,32,13,2 · 75 Re 186.2	2,8,18,32,14,2 · 76 Os 190.2	2,8,18,32,15,2 · 77 Ir 192.2
7	2,8,18,32,18,8,1 · 87 Fr (223)	2,8,18,32,18,8,2 · 88 Ra (226)	89–103 **							

* Lanthanide Series

| 2,8,18,18,9,2 · 57 La 138.91 | 2,8,18,20,8,2 · 58 Ce 140.12 | 2,8,18,21,8,2 · 59 Pr 140.907 | 2,8,18,22,8,2 · 60 Nd 144.24 | 2,8,18,23,8,2 · 61 Pm (147) | 2,8,18,24,8,2 · 62 Sm 150.35 |

** Actinide Series

| 2,8,18,32,18,9,2 · 89 Ac (227) | 2,8,18,32,18,10,2 · 90 Th 232.038 | 2,8,18,32,20,9,2 · 91 Pa (231) | 2,8,18,32,21,9,2 · 92 U 238.03 | 2,8,18,32,22,9,2 · 93 Np (237) | 2,8,18,32,23,9,2 · 94 Pu (242) |

the Elements

		III	IV	V	VI	VII	O
						2	2 He 4.0026
		2,3 5 B 10.811	2,4 6 C 12.01115	2,5 7 N 14.0067	2,6 8 O 15.9994	2,7 9 F 18.9984	2,8 10 Ne 20.183
		2,8,3 13 Al 26.9815	2,8,4 14 Si 28.086	2,8,5 15 P 30.9738	2,8,6 16 S 32.064	2,8,7 17 Cl 35.453	2,8,8 18 Ar 39.948
2,8,16,2 28 Ni 58.71	2,8,18,1 29 Cu 63.546	2,8,18,2 30 Zn 65.37	2,8,18,3 31 Ga 69.72	2,8,18,4 32 Ge 72.59	2,8,18,5 33 As 74.9216	2,8,18,6 34 Se 78.96	2,8,18,7 35 Br 79.904	2,8,18,8 36 Kr 83.80
2,8,18,18 46 Pd 106.4	2,8,18,18,1 47 Ag 107.868	2,8,18,18,2 48 Cd 112.40	2,8,18,18,3 49 In 114.82	2,8,18,18,4 50 Sn 118.69	2,8,18,18,5 51 Sb 121.75	2,8,18,18,6 52 Te 127.60	2,8,18,18,7 53 I 126.9044	2,8,18,18,8 54 Xe 131.30
2,8,18,32,18,1 78 Pt 195.09	2,8,18,32,18,1 79 Au 196.967	2,8,18,32,18,2 80 Hg 200.59	2,8,18,32,18,3 81 Tl 204.37	2,8,18,32,18,4 82 Pb 207.19	2,8,18,32,18,5 83 Bi 208.980	2,8,18,32,18,6 84 Po (210)	2,8,18,32,18,7 85 At (210)	2,8,18,32,18,8 86 Rn (222)

2,8,18,25,9,2 63 Eu 151.96	2,8,18,27,8,2 64 Gd 157.25	2,8,18,28,8,2 65 Tb 158.924	2,8,18,29,8,2 66 Dy 162.50	2,8,18,30,8,2 67 Ho 164.930	2,8,18,31,8,2 68 Er 167.26	2,8,18,31,8,2 69 Tm 168.934	2,8,18,32,8,2 70 Yb 173.04	2,8,18,32,9,2 71 Lu 174.97
2,8,18,32,25,9,2 95 Am (243)	2,8,18,32,26,9,2 96 Cm (247)	2,8,18,32,27,9,2 97 Bk (247)	2,8,18,32,28,9,2 98 Cf (249)	2,8,18,32,29,9,2 99 Es (254)	2,8,18,32,29,9,2 100 Fm (253)	2,8,18,32,30,9,2 101 Md (256)	2,8,18,32,31,9,2 102 No (254?)	2,8,18,32,32,9,2 103 Lw† (257)

Atomic weights are based on carbon-12; values in parentheses are for the most stable or the most familiar isotope.
† Symbol is unofficial.

FIGURE 13–4

The periodic system of the elements.

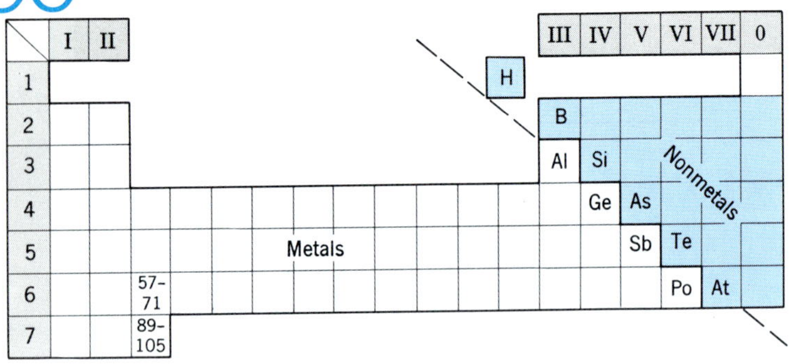

13 THE PERIODIC SYSTEM AND ELECTRONIC CONFIGURATIONS

FIGURE 13-5
Locations of the metals and nonmetals in the periodic system of the elements.

calculate the probability of finding that electron at any point in space (with reference to the nucleus). At some points the probability is zero; the electron is never there. At other points the probability is high. We can imagine making calculations of these probabilities at innumerable points in space for a given energy state. We can then begin to visualize something like a cloud that is thin in some places and thick in others and tapers off essentially to nothing as we go far from the nucleus. Where probabilities are high of finding an electron in a particular energy state, the cloud is thickest. We might even throw an "envelope" around the part of the cloud that will enclose a region having at least 90 percent probability of having the electron all of the time. That, roughly, is how we get the pictures of *atomic orbitals* introduced in Chapter 9 (see Figure 9-8).

An atomic orbital is a uniquely shaped volume of the space near an atomic nucleus in which an electron having a particular allowed energy might be. Several aspects of the idea of an atomic orbital must be emphasized. First, the orbital is a particular portion of space; it does *not* have to have an electron in it. An orbital is a certain region in all the space enclosing an atomic nucleus. We think of an orbital as having a definite shape. Each shape arises from wrapping an imaginary envelope about just that much of the space whose chances are arbitrarily high (e.g., at least 80–90 percent) of having an electron of a particular set of quantum numbers—if an electron is to have such a set at all and be associated with that nucleus. These shapes have been calculated with great precision for only one very simple system: one proton and one electron, the hydrogen atom. *Exact* calculations for a system of one proton and two electrons have not been possible. Approximate solutions, as contrasted with exact solutions, have been worked out. In order to apply what has been found to be exact for the hydrogen atom to more complicated systems, we must *assume* that the orbitals available for electrons in other systems are similar to the hydrogen orbitals. Over the years great confidence has been built in this assumption because it works so well. Thus, we use hydrogen-like atomic orbitals for more complicated atoms.

Since *electronic configurations* of atoms more complex than hydrogen are based on information we have about hydrogen itself, let us review the nature of hydrogen's atomic orbitals. At the first *energy level*, number 1 or *K*-shell, there is only one orbital, and its "probability envelope" encloses a sphere whose center is the nucleus (refer to Figure 9-8). The envelope is spherically symmetrical. Whatever the probability is of finding an electron at a specific distance *and direction* from the nucleus, it will be the same as the probability of finding an electron at another spot at the same distance but in any other direction. Any orbital of this symmetry is called an *s* orbital.[1] At level one, it is specifically the 1*s* orbital.

At the next energy level, number 2 or the *L*-shell, there are four orbitals. An electron can have the next highest allowed energy in any one of four ways. Four equations exist from which the same value of energy is calculated but from which four somewhat differently shaped and positioned orbitals can be drawn. One of them does have the spherical symmetry we described earlier for the lone orbital at the first level. Therefore, it is an *s*-orbital, but at energy level 2 it is the 2*s* orbital (refer to Figure 9-8). The other three orbitals at level 2 have the shapes shown somewhat simplified in Figure 13-6. These hourglass shapes are identical, but their main axes are mutually perpendicular. An orbital with this hourglass shape having symmetry around only one axis is called a *p*-orbital; to give the three 2*p* orbitals separate names, we call them $2p_x$, $2p_y$, and $2p_z$ orbitals (*x*, *y*, and *z* corresponding, of course, to the three mutually perpendicular axes).

At the third energy level, there are nine orbitals, nine regions where an electron can have the energy of this level, and nine equations all leading to the calculation of the same energy. Each

---

[1] We shall encounter the letters *s*, *p*, *d*, and *f* as names for certain kinds of orbitals. These symbols stand for the words "sharp," "principal," "diffuse," and "fine"—words that have particular meaning to spectroscopists. What that meaning is will not be explained, for it requires greater knowledge of the subject than we need for our purposes.

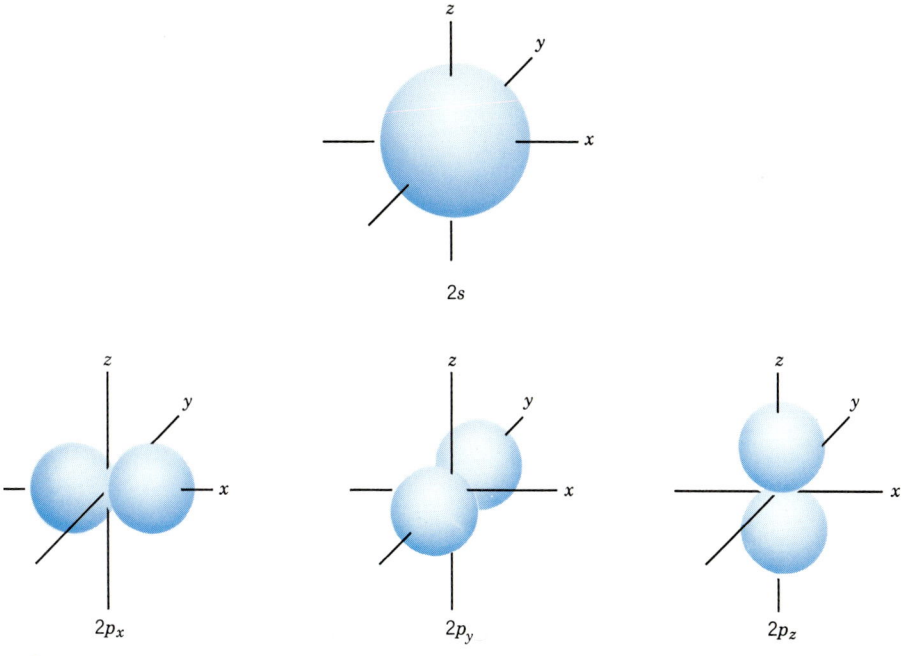

13.9 SPLITTING OF PRINCIPAL ENERGY LEVELS IN ATOMS OTHER THAN HYDROGEN

FIGURE 13-6
Orbitals at principal energy level 2 (the *L*-shell). For the hydrogen atom the electron may be at the second level in any one of four ways, in the 2s orbital or in one of the mutually perpendicular 2p orbitals.

FIGURE 13-7
Shapes of the 3d orbitals and their orientations in relation to each other and the same set of axis.

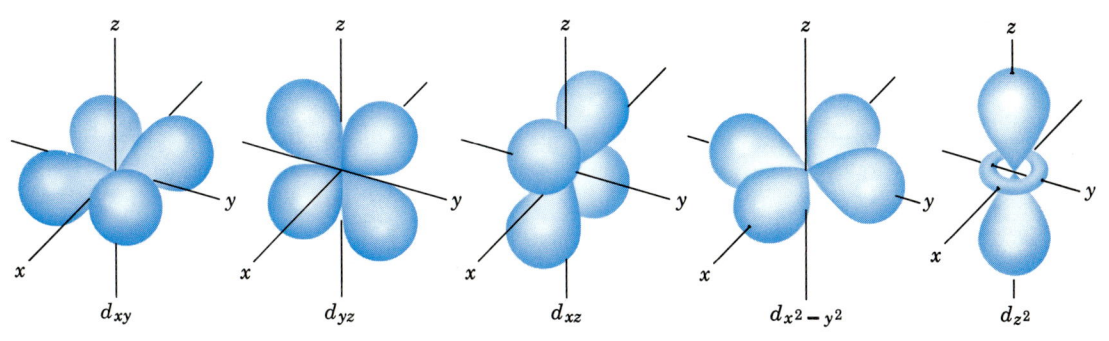

equation, however, leads to the calculation of a particular orbital shape. One orbital, the 3s, is spherically symmetrical; three have the symmetry and projections into space of the *p* orbitals—$3p_x$, $3p_y$, and $3p_z$ (they have the *symmetry*, but the actual shapes are somewhat different). The remaining five, designated as *d* orbitals, have the shapes and orientations in space shown in Figure 13-7.

At the fourth energy level there are sixteen orbitals—4s, $4p_x$, $4p_y$, and $4p_z$—plus five orbitals of the *d* type, plus seven orbitals designated as *f* orbitals. The orbitals of the *d* and *f* types will scarcely be encountered in our future study. These orbitals are important to the chemistry of a large number of metallic elements.

13.9 SPLITTING OF PRINCIPAL ENERGY LEVELS IN ATOMS OTHER THAN HYDROGEN

Calculations for the hydrogen atom result in the orbitals described in Section 13.8. We shall now introduce a necessary modification of this theory so that we can use these orbitals with atoms having nuclei of higher charge and more electrons. For hydrogen and hydrogen alone, the four orbitals at the second energy level are of *equal* energy (and then only in the absence of a magnetic field). For all other atoms the 2s orbital is of slightly less energy than any one of the 2p orbitals. It is as if a small splitting of the second

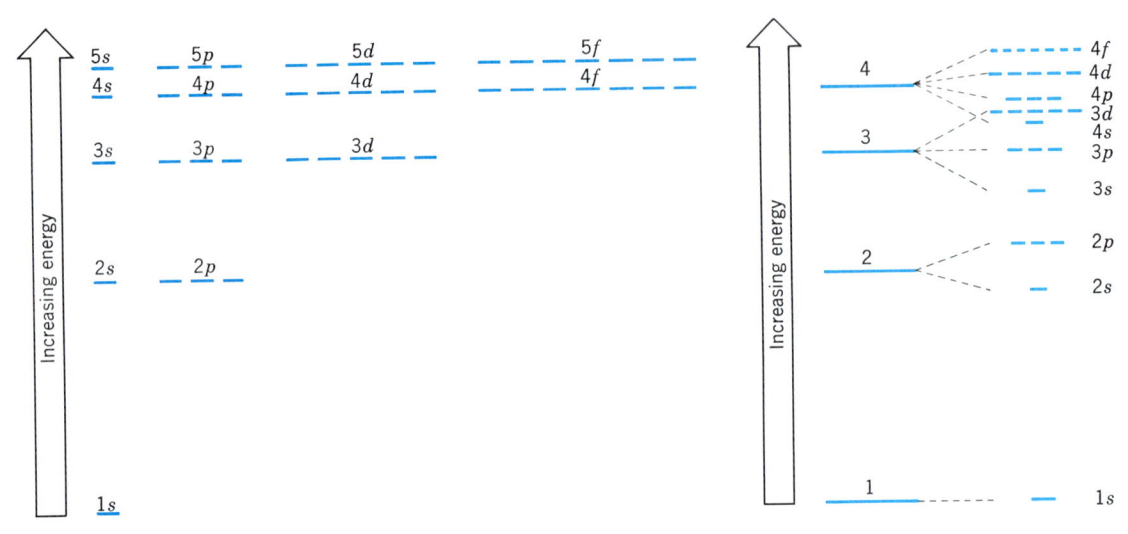

13 THE PERIODIC SYSTEM AND ELECTRONIC CONFIGURATIONS

FIGURE 13-8

Energy levels. (a) The relative energies of the atomic orbitals of the hydrogen atom. Note that the sublevels are not "split," and the levels are therefore said to be degenerate. (b) Relative energies of atomic orbitals when splitting of the principal energy levels occurs. The shapes of the orbitals (not shown here) remain basically as they are for hydrogen, but their relative energies are changed. (The splitting of the fifth level is not shown. No inferences should be made about the relative separations of the sublevels, for the figure implies no quantitative information.)

energy level occurs, and in atoms more complicated than hydrogen there are two sublevels, the s and the p. Sublevel s consists of just one orbital, the $2s$, and sublevel p has three orbitals, $2p_x$, $2p_y$, and $2p_z$. At the third main energy level, there are three sublevels, one orbital at the s, three at the p sublevel, and five at the d sublevel. The fourth energy level is split into four sublevels. Figure 13-8 illustrates energy relations among the sublevels. (For hydrogen, this splitting occurs only if the hydrogen atoms are subjected to an external magnetic field.) The reader may note that we have not explained how these splittings occur but have simply stated that they do. Generally speaking, energy levels split because of repulsions between electrons when there are many in a system.

To summarize our model for studying how electrons are distributed about a nucleus in a particular atom, we have a nucleus containing all (or virtually all) the mass of an atom and all the positive charge. Surrounding this core we have uniquely shaped volumes of space or orbitals in which electrons are allowed to be. Only certain allowed values of energy are permitted for the electrons. We have not given actual values to these energies (e.g., stating them in joules or calories), but we have given them index numbers, the principal quantum numbers, and letters s, p, d, and f. (If and when we need actual values of energy, we shall introduce them.) Our next step is to work from the simplest kind of atom, hydrogen, to the progressively more complex and examine just how electrons are distributed in them. To do this systematically, we conceive of a step-by-step *aufbau* (German, "building up").

13.10 AUFBAU RULES

The most stable form of the hydrogen atom is one in which the electron is in the lowest allowed energy level, the $1s$ state. If it were in the $2s$ state, it could and eventually would jump to the $1s$ state and deliver a quantum of electromagnetic energy to the surroundings. Thus, the first *aufbau* rule for working out the distributions of electrons about a nucleus is that electrons are "fed" one at a time, first into the allowed orbitals of lowest

13.11 ELECTRONIC CONFIGURATIONS

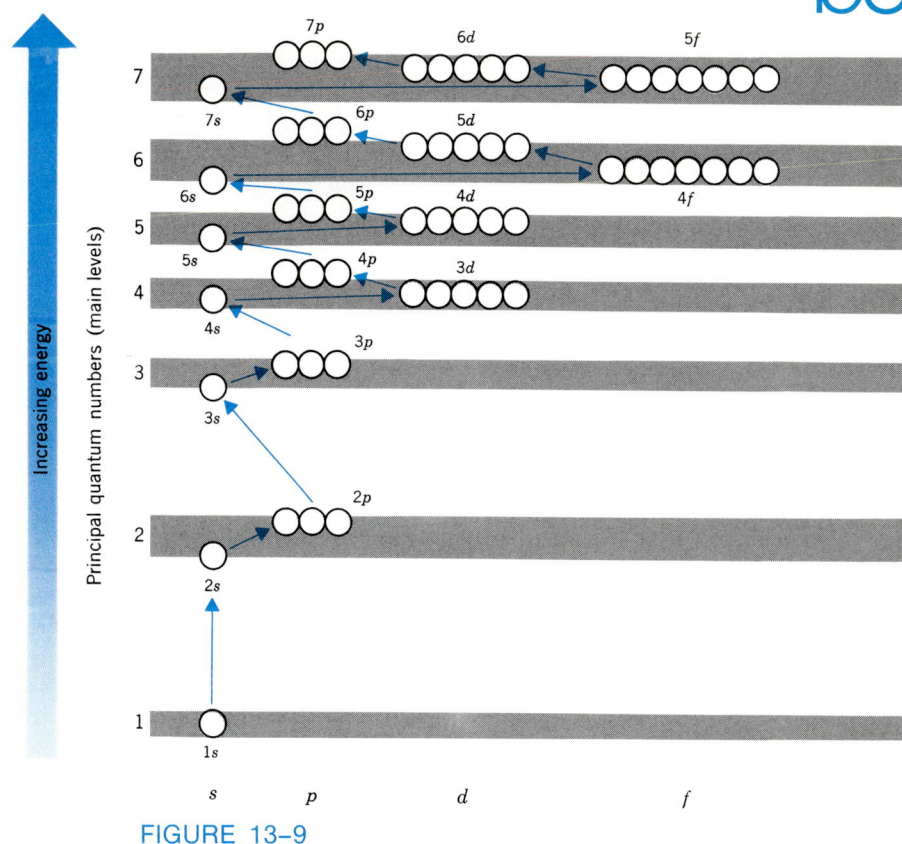

FIGURE 13-9
Beginning with the 1s-orbital, the general order of filling successive orbitals is shown by the paths of the arrows.

energy and then successively up through higher levels as required. The order of filling, specified in Figure 13-9, follows the order of relative energies of the sublevels.

The second aufbau rule is the Pauli exclusion principle, which we studied in Section 9.16. An orbital can hold at most only two electrons and these only if they are spinning in opposite directions. The Pauli principle puts the limits on the number of electrons that can be a part of a given principal energy level or any sublevel. These limits are summarized in Table 13.3.

13.11 ELECTRONIC CONFIGURATIONS

When the electron of a hydrogen atom is in a 1s orbital, we say that the atom is in its *ground state*—its most stable state. If the atom absorbs the right amount of energy, the electron may be moved to another orbital of higher energy. Now the atom is in one of its many possible *excited states*. When we describe the electronic configuration of an atom, we mean the distribution of the atom's electrons in the ground state. One fundamentally unique feature of each element is the *electronic configuration* of its atoms. We shall look at them all. They are the key to understanding not only the periodic law but also the chemical properties of the elements.

We shall use boxes to represent orbitals. The higher up the box, the higher the energy an electron has when it is in that orbital. To indicate the two ways electrons can spin, clockwise or counterclockwise, we shall use arrows pointing in opposite directions. By this scheme the electronic configuration of the hydrogen atom, atomic number 1, is simply

In hydrogen atoms the 2s and 2p "sublevels" correspond to the same energy.

H, condensed symbol: 1s

The condensed symbol for hydrogen is the customary way of writing its electronic configuration. The pattern will become clearer as additional examples are developed. (Some authors express 1s as $1s^1$, a practice we shall not follow.)

The element of atomic number 2 is helium. Having a nuclear charge of 2+, its atoms must each have two electrons. According to the aufbau rules, its electronic configuration is

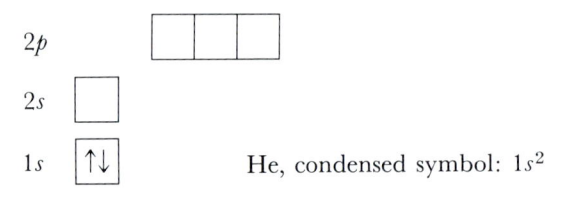

2p [ ][ ][ ]

2s [ ]

1s [↑↓]   He, condensed symbol: $1s^2$

The condensed symbol says that there are two electrons (the superscript) in the 1s state.

Because the 1s state is now filled, the next electron will have to go into the next highest orbital, the 2s. The electronic configuration of lithium, atomic number 3, is represented as

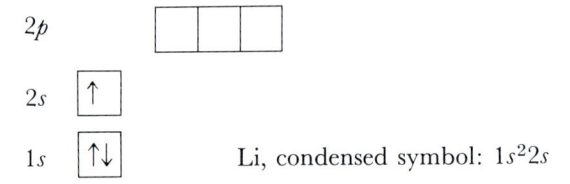

2p [ ][ ][ ]

2s [↑]

1s [↑↓]   Li, condensed symbol: $1s^2 2s$

Carrying this development to atoms of higher and higher atomic number, in accordance with the aufbau rules and the order of filling orbitals (Figure 13–9), the following electronic configurations show how the second energy level is filled.

Beryllium (atomic number 4)

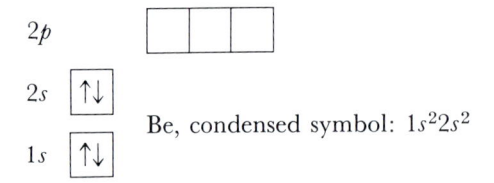

2p [ ][ ][ ]

2s [↑↓]

1s [↑↓]   Be, condensed symbol: $1s^2 2s^2$

Boron (atomic number 5)

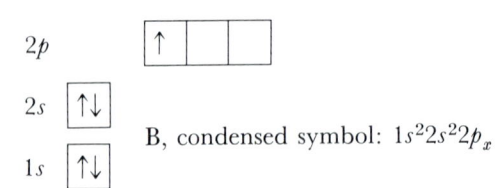

2p [↑][ ][ ]

2s [↑↓]

1s [↑↓]   B, condensed symbol: $1s^2 2s^2 2p_x$

Carbon (atomic number 6)

2p [↑][↑][ ]

2s [↑↓]

1s [↑↓]   C, condensed symbol: $1s^2 2s^2 2p_x 2p_y$

With carbon, another aufbau rule—*Hund's rule*—is applied. It says that if empty orbitals of the *same* energy are available, electrons will be distributed as evenly among them as possible. Carbon is therefore not shown as $1s^2 2s^2 2p_x^2$. The sixth and last electron is not put into the same p-orbital as the fifth but in a different 2p-orbital. In this arrangement, the sixth electron can be farther away from the fifth electron, repulsions between the two will be less, and the system can be more stable. The Pauli exclusion principle *permits* two electrons (and no more) in the same space orbital, but it does not suggest that two paired electrons in such a circumstance have some special, extra stability. They do not. Paired electrons are more stable in the same space orbital only if the second electron of the pair would otherwise be in an orbital of higher energy. Hund's rule states that the second electron will go to a different orbital of the same energy if one is vacant. We can now proceed to atoms of higher atomic number.

Nitrogen (atomic number 7)

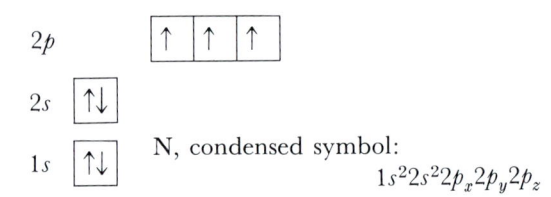

2p [↑][↑][↑]

2s [↑↓]

1s [↑↓]   N, condensed symbol:
                  $1s^2 2s^2 2p_x 2p_y 2p_z$

Oxygen (atomic number 8)

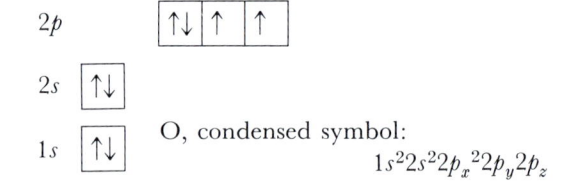

2p [↑↓][↑][↑]

2s [↑↓]

1s [↑↓]   O, condensed symbol:
                  $1s^2 2s^2 2p_x^2 2p_y 2p_z$

Note that the eighth electron is put into a 2p-orbital with a spin opposite that of the electron already there. Although this arrangement means some crowding of like-charged particles, it is more stable than if the eighth electron were placed in the empty but higher-energy 3s-orbital.

Fluorine (atomic number 9)

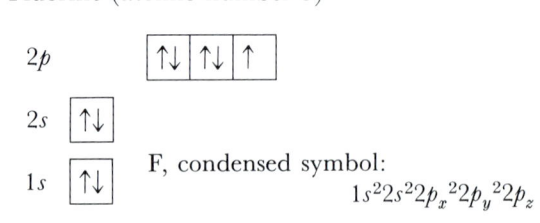

2p [↑↓][↑↓][↑]

2s [↑↓]

1s [↑↓]   F, condensed symbol:
                  $1s^2 2s^2 2p_x^2 2p_y^2 2p_z$

Neon (atomic number 10)

2p [↑↓][↑↓][↑↓]

2s [↑↓]

1s [↑↓]   Ne, condensed symbol:
$1s^2 2s^2 2p_x^2 2p_y^2 2p_z^2$

With this atom the orbitals in the second main level are filled.

For the element of atomic number 11, sodium, we shall have to start using orbitals of the third principal level.

Sodium (atomic number 11)

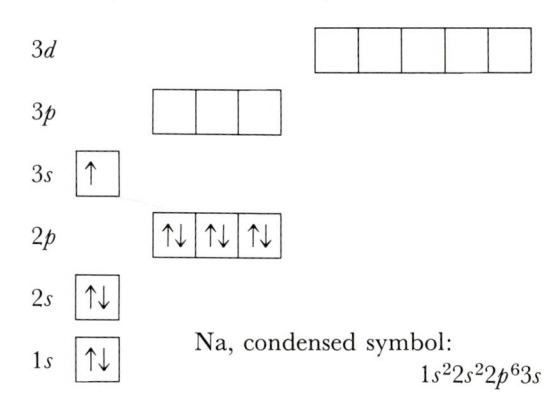

3d [ ][ ][ ][ ][ ]

3p [ ][ ][ ]

3s [↑]

2p [↑↓][↑↓][↑↓]

2s [↑↓]

1s [↑↓]   Na, condensed symbol:
$1s^2 2s^2 2p^6 3s$

For convenience, we switch now to tabular organizations of the data (see Table 13.4). When we come to the element of atomic number 19, we see in Figure 13–8 that the 4s-orbital provides a lower-energy location for an electron than the 3d-orbitals. Hence electron 19, and electron 20 when we get to the element of atomic number 20, will go into the 4s state. Only in elements of atomic number 21 and above are the 3d states filled. Table 13.4 summarizes the electronic configurations of the elements. It may be noted that a few exceptions to the aufbau rules occur; sometimes the orbitals do not fill up in the regular way predicted by these rules. These exceptions will be discussed later.

13.12 TRANSITION ELEMENTS

We saw earlier that the horizontal rows, called periods, in the periodic table are not of equal length.

The first period is very short, and hydrogen is not placed with any particular family. The second and third periods have eight elements each, but the rest of the periods are longer. The shaded portion of the chart in Figure 13–10 contains elements classified as the transition elements. Why it is reasonable for the periods to lengthen may be seen by referring to the electronic configurations of some of these elements. Those for elements 20 to 31 are given in Table 13.5. (Elements 20 and 31 are included for reference; they are not properly transition elements themselves.) Because the 4s state is of slightly lower energy than the 3d state, the configuration of calcium shows an empty 3d sublevel and two electrons in its 4s sublevel. The 4s sublevel is now full. Because the 4p level is of higher energy than the 3d, electrons next go into the 3d sublevel, one by one, from scandium (21) to zinc (30). Apparently having a d sublevel exactly half filled, d [↑][↑][↑][↑][↑], represents a special condition of stability, for chromium (24) has at its outermost sublevels a distribution of $3d^5 4s$ instead of $3d^4 4s^2$—a configuration that would make the series from 21 to 30 look more uniform. Similarly, in copper (29), an entirely filled level-3 instead of a filled 4s sublevel is the more stable arrangement. Copper has, at its outer fringes, a distribution of $3d^{10} 4s$ instead of $3d^9 4s^2$. In summary, for the transition metals in the fourth period of the periodic table, the 3d sublevel fills while the 4s level has one or two electrons.

Electronic configurations for the transition elements of the fifth period are given in Table 13.6. Again, we have an inner sublevel filling, the 4d, while the outer sublevel, the 5s, has one, two, or no electrons. The 4f sublevel remains empty until the 5p and 6s sublevels fill up (elements 49 through 56). Then the 4f level starts filling again, and the elements with these configurations comprise the *lanthanide series,* or "rare earths," named after the first member lanthanum (57). We note here that the order of filling of sublevels corresponds quite closely with the order of increasing energy summarized in Figure 13–8. We mentioned earlier that the lanthanide series is usually placed outside the main body of the periodic table because it will not conveniently fit into the pattern that vertical columns consist of families. In the other similar series, the *actinide series,* or the "transuranium elements" 89 to 103, the following sublevels near the outer regions are filled: the 5s, 5p, and 5d, the 6s and 6p, and the 7s (see Table 13.4). At element number 88 (radium), the 5f and the 6d and 6f are empty. In the actinide series the 5f sublevel fills while the 6d sublevel holds one, two, or no electrons.

In general, the transition elements all have one, two, or no electrons in the s orbitals while the d or f orbitals at lower numbered levels are filling.

13 THE PERIODIC SYSTEM AND ELECTRONIC CONFIGURATIONS

TABLE 13.4
ELECTRONIC CONFIGURATIONS OF THE ELEMENTS

Element	1s	2s	2p	3s	3p	3d	4s	4p	4d	4f	5s	5p	5d	5f
1 H	1													
2 He	2													
3 Li	2	1												
4 Be	2	2												
5 B	2	2	1											
6 C	2	2	2											
7 N	2	2	3											
8 O	2	2	4											
9 F	2	2	5											
10 Ne	2	2	6											
11 Na	2	2	6	1										
12 Mg	2	2	6	2										
13 Al	2	2	6	2	1									
14 Si	2	2	6	2	2									
15 P	2	2	6	2	3									
16 S	2	2	6	2	4									
17 Cl	2	2	6	2	5									
18 Ar	2	2	6	2	6									
19 K	2	2	6	2	6		1							
20 Ca	2	2	6	2	6		2							
21 Sc	2	2	6	2	6	1	2							
22 Ti	2	2	6	2	6	2	2							
23 V	2	2	6	2	6	3	2							
24 Cr	2	2	6	2	6	5	1							
25 Mn	2	2	6	2	6	5	2							
26 Fe	2	2	6	2	6	6	2							
27 Co	2	2	6	2	6	7	2							
28 Ni	2	2	6	2	6	8	2							
29 Cu	2	2	6	2	6	10	1							
30 Zn	2	2	6	2	6	10	2							
31 Ga	2	2	6	2	6	10	2	1						
32 Ge	2	2	6	2	6	10	2	2						
33 As	2	2	6	2	6	10	2	3						
34 Se	2	2	6	2	6	10	2	4						
35 Br	2	2	6	2	6	10	2	5						
36 Kr	2	2	6	2	6	10	2	6						
37 Rb	2	2	6	2	6	10	2	6			1			
38 Sr	2	2	6	2	6	10	2	6			2			
39 Y	2	2	6	2	6	10	2	6	1		2			
40 Zr	2	2	6	2	6	10	2	6	2		2			
41 Nb	2	2	6	2	6	10	2	6	4		1			
42 Mo	2	2	6	2	6	10	2	6	5		1			
43 Tc	2	2	6	2	6	10	2	6	6		1			
44 Ru	2	2	6	2	6	10	2	6	7		1			
45 Rh	2	2	6	2	6	10	2	6	8		1			
46 Pd	2	2	6	2	6	10	2	6	10		1			
47 Ag	2	2	6	2	6	10	2	6	10		1			
48 Cd	2	2	6	2	6	10	2	6	10		2			
49 In	2	2	6	2	6	10	2	6	10		2	1		
50 Sn	2	2	6	2	6	10	2	6	10		2	2		
51 Sb	2	2	6	2	6	10	2	6	10		2	3		
52 Te	2	2	6	2	6	10	2	6	10		2	4		
53 I	2	2	6	2	6	10	2	6	10		2	5		
54 Xe	2	2	6	2	6	10	2	6	10		2	6		

TABLE 13.4
ELECTRONIC CONFIGURATIONS OF THE ELEMENTS (continued)

Element	K	L	M	4s	4p	4d	4f	5s	5p	5d	5f	6s	6p	6d	7s
55 Cs	2	8	18	2	6	10		2	6			1			
56 Ba	2	8	18	2	6	10		2	6			2			
57 La	2	8	18	2	6	10		2	6	1		2			
58 Ce	2	8	18	2	6	10	2	2	6			2			
59 Pr	2	8	18	2	6	10	3	2	6			2			
60 Nd	2	8	18	2	6	10	4	2	6			2			
61 Pm	2	8	18	2	6	10	5	2	6			2			
62 Sm	2	8	18	2	6	10	6	2	6			2			
63 Eu	2	8	18	2	6	10	7	2	6			2			
64 Gd	2	8	18	2	6	10	7	2	6	1		2			
65 Tb	2	8	18	2	6	10	9	2	6			2			
66 Dy	2	8	18	2	6	10	10	2	6			2			
67 Ho	2	8	18	2	6	10	11	2	6			2			
68 Er	2	8	18	2	6	10	12	2	6			2			
69 Tm	2	8	18	2	6	10	13	2	6			2			
70 Yb	2	8	18	2	6	10	14	2	6			2			
71 Lu	2	8	18	2	6	10	14	2	6	1		2			
72 Hf	2	8	18	2	6	10	14	2	6	2		2			
73 Ta	2	8	18	2	6	10	14	2	6	3		2			
74 W	2	8	18	2	6	10	14	2	6	4		2			
75 Re	2	8	18	2	6	10	14	2	6	5		2			
76 Os	2	8	18	2	6	10	14	2	6	6		2			
77 Ir	2	8	18	2	6	10	14	2	6	7		2			
78 Pt	2	8	18	2	6	10	14	2	6	9		1			
79 Au	2	8	18	2	6	10	14	2	6	10		1			
80 Hg	2	8	18	2	6	10	14	2	6	10		2			
81 Tl	2	8	18	2	6	10	14	2	6	10		2	1		
82 Pb	2	8	18	2	6	10	14	2	6	10		2	2		
83 Bi	2	8	18	2	6	10	14	2	6	10		2	3		
84 Po	2	8	18	2	6	10	14	2	6	10		2	4		
85 At	2	8	18	2	6	10	14	2	6	10		2	5		
86 Rn	2	8	18	2	6	10	14	2	6	10		2	6		
87 Fr	2	8	18	2	6	10	14	2	6	10		2	6		1
88 Ra	2	8	18	2	6	10	14	2	6	10		2	6		2
89 Ac	2	8	18	2	6	10	14	2	6	10		2	6	1	2
90 Th	2	8	18	2	6	10	14	2	6	10		2	6	2	2
91 Pa	2	8	18	2	6	10	14	2	6	10	2	2	6	1	2
92 U	2	8	18	2	6	10	14	2	6	10	3	2	6	1	2
93 Np	2	8	18	2	6	10	14	2	6	10	5	2	6		2
94 Pu	2	8	18	2	6	10	14	2	6	10	6	2	6		2
95 Am	2	8	18	2	6	10	14	2	6	10	7	2	6		2
96 Cm	2	8	18	2	6	10	14	2	6	10	7	2	6	1	2
97 Bk	2	8	18	2	6	10	14	2	6	10	8	2	6	1	2
98 Cf	2	8	18	2	6	10	14	2	6	10	10	2	6		2
99 Es	2	8	18	2	6	10	14	2	6	10	11	2	6		2
100 Fm	2	8	18	2	6	10	14	2	6	10	12	2	6		2
101 Md	2	8	18	2	6	10	14	2	6	10	13	2	6		2
102 No	2	8	18	2	6	10	14	2	6	10	14	2	6		2
103 Lw	2	8	18	2	6	10	14	2	6	10	14	2	6	1	2

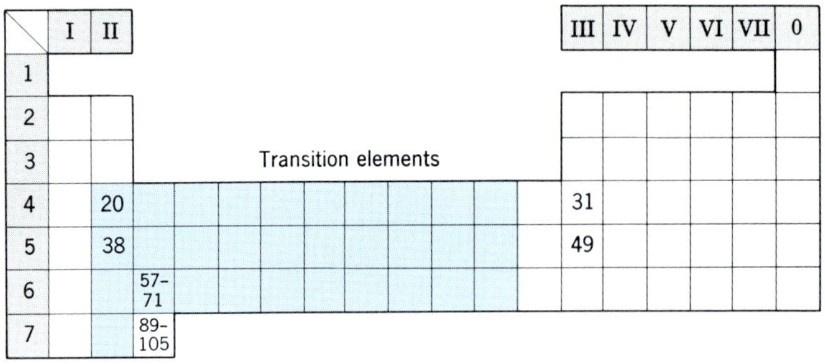

FIGURE 13–10
Location of the transition elements in the periodic system.

TABLE 13.5
TRANSITION ELEMENTS, FOURTH PERIOD

Atomic Number	Element: Name	Element: Symbol	Electronic Configuration: 1 s	2 s	2 p	3 s	3 p	3 d	4 s	4 p	4 d	4 f
20	Calcium	Ca	2	2	6	2	6		2			
21	Scandium	Sc	2	2	6	2	6	1	2			
22	Titanium	Ti	2	2	6	2	6	2	2			
23	Vanadium	V	2	2	6	2	6	3	2			
24	Chromium	Cr	2	2	6	2	6	5	1			
25	Manganese	Mn	2	2	6	2	6	5	2			
26	Iron	Fe	2	2	6	2	6	6	2			
27	Cobalt	Co	2	2	6	2	6	7	2			
28	Nickel	Ni	2	2	6	2	6	8	2			
29	Copper	Cu	2	2	6	2	6	10	1			
30	Zinc	Zn	2	2	6	2	6	10	2			
31	Gallium	Ga	2	2	6	2	6	10	2	1		

TABLE 13.6
TRANSITION ELEMENTS, FIFTH PERIOD

Atomic Number	Element: Name	Element: Symbol	Electronic Configuration: 1 s	2 s	2 p	3 s	3 p	3 d	4 s	4 p	4 d	4 f	5 s	5 p	5 d	5 f
38	Strontium	Sr	2	2	6	2	6	10	2	6			2			
39	Yttrium	Y	Krypton configuration for all inner sublevels of these elements								1		2			
40	Zirconium	Zr									2		2			
41	Niobium	Nb									4		1			
42	Molybdenum	Mo									5		1			
43	Technitium	Tc									6		1			
44	Ruthenium	Ru									7		1			
45	Rhodium	Rh									8		1			
46	Palladium	Pd									10					
47	Silver	Ag									10		1			
48	Cadmium	Cd									10		2			
49	Indium	In									10		2	1		

13.13 FAMILIES OF ELEMENTS—OUTSIDE SHELL CONFIGURATIONS

We now ask why the elements that group into families in the periodic table have similar chemical properties. In answer, let us look at the electronic configurations of the atoms of four important families given in Table 13.7. Parts of the table are shaded to emphasize the most important single feature of the groups. *Within* each family the numbers of electrons in the outside shells (shaded) are identical. (The exception in the table is in group 0, the noble gases, where the outside shell of the first member, helium, has only 2. It could not have more because its outside shell is the K-shell.) *Between* each family the numbers of electrons in the outside shells are different. In the next chapter we shall build on these simple facts about our world to explain much about the driving forces for chemical changes and to explain how various chemical compounds must have definite formulas.

On grounds of physical properties, we grouped elements into the classes of metals, nonmetals, and metalloids. Atoms of metallic elements have 1, 2, or 3 electrons in their highest numbered, occupied shell (we refer to the principal quantum number). In most metals there are 2 or 1 electrons in that level, as the data of Tables 13.5 and 13.6 show. See also the metals in groups I and II, Table 13.7. (A few metals near the bottom of the chart and close to the borderline with the nonmetals have 4 electrons in their highest numbered, occupied shell.)

Atoms of the nonmetals have from 4 to 8 electrons in their highest numbered, occupied shells. Groups 7 and 0 in Table 13.7 illustrate this.

13.14 ELECTRONIC CONFIGURATIONS AND PERIODICITY

The periodic law is a fact about our world. It does not explain itself, but it is a wondrous part of the human condition to ask "why?" or "how?" "Why are properties a periodic function (instead of some other function) of atomic number?"

Correlations are not themselves explanations, but they are first steps toward them. What we now may correlate with the periodic law is the periodic changes in electronic configurations as we move from low to high atomic numbers. Each period represents the gradual filling of a particular set of orbitals until we reach an atom of a noble gas with eight outside-shell electrons.

TABLE 13.7
ELECTRONIC CONFIGURATIONS OF ATOMS IN SEVERAL COMMON FAMILIES

Family	Element	Atomic Number	K	L	M	N	O	P	Q
Group 0—noble gases	Helium	2	2						
	Neon	10	2	8					
	Argon	18	2	8	8				
	Krypton	36	2	8	18	8			
	Xenon	54	2	8	18	18	8		
	Radon	86	2	8	18	32	18	8	
Group I—alkali metals	Lithium	3	2	1					
	Sodium	11	2	8	1				
	Potassium	19	2	8	8	1			
	Rubidium	37	2	8	18	8	1		
	Cesium	55	2	8	18	18	8	1	
	Francium	87	2	8	18	32	18	8	1
Group II—alkaline earth metals	Beryllium	4	2	2					
	Magnesium	12	2	8	2				
	Calcium	20	2	8	8	2			
	Strontium	38	2	8	18	8	2		
	Barium	56	2	8	18	18	8	2	
	Radium	88	2	8	18	32	18	8	2
Group VII—halogens	Fluorine	9	2	7					
	Chlorine	17	2	8	7				
	Bromine	35	2	8	18	7			
	Iodine	53	2	8	18	18	7		
	Astatine	85	2	8	18	32	18	7	

13 THE PERIODIC SYSTEM AND ELECTRONIC CONFIGURATIONS

(We note again the exception of helium, whose outside shell, the K-shell, can hold only two electrons.) Then we start over again; we begin to fill a new set of orbitals. What all of this must mean, at least so we reason, is that *properties are intimately related to electronic configurations*. It is not so much the atomic number that determines properties; it is the electronic configuration. Of course, the latter indirectly relates to the former. The atomic number is the same as the charge on the nucleus. Because an atom, by definition, is electrically neutral, it must have electrons in a quantity identical to the atomic number. Given the available orbitals and the aufbau rules, once you pick a certain number of electrons you automatically will have a particular electronic configuration (of the ground state). Anything periodic in electronic configuration means many things periodic in properties.

We have not explained *why* nature is this way—but simply *how* nature is made.

SUMMARY

Periodic functions. Many chemical and physical properties of the elements are periodic functions of their atomic numbers (the periodic law). This law made possible the prediction of elements that had yet to be discovered.

Periodic table. The elements in a chemical family are in vertical columns called groups in the periodic table. The periods are horizontal rows. There are eight main groups, three series of transition elements and two families outside of the pattern of vertical grouping—the lanthanides and actinides. The number of a main group corresponds to the number of electrons in the outside shell of the atoms of the elements in that group.

Transition elements. The filling of *d*- or *f*-orbitals at levels lower than the highest numbered occupied orbitals occurs as one moves across the transition elements in the periodic chart. All are metals, and like all metals their atoms have 1, 2, or 3 electrons in their orbitals of highest occupied principal quantum number.

Nonmetals and metals. The nonmetallic elements are clustered in the upper right corner of the periodic chart; the metals are spread over the remainder. Nonmetals have atoms with 4, 5, 6, 7, or 8 outside shell electrons.

Electronic configurations. Aufbau rules enable us to construct the electronic configuration of an atom of an element, given only the atomic number. The first rule is to place electrons in available orbitals of as low energy as possible. The second, the Pauli exclusion principle, states that each orbital may hold only two electrons (and if two, then with opposite spin). The third rule, Hund's rule, tells us that if empty orbitals of the same energy are available, electrons will be distributed as evenly among them as possible. The electronic configuration of the atoms of an element as much as any other factor determines its chemical properties.

TERMS AND CONCEPTS

periodic property
valence
chemical family
periodic law
periodic table
period
group
transition element
Bohr model
electronic configuration
energy level
orbital
aufbau rules
noble gas
ground state
excited state
Hund's rule
Pauli exclusion principle
actinide
lanthanide
rare earth

QUESTIONS AND EXERCISES

1. Describe in general terms how orbitals for atoms other than hydrogen are like those of hydrogen and how they are different.

2. How many protons must the nucleus of an atom having the electronic configuration $1s^2 2s^2 2p^6 3s^2 3p^6 3d^5 4s$ contain?

3. What is its atomic number?

4. Still referring to the configuration in question 2, is this element more likely be a transition element or a member of another group? Explain. (You should be able to handle such a question without looking at a periodic table.)

5. Without referring to tables or charts, you should be able to construct electronic configurations for any element from 1 to 20. (a) Using condensed symbolism (see question 2), write electronic configurations for elements of atomic number 3, 7, 5, 19, 14, 10, 20, 9, 5, 17, and 13. (b) Without referring to tables or charts, predict on the basis of electronic configurations just written for part (a) whether the element in question is more likely to be a metal or a nonmetal. Explain.

6. Write the electronic configuration, and this time include the composition of the nucleus, for a reasonable isotope of an element of atomic number 12 and atomic mass 25.

	7	8	9	10
	15 W	16 X	17 Y	18 Z
	33	34	35	36

7. The accompanying diagram shows a section of a period chart. Hypothetical atomic symbols have been used. You should be able to supply answers to the following without having to refer to an actual period chart (the numbers are atomic numbers).
(a) Write the atomic number of an element in the same family as Y. (b) Write the atomic number of an element in the same period as W. (c) Are the elements shown more likely to be metals or nonmetals? (d) If there are six electrons in the outside shell (highest occupied principal energy level) of X, how many would the outside shells of elements 8, 15, and 36 hold? (You should be able to deduce the answers without constructing electronic configurations.)

8. An element that is a gas at room temperature and pressure will appear where in the periodic table—the upper right or the lower left corner?

9. Why is the lanthanide series set outside the periodic chart?

CHAPTER 14
COMPOUNDS, CHEMICAL BONDS, AND CHEMICAL ENERGY

14.1 THE IDEA OF A CHEMICAL BOND

"How in the world do things stick together?" The question implies neither despair nor jest. It quite literally is a fundamental question about our world—about our entire universe, in fact.

"I can understand how I stick to the earth, but how can the Australians do so at the same time?" We answer that question in terms of the concept of a gravitational force. We do not explain gravity; we describe it. "Gravity" is the name we give to a universal human experience, and naming something is not necessarily the same thing as explaining it. Gravity is part of the "givenness" of the universe; it is there. We do not understand "how come," but we do recognize it and we can measure it. We understand the laws of gravity and gravitational forces well enough to be able to make some predictions about how rapidly falling bodies move and where they will land. We put men at particular places on the moon with that knowledge, and we confidently designate targets for space probes that we send to outer planets.

One way by which things stick together, then, is by a gravitational force. Outside an atom's nucleus, where special nuclear forces hold things together, there are two other important natural forces: magnetic and electric. Because nuclei and electrons are electrically charged, we may expect to find that electrical forces are responsible for holding chemical substances together.

You may not have given it much thought, but substances do stick together. Your skin keeps your insides in. Whatever skin is, it is made from atoms of various elements. Atoms, however, are electrically neutral. How can they stick in the form of skin or (with other kinds of atoms) in the form of a mountain, or a bridge, or anything else?

Whenever any two things stick together, we say they are bonded, and we may call the "something" that makes them stick a *bond*. Our task in this chapter is to explain how neutral atoms can develop powerful forces of attraction called *chemical bonds*. When we have done that we shall have made huge strides toward understanding our universe.

14.2 KINDS OF CHEMICAL BONDS

Atoms consist of atomic nuclei and electrons. Because like charges repel, nuclei repel each other. Electrons repel each other also. Forces of repulsion must therefore exist in atoms, and they work against letting anything stick to anything else. There are forces of attraction in atoms, too. Nuclei are attracted to electrons, and this attraction will be the chief means of explaining the chemical bond. Somehow two or more atoms can reorganize their nuclei and electrons into a new arrangement that lets forces of attraction win out over forces of repulsion in the new particle.

Chemists, who have studied thousands of substances, recognize three principal ways by which atoms reorganize their nuclei and electrons into new, stable units. The bonds arising from these three ways are named the ionic bond, the covalent bond, and the metallic bond. We shall study them in that order after we look again at a feature of the elements—and the electronic configurations of their atoms—that now is especially important.

14.3 THE SIGNIFICANCE OF AN OUTER OCTET

Table 13.7 contained electronic configurations of the atoms of members of several families of elements. Details about sublevels were omitted in order to dramatize how the outside levels of atoms in the same family were identical (insofar as the atom's outside level was at energy level 2 or higher). We now take note of three facts about the world of elements and a guideline that we may make from these facts.

The noble-gas elements have outer *octets;* their outside levels hold 8 electrons (except for helium with an outside level of 2 electrons, which is a filled first-level). That is the first fact. The second fact is the unusual chemical unreactivity of noble-gas elements. Helium, neon, argon, and krypton form no known compounds with any other element. Xenon forms a few compounds with fluorine and oxygen, which are two of the most reactive nonmetallic elements. Radon is expected to form similar compounds. The general unreactivity of these noble-gas elements is in striking contrast to the reactivity of the other elements that form millions of chemical compounds. Even more striking is the third fact: in most changes of the other elements into compounds, their atoms cooperate in ways that lead to outer octets. We shall see how in the next sections. We simply make the point here, based on what chemists have observed, that an outer octet is a condition of unusual chemical stability. Atoms that do not have outer octets will (given the opportunity) take, shed, or share electrons to acquire one (or to acquire an outer, filled first level of two electrons). From these facts we draw the following guide or rule, called the *octet rule*, which holds especially well through elements of atomic number 20. The octet rule states that rearrangements of the electrons relative to nuclei of reacting atoms of elements usually generate particles having outer octets of electrons (or where level 1 is the outer level, then an outer pair). One way that certain atoms acquire outer octets is by a process of electron transfer that leads to the first type of chemical bond mentioned earlier, the ionic bond.

14.4 IONIC BOND

The elements sodium and chlorine are not stable in each other's presence. When they are mixed, the following changes in electronic configurations occur among billions and billions of their atoms:

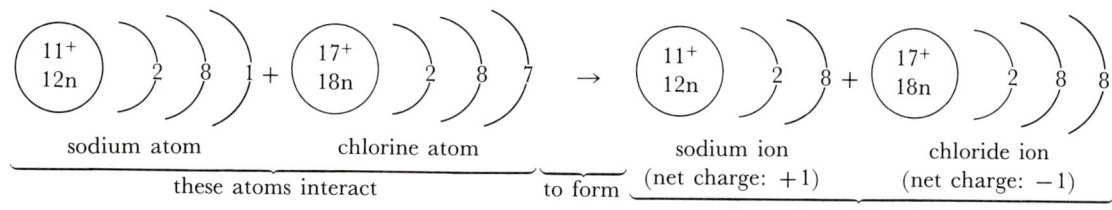

$$\underbrace{\begin{pmatrix}11^+\\12n\end{pmatrix}\;2\;8\;1 \;+\; \begin{pmatrix}17^+\\18n\end{pmatrix}\;2\;8\;7}_{\substack{\text{sodium atom}\qquad\qquad\text{chlorine atom}\\ \text{these atoms interact}}} \;\underset{\text{to form}}{\longrightarrow}\; \underbrace{\begin{pmatrix}11^+\\12n\end{pmatrix}\;2\;8 \;+\; \begin{pmatrix}17^+\\18n\end{pmatrix}\;2\;8\;8}_{\substack{\text{sodium ion}\qquad\qquad\text{chloride ion}\\ \text{(net charge: }+1)\qquad\text{(net charge: }-1)\\ \text{charged particles that in each}\\ \text{other's presence are very stable}}}$$

This event is an example of a *chemical reaction* as seen at an "atom's eye view." A chemical reaction is any event in which there is a net rearrangement or redistribution of electrons and nuclei relative to each other. Physical changes, which we have discussed already, leave electronic configurations essentially unchanged. That is the fundamental difference between these two major kinds of changes in our world.

In this example, sodium atoms transfer one electron each to chlorine atoms.[1] The new parti-

---

[1] We are glossing over the fact that elemental chlorine exists not as separate atoms but as diatomic "clusters" or molecules with the formula Cl_2. This complication, however, does not alter the nature and outcome of the overall electron-transfer event illustrated here.

cles still having sodium nuclei are now positively charged. The other new particles with chlorine nuclei are negatively charged. *They are no longer atoms.* They are examples of new kinds of particles called *ions*. Ions are positively or negatively charged particles at the atomic (or molecular) level in size. Nature is full of them. We have to learn how they are named and symbolized.

14.5
SYMBOLS AND NAMES FOR IONS

Several important ions derived from single atoms are given in Table 14.1. (Other kinds of ions that involve clusters of atoms are also known; see Section 14.17.) Note in Table 14.1 that the positively charged ions are named simply after the corresponding elements. The names of the negatively charged ions in the Table end in *-ide*, with the prefixes taken from the names of their elements.

14.6
IONIC COMPOUNDS

We cannot do an experiment involving just one atom of sodium and one of chlorine. Any seeable and weighable sample of either will contain billions and billions of atoms. Hence, when we mix sodium and chlorine in an actual experiment, a storm of ions forms—billions and billions of each kind. Sodium ions, of course, will stay as far apart from each other as they can. Like charges repel. Chloride ions will likewise stay away from each other, for the same reason. Sodium ions, however, cannot help but be attracted to chloride ions. They cannot help anything, and unlike charges attract. That is why sodium ions and chloride ions quickly begin to stack together in such a way that oppositely charged ions are nearest neighbors and like charged ions are farther away from each other. The result is a crystal of a new substance, sodium chloride or ordinary table salt. (Actually, many small crystals of salt form.) The arrangement of ions is illustrated in Figure 14–1.

In crystalline sodium chloride we do not have whole, intact *atoms* anymore. We have the pieces of sodium and chlorine atoms, their nuclei and electrons reorganized into ions that are oppositely charged. That is how they stick together. The force of attraction between oppositely charged ions in a crystal is called the *ionic bond*. Substances held together by ionic bonds are called *ionic compounds*. An ionic compound is one

TABLE 14.1
SOME IMPORTANT IONS[a]

Group	Element	Symbol for Neutral Atom	Symbol for Its Common Ion	Name of Ion
I	Lithium	Li	Li^+	Lithium ion
	Sodium	Na	Na^+	Sodium ion
	Potassium	K	K^+	Potassium ion
II	Magnesium	Mg	Mg^{2+}	Magnesium ion
	Calcium	Ca	Ca^{2+}	Calcium ion
	Barium	Ba	Ba^{2+}	Barium ion
III	Aluminum	Al	Al^{3+}	Aluminum ion
VI	Oxygen	O	O^{2-}	Oxide ion
	Sulfur	S	S^{2-}	Sulfide ion
VII	Fluorine	F	F^-	Fluoride ion
	Chlorine	Cl	Cl^-	Chloride ion
	Bromine	Br	Br^-	Bromide ion
	Iodine	I	I^-	Iodide ion
Transition elements	Silver	Ag	Ag^+	Silver ion
	Zinc	Zn	Zn^{2+}	Zinc ion
	Copper	Cu	Cu^+	Copper(I) ion (cuprous ion)[b]
			Cu^{2+}	Copper(II) ion (cupric ion)
	Iron	Fe	Fe^{2+}	Iron(II) ion (ferrous ion)
			Fe^{3+}	Iron(III) ion (ferric ion)

[a] Other common ions, which are derived from more than one element, are listed in Table 14.4.
[b] Names in parentheses represent older practice.

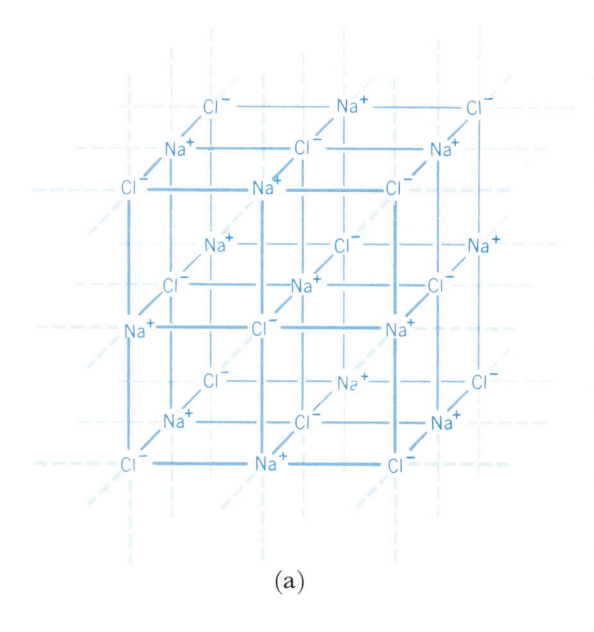

FIGURE 14-1
(a) The arrangement of ions in sodium chloride (table salt). (b) The crystal acquires its geometry from the geometry at the ionic level. (The Smithsonian Institution)

that consists of an orderly aggregation of oppositely charged ions assembled in a definite ratio that assures that the crystal will have overall electrical neutrality. Examples include, besides table salt, sodium bicarbonate in baking soda, sodium hydroxide in lye (and some drain cleaners), sodium tripolyphosphate (the "phosphate" in many detergents), sodium propionate (a preservative in commercial bread), and silver nitrate in photographic film.

14.7 SYMBOLS FOR IONIC COMPOUNDS

Formulas of ionic compounds are written to show what ions are present and in what ratio they occur by a combination of atomic symbols and subscripts following those symbols. A subscript number 1 is always understood; it does not appear in the formula. Thus, the symbol for sodium chloride is NaCl. Magnesium ions and chloride ions aggregate in a ratio of $1Mg^{2+}$ to $2Cl^-$. They cannot assemble in any other ratio, since opposite charges must cancel each other to preserve electrical neutrality for the substance.

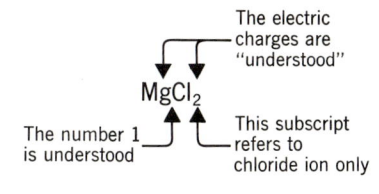

MgCl$_2$
- The number 1 is understood
- The electric charges are "understood"
- This subscript refers to chloride ion only

The $+2$ charge on one magnesium ion is neutralized by the -2 charge resulting from having two chloride ions, each with a charge of -1. Table 14.2 gives other examples of writing formulas of ionic compounds.

14.8 IONS AND THE OCTET RULE

The Group I elements in the periodic table have atoms whose highest occupied main energy levels have one electron. Sodium is an example. Like sodium, all other Group I elements form ions with charges of $+1$. Like sodium atoms, these other atoms in Group I can establish outer octets (for lithium, an outer pair in level 1) simply by giving up one electron. This will happen any time we mix one of these elements with an element whose atoms can accept electrons. Elements in groups VI and VII have such atoms. Their atoms have, respectively, 6 and 7 outside level electrons. Chlorine, in Group VII, is an example we have already used. With seven electrons, just one electron shy of an outer octet, it is poised to pick up the eighth electron and become an ion, Cl^-.

In general, elements whose atoms have 1, 2, or 3 electrons in their outside levels can lose electrons to suitable acceptors. *These elements are all metals,* except for hydrogen and helium. When atoms of metals form ions they generally are ions having positive charges of 1, 2, or 3—a number

TABLE 14.2
CONSTRUCTING FORMULAS AND NAMES OF IONIC COMPOUNDS

Compound Name	Constituent Ions	Ratio of Aggregation (to ensure neutrality)	Compound Formula
Magnesium bromide	Mg^{2+} and Br^-	1(++):2(−)	$MgBr_2$
Calcium oxide	Ca^{2+} and O^{2-}	1(++):1(=)	CaO
Sodium oxide	Na^+ and O^{2-}	2(+):1(=)	Na_2O
Iron(III) oxide (ferric oxide)	Fe^{3+} and O^{2-}	2(+++):3(=)	Fe_2O_3
Copper(I) sulfide (cuprous sulfide)	Cu^+ and S^{2-}	2(+):1(=)	Cu_2S

corresponding to electrons that must be lost in order for the particle to have a new outside level with an octet.

Elements whose atoms have 6 or 7 electrons in their outside levels can accept electrons from metallic atoms. It is, in terms of energy costs, much easier for atoms of Group 6 and 7 elements to acquire outer octets by gaining 2 or 1 electrons than it is for them to lose 6 or 7 electrons. Table 14.3 displays the simple electronic configurations of atoms and ions of common elements to show how the octet rule helps us predict what kinds of ions these elements can form.

14.9
IONIC COMPOUNDS AND LAW OF DEFINITE PROPORTIONS

Chemical compounds, like elements, are electrically neutral substances. Hence, if a compound is made of electrically charged particles called ions, the ions must assemble in a definite ratio. The opposite charges must cancel or the substance could not be electrically neutral. Because ions have fixed, given weights, a definite ratio by ions inevitably means a definite ratio by weight. This explains the law of definite proportions and the other laws of chemical combination, which were discovered before people were sure that atoms exist.

14.10
MOLECULAR SUBSTANCES

Atoms of elements in Groups IV and V seldom change to ions. Yet these elements occur as part of nature's most abundant and important substances. The element silicon, for example, is found in chemically combined forms in sands and silicates that are major parts of the earth's crust. The element carbon (Group IV) is chemi-

TABLE 14.3
CONFIGURATIONS OF IONS AND NOBLE GAS STRUCTURES

Group	Common Element	Nuclear Charge (Atomic Number)	Atoms Electronic Configurations 1 2 3 4 5 6	Ion	Ions Electronic Configurations 1 2 3 4 5 6	Comparable Noble Gas
I Alkali metals	Li	3+	2 1	Li^+	2	Helium
	Na	11+	2 8 1	Na^+	2 8	Neon
	K	19+	2 8 8 1	K^+	2 8 8	Argon
II Alkaline earth metals	Mg	12+	2 8 2	Mg^{2+}	2 8	Neon
	Ca	20+	2 8 8 2	Ca^{2+}	2 8 8	Argon
	Ba	56+	2 8 18 18 8 2	Ba^{2+}	2 8 18 18 8	Xenon
VI Oxygen family	O	8+	2 6	O^{2-}	2 8	Neon
	S	16+	2 8 6	S^{2-}	2 8 8	Argon
VII Halogens	F	9+	2 7	F^-	2 8	Neon
	Cl	17+	2 8 7	Cl^-	2 8 8	Argon
	Br	35+	2 8 18 7	Br^-	2 8 18 8	Krypton
	I	53+	2 8 18 18 7	I^-	2 8 18 18 8	Xenon
O Noble gases	He	2+	2			
	Ne	10+	2 8	The noble gases do not form stable ions		
	Ar	18+	2 8 8			
	Kr	36+	2 8 18 8			
	Xe	54+	2 8 18 18 8			

cally combined with hydrogen, nitrogen, and oxygen in all living things, and in many plastics and drugs and dyes and fabrics. Carbon atoms furnish the molecular frameworks of all organic substances. These atoms are particularly unique among all the elements. Only carbon atoms can become strongly bonded to each other over and over again in chains, rings, cages, and lattices and still retain combining power for atoms of other elements. To understand how atomic nuclei in these substances are kept from repelling each other to great distances, we must study the second kind of chemical bond, the covalent bond.

14.11 DIATOMIC MOLECULES

The covalent bond is found in some of nature's simplest substances—the reactive, gaseous elements: hydrogen, oxygen, fluorine, chlorine, and nitrogen. The formula of chlorine, as presented but not explained earlier in this chapter, is Cl_2, not Cl. The smallest, stable, representative sample of this element is not an atom. Instead, it is a particle about twice the size of an atom, consisting of two chlorine nuclei and enough electrons (34) to make a small *compound particle* that is stable and electrically neutral. It is called a *molecule* (Latin, "little mass"). The word *molecule* is generally used for any electrically neutral, non-ionic particle consisting of two or more nuclei and enough electrons to ensure electrical neutrality; it is the smallest stable sample of the substance. A substance made up of essentially identical molecules is called a *molecular substance*.

14.12 STRUCTURAL FORMULAS

A molecule such as Cl_2 may be represented by Cl-Cl. There is no ambiguity here. The two chlorines must be bound to each other. A triatomic molecule such as water, H_2O, however, could be represented as H-H-O or H-O-H. The second, H-O-H, is correct. Chemists know that the hydrogens are attached individually to the oxygen. Symbols such as Cl-Cl or H-O-H, wherein straight lines connecting atomic symbols are used to show the general arrangement of atoms in a molecule, are called *structural formulas* or, simply, *structures*.

14.13 LEWIS THEORY

The concept of an ionic bond is inadequate for explaining how a molecule such as Cl-Cl can be stable. Electron transfers between *like* atoms are unlikely. Chlorine does not consist of an orderly aggregation of Cl^+ and Cl^- ions. Energetically, the cost of forming Cl^+, which would have an outer sextet rather than an outer octet, is too high.

To explain how a molecule such as Cl-Cl held together, G. N. Lewis proposed in 1916 that the outside levels of certain atoms can interpenetrate to the extent that two electrons could be held in common by two atoms. Such shared electrons, not belonging exclusively to either atom, would have to be counted toward completing the outer octets of both atoms. Lewis represented outer level electrons by dots, which he grouped as much as possible in pairs around the atomic symbol. (This symbol now stood for all the rest of the atom, which Lewis called the atomic *kernel*.) Thus, in the Lewis "electron-dot" symbolism, two chlorine atoms were written as shown:

$$:\ddot{Cl}\cdot \rightleftarrows \cdot\ddot{Cl}: \rightarrow :\ddot{Cl}:\ddot{Cl}: \quad \text{shared pair of electrons}$$

two chlorine atoms   chlorine molecule

The chlorine molecule was said to result from a pairing and a sharing of two electrons. The straight lines, which for years before Lewis had appeared in structures, were now identified as representing shared pairs of electrons. Irving Langmuir coined the name *covalent bond* for them. Typically, the atoms that form covalent bonds are found among the nonmetals.

14.14 COVALENT-BONDING CAPACITY

Atoms that are able to form covalent bonds generally form as many as the number of electrons needed to complete outer octets (or, in the case of hydrogen, an outer filled first-level of two electrons). A carbon atom has four electrons in its outside level. It requires four more to complete its octet. To acquire them by a sharing process, it must form four bonds. The methane molecule, shown below, has shared pairs of electrons between the central carbon atom and each of the four hydrogen atoms.

$$H\cdot\cdot\overset{H}{\underset{H}{C}}\cdot\cdot H \rightarrow H:\overset{H}{\underset{H}{C}}:H \quad \text{or} \quad H-\overset{H}{\underset{H}{\overset{|}{C}}}-H$$

atomic parts for methane: one carbon atom and four hydrogen atoms   Methane, CH_4 (boiling point $-161.5°$ C)

By sharing a pair of electrons with carbon, each hydrogen fulfills its condition of stability, a first-level of two electrons. Its covalent-bonding capacity, called simply its *covalence,* is just one. Carbon's covalence is four. A nitrogen atom has five outside level electrons. To acquire an outer octet it must have three more by a sharing process. Nitrogen, therefore, can form three covalent bonds. Its covalence is three. Oxygen's covalence is two; its atoms have six electrons in their outside levels. All the halogens of Group VII, whose atoms have seven outside level electrons, have a common covalence number of one. The following series of structures of known compounds illustrate these facts. Shared pairs are shown by straight lines; unshared pairs in the outside levels are represented by dots, a practice usually not followed unless the condition of satisfied outer octets has to be stressed.

$$H-\underset{\underset{H}{|}}{\overset{\overset{H}{|}}{C}}-H \qquad H-\underset{..}{\overset{\overset{H}{|}}{N}}-H$$

Methane   Ammonia

$$H-\underset{..}{\overset{..}{O}}-H \qquad H-\underset{..}{\overset{..}{Cl}}:$$

Water   Hydrogen chloride

14.15
MULTIPLE BONDS

Two atoms may share more than one pair of electrons. Double and triple bonds are common, as illustrated by the structures of ethylene, carbon dioxide, acetylene, and nitrogen. All structures meet the requirements of the octet theory.

$$\underset{H}{\overset{H}{\diagdown}}C=C\underset{H}{\overset{H}{\diagup}} \qquad \overset{..}{\underset{..}{O}}=C=\overset{..}{\underset{..}{O}}$$

Ethylene   Carbon dioxide
(boiling point $-104°$ C)   (sublimes $-79°$ C)

$$H-C\equiv C-H \qquad :N\equiv N:$$

Acetylene   Nitrogen
(boiling point $-84°$ C)   (boiling point $-196°$ C)

14.16
MOLECULAR ORBITALS

Since the advent of quantum chemistry in the late 1920s and early 1930s, the covalent bond as seen by Lewis has been reinterpreted. In one theory, shared pairs of electrons are said to exist in regions of space, called *molecular orbitals,* that encompass two or more atomic nuclei. These regions are formed by the partial merging or overlapping of the atomic orbitals containing the electrons that became shared. The simplest example of a molecular orbital is that of the hydrogen molecule illustrated in Figure 14–2. The shared electrons in a hydrogen molecule can be thought of as originating from $1s$ atomic orbitals of isolated hydrogen atoms. Each atomic orbital had one electron. These two atomic orbitals (from two hydrogen atoms) become partially overlapped so as to form a new space enclosing the two nuclei and containing the two electrons. The electrons move with respect to both nuclei. Since they are more frequently *between* the two nuclei than elsewhere, the positively charged nuclei are shielded from each other. They do not repel each other to great distances, and the system is very stable. In fact, this way of organizing two electrons and two protons contains 103 kcal/mole-H_2 *less* energy than the same particles organized as separated atoms. It takes 103 kcal/mole to split hydrogen molecules into hydrogen atoms. That gives us a measure of how stable the molecular arrangement is compared to the separated atoms.

Elemental fluorine, F_2, illustrates overlap of atomic p-orbitals, as shown in Figure 14–3. The overlap of an atomic s-orbital with a p-orbital is illustrated in Figure 14–4 for the case of hydrogen fluoride, H-F.

14.17
POLYATOMIC IONS

It is not necessary that the overlapping atomic orbitals contribute one electron each to the shared pair. One of the atomic orbitals may bring in both electrons, and the other may be empty. The important requirement is that molecular orbitals may contain only two electrons each. Where these electrons originate is less important. Two very important polyatomic ions illustrate these ideas, the ammonium ion NH_4^+ and the hydronium ion H_3O^+. The names, formulas, and electric charges of several other polyatomic ions are listed in Table 14.4. These ions are small, molecular-sized clusters of nuclei and electrons held together by covalent bonds, but they are electrically charged. The sum of the nuclear charges is not canceled by the sum of the

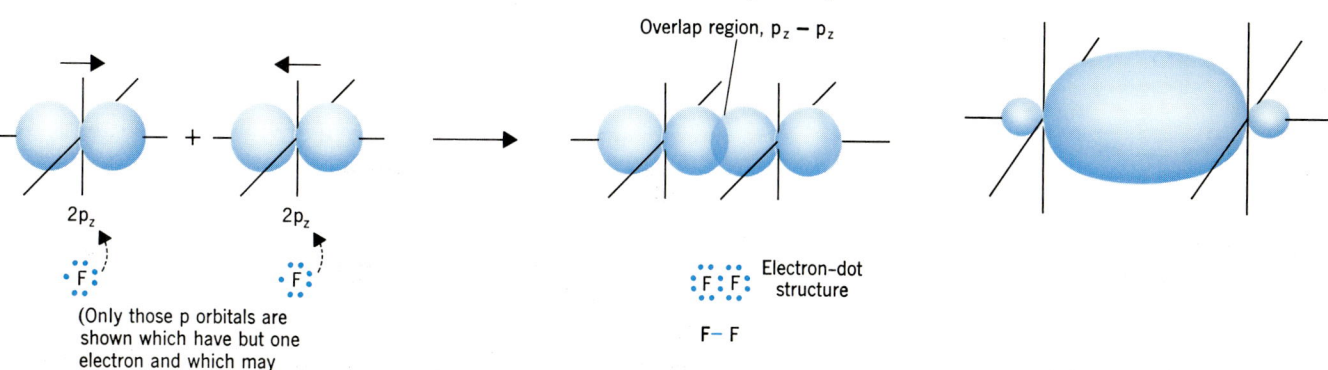

14.17 POLYATOMIC IONS

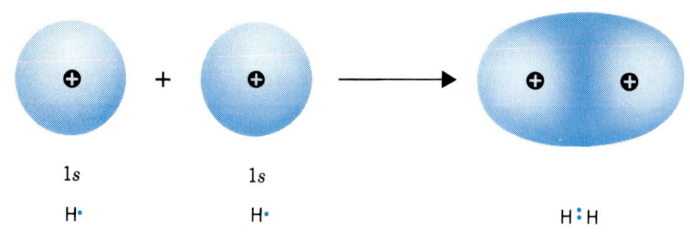

FIGURE 14-2
The formation of the covalent bond in H₂ is shown as the overlap or the interpenetration of the 1s atomic orbitals of the hydrogen atoms. The relative darkness of the shading in the structure of the hydrogen molecule gives a qualitative idea of where the electron cloud is densest, between the two nuclei. Only the bonding orbital of H₂ is described.

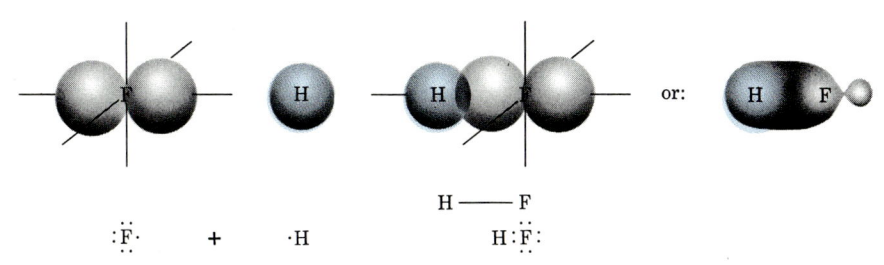

FIGURE 14-3
A bond that is symmetrical about the bonding axis may also form from overlap between a *p*-orbital and an *s*-orbital, as illustrated here for hydrogen fluoride, H—F. Note that the maximum electron density does not fall halfway between the two nuclei. This maximum is shifted toward the fluorine end of the bond, and the electron density in the immediate vicinity of the nucleus of hydrogen is *less* than it was in the hydrogen atom. Similarly, the total electron density in the vicinity of the nucleus of fluorine is greater than in the fluorine atom. The H—F molecule is therefore polar as described in the text.

FIGURE 14-4
The overlap of the two p_z orbitals of two fluorine atoms gives a bond that is symmetrical about the bonding axis. Thus, *p—p* overlap of this kind also gives rise to a single bond. The shape of the resultant molecular orbital is oversimplified. A better model would look more like the following:
In other words, the electron density is more concentrated between the nuclei than is indicated in the simplified picture.

(Only those p orbitals are shown which have but one electron and which may overlap.)

TABLE 14.4
POLYATOMIC IONS

Name of Ion	Formula	Electrovalence of Polyatomic Ion
Hydroxide ion	OH^-	$1-$
Carbonate ion	CO_3^{2-}	$2-$
Bicarbonate ion	HCO_3^-	$1-$
Sulfate ion	SO_4^{2-}	$2-$
Bisulfate ion (or hydrogen sulfate ion)	HSO_4^-	$1-$
Nitrate ion	NO_3^-	$1-$
Nitrite ion	NO_2^-	$1-$
Phosphate ion	PO_4^{3-}	$3-$
Monohydrogen phosphate ion	HPO_4^{2-}	$2-$
Dihydrogen phosphate ion	$H_2PO_4^-$	$1-$
Cyanide ion	CN^-	$1-$
Permanganate ion	MnO_4^-	$1-$
Chromate ion	CrO_4^{2-}	$2-$
Dichromate ion	$Cr_2O_7^{2-}$	$2-$
Sulfite ion	SO_3^{2-}	$2-$
Bisulfite ion (or hydrogen sulfite ion)	HSO_3^-	$1-$
Ammonium ion	NH_4^+	$1+$

electrons after the cluster has formed and octets have been established.

14.18 COORDINATE COVALENT BOND

As illustrated below, the ammonia molecule has an unshared pair of electrons in its outside shell. The hydrogen ion, H^+, is nothing more than a bare proton that has an empty $1s$ atomic orbital. A molecular orbital involving ammonia's unshared pair of electrons can form so as to enclose the proton, and this produces the ammonium ion. It has a net charge of $+1$ because a particle having that charge, a proton, has been added to a neutral particle. The ammonia molecule is said to have *coordinated* the proton.

The new bond formed this way is often called a *coordinate covalent bond*, but all four N-H bonds in ammonia are equivalent, and the distinction is unimportant.

A water molecule has two pairs of unshared electrons in its outside shell. One of its important chemical properties is that it can coordinate with one proton to form the hydronium ion, the species responsible for most of the chemical properties of aqueous solutions of acids.

Our study of the molecular orbital theory of covalent bonds has dealt only with single bonds. To develop a molecular orbital theory of double and triple bonds is beyond the scope of our study. The second and third bonds in double and triple bonds, like the first, involve molecular orbitals, and they form by the overlap of atomic orbitals. But the atomic orbitals are not the simple s- or p- types with which we are familiar. They are "hybrid" types, and our future needs do not require this development. Most textbooks in general or organic chemistry present it.

14.19 POLAR MOLECULES AND THE PROPERTIES OF MOLECULAR SUBSTANCES

Sugar, a crystalline material, readily dissolves in water. But the solution will not conduct electricity. Neither will melted sugar. Sugar is not an ionic substance. It is molecular, and its molecules are electrically neutral. Yet they stick to each other so strongly in the crystal that efforts to render them mobile by melting sugar usually cause some decomposition (caramelizing). This means that covalent bonds *within* the molecules start to break leading to colored substances before individual molecules become free enough from each other to move around in the liquid state. Sugar molecules, of course, are huge; their molecular formula is $C_{12}H_{22}O_{11}$. Water molecules are very small. Still, they must also stick to each other, because water at room temperature is a liquid, not a gas. But apparently they do not stick so strongly that slippage is impossible.

The question raised by these examples is how can electrically neutral molecules stick together? Our understanding of virtually all the chemical and physical properties of organic chemicals, of proteins, carbohydrates, fats, enzymes, and genes depends on finding the answer.

14.20 ELECTRICAL DIPOLES IN MOLECULES

Just as we may talk about a center of gravity of some oddly shaped mass, we may also say that an oddly shaped molecule has a "center of density of positive charge," or a "center of density of negative charge." There will be some geometric point within a molecule at which all the nuclear plus charge will appear to act as if it were centered at that point. The electrons swarming about these nuclei will likewise have a "center of density of negative charge." *If the center of positive charge density and the center of negative charge density are not in the same place, the molecule will have an electric dipole exactly analogous to a magnetic dipole of a magnet.* Just as it is possible for a piece of iron to be magnetically polar, it is possible for a molecule to be electrically polar. Just as it is possible for two properly lined up magnets to stick together, it is possible for two properly oriented polar molecules to stick. Figure 14-5 shows a molecule's-eye view of a crystal of a polar, covalent substance. Each molecule is represented as having a site of partial or fractional positive charge and somewhere else a site of partial negative charge. (The lower-case Greek letter *delta*, δ, is the symbol we use for "partial" or "fractional.") These partial charges are equal but opposite, and the molecule as a whole is electrically neutral. Electrically neutral molecules can therefore stick together if they are polar. We learned in this section what it means to be polar. We next ask, "how?" How can molecules be polar?

14.21 POLAR MOLECULES ARISE FROM POLAR BONDS

For a molecule to be polar, its individual bonds must be polar. A bond will be polar if the bonding electrons are not *equally* shared between the two atoms. At the fluorine end of a hydrogen fluoride molecule, H-F, there is a nuclear charge of +9; at the hydrogen end, a nuclear charge of only +1. The shared pair is drawn slightly toward the fluorine because its nucleus has the larger plus charge. We say that fluorine is more *electronegative* than hydrogen; it draws electrons better than hydrogen (see Section 14.22). Because of this drift in electron density away from the hydrogen end, the shared electrons do not completely cancel the +1 charge there. The hydrogen end of a molecule of H-F therefore bears a partial *positive* charge. At the other end, the electron density is a bit greater than needed exactly to cancel the +9 nuclear charge, and a partial negative charge exists there as a result. This means that at the ends of the bond we may write a δ+ and δ−, thus

$$\overset{\delta+}{\text{H}}—\overset{\delta-}{\text{F}}$$

Hydrogen fluoride—a polar molecule

In polyatomic molecules, the overall polarity is the result of all the individual bond polarities. A molecule's polarity is the vector sum of individual bond polarities. Thus, it is not enough for a molecule's bonds to be polar in order to have a polar molecule. These bonds must be pointing in

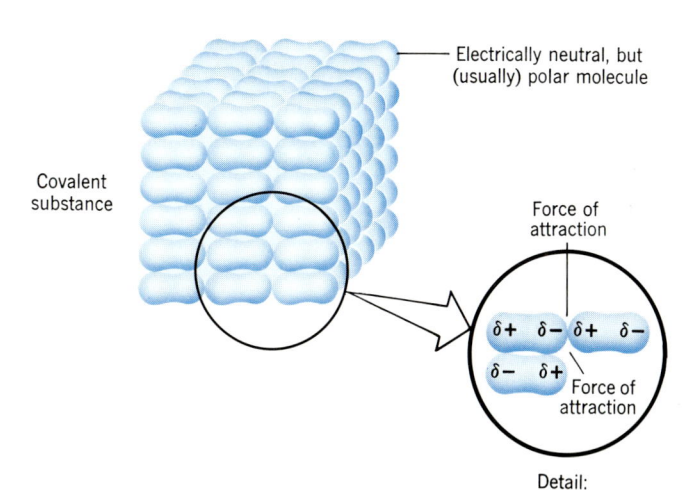

FIGURE 14-5

Electrically neutral molecules may stick to each other if they are polar, as illustrated here. Only an orderly aggregation would minimize the forces of repulsion between sites of like partial charge and maximize the forces of attraction between the unlike partial charges. If the net force of attraction is strong enough, the substance will be a solid at room temperature. If it is relatively weaker, the substance will be a liquid; if virtually zero, a gas. The shape of the polar molecule will be important in determining the magnitude of the net force of attraction. Molecules with highly irregular shapes may not be able to fit close enough together to take advantage of possible forces of attraction. Irregularly shaped molecules are often encountered among organic substances.

directions within the molecule that will let the molecule have a resultant polarity. We shall see how covalent bonds have definite directions after we study how such bonds can be polar.

14.22 ELECTRONEGATIVITIES

An element whose atoms in a molecule can attract shared pairs of electrons to themselves is called an *electronegative* element. Its atoms have a tendency to have partial negative charges in polar molecules. There are wide differences among the elements in electronegativity, as shown by the data in Figure 14–6. The numbers are relative values, and fluorine has the highest relative electronegativity of all the elements. The metals, which we know have little tendency to acquire negative charge, have the lowest values. The relative electronegativity of an element is a measure of the ability of its covalently bound atoms to draw electron density toward themselves along the bond. The greater the *difference* in electronegativity of two atoms joined by a covalent bond, the more polar will that bond be. Thus, an oxygen-to-hydrogen bond, as in water, will be more polar than a nitrogen-to-hydrogen bond, as in ammonia. When the difference in electronegativity of two elements is extreme, as for NaCl, the polarity is so great that the bond is ionic rather than covalent. The $\delta+$ has become a full $+1$ charge; the $\delta-$, a full -1 charge.

14.23 GEOMETRIES OF MOLECULES

Oxygen is much more electronegative than carbon. Therefore, in the carbon dioxide molecule, O=C=O, we should expect the bonds to be quite polar. Yet the molecule as a whole is nonpolar. The reason is that the molecule is linear; the three nuclei lie on a straight line. The individual bond polarities point in exactly opposite directions and their effects cancel each other.

Oxygen is also much more electronegative than hydrogen. Hence, in the water molecule, H-O-H, we should expect the two bonds to be polar. They are, and in this example their polarities do not cancel because the water molecule is not linear, which prompts us to raise an enduring question, "how come?"

14.24 BOND ANGLES

By using x-ray techniques, bond angles within molecules can be measured. The water molecule has a bond angle of 104.5 degrees. In a molecule of hydrogen sulfide it is 92 degrees.

Water (104.5°)   Hydrogen sulfide (92°)

Oxygen and sulfur are in the same family in the periodic system, Group VI. Their atoms have identical outside shell configurations.

Oxygen, O, atomic number 8
$$1s^2 2s^2 2p_x^2 2p_y 2p_z$$
Sulfur, S, atomic number 16
$$1s^2 2s^2 2p_x^2 2p_y^2 2p_z^2 3s^2 3p_x^2 3p_y 3p_z$$

In each case, two p-orbitals in an outside shell possess one electron each. These can overlap with 1s orbitals from hydrogen atoms to form the covalent bonds in water and hydrogen sulfide. Of course, p-orbitals are perpendicular to each other. When hydrogens come up to two of them in either oxygen or sulfur to develop covalent bonds, the resulting bond angle should be 90 degrees, as illustrated for water in Figure 14–7. The fact that the angle is much larger than 90 degrees in a water molecule is explained (in this theory) by the mutual repulsion between the two hydrogen nuclei. In bonding to oxygen in water they have been brought close together, but the strain is relieved by some spreading. In hydrogen sulfide, where sulfur's p-orbitals are farther away from the nucleus of the sulfur atom (third energy level, instead of the second, as in oxygen), the hydrogens can form their bonds without their nuclei coming as close to each other. There is therefore much less spreading, and the bond angle (92 degrees) is very close to the predicted angle.

The shape of a molecule depends on its composition and its bond angles. It is determined by the requirement that the molecule have a minimum internal energy. It achieves this when its molecular orbitals correspond to maximum overlaps of atomic orbitals consistent with mini-

H 2.20						
Li 0.97	Be 1.47	B 2.01	C 2.50	N 3.07	O 3.50	F 4.10
Na 1.01	Mg 1.23	Al 1.47	Si 1.74	P 2.06	S 2.44	Cl 2.83
					Se 2.48	Br 2.74
					Te 2.01	I 2.21

FIGURE 14–6
Electronegativities for some elements.

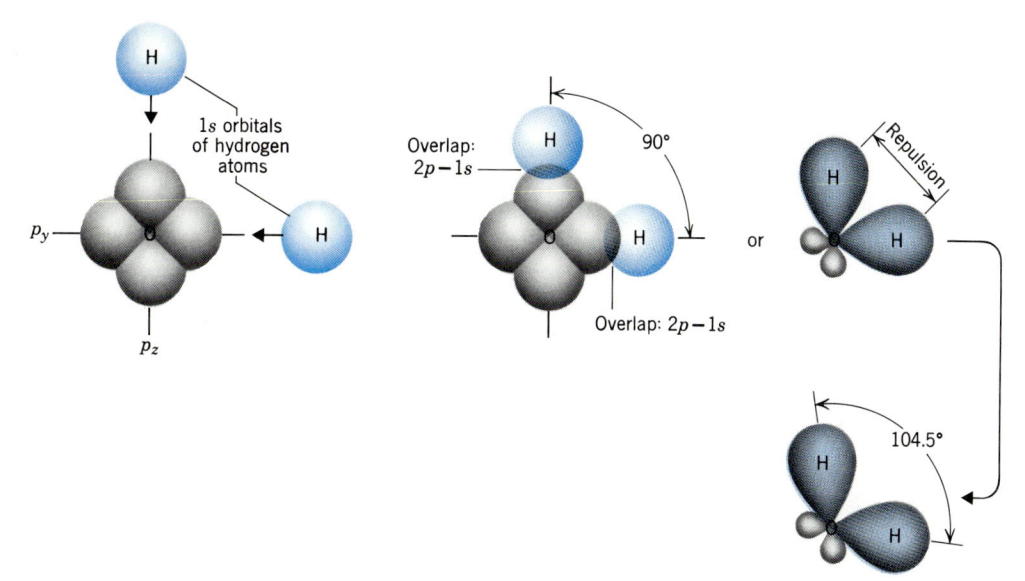

FIGURE 14-7
Molecular orbital picture of the water molecule. According to one model, the bonds are created from the overlap of *p*-orbitals in oxygen with *s*-orbitals provided by the hydrogens. Then repulsions between the hydrogen nuclei and repulsions between the electron clouds of the new molecular orbitals spread the bond angle from the "predicted" value of 90 degrees to the actual value of 104.5 degrees. (From J. R. Holum, *Elements of General and Biological Chemistry*, 4th edition, 1975, John Wiley & Sons, Inc., New York.)

mum repulsions between nuclei and between the orbitals containing unshared electrons.

14.25
THE METALLIC BOND

All the metallic elements share two features that are important to understanding how atoms of metals can stick very tightly to each other and still be in the solid state, conduct electricity, and be capable of being hammered into sheets and drawn into wires. All the metals have relatively low ionizaion energies. It takes relatively little energy to force an electron out of an atom of a metal—in all cases less than 250 kcal/mole. Another common feature is that most metals have only 1–3 electrons in their highest occupied principal energy levels. This means that at these levels, atoms of metals have many unfilled and unoccupied atomic orbitals. These orbitals can overlap with those of neighbors on several sides. Valence shell electrons are shared among several atoms. Large, multicentered molecular orbitals form that link atoms to neighbors throughout the entire solid. The outside shell electrons of the metallic atoms move more or less freely in these, accounting for the fact that metals are good electrical conductors. It is as if in a metal there were a collection of metallic ions immersed in a sea of outside shell electrons, and this model for a metal is sometimes called the *electron-sea model* or the *electron-gas model*.

The fact that metals can be hammered and drawn suggests that the inner cores, the atomic kernels Lewis discussed, can shift around without seriously affecting the opportunities for forming the multicentered molecular orbitals.

The electron-sea model of metals is misleading in one important way. It suggests that the nuclei bob around and drift here and there. On the contrary, the nuclei and their inner shell electrons are in very definite positions relative to each other. They are so regularly spaced that they bring to mind the regularity and order found in crystals of an ionic compound.

14 COMPOUNDS, CHEMICAL BONDS, AND CHEMICAL ENERGY

(a)

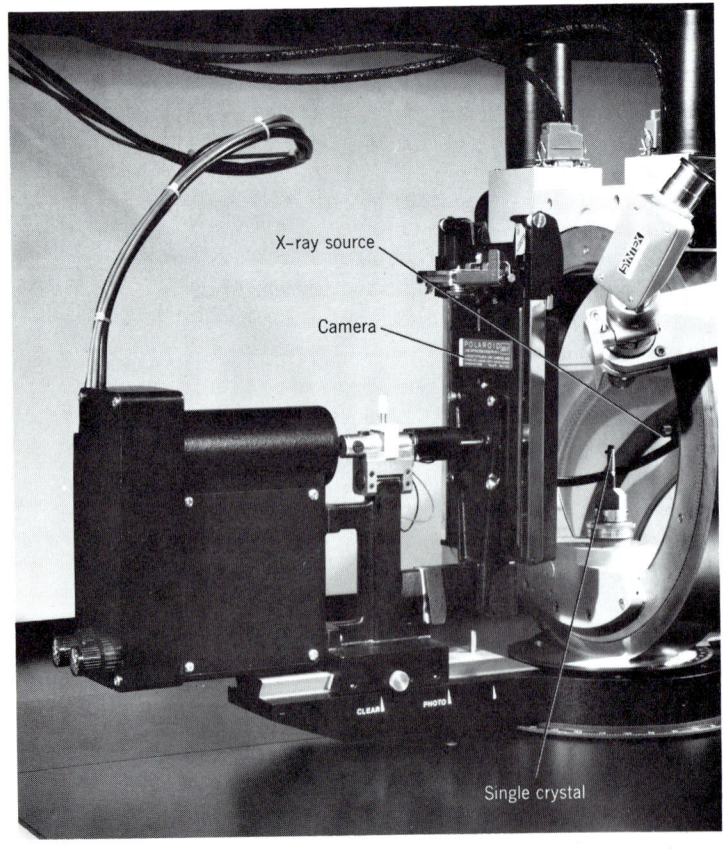

(b)

FIGURE 14-8

Single-crystal x-ray diffractometer. (a) Computer-controlled diffractometer. Circular device (goniometer) allows rotation of the crystal around four axes independently. Readouts from the detector are digitized and fed to the computer at left. (b) A Polaroid camera is used to obtain a preliminary diffraction pattern (Figure 14-9) before actual data collection begins. (Courtesy Syntex Analytical Instruments, Inc.)

14.26
CRYSTAL STRUCTURE

Whether in rock shops, museums, or color photos we have all admired some of nature's beautiful crystals. Given the right growing conditions, most of nature's solid, pure substances will form crystals. (Those solids that do not are termed *amorphous solids;* glass is one example; cold tar is another.) Crystalline solids, whether they are elements, ionic compounds, or molecular compounds, have regular shapes because their constituent particles—atoms, ions, or molecules—stack together in very regular patterns resembling latticework. These patterns are discovered through the use of x-rays (see Figure 14-8). When a crystal is held in an x-ray beam, most of the x-rays go right on through. Some, however, diffract at various angles, and these x-rays are allowed to impinge on photographic film producing lines. The data about these lines—their relative locations and intensities—are fed into computers programmed to analyze them in terms of crystal structure. What diffracts the x-rays are the atomic nuclei present in the crystal. Only if nuclei are arranged in an orderly way will diffraction occur to give regularly spaced lines or dots on the film (see Figure 14-9). Amorphous solids do not give regular x-ray diffraction patterns. Some of the common types of crystal forms are illustrated in Figure 14-10. In ionic compounds, oppositely charged ions stack together. We saw an example earlier in Figure 14-1. In crystals of molecular compounds, polar molecules get themselves arranged so that forces of attraction can have the advantage over forces of repulsion, as we saw in Figure 14-5. These forces originate from the partial positive and negative charges each molecule has if it is polar. Generally, the larger the partial charges, the more strongly is the molecular crystal held together. Seldom, however, are the forces of attraction as strong in a molecular crystal as in an ionic crystal. That is why it takes a much higher

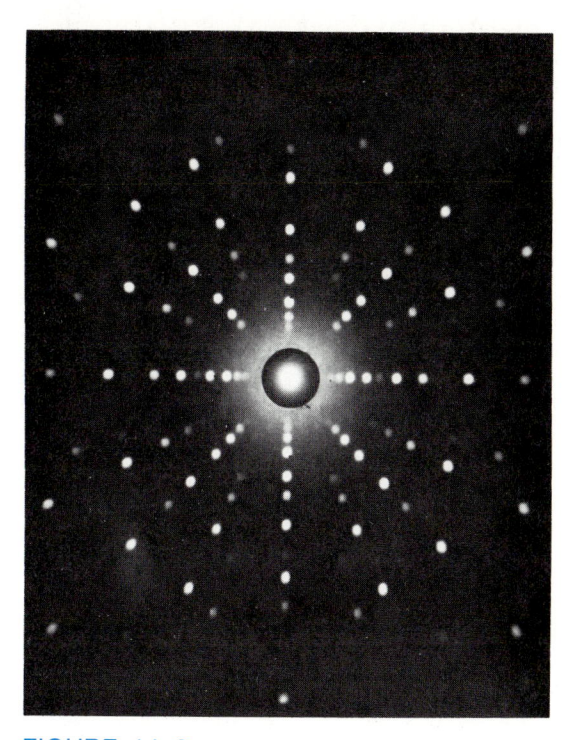

14.27 ENERGY FEATURES OF CHEMICAL CHANGES

FIGURE 14-9
X-ray diffraction pattern of a single crystal of niobium diboride, a synthetic ceramic material. (Courtesy The Polaroid Corporation.)

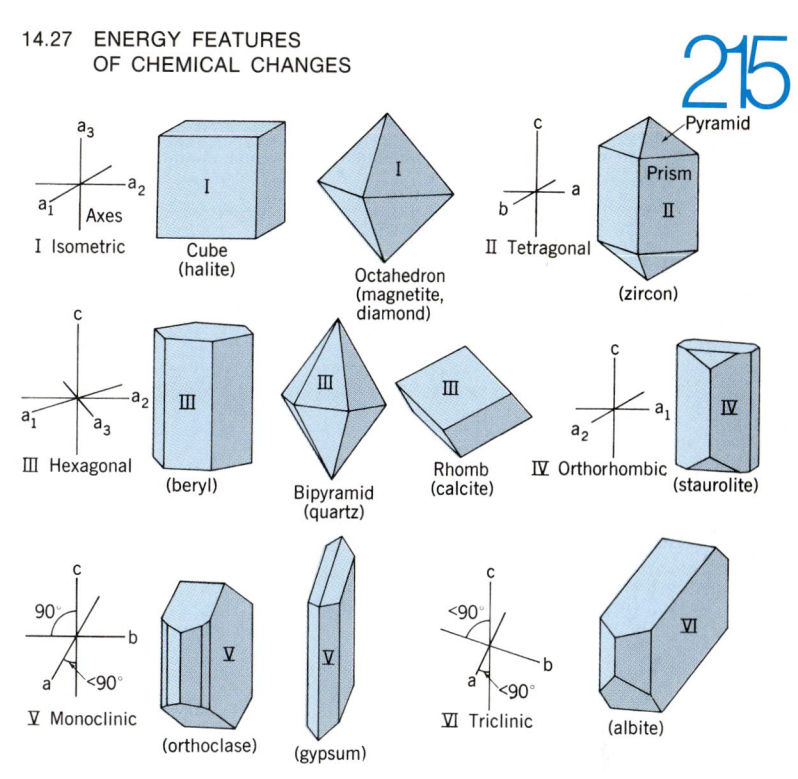

FIGURE 14-10
The six crystal systems differ in the relative lengths of the axes and the angles at which they meet. These are shown here together with sketches of representative external crystal forms. (From A. N. Strahler, *The Earth Sciences*, 3rd edition, 1977. Harper & Row, New York. Used by permission.)

temperature to melt an ionic compound than a molecular compound. Most ionic substances melt well above 350°C; most molecular substances melt below 300°C. Some molecular substances do not really consist of discrete molecules at all. Instead, they consist almost of one huge, continuous molecule in which covalent or polar covalent bonds hold much of the system together. Diamond and graphite are two important examples, as seen in Figure 14-11. The several varieties of silicates, major components of the earth's crust, are others.

14.27 ENERGY FEATURES OF CHEMICAL CHANGES

If two atoms, molecules, or ions are to react with each other they must collide. This seems obvious, and it implies some very important practical procedures for carrying out chemical reactions. To have a chemical change occur smoothly, the chemist not only arranges for the atoms or molecules or ions of the chemicals to meet each other by mixing them together. The chemist also controls many experimental aspects of the "meet-ing." How crowded are the intermingling, bumping-together particles? This is a question of concentration. The more concentrated the particles, the more likely are they to collide and react. Concentration, therefore, is one important factor in how fast chemicals react.

How violently are the bumping-together particles colliding with each other? The more violent these collisions, on the average, the more frequently will the collisions be effective. By *effective* we mean in the sense that the change proceeds and that the particles become those of the products. Except for the most reactive substances, mere taps and nudges between particles are not enough to get them to change chemically.

Why not? These reacting particles are electron clouds within which nuclei are buried. Electron clouds tend to repel each other. Yet, for many chemical changes to take place, electron clouds from the reacting particles must interpenetrate or merge somewhat. Then electrons and nuclei can reorganize themselves into the arrangements of the product particles. To get electron clouds to interpenetrate they must not only be on a collision course, they must also

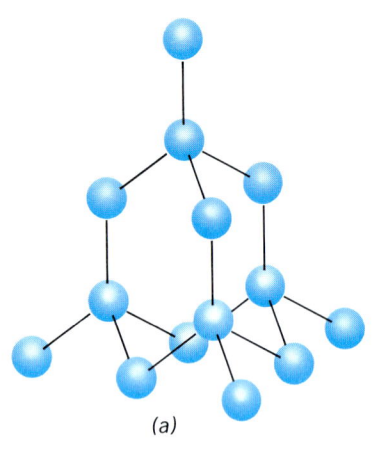

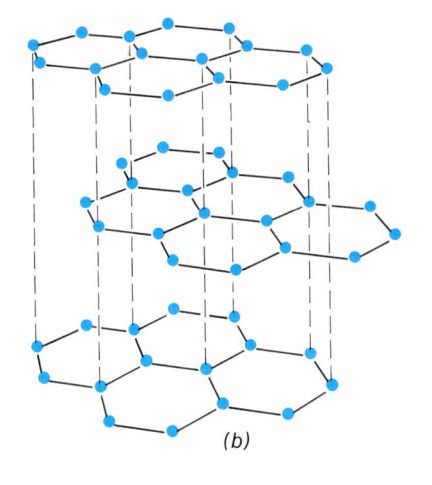

14 COMPOUNDS, CHEMICAL BONDS, AND CHEMICAL ENERGY

FIGURE 14-11
The arrangement of atoms in (a) diamond and (b) graphite.

come at each other quite energetically. The collision has to be more or less violent or it will not be effective in leading to products. Let us study this by the example of burning coal.

14.28
SPONTANEOUS EVENTS AND HEAT

Coal is mostly carbon, but we know that we can store coal in the open air almost indefinitely. Coal does not spontaneously combine with the oxygen in the air. At normal outdoor and room temperatures, the oxygen molecules in air almost never collide violently enough with carbon atoms on the surface of the coal to launch the peculiar rearrangement of their electrons and nuclei that changes these particles into a molecule of carbon dioxide. Even though the arrangement of electrons and nuclei in the carbon dioxide molecule is more stable than the original, separate arrangements of the carbon atom and the oxygen molecule, the chemical change is not spontaneous at room temperature.

To get this change to occur we do something very commonplace and familiar. We ignite the coal. We hold a match or some other igniting device next to the surface of the coal. Heat makes particles move more vigorously, as we learned in Section 4.10. Atoms of coal begin to vibrate more and more violently; oxygen molecules hit them more violently. All of a sudden so many carbon atoms and oxygen molecules are changing to carbon dioxide that the heat this change produces helps to ignite unchanged coal nearby. The change is called exothermic (*exo-* "out"; *-thermic* "heat"), because heat is released.

Because this event is so common, we no longer think about how astonishing it is. We began with coal (or paper or wood) in air at room temperature. These are materials we can handle safely and put wherever it suits our convenience. With a little nudge from an igniting device, however, we suddenly have the proverbial blazing inferno. Now the materials had best be in a stove or fireplace. It appears that we are creating energy! Where did all that heat come from? The carbon dioxide molecules evidently are moving with a collective violence and energy much greater than the collective energy of the moving and vibrating oxygen molecules and carbon atoms. That is why it is hot in the vicinity of the blaze. The product molecules are emerging with great collective violence.

Before the collisions that led to carbon dioxide there was energy of a potential nature within the structure of the coal and the oxygen, within their chemical bonds. Because there is less of this potential energy in carbon dioxide molecules, the difference appears in other forms of energy—mostly heat but some light and sound in the burning of coal. The potential energy within chemicals, called chemical energy, thus may emerge as heat in some reactions, as electricity in others (see Chapter 5, Section 8), as sound (in an explosion of chemicals, for example), and as light. Often two or more of these forms of energy are released in a chemical change.

14.29
ENERGY OF ACTIVATION

When we ignite coal we are doing something analogous to pushing a boulder up to the brink

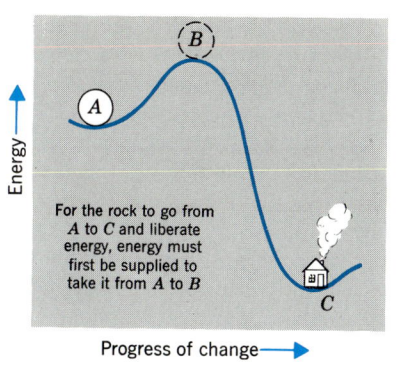

FIGURE 14-12

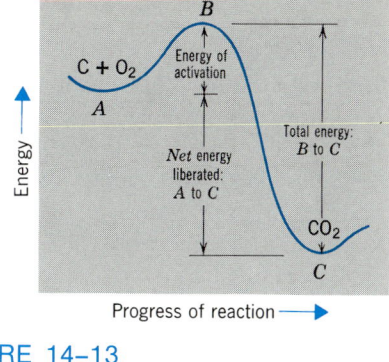

FIGURE 14-13

of a hill or cliff. Figure 14-12 illustrates this. We know from experience that if we could get the rock to point B it would spontaneously go to C. Point C is a more stable place for the rock than B, and spontaneous events in nature tend always to go in the direction of increasing stability under a given set of conditions. C is also at a lower energy level than A, but the change from A *directly* to C does not happen. Boosting the rock to B is analogous to igniting the carbon, as seen in Figure 14-13. We invest a little energy (A to B is uphill) and get a whole lot more back. The energy we must invest to get the reaction going is called the *energy of activation*.

14.30 CATALYSTS

Catalysts are substances that, when present in trace concentrations during a reaction, accelerate the rate of the reaction. Yet the catalyst itself is generally not permanently changed by the reaction; it can usually be recovered (and used again, sometimes).

Catalysts work by reducing the energy of activation for the chemical reaction they affect. It is as if a catalyst opens a different route from A to C in Figure 14-13, a route that might be analogous to a tunnel under the energy hill. Exactly how a catalyst brings this about depends on the catalyst and the reaction. We cannot write a general explanation. All we may say is that catalysts do accelerate reactions (by definition) and that they do lower energies of activation by providing for different routes from reactants to products.

The importance of catalysts in our daily lives simply cannot be overstated. Almost without exception the reactions within the cells of living systems require catalysts. Life as we know it would be virtually unimaginable without these special catalysts. In living things they are called *enzymes*, and without enzymes the chemical events of life would be much too slow or would not go at all.

In the chemical industry catalysts are employed in almost every single chemical process whereby a drug, a dye, a plastic, or anything else, including synthetic gasoline, is made by a chemical reaction. Without catalysts, these reac-

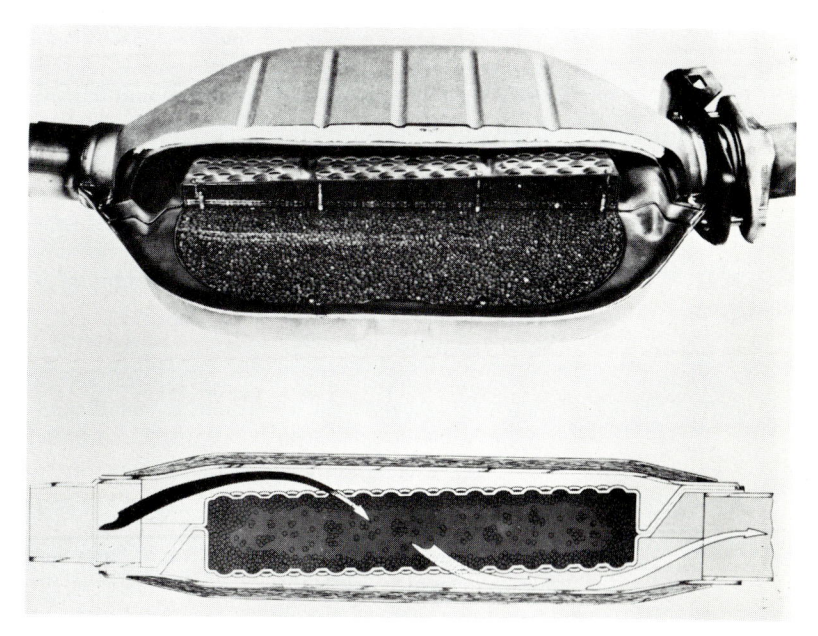

FIGURE 14-14
Catalytic muffler. Top, cutaway of oxidizing catalytic converter developed for use in light-duty General Motors vehicles to meet automotive emissions regulations under the Federal Clean Air Act. As exhaust gases flow through the converter (bottom), ceramic granules containing a platinum catalyst aid the oxidation of carbon monoxide to carbon dioxide and unburned hydrocarbons to carbon dioxide and water. Lead-free gasoline must be used because lead compounds in the exhaust would poison the catalyst. (Courtesy AC Spark Plug Division of General Motors.)

tions either would not go as desired or would cost so much energy as to make them unprofitable.

The catalyst in a catalytic muffler (Figure 14–14) is a specific example of a catalyst at work. It helps to change pollutants in the exhaust gases to less harmful compounds. Unhappily, the best catalyst is made from platinum, a rare and very expensive metal and one easily "poisoned" or inactivated by compounds of lead that are in the exhaust when leaded gasoline is used.

SUMMARY

Chemical bonds are generally three major kinds of holding forces that keep subatomic particles together in substances. When neutral atoms react chemically, their electrons and atomic nuclei become reorganized. Old electronic configurations change into new ones that provide binding forces.

Ionic bond. When a metal and a nonmetal react, electrons transfer from metallic atoms to nonmetallic atoms to produce oppositely charged ions that attract each other. The force of attraction in this instance is called an ionic bond.

Covalent bond. Often, reactive atoms that cannot engage in electron transfer will cooperate and share one or more pairs of electrons between them. Sharing occurs by means of a molecular orbital formed when two atomic orbitals partly merge or overlap.

Metallic bond. Multicentered molecular orbitals between metal atoms contain shared electrons that are very mobile and account for the electrical conductivity of metals. The atomic nuclei are not mobile. Their fixed arrangements in metals (and in any crystalline substance) can be determined by x-ray diffraction analysis.

Octet rule. In a great number of compounds, the proportions of the elements chemically combined fit this rule: rearrangements of electrons and nuclei of reacting atoms of elements usually generate particles whose nuclei have outer octets of electrons. If level 1 is the outer level, then the generated particles will have an outer pair of electrons. The rule works especially well with the first 20 elements of the periodic table and their compounds.

Dipole-dipole interactions. Electrically neutral molecules will stick to each other if they are electrically polar—if they possess dipoles. A net molecular dipole is the result of polar bonds whose polarities do not cancel each other. Polar bonds arise when one atom of a pair joined by a covalent bond is more electronegative than the other, when it can draw electron density toward itself in the bond and acquire more than its fair (50:50) share.

Rates of chemical reactions. Three factors have particular influence on how rapidly a chemical reaction will go: (1) the temperature of the reacting mixture, (2) the concentration of the reactants, and (3) the presence of a catalyst.

Energy of activation is a fourth factor affecting the rate of a reaction, but once the reaction is selected, only the discovery of a catalyst can affect it. The energy of activation of a reaction is the minimum energy that must be present in a collision for the two particles to be able to get their nuclei and electrons rearranged into product particles. A catalyst is any substance that in relatively trace amounts will speed up a reaction without itself being permanently changed. Normally it is temporarily changed as it provides an alternative route to products by a lower energy hill—a lower energy of activation. Normally, reactions with very high energies of activation are very slow even if the change is, overall, very exothermic. A reaction with a low energy of activation will be rapid and difficult to control.

TERMS AND CONCEPTS

bond, chemical
octet
octet rule
chemical reaction
ion
ionic compound
ionic bond
Lewis octet theory
molecular compound
molecule
structural formula
structure
covalent bond
covalence
molecular orbital
polyatomic ion
coordinate covalent bond
polar molecule
electronegativity
bond angle
metallic bond
electron-gas model
exothermic change
endothermic change
energy of activation
catalysis
catalyst
enzyme

QUESTIONS AND EXERCISES

1. The magnesium atom (atomic number 12) has two electrons in its outside shell. Can this atom become a reasonably stable ion? If so, what is the charge on the ion? If not, why not?

2. Bromine atoms (atomic number 35) have seven electrons in their outside shells. Can the bromine atom become an ion and if so, what will be its electric charge?

3. The aluminum ion has the formula Al^{3+}; the fluoride ion, F^-. Write the formula of their combination, aluminum fluoride.

4. Using information found in Table 14.1 about charges on ions write formulas for compounds between each of the following pairs: (a) calcium and oxygen, (b) potassium and sulfur, (c) magnesium and oxygen.

5. In what specific structural ways are potassium atoms and potassium ions alike and in what ways are they different?

6. Atom X has atomic number 9 and atom Y has atomic number 11. Write the electronic configurations of each. Predict from this information alone if X and Y can react to form any *ionic* compound.

7. A compound of the hypothetical formula YZ is a liquid at room temperature and pressure. The bond between Y and Z is therefore most likely what kind of bond—ionic, covalent or metallic? How do you know?

8. The compound YZ (problem 7) had a very high boiling point. What does this fact indicate about the particles in this compound?

9. What are the chief differences between a molecule and an ion? Between an atom and an ion? Between an atom and a molecule?

10. What are the three major kinds of "holding forces" in nature?

11. What is structurally true about *all* the chemically reactive elements? About the chemically unreactive elements?

12. Sodium (atomic number 11) reacts much more vigorously than magnesium (atomic number 12) with chlorine. What is it about sodium atoms as compared to magnesium atoms that might account for this greater reactivity?

13. Although fluorine, F_2, has the higher formula weight (38) than hydrogen fluoride, HF (formula weight 20), it has a much lower boiling point ($-188\,°C$ for F_2 and $19.5\,°C$ for HF). Explain this in terms of what we have studied in this chapter.

14. Concerning problem 13, what has the formula weight to do with boiling point?

15. Suppose we have a chemical reaction that may be expressed simply as $R \rightarrow P$. (a) What is true about the relative chemical energies of R and P if the reaction is exothermic? (b) Which would be more important in determining how fast R changes to P under a given set of conditions of temperature and concentration: (1) the energy of activation for the reaction is unusually low? or (2) P is at a much lower chemical energy level than R?

16. Why isn't the chemical symbol for the catalyst included among those of the reactants in a chemical equation?

17. What is the general name for catalysts found in living things?

18. What is the reason for needing a catalyst in a catalytic muffler?

CHAPTER 15
THE EARTH'S SUBSTANCES—INORGANIC MATERIALS

15.1 INORGANIC AND ORGANIC SUBSTANCES

The word *organic*, related to the word *organism*, implies a substance coming exclusively from plants and animals. This was once the basis for classifying matter into the two very broad classes: organic and inorganic (the prefix *in-* means "not" in the word *inorganic*). Substances obtained only from living things or from the products of their decay were classified as organic. Anything else was inorganic. Although plants and animals contain considerable water, this compound, of course, is obtained from nonliving sources, too. Therefore, it was classified as inorganic. Scientists eventually found that organic compounds seemed always to be compounds of carbon together with a few other elements such as hydrogen, oxygen, nitrogen, and sulfur. By the middle of the nineteenth century, chemists could make many organic compounds from minerals. They discovered—to their great surprise, actually—that organisms are not essential as sources of organic compounds. Because of these developments we now define organic compounds in a broad way. An *organic compound* is any compound of carbon, synthetic or naturally occurring, except carbonates, bicarbonates, and cyanides.

We shall devote later chapters to organic compounds. In this chapter we survey the principal kinds of *inorganic compounds* and those important properties that most directly affect human life and well-being on our planet. Many of them are of particular interest only when they are in a solution—that is, they are dissolved in water or something else. For that reason we turn first to a study of solutions, solutions being a special kind of mixture. In Section 12.7 we learned that when we use the law of definite proportions as a basis for classifying matter, mixtures join elements and compounds as the three broad classes of matter. The relations between these three are sketched in general terms in Figure 15–1.

15.2 HOMOGENEOUS MIXTURES

Something that is homogeneous has a uniform structure and composition throughout. Virtually all milk sold in stores and restaurants is homogenized milk. When you buy paint you mix it carefully to make it homogeneous. With or without sugar, tea is homogeneous. You can buy skim milk, 2 percent milk, or 3 percent milk. You can buy red paint of widely varying "redness"; it depends on how much dye as well as what kind of dye is used. You can make tea as sweet tasting as you please. Because milk or paint or tea can be formulated with different compositions and because the substances in these can be separated by physical means, we know they cannot be called pure compounds. They are mixtures. When paint is left to stand for a long period the pigments settle to the bottom of the can. In the days before homogenized milk, the cream in natural milk used to rise to the top of the bottle. The sugar in a cup of hot tea, however, never settles

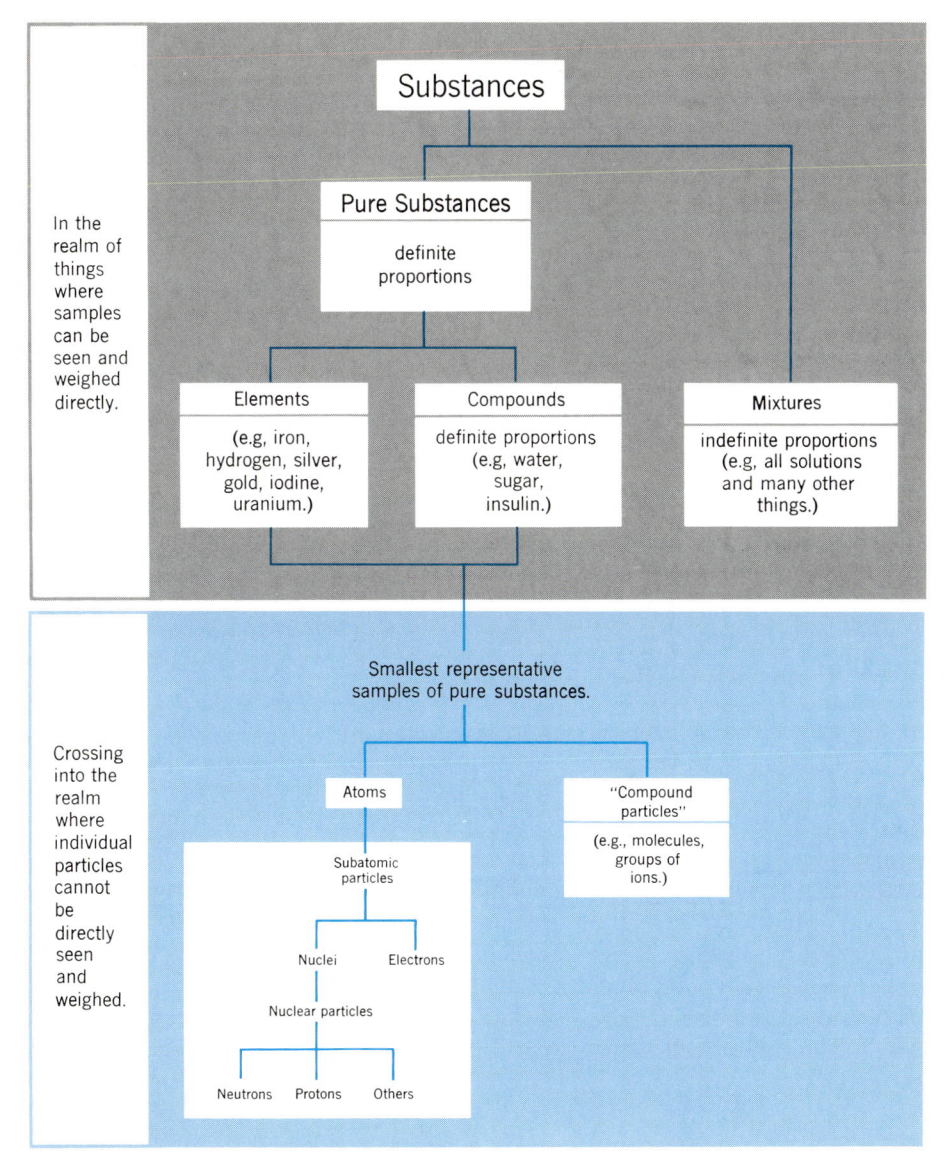

FIGURE 15-1
A scheme for classifying matter. (From J. R. Holum. *Elements of General and Biological Chemistry,* 4th edition. 1975. John Wiley & Sons, Inc., New York.)

out. Why? What determines the stability with respect to settling of a homogeneous mixture? The answer is the size of the particles blended and mixed together. At the very smallest level of particle size, we have what scientists call the *solution.*

15.3 SOLUTIONS

The solution is a homogeneous mixture of two or more substances that have subdivided into ionic and molecular-sized particles. When the blended and intermingled particles are ions or small molecules, they are small enough to be kept indefinitely homogeneous by the buffeting actions of their constant motions. If the particles are larger, we have something else, not a true solution.

The eight possible kinds of solutions are described in Table 15.1. The component normally present in the highest relative amount is often called the *solvent*. In actual practice, this word is normally used in connection with some liquid, such as water or alcohol or gasoline. The substance dissolved in the solvent is called the *solute*.

Ions and small molecules that can be in a true solution generally have diameters in the range of 0.05 to 0.25 nanometers (1 nanometer = 1 nm = 1×10^{-9} meter). That is too small to be

15 THE EARTH'S SUBSTANCES—INORGANIC MATERIALS

TABLE 15.1
KINDS OF SOLUTIONS

Kinds	Common Examples
Gas in a liquid	Carbonated beverages (carbon dioxide in water)
Liquid in a liquid	Vinegar (acetic acid in water)
Solid in a liquid	Sugar in water
Gas in a gas	Air
Liquid in a gas	Humid air
Solid in a gas	(Not considered possible; instead, this is a colloidal dispersion)
Gas in a solid	Alloy of palladium and hydrogen
Liquid in a solid	Benzene in rubber (e.g., rubber cement)
Solid in a solid	Carbon in iron (steel)

seen even with the aid of an electron microscope and is too small to be trapped by even the best filters. Because the particles are so small, the force of gravity that urges them to settle is defeated by the chaotic bumping around they get from innumerable, random collisions in the solution.

Particles on the order of 10,000 nanometers in average diameter—particles such as those making up very fine clay or talcum powder—are finally large enough to settle under the influence of gravity from a suspension in water, provided the water is not stirred or otherwise turbulent. In erosion of landforms by wind or water, the smallest particles are carried the farthest. They can be trapped by filters, which they soon plug. Particles in true solution are so small that they readily pass through filters together with molecules of the solvent.

15.4 COLLOIDAL DISPERSIONS

Between the solution and the suspension is a third kind of homogeneous mixture, the colloidal dispersion, in which the dispersed particles may be clusters of several hundred or a few thousand atomic, ionic, or molecular-sized particles. In other examples, instead of clusters of small molecules, we may be dealing with enormous molecules such as those found in starch, most proteins, and plastics. These substances consist of molecules with formula weights of several hundred thousand or a few million. At this molecular size (or cluster size), at least one dimension will be between 100 and 10,000 nm. Matter whose particles are this large is described as being in the *colloidal state*. (*Colloidal* comes from the Greek for "glue-like." Common glues contain particles in the colloidal state.)

When matter of colloidal size is dispersed in some medium—a dispersing medium such as air or water or other liquid—the mixture is called a *colloidal dispersion* instead of a solution. Examples are given in Table 15.2. The particles in a colloidal dispersion, in contrast to those in a solution, are large enough to reflect and scatter light. Therefore, colloidal dispersions sometimes have a milky, cloudy appearance. This effect is called the *Tyndall effect*, and you have seen it whenever you have seen light "streaming" through a forest canopy (Figure 15-2) or through a window into a darkened room with smoky air.

Colloidal particles are usually not large enough to be trapped by filters. The material that is used to make the walls of a blood vessel is like a superfilter, however. It cannot let large, colloidal-size molecules through. One of the many purposes of digesting food is to break up its big molecules into much smaller ones. These can get out of the digestive tract into the bloodstream and into cells needing them. Thus, the relative sizes of particles dissolved or dispersed in body fluids are matters of life and death.

Under ordinary gravitational forces, even large colloidal particles settle so slowly that they give the appearance of not being affected by gravity. Smaller colloidal particles successfully resist settling by a combination of thermal convection currents and the constant buffeting they receive from ions and molecules in the medium. These erratic motions of small colloidal particles can be seen as moving scintillations of light when a colloidal dispersion is viewed through a microscope. The phenomenon is called the *Brownian movement* after the English botanist, Robert Brown, who made the first detailed study of it.

15.5 SKY COLORS

If you are near a window on a reasonably clear day, fix your eyes on some part of the sky that is well away from the sun. If there were nothing in the earth's atmosphere to reflect or to scatter light in all directions, including back to your eyes,

TABLE 15.2
COLLOIDAL SYSTEMS

Type	Dispersed Phase[a]	Dispersing Medium[b]	Common Examples
Foam	Gas	Liquid	Suds, whipped cream
Solid foam	Gas	Solid	Pumice, marshmallow
Liquid aerosol	Liquid	Gas	Mist, fog, clouds
Emulsion	Liquid	Liquid	Cream, mayonnaise, milk
Solid emulsion	Liquid	Solid	Butter, cheese
Smoke	Solid	Gas	Dust in air
Sol	Solid	Liquid	Starch in water, jellies,[c] paints
Solid sol	Solid	Solid	Black diamonds, pearls, opals, alloys

[a]The colloidal particles constitute the dispersed phase.
[b]The continuous matter into which the colloidal particles are scattered is called the *dispersion medium*.
[c]Sols that adopt a semisolid, semirigid form (e.g., gelatin desserts, fruit jellies) are called *gels*.

the sky at that point would be completely black. Instead, it is blue. Why isn't it white, the "color" of the light radiated by the sun? The sun's white light consists, of course, of light frequencies corresponding to all the colors of the visible spectrum, from the longer-wavelength, lower-frequency red light to the short-wavelength, higher-frequency blue light. The extent to which light is scattered by particles in the air depends on its frequency. The higher the frequency, the more intensely is it scattered. According to the *Rayleigh scattering law* (after the third Lord Rayleigh, an English physicist, 1842–1919), the intensity of light scattered away from its initial direction varies directly as the *fourth* power of the frequency. Thus, of all the radiations streaming from the sun toward that spot in the sky that you are viewing, the blue portions are most intensely scattered back, and the red portions go on. If you view the sky near the sun at sunset or sunrise, it may appear one of several shades of yellow or red. Light at the red end of the spectrum gets through, but the blue components and others at that end of the spectrum are almost completely scattered as they pass to your view through the smoke- and dust-filled lower atmosphere.

15.6 PROPERTIES OF WATER

Water's abundance and its dissolving powers for chemicals are major facts of life. Life as we know it cannot be imagined without it. Space surveys of other planets nearly always include efforts to detect water's presence. Where water is shown to be absent, life is known to be absent.

Water makes chemical reactions possible between dissolved substances that otherwise are solids and therefore cannot intermingle. By dissolving them, water gives their ions or molecules

FIGURE 15-2
Tyndall effect as sunlight filters through the fog in the redwoods of California's Del Norte State Park. (U.S. Forest Service. Photo by L. J. Prator.)

15 THE EARTH'S
SUBSTANCES—INORGANIC MATERIALS

the mobility they lack in the solid state. In living systems, water transports chemicals to and from cells and holds them in solution while they are in various tissues. Outside the body, water is a servant or a master wherever we turn. Where it is either absent or frozen, there can be no agriculture and, hence, no permanent civilization. Feuds, skirmishes, battles, and wars—both physical and legal—have been waged over water rights. Water is also one of the great agents of erosion. It acts not just by physical wearing and transporting eroded materials but also by chemical reaction.

One of the most remarkable features of water is its dissolving power. For a great number of ionic substances, water can overcome ionic bonds in the crystalline lattices and cause the crystalline structure to disintegrate. What is the secret of water's ability and why do some things dissolve in water and others do not?

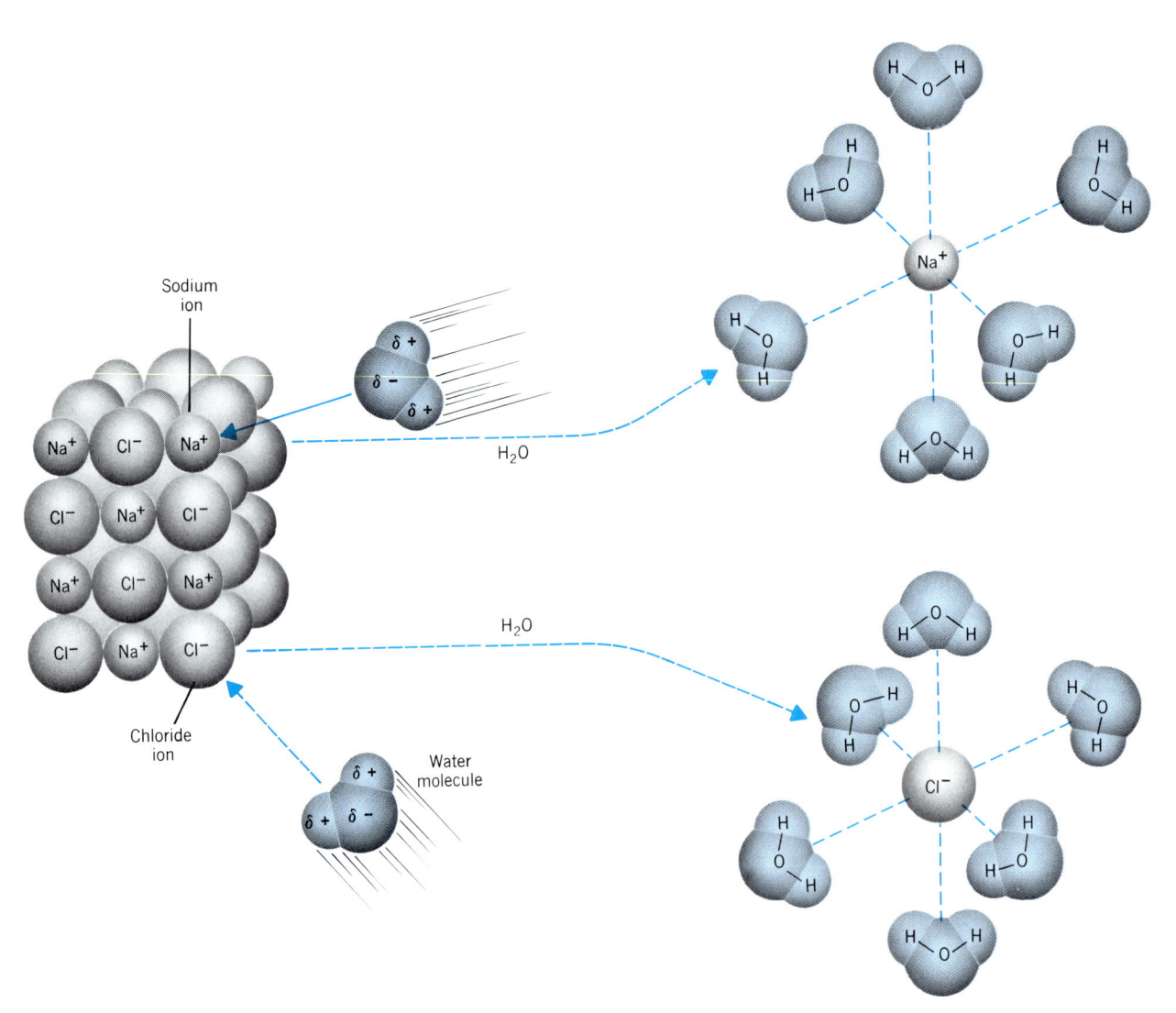

FIGURE 15-3
Factors making possible the dissolution of sodium chloride in water. On the left, polar water molecules are seen bombarding the surface of the crystal. A combination of mechanical bumping and electric forces of attraction dislodges ions on the surface and the edges of the crystal, and they go into solution. As they do so, on the right, they become surrounded by water molecules. Hydrated forms of the sodium and chloride ions are shown. (From J. R. Holum, *Elements of General and Biological Chemistry*, 4th edition, 1975, John Wiley & Sons, Inc., New York.)

15.7 HYDRATION

Partial charges are believed to be distributed in water molecules as follows:

$$\overset{\delta+}{H}\diagdown\underset{\delta-}{O}\diagup\overset{\delta+}{H}$$

Water molecules are very polar. Even though ions have outer octets, they are not very stable unless they are surrounded by ions of opposite charge as in a crystalline lattice. Ions must have something of opposite charge near by to be reasonably stable. A positive ion is more stable if its nearest neighbors are negatively charged, *but these neighbors need not carry whole negative charges. Partial* charges will also sometimes work, and water molecules have them. Figure 15–3 illustrates how water exerts its dissolving action on an ionic substance, such as sodium chloride. Ions that have become surrounded by water molecules are said to be *hydrated*. To break up a crystalline lattice of ions in air by heat alone, we have to provide considerable energy at a high temperature. In water, however, this energy cost is much less because water molecules, unlike air molecules, can surround the dissolved ions. The effectiveness of hydration in helping to dissolve substances varies widely, and each substance has its own particular solubility in water, as illustrated by data in Table 15.3. For a glossary of terms and a summary of some ways to express solubilities and concentrations, see Table 15.4.

15.8 HYDROGEN BONDS

While water is a good solvent for ionic compounds and polar molecular compounds, it does not dissolve everything (luckily!). Water molecules, after all, stick to each other, too. A site of partial positive charge on one water molecule cannot help but be attracted to a site of partial negative charge on another, as illustrated in Figure 15–4. These sites are so exposed that they can get quite close to each other. The forces of attraction between water molecules are consequently quite strong, so strong that we even say that bonds exist between them. Called *hydrogen bonds*, they are represented by dotted lines in Figure 15–4. They must not be confused with covalent bonds or with ionic bonds, although they are weak cousins of the latter. *Hydrogen bond* is the name for a force of attraction between *partial* charges in which the partial postive charge is on a hydrogen attached to a strongly electronegative element (e.g., fluorine, oxygen, or nitrogen).

When solute particles enter water, they disturb hydrogen bonds. To break any kind of bond costs energy, and this cost is repaid if the solute particles—ions or *polar* molecules—provide alternate places, substantially equivalent, to which water's partial charges can be attracted (see Figure 15–5). Thus, polar covalent substances whose molecules contain oxygen or nitrogen atoms are much more soluble in water than those which do not (see Figure 15–5). The only gas that is appreciably soluble in water (without reacting with it to any significant ex-

TABLE 15.3
SOLUBILITIES OF SOME SUBSTANCES IN WATER

Solute	Solubilities (g/100g water)			
	0°C	20°C	50°C	100°C
Sodium chloride, NaCl	35.7	36.0	37.0	39.8
Sodium hydroxide, NaOH	42	109	145	347
Barium sulfate, BaSO$_4$	0.000115	0.00024	0.00034	0.00041
Calcium hydroxide, Ca(OH)$_2$	0.185	0.165	0.128	0.077
Oxygen, O$_2$	0.0068	0.0043	0.0027	0
Carbon dioxide, CO$_2$	0.335	0.169	0.076	0
Nitrogen, N$_2$	0.0029	0.0019	0.0012	0
Ammonia, NH$_3$	89.5	53.1	—	—

TABLE 15.4
SOME COMMON WAYS OF EXPRESSING CONCENTRATIONS OF SOLUTIONS

Type	Definition	Example of Label	Translation of the Label
Quantitative			
Weight-volume percent (weight per unit volume)	The number of grams of solute per 100 ml of solution	10% sugar	The *ratio* of solute to solution is 10 grams sugar in every 100 ml of solution
Molar concentration (moles solute per liter solution)	The *molar concentration* of a solution is the number of moles of solute per liter of solution. Molar concentration = M = $\frac{\text{moles solute}}{\text{liters solution}}$	1.0M NaCl	The bottle with this label contains sodium chloride in water in the ratio of one mole sodium chloride per liter of solution
Qualitative Expression			
Saturated solution	One in which the solvent has dissolved the maximum amount of the solute it can at that temperature. Usually, undissolved solute will be at the bottom of the container, and undissolved solute and dissolved solute will be in dynamic equilibrium	Concentrated solution	A solution that contains a relatively high percent of solute. (Not to be confused with "saturated." A saturated solution of a solute with a very low solubility would not be described as concentrated)
		Dilute solution	A solution that contains a low percent of solute
Unsaturated solution	A solution that is less than saturated		

tent) is ammonia (see Table 15.2). Hydrogen bonds can form between molecules of water and ammonia as illustrated.

$$\underset{\text{Hydrogen bond}}{H-O-H \overset{\delta+ \quad \delta-}{\cdots} N\begin{array}{c}H\\|\\H\end{array}H}$$

15.9 THE IONIZATION OF WATER

According to the kinetic-molecular theory, water molecules in liquid water hit each other all the time. The energies of collision vary from mere nudges to some of great violence. Some are low-energy collisions; others have very high energy. Figure 15-6 shows two molecules of water on a collision course. The hydrogen of one will hit the oxygen of the other. If the collision is violent enough, the proton caught in the middle will momentarily be as close to one oxygen as the other. On the rebound, it could stay with the oxygen it hit, leaving *both* of the shared electrons on the first oxygen. It will get a share in a new pair on the other oxygen. This violent collision produces two ions, a hydronium ion, H_3O^+, and a hydroxide ion, ^-OH.

These ions are not very stable in each other's presence, but they will not find each other (or

15.9 THE IONIZATION OF WATER

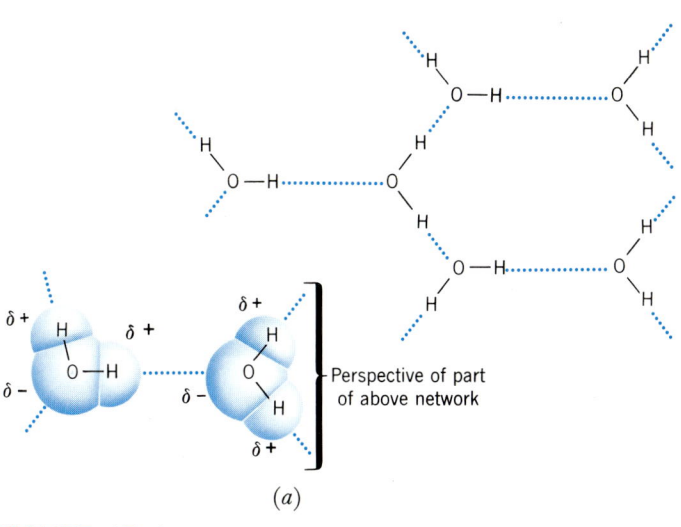

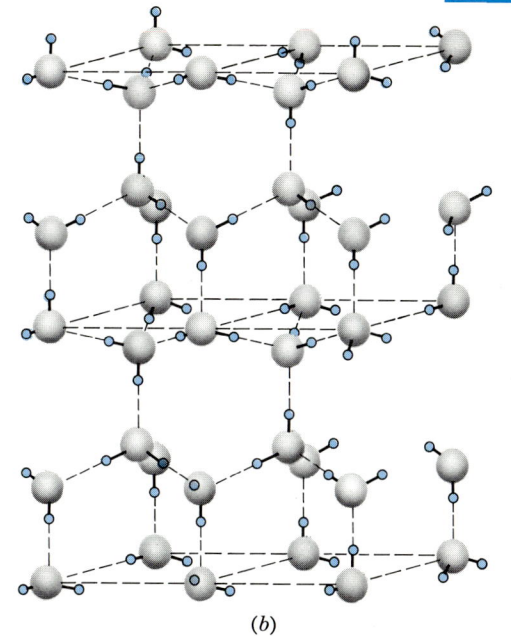

FIGURE 15-4
Forces of attraction between polar water molecules are strong enough that bonds, hydrogen bonds—represented in part *a* by dotted lines—are said to exist between them. The high degree of order implied in part *a* does not apply to liquid water. When water freezes to ordinary ice, however, the molecules form an intricate network of "cages" with hydrogen bonds furnishing the forces of attraction to hold the crystalline network in part *b* together. To build this lattice network, more volume is required than for the liquid state, and ice is less dense than water. (Crystal structure of ice from Linus Pauling, *The Nature of the Chemical Bond*, 3rd edition, © 1960 by Cornell University Press, Ithaca, N.Y. Used by permission of Cornell University Press.)

FIGURE 15-5
Molecules of covalent compounds can be dissolved in water if they are polar enough, particularly if they have structural features that make possible hydrogen bonds.

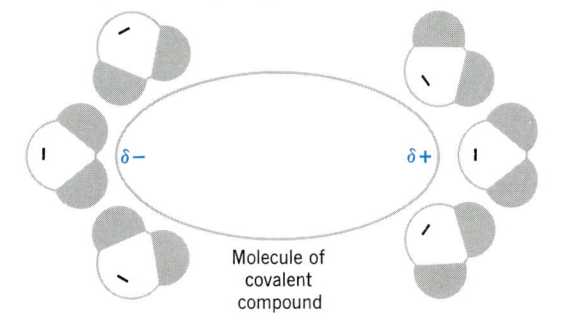

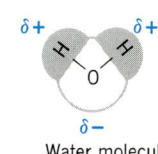

FIGURE 15-6
A collision of sufficient energy and proper orientation between two water molecules will produce a hydronium ion and a hydroxide ion. These are immediately hydrated (not shown). At room temperature, the fraction of these successful collisions is evidently very low for at any instant only about 10^{-7} percent of all water molecules have ionized.

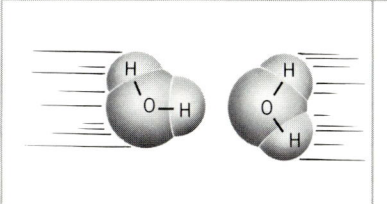

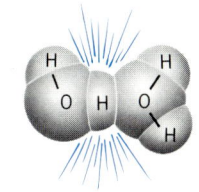

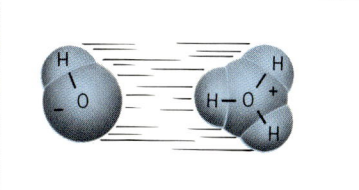

others produced elsewhere in water) right away. When they do, a low-energy collision is all that is needed to transfer the proton right back again.

$$H_3O^+ + {}^-\!:\!O\!-\!H \rightarrow H_2O + H\!-\!O\!-\!H$$

Therefore, at any given instant in a sample of pure water, there will be only a very small concentration of hydronium ions and hydroxide ions. Water and any aqueous solution in which the concentrations of these ions are equal are said to be *neutral*. If we add anything to water that makes these concentrations unequal, we have on our hands either an *acid* or a *base*. We shall say more about these later; that is where we are leading.

At room temperature (25°C), there is only 1×10^{-7} mole of hydronium ion (or hydroxide ion) per liter of water. (This amounts roughly to one hydronium ion for every 555 million molecules of water.)

15.10
DYNAMIC EQUILIBRIUM

The equation for the ionization of the water is usually written with oppositely pointing arrows to show that the reaction goes forward and reverse:

$$2H_2O \rightleftharpoons H_3O^+ + {}^-OH$$

The longer arrow points toward the species present in the higher concentration—the "favored" species.

The situation in water is one in which there is no *net* change in the concentration of any species over a period of time. Even though the forward and the reverse reactions constantly take place, they do so at *equal* rates, and a condition of *dynamic equilibrium* exists—much coming and going but no net change.

15.11
IONIZATION CONSTANT OF WATER

At any given temperature, the product of the concentrations of hydrogen ion and hydroxide ion is a constant. If the temperature is 25°C, and if the units are moles per liter, then[1]:

$$[H^+][OH^-] = (1 \times 10^{-7})(1 \times 10^{-7})$$
$$= 1 \times 10^{-14}$$

[1] Brackets, [ ], around a chemical formula signify units of *moles/liter* (see Table 15.4).

(It is standard practice to abbreviate the symbol for the hydronium ion to H^+ and to call it a hydrogen ion.) This constant is called the ionization constant of water. Its value changes with temperature, being higher at higher temperatures because violent, ion-producing collisions occur more frequently. At a given temperature, however, it is a constant *regardless of the solutes present*. Thus, if some chemical were added to water to make the concentration of hydronium ion 1×10^{-4} mole/liter, then the concentration of hydroxide ion (at 25°C) will adjust to a value of 1×10^{-10} mole/liter. We may calculate that as follows:

$$[H^+][{}^-OH] = 1 \times 10^{-14} \text{ (at 25°C)}$$
Hence,
$$(1 \times 10^{-4})(1 \times 10^{-10}) = 1 \times 10^{-14}$$

The adjustment in concentration of hydroxide ion occurs because at a higher hydronium ion concentration, hydroxide ions have more frequent opportunities to pick hydrogen ions (H^+) and revert to water molecules. Consequently, the concentration of hydroxide ions will diminish.

15.12
LE CHÂTELIER'S PRINCIPLE

One of the important facts about our world is summarized in a statement credited to Henry Louis Le Châtelier (1850–1936), a French inorganic chemist:

If a system is in stable equilibrium and something happens to change the conditions, the system will shift in whatever way will most restore equilibrium.

Pure water represents a stable equilibrium with equal (but very small) concentrations of hydronium and hydroxide ions. If we add a chemical to this system that can furnish hydronium ions but not hydroxide ions, we are "changing the conditions." The system changes or shifts; some of the hydroxide ions react with the added hydronium ions until the rates of the forward and reverse reactions in the ionization of water are again equal. At this point we now have new concentrations of H_3O^+ and OH^-, but the number we get when we multiply the values of these two (expressed in moles per liter) is still, at room temperature, 1×10^{-14}. Note that we do not restore the original *concentrations; we restore a condition of equilibrium*—a condition of no further net change in concentrations.

Another example of the working of Le Châtelier's principle occurs if we simply add heat to an ice slush. In this case, equilibrium means the amount of water present as ice does not change; the amount present as liquid does not change either, and the temperature is a constant 0°C. If we heat the slush we place a stress on one of the conditions, the temperature. The equilibrium shifts in response to the stress. It absorbs the heat using it to melt some of the ice. Assuming we do not overheat the system, melting will stop eventually. We no longer have the original amounts of ice and liquid water. We now have less water present as ice and more present as liquid, but we once again have equilibrium—a condition of no further net change in the amounts of solid and liquid. If we had cooled the slush, some liquid would have changed to ice. The equilibrium would have shifted in the opposite direction in response to an opposite kind of stress.

We turn next to substances that can make the concentrations of hydronium and hydroxide ions in water unequal: acids and bases.

15.13
ACIDS

The word *acid* generates fear. Acids can corrode and dissolve such common materials as cars and garbage cans and teeth. The dastardly Baron Gruner in Arthur Conan Doyle's "The Adventure of the Illustrious Client" was foiled in his murderous lunge for Sherlock Holmes when Kitty Winter hurled a vial of vitriol (sulfuric acid) into the baron's face. In TV ads, which often create the very nausea they would cure, we learn of the horrors that await those who do not promptly take this or that "antacid."

It's time someone spoke up for acids. They are neither good nor bad. Like fire they may be servants or menaces. Without acids neither your life nor civilization would be possible! Sulfuric acid, H_2SO_4, is by far the largest-volume industrial chemical used in American industry; annual production is over 32,000 tons. It is used in processing steel and in manufacturing fertilizers, batteries, dyes, explosives, drugs, and plastics. Your stomach juices include a low concentration of hydrochloric acid, and without it you could not efficiently digest meat or other protein products. Citric acid gives the pleasant tart taste to citrus fruits and many soft drinks. Ascorbic acid is vitamin C. Nicotinic acid is one of the B-vitamins. Acetic acid is present in vinegar and any salad dressing using vinegar. Why is it that all these things are classified as acids?

15.14
ARRHENIUS THEORY OF ACIDS AND BASES

The simplest definitions of acids and bases were first proposed by a young Swedish chemist, Svante August Arrhenius (1859–1927). His theory was part of his doctoral thesis, and his examiners (according to legend) thought so poorly of his ideas about ions that they gave him the lowest passing grade. He went on to win a Nobel Prize in 1903.

TABLE 15.5
COMMON ARRHENIUS ACIDS

Name	Formula	Percent Ionization in 0.1M Solutions
Strong acids		
Perchloric acid	$HClO_4$	~100
Hydrochloric acid	HCl	~100
Hydrobromic acid	HBr	~100
Hydriodic acid	HI	~100
Nitric acid	HNO_3	~100
Sulfuric acid[a]	H_2SO_4	60 (in 0.05M solution)
Moderate acid		
Phosphoric acid	H_3PO_4	27
Weak acids		
Acetic acid	$CH_3\overset{O}{\overset{\|}{C}}O{-}H$	1.3
Carbonic acid	H_2CO_3	0.2
Boric acid	H_3BO_3	0.01

[a]Concentrated sulfuric acid is a particularly dangerous chemical not only because it is a strong acid but also because it is a powerful dehydrating agent. This action generates considerable heat very quickly, and as the temperature rises other reactions are accelerated. Moreover, it is a thick, viscous liquid that does not wash away from skin or fabric at once.

The theory applies to acids and bases in water, not in any other solvent. According to Arrhenius (and we shall modernize his views a trifle), an acid is any substance that can produce hydronium ions in water. A base is any substance that can release hydroxide ions in water. Several common examples of acids are given in Table 15.5. Table 15.6 lists a few common bases. The reason, therefore, that such diverse things as ascorbic acid and sulfuric acid are called acids is that in water thy both release hydronium ions. Sulfuric acid, however, is called a strong acid because a large percent of its molecules will break up into ions in water. Ascorbic acid, acetic acid, nicotinic acid, and citric acid (all organic acids, incidentally) break up or ionize to only a very small percent. Therefore, we call them weak acids.

15.15 COMMON PROPERTIES OF ACIDS

These properties are really properties of the hydronium ion in water.

1. Acids have a sour taste.
2. Acids react with several metals, rapidly with some and very slowly with others. With still others (the precious metals, gold, platinum, silver, and copper), hydronium ions give no reactions. Lead reacts so slowly that sulfuric acid can be stored in lead (or glass) containers. Zinc is a metal that reacts with sulfuric acid rather vigorously:

$$Zn + H_2SO_4 \rightarrow ZnSO_4 + H_2$$
Zinc / Sulfuric acid / Zinc sulfate / Hydrogen

Actually, when a piece of zinc is placed in a dilute solution of sulfuric acid, atoms of zinc are attacked by hydrogen ions (and we now use the handy abbreviation of hydronium ion; we use H^+ to stand for H_3O^+).

$$Zn + 2H^+ \rightarrow Zn^{2+} + H_2$$
Zinc atom / Hydrogen ions / Zinc ion / Hydrogen

This is a new kind of equation, an *ionic equation*. The equation is balanced in two ways. First, because all nuclei on the left are found somewhere on the right of the arrow, we have a "material balance." Moreover, the net charge on the left (+2) is exactly balanced by a net charge of +2 on the right, and we have an "electrical balance." No equation in chemistry is balanced unless it has both a material and an electrical balance. The helpful feature of an ionic equation is that it excludes all those particles that are present (e.g., SO_4^{2-} ions) as "spectators" but that are not involved in any chemical change. (The name *ionic equation* is a trifle misleading. Obviously, not all the formulas shown are of ions.)

Iron, aluminum, and magnesium are other metals that react with hydrogen ions. Iron does so slowly but surely, and as it reacts it corrodes. In tin cans, steel (an alloy of iron) is coated with a thin layer of a much less reactive metal, tin, in order that any food in the tin can that might be slightly acidic will not eat its way out. Many alloys of iron—many special steels—are exceptionally resistant to acids.

3. Acids react with carbonates and bicarbonates. Here is the chemistry of the antacid action of bicarbonate of soda, $NaHCO_3$.

TABLE 15.6
COMMON ARRHENIUS BASES

Name	Formula	Solubility in Water[a]	Percent Dissociation
Strong bases			
Sodium hydroxide	NaOH	109	91 (in 0.1M solution)
Potassium hydroxide	KOH	112	91 (in 0.1M solution)
Calcium hydroxide (an aqueous solution is called "limewater")	$Ca(OH)_2$	0.165	100 (saturated solution)
Magnesium hydroxide (a slurry in water is called "milk of magnesia")	$Mg(OH)_2$	0.0009	100 (saturated solution)
Weak base			
Ammonium hydroxide	$NH_{3(aq)}$ ("NH_4OH")	89.9 grams NH_3 at 0°C	1.3 (at 18°C)

[a] Solubilities are in grams of solute per 100 grams of water at 20°C except where noted otherwise.

15.15 COMMON PROPERTIES OF ACIDS

$$\text{HCl} + \text{NaHCO}_3 \rightarrow$$
Hydrochloric acid (as in "stomach acid"), Sodium bicarbonate

$$\text{CO}_2 + \text{H}_2\text{O} + \text{NaCl}$$
Carbon dioxide, Water, Sodium chloride

The fizzing action as certain antacid tablets (e.g., AlkaSeltzer) or soft drink tabs (e.g., Fizzies) dissolve in water is caused by carbon dioxide, the gas released in this reaction. The acid in these tablets is citric acid, a solid, organic acid. As long as the citric acid and sodium bicarbonate powders are kept dry, their ions and molecules cannot intermingle and react. When dropped into water, these solid mixtures promptly dissolve and then they can intermingle and react.

Calcium carbonate is found in nature both as limestone and marble. It unites quartz particles in sandstone. These substances have been widely used as building materials, with marble reserved for the finest effects and for statues and monuments. Buildings and art objects made of these substances suffer heavily in areas of air pollution, because moist polluted air contains traces of sulfuric and sulfurous acid (H_2SO_3) (see Figure 15-7). Sulfuric acid attacks calcium carbonate as follows:

$$\text{H}_2\text{SO}_4 + \text{CaCO}_3 \rightarrow$$
Sulfuric acid, Calcium carbonate

$$\text{CaSO}_4 + \text{CO}_2 + \text{H}_2\text{O}$$
Calcium sulfate, Carbon dioxide, Water

The calcium sulfate leaches away as damage to the building material continues.

4. Acids react with metal hydroxides.

$$\text{HCl} + \text{NaOH} \rightarrow$$
Hydrochloric acid, Sodium hydroxide

$$\text{NaCl} + \text{H}_2\text{O}$$
Sodium chloride, Water

FIGURE 15-7
Stone decay in the polluted atmosphere of Germany's Rhein-Ruhr. This statue of porous Baumberg sandstone decorates the Herten Castle near Recklinghausen, Westphalia, West Germany. (Left) Appearance in 1908 after 206 years; (right) appearance in 1968 after only 60 more years. (Photos by Schmidt-Thomsen, Landesdenkmalamt, Westfalen-Lippe, Münster, West Germany.)

This reaction is called *neutralization*—acid-base neutralization. It actually occurs simply between the hydroxide ion and the hydrogen ion:

$$H^+ + OH^- \to H_2O$$

Any chemical event that takes hydrogen ions out of water and ties them up in some way can be called neutralization. Thus, carbonates and bicarbonates also neutralize acids.

5. The hydronium ion is a powerful catalyst in many reactions affecting the health of living things or the safety of other materials.
6. Acids turn blue litmus red. Litmus is a naturally occurring dye that changes color from blue to red in acid and from red to blue in base. In a neutral solution it will be of intermediate color. Litmus is usually sold as "litmus paper," strips of porous paper impregnated with the dye. Chemists have several dyes called *indicators* that they use to do what their name implies—indicate the acidity or basicity of a solution. Several are listed in Table 15.7. The most sensitive test is done with an instrument called a pH meter.

15.16
THE pH CONCEPT

Solutions in which the concentrations of hydronium ions are in the range of 10^{-1} to 10^{-14} mole/liter are so commonly encountered that chemists have adopted a shorthand way of describing their concentrations—the *pH* of the solutions. The pH of a solution is the negative power to which the number 10 must be raised to describe the hydrogen ion concentration in moles per liter.

$$[H^+] = 1 \times 10^{-pH}$$

If the hydrogen ion concentration of a solution is 1×10^{-4}, then the pH of the solution is 4. A solution with a pH of 8 would have a concentration of hydrogen ion of 1×10^{-8} mole/liter. If the pH is 7.4, then the hydrogen ion concentration is $1 \times 10^{-7.4}$ mole/liter. This number could be expressed as $10^{0.6} \times 10^{-8}$, but what does $10^{0.6}$ mean? It is somewhere between 10^0 (which is 1) and 10^1 (which is 10). To determine it more exactly we would have to use logarithms, a mathematical concept not assumed to be known by the users of this text.[2] We shall therefore have to be content with learning some simple relations between values of pH and conditions of acidic,

basic, or neutral for solutions. A neutral solution is easy. Its pH must be 7 at room temperature (why?). Values of pH less than 7 correspond to acidic solutions; values greater than 7 are for basic solutions.

The pH meter (Figure 15–8) measures the pH of a dilute solution quickly and accurately. One place requiring this is the clinical laboratory of a hospital and certain of its emergency rooms. The pH of your blood is normally about 7.35. If it falls too close to 7, you will die. If it rises upward to 7.9, you likewise will die. (Only at a blood pH close to 7.35 does the blood effectively transport oxygen.) These changes in pH represent extremely small changes in hydrogen ion concentration. If you were to add one-tenth of a drop of concentrated hydrochloric acid to a liter of pure water (pH 7) you probably would not be able to detect any change in the taste of the water. Yet the pH changes by about 0.3 units, enough of a change to kill you if it happened to your blood. Fortunately, the blood carries a number of chemicals, including the bicarbonate ion, to help protect it against changes in its pH. These chemicals stand ready to neutralize any acid that might enter the blood, and chemical reactions in cells of the body generate organic acids all the time. Substances that protect a solution such as blood against any change in pH are called *buffers*.

15.17
BASES

The most common strong *base* is sodium hydroxide (lye, soda lye, caustic soda). It is a base because it releases *hydroxide ions* in water. It is a strong base because the *percent* ionization is high. Another strong base is potassium hydroxide, or caustic potash. Weak bases include magnesium hydroxide (milk of magnesia) and aqueous ammonia (ammonium hydroxide). Bases in water share the following properties:

1. They are bitter to the taste, and they feel slippery or soapy between the fingers.
2. They turn red litmus blue.
3. They react with acids (see Section 15.15).

---

[2] For those who are familiar with logarithms, pH is better defined by the equation

$$pH = \log \frac{1}{[H^+]} = -\log [H^+]$$

Remember that brackets, [ ], signify the units of moles per liter.

15.17 BASES

TABLE 15.7
COLORS OF SOME COMMON INDICATORS AT VARIOUS pH RANGES AND APPROXIMATE pH VALUES FOR SOME COMMON SUBSTANCES AT ROOM TEMPERATURE

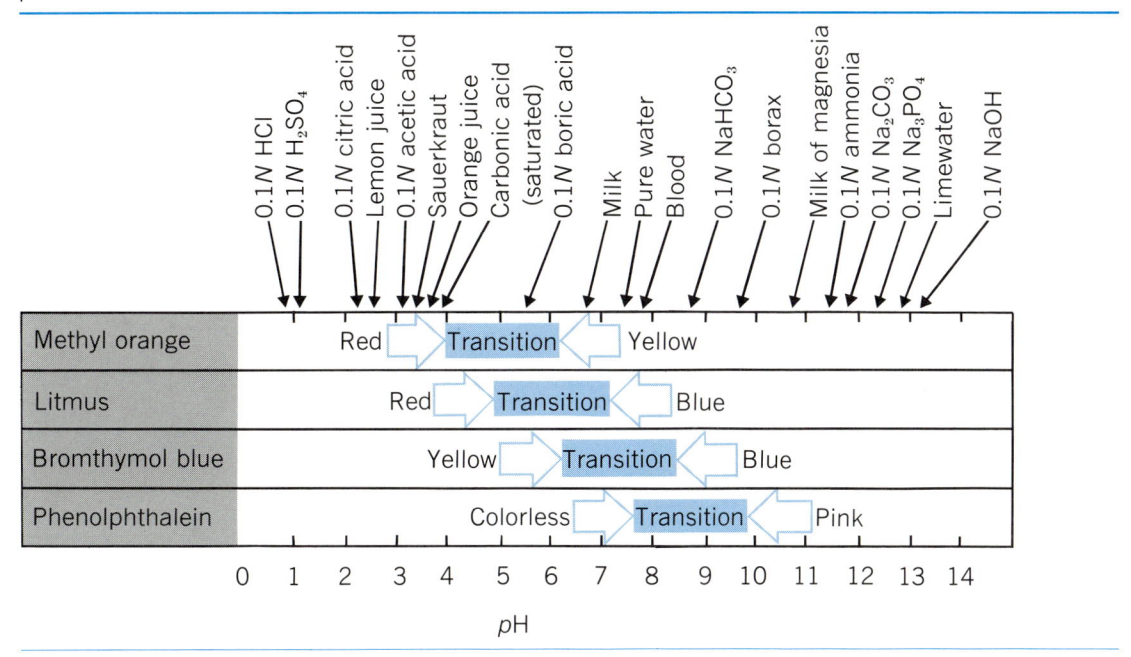

Substances (by approximate pH): 0.1N HCl, 0.1N H$_2$SO$_4$, 0.1N citric acid, Lemon juice, 0.1N acetic acid, Sauerkraut, Orange juice, Carbonic acid (saturated), 0.1N boric acid, Milk, Pure water, Blood, 0.1N NaHCO$_3$, 0.1N borax, Milk of magnesia, 0.1N ammonia, 0.1N Na$_2$CO$_3$, 0.1N Na$_3$PO$_4$, Limewater, 0.1N NaOH

Indicator	Colors across pH 0–14
Methyl orange	Red → Transition ← Yellow
Litmus	Red → Transition ← Blue
Bromthymol blue	Yellow → Transition ← Blue
Phenolphthalein	Colorless → Transition ← Pink

pH scale: 0 1 2 3 4 5 6 7 8 9 10 11 12 13 14

(Adapted from J. R. Holum. *Elements of General and Biological Chemistry,* 4th edition. 1975. John Wiley & Sons, Inc., New York.)

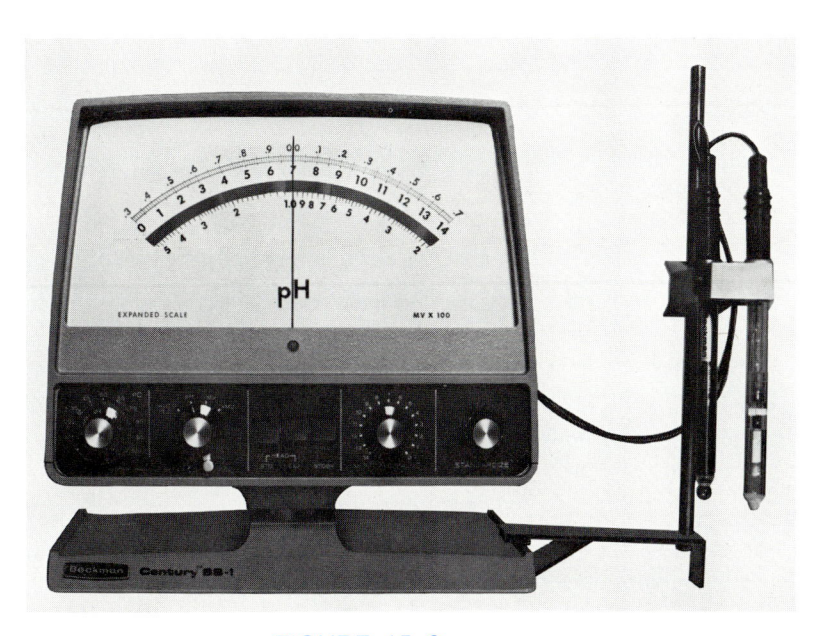

FIGURE 15–8
pH meter. The electrodes, on the right, are dipped into the solution and the pH of the solution is read on the dial. (Courtesy Beckman Instruments.)

TABLE 15.8
IMPORTANT SALTS AND THEIR USES

Formula and Name	
$(NH_4)_2CO_3$ Ammonium carbonate	Aromatic spirits of ammonia, "smelling salts" (together with ammonium hydroxide and the aromatic oils of nutmeg, lemon, and lavender)
$BaSO_4$ Barium sulfate (barite)	White pigment for rubber goods, paper, oilcloth, linoleum; used in x-raying gastrointestinal tract ("barium cocktail")
$(CaSO_4)_2 \cdot H_2O$ Calcium sulfate (plaster of paris)	Plaster casts; wall stucco; wall plaster
$MgSO_4 \cdot 7H_2O$ Magnesium sulfate (epsom salt)	Purgative; tanning and dyeing
$HgCl$ Mercury(I) chloride (mercurous chloride; calomel)	Purgative; diuretic
$HgCl_2$ Mercury(II) chloride; (mercuric chloride; corrosive sublimate)	Disinfectant, in dilute solutions, for hands and instruments that cannot be boiled; poisonous
$KMnO_4$ Potassium permanganate	In 0.02% solution—disinfectant for irrigation of urinary tract, vagina, and infected wounds In 1% solution—for local treatment of athlete's foot and poison ivy
$AgNO_3$ Silver nitrate (lunar caustic)	Antiseptic and germicide (used in eyes of infants to prevent gonorrheal conjunctivitis); photographic film sensitizer; indelible ink; hair dyes
$NaHCO_3$ Sodium bicarbonate (baking soda)	Baking powders, effervescent salts; stomach antacids; fire extinguishers
$Na_2CO_3 \cdot 10H_2O$ Sodium carbonate (soda ash; sal soda; washing soda)	Water softener; used in soap and glass manufacture
SnF_2 Tin(II) fluoride (stannous fluoride; Fluorostan)	Toothpaste additive to combat dental caries

4. They are often powerful catalysts or promoters of reactions affecting the health of living things or the safety of other substances.

Sodium hydroxide is used in industry to make soap, dyes, drugs, plastics, and other materials. In the house and apartment, it may be found in certain drain cleaners and oven cleaners where warnings on labels must be followed very carefully. Sodium hydroxide is truly a caustic material that can, in concentrated form, act rapidly to destroy tissue (skin or eyes, for example).

15.18 SALTS

Substances most frequently classified as *salts* are ionic compounds whose positive ions are from an Arrhenius base and whose negative ions are from an Arrhenius acid. A number of examples are listed in Table 15.8.

It is difficult to generalize about salts. Unlike acids or bases, there is no one ion common to all salts. At room temperature, salts are solids, but beyond that generalization their properties depend on the specific ions present.

TABLE 15.9
SOME COMMON HYDRATES

Formulas	Names	Decomposition Modes and Temperatures[a]	Uses
$(CaSO_4)_2 \cdot H_2O$	Calcium sulfate sesquihydrate (plaster of paris)	$-\frac{1}{2}H_2O$ (163°C)	Casts, molds
$CaSO_4 \cdot 2H_2O$	Calcium sulfate dihydrate (gypsum)	$-2H_2O$ (163°C)	Casts, molds, wallboard
$CuSO_4 \cdot 5H_2O$	Copper(II) sulfate pentahydrate (blue vitriol)	$-5H_2O$ (150°C)	Insecticide
$MgSO_4 \cdot 7H_2O$	Magnesium sulfate heptahydrate	$-6H_2O$ (150°C)	Cathartic in medicine
		$-7H_2O$ (200°C)	Used in dyeing and tanning
$Na_2B_4O_7 \cdot 10H_2O$	Sodium tetraborate decahydrate (borax)	$-H_2O$ (33.5°C)	Water softener
$Na_2SO_4 \cdot 10H_2O$	Sodium sulfate decahydrate (Glauber's salt)	$-10H_2O$ (100°C)	Cathartic
$Na_2S_2O_3 \cdot 5H_2O$	Sodium thiosulfate pentahydrate	$-5H_2O$ (100°C)	Photographic developing

[a] Loss of water is indicated by the minus sign before the symbol, and the loss occurs at the temperature given in parentheses.

TABLE 15.10
SOLUBILITY RULES FOR COMMON SALTS

All sodium, potassium, and ammonium salts are soluble, regardless of the counter ion[a]
All nitrates and acetates are soluble, regardless of the counter ion
All chlorides, except those of lead, silver, and mercury(I), are soluble
Salts not in the above categories are generally insoluble or, at best, only slightly soluble[b]

There are exceptions to these rules, but we will not be wrong very often in applying the rules.

[a] The *counter ion* is the oppositely charged ion present together with the named ions.
[b] *Soluble* means at least to the extent of a 3 percent solution.

15.19 HYDRATES

Many salts crystallize from water in such a way that hydrated water molecules stay with the ions. Salts that contain intact molecules of water in definite proportions are called *hydrates*. Several examples are given in Table 15.9. Notice that their formulas are written showing that water molecules are present intact. The water of hydration can be expelled from hydrates by heating them, and the residue is called the *anhydrous* form of the salt. Anhydrous calcium chloride will combine with water in moist air, and it is a common chemical dessicant or drying agent. When plaster of paris is mixed with water, gypsum forms—a hydrate with a higher proportion of water. Accompanying this change is a small increase in volume and the material hardness. That is why it makes a good casting material. As it hardens it expands into small nooks and crannies of the mold.

15.20 SOLUBILITIES OF SALTS

Solubilities of salts vary widely. Generalizations about their solubilities are summarized in Table 15.10. We can say, based roughly on an arbitrary decision, that a salt is soluble if it will dissolve to the extent of 3 grams per 100 cm² of water.

15.21 THE COMMON REACTIONS OF IONS

The common reactions of solutions of Arrhenius acids, bases, and salts are of the following types.

1. A new, water-insoluble substance forms.
2. A gas is produced that largely bubbles out of solution.
3. A new, un-ionized, water-soluble molecule is produced from two oppositely charged ions.

The first is illustrated by the change that occurs when separate solutions of sodium chloride and silver nitrate are mixed. If no change happened, the new solution would contain the ions: Na^+, Cl^-, Ag^+, and NO_3^-. But one combination of these oppositely charged ions, AgCl, represents a water-insoluble salt. (In Table 15.10, "all chlorides are soluble *except* those of lead, silver, and mercury(I).") The following event will therefore take place, leaving sodium ions and chloride ions behind in solution.

$$Ag^+ + Cl^- \longrightarrow AgCl\downarrow$$

The second type of reaction, above, has been illustrated by the behavior of carbonates and bicarbonates toward acids. Carbon dioxide is produced.

The third type of reaction, above, has been illustrated by the reaction between a hydrogen ion and a hydroxide ion to produce a water molecule.

15.22 COORDINATION COMPOUNDS

Many metallic ions, especially those of the transition metals, can form bonds to neutral molecules or to nonmetallic ions without necessarily becoming insoluble in water. The molecules or ions that tie up metal ions are called *ligands* (from *ligare*, Latin for "to bind or tie"). Water is an example of a ligand whenever it hydrates ions. The chemical event of tying up a metallic ion in solution—forming a coordinate covalent bond to it—is called *coordination*. The products, when isolated in a pure state, are called *coordination compounds*. The ligand donates an unshared pair of electrons to share with the positively charged metal ion as the bond is formed. Unshared electrons on nitrogens are common in molecules of living systems. The importance of vitamins and minerals in our diets is well known. The minerals we need include trace amounts of such metallic ions as Fe^{2+}, Cu^{2+}, Co^{2+}, Zn^{2+}, Mn^{2+}, Mg^{2+}, and Mo^{2+}—that is, ions from iron, copper, cobalt, zinc, manganese, magnesium, and molybdenum (the list is only partially complete). For example, the iron(II) ion, Fe^{2+}, is essential to the structure of hemin, the red-colored, nonprotein portion of hemoglobin in red blood cells. The iron coordinates with unshared pairs of electrons on nitrogens in the center of this complex molecule. Chlorophyll in all green plants is a metal coordination compound containing the magnesium ion. Many systems of enzymes are also metal coordination compounds, and these systems are primarily the ones that need most of the trace minerals essential to health. (The enzymes are catalysts essential for all chemical processes in living things.)

Some of the most powerful, rapid-acting poisons consist of ligands, which are strong binders of trace metallic ions in living systems. By tying up these ions, they turn enzyme-catalyzed reactions off. The cyanide ion acts on trace ions needed by cells to use molecular oxygen. Carbon monoxide, CO, is a strong ligand for iron(II) ions, Fe^{2+}. By coordinating with them they make it impossible for hemoglobin to transport oxygen, a much weaker ligand.

Many modern detergents contain the diphosphate and the triphosphate ion. These are strong ligands for the metallic ions responsible for hard water—Ca^{2+} and Mg^{2+} (and sometimes Fe^{2+} or Fe^{3+}). The abundance of calcium and magnesium ions in the substances of rocks and minerals means that some of these ions will be leached by ground water as it percolates into wells and springs. These ions form water-insoluble precipitates with the negative ions of common soap, and that is why water containing these ions is called "hard."

$$\begin{array}{cc} \text{Diphosphate ion} & \text{Triphosphate ion} \end{array}$$

The diphosphate and the triphosphate (or "tripolyphosphate" ion), shown above in their fully ionized forms, can coordinate or "sequester" the metallic ions in hard water *without forming precipitates* and without generating excessive foam.

SUMMARY

Types of compounds. Two broad classes of compounds are the inorganic and the organic. Organic compounds are compounds of carbon (except carbonates, bicarbonates, and cyanides). Inorganic compounds are all others, although there are many on the borderline, compounds involving metals and organic substances. Virtually all organic compounds are molecular substances. Most inorganic compounds are ionic substances. Among the inorganic and organic compounds, other useful (and overlapping) classes are acids, bases and salts.

Types of mixtures. Properties of homogeneous mixtures—solutions and colloidal dispersions—depend in part on the materials mixed together and in part on the sizes of their particles. The third kind of mixture, usually less homogeneous, is the suspension. In true solutions, the dissolved particles are small ions or molecules. In colloidal dispersions, they are larger—either large molecules or aggregations of smaller ones. In suspensions the suspended particles are large enough eventually to settle out or be trapped in a filter.

Water undergoes a small amount of self-ionization; at equilibrium at room temperature, the concentration of hydrogen ions equals the concentration of hydroxide ions, or 1×10^{-7} mole per liter. The product of the two concentrations, called the ionization constant of water is, at 25°C, 1×10^{-14}. The high polarity of water aids it in dissolving ionic compounds and the more polar molecular compounds.

Hydrogen bond. The force of attraction between the partially positively charged hydrogen of one molecule of water and the partially negatively charged oxygen of another molecule is strong enough to be named—the hydrogen bond. Its existence helps explain the unusually high boiling and melting points of water and its exceptional thermal properties.

Le Châtelier's principle gives us a way to predict how certain things will change. It says that if a system is in a stable equilibrium and something happens to change the conditions, the system will shift in whatever way will most restore equilibrium. One equilibrium described in this chapter is that of hydrogen and hydroxide ions with water in pure water.

Acids make the concentration of hydrogen ions greater than hydroxide ions in water, and those that can undergo an especially high percent breakdown into hydrogen ions are strong acids. In brief, acids release hydrogen ions in water. They also give it a tart taste; will change litmus from blue to red; react with metals to change them to salts (and release hydrogen); react with carbonates or bicarbonates to make salts, water, and carbon dioxide; and react with metal hydroxides also to form salts and water.

Bases produce a higher concentration of hydroxide ions than hydrogen ions in water. A base in water will change red litmus to blue; have a bitter taste and a slippery feel; and neutralize acids.

Salts are ionic compounds whose positive ions are anything but hydrogen ions and whose negative ions are anything but hydroxide ions. When they dissolve in water they generally break up into their ions. Water handles this especially well because its polar molecules can surround charged particles shielding them from each other. Certain salts called hydrates crystallize together with definite proportions of water. Coordination compounds (or a coordination ion) are special kinds of compounds resulting from the binding together of a neutral molecule with an ion.

pH concept. We estimate the concentration of hydrogen ions in water either by indicator dyes (e.g., litmus) or by a pH meter. The pH is the negative power to which 10 must be raised to express the hydrogen ion concentration in moles per liter:

$$[H_3O^+] = 1 \times 10^{-pH}$$

The higher the concentration of hydrogen ion (or, more accurately, hydronium ion, H_3O^+) the lower the pH. The pH of pure water at room temperature is 7.

TERMS AND CONCEPTS

organic compound
inorganic compound
homogeneous mixture
solution, true
solvent
solute
colloidal state
emulsion
colloidal dispersion
gel
sol
suspension
Tyndall effect
Brownian movement
Rayleigh scattering law
hydration
hydrogen bond
equilibrium, dynamic
Le Châtelier's principle
Arrhenius theory
acid
hydronium ion
hydroxide ion
neutral solution
neutralization
indicator
pH
base
salt
hydrate
solubility
coordination compound
ligand
concentration
parts per million

parts per billion
percent concentration
unsaturated solution
saturated solution
molar

PROBLEMS AND EXERCISES

1. What can colloidal-sized particles dispersed in water or air do that smaller particles of a true solution cannot do?

2. What can colloidal-sized particles do that those of a larger size in a suspension cannot do?

3. What is one of the many reasons why the digestive system must break up huge molecules found in many foods?

4. Why does a sunset often have colors we do not see at noontime?

5. Pure water is really transparent to all colors. Suggest a reason why the ocean is blue.

6. How is it that a cup of water at room temperature can break up a salt crystal when the action of heat alone on the crystal does nothing to it until you get to unusually high temperatures? (What holds the crystal together in the face of heat and why doesn't it work in the face of water?)

7. When you heat water you increase the frequency with which its molecules hit each other, and you increase the average energy or violence of the collisions. We learned that at room temperature the concentration of hydrogen ions and hydroxide ions in pure water is for each 1×10^{-7} moles per liter. (a) What must happen to the size of this number if you heat water? (b) Why?

8. Suppose you change the temperature of pure water in such a way that the concentration of hydrogen ions changes to 5×10^{-7} moles. (a) What *must* be the new value of the concentration of hydroxide ions? (b) Why must it change to this new value? (c) At this new temperature, what will then be the value of the ionization constant of water.

9. Hydrochloric acid attacks potassium bicarbonate ($KHCO_3$) in the same way it reacts with sodium bicarbonate (Section 15.15). Write the full equation for that reaction.

10. A particular water pollutant changes the pH of the water in a nearby stream to 5. What effect would this have on the limestone used in the piers of a nearby bridge? Why?

11. What happens chemically when an acid is neutralized?

12. When dissolved in water, sodium nitrate is dissociated into separate sodium ions (Na^+) and nitrate ions (NO_3^-). In water, potassium chloride is similarly dissociated into potassium ions (K^+) and chloride ions (Cl^-). Suppose you have in one test tube a solution of sodium nitrate and in another test tube a solution of potassium chloride. Then you pour one into the other. Do the various combinations of oppositely charged ions leave each other alone, or does one combination become an insoluble salt and precipitate? (Write the formulas of the possible combinations and ask yourself if each one represents a salt that is soluble in water according to the solubility rules of Table 15.10.)

13. Repeat question 12 using the mixing of solutions in water of each of the following pairs of compounds: (a) sodium sulfate, Na_2SO_4, and barium nitrate, $Ba(NO_3)_2$. (b) sodium carbonate, Na_2CO_3, and calcium chloride, $CaCl_2$.

14. Washing soda is the decahydrate of sodium carbonate, $Na_2CO_3 \cdot 10H_2O$. (a) What ions separate (dissociate) when this is dissolved in water? (Ignore the small self-ionization of water.) (b) Which of these ions will combine with a magnesium ion (Mg^{2+}) or a calcium ion (Ca^{2+}) and form an insoluble precipitate? (c) We learned in Section 15.22 that ground water that contains calcium or magnesium ions is called hard water. (It is "hard" to use it for washing things with ordinary soap.) If we take out these ions or in any way tie them up, the water will then be soft water. The question now is this. Does washing soda, when added to hard water, soften it? If so, write an equation for the action.

CHAPTER 16
NONMETALLIC INORGANIC SUBSTANCES

16.1 SPHERES OF ACTION

Our natural world has many parts or "spheres of action"—the biosphere, lithosphere, hydrosphere, and atmosphere. While they are familiar and easily recognized, no simple dividing line exists between them. We use them because they are helpful points of departure for our study.

The *biosphere* is that part of the world where living things exist, and life is almost everywhere. The biosphere intermingles with the *lithosphere,* the solid earth. Living things are in the *hydrosphere,* the waters of the planet, wherever water occurs. There are also living things in the *atmosphere,* the gaseous envelope of the earth.

Nature uses nonmetallic elements almost entirely to make the materials of the biosphere, the atmosphere, and the hydrosphere. The metallic elements that are present in these regions are only in low concentrations as their ions. Most of the substances in the biosphere are organic compounds or mixtures of them. They are important enough to merit separate chapters—18 and 19. Nonmetallic inorganic substances are found combined with metals in the lithosphere as various rocks, minerals, and ores. These, too, merit their own chapters—17 and 23. In this chapter we concentrate mainly on nonmetallic elements and compounds that are in the atmosphere whether as substances needed for life or harmful to life. They enter and leave the atmosphere in various ways and are parts of major natural cycles, three of which we shall study—the oxygen, nitrogen, and carbon cycles.

16.2 THE ATMOSPHERE

The earth's atmosphere is a mixture made up almost entirely of oxygen and nitrogen but with trace concentrations of several other elements and gaseous compounds. The forces that create winds, storms, rainfalls, and the weather in general will be the subject of parts of Chapter 22. We look only at the chemicals in the atmosphere here. Nitrogen (N_2) makes up 79 percent of the volume of air. It dilutes oxygen, which is 21 percent of air and is highly reactive, to the concentration to which life has adapted.

16.3 THE OXYGEN FAMILY

The members of the *oxygen family* and some of their compounds are listed and described in Table 16.1. All the atoms in the oxygen family have outside energy shells of six electrons. Therefore, these elements have similar bond-forming abilities (valences), which explains how they form compounds with similar formulas. Our study of this family will be limited mostly to oxygen, sulfur, and some oxides.

16.4 OXYGEN OCCURRENCE

Oxygen is the most abundant element on our planet. Its nuclei account for almost 50 percent of the estimated weight of the earth's crust, and they are about one-fourth of the nuclei in living

TABLE 16.1
THE OXYGEN FAMILY—GROUP 6

Element	Symbol	Melting Point (°C)	Boiling Point (°C)	Appearance (at room temperature and atmospheric pressure)
Oxygen	O	−218	−183	Colorless gas
Sulfur	S	113[a]	445	Yellow, brittle solid
Selenium	Se	217	685	Bluish-gray metal
Tellurium	Te	452	1390	Silvery-white metal
Polonium	Po	254	962	Intensely radioactive metal (an emitter of α- and γ-rays); a half gram sample in a capsule would heat to 500°C

Some Compounds

Hydrides Formula	Boiling Point (°C)	Oxides Formula	Boiling Point (°C)
H_2O	100	O_3 (ozone)	−112
		SO_2	−10
H_2S	−61	SO_3	−45
		SeO_2	melting point > 300 (sublimes)
H_2Se	−42		
		SeO_3	melting point 118
H_2Te	−2	TeO	decomposes
		TeO_3	decomposes
—	—	PoO_2	decomposes

[a]Elemental sulfur exists in more than one crystalline form. In rhombic crystals (α-form) it melts at 113°C; in monoclinic needles (β-form) it melts at 119°C; and in a noncrystalline, amorphous form it softens and melts at about 120°C.

things. The world's lakes and oceans are nearly 90 percent oxygen by weight. In clean air over the oceans between latitudes 50 degrees north and 60 degrees south the oxygen content is 20.946 percent by volume. Essentially no change in the percentage of oxygen in the atmosphere has been detected since 1910. Our planet's supply of oxygen is huge. Geochemist Wallace Broecker (Columbia University) estimates that in the air above each square meter of the earth's surface are 60,000 moles of oxygen.

16.5
OXYGEN CYCLE

The constancy of the oxygen in the atmosphere even while it is used and remade means that a grand chemical cycle exists in nature, an *oxygen cycle*. We know that oxygen is constantly being consumed by respiration, decay, and combustion. It is being regenerated in green plants by photosynthesis.

Living things use oxygen in their *respiration*, the set of events by which they remove oxygen from the surroundings, put it into circulation to cells needing it, and exchange it for waste products.

Decay is another process using oxygen. When living things die they are slowly broken down, with the aid of microorganisms that use oxygen, to simple chemicals—carbon dioxide, water, minerals, and certain nitrogen compounds. Even some rocks undergo decay as oxygen slowly attacks their minerals. Many metals exposed to the air are attacked by oxygen. The rusting of iron is the change of iron to a mixture of various iron oxides.

Besides respiration and decay there is a third way in which oxygen is consumed, the familiar process of *combustion*. The distinction between combustion and decay is a matter of speed. Combustion is very rapid and heat is produced quickly. Decay is much slower and the heat produced escapes as rapidly as it is generated. When it cannot escape, the heat may build up enough to raise the material's temperature to its kindling temperature, the temperature at which it ignites in air. Spontaneous combustion then occurs; this often happens in piles of oily rags or in haylofts of barns.

Except for the interactions of oxygen with metals or minerals, respiration, decay, and combustion not only consume oxygen, they also produce carbon dioxide, CO_2. With the aid of solar

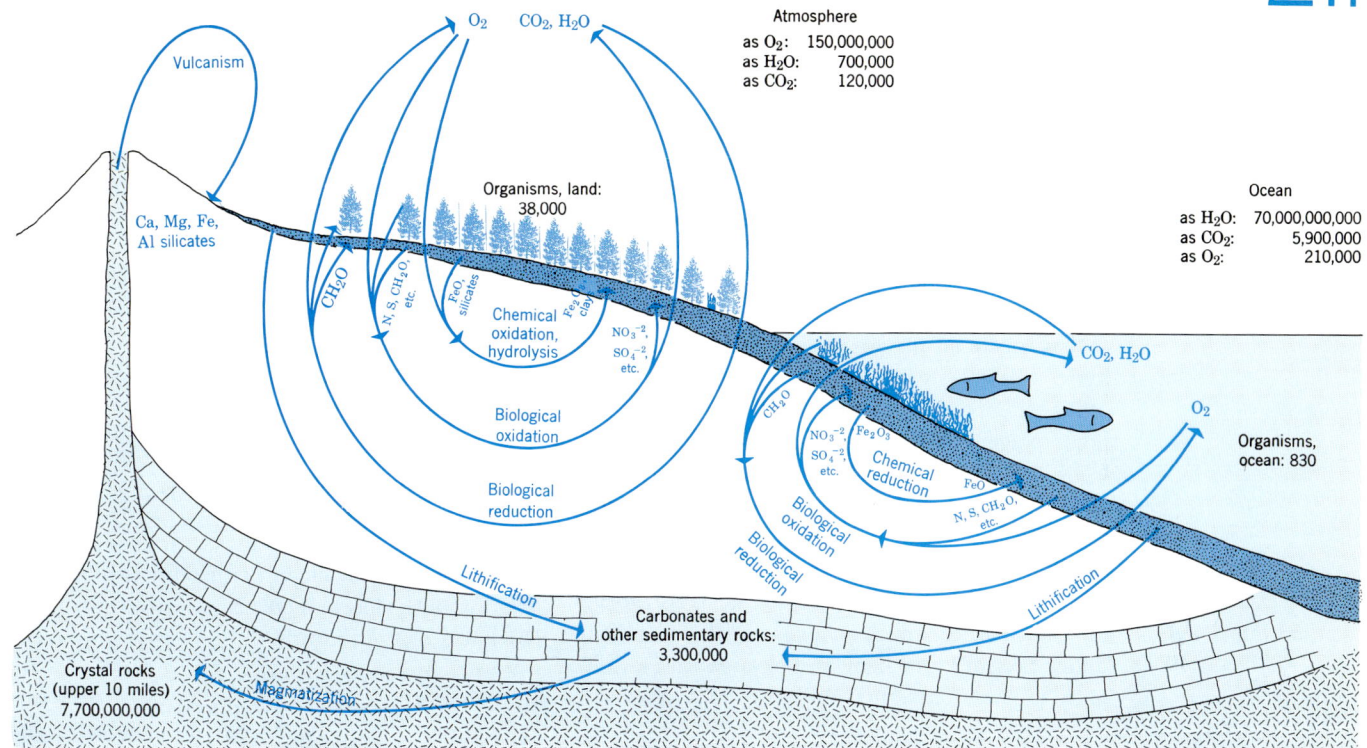

FIGURE 16-1
The oxygen cycle. The figures on oxygen supplies in various places of the system are in units of 10^{12} moles. (From A. N. Strahler and A. H. Strahler, *Introduction to Environmental Science*, 1974, Hamilton Publishing Co., Santa Barbara, Calif. Used by permission.)

energy and chlorophyll, a green pigment in most plants, plant cells use carbon dioxide, water, and minerals to make substances the plant needs and we sometimes use. The plant's use of solar energy to make plant materials is called *photosynthesis*, a subject we shall study in more detail in Section 19.1. What is of interest here is that photosynthesis also produces oxygen. Respiration, decay, and combustion continuously consume oxygen; photosynthesis continuously regenerates it. These events make up the bulk of the oxygen cycle in nature, sketched in Figure 16-1.

According to Broecker, the plants of the world, including oceanic plants such as algae, produce each year an average of eight moles of oxygen per square meter of earth surface (one mole of oxygen weighs 32 grams). Virtually the entire production is eventually used up again by decay, respiration, and combustion. Broecker estimates that for every 15 million moles of oxygen in the atmosphere, one year's *net* oxygen production is not over one mole. There is a nearly perfect balance, then, in oxygen production and consumption.

16.6 OXIDATION

Oxygen is an *oxidizing agent*. Any substance that it successfully attacks is said to be *oxidized*. The following equations illustrate oxidation as here defined:

$$C + O_2 \xrightarrow{\text{combustion}} CO_2 + \text{Energy}$$

Carbon (as in coal)   Oxygen   Carbon dioxide

$$4Fe + 3O_2 \xrightarrow{\text{rusting}} 2Fe_2O_3 + \text{Energy}$$

Iron   Iron oxide

$$S + O_2 \longrightarrow SO_2 + \text{Energy}$$

Sulfur (as in some coal and oil)   Sulfur dioxide

It is convenient, however, to consider oxidation in a broader way. In the above equations, the substances being oxidized—carbon, iron, and sulfur—all gain oxygen atoms. An attack by oxygen, however, does not always cause that and yet the reaction is still an oxidation. Oxygen attacks methyl alcohol; for example,

$$2H-\underset{H}{\overset{H}{\underset{|}{C}}}-O-H + O_2 \rightarrow 2\ \underset{H}{\overset{H}{>}}C=O + 2H_2O$$

Methyl alcohol   Formaldehyde

Molecules of methyl alcohol are changed by loss of hydrogen atoms to molecules of formaldehyde, a raw material for certain plastics. Many other examples could be shown that illustrate how oxygen can strip some hydrogen atoms from another molecule. Oxidation, therefore, is an event in which the oxidized molecules may gain oxygen or lose hydrogen. The opposite events—loss of oxygen or gain of hydrogen—are called *reduction*. It is possible in the laboratory, for example, to mix iron oxide with hydrogen under pressure and heat and change iron oxide back to iron. The equation is:

$$Fe_2O_3 + 3H_2 \xrightarrow{\text{heat, pressure}} 2Fe + 3H_2O$$

Iron oxide   Hydrogen   Iron

The iron oxide has been reduced—it has lost oxygen. The hydrogen has been oxidized—it has gained oxygen. Even the simple but highly explosive reaction of oxygen with hydrogen illustrates these definitions.

$$2H_2 + O_2 \rightarrow 2H_2O + \text{Energy}$$

Hydrogen is oxidized—it gains oxygen; oxygen is reduced—it gains hydrogen.

All oxidations simultaneously involve reductions. We saw this in the reaction of hydrogen with oxygen, given above. You cannot oxidize something without reducing something else, and this implies that something *transfers* from one chemical to another in an oxidation-reduction event. At the most fundamental level, electrons are the things that transfer. Electron-transfer events include some of the most important in our world, including respiration, decay, combustion, and photosynthesis. In the oxidation (rusting) of iron, neutral atoms of iron, Fe, lose electrons and become positively charged ions of iron, Fe^{3+}. Oxygen atoms gain these electrons and become oxide ions, O^{2-}. Wholesale transfer of electrons occurs, but it need not be that dramatic to be oxidation-reduction. It need not be a gain of whole electrons; it can be simply a gain in electron density or a gain in relative control over electrons. In the oxidation of sulfur (above) molecules of sulfur dioxide are produced. Nonpolar atoms of sulfur and nonpolar molecules of oxygen are changed to molecules of sulfur dioxide in which each oxygen, being more electronegative than sulfur, has a fractional or partial negative charge:

$$\overset{\delta-}{O}=\underset{\delta+}{S}=\overset{\delta-}{O}$$

Sulfur atoms lose some control of electron density; oxygen atoms gain some control of electron density. Sulfur atoms are oxidized and oxygens are reduced. Often it is not easy to apply just one very general definition to find out if oxidation and reduction are occurring in a reaction. That is why scientists commonly use the following set of definitions to apply to a particular reaction until a decision is made:

Oxidation: A particle is oxidized if it
(a) obviously loses electrons.
(b) gains oxygen atoms.
(c) loses hydrogen atoms.

Reduction: A particle is reduced if it
(a) obviously gains electrons.
(b) loses oxygen atoms.
(c) gains hydrogen atoms.

16.7 OZONE

Oxygen is not the only oxidizing agent in air. In smog some oxygen is converted to the most powerful oxidizing agent in our lower atmosphere, *ozone*, O_3. Ozone (from the Greek *ozein*, "to smell") is responsible for the peculiar odor sometimes noticed near electrical machines and after severe electrical storms. Electrical discharges convert small amounts of oxygen in air to what might be called the oxide of oxygen, ozone. Ozone's odor can be detected by most people at a level of 0.02 ppm (1 ppm = 1 part per million parts). Ozone is a friend or an enemy depending on where it is. In the stratosphere ozone performs a service essential to life on earth.

16.8 OZONE CYCLE IN THE STRATOSPHERE

In the stratosphere, the region of the atmosphere roughly 7 to 30 miles up, high-energy ultraviolet radiation from the sun encounters oxygen molecules. It starts reactions that change oxygen to ozone. The light energy first splits oxygen molecules to atoms. When light energy breaks a chemical bond the event is called photolysis.

$$O_2 + \text{Ultraviolet light} \rightarrow O + O$$

Oxygen molecule   Oxygen atoms

The oxygen atoms combine with unchanged oxygen molecules to form ozone.

$$O_2 + O \rightarrow O_3 + \text{Heat}$$

This collision must occur on the surface of some other particle, which we may symbolize by M. Particles of M shoot away after the collision, and it is their increased energy that constitutes the liberated heat. Without the involvement of these particles the extra energy, lodged in the newly formed ozone molecules, would sunder them again.

The ozone concentration does not simply become higher and higher, however. There are two major mechanisms for breaking ozone molecules down again, all natural. Probably the chief means is by interaction with nitric oxide, NO, which exists naturally in the stratosphere. Another important mechanism is through interaction between high energy ultraviolet rays and ozone.

$$O_3 + \text{Ultraviolet light} \rightarrow O_2 + O + \text{Heat}$$

Then
$$O_3 + O \rightarrow 2O_2 + \text{Heat}$$

Net effect
$$2O_3 + \text{Ultraviolet light} \rightarrow 3O_2 + \text{Heat}$$

The natural nitric oxide cycle for interacting with ozone is essentially identical to one that appears to be at work at ground level in photochemical smog, and we shall study it in Sections 16.18 and 16.19.

The net effect of these stratospheric reactions is an *ozone cycle*, as illustrated in Figure 16–2. We may add up the equations as follows, and when we do, canceling all that appear equally on opposite sides of the arrow, we reach a very simple but important result—we change ultraviolet (UV) rays to heat.

$$O_2 + UV \rightarrow O + O$$
$$2M + 2O_2 + 2O \rightarrow 2O_3 + 2M \text{ (heated)}$$
$$O_3 + UV \rightarrow O_2 + O + \text{Heat}$$
$$O_3 + O \rightarrow 2O_2 + \text{Heat}$$

Net effect
$$UV \rightarrow \text{Heat}$$

Only the high-energy ultraviolet rays participate—those having wavelengths below 242 nm. The ozone cycle converts these rays to heat, which warms the stratosphere and they do not reach earth to harm life.

Life as we know it would be impossible without the ozone layer of the stratosphere. As long as the ozone remains there, it is our friend. We have done something here on earth, however, that may reduce our stratospheric ozone shield.

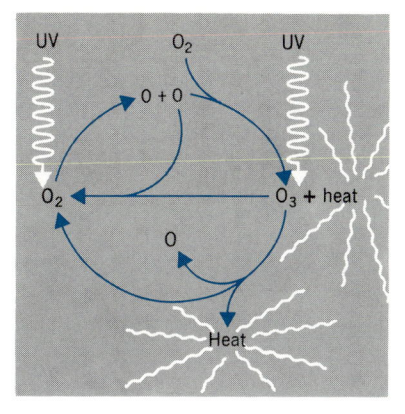

FIGURE 16–2
The ozone cycle in the stratosphere that, on balance, converts lethal ultraviolet rays from the sun into heat.

16.9
STRATOSPHERIC OZONE AND AEROSOLS

The principal chemicals used until 1978 as refrigerant gases and as propellants in *aerosol* spray cans are two members of a family of compounds known as *Freons*[1]—Freon 11, $CFCl_3$ (boiling point 23.7°C) and Freon 12, CF_2Cl_2 (boiling point $-29.8°C$). Both are gases at normal temperatures and pressures, but in refrigerant systems and in aerosol cans they exist mostly as liquids under pressure. When the can's valve is opened they change to gases that force the material out of the can—synthetic whipped cream, shaving cream, toothpaste, hair spray, paint, even cheese spread. Advantages of these Freons are that they are essentially tasteless and odorless, and they do not appear to react with anything in our low-altitude environment; they do not corrode metals. Over two billion pounds of these two Freons were manufactured in the world in 1975. If these substances react with nothing, where do they go? If there is no natural "sink" for Freons, what happens to them? One answer is that some Freons migrate into the stratosphere where they encounter high-energy ultraviolet rays. From studies on earth's surface, it is known that these rays can crack Freon molecules. The primary reactions are

$$UV + CFCl_3 \rightarrow CFCl_2 + Cl$$
$$UV + CF_2Cl_2 \rightarrow CF_2Cl + Cl$$

In other words, carbon-chlorine bonds are broken and chlorine atoms are produced along with

---

[1] Freon is a registered trademark.

other unstable fragments. It is also known that chlorine atoms can attack ozone:

$$\underset{\text{Chlorine atom}}{Cl} + \underset{\text{Ozone}}{O_3} \rightarrow \underset{\text{Chlorine monoxide}}{ClO} + \underset{\text{Oxygen}}{O_2}$$

Chlorine monoxide, in turn, can react with oxygen atoms (needed to replenish the lost ozone) and generate another chlorine atom

$$ClO + O \rightarrow Cl + O_2$$

The net effect, if it all happens this way in the stratosphere, is another mechanism for destroying ozone. In the early 1970s there was considerable alarm about the possibility that our continued use of Freons as aerosol propellants and refrigerants would jeopardize our ozone shield. Estimates varied, but there was fairly wide agreement among scientists that in the United States population a yearly average of 1 percent ozone reduction would cause 8000 additional skin cancer cases per year. The debate was not over the relation between high-energy ultraviolet rays, ozone, and skin cancer. Instead, it concerned the possible relation between the use of Freon down here and the reduction of ozone up there. Were the Freons doing up there what experiments under quite different conditions suggested they might? What of other sources of chlorine atoms—natural sources? How good are our models of how things circulate into the stratosphere? A total ban on Freons by the United States alone would still leave the Freon production of the rest of the world, roughly half of the world total. Moreover, Freons are so essential to present-day refrigeration that a total ban would have seriously affected how we transport and store perishable food. It was a classic public debate over risks versus benefits. The benefits went to all the people in the form of safe, efficient refrigeration without which our present ways of handling and storing food would be difficult. Many people also saw benefits in the convenience of the aerosol can. The risks also were distributed to all the people, including many not yet born, in the form of a possible rise in the rate of skin cancer. Efforts to ban the use of Freons in aerosol cans were intense in the mid-1970s even as their use in closed-cycle systems (refrigerators) came to be more widely appreciated.

16.10 EFFECTS OF OZONE ON LIFE

Ozone, which poses different problems where life exists, is present in certain smogs. Exposure to ozone on a continuous basis at a concentration of only 1 ppm (parts per million) is hazardous to health. Only the most severe smog episodes raise the ozone level to that value, but even at half that value, 0.5 ppm, vigorous activity that results in increased breathing should be reduced.

Plants are very sensitive to ozone with some being afflicted by an ozone level of only 0.03 ppm for 8 hours. As is so common with pollutants, one will *potentiate* the other—make the other more potent. Thus, when sulfur dioxide is present, ozone causes injury to plants more rapidly. Because of ozone some areas of southern California will no longer sustain crops of leafy vegetables. Whole stands of pines in the surrounding mountains are dying (see Figure 16–3). Crops of many vegetables and grains suffer at least 20 percent injury with an 8-hour exposure at 0.05–0.15 ppm ozone.

A few cities use ozone instead of chlorine to purify and deodorize their water supplies and to treat sewage. Ozone is highly reactive toward organic substances including, of course, disease and odor-causing microbes.

16.11 AIR TEMPERATURE INVERSIONS AND SMOG

The word *smog*, a contraction of smoke-fog, was coined in 1911 by Harold Des Voeux some time after a dreadful air pollution disaster killed roughly a thousand people in Glasgow, Scotland. Today the word *smog* means polluted air in which visibility is reduced and which contains eye and lung irritants. There are two types, the coal-fire, or simply coal smog and photochemical smog. They share some features, but there are important differences. For either to occur there must be an *air temperature inversion*.

Normally the atmosphere becomes cooler as you go to higher altitudes. In this circumstance, air near the ground, being warmer and therefore less dense, naturally tends to rise carrying pollutants to higher altitudes where they are dispersed. In an air temperature inversion air near the ground is *colder* than air above it (see Figure 16–4). The colder, more dense air therefore stays right near the ground, along with its pollutants.

One way by which an inversion occurs involves the radiation of earth heat to outer space in hours of darkness when the ground loses heat more rapidly than it gets it. The air closest to the ground therefore becomes colder and more dense than air higher up. In other words an inversion has been established. If the next day is relatively windless and especially if it is wintertime and the earth with the air nearest it does not become warmed enough, the inversion is not removed and on the following night it simply becomes

16.12 POLLUTANTS IN SMOG

FIGURE 16-3
Effect of ozone on vegetation. Left, a healthy ponderosa pine in the San Bernardino National Forest, Southern California. Right, the same tree ten years later, dying of smog. (Courtesy U.S. Forest Service, San Bernardino National Forest.)

worse. That is why the worst of the coal smog episodes occur in the months of November through January, months when home-heating units and power plants fueled by coal or heavy oil are especially active in emitting pollutants. It takes the return of winds to terminate the inversion and disperse the pollutants. The city of London was particularly plagued by coal-fire smogs, but strict controls on the use of coal have largely solved the problem.

An inversion will also occur if a cold, more dense mass of air moves into a region resting in a basinlike landform (e.g., the Los Angeles basin or Denver). Being more dense, the cold air mass will sink toward ground level displacing warmer, less dense air upward. A cool, dense air mass may also "slide" down the slopes of surrounding mountains or be generated out over the ocean. Either event establishes an inversion.

16.12 POLLUTANTS IN SMOG

The two types of smogs are compared and contrasted in Table 16.2. *Photochemical smog* is so named because many of its pollutants are produced by sun-powered chemical changes that take place among the primary pollutants—oxides of nitrogen, unburned hydrocarbons, sulfur oxides, particulates, and carbon monoxide. Ozone, in particular, is produced by photochemical changes, and the ozone concentration

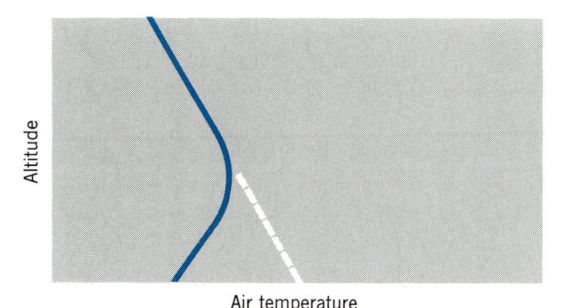

FIGURE 16-4
Air temperature inversion. The dashed line that joins the solid line indicates the normal lapse rate, which is the change in temperature with altitude. The solid line shows the lapse rate during an air-temperature inversion. Normally, the line slopes to the left. In an inversion it starts to slope to the right, but then it "inverts" and switches to the left. The layer of air at the altitude of the inversion is often called the inversion layer or the "lid."

TABLE 16.2
MAJOR TYPES OF SMOG

Characteristics	Coal Smog	Photochemical Smog
When peak intensities occur	Early in the morning, wintertime	Around noontime
Ambient temperature	30 to 40°F	75 to 90°F
Humidity	Humid and foggy	Low humidity
Thermal inversion	Close to ground	Overhead (varies)
Causes irritation principally to	Bronchia and lungs (also eyes)	Eyes (but also bronchia, lungs)
Chief irritants	Soot and other particulates; sulfur oxides	Ozone; PAN; aldehydes; oxides of nitrogen. Also sulfur dioxide and particulates (carbon monoxide, which has no odor, is also present)
Major source of irritants	Coal fires	Vehicular traffic
Other effects	Severe haze	Haze; severe damage to crops, pines, ornamentals; cracking of rubber
General chemical feature	Reducing atmosphere	Oxidizing atmosphere

or "ozone level" is one criterion of severity of photochemical smogs. Besides ozone, the principal irritants in photochemical smog are a family of compounds collectively labeled PAN. Photochemical smog is the type that plagues cities such as Tokyo and Los Angeles.

The death-dealing pollutants in *coal smogs* are oxides of sulfur and particulate matter (these also occur in photochemical smogs). Both are produced by burning sulfur-rich fuels, particularly coal. Until London authorities severely restricted the use of coal, that city experienced one smog disaster after another (coal smog is commonly called London-type smog). In the most infamous smog of all time, London's great killer smog of 1952, 4000 more people died than normal for the period. Traffic came to a halt because drivers could not see beyond their windshields!

16.13
SULFUR OCCURRENCE

Sulfur, a brittle, yellow, nonmetallic element just below oxygen in the periodic table, was known in ancient times as *brimstone* ("burning stone"). It ranks fourteenth in abundance at 0.052 percent in the earth's crust. Large deposits of native sulfur occur in Louisiana, Texas, Sicily, and a few other places (Figure 16–5). Several sulfur-containing compounds are among the minerals and ores of the earth—substances such as those described in Table 16.3. Sulfuric acid, H_2SO_4, is the most important compound of sulfur commercially. A country's consumption of sulfuric acid is regarded by some economists to be as valid a gauge of its economic activity and prosperity as its steel production. About 85 percent of the sulfur produced in the United States is converted to sulfuric acid, of which 50 percent goes into fertilizers.

Sulfur is a minor but essential constituent in nearly all proteins. The proteins of hair, fur, and feathers are particularly rich in combined sulfur.

To our great discomfiture, sulfur compounds are present in both coal and crude oil, and when these fuels are burned their sulfur is converted to sulfur dioxide and traces of sulfur trioxide. Middle Eastern crude is 3.5 to 5 percent sulfur; Alaskan oil, low in sulfur, is referred to as "sweet" crude. Soft coal has from 0.5 to 5 percent sulfur; 40–60 percent of this sulfur is present as pyrite (iron sulfide), the rest as organic sulfur. Both types can be removed but at a cost. Failure to remove the sulfur is also expensive, but in a different way as we shall next see.

16.14
THE SULFUR OXIDES: SO_2, SO_3

When sulfur burns in air or when sulfide ores are roasted in air, it is largely *sulfur dioxide* that is produced. The trioxide of sulfur also forms, but in much, much smaller amounts. Normally it is made by oxidation of sulfur dioxide, and as we shall see, the needed catalysts sometimes occur in smoggy air.

Both oxides of sulfur have sharp, pungent, irritating odors. However, the sulfur trioxide level in polluted air is normally too low to be smelled. (It is high enough, however, to be important in corroding metals and building materials.) Most people can taste the sulfur dioxide in air when it is at a level or concentration of 0.3 to 1 ppm. Its irritating odor becomes obvious at only 3 ppm. The irritant properties of sulfur oxides are, however, enhanced two to three times by the presence of the suspended solids in polluted air called *particulates*. Fortunately for human comfort, SO_2 levels almost never exceed 1 ppm except near power plants, refineries, and smelters where it can go over 3 ppm. (London's terrible smog of 1952 had an SO_2 level of 1.34 ppm, mean daily average—ten times normal.)[2]

Sulfur dioxide damages and destroys vegetation. This is particularly evident near nickel and copper smelters where the roasting of sulfide ores of these metals releases sulfur dioxide.

Perhaps the worst property of sulfur dioxide is that in the presence of sunlight and other pollutants in photochemical smog (ozone, hydrocarbons,[3] nitrogen oxides, and particulates), sulfur dioxide is oxidized to sulfur trioxide. The change is particularly rapid when the SO_2 is dissolved in microdroplets of water, and in water sulfur trioxide changes to the highly corrosive sulfuric acid mists mentioned earlier.

$$SO_2 + O_2 \xrightarrow[\text{other pollutants}]{\text{sunlight}} SO_3 \xrightarrow{H_2O} H_2SO_4$$

Much of the damage to materials and property caused by the sulfur compounds of polluted air is the result of the actions of sulfuric acid that forms indirectly from sulfur dioxide. (Sulfur dioxide also reacts slightly with water. Sulfurous acid, H_2SO_3, forms, but it is a relatively weak acid.)

When humid air is polluted with sulfur oxides it can be more corrosive than seawater. Some of our best building stones—marble, sandstone, slate, limstone—are attacked by the acids in polluted air (Section 15.15). Some of our finest buildings, cathedrals, and monuments are suffering serious deterioration. Cleopatra's Needle, a famous stone obelisk now in London, suffered less decay in the 3000 or more years it stood in Alexandria, Egypt, than it has endured in the some 80 years it has stood in the humid, smoggy, acid air of London. Venice has for years suffered from the acidic "fallout" of mainland chemical plants, and its famous buildings, statues, and paintings are decaying at an alarming rate.

[2] The concentration of particulates was 4500 micrograms per cubic meter (mean daily average)—also about ten times normal.

[3] Hydrocarbons are compounds whose molecules are made exclusively of carbon and hydrogen and that comprise gasoline and fuel oil. They will be studied in more detail in Chapter 19. Octane, present in gasoline, has the molecular formula C_8H_{18}.

FIGURE 16–5
Underground deposits of sulfur are mined by pumping hot steam into a pipe leading to the sulfur. The heat melts the sulfur, which is forced out in the molten state onto the surface where it cools in huge fenced areas. The power shovel on the right in this photograph is loading sulfur onto boxcars. (Courtesy Texasgulf.)

TABLE 16.3
IMPORTANT COMPOUNDS OF SULFUR

Sulfide ores	Pyrite	FeS_2
	Chalcopyrite	$CuFeS_2$
	Galena	PbS
	Sphalerite	ZnS
Oxides	Sulfur dioxide	SO_2
	Sulfur trioxide	SO_3
Acids	Sulfuric acid	H_2SO_4
	Sulfurous acid (unstable)	H_2SO_3
	Hydrogen sulfide	H_2S
Sulfates	Compounds derived from sulfuric acid having either the hydrogen sulfate ion, HSO_4^-, or the sulfate ion, SO_4^{2-}. Examples: gypsum $CaSO_4 \cdot 2H_2O$ barite $BaSO_4$	
Sulfites	Compounds derived from sulfurous acid having either the hydrogen sulfite ion, HSO_3^-, or the sulfite ion, SO_3^{2-}. Examples: sodium hydrogen sulfite $NaHSO_3$ sodium sulfite Na_2SO_3	

TABLE 16.4
THE NITROGEN FAMILY—GROUP 7

Name	Symbol	M. Pt. (°C)	B. Pt. (°C)	Important Types of Compounds
Nitrogen	N_2	−210	−196	Nitrates (fertilizers, explosives) Oxides (air pollutants) Ammonia (fertilizer)
Phosphorus	P	44	281	Phosphates and polyphosphates (detergents, fertilizers)
Arsenic	As	815	610	Arsenates (pesticides)
Antimony	Sb	631	1440	Lead-antimony mixtures (alloys) for storage batteries and type metal
Bismuth	Bi	271	1420	In mixtures with other metals; low melting alloys for automatic fire alarms and sprinkler systems

Scandinavia receives sulfur oxides from great industrial regions in faraway Germany (Ruhr Valley) and England (the Midlands).

As long as we continue to use petroleum and coal for fuel we shall have to contend with sulfur oxides. Most coal east of the Mississippi River used to generate electricity has an average of 2.5 percent sulfur, about half in the form of iron pyrite (Table 16.3) and the rest as organic sulfur compounds. Only the pyrites can be removed easily (i.e., at relatively low cost). Western coal and lignite is quite low in sulfur. Sulfur can be removed from petroleum, but also at considerable cost. Where sulfur is not removed before combustion, sulfur dioxide can be removed from the smokestack gases. In one approach, these gases are allowed to interact with wet limestone (mostly calcium carbonate). Sulfites and sulfates of calcium form, displacing carbon dioxide, which is sent out into the atmosphere.

16.15
NITROGEN AND THE NITROGEN FAMILY

The members of the *nitrogen family* are given in Table 16.4. We shall study in detail only nitrogen.

Nitrogen, a gas that we cannot see, smell, or taste, is present in air at a concentration of about 79 percent (by volume). In its structure ($:N\equiv N:$) each nitrogen has an outer octet, and under ordinary conditions nitrogen reacts with almost no other substances. We need it in air to dilute the oxygen, and we also need it as a source of compounds of nitrogen for plants.

Nitrogen is also a raw material in chemical manufacture. By a process discovered by a German chemist, Fritz Haber (1868–1934), nitrogen and hydrogen combine at high pressure and temperature in the presence of a special iron catalyst to form ammonia. For this discovery Haber won the 1918 Nobel Prize in chemistry.

$$N_2 + 3H_2 \xrightarrow[\text{pressure}]{\text{heat}} 2NH_3$$

Nitrogen   Hydrogen   Ammonia

It takes energy to develop both the high pressure and the high temperature. It also takes energy to prepare hydrogen and nitrogen of sufficient purity. This cost in energy translates the world fuel problem into a food problem as well. High-yield agriculture requires fertilizers, and ammonia is the basis for virtually all nitrogen fertilizers. Thus, the bottleneck for nitrogen fertilizers became, after the oil embargo of 1973, the availability of energy. Ammonia is itself a fertilizer, being sold as liquified ammonia in pressure tanks and injected directly into the soil. Ammonia is also oxidized to nitric acid, HNO_3, which is used to make another form of nitrogen fertilizer, ammonium nitrate, NH_4NO_3. Still another nitrogen fertilizer, urea, one that acts relatively slowly, is also made from ammonia.

Table 16.5 provides a summary of the principal types of nitrogen compounds. Ammonium salts, nitrites, and nitrates all are involved in the lives of plants and, therefore (indirectly), animals. When plants and animals die and decompose, their complex organic nitrogen materials are broken down and returned to the soil and the atmosphere. We shall look briefly at the nitrogen cycle and then at two oxides of nitrogen, NO and NO_2, that are important pollutants in smog.

16.16
THE NITROGEN CYCLE

Nature and humans shuttle nitrogen nuclei among all their "spheres." Considering the extreme conditions of temperature and pressure

TABLE 16.5
IMPORTANT COMPOUNDS OF NITROGEN

Inorganic Compounds			Organic Compounds	
Oxides	Nitrogen(I) oxide (nitrous oxide,[a] laughing gas)	N_2O	Amines	Organic derivatives of ammonia. Example: ethylamine $CH_3CH_2NH_2$
	Nitrogen(II) oxide (nitric oxide[a])	NO	Amino acids	Building blocks of all proteins. One example of about 24: glycine $NH_2CH_2CO_2H$
	Nitrogen dioxide	NO_2		
	Dinitrogen pentoxide	N_2O_5		
Ammonia and ammonium salts	Ammonia	NH_3	Proteins	Including all enzymes and many hormones
	Ammonium nitrate	NH_4NO_3	Nucleic acids	Chemicals of genes
Acids	Nitrous acid (unstable)	HNO_2		
	Nitric acid	HNO_3		
Nitrites	(Compounds derived from nitrous acid having the nitrite ion, NO_2^-.) Example: sodium nitrite	$NaNO_2$		
Nitrates	(Compounds derived from nitric acid having the nitrate ion, NO_3^-.) Example: sodium nitrate	$NaNO_3$		

[a] These are the commonly used names.

needed by the Haber process to reduce nitrogen to ammonia, it is a remarkable tribute to the service catalysts give that soil bacteria can "fix" atmospheric nitrogen at soil temperature and atmospheric pressure. They have catalysts—enzymes—that accomplish *nitrogen fixation*:

$$N_2 + H_2 \rightarrow NH_3 \text{ (or } NH_4^+)$$

(We shall not balance this equation; we show only reactants and products.) Nitrogen fixation is a primary event in our world because plants need nitrogen in the form of ammonia or close relatives to make the amino acids needed to put together plant proteins (as discussed further in Chapter 19). Animals, in turn, cannot use nitrogen in any inorganic form to make their proteins. Animals must get basic amino acids for their proteins from plants or from other animals that feed on plants. To continue the cycle, decomposer organisms use the proteins and amino acids of plants and animals when they die, and change organic nitrogen back to inorganic forms such as ammonia.

Some soil bacteria oxidize ammonia to nitrite ion, NO_2^-.

$$4NH_3 + 7O_2 \xrightarrow{\text{nitrosomonas bacteria}} 4NO_2^- + 6H_2O$$

In a reaction called *nitrification*, *nitrobacter* (another soil microorganism) oxidizes nitrite ion to nitrate ion, NO_3^-.

$$4NH_3 + 9O_2 \xrightarrow{\text{nitrobacter}} 4NO_3^- + 6H_2O$$

Nitrate and nitrite ions, generated within the soil or added to the soil as chemical fertilizers, can be reduced to ammonia. Some, however, are reduced all the way back to molecular nitrogen and returned to the atmosphere. This reaction is known as *denitrification*. There are, therefore, several events involving the soil and the use of nitrogen. The flow of nitrogen nuclei forms another great chemical cycle in nature, the nitrogen cycle shown in Figure 16–6.

16.17
THE NITROGEN OXIDES: NO, NO_2

When fuels and air are mixed and burned in the cylinders of automobiles or in stationary furnaces (e.g., power plants), very high temperatures result. Under these conditions, even the air's normally inert nitrogen will react with the air's oxygen, and the waste gases therefore include nitric oxide, NO. There are several ways by which some nitric oxide is oxidized to nitrogn dioxide. One involves complex interaction with hydrocarbons (as in automobile exhaust). Another uses ozone. In the moments when hot exhaust gases are mixed with and diluted by air, some of the nitric oxide is further changed by

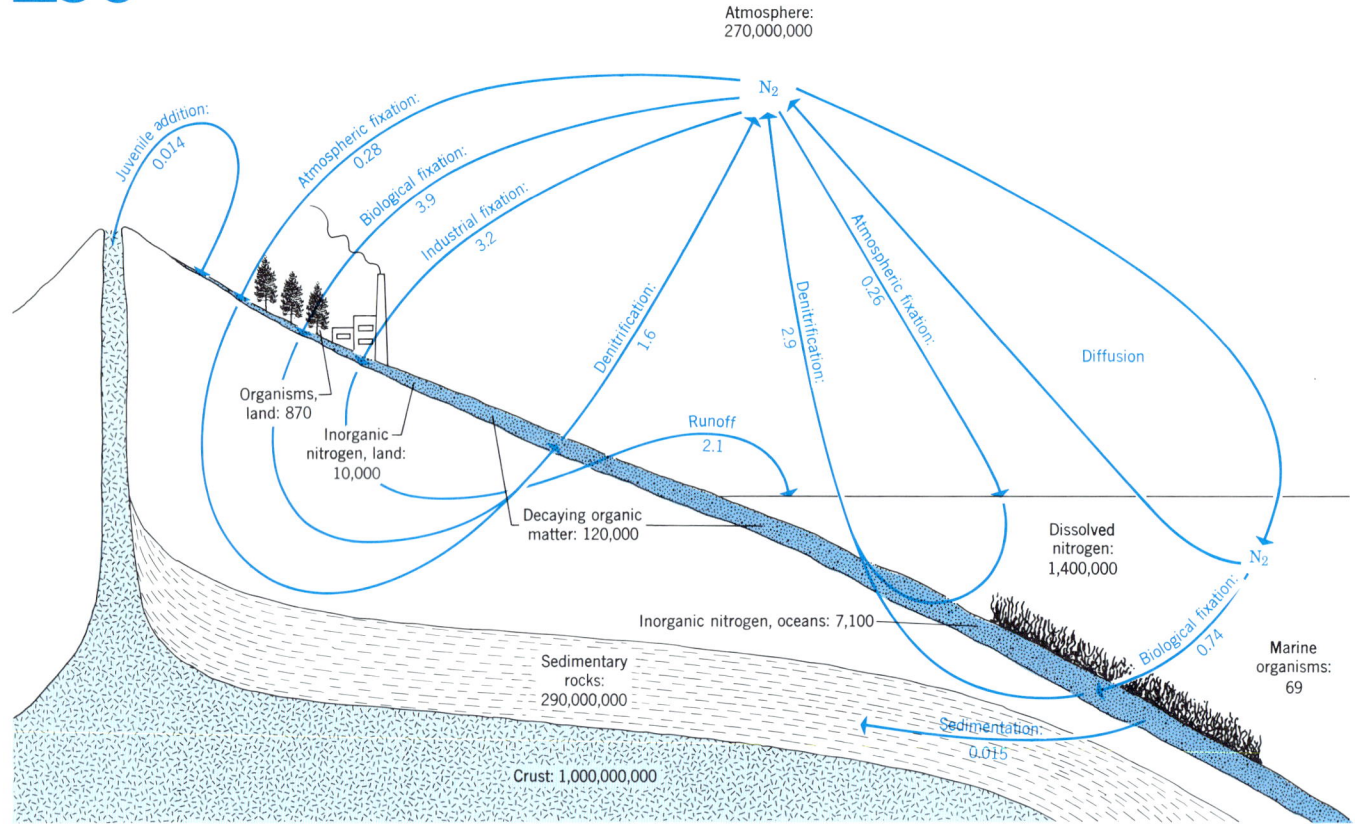

FIGURE 16-6
The nitrogen cycle. The figures on the curved arrows represent the flows of nitrogen in units of 10^{12} moles per year from one "pool" or sphere to another. The other figures represent the approximate amounts of nitrogen in each sphere in units of 10^{12} moles. (From A. N. Strahler and A. H. Strahler, *Introduction to Environmental Science*, 1974, Hamilton Publishing Company, Santa Barbara. Used by permission.)

oxygen to nitrogen dioxide, NO_2. These two oxides are often referred to as the *nitrogen oxides* and given the general formula, NO_x.

$$N_2 + O_2 \rightarrow 2NO$$
$$2NO + O_2 \rightarrow 2NO_2$$

The more villainous of the two is nitrogen dioxide because it is responsible for the production of ozone in the lower atmosphere. It is also the chief cause of the yellowish-brown color normally assumed by smog over an urban area. Traffic accounts for about 40 percent and stationary furnaces about 50 percent of all nitrogen oxides emitted into the air.

The combined NO_x concentration in severe smog has rarely gone above 3 ppm. One of the remarkable features of smog is the misery caused by such low concentrations of pollutants.

16.18
THE NITROGEN DIOXIDE–OZONE CYCLE—OZONE IN SMOG

Lower-energy ultraviolet rays filter through the stratospheric ozone screen. The nitrogen dioxide molecule is just weakly enough held together that these rays can break it apart to nitric oxide and oxygen atoms.

$$NO_2 \xrightarrow{\text{lower-energy UV rays}} NO + O$$

Nitrogen dioxide → Nitric oxide + Atomic oxygen

The oxygen atoms collide with and combine with molecular oxygen to form ozone.

$$O + O_2 \rightarrow O_3$$

The newly formed ozone can attack nitric oxide and thereby both are removed.

$$O_3 + NO \rightarrow NO_2 + O_2$$

If we simply add the last three equations, canceling things on opposite sides of arrows, everything cancels—except the low-energy ultraviolet light. The net effect of this cycle (Figure 16–7) is apparently the simple conversion of ultraviolet light to heat.

The cycle (Figure 16–7) does not account for the levels of ozone in photochemical smog. See, however, Figure 16–8. Smog also contains unburned or partly burned hydrocarbons from gasoline and diesel fuels. Atomic oxygen, generated by sunlight's lower-energy ultraviolet rays, reacts very rapidly with hydrocarbons, oxidizing some to very reactive partly oxidized hydrocarbons (Figure 16–8). These can change nitric oxide (NO) to nitrogen dioxide (NO_2). This chemical mischief means there is less nitric oxide, not enough to react with and destroy newly formed ozone. Ozone slowly accumulates, and that is how we get ozone in smog.

16.19 NITRIC OXIDE, OZONE, AND THE SST

Supersonic transports (SSTs) are airliners designed to fly in the stratosphere faster than sound. Their engines, like any combustion engine, produce nitric oxide, NO. Nitric oxide reacts with ozone (Section 16.18) producing oxygen and nitrogen dioxide, NO_2. For each kilogram of fuel burned, about 18 g NO_2 is produced. If SST flights inject substantially more nitric oxide into the stratosphere than naturally exists, the rate of ozone destruction would exceed the rate at which it forms. Even though there are huge day-to-day variations in the concentration of ozone at any one point in the stratosphere,

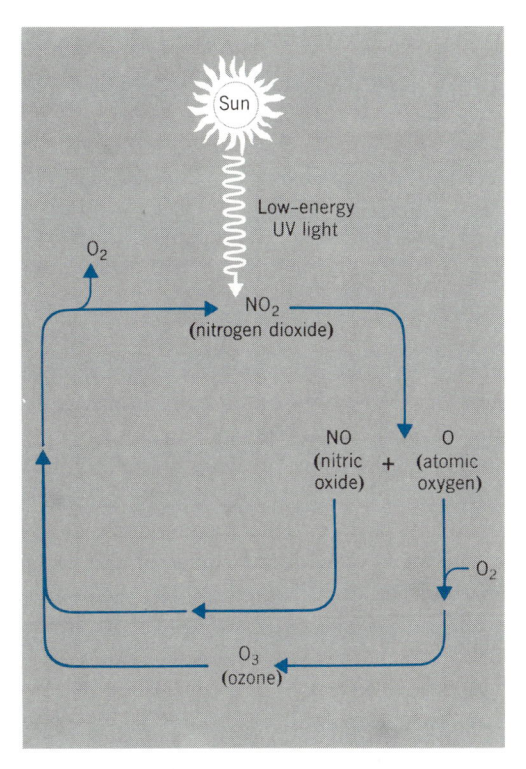

FIGURE 16–7
The nitrogen dioxide-ozone cycle. Nitric oxide produced by the UV-energized decomposition of nitrogen dioxide will eventually destroy the ozone produced from the atomic oxygen made in the same decomposition. The arrows are shown of equal thickness to indicate this balance. However, the balance is upset if hydrocarbons are present (Figure 16–8). (From J. R. Holum, *Elements of General and Biological Chemistry*, 4th edition, 1975, John Wiley & Sons, Inc., New York.)

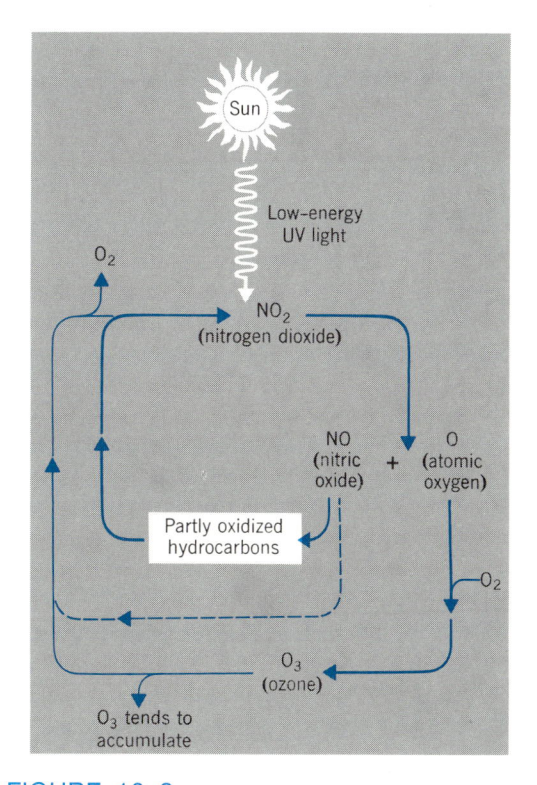

FIGURE 16–8
The nitrogen dioxide-ozone cycle when hydrocarbons are present. If oxidation products of hydrocarbons react with newly formed nitric oxide, then newly formed ozone (thicker arrow) cannot as readily be taken out (thinner arrow) and it can accumulate to the levels observed in severe photochemical smog. (From J. R. Holum, *Elements of General and Biological Chemistry*, 4th edition, 1975, John Wiley & Sons, Inc., New York.)

16 NONMETALLIC INORGANIC SUBSTANCES

and variations that depend on latitude, the long-term result would be a reduction in our ozone shield (Section 16.8), and this would cause an increase in the incidence of skin cancer. On economic grounds the U.S. Congress has thus far prevented the development of a United States SST fleet. The number of Concordes and Tupolevs, the British-French and Russian versions of SSTs, presently in operation is low (there were about 30 in the mid 1970s). However, at that time the Environmental Protection Agency believed there might be as many as 375 SSTs (mostly Concordes) by 1990. If that were to happen, there would be an increase of 22,500 skin cancer cases per year by 1990 in the United States, based on a conclusion of the National Academy of Sciences–National Research Council that each SST could be responsible for 60 skin cancer cases per year. This potentially very serious matter is the subject of intensive research sponsored by various agencies of the government.

16.20 THE CARBON FAMILY

The members of the *carbon family* are listed in Table 16.6 together with some of their properties. We shall concentrate our study primarily on carbon and its two oxides. Silicon and its various oxides will be discussed in Chapter 23.

16.21 CARBON

Pure carbon may exist in principally three forms—diamonds, graphite, and varieties of carbon black. In hardness these represent a range from one of the hardest-to-scratch substances known, the diamond, to some of the easiest. These differences are attributed to different crystal structures, which we illustrated in Figure 14-11. In a diamond, bonds extend from atom to atom, four single bonds from each, each bond equivalent to each other and very strong. In graphite the carbon atoms are arranged in sheets, and it is relatively easy for slippage between these sheets to occur. Hence, graphite makes an excellent dry lubricant; it is even slippery to the touch. Carbon black is roughly like ultrapowdered graphite.

Carbon is a constituent of the molecules of all organic compounds, including, of course, the hydrocarbons found in gasoline and oil. When these are burned in cars, trucks, heating plants, and power plants two oxides of carbon are produced (besides water).

$$\text{Hydrocarbons} + (O_2) \xrightarrow{\text{if combustion is incomplete}} CO_2 + H_2O + CO$$

(Compounds of carbon and hydrogen)

Carbon dioxide (mostly)   Carbon monoxide (traces)

16.22 CARBON MONOXIDE

Carbon monoxide, CO, is a colorless, odorless, tasteless gas that forms in small amounts when just about anything burns. It is extremely poisonous and it acts without warning. Auto exhaust and poorly vented heating systems are the most dangerous sources of carbon monoxide, which annually kills about 1400 people in the United States, mostly from space heaters. The idling engine of a small car in a small garage produces a lethal concentration of carbon monoxide in less than 10 minutes. Carbon monoxide levels in major, smog-ridden cities average from 1 to 10 ppm, but they may reach 100 ppm in tunnels, garages, and in densely packed, stalled traffic.

Carbon monoxide acts as a poison by combining strongly with molecules of the oxygen-carrying red pigment in blood, hemoglobin, which then cannot carry oxygen from lungs to cells.

Exposure to a CO level of 120 ppm for an hour, not impossible in smoke-filled rooms, causes headache, dizziness, and dullness. Inhaled cigarette smoke has 400 ppm carbon monoxide, and the smoke from the tip is as high as 20,000–40,000 ppm (2–5 percent). It is one factor in the lower average birth weights of babies born of mothers who smoke.

16.23 CARBON DIOXIDE

The principal natural sources of carbon dioxide, CO_2, are the chemical reactions that all plants and animals undergo when they die and decay

TABLE 16.6
THE CARBON FAMILY—GROUP 4

Name	Symbol	Density (g/cc; 20°C)	Melting Point (°C)	Boiling Point (°C)
Carbon	C	3.51 (diamond)	3570	3470 (sublimes)
Silicon	Si	2.33	1414	2355
Germanium	Ge	5.36	959	2700
Tin	Sn	7.29	232	2275
Lead	Pb	11.34	327	1750

as well as the reactions of respiration while they are alive.

If the carbon dioxide is cooled and compressed, it can be changed into the solid form known as *dry ice,* a solid that vaporizes in air without melting. A familiar use of carbon dioxide is the CO_2 fire extinguisher.

Carbon dioxide is not classified as a poison, but it will not support life. If it is unknowingly allowed to displace oxygen from a room or some other space (e.g., a silo), living things in such places will die.

The principal fates of carbon dioxide in nature are the photosynthetic production of plant materials and the formation of carbonate rocks, limestones and dolomites. Until we began to burn fossil fuels at the increasing rates we now use, the concentration in air was quite steady. By our current rate of burning coal and petroleum, we add carbon dioxide to the atmosphere more rapidly than it is removed. In the last 100 years the level of carbon dioxide in the atmosphere has increased from 290 ppm to 320 ppm. About 20 percent of this increase occurred in the 1960s alone. This trend created some alarm because the carbon dioxide level in the atmosphere is one factor that helps the earth's environment stay as warm as it does (see Section 21.16). Too much warming, however, would cause the glaciers and icecaps of the world to melt, which would raise the world ocean level enough to inundate all coastal cities.

The oceans help to regulate atmospheric carbon dioxide because of the following equilibrium:

$$Ca^{2+} + CO_2 + H_2O \rightleftharpoons CaCO_3\downarrow + 2H^+$$

The ocean is saturated with respect to calcium carbonate ($CaCO_3$), but more calcium ions constantly enter the oceans by way of the rivers of the world. Calcium ions absorb carbon dioxide from the air above the oceans, which helps to control the carbon dioxide level of the atmosphere. It is a very slow process, however. The rate is so low that it will take several centuries for the oceans and the atmosphere to come again into equilibrium with respect to carbon dioxide.

16.24 CARBON CYCLE

Carbon atoms move, in combined forms, through all the spheres of our planet—the lithosphere as carbonate sediments and rocks, the atmosphere as carbon dioxide, the hydrosphere as dissolved carbon dioxide and carbonate or bicarbonate ions, and the biosphere as organic compounds as well as carbonates and carbon dioxide. Respiration, decay, and combustion produce it; photosynthesis consumes it. Thus, there is another chemical cycle in nature, the *carbon cycle,* illustrated in Figure 16-9.

SUMMARY

Oxygen family. Oxygen, 21 percent by volume of air, is in the same family as sulfur, selenium, tellurium, and polonium. All form hydrides of the same general formula, H_2X and oxides of the form XO_2 (as well as other oxides). The oxide of oxygen, ozone, exists normally in the stratosphere where it interacts with incoming high-energy ultraviolet rays to convert them to heat. Two substances that may imperil the ozone shield are Freons and nitric oxide. A reduction in the ozone shield will cause an increase in skin cancer. Oxides of sulfur, SO_2 and SO_3, form when sulfur-bearing fuels such as certain coals and crude oils are burned in air. They are primary air pollutants in industrial areas.

Oxygen cycle. Atoms of the earth's most abundant element, oxygen, cycle through all spheres—atmosphere, lithosphere, hydrosphere, and biosphere—chiefly through respiration, decay, combustion, and photosynthesis. The cycle is intimately entwined with the carbon cycle.

Nitrogen family. Nitrogen, 79 percent by volume of air, is in the same family as phosphorus, arsenic, antimony, and bismuth. Two of its oxides, NO_2 and NO, are produced in high-temperature internal combustion engines of cars, trucks and jet planes as well as stationary power plants that burn coal or oil. Nitrogen dioxide (NO_2) is involved in the production of ozone in smog.

Nitrogen cycle. The movement of nitrogen atoms in nature depends on several activities of soil microorganisms. Some help the soil fix nitrogen—change it from gaseous nitrogen to ammonia or nitrite or nitrate ions. Others can oxidize ammonia to nitrite ion. Nitrifying bacteria oxidize nitrite ion to nitrate ion—nitrification. Denitrification—reduction of nitrate and nitrite ions to nitrogen—is accomplished by still other organisms. Plants take soil minerals and ammonia to make amino acids for making plant proteins. Animals get their proteins either from plants or from animals that feed on plants. To close the cycle, plants and animals die and decay, releasing nitrogen from amino acids in various forms.

Carbon family. Carbon, which forms the molecular skeletons of all organic compounds, is in the same family as silicon, germanium, tin, and lead. Carbon dioxide is good for making plants; silicon dioxide for making rocks. Carbon monoxide is good for nothing except as an industrial chemical or an intermediate. It is a powerful poison produced when gasoline or coal is burned in insufficient oxygen to oxidize carbon monoxide fully to carbon dioxide.

Carbon cycle. Carbon cycles through the biosphere as part of organic compounds made in photosynthesis and carbon dioxide released by respiration, decay,

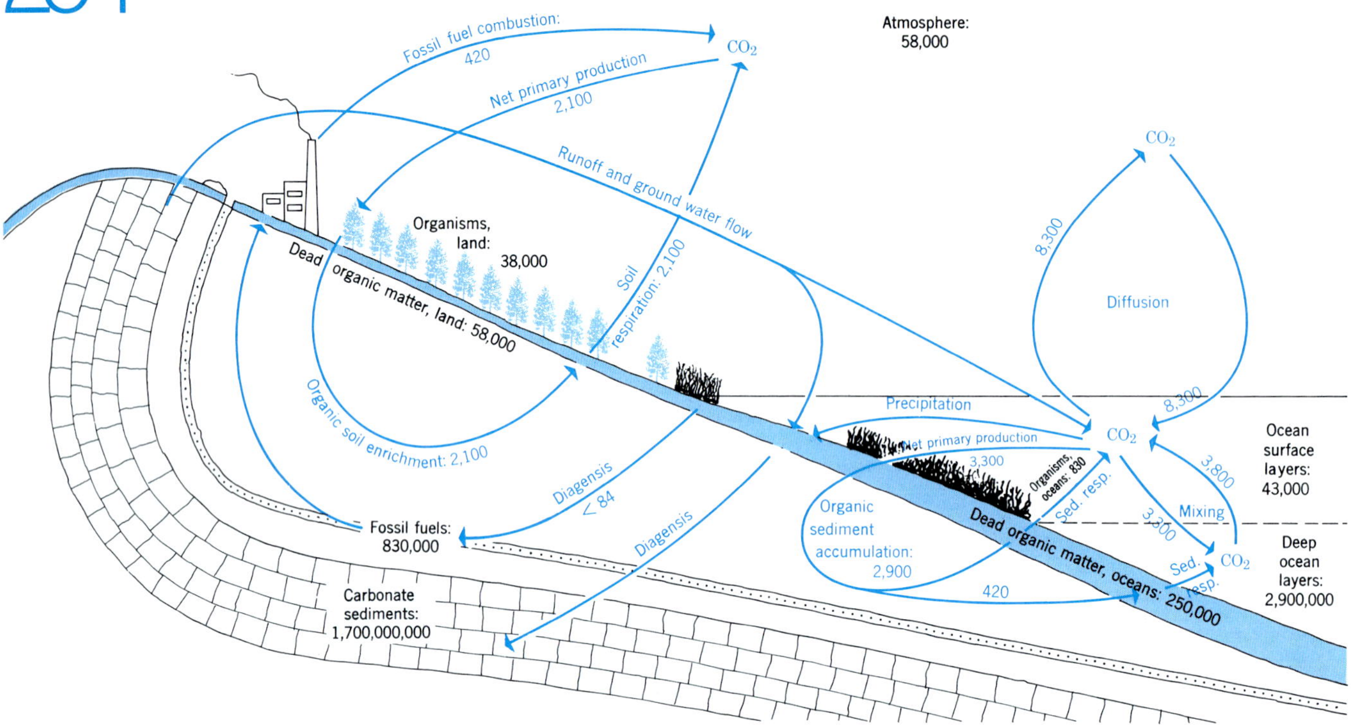

FIGURE 16-9
The carbon cycle in nature. The figures indicate estimates of the amounts of carbon in various "pools" in units of 10^{12} moles per year. (From A. N. Strahler and A. H. Strahler, *Introduction to Environmental Science*, 1974, Hamilton Publishing Company, Santa Barbara. Used by permission.)

and combustion. Some carbon dioxide moves in and out of storage as carbonate rocks and sediments.

Smog. A smog episode develops during an air-temperature inversion in which air nearer the ground is cooler and denser than air above it. This inversion traps air pollutants. In coal smog, the principal pollutants are sulfur oxides and particulates produced from burning coal. In photochemical smog, the pollutants are chiefly nitrogen oxides, and eye or lung irritants such as ozone and PAN. They arise from complex reactions involving solar (photochemical) energy. Carbon dioxide and carbon monoxide are also present.

TERMS AND CONCEPTS

biosphere
lithosphere
hydrosphere
atmosphere
oxygen family
oxygen cycle
respiration
decay
combustion
photosynthesis
oxidation
oxidizing agent
reduction
reducing agent
pollutant
ozone
ozone cycle
Freon
aerosol
potentiation
smog
air temperature inversion
photochemical smog
coal smog
sulfur oxides
particulates
nitrogen family
nitrogen cycle
nitrogen fixation
nitrification
denitrification
nitrogen oxides
nitrogen dioxide-ozone cycle
carbon family
carbon cycle

QUESTIONS AND EXERCISES

1. Explain how members of the oxygen family have hydrides with the same kind of formula.

2. Photosynthesis puts oxygen into the air, and yet the concentration of oxygen in air does not change. Explain.

3. How do the oceans help to maintain a fairly even level of carbon dioxide in the atmosphere?

4. How does the ozone cycle in the stratosphere convert the sun's high-energy ultraviolet rays into heat?

5. Not much ozone can accumulate in smog unless hydrocarbons are also present. How is this explained?

6. By what sequence of reactions does the combustion of high-sulfur fuels produce acid mists in smog?

7. How do the nitrogen oxides in smog, NO_2 and NO, arise?

8. What are many scientists worried about (a) continued use of Freons in aerosol cans? (b) stratospheric flights by fleets of SSTs?

9. In what ways do photochemical smog and coal smog differ?

10. Which is the more dangerous air pollutant, SO_2 or SO_3? Why?

11. Describe the chemistry of the deterioration of marble or sandstone building material and statues by polluted air.

12. What are the principal forms of carbon and how do they differ structurally?

CHAPTER 17
METAL REFINING

17.1 METALS, ORES, AND RESOURCES

The machines and appliances of the home and of commerce and industry, the vehicles of transport—ships, trucks, air carriers, trains—the bridges that span rivers and roads, the carriers of electrical power and of water and wastes, and the strength-giving elements of virtually all major buildings, factories, and freeways are all made mostly or entirely of *metals*. At one stage or another, the production of nearly every industrially important metal involves chemistry in action. We shall see how in this chapter.

Except for gold, some silver, and some copper, metals occur in nature as chemical compounds known as *minerals*. These minerals are usually intimately mixed with other minerals such as various silicates. A mineral is a solid, homogeneous chemical element or compound produced by processes in nature, possessed of a definite chemical composition (or range of composition), and characterized by a definite crystalline structure and a set of particular chemical properties. An *ore* is a mixture of minerals from which a metal can be obtained at a profit. What is or is not an ore, therefore, depends not just on the presence of the desired metal but also on a demand for that metal in the marketplace at a price that makes it worthwhile to do all the work to extract it. As the price of gold rises, more and more abandoned gold mines are reinvestigated. As the price of copper increases, poorer and poorer grades of copper-bearing ores change in status from "neutral stuff" to minable ores. As veins of iron-rich ores in Minnesota run out, low-grade taconite becomes an ore of great value.

17.2 IRON AND STEEL

Over 90 percent of all metal refined in the world is *iron*, nearly all of which goes into steel. The principal iron minerals are hematite (Fe_2O_3), magnetite (Fe_3O_4), limonite (mostly $2Fe_2O_3 \cdot 3H_2O$), and siderite ($FeCO_3$). The major United States deposits occur in northern Minnesota where, until the 1960s, the workable ore deposits ran over 50 percent in contained iron. As the principal iron ranges of Minnesota became depleted, scientists at the University of Minnesota developed methods to use a lower-grade material, *taconite,* containing about 23 percent iron. The iron minerals in Minnesota taconite—mostly small crystals of magnetite, Fe_3O_4—are mixed with several kinds of silicate minerals in a very hard type of rock. To extract the magnetite, the taconite rock is crushed to a fine powder. Powerful magnets than pull the magnetite away from the silicates. Finely powdered magnetite cannot be used directly in blast furnaces because its texture and magnetic properties do not permit efficient operation. Therefore, the powder is mixed with a clay binder, some hard coal, and moisture and baked in a rotating drum. The magnetite changes to nonmagnetic hematite, Fe_2O_3. Golfball-sized spheres emerge containing 63 percent iron. The principal operations are shown in Figure 17–1.

17.3 THE BLAST FURNACE

The *blast furnace* is a huge tower made of steel and lined with refractory bricks made of a special silica and clay that will withstand very high temperatures. In operation the furnace is loaded with a mixture of iron ore, coke, and limestone (see Figure 17-2).

Coke, a porous material that is nearly all carbon, is made by heating coal to drive off moisture and other volatile materials (Section 20.8). About 13 percent of the 1973 United States production of soft coal went into the manufacture of coke.

Limestone (mostly $CaCO_3$) is used in the blast furnace to combine with impurities in the iron ore and form a molten slag. (Without limestone, the silicate impurities would not form a *molten* slag and hence would be harder to remove.)

Once the furnace is charged, blasts of hot air or air enriched with oxygen are shot into the furnace through ports near the bottom. Some of the coke burns, producing both intense heat and carbon monoxide (CO), which reduces the iron oxides to molten iron.

$$3CO + Fe_2O_3 \rightarrow 2Fe + 3CO_2$$

Some carbon dissolves in the iron to form an alloy (Section 11.13) called *pig iron* that has a lower melting point (1130°C) than pure iron (1540°C). Molten pig iron collects in a pool on the bottom of the blast furnace. The less dense *slag* forms a second molten layer above it. At appropriate times first the slag and then the iron are tapped and drained from the furnace (Figure 17-3). The slag, mostly calcium aluminum silicate, may be used to make portland cement, or it may be broken and used like gravel in making concrete. If the pig iron is not to be made immediately into steel, it is poured into molds and allowed to harden.

The blast furnace operates continuously. Raw materials are continuously fed into the top, and they undergo their changes as they work their way down toward the bottom.

17.4 STEEL MAKING

Pig iron, while very hard, is also too brittle for most uses. Its chief impurities, besides carbon, are manganese, phosphorus, and silicon left over from the original ore and some sulfur from the coke. To remove these impurities, the method in greatest use today is the *basic oxygen process* or BOP (Figure 17-4). In principle it is very simple. A jet of nearly pure oxygen is blown directly into the molten pig iron to oxidize the impurities and form a slag. (In an older method still used, the Bessemer converter, oxygen is supplied by a blast of air. The principle is still the same.) Scrap iron becomes an important commodity at this stage. It can be melted together with pig iron prior to the BOP operation.

The new molten steel, which might be mixed with any of the several alloying metals (Section 11.13), is cast into molds or slabs for further milling into semifinished and finished articles. The entire operation from taconite to steel is outlined in Figure 17-5.

The slabs of new steel are hot, and they react at their surfaces with oxygen in air to form oxide coatings that are removed by acid in an operation called pickling. The slabs are moved continuously through pickling tanks containing sulfuric acid or hydrochloric acid, which dissolves the oxide.

For some uses steel plate must be coated with tin, which is much less readily attacked by either air or weak acids. To ensure that the tin is deposited uniformly but not too thickly on the steel, the *tinplate* industry uses an electrolytic method. Since the basic principle is the same as that used to refine many metals, we shall study that next.

17.5 ELECTROCHEMISTRY

We generally think of an oxidation-reduction reaction as a spontaneous event—one that also releases energy, usually heat. In the blast furnace, carbon is oxidized; oxygen is reduced. Iron oxides are reduced to iron, and carbon monoxide is oxidized to carbon dioxide. These are electron-transfer events, and the electrons pass directly from one chemical to another.

In the chemical storage battery (Section 5.11), the chemicals are packaged in an arrangement that forces the transferring electrons to pass through an external circuit. In this way, the spontaneous chemical reaction occurring inside the battery draws electrons through the external circuit where useful work is done. The battery produces direct current. What happens when we force direct current through chemicals? That is one of the questions studied in the branch of chemistry called *electrochemistry*.

Suppose we have dissolved some copper(II) chloride, $CuCl_2$, in water contained in a glass jar or beaker (see Figure 17-6). To simplify matters, we assume that this salt breaks up entirely into copper ions, Cu^{2+}, and chloride ions, Cl^-. Now suppose we dip two platinum wires called *electrodes* into the solution. We use platinum because it is a good conductor of electricity that does not react chemically with our reagents. We next

17 METAL REFINING

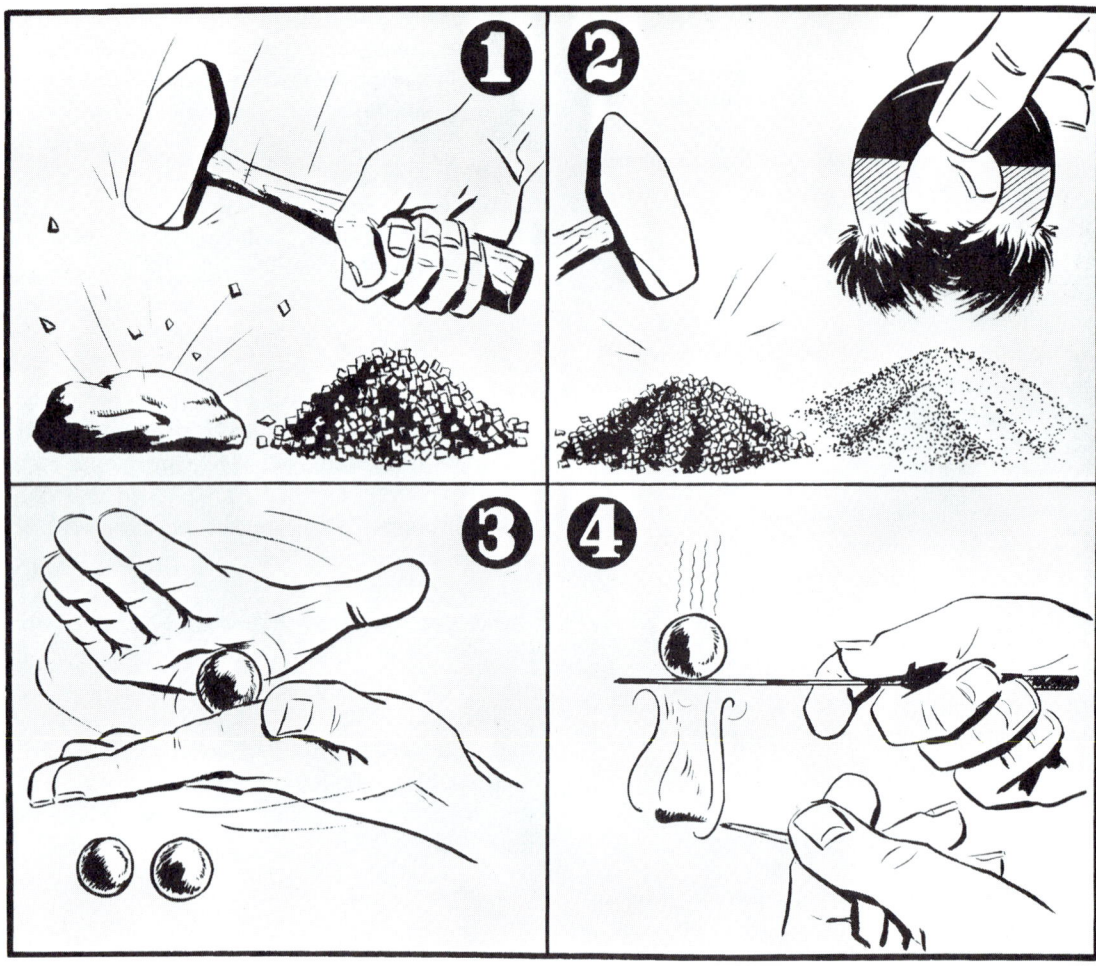

FIGURE 17-1

Taconite—from hard rock to iron-rich pellets ready for the steel mill. The four main steps in processing taconite are shown by the artist's sketches in part a. The exceptionally hard taconite rock must be broken and crushed (1). Particles are separated as much as possible from a sandlike material by magnetic separation (2). The iron ore is moistened, mixed with a binder and rolled into balls (3). Then the pellets are fired to make them strong enough to withstand rough handling (4).

Part b is a cutaway drawing of the taconite crusher. Rod and ball mills (not shown) are used to convert the crushed rock into a fine powder before magnetic separation is used. Part c is a cutaway drawing of a magnetic separator. Vacuum disk filtration will next remove 90 percent of the water from the concentrate, now 65 percent iron. Part d is a photo of one of the huge pelletizing furnaces for firing the pellets, which move slowly through the heated area on continuous grates. They then are cooled and stockpiled for shipment. Part e shows finished taconite pellets. (a, b, d, and e Courtesy Reserve Mining Company; c courtesy Dings Co.—Magnetic Group.)

17.5 ELECTROCHEMISTRY

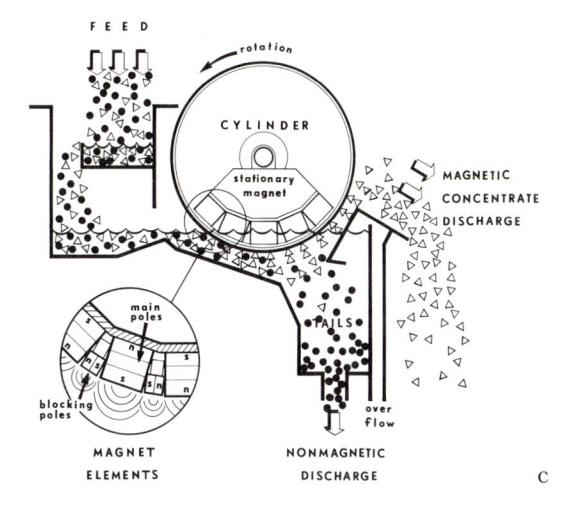

FIGURE 17-2
Blast furnace for converting iron ore into pig iron. A series of chemical reactions occurs in the furnace, as indicated. The furnace is operated continuously, and molten iron is tapped every five or six hours. (Bethlehem Steel Corporation.)

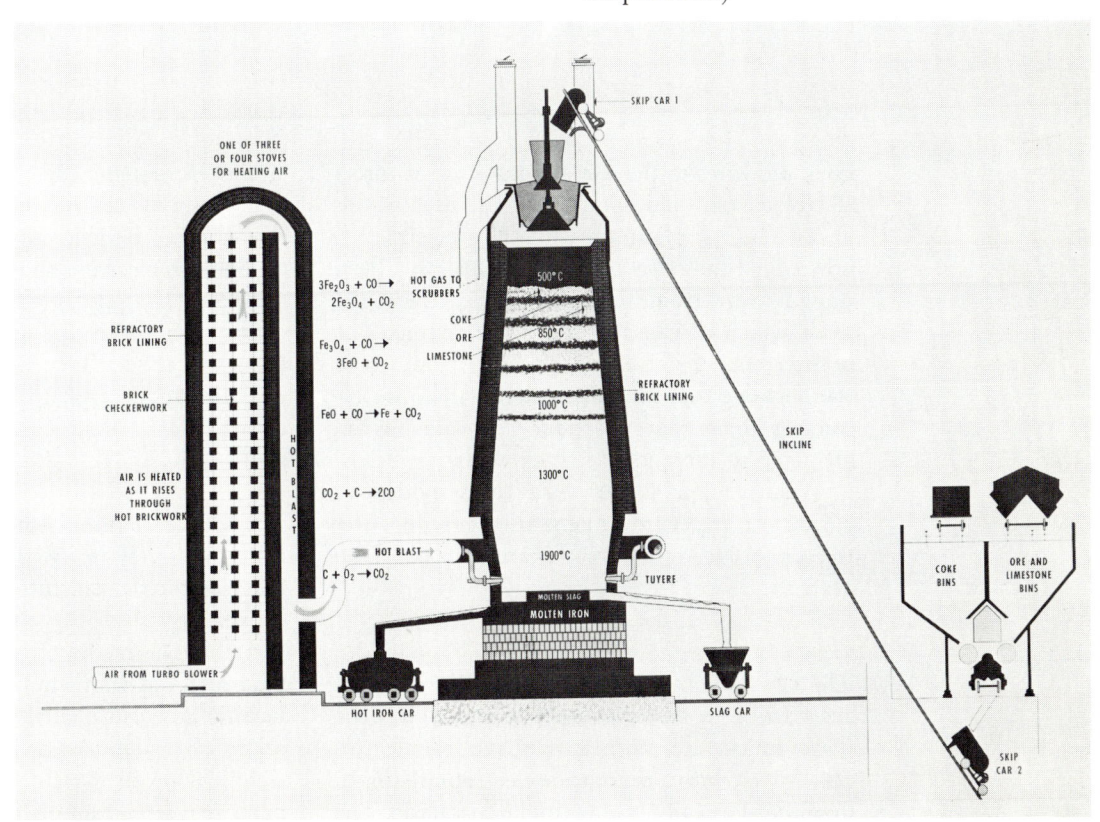

17 METAL REFINING

FIGURE 17-3
Molten iron rushes from a blast furnace through a series of clay-lined runners in a receptacle called a "submarine." Most of this iron will be charged, still molten, into a basic oxygen furnace for refinement into steel. (Bethlehem Steel Corporation.)

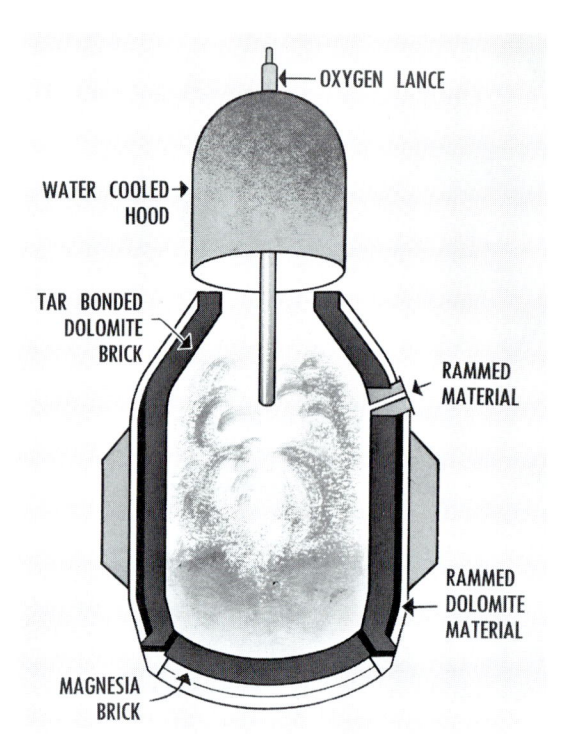

FIGURE 17-4
Basic oxygen process (BOP). Shown here in an artist's sketch is the furnace for this process. The pot is charged with molten iron, iron and steel scrap, and some iron ore. A lance is lowered into the molten mass, and oxygen is blown down into the mixture to oxidize impurities. Newer BOP furnaces employ a tube for oxygen entering from the bottom, and oxygen is blown upward through the molten mass. (Bethlehem Steel Corporation.)

hook the wires to the leads of a source of direct electric current that pumps a surplus of electrons at once to one platinum wire. This electrode is now called the *cathode*. The electrons have been drawn away from the other wire, called the *anode*, which now carries a positive charge. The positive copper ions in solution are attracted to the negative cathode. They cannot help be attracted; unlike charges attract. Chloride ions are attracted to the anode.

When a copper ion arrives at the cathode, two electrons transfer from the platinum wire, enter the copper ion, and thereby change the ion to an atom.

$$Cu^{2+} + 2e^- \to Cu$$

The new atom of copper sticks to the wire. We are starting to *electroplate* copper metal onto the platinum surface. With a silver salt, we electroplate silver; with a tin salt, we electroplate tin. In fact, that basically is how tinplate is made. The steel to be plated is used as one electrode, and tin in a dissolved state is forced onto the steel by means of an electric current.

The two electrons that went to Cu^{2+} to make one atom of copper, Cu, must be replaced. They come from the direct current that takes them from the anode. Two chloride ions in solution near the anode give up an electron each to the anode.

$$2Cl^- \to Cl_2 + 2e^-$$

The source of direct current pumps these electrons into the system and makes a net delivery of two electrons to the cathode. For every atom of copper deposited two atoms of chlorine form, and these immediately combine to form one molecule, Cl_2, which bubbles out of the solution. Passing an electric current through a solution is called *electrolysis*. We have left many important technical problems out of our discussion. Our purpose is served simply by a general description

17.5 ELECTROCHEMISTRY

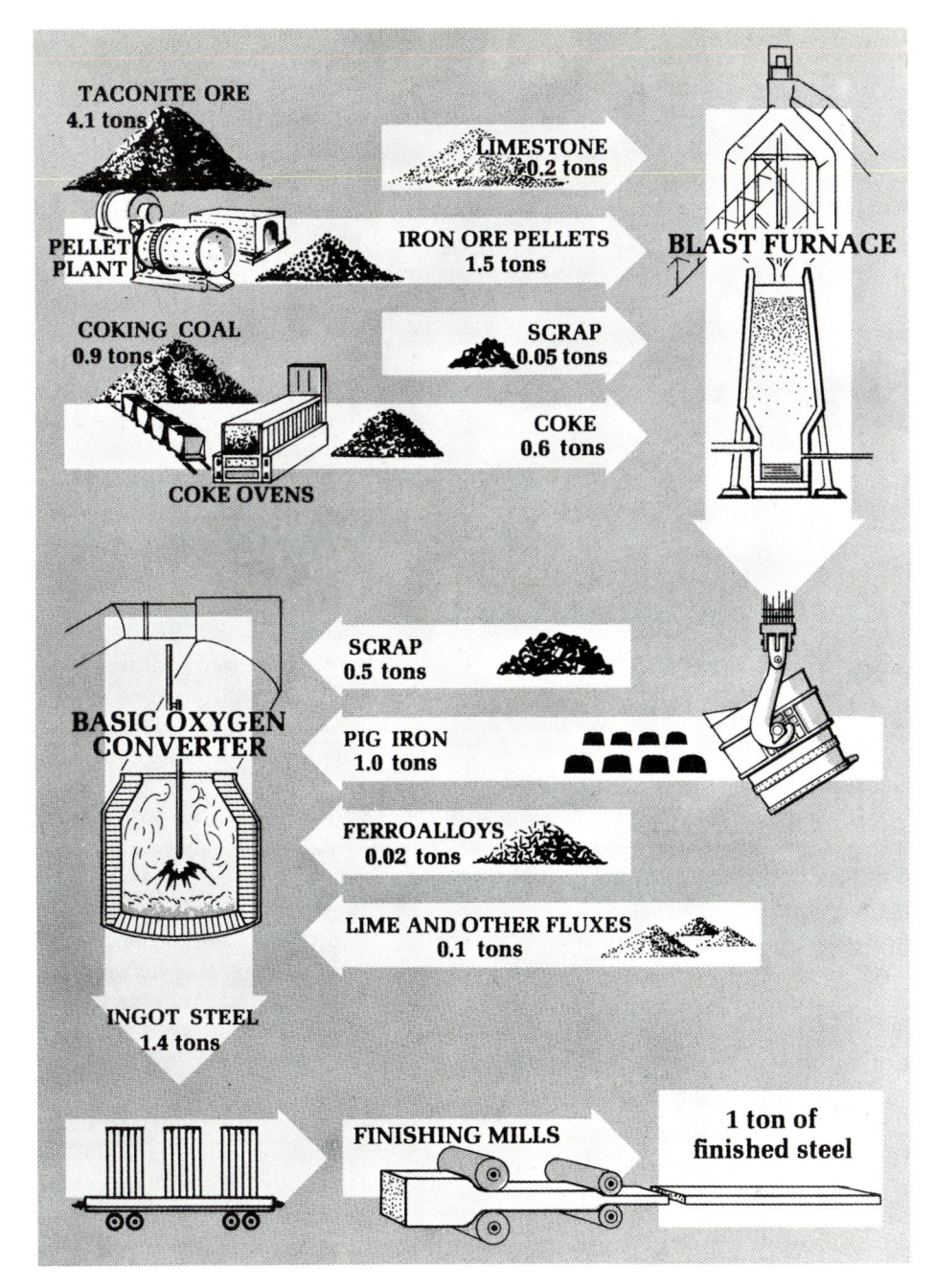

FIGURE 17-5
From taconite to steel—via the blast furnace, the basic oxygen converter, and finishing mills—it takes over five tons of raw materials per ton of steel. (From J. F. McDivitt and G. Manners, *Minerals and Men*, 1974, Published for Resources for the Future by The Johns Hopkins Press. Used by permission.)

of how electroplating works. With many variations, this method is how aluminum, magnesium, zinc, sodium, nickel, copper and a few other metals are refined.

Sometimes the ion that migrates to an electrode becomes involved in a secondary chemical reaction and produces a product that is not obvious from the formula of the chemicals. Sometimes solvent molecules react. If dilute hydrochloric acid is electrolyzed, for example, the hydrogen ions that migrate to the cathode are changed to hydrogen atoms and thence, immediately, to molecules of hydrogen gas H_2, which bubble out. At the anode, however, we do not

17 METAL REFINING

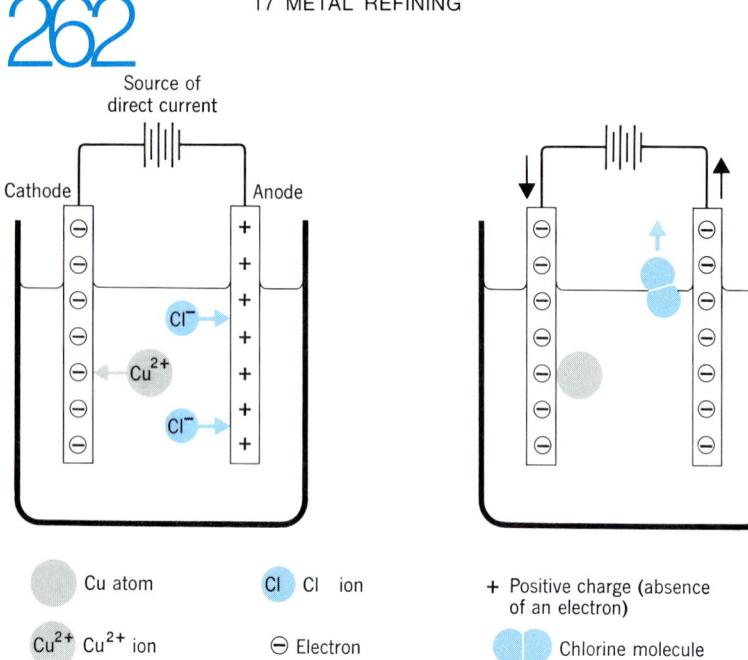

FIGURE 17-6
Basic events in electroplating copper from a solution of copper(II) chloride. (a) Ions migrate to electrodes. Copper ions move to the cathode; chloride ions to the anode. (b) The cathode delivers electrons to copper ions, which change to copper atoms and stick on the electrode. The anode removes electrons from chloride ions, which change to molecules of chlorine that bubble out.

generate chlorine from chloride ions. Energetically, it is easier for a different reaction to occur. Water molecules are changed to oxygen molecules and hydrogen ions are released.

$$\text{At the anode } 2H_2O \rightarrow O_2 + 4H^+ + 4e^-$$
$$\text{At the cathode } 4H^+ + 4e^- \rightarrow 2H_2$$
$$\text{The sum } 2H_2O + \text{energy} \rightarrow 2H_2 + O_2$$

This is one industrial synthesis of hydrogen and oxygen. It may have an important use in the future as a possible solution to the disappearance of natural gas reserves. The basic idea is to use nuclear or solar energy near the ocean to electrolyze ocean water and then distribute the hydrogen through the pipeline networks now used for natural gas. The technical problems are, it must be said, awesome. Considerable thought and some research have been given to it, and quite possibly a hydrogen energy economy will emerge in the twenty-first century.

17.6 ALUMINUM

Aluminum is the most abundant metallic element and the third most abundant of all elements found in the earth's crust. Out of every hundred atoms that make up crustal material, roughly 63 are atoms of oxygen, 21 are silicon, 6 are aluminum, and 2 are iron—all being in chemically combined forms together with small amounts of other elements. In spite of how widely it occurs, aluminum was not discovered until 1827, and it remained a costly laboratory curiosity until 1886. In that year Charles Martin Hall (United States) and P. L. T. Héroult (France) discovered how to extract aluminum from aluminum oxide by electrolysis. The price of aluminum fell from several dollars to a few pennies a pound, and extensive exploitation of this remarkable metal began.

The principal aluminum ore is *bauxite*, a mixture of various hydrated oxides of aluminum ($Al_2O_3 \cdot xH_2O$; x varies) together with more or less clay, iron oxide, silica, and other impurities. The largest deposits being mined today are in Australia and Jamaica, but huge deposits are known to exist in Africa and China.

Bauxite is processed to produce a relatively pure aluminum oxide (alumina, Al_2O_3), which is then reduced to aluminum electrolytically. Hall and Héroult solved the chief problem in aluminum refining, a solvent for the alumina while it is electrolyzed. Alumina is insoluble in water. It does dissolve in acids, but when these solutions are electrolyzed, hydrogen, not aluminum, is produced at the cathode. Simply passing electricity through molten alumina is uneconomical because it takes over 2000° to melt this oxide. Hall found that molten cryolite (Na_3AlF_6, melting point 1000°C) dissolves alumina, and that the metal is deposited on the cathode when direct current is passed through this hot solution. Cryolite occurs very rarely in nature. Minable quantities exist only at Ivigtut in southwestern Greenland. However, cryolite can be made synthetically from fluospar (mostly CaF_2), and all cryolite used today is prepared from this more abundant mineral.

Molten alumina-cryolite is held in a large pot lined with carbon, which serves as the cathode (Figure 17-7). The anode is made of carbon rods or blocks that are lowered into the mixture. Aluminum ions, Al^{3+}, acquire electrons at the cathode, and molten aluminum collects at the bottom of the pot.

$$Al^{3+} + 3e^- \rightarrow Al$$

At the anode oxide ions react with the carbon as they deliver electrons. (As a result the carbon anodes disappear and must be replaced.)

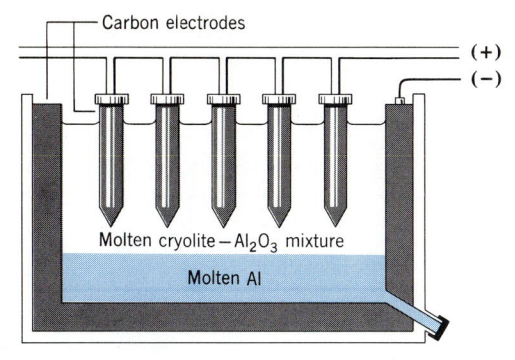

FIGURE 17-7
Production of aluminum by the Hall process. (From J. E. Brady and G. E. Humiston, *General Chemistry: Principles and Structure*, 1975. John Wiley & Sons, New York. Used by permission.)

$$2O^{2-} + C \rightarrow CO_2 + 4e^-$$

Multiplying the first equation by 4 and the second by 3 before adding them, the net electrolytic reaction is

$$4Al^{3+} + 6O^{2-} + 3C \xrightarrow{electricity} 4Al + 3CO_2$$

The carbon dioxide escapes; leaving with it, unfortunately, are large amounts of hydrogen fluoride (HF) and fluorine (F_2), both of which are air pollutants in the vicinities of aluminum plants. They damage fruit crops and forests, and cause serious bone changes in animals.

It takes about 4 to 6 tons of bauxite to make 2 tons of alumina from which 1 ton of aluminum metal is obtained. Electricity is the principal expense, however, because production of one ton of aluminum metal from two tons of alumina consumes about 15,000 kilowatt-hours of electricity. The United States aluminum industry uses each *day* more electric energy than a small city of 100,000 people might consume in one *year*. The daily electricity used by the aluminum industry is roughly what is used domestically by five metropolitan areas the size of Chicago. Understandably, aluminum refineries are located where electricity is the cheapest. In the United States that is largely in the Tennessee Valley, near Niagara Falls, and in the Pacific Northwest where the Columbia River is harnessed by several major hydroelectric dams. In 1970 the United States accounted for 37 percent of all the aluminum produced in the world.

Because pure aluminum is too soft for most applications, it is alloyed with various metals such as silicon, copper, magnesium, and manganese. Even though aluminum oxidizes in air to aluminum oxide, aluminum does not "rust" away. The oxide forms a tough but almost invisible coating that protects the metal beneath from further oxidation. The metals and its alloys are good conductors of heat and electricity. The low density of aluminum alloys together with their great strength make them ideal for building materials, transportation equipment, containers, cans, and other packaging (Figure 17-8). Aluminum is now second only to iron as the most used metal in the United States. As the cost of energy climbs, increasing interest is shown in recycling aluminum (Figure 17-9).

17.7 COPPER

Copper occurs in the earth's crust in several forms, particularly as sulfides and oxides often mixed with similar compounds of nickel, iron, and arsenic. Principal deposits of mined copper ores are in the Republic of Zaire in Africa, the western United States (Montana, Utah, and Arizona), Chile, and Ontario and Quebec, Canada (Figure 17-10). The metal's high ability to conduct electricity and heat, and its exceptional resistance to corrosion, were properties that kept it until 1970 as the world's second most used metal. Aluminum has since eased copper into

FIGURE 17-8
Aluminum ingot weighing 10,000 pounds is maneuvered into place at the hot rolling line where it will be rolled into thin plate or sheet or foil. (Courtesy Kaiser Aluminum and Chemical Corporation.)

264   17 METAL REFINING

(a)

(b)

(c)

FIGURE 17-9
Recycling aluminum. (a) Aluminum beverage cans are delivered to a hopper from which they move on a belt carrier to a shredder. A magnetic separator acts to kick away steel cans. (b) In the shredder the cans are reduced to dime-size chips that are fed into a remelt furnace. Enamel and other foreign matter is burned off. (c) From the furnace the metal flows into molds to form reclaimed scrap ingots. (d) At a fabricating plant the ingots may go into a rolling mill to make sheets of aluminum that might be used to make aluminum cans. (Courtesy Reynolds Metals Company.)

(d)

third place. Although aluminum is only about 60 percent as efficient as copper in conducting electricity, its lower density and lower cost have made it more economical than copper for some electrical applications. Those uses, however, continue to consume about one-half of the copper produced. Other major uses are in corrosion-resistant fittings and devices made of bronze (copper-tin alloy) and brass (copper-zinc alloy).

The typical copper ore in the United States contains less than one percent copper, and several complicated operations are required to obtain the copper. The principal stages are illustrated in Figure 17–11. To concentrate the copper-bearing minerals as much as possible, the ore is crushed and ground to a fine dust, which is stirred in water containing special chemicals. These chemicals cause the water to foam and froth as air is bubbled through the mixture. Copper mineral particles stick to the foam while silica rock and other debris sink. Periodically the froth is skimmed, and its load of mineral particles is separated. The operation is called froth *flotation* (Figure 17–12). By choosing the foaming chemicals properly, first one mineral and then another can be separated. Flotation does not work perfectly, but about 95 percent of the copper minerals in the original ore may be concentrated into a mass having only 10 to 20 percent of the original weight. If the mineral is mostly sulfide, the material obtained by flotation is roasted. Roasting brings about chemical changes that adjust the sulfur content of the minerals. Air reacts with the hot concentrate to change some of the copper sulfides to copper oxides. The sulfur is changed mostly to sulfur dioxide. The sulfur dioxide released into the air around copper refineries kills almost all vegetation in the vicinity. To minimize this, enormous smokestacks are erected by some companies to try to mix the sulfur dioxide with more air before it gets back to the ground. It is an immense air pollution problem that becomes a water pollution problem in lakes even hundreds of miles downwind where sulfur oxides dissolve, make the lakes slightly more acidic, and harm the fish.

The product of roasting is a mixture of sulfides and oxides of copper and its impurities (iron and siliceous matter plus traces of gold and silver). The next steps, converting and refining, produce a copper pure enough for many uses but that contains some gold and silver. These are often present in enough concentration that it is economical to remove them. Thus, the last stage of copper refining is electrolytic.

In *electrolytic refining,* copper is used both as anode and cathode (Figure 17–13). The anode is the impure copper. The plating solution contains dissolved copper—copper sulfate, $CuSO_4$.

FIGURE 17–10
The Bingham Canyon copper mine, Utah—one of the largest open-pit mines in the world. Note the new tunnel at the bottom of the pit. (Courtesy Kennecott Copper Corporation.)

As direct current moves through the system, copper ions (Cu^{2+}) in solution plate out at the cathode, and at the anode copper atoms change to copper ions leaving behind two electrons each.

Cathode reaction $Cu^{2+} + 2e^- \rightarrow$ Cu
                                         Purified
                                         copper

Anode reaction   Cu $\rightarrow Cu^{2+} + 2e^-$
                   Impure
                   copper

Net change   Cu $\xrightarrow{\text{electricity}}$ Cu
               Impure   Pure
               copper   copper

As impure copper anodes dissolve, their impurities settle into the bottom "slime" of the electrolytic cell. Silver and gold can be recovered from this material.

17.8
LEAD

The principal lead mineral in lead ore is galena, PbS. Often the ore includes zinc sulfide (ZnS) as well. The state of Missouri is the main source of lead ore in the United States, but lead is also obtained from several western states as a product of the refining of precious metals.

Much of the separation of lead minerals from zinc minerals is done by selective flotation (Section 17.7). The lead sulfide is then roasted to change it to lead oxide.

17 METAL REFINING

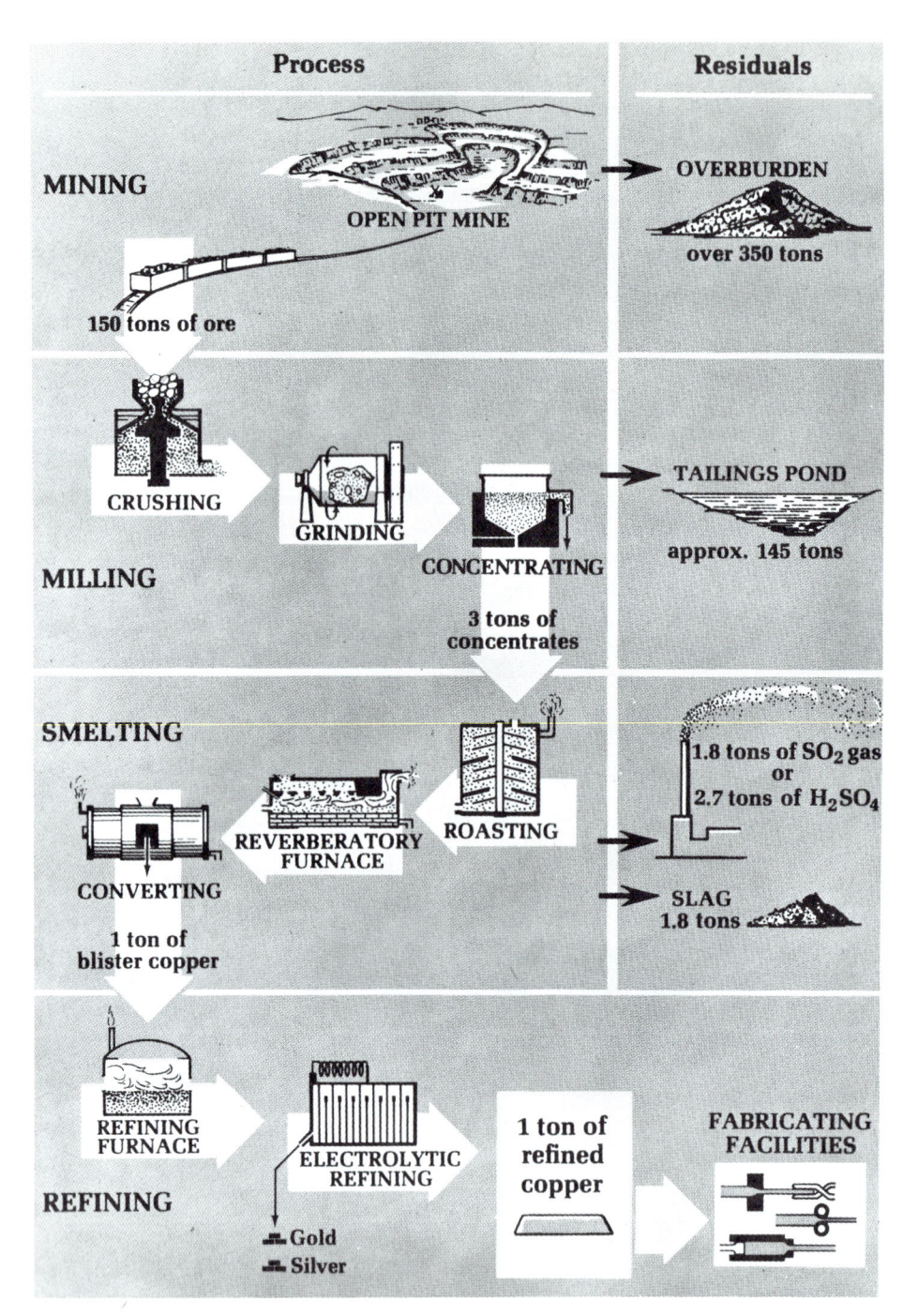

Process	Residuals
MINING — OPEN PIT MINE	OVERBURDEN over 350 tons
150 tons of ore	
MILLING — CRUSHING, GRINDING, CONCENTRATING	TAILINGS POND approx. 145 tons
3 tons of concentrates	
SMELTING — ROASTING, REVERBERATORY FURNACE, CONVERTING	1.8 tons of SO₂ gas or 2.7 tons of H₂SO₄; SLAG 1.8 tons
1 ton of blister copper	
REFINING — REFINING FURNACE, ELECTROLYTIC REFINING (Gold, Silver)	
1 ton of refined copper → FABRICATING FACILITIES	

FIGURE 17–11
From copper ore to final products via milling, smelting, and electrolytic refining. (From J. F. McDivitt and G. Manners, *Minerals and Men*, 1974, Published for Resources for the Future by The Johns Hopkins Press. Used by permission.)

$$2PbS + 3O_2 \xrightarrow{air} 2PbO + 2SO_2$$
Lead sulfide — Oxygen — Lead oxide — Sulfur dioxide

The oxide is then reduced by the use of coke in a manner similar to that used in an iron blast furnace. Greatly simplified, the equation is

$$2PbO + C \rightarrow 2Pb + CO_2$$
Lead oxide — Carbon (coke) — Lead — Carbon dioxide

Zinc oxide, if present, is also reduced. Zinc metal, however, boils at 907°C; lead at 1744°C. Zinc, therefore, emerges as a vapor that is car-

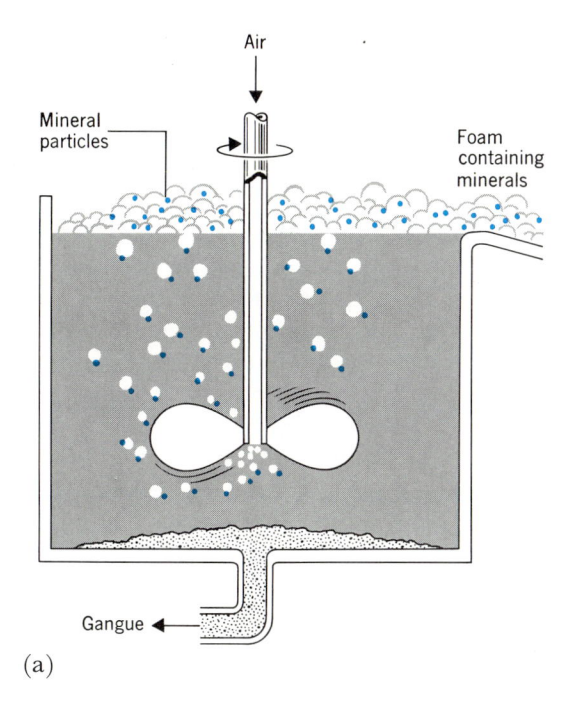

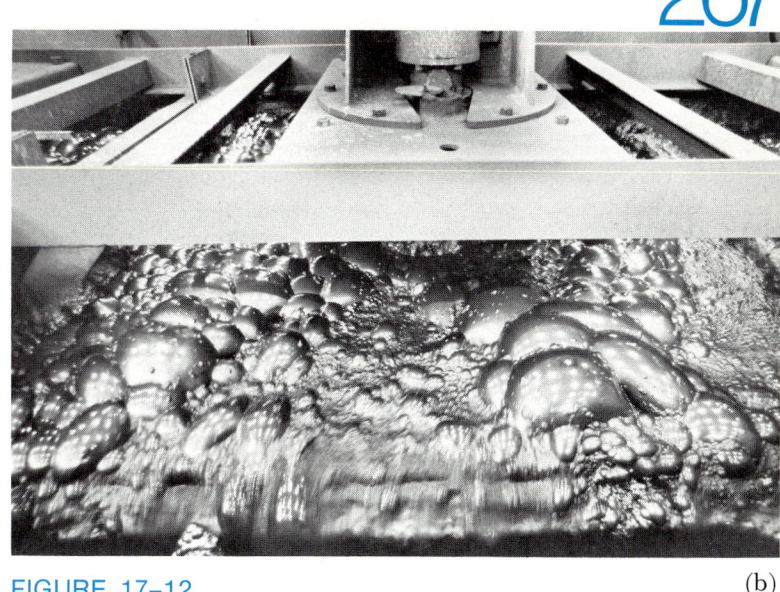

17.8 LEAD

FIGURE 17-12
Froth flotation. (a) An artist's sketch of how froth flotation generates a foam that catches mineral particles and floats them over the edge of the tank while waste sand settles. (b) Photo of the actual operation. (Sketch from J. E. Brady and G. Humiston, *General Chemistry: Principles and Structure*. 1975. John Wiley & Sons, New York. Used by permission. Photo courtesy of Lead Industries Association, Inc.)

ried away to collecting pots. Lead is further refined electrolytically in a manner very similar to that used in copper refining.

The principal use of lead is in making lead storage batteries for the transportation industry, which takes slightly over 40 percent of United States lead production. Most battery lead is recycled. The second major use of lead is to make tetraethyl lead and other lead additives for improving the octane rating of gasoline. The oldest use of lead, one dating back to early Roman times, was in making pipes, fixtures, and seals for plumbing. The symbol for lead, Pb, comes from *plumbum*, a Latin word from which we get our word *plumber*. Lead, a soft metal that is easy to work and that resists corrosion, melts at a low enough temperature, 327.5°C, to make it easy to liquify. When poured into molds or around pipe joints, it does not shrink when it cools, making it an ideal sealant. Unhappily, lead in the form of lead ions, Pb^{2+}, is a very dangerous poison, because these ions react with certain enzymes in the human body to inactivate them. Areas of the central nervous system are hit particularly hard, especially in infants. If substances of slight acidity such as fruit juices are allowed to stand for a period of time in lead bowls or goblets, enough lead can dissolve to injure infants and young children who later drink the liquid. Lead salts were once widely used to make paint pigments. In old tenements flaked paint chips are sometimes eaten by toddlers, who eventually develop the symptoms of lead poisoning.

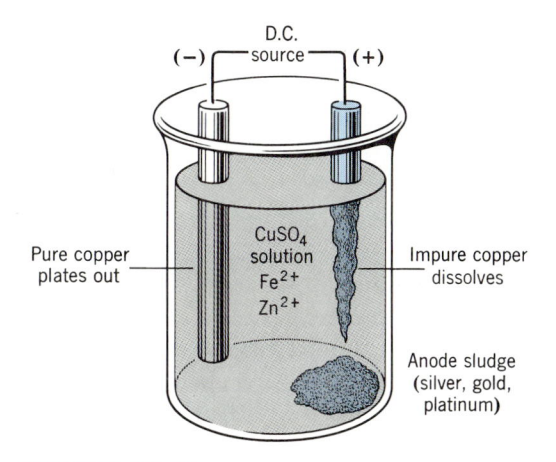

FIGURE 17-13
Purification of copper by electrolysis. (From J. E. Brady and G. E. Humiston, *General Chemistry: Principles and Structure*. 1975. John Wiley & Sons, New York. Used by permission.)

17.9
ZINC

The chief ore of zinc is called sphalerite, ZnS, a sulfide ore that usually occurs with galena, the chief lead ore. Both lead and zinc are often refined as parts of the same set of operations.

Zinc is used principally to plate steel to protect it against oxygen and moisture in the air. The product, called "galvanized steel," is protected by two mechanisms. First, the zinc coat provides a mechanical barrier. Where it covers the steel completely, nothing can get through to the steel to corrode it. Second, it works by *galvanic action*.

17.10
GALVANIC ACTION

Suppose the zinc-coated steel develops a pit or crack that exposes some steel and then becomes moistened by water with enough electrolyte (dissolved salts) to conduct electricity. The conditions are now ripe for this special type of corrosion—galvanic action. When two different metals are in contact with each other and with a solution that is a conductor, the more chemically active of the two metals will preferentially oxidize. By the "more chemically active" metal, we mean a greater tendency for its atoms to change to ions, giving up electrons. Zinc is more active in this sense than iron. We say it is more *electropositive* than iron; it has the greater tendency to change to a more positive form, in this case the positive ion. Zinc atoms give up electrons that travel through the steel (usually the shortest possible distance) to the moisture. Something in the water, usually hydrogen ions, accepts the electrons and is reduced. The zinc coating slowly dissolves but nothing happens to the steel. It is protected. The phenomenon was discovered by an Italian physicist, Luigi Galvani (1737–1798), after whom galvanized steel is named. The steel article to be galvanized is either dipped in molten zinc or the zinc is applied by an electrolytic process. Zinc plates or magnesium blocks (Figure 17–14) are sometimes bolted to the steel hulls of ships or to underground steel tanks and pipelines to protect them from corrosion. Thus, it is unnecessary for the entire steel surface to be plated to give it protection by galvanic action.

Another interesting example of galvanic action occurs when iron nails are used to attach aluminum objects such as storm windows or screens having aluminum frames. Aluminum is more electropositive than iron. Aluminum metal closest to the surface of the iron nails corrodes more rapidly than elsewhere because of galvanic action, and objects fasted by the nails loosen. The "rule" is "always use aluminum nails with aluminum objects."

FIGURE 17–14
Twin ribbons of zinc wire, laid into a ditch alongside a portion of the trans Alaska pipeline will help to prevent electrochemical corrosion of the pipe. Called anodes, the diamond-shaped ribbons of zinc wire, about 0.5 inch in diameter, are connected to the pipe at 500- and 1000-foot intervals, completing an electrical circuit caused by chemical reaction between the iron in the pipe and corrosive agents in the soil. The zinc acts sacrificially as it corrodes away instead of the steel in the pipe. (Courtesy Alyeska Pipeline Service Company.)

SUMMARY

Ores and minerals. Nearly all commercially important metals occur in nature as chemical compounds called minerals that are intermixed with other minerals of little or no commercial value. An ore is a mixture of minerals from which one or more metals can be extracted at a profit.

Iron refining. Iron minerals are usually oxides. Taconite, a particularly hard rock, contains a magnetic iron mineral called magnetite. In getting iron from taconite, the magnetite is separated magnetically from the crushed ore and shaped into pellets as it is heated to convert the magnetite to hematite. In this form the refined ore together with coke and limestone are heated in a blast furnace where carbon monoxide reduces iron oxides to molten iron. Limestone gathers up impurities and forms a molten slag that is separated. For further refining, crude pig iron is mixed with scrap metal, heated, and subjected in the basic oxygen process (BOP) to a jet of oxygen that oxidizes most impurities. The product is steel.

Refining by electrolysis. When direct current is passed through a solution containing the ions of a desired metal, the ions change at the cathode to atoms and the metal deposits. Electroplating of tin or zinc onto steel can be done this way.

Aluminum refining. Bauxite, the chief ore of aluminum, is processed to make alumina, Al_2O_3, which is dissolved in molten chyrolite and subjected to electrolytic refining. Next to iron, aluminum is the most used metal in the United States.

Flotation. Without froth flotation several metals, including copper and nickel, would probably not be available except at exorbitant prices. By bubbling air through a slurry of ore in water containing special frothing chemicals, the particles of the desired mineral are carried onto the surfaces of the bubbles and can be skimmed away from the crushed rock, which settles out. In this way ores of very low concentration in the desired metal can be refined at relatively low cost.

Sulfide ores. Ores of copper, nickel, lead, and zinc contain these metals chiefly as their sulfides instead of their oxides. Such ores are generally roasted to adjust the sulfur content by changing some sulfides to oxides. Sulfur dioxide becomes an air pollutant in this operation. In the final stages crude copper, lead, and nickel can be purified electrolytically. Zinc is a metal that can be distilled economically.

Galvanic action. Two metals in contact with each other and a water solution of any conducting materials (e.g., salt) will interact, and one of the metals tends to dissolve or become oxidized. Galvanic action is this selective corrosion of the more active of the pair of metals. Atoms of the more active metal change to ions; the released electrons go through the less active metal to reduce something dissolved in the water. The less active metal is simply a conductor; it does not corrode. Hence, steel items that must be in salt water (such as ships) can be protected in part by fixing zinc (a more active metal than iron) to them.

TERMS AND CONCEPTS

mineral
ore
taconite
blast furnace
slag
pig iron
basic oxygen process
steel
tinplate
electrochemistry
electrode
anode
cathode
electroplating
electrolysis
electrolytic refining
froth flotation
galvanic action
electropositivity

QUESTIONS AND EXERCISES

1. Any clay deposit in the world contains aluminum. Suggest reasons why clay is not classified as an aluminum ore.

2. What physical property of magnetite is exploited to separate this mineral from crushed taconite ore?

3. What is the function of limestone in the operation of a blast furnace to make pig iron?

4. The partial combustion of coke in the iron blast furnace serves what two purposes?

5. What are four impurities in pig iron?

6. What does the basic oxygen process do to change pig iron to steel?

7. What is the purpose of "pickling" in processing steel and how is it done?

8. Metallic ions are sometimes called *cations* because they migrate to the cathode during electrolysis. What kind of electric charge does a cathode have?

9. Anions that migrate to the anode during electrolysis are sometimes called *anions*. What electric charge do these ions bear? What is the electric charge on the anode?

10. Write the equation for the reaction occurring at the cathode when a solution of copper chloride, $CuCl_2$, is electrolyzed. (Treat electrons as a reagent having the symbol "e^-.")

11. Write the equation for the reaction at the anode when $CuCl_2$ is electrolyzed.

12. In the future, nuclear or solar energy might be converted to electricity and "shipped" by transmission wires or it might be used to generate hydrogen which would be shipped through pipelines. What would be used as the source of hydrogen?

13. Why cannot aluminum be made by electrolyzing some aluminum salt dissolved in water or in dilute acid?

14. What makes it uneconomic to electrolyze molten alumina (Al_2O_3) to make aluminum metal?

15. Why are the major aluminum refineries situated near the Columbia River, the Tennessee Valley, and Niagara Falls?

16. What air pollutant is produced during aluminum refining?

17. What is flotation as used in refining certain metals? In general terms, how does it work?

18. In the electrolytic refining of copper, impure copper is used as the anode, copper sulfate ($CuSO_4$) is in the solution between the electrodes,

17 METAL REFINING

and copper is used as the cathode. Explain how the impure copper on the anode gets to the cathode in a purer state.

19. Write the equation for the change in lead sulfide when it is roasted.

20. Write the equation for the change in lead oxide when it is reduced to lead.

21. How does ionized lead affect a person's health?

22. What is meant by *galvanic action?*

23. How does zinc protect iron by galvanic action?

24. Why are zinc plates attached to ship hulls?

CHAPTER 18
THE EARTH'S SUBSTANCES—ORGANIC COMPOUNDS

18.1 THE ORGANIC REALM

Organic compounds are invariably compounds of carbon. They occur so widely in nature that by far the majority of all compounds are organic. The food you had for breakfast, the clothes you are wearing, the paper in this book, the drugs you take when you are sick, the vitamins that help keep you well, the plastics in almost every room of a house, the gasoline and oil in the transportation you probably took to school—all these are organic substances or mixtures of them.

Within your eyes scanning this page, in your brain processing what you see, in your heart and lungs silently busy with their vital tasks, in and among all these parts, down at the subcellular level, there are countless organic molecules. All of life has a molecular basis. So, too, does all of disease, and most of the molecular types are organic.

18.2 THE ARCHITECTURE OF ORGANIC MOLECULES

There is no other element like carbon, none that forms the basis, the molecular skeletons of as many and as wide a variety of substances. One reason carbon is unique is that its atoms form strong covalent bonds *to each other* while still being able to hold atoms of many other elements by covalent bonds. The molecules of virtually all organic compounds, include, besides carbon, bound atoms of hydrogen. Most also include bound atoms of oxygen; many include nitrogen, or one of the halogens, or sulfur or phosphorus.

18.3 CARBON CHAINS AND RINGS

In support of the octet rule (Section 14.3), a carbon atom will have four covalent bonds going from it. The structural formulas of Table 18.1 are just a few examples of the many ways in which these bonds are used. In some the carbon "skeleton" consists of one carbon following another. We call this a *straight chain*. There is in principal no upper limit to the length of a chain. In others we have *branched chains;* a carbon atom can directly bind as many as four other carbons. The skeleton of carbon atoms can close into a *carbon ring*. As few as three carbons can make up a ring, and there is in principal no upper limit to the size of a ring. Often we find one molecule containing more than one ring; molecules of a compound may be bicyclic, tricyclic, and polycyclic. As you study Table 18.1, you should pay close attention to the ways used to write convenient symbols of structural formulas.

18.4 STRUCTURAL ISOMERS AND ISOMERISM

The unique ability of atoms of carbon to join together to form long chains and large ring systems is the main reason there are so many organic compounds. Another reason, the phenomenon of *isomerism,* is directly related to this.

TABLE 18.1
SOME EXAMPLES OF ORGANIC COMPOUNDS

Full Structural Formulas (with names and an occurrence)	Condensed Structures
Some Straight-Chain Molecules	
Methane (main component of natural gas) — $H-C(H)(H)-H$	CH_4 All single bonds between carbon and hydrogen may be "understood"
Hexane (a minor component of gasoline) — H—C—C—C—C—C—C—H with H's	$CH_3-CH_2-CH_2-CH_2-CH_2-CH_3$ or $CH_3CH_2CH_2CH_2CH_2CH_3$ Single bonds between carbons that appear on a horizontal line may be "understood"
Ethyl alcohol (grain alcohol) — H—C—C—O—H with H's	CH_3-CH_2-O-H or CH_3CH_2OH Again, single bonds that would otherwise appear on a horizontal line may be "understood." Note that oxygen always has a covalence of 2—two bonds from each
Ethylamine (a fishy odor) — H—C—C—C—C—N—H with H's	$CH_3-CH_2-CH_2-CH_2-NH_2$ or $CH_3CH_2CH_2CH_2NH_2$ Nitrogen has a simple covalence of 3—three bonds from each nitrogen
Ethyl mercaptan (part of a skunk's defensive fluid) — H—C—C—C—C—S—H with H's	$CH_3CH_2CH_2CH_2SH$ Sulfur, in the same family as oxygen, has a covalence number of 2
A Branch-Chain Compound	
Isooctane (the standard for for 100 octane gasoline)	$CH_3-\underset{CH_3}{\overset{CH_3}{C}}-CH(CH_3)-CH_3$ or $(CH_3)_3CCH(CH_3)_2$ Usually, the second condensed structure is not used. In other words, "branches" on a main chain are shown sticking up or down and vertical bonds are shown; they are not "understood"
Some Cyclic Compounds	
Cyclopropane (an anesthetic)	CH_2-CH_2 / CH_2 or △ Ring compounds are usually represented simply by the corresponding geometric figure, a triangle for a three-membered ring. It is "understood" that at each corner there is a carbon atom holding as many hydrogen atoms as required to fill out carbon's tetravalence

TABLE 18.1 (continued)

Full Structural Formulas (with names and an occurrence) | **Condensed Structures**

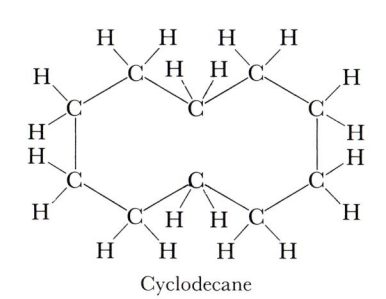

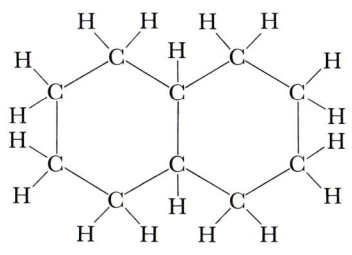

Cyclodecane

Decalin (a bicyclic compound)

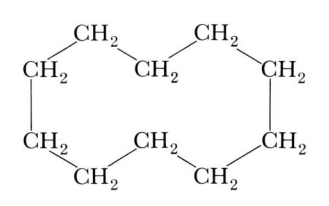

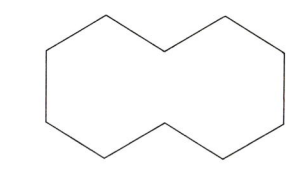

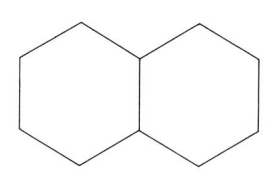

Each line is a carbon-carbon bond. Where only two lines meet at a corner, there must be two hydrogens bound to the carbon "understood" to be at that corner. Where three lines meet, there can be only one hydrogen

---

Isomerism is the existence of compounds having identical molecular formulas but different structures. Compounds having different structures for the same molecular formula are called *isomers* (after the Greek, *iso*, "the same" and *-meros*, "parts").

Butane, **1**, and isobutane, **2**, are isomers. They are truly different compounds, as is evident by their physical properties, but they have identical molecular formulas, C_4H_{10}. Hence, the pieces (atoms) used to make the molecule are not as important as the way in which the pieces are organized. In butane the carbons form a straight chain; in isobutane, a branched chain.

$$CH_3-CH_2-CH_2-CH_3 \qquad CH_3-\underset{\underset{CH_3}{|}}{CH}-CH_3$$

1 butane
boiling point −0.5°C
melting point −138.4°C

2 isobutane
boiling point −11.7°C
melting point −159.6°C

The differences in properties between isomers can sometimes be very great. Ethyl alcohol, **3**, and dimethyl ether, **4**, share the molecular formula C_2H_6O (see Figure 18-1). At room temperature and pressure, ethyl alcohol is a liquid, and dimethyl ether is a gas; their chemical properties are very different.

CH_3CH_2OH
3 ethyl alcohol
boiling point 78.5°C
melting point −117°C
density 0.789 g/cm³

CH_3-O-CH_3
4 dimethyl ether
boiling point −24°C
melting point −138.5°C
density 2 g/liter

Ethyl alcohol is changed by hot, concentrated hydriodic acid, H-I, to a two-carbon compound, ethyl iodide, CH_3CH_2I. Dimethyl ether is broken by hydriodic acid into two molecules of a one-carbon compound, methyl iodide, CH_3I.

$$CH_3CH_2OH + HI \xrightarrow[water]{} CH_3CH_2I + H_2O$$

$$CH_3OCH_3 + 2HI \xrightarrow[water]{} 2CH_3I + H_2O$$

These reactions are evidence that in molecules of ethyl alcohol the carbons are joined to each other (see **3**) and in those of dimethyl ether they are not (see **4**).

18 THE EARTH'S SUBSTANCES—ORGANIC COMPOUNDS

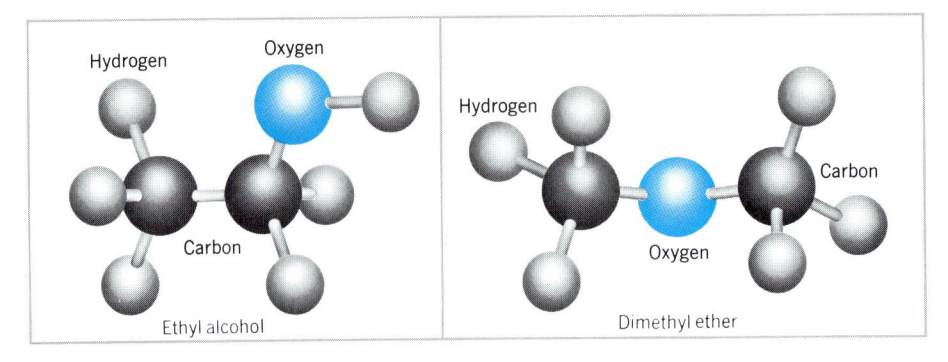

FIGURE 18–1
The "ball-and-stick" models of two isomers.

Ethyl alcohol reacts with sodium metal quite vigorously, and dimethyl ether does not react at all. Ethyl alcohol changes to a two-carbon salt-like compound of sodium, sodium ethoxide, CH_3CH_2ONa; hydrogen gas is released. One of

$$2CH_3CH_2OH + 2Na \longrightarrow 2CH_3CH_2ONa + H_2$$

the six hydrogens in ethyl alcohol evidently is joined to oxygen, but in dimethyl ether there is no hydrogen-oxygen bond.

18.5 FLEXING, TWISTING, ROTATING MOLECULES

Groups that are joined by single, covalent bonds can rotate relative to each other about that bond. In Section 14.16 we studied how a covalent bond arises from the partial merger or interpenetration of two atomic orbitals. The orbitals overlap, and two electrons, the shared pair, go into the new space called a molecular orbital. The amount of overlap is not affected very much

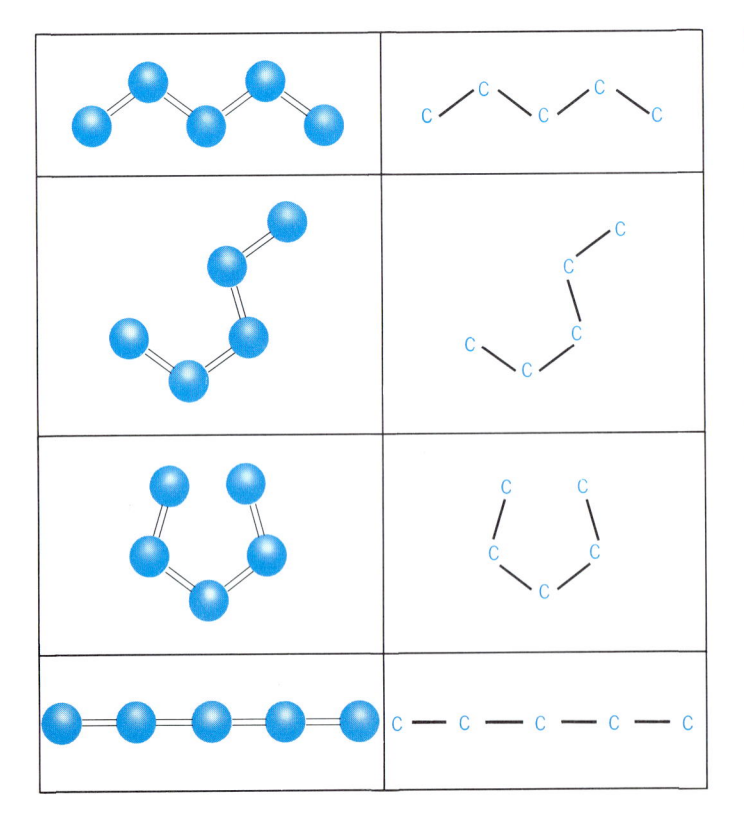

FIGURE 18–2
The carbon skeleton of pentane in various conformations. Only the carbon-carbon bonds are shown here. There are many other possible conformations. The one at the top and the one at the bottom are really the same, the bottom conformation simply being an edge-on view of the one at the top. The most populous conformation in a sample of pentane will be the one in which parts of the molecule get as far away from each other as they can. (From J. R. Holum. *Elements of General and Biological Chemistry*, 4th edition. 1975. John Wiley & Sons, Inc., New York.)

by internal rotations of the groups joined together. Hence, these rotations occur at virtually no expense in energy. That is why they are called *free rotations*.

The phenomenon of free rotation about single bonds means that a sample of an organic compound will have molecules in an almost infinite variety of shapes. Figure 18–2 illustrates just a few that are possible for molecules of pentane, a compound found in petroleum. The physical and chemical properties we measure for pentane are therefore the average effect of molecules in all these conformations on whatever instruments or chemicals used to study the property. Granting that molecules flex and twist, we then select just one conformation—the one easiest to set in type or write—when we represent a molecule's structure on paper or the blackboard.

18.6 RESTRICTED ROTATION AT DOUBLE BONDS AND IN RINGS

Groups joined by a double bond cannot rotate with respect to each other. The difference between the single and double bond is analogous to the difference between joining two boards by one nail or two. The boards can spin when attached to each other by one nail, but a second nail freezes this action. Molecules of ring compounds likewise have little or no flexibility compared to those of open-chain compounds.

18.7 GEOMETRIC ISOMERISM

Restricted rotations at double bonds or in rings make *geometric isomerism* possible, a type of isomerism in which molecules of the isomers differ only in internal geometries, not in basic atom-to-atom sequences. Examples of geometric isomers are shown in Figure 18–3.

18.8 MOLECULAR GEOMETRY AT CARBON

In Section 14.24 we learned that covalent bonds have direction to them. The angle between the bonds from oxygen to the hydrogens in a water molecule is 105 degrees. In like manner, when a carbon holds four groups—when it has four *single* bonds from it—there are definite angles between them. The four bonds in methane, the simplest and most symmetrical of all organic molecules, point outward in such a way that if we connect their ends by lines we have a regular tetrahedron, as seen in Figure 18–4. The carbon of the methane molecule is at the exact center of this tetrahedron, and the angle between any two bonds is 109 degrees 28 minutes. We call any carbon having four single bonds a *tetrahedral carbon*.

When a carbon has a double bond going from it, while it will have a total of four bonds, it holds only three atoms or groups. It is using two bonds to hold one of the groups, as seen in the examples of the 2-butenes of Figure 18–4. The atoms directly attached to such a carbon are now all in the same plane, and the bond angle is approximately 120 degrees.

One of the profound consequences of a tetrahedral carbon is that it allows nature its most subtle way of having isomers, which we shall study in the next section.

18.9 OPTICAL ISOMERISM— LEFT- AND RIGHT- HANDED MOLECULES

In many respects, the most interesting way in which substances can be isomers occurs when their molecules possess a "handedness" analogous to our left and right hands. We shall consider some features of "handedness" first and then see what that means when a molecule has it.

Your left and right hands are very similar. You could say, disregarding wrinkles and fingerprints, that they have the same "molecular" structure, where the center of the palm is one atom and the fingers are attached groups. In each hand there are the same groups and the same "bonds" to groups, and they point in the same relative directions. The distance from the thumb to little finger (or to any other finger you wish to select) is the same in each hand. So alike are they we are tempted to say the hands are identical, not isomers. Yet the two hands are different. You find that out if you try to fit a left hand glove to your right hand.

Now hold the two hands with palms facing each other. Imagine that there is a mirror between them. Think of the left hand as the object facing the mirror. The right hand then becomes the image in the mirror. Left and right hands are related as object to mirror image. That is how very closely they are alike. And yet you cannot comfortably fit a left hand into a right glove. The two hands are different, and this brings us to the ultimate test for deciding if two things—two hands or two molecules—are identical while having an object-to-mirror image relation. That test is *superimposability*.

Let your left and right hands face each other again. Now bring them together so that they

18 THE EARTH'S SUBSTANCES—
ORGANIC COMPOUNDS

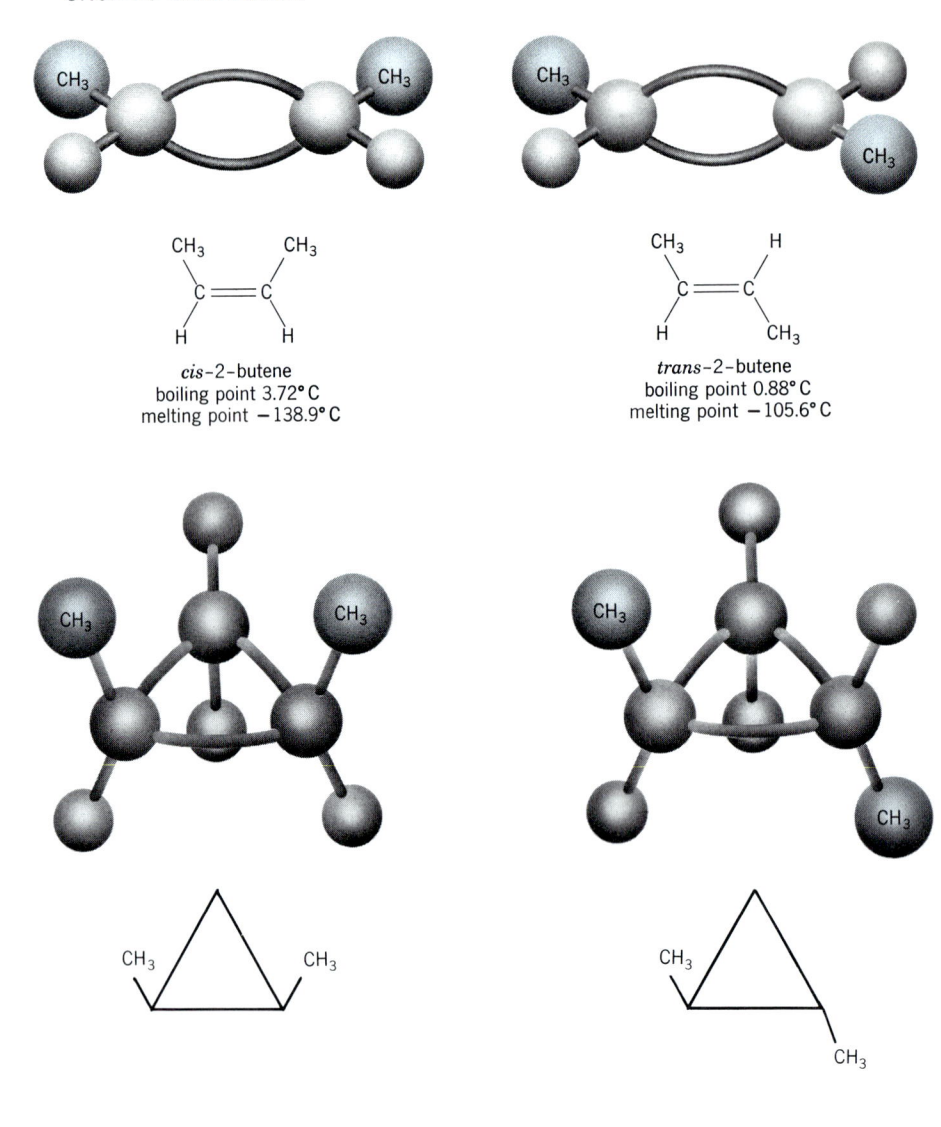

cis-2-butene
boiling point 3.72°C
melting point −138.9°C

trans-2-butene
boiling point 0.88°C
melting point −105.6°C

cis-1,2, Dimethylcyclopropane
bp 37°C
mp −141°C

trans-1,2, Dimethylcyclopropane
bp 28°C
mp −150°C

FIGURE 18-3

Geometric isomers. Restricted rotation at double bonds and within rings makes possible geometric isomerism. *Cis* means that two groups are on the same side; *trans*, on the opposite side.

touch, thumbs touching thumbs, palms touching palms. What you are now doing is simply "imposing" them. Superimposability is something more. Superimposing the two hands can only be done mentally (which is why it is called *superimposing*). Pretend you blend the right hand right through the left to make it, in your mind, completely inside the left hand. Do all parts of the right hand now blend and perfectly match all parts of the left? The answer is "No." The fingers may *seem* to be satisfactorily blending, but the palms are not blending and matching at all. The palm of one hand is on the back side of the other. If you try to get the palms to "blend" by turning one hand over, the fingers do not superimpose. The left and the right hands, although related as object to mirror image, do not superimpose (Figure 18–5). Therefore, they are different; they do possess (what we have known all along) a handedness. Only if two objects superimpose may we call them identical.

Try the same thought experiment with some regular objects such as two thumbtacks. You can line them up with a make-believe mirror be-

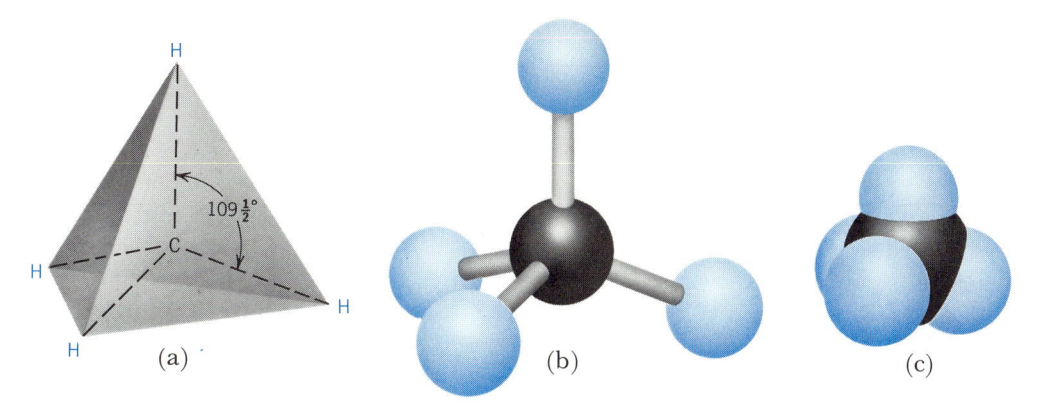

18.9 OPTICAL ISOMERISM—LEFT- AND RIGHT-HANDED MOLECULES

FIGURE 18-4
Tetrahedral carbon atom. (a) Dotted lines show the four covalent bonds of carbon that are directed toward the corners of a regular tetrahedron. (b) Common "ball and stick" model of methane. (c) Scale model of methane, a model that attempts to indicate the relative volumes occupied by the various "electron clouds" in the molecule.

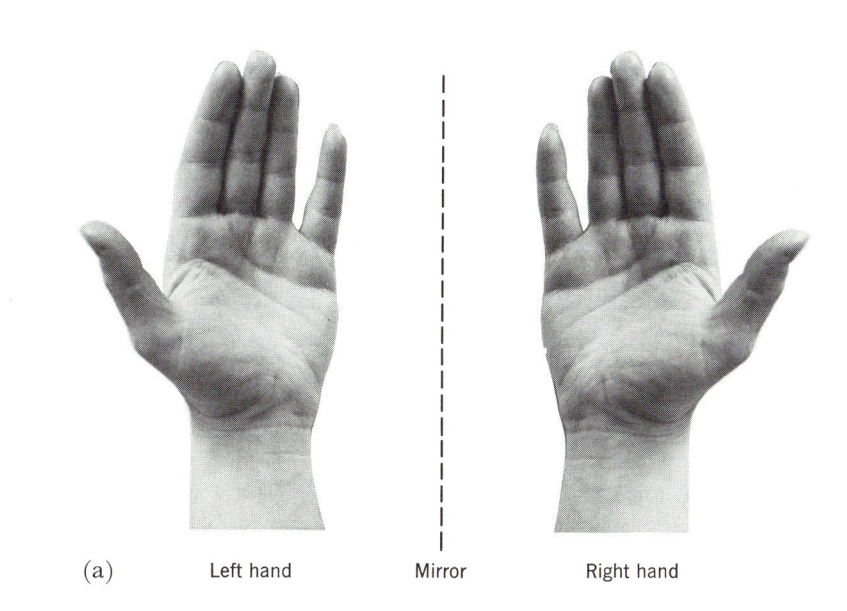

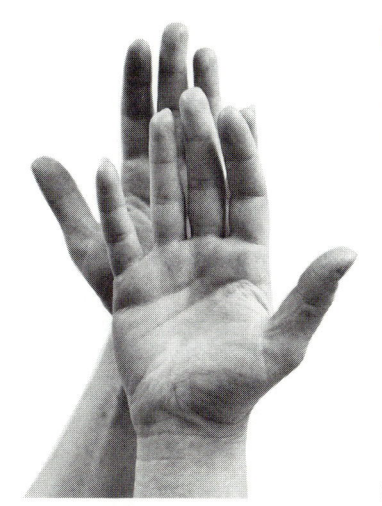

(a) Left hand   Mirror   Right hand

FIGURE 18-5
Your left hand and your right hand are related as object-to-mirror image (a) but they do not superimpose (b). (K. T. Bendo)

tween them and have one appear as the mirror image of the other. And you can also image the operation of superimposition. The two do superimpose; they do not possess a handedness. They are, in fact, identical and not "isomers."

Now turn to Figure 18–6, where two molecules have identical structures—identical atom-to-atom sequences. They are models of the molecules in a compound known as asparagine, **5**, with a molecular formula of $C_4H_8N_2O_3$. Each model has a handedness.

$$H-O-\overset{\overset{O}{\|}}{\underset{1}{C}}-\underset{2}{\underset{|}{CH}}-\underset{3}{CH_2}-\overset{\overset{O}{\|}}{\underset{4}{C}}-OH$$
$$\text{NH}_2$$

5 asparagine

Pretend you view each in the way the eyes are looking in the figure—that is, down the bond from carbon to hydrogen. Now note carefully that if your eye is to sweep from group *g* then to *e* and then to *f*, *in that order* of *g* to *e* to *f*, it must sweep clockwise in one model and counterclockwise in the other. It is the same with your two hands. Hold them with palms facing you. Let your eyes now sweep from thumb to middle finger to little finger *in that order*. With the left hand you sweep clockwise (or left to right) and with the right hand you sweep counterclockwise (or right to left). The two asparagine models were made up with this kind of difference in handedness at the carbon we numbered 2 in the structure **5**. It happened to be a carbon to which *four different* groups were joined, not just three but four groups, and not just four groups but four different groups. (At carbon number 1 in **5**, there are only three atoms or groups: OH, =O, and the third, more complicated group. At carbon number 3 there are four groups or atoms, but two are identical, the two hydrogens. At carbon number 4, we again have only three groups or atoms. Only carbon 2 has four different groups.)

The remarkable facts about the two asparagines are these (and we have selected asparagine only as one of thousands—hundreds of thousands—of possible examples in nature). They exist; there are two forms of asparagine. One occurs in the juice of asparagus and it has a bitter taste. The other can be isolated from an herb called vetch, and it has a sweet taste. Taste is a chemical sense. The sensation of taste begins with a chemical reaction between molecules of what you are tasting and receptor molecules in the taste buds. These receptor molecules also have a handedness, *but only one form is present* in the taste bud (it is as if you had a box of only right-hand gloves). Asparagine molecules of one handedness generate one kind of chemical reaction in the taste buds, a reaction eventually producing the sensation of bitterness. Those asparagine molecules of the other handedness lead to the sensation of sweetness. The two types of asparagine are called *optical isomers* of each other. Optical isomers are compounds whose molecules are related as object to mirror image but which cannot be superimposed. There are other differences in properties between the two asparagines. We have described only one—a difference in chemistry involving taste—in order to establish the fact that these two are different compounds. Yet they have the same molecular formula, $C_4H_8N_2O_3$, and they also have the same basic structures. They differ only in handedness at one carbon.

The technical term for handedness is *chirality* (it comes from the Greek, *cheir*, meaning "hand"). The asparagine molecules are chiral molecules; they are related as object to mirror image but cannot be superimposed. Any molecule having *one* carbon holding four different groups will be chiral. (Where there are two or more such carbons in one molecule, the situation becomes more complicated. We shall not study that, except to state that there will always be at least some forms that will be chiral.)

Virtually all of the molecules of nutrients in the food you eat are chiral and of only one handedness or chirality. If you were to eat synthetic food whose molecules were identical except for their being of opposite handedness you could not survive. It would be as if you ate no food at all! That is a measure of how important is the handedness of molecules. The reason is that the enzymes you have, the catalysts needed to process food molecules, are also of one handedness. They can act *only* on food molecules of the correct (one might even say "matching") handedness, just as a right hand "reacts" smoothly only with a right hand glove.

The handedness of the "building block" molecules used by some forms of life can lead to some beautiful chiral forms in nature. Certain seashells have a spiral shape that is chiral; a model of the mirror image of the shell could not be superimposed with it. Another analogy is a wood screw. A right-handed screw is one you turn clockwise, to the right, to seat it. You would turn the mirror image oppositely. The two are related as object to mirror image, which cannot be superimposed. See Figure 18–7. Some vines will twist in one direction but not in another as they climb and grow.

Where molecules of nature's organic substances can have a handedness, nearly always only one form (one "hand") actually occurs. It is as if during chemical and biological evolution, nature by some random accident selected for

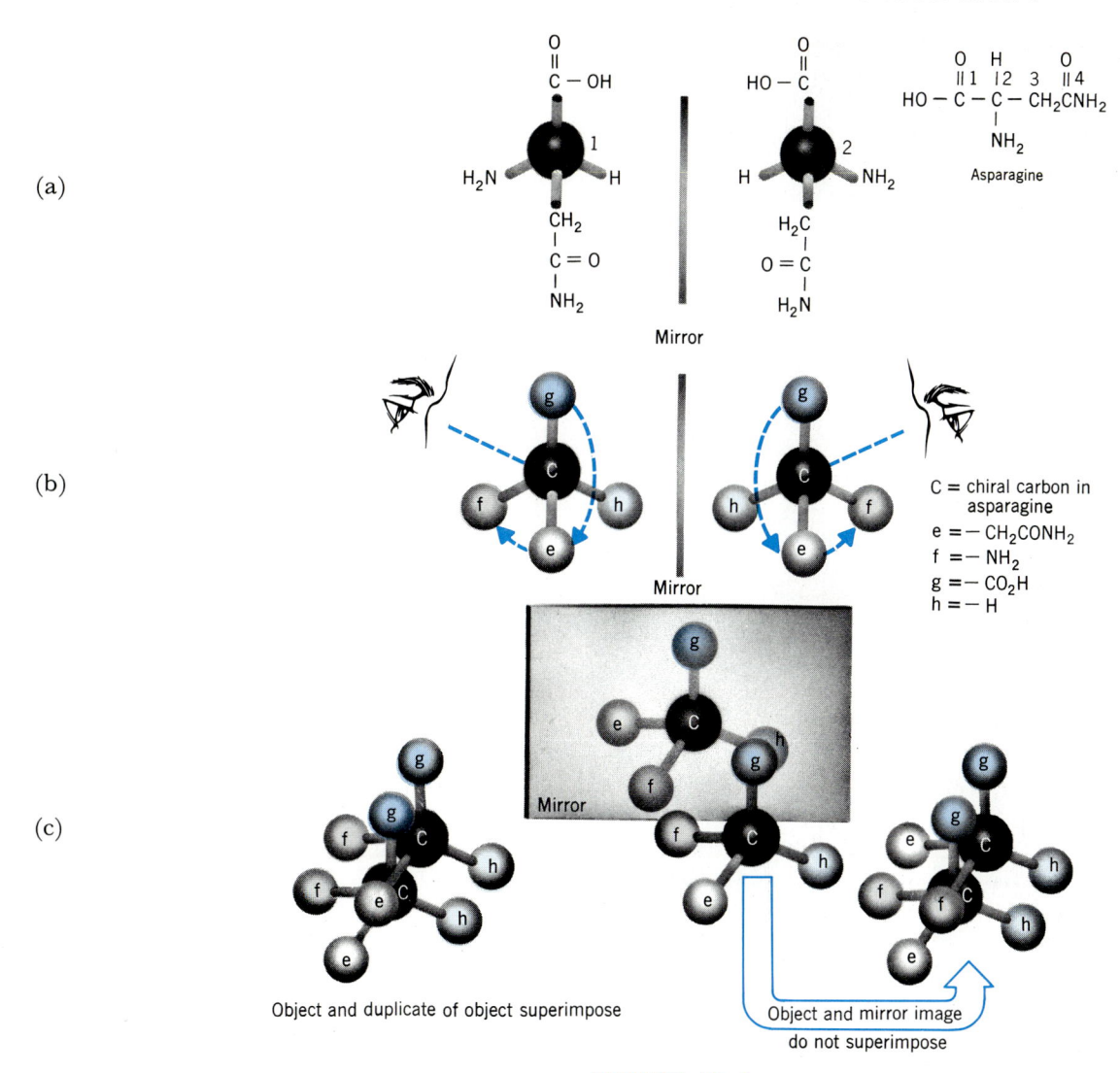

18.10 FAMILIES OF ORGANIC COMPOUNDS—FUNCTIONAL GROUPS

FIGURE 18-6
The two forms of asparagine are shown in part (a), but it is simpler to use the structures of part (b) to study the internal configurations of these two. In part (c) the attempt to superimpose object and mirror image fails as groups (f) and (e) do not coincide. However, an object and its exact duplicate, not its mirror image, do superimpose, as seen on the left in part (c). (From J. R. Holum, *Organic Chemistry: A Brief Course,* 1975, John Wiley & Sons, Inc., New York.)

each such substance just one of the possible forms. Perhaps it was the chance formation of a catalyst having one kind of handedness that led to the production of molecules of just one kind of handedness and, in turn, inhibited the generation of the opposite form of the original catalyst. One can only speculate, but the handedness of most organic molecules in nature is not just an important fact about the world we live in, it is one of the most interesting and thought-provoking. It is one of nature's "tremendous trifles."

18.10 FAMILIES OF ORGANIC COMPOUNDS— FUNCTIONAL GROUPS

In Section 18.4 we learned that molecules of ethyl alcohol have the structure CH_3CH_2OH and that they react with hydriodic acid. However, in this reaction only one part of the molecule functioned chemically, the -OH group. The ethyl group, CH_3CH_2-, rode out the chemical

18 THE EARTH'S SUBSTANCES—ORGANIC COMPOUNDS

(a)

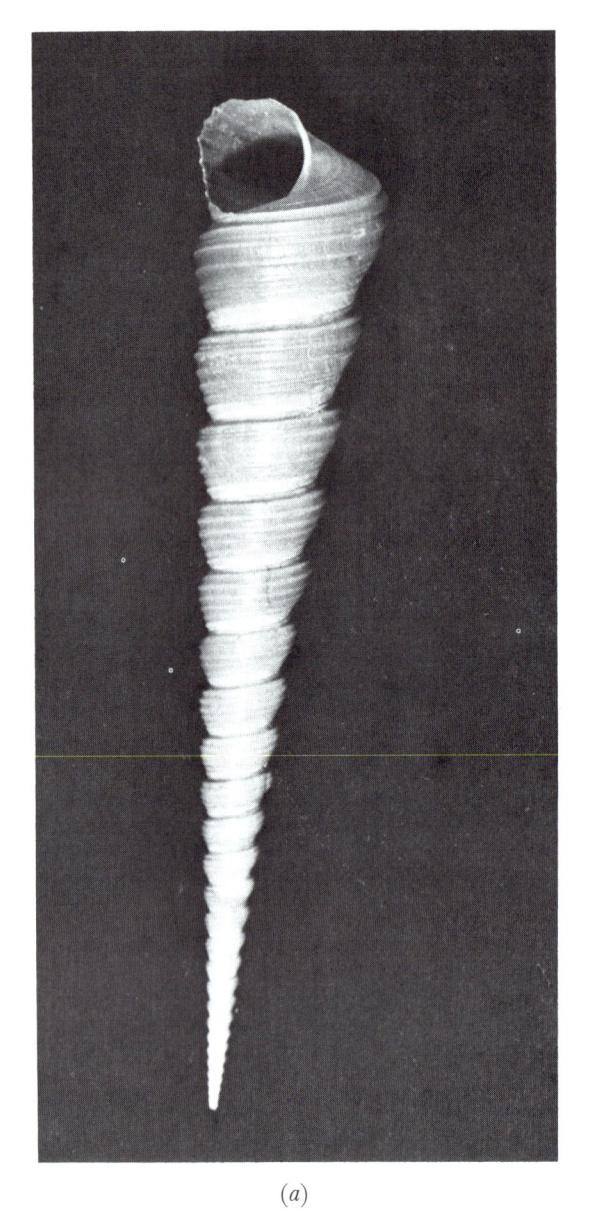

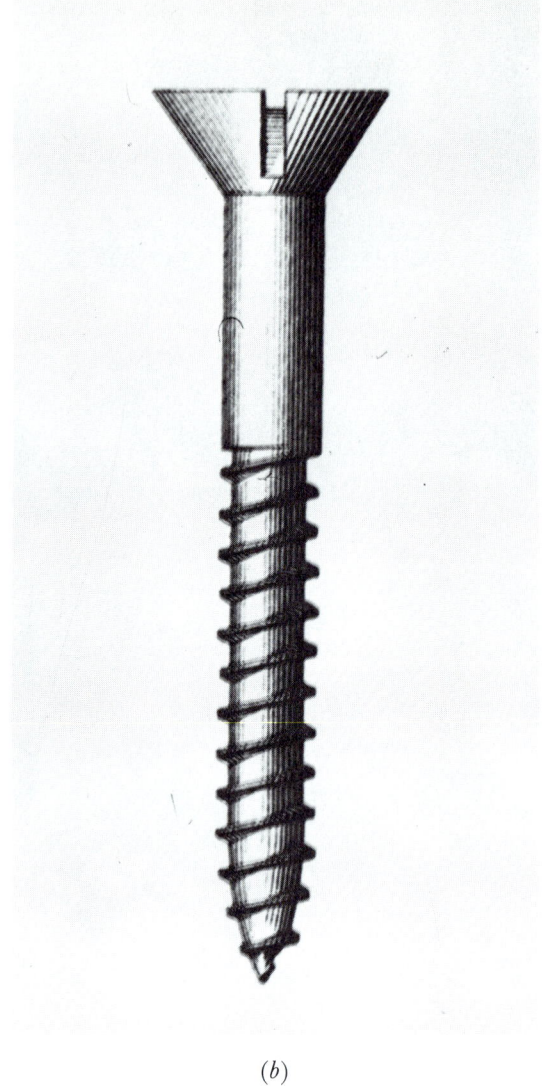

(b)

FIGURE 18–7
Handedness or chirality in nature and in manufactured objects. (a) The turritella snails, tropical relatives of common periwinkles, have right-handed twists. In fact, all the sea shells that have any twist to them at all, listed in *Sea Shells of the World* by R. T. Abbott (Golden Press, New York), have a right-handed twist. (Photo: Carlton Ray/Photo Researchers.) (b) The common wood screw used in the United States, Canada, and Mexico, has a right-handed twist. (Photo: Brown Brothers.)

change untouched. It did not function chemically; it is a nonfunctional group. We call the -OH group a *functional group,* and its particular name is *alcohol group*. Any substance whose molecules contain an -OH group attached to a carbon that, in turn, has only single bonds going from it will behave more or less like ethyl alcohol toward hydriodic acid. Here are some examples.

CH_3OH + HI → CH_3I + H_2O
Methyl alcohol

$CH_3CH_2CH_2CH_2OH$ + HI →
Butyl alcohol

$CH_3CH_2CH_2CH_2I$ + H_2O
Butyl iodide

CH$_3$CH$_2$CH$_2$CH$_2$CH$_2$CH$_2$CH$_2$CH$_2$OH + HI
Octyl alcohol

→ CH$_3$CH$_2$CH$_2$CH$_2$CH$_2$CH$_2$CH$_2$CH$_2$I + H$_2$O
Octyl iodide

In each case the -OH is replaced by -I and a molecule of water is produced.

There are many kinds of functional groups among the molecules of organic compounds. Because molecules having identical functional groups but differing nonfunctional groups have similar chemical properties, we have a simple way of classifying organic compounds into just a few families (see Table 18.2). A few will be described next. The simplest type is the family of alkanes whose molecules have no functional group. Alkanes belong to the hydrocarbon family.

18.11 FAMILIES OF HYDROCARBONS

Hydrocarbons are substances whose molecules contain only carbon and hydrogen. The carbon skeletons may be in open chains or in rings. If all the bonds are single, we say that the substance is *saturated*. Double and triple bonds in varying numbers and combinations may be present, and substances having them are *unsaturated*. The principal families of hydrocarbons are outlined in Figure 18-8. The two main families are the *aliphatic* and *aromatic* hydrocarbons. *Aromatic* originally denoted occurrence among fragrant substances, such as oil of wintergreen. It now is a label for any compound having a benzene ring system or that has benzenelike chemical properties. *Aliphatic* (Greek, *aleiphar*, "oil") designates nonaromatic compounds.

TABLE 18.2
SOME IMPORTANT CLASSES OF ORGANIC COMPOUNDS

Class	Characteristic Structural Feature
Hydrocarbons	Composed only of carbon and hydrogen, with many subclasses according to whether single, double, or triple bonds are present
	Alkanes: all single bonds (e.g., CH$_3$CH$_3$)
	Alkenes: at least one double bond (e.g., CH$_2$=CH$_2$)
	Alkynes: at least one triple bond (e.g., H—C≡C—H)
	Aromatic: at least one benzenoid ring system
Alcohols	Contain the —OH group, as in CH$_3$CH$_2$—OH
Ethers	Contain C—O—C system as in CH$_3$—O—CH$_3$
Aldehydes	Contain the —C(=O)—H group as in CH$_3$—C(=O)—H
Ketones	Contain the C—C(=O)—C group as in CH$_3$—C(=O)—CH$_3$
Carboxylic acids	Contain the —C(=O)—OH system as in CH$_3$—C(=O)—OH
Esters of carboxylic acids	Contain the —C(=O)—O—C system as in CH$_3$—C(=O)—O—CH$_2$CH$_3$
Amines	Contain trivalent nitrogen, where all covalent bonds are single bonds as in CH$_3$—NH$_2$
Amides	Contain the grouping —C(=O)—N as in CH$_3$—C(=O)—NH$_2$

The structures of the ten smallest "straight-chain" alkanes are given in Table 18.3. From the C$_4$ alkane, butane, and on, isomerism is possible. There are two butanes, C$_4$H$_{10}$, three pentanes, C$_5$H$_{12}$, five hexanes, C$_6$H$_{14}$. The number of substances in a set of isomers increases very rapidly with the carbon content. Decane, C$_{10}$H$_{22}$, exists

TABLE 18.3
STRAIGHT CHAIN ALKANES

Number of Carbon Atoms	Name[a]	Molecular Formula	Structural Formula	Boiling Point (°C)
1	Methane	CH$_4$	CH$_4$	−162
2	Ethane	C$_2$H$_6$	CH$_3$CH$_3$	−88.6
3	Propane	C$_3$H$_8$	CH$_3$CH$_2$CH$_3$	−42
4	n-Butane	C$_4$H$_{10}$	CH$_3$CH$_2$CH$_2$CH$_3$	0
5	n-Pentane	C$_5$H$_{12}$	CH$_3$CH$_2$CH$_2$CH$_2$CH$_3$	36
6	n-Hexane	C$_6$H$_{14}$	CH$_3$CH$_2$CH$_2$CH$_2$CH$_2$CH$_3$	69
7	n-Heptane	C$_7$H$_{16}$	CH$_3$CH$_2$CH$_2$CH$_2$CH$_2$CH$_2$CH$_3$	98
8	n-Octane	C$_8$H$_{18}$	CH$_3$CH$_2$CH$_2$CH$_2$CH$_2$CH$_2$CH$_2$CH$_3$	126
9	n-Nonane	C$_9$H$_{20}$	CH$_3$CH$_2$CH$_2$CH$_2$CH$_2$CH$_2$CH$_2$CH$_2$CH$_3$	151
10	n-Decane	C$_{10}$H$_{22}$	CH$_3$CH$_2$CH$_2$CH$_2$CH$_2$CH$_2$CH$_2$CH$_2$CH$_2$CH$_3$	174

[a] These are the *common* names. Among the common names *n-* stands for *normal*. From butane and up, the alkanes can exist as sets of isomers, and *normal* specifies the straight-chain member of a set of isomers. Where isomers are impossible (the first three), *n-* is omitted. The *formal* names for the straight-chain alkanes shown here are the same as the common names except that the *n-* designations are omitted.

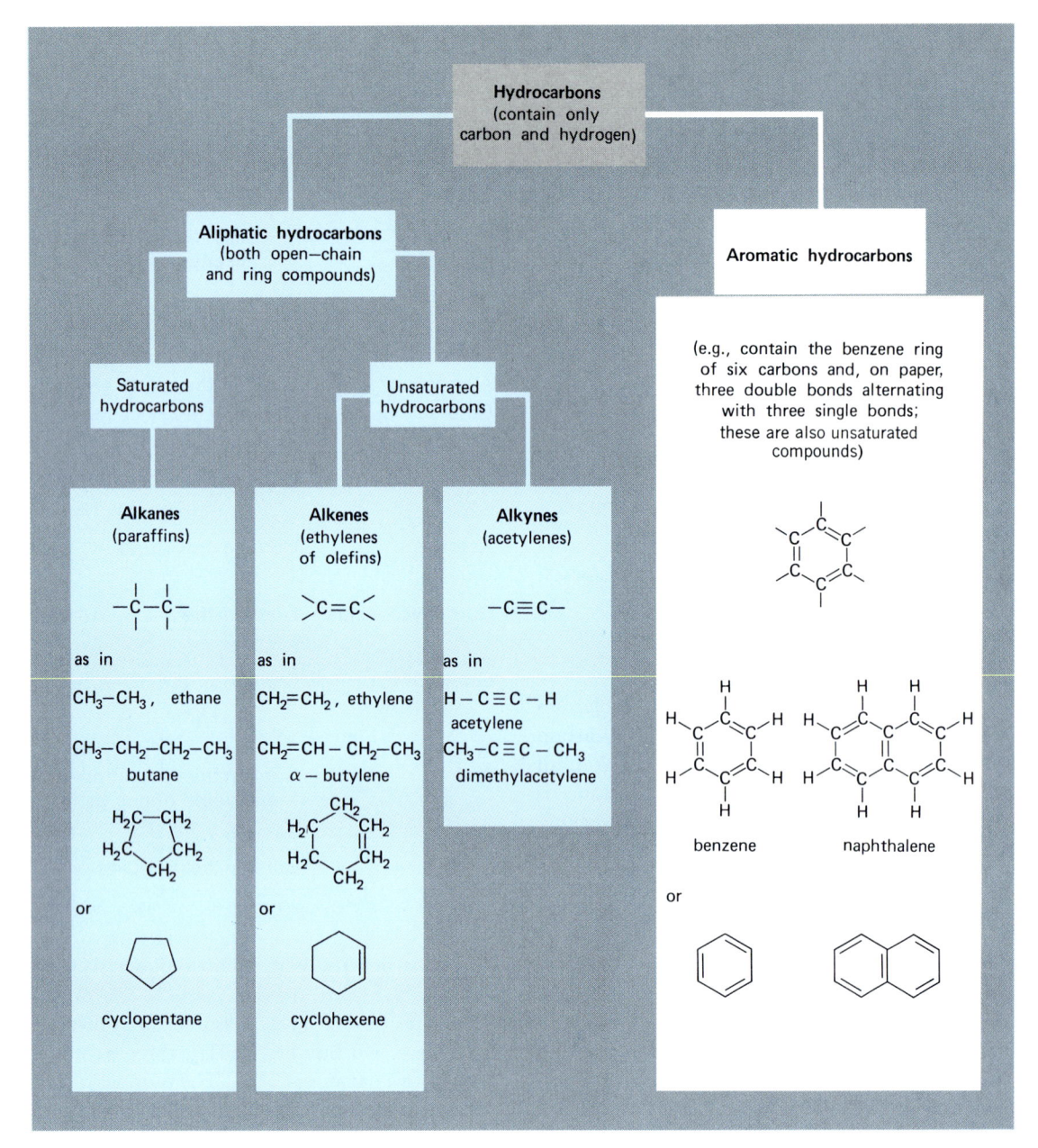

FIGURE 18-8
Principal families of hydrocarbons.

in 75 isomeric forms; $C_{20}H_{42}$ has 366,319 isomers; $C_{40}H_{42}$ has 6.25×10^{13}.

Repesentative unsaturated hydrocarbons are listed in Table 18.4.

18.12
NAMES

The name ending for all alkanes is *-ane;* for all alkenes, *-ene;* and for all alkynes, *-yne.* Prefixes to these endings usually indicate the total number of carbons in the molecule. Thus *pent-* in *pentane* signifies five carbons.

Each of the few million organic compounds must have its own name. Each name must be translatable into just one structure. These minimum requirements for communication in organic chemistry make the naming of compounds an important field for specialists. Our needs, however, will not be great enough to require us to study how to name organic compounds.

18.13
PHYSICAL PROPERTIES OF HYDROCARBONS

Carbon and hydrogen have nearly the same relative electronegativities (Section 14.22). None of the bonds in molecules of hydrocarbons is very polar, and hydrocarbons are therefore the least polar of all organic compounds. As a result, none is soluble in water.

18.14 CHEMICAL PROPERTIES OF HYDROCARBONS

Oxygen is the only common chemical that will attack alkanes (fluorine, chlorine, bromine and hot nitric acid will, too). The reaction with oxygen is combustion, and the major use of alkanes is to supply energy by combustion. Natural gas is mostly methane. Gasoline is a complex mixture mostly of various alkanes whose molecules have from five to ten carbons. The chemical products of combustion are carbon dioxide and water, provided enough oxygen is available. Otherwise, carbon monoxide, a dangerous poison, will be produced as well. The equation for the combustion of propane is

$$CH_3CH_2CH_3 + O_2 \rightarrow$$
Propane
$$3CO_2 + 4H_2O + 531 \text{ kcal/mole propane}$$

All other hydrocarbons—indeed, all other organic compounds—will burn. (Only a few, such as carbon tetrachloride, are exceptions.)

Alkenes are hydrocarbons whose molecules contain carbon-carbon double bonds, an important functional group. Alkenes react with many substances. Ethylene, for example, will add water, provided an acid catalyst is present.

$$CH_2=CH_2 + H-OH \xrightarrow{acid} CH_3-CH_2-OH$$
Ethylene   Ethyl alcohol

This reaction is one of the ways we use a chemical that we can obtain from petroleum to make another chemical that is important in chemical manufacture.

The carbon-carbon double bond will also add hydrogen.

$$CH_2=CH_2 + H_2 \xrightarrow[\text{heat, high pressure}]{\text{platinum catalyst}} CH_3-CH_3$$

One of the steps in making oleomargarine is this kind of addition of hydrogen to carbon-carbon double bonds in molecules found in vegetable oils (e.g., corn oil). The vegetable oils are called "polyunsaturated" because their molecules have several double bonds. The addition of hydrogen changes them to substances having more of the consistency of butter—a "saturated" animal fat.

Molecules of alkenes can even be made to react with themselves in the presence of special catalysts. The products are called *polymers* (from the Greek *poly,* "many," and *-meros,* "parts"). The starting materials, the alkenes, are called *monomers* ("one part"). Polymers are substances whose molecules are extremely large and are made of

TABLE 18.4
SOME UNSATURATED HYDROCARBONS

Name[a]	Structure	Boiling Point (°C)
Alkenes		
Ethylene	$CH_2=CH_2$	−104
Propylene	$CH_2=CHCH_3$	−47
α-Butylene	$CH_2=CHCH_2CH_3$	−5
cis-α-Butylene	$\begin{array}{cc} CH_3 & CH_3 \\ \diagdown & \diagup \\ CH=CH & \end{array}$	+3.72
trans-β-Butylene	$\begin{array}{cc} CH_3 & \\ \diagdown & \\ CH=CH & \\ & \diagdown \\ & CH_3 \end{array}$	+0.88
Isobutylene	$\begin{array}{c} CH_3 \\ \diagdown \\ CH=CH_2 \\ \diagup \\ CH_3 \end{array}$	−6
Dienes (alkenes with two double bonds)		
Butadiene	$CH_2=CH-CH=CH_2$	−3
Isoprene	$\begin{array}{c} CH_3 \\ | \\ CH_2=C-CH=CH_2 \end{array}$	+34
Alkynes		
Acetylene	$H-C\equiv C-H$	−84
Methylacetylene	$H-C\equiv C-CH_3$	−23
Aromatic Hydrocarbons		
Benzene	(benzene ring)	+80
Toluene	(benzene ring)—CH_3	+111
Naphthalene	(naphthalene ring)	+218 melting point +80

[a] Common system of naming.

repeating units furnished by molecules of the monomer. We shall survey the field of polymers in Chapter 20, but the terms *polymer* and *monomer* need introduction here. The polymerization of ethylene may be written as follows:

$$n CH_2=CH_2 \xrightarrow{\text{catalyst}} (CH_2-CH_2)_n$$
Ethylene   Polyethylene
(a monomer)   (a polymer)

The catalyst makes it possible for bonds to reorganize with one of the two bonds of each double bond moving out, pivoting around, to form a new covalent bond to a carbon of a neighboring molecule. (Pieces of the catalyst eventually become attached to ends of the polymer chain.)

$$+ CH_2{=}CH_2 + CH_2{=}CH_2 + CH_2{=}CH_2 + \rightarrow$$
$$—CH_2—CH_2—CH_2—CH_2—CH_2—CH_2—$$

Aromatic hydrocarbons such as benzene are very resistant to addition reactions in spite of being unsaturated. In fact, structure **6**, while it fits rules of covalence, is too misleading.

6 benzene (traditional structure)   **7** benzene (modern representation)

Chemists generally write a circle within the six-membered ring, **7**, instead of writing three double bonds, in order to get away from any suggestion that benzene gives reactions like those of ethylene. The difference with benzene is that some of the bonding electrons actually do circulate around the ring, as the circle in **7** implies. They move in a closed circuit, and this imparts great chemical resistance to any change that would break the circuit. A typical reaction of benzene, one used on a large scale by the chemical industry, is its reaction with hot nitric acid:

Benzene + Nitric acid (H + HO)—NO$_2$ $\xrightarrow{\text{heat}}$

Nitrobenzene —NO$_2$ + H$_2$O

All that has happened is that one of the hydrogens on the ring of the benzene molecule has been replaced by a nitro group, -NO$_2$. Hence, instead of addition we have substitution. (Nitric acid would virtually destroy ethylene under the same conditions.)

18.15
ALCOHOLS, PHENOLS, AND ETHERS

Alcohols are compounds whose molecules contain an -OH group joined to a carbon from which extend only single bonds. We can think of them as waterlike in that one hydrogen in a water molecule has been replaced by an alkanelike group. Or we may think of them as alkanelike in that one (or more) of the hydrogens of an alkane has been replaced by an -OH group. Being somewhat waterlike, alcohol molecules are much more polar than those of alkanes. The simple alcohols of one to four carbons dissolve in water. Yet they are not so polar that they do not dissolve in many nonpolar solvents such as carbon tetrachloride or gasoline.

Phenols are compounds whose molecules have an -OH group attached to a carbon of a benzene ring.

CH$_3$OH   CH$_3$CH$_2$OH   CH$_3$CH$_2$CH$_2$OH
Methyl   Ethyl   Propyl
alcohol   alcohol   alcohol

CH$_2$—CH$_2$ —OH
 | |
OH   OH
Ethylene glycol   Phenol

Methyl alcohol is wood alcohol. It blinds or kills if taken internally. Ethyl alcohol is the "alcohol" of alcoholic beverages. Ethylene glycol is used in permanent antifreeze. *Phenol* was the first chemical germ-killer or antiseptic used by the British surgeon Joseph Lister (1827–1912) to prevent infection in surgical wounds. (Since then, less potent but still effective materials have been found. Phenol in undiluted form as first used by Lister can cause burns.) Phenols, unlike alcohols, are weak acids. The alcohol group occurs in molecules of all carbohydrates and proteins.

When we replace one hydrogen of a water molecule by an alkanelike group we get an alcohol. When we replace both hydrogens, we obtain a member of the *ether* family of organic compounds. For example, diethyl ether may be made as follows:

$$CH_3CH_2—OH + H—OCH_2CH_3 \xrightarrow[\text{high temperature}]{H^+ \text{ catalyst}}$$
Ethyl alcohol

$$CH_3CH_2—O—CH_2CH_3 + H_2O$$
Diethyl ether

This ether is a common anesthetic.

Ethers resemble alkanes in chemical properties. Like alkanes, ethers have no hydrogens attached to oxygen, just to carbon. Because of the oxygen, however, ethers are somewhat more polar than alkanes and a trifle more soluble in water therefore.

18.16 AMINES

Amines are ammonia-like compounds whose molecules contain alkanelike groups joined to nitrogen.

CH_3NH_2 CH_3NHCH_3 $CH_3CH_2NH_2$
Methylamine Dimethylamine Ethylamine

$CH_3CH_2-\underset{\underset{CH_2CH_3}{|}}{N}-CH_2CH_3$ Ph$-NH_2$

Triethylamine   Aniline

The simple amines even have ammonia-like odors. Like ammonia they will neutralize acid; for example,

CH_3NH_2 + HCl →
Methylamine Hydrochloric acid

$CH_3NH_3^+$ + Cl^-
Methylammonium ion Chloride ion

The amino group occurs in all proteins and in a large family of drugs known as the *alkaloids* ("alkalilike"). Aniline is used to make a large family of dyes known as the aniline dyes.

18.17 CARBOXYLIC ACIDS

The functional group present in molecules of *carboxylic acids* is called the carboxyl group. It has a structure $-\overset{\overset{O}{\|}}{C}-OH$ that includes a hydroxyl group attached at a carbon-oxygen double bond.

$CH_3\overset{\overset{O}{\|}}{C}-OH$ $CH_3CH_2CH_2\overset{\overset{O}{\|}}{C}-OH$
Acetic acid   Butyric acid
(in vinegar)   (in rancid butter)

$HO-\overset{\overset{O}{\|}}{C}-\overset{\overset{O}{\|}}{C}-OH$
Oxalic acid
(present in rhubarb)

$HO-\overset{\overset{O}{\|}}{C}-\underset{\underset{CH_2\overset{\overset{O}{\|}}{C}-OH}{|}}{\underset{|}{\overset{|}{C}}}-OH$
$\overset{|}{CH_2\overset{\overset{O}{\|}}{C}-OH}$
Citric acid
(present in citrus fruits)

The hydrogen on oxygen in the carboxyl group is much more acidic than when it is in ordinary alcohols. Carboxylic acids neutralize bases.

$CH_3\overset{\overset{O}{\|}}{C}-OH$ + NaOH → $CH_3\overset{\overset{O}{\|}}{C}-ONa$ + H_2O
Acetic acid Sodium hydroxide Sodium acetate Water

The tart tastes of rhubarb, sour apples, lemons, limes, and other citrus fruits are caused by various organic acids.

The carboxyl group is present in all proteins and is involved in the chemical changes in the body in which carbohydrates and fats or oils are digested and used. The group arises principally when certain alcohol groups are attacked by oxygen or other oxidizing agents. Wine, for example, changes to vinegar by a reaction involving oxygen in the air.

CH_3CH_2OH + O_2 → $CH_3\overset{\overset{O}{\|}}{C}-OH$ + H_2O
Ethyl alcohol   Acetic acid
(in wine)   (in vinegar)

18.18 ESTERS

Esters are made from the splitting out of water between molecules of a carboxylic acid and an alcohol. For example,

$CH_3-\overset{\overset{O}{\|}}{C}-(OH + H)-O-CH_2CH_3 \xrightarrow[\text{heat}]{H^+ \text{ catalyst}}$
Acetic acid   Ethyl alcohol

$CH_3-\overset{\overset{O}{\|}}{C}-O-CH_2CH_3 + H_2O$
Ethyl acetate

This reaction can be made to go in the opposite direction if a large excess of water is added to an ester and the mixture is heated. In other words, one of the reactions of esters, one of the most important, is their hydrolysis, their reaction with water. This kind of reaction occurs when you digest the fats and oils in your diet.

A number of esters have very pleasant flavors and fragrances.

$CH_3\overset{\overset{O}{\|}}{C}-O-CH_2CH_2CH_2CH_2CH_2CH_2CH_2CH_3$
Octyl acetate (oranges)

$CH_3CH_2CH_2\overset{\overset{O}{\|}}{C}-O-CH_2CH_3$
Ethyl butyrate
(pineapples)

18 THE EARTH'S SUBSTANCES—
ORGANIC COMPOUNDS

Methyl salicylate
(oil of wintergreen)

Many plastics including Dacron are polyesters, as we shall study in Section 20.17.

18.19 AMIDES

Carboxylic acids can be made to react with ammonia or certain amines to form compounds known as *amides*. For example,

$$CH_3-C(=O)-(OH + H)-NH_2$$
Acetic acid   Ammonia

$$\xrightarrow{heat} CH_3-C(=O)-NH_2 + H_2O$$
Acetamide   Water

This kind of reaction, the splitting out of water between a carboxyl group and an amine, is how living things make proteins. When we digest proteins, we bring about the exact reverse reaction, the action of water on an amide to regenerate the original acid and amine.

Proteins are polyamides of a rather complex nature, as we shall study in Section 19.14. Polyamides of simpler types are important plastics and synthetic fibers. Nylon is the principal example (Section 20.16).

18.20 ALDEHYDES AND KETONES

Molecules of *aldehydes* have one of nature's most easily oxidized functional groups, the aldehyde group, $-\overset{O}{\overset{\|}{C}}-H$.

$$H-\overset{O}{\overset{\|}{C}}-H \qquad CH_3CH_2CH_2\overset{O}{\overset{\|}{C}}-H$$
Formaldehyde   Butyraldehyde

Vanillin

Formaldehyde in one form or another is used as a disinfectant. Butyraldehyde happens to have a dreadful odor—the "essence of dirty socks or stale locker rooms." Vanillin, on the other hand, provides the delightful fragrance and taste to vanilla extract, an important flavoring agent in cooking. The aldehyde group occurs widely in molecules of carbohydrates.

The keto group is simply the carbon-oxygen double bond joined to two other carbons, as in a very common solvent, acetone.

$$CH_3-\overset{O}{\overset{\|}{C}}-CH_3$$
Acetone

This functional group is present in fructose, a carbohydrate.

SUMMARY

Structural isomers. Because carbon can bond to its own kind in many kinds of chains, branched chains, and rings, many compounds of carbon have identical molecular formulas but different structures. These isomers differ in nucleus-to-nucleus sequences. Differences between structures based only on temporary flexings of a chain are not responsible for isomers. In open-chain compounds there is generally unrestricted rotation about single bonds.

Geometric isomers. Ring structures and those structures having carbon-carbon double bonds give rise to isomers that differ not in basic nucleus-to-nucleus sequences but in geometry at the double bond or on the ring.

Optical isomers. Still another way compounds can be different while being closely similar and sharing identical molecular formulas is in their molecular chiralities or handednesses. A chiral center in a molecule is a point, usually a carbon atom, from which four different groups radiate either in a clockwise or in a counterclockwise direction. Optical isomers are very common among biochemical compounds, and usually nature's supply of a given optical isomer is solely of molecules of just one of the possible chiralities.

Functional groups. Families of organic compounds are organized according to characteristic functional groups. The carbon-carbon double bond characterizes alkenes; the benzene ring, aromatic compounds; the hydroxyl group, alcohols; and the carbonyl group is in aldehydes, ketones, carboxylic acids, esters, and amides. Structural features are summarized in Table 18.2.

Groups that neutralize. Amines are organic compounds with amino groups that can neutralize acids. Carboxylic acids can neutralize alkalies.

Groups that split out water. Three families (that we studied) can be made from others simply by splitting out a water molecule between two other molecules. Ethers can be made from two molecules of an alcohol; esters from an alcohol and a carboxylic acid; and amides from ammonia (or an amine) and a carboxylic acid.

Groups that react with water. Two families can be made to react with water. They can be hydrolyzed. Esters are hydrolyzed to acids plus alcohols, and amides are hydrolyzed to acids plus ammonia (or amines).

Nonfunctional groups are alkanelike, consisting only of carbon and hydrogen and nothing but single bonds, and are almost entirely nonpolar. In fact, hydrocarbons in general are nonpolar and are insoluble in water.

TERMS AND CONCEPTS

carbon chain
carbon ring
branched chain
isomerism
isomers
free rotation
geometric isomers
optical isomers
optical isomerism
superimposibility
chirality
functional group
nonfunctional group
saturated compound
unsaturated compound
aliphatic compound
aromatic compound
hydrocarbon
alkane
alkene
alkyne
monomer
polymer
polymerization
alcohol
ether
phenol
amine
alkaloid
carbonyl group
carboxylic acid
ester
amide
aldehyde
ketone

QUESTIONS AND EXERCISES

1. Consider the two structures $CH_3CH_2CH_2OH$ and $CH_3CH_2CH_2CH_2OH$. (a) What is the molecular formula of each? (b) Are they organic compounds? Why? (c) Are they isomers? Why? (d) To what family do they belong?

2. Consider the two structures CH_3CHCH_3 and
$\qquad\qquad\qquad\qquad\qquad\ \ \ |$
$\qquad\qquad\qquad\qquad\qquad\ OH$
$CH_3CH_2CH_2OH$. (a) What is the molecular formula of each? (b) Are they isomers? Why? (c) Will they be more soluble or less soluble in water than $CH_3CH_2CH_3$, propane? Explain.

3. Write an equation for the complete combustion of methane, CH_4.

4. Consider the structure $CH_3-CH-CH_2CH_3$.
$\qquad\qquad\qquad\qquad\qquad\qquad\ \ \ |$
$\qquad\qquad\qquad\qquad\qquad\qquad\ OH$
(a) Would it be possible to make "ball-and-stick" models of this compound that were related as object to mirror image? (b) Would object and mirror image superimpose? (c) What kind of isomers are the two?

5. Chemists do not like to use a symbol for benzene that shows three double bonds within the six-membered ring. Instead, they prefer a circle within the ring. Why?

6. What is the basis for deciding if a compound is an aromatic compound or an aliphatic compound?

7. Write the structures of three isomers of C_5H_{12}, the isomeric pentanes.

8. Assign each of the following compounds to their correct chemical families:

(a) $CH_3CH=CH_2$

(b) $HO-\overset{\overset{\displaystyle O}{\|}}{C}-CH_3$

(c) $NH_2-CH_2CH_2CH_3$

(d) $H-\overset{\overset{\displaystyle O}{\|}}{C}-CH_2CH_3$

(e) $NH_2-\overset{\overset{\displaystyle O}{\|}}{C}-CH_3$

(f) $CH_3-O-\overset{\overset{\displaystyle O}{\|}}{C}-CH_3$

(g) benzene ring$-CH_3$

(h) $HO-CH_2CH_2CH_3$

(i) CH_3-O-CH_3

(j) $CH_3\overset{\overset{\displaystyle O}{\|}}{C}CH_2CH_3$

9. Which compound in question 8 could neutralize hydrochloric acid? Write an equation.

10. Which compound in question 8 could neutralize sodium hydroxide? Write an equation.

11. Which compound would be split apart by the action of water?

12. Which compound would combine with water to form an alcohol?

13. Which compound is aromatic?

14. Which two would be the least soluble in water?

18 THE EARTH'S SUBSTANCES— ORGANIC COMPOUNDS

15. Which compounds would burn?

16. Molecules of methyl alcohol CH_3OH and ethane CH_3CH_3 have virtually the same weights (the formula weight of methyl alcohol is 32; of ethane, 30). Methyl alcohol, boiling at 64.5°C, boils 153°C higher than ethane, whose boiling point is −88.6°C. What this means is that molecules of methyl alcohol stick to each other and resist the change to a vapor far more strongly than do molecules of ethane. What makes methyl alcohol molecules stick to each other?

17. In the light of problem 16, predict which would have the higher boiling point, $CH_3CH_2CH_2OH$, propyl alcohol, or $HOCH_2CH_2OH$, ethylene glycol.

CHAPTER 19
ORGANIC COMPOUNDS OF THE BIOSPHERE

19.1 PHOTOSYNTHESIS

In this chapter we shall introduce the principal kinds of organic compounds that make up the bulk of living things, materials that are used to make cells and tissues and to furnish them with the chemical energy needed for life. That chemical energy comes originally from the sun. Solar energy pours into our planet, free of charge, at a rate of about 2.2×10^{25} joules per year. The clouds reflect approximately a third back to space. Just about all of the rest is also eventually radiated back to space. Otherwise the planet would become hotter and hotter, but we use some of it first. Its warmth keeps most of the planet inhabitable, and about 0.04 percent of the received solar energy is used by green plants from microscopic, marine phytoplankton to towering forest giants.

Green plants contain a pigment called *chlorophyll* that can capture and temporarily store solar energy from the visible parts of the spectrum. Besides chlorophyll, plant cells have several catalysts—enzymes—and they help take the energy-poor molecules of carbon dioxide and water plus certain minerals and move them up an energy "hill" to make energy-rich compounds that the plant needs (Figure 19–1). These changes are called *photosynthesis*. The overall result of the series of reactions may be expressed by

$$6CO_2 + 6H_2O + \text{energy} \xrightarrow[\substack{\text{chlorophyll} \\ \text{(several steps)}}]{\text{plant enzymes}} \underset{\substack{\text{a fundamental} \\ \text{unit in carbohydrates}}}{C_6H_{12}O_6} + 6O_2$$

Certain amino acids, building blocks of proteins, are also produced. These, together with carbohydrates and oxygen, are the primary products of photosynthesis. Plants use them for their own needs. Many plants put proteins, lipids, and starches into their seeds to nourish new sprouts, but animals and people often intervene and divert seeds, tubers, and fruits to their own needs. Plants also specialize in making different kinds of secondary products. These include oils and waxes, steroids, alkaloids, pigments, cellulose fibers, gums and pectins, and resins and rubber, to say nothing of wood for lumber.

Swedish meterorologist Bert Bolin has estimated that between 20 and 30 billion tons of carbon dioxide are fixed via photosynthesis each year by land plants and about 40 billion tons (probably more) by microscopic organisms, such as the phytoplankton, in the oceans. The earth's forests make about 15 billion tons of wood each year. The death and decay of plants and animals, as well as their normal respiration while living, converts photosynthetically fixed carbon and atmospheric oxygen back to carbon dioxide and water. An outline of these relations is in

19 ORGANIC COMPOUNDS OF THE BIOSPHERE

FIGURE 19-1
Young sugar beets growing in Colorado. Each sugar beet grows to a weight of about two pounds as its giant leaves soak up sunlight and use its energy for photosynthesis. Each beet yields about 14 teaspoons of pure white sugar. (Grant Heilman.)

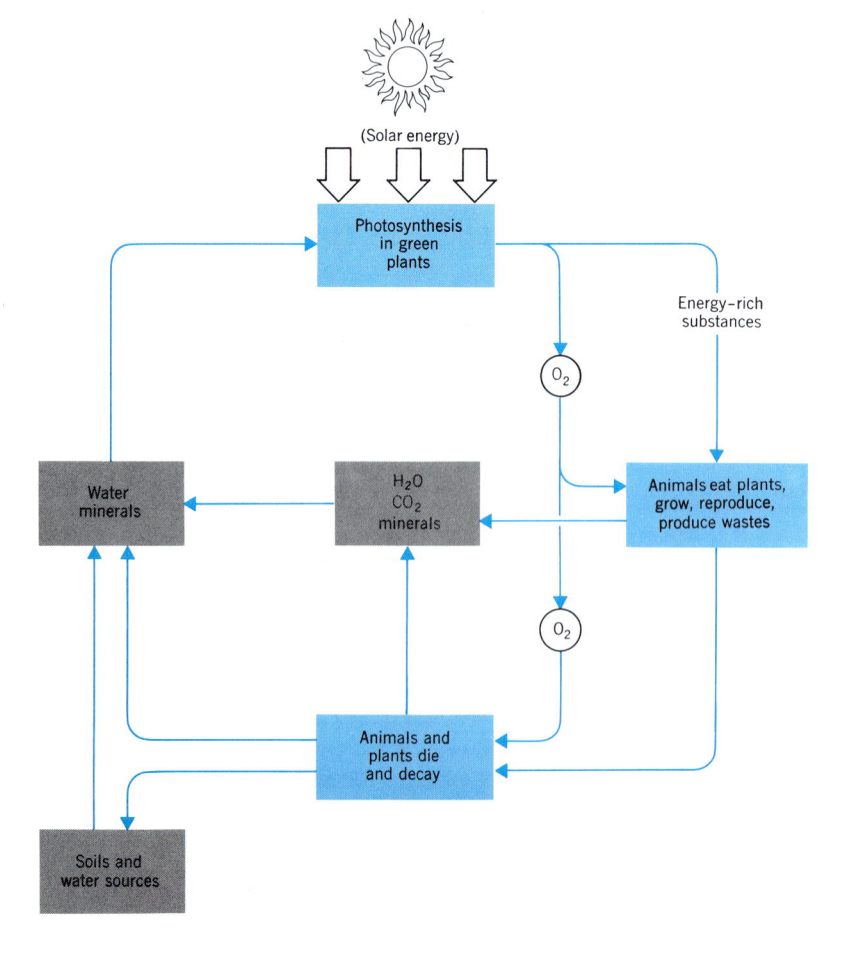

Figure 19-2. We are a part of the cycles of nature. The first type of plant material we shall study is the carbohydrates.

19.2 CARBOHYDRATES

If you have a sweet tooth, you have a weakness for *carbohydrates*, for this family includes common table sugar. It also includes starch, dextrose (glucose), milk sugar (lactose), and other sugars—all of which are digestible—plus something humans cannot digest, cellulose, the chief substance of a cotton fiber.

Carbohydrates are polyhydroxyaldehydes, or polyhydroxyketones, or substances that yield these by hydrolysis. Carbohydrates that cannot be broken down by water are called simple sugars or *monosaccharides*. Glucose is the most common example. Molecules of *disaccharides* are capable of being hydrolyzed to two monosaccharide units, and they include sucrose (table sugar), lactose (milk sugar), and maltose (malt sugar). Starch and cellulose, two common *polysaccharides*, are polymers of glucose. Glucose is by far the most important carbohydrate. It may well be that glucose units are the most abundant organic species in the world considering their importance in cellulose and starch. The following "word equations" reveal in a different way how widely occurring glucose units are.

Disaccharides
$$\text{Maltose} + H_2O \rightarrow \text{glucose} + \text{glucose}$$
$$\text{Lactose} + H_2O \rightarrow \text{glucose} + \text{galactose}$$
$$\text{Sucrose} + H_2O \rightarrow \text{glucose} + \text{fructose}$$

Polysaccharides
$$\text{Starch} + x\,H_2O \rightarrow x\,\text{glucose}$$
$$\text{Glycogen} + x\,H_2O \rightarrow x\,\text{glucose}$$
(animal starch)
$$\text{Cellulose} + x\,H_2O \rightarrow x\,\text{glucose}$$

By concentrating our attention on glucose and its derivatives, we can survey much of the field of carbohydrates.

FIGURE 19-2
The flux of energy from sun to earth is rendered most useful by means of photosynthesis. Shown here are the principal cycles necessary to life and powered by solar energy—the carbon dioxide, oxygen, and mineral cycles. (From J. R. Holum, *Elements of General and Biological Chemistry*, 4th edition, 1975, John Wiley & Sons, Inc., New York.)

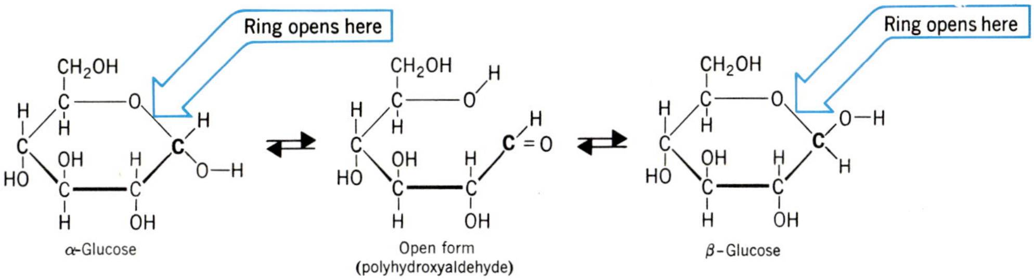

19.5 LACTOSE (MILK SUGAR)

FIGURE 19-3
Naturally occurring glucose can exist in each of these interconvertible forms. In aqueous solution, all three are present in a dynamic equilibrium with the "open form" present to an extent of less than 0.5 percent.

19.3
GLUCOSE (DEXTROSE, GRAPE SUGAR, BLOOD SUGAR)

Glucose occurs in an uncombined form in sweet fruits, especially ripe grapes. It is also found in circulation in the bloodstream, and it serves as one of the major sources of chemical energy in living systems.

The formula for glucose is $C_6H_{12}O_6$. Its structure is complicated by the fact that a glucose molecule can exist in three readily interconvertible forms, shown in Figure 19-3. In aqueous solution, all three forms are present in dynamic equilibrium. The only structural differences between alpha- and beta-glucose (α- and β-glucose) is the upward or downward projection of the -OH group attached to the boldfaced carbon. This difference may seem trivial, but it will make the difference between starch (which we can digest) and cellulose (which we cannot digest).

Convenient condensed structures for the glucose forms are as follows, wherein the -CH$_2$OH group and the -OH groups attached to the ring are indicated simply by lines. (As is usual in such condensations, hydrogens are mostly "understood.")

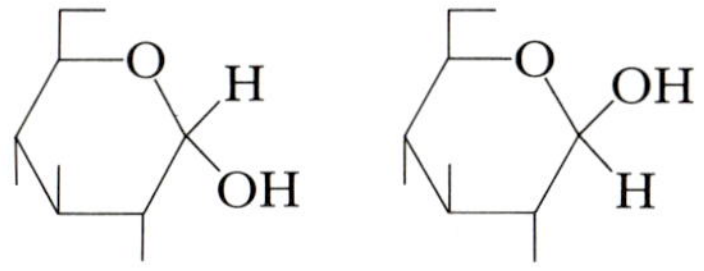

α-glucose   β-glucose

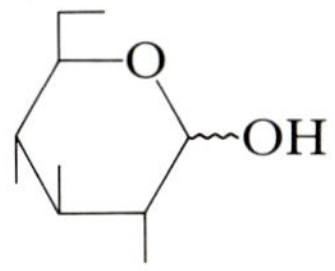

(may be used when α and β designations are unimportant)

Although glucose molecules are large, the several hydroxyl groups make them very soluble in water and insoluble in the common nonpolar solvents.

19.4
MALTOSE (MALT SUGAR)

Maltose is present in germinating grains; it does not occur abundantly in the free state in nature. Maltose can be prepared from starch, and along with glucose and dextrins it is present in corn syrup.

Maltose consists of two glucose units linked by an oxygen "bridge." Its formation could be visualized as the splitting out of a water molecule between two -OH groups.

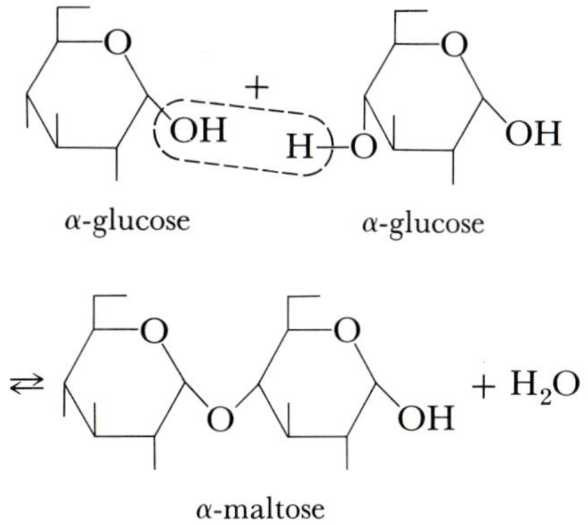

α-glucose   α-glucose

α-maltose + H$_2$O

19.5
LACTOSE (MILK SUGAR)

Lactose is the principal carbohydrate found in milk—4-6 percent in cow's milk; 5-8 percent in human milk. It is also a by-product in the manufacture of cheese. Its structure and its relation to its monosaccharide units are shown by the following equation:

19 ORGANIC COMPOUNDS OF THE BIOSPHERE

β-galactose   Glucose (either α, or β)

Lactose

19.6
SUCROSE (CANE SUGAR, BEET SUGAR, TABLE SUGAR)

Juice that can be expressed from sugar cane contains about 14 percent sucrose. Lime (mostly calcium oxide) is added to this juice to precipitate proteins, and the clear, remaining liquid is evaporated until a semisolid mass remains. The liquid portion (blackstrap molasses) is removed and the remaining raw sugar is now about 95 percent pure. Odor- and color-causing contaminants are removed, and the resulting white sucrose is probably the purest single organic compound known that is so widely and inexpensively sold. Much of our supply now is obtained from sugar beets.

α-glucose   Fructose

Sucrose

19.7
STARCH

Starch is the form in which plants store glucose units for eventual use for chemical energy and other processes required for life. Starch consists of two types of molecules with very similar features—amylose and amylopectin, shown in Figure 19–4. Both types are polymers of the α-form of glucose with each glucose unit hooked into the polymer by oxygen bridges. The digestion of starch is simply hydrolysis at these bridges, and animals have the necessary enzymes in their digestive juices. In humans, the principal starch-splitting enzyme is found in saliva.

19.8
GLYCOGEN

Animals, including humans, can also store glucose units in the form of a polymer of α-glucose called glycogen. Structurally, glycogen is very much like amylopectin. Between meals or during longer periods of fasting, people and animals withdraw glucose units from glycogen reserves to resupply the glucose used up from the bloodstream. These glycogen reserves in animals, however, are not extensive. In humans they cannot be expected to last for more than a few hours of fasting.

19.9
CELLULOSE

Cellulose is also a high-formula-weight polymer of glucose, but with a difference. Where starch and glycogen are derived from α-glucose, cellulose is a polymer of β-glucose. Figure 19–5 shows its main structural features. It resembles amylose, one of the kinds of molecules in starch. We have here another of the very interesting "tremendous trifles" in nature. Amylose is digestible; cellulose is not. Yet the only difference between them is the direction of the oxygen bridge between their glucose units. *Digestion* is nothing more than hydrolysis, a reaction of water with food molecules. Whether or not something can be digested depends on the presence of the right digestive enzyme or catalyst. Humans and other meat-eating animals do not have an enzyme for digesting cellulose. Many microorganims do. Some varieties are found in the stomachs of cows. They act on cellulose in hay, grass, and fodder in the cow's diet to convert it to small molecules the cows then use.

Cellulose makes up the cell membranes of the higher plants and gives them their rigidity. Cotton fiber (Figure 19–6) is almost 98 percent cellulose, and each molecule in cotton has 2000–9000 glucose units. These extraordinarily long molecules lie side by side in the fiber, overlapping each other, and sticking together by hydrogen bonds. Rayon is cellulose that has been processed chemically to give it more desirable properties as a fabric.

FIGURE 19-4
The basic structural features of the types of polymers of α-glucose found in starches. Glycogen, which is to animals what starch is to plants—a reserve source of energy—resembles amylopectin except that the branching may be much more extensive.

19.10
LIPIDS

The animal fats such as butter, lard, and tallow, and the vegetable oils such as corn oil, olive oil, peanut oil, and linseed oil, all belong to a family of natural products known as the *lipids*. They are insoluble in water and are generally soluble in the less-polar solvents—the "fat solvents" such as benzene, carbon tetrachloride, and ether. For a substance to belong to the lipid family, it need only be extractable from plant and animal sources by these solvents. Lipids therefore are defined according to an operation used to obtain them rather than according to a particular structural feature. Structure determines property, however, and while there is a great variety of lipids, all have one feature in common. Their molecules are substantially hydrocarbon-like. Only one type makes up a significant portion of an average daily diet—the simple lipids or triglycerides or animal fats and vegetable oils.

19.11
ANIMAL FATS AND VEGETABLE OILS

The family of *animal fats* and *vegetable oils* constitutes the most abundant type of lipid found in plants and animals. Any particular member—corn oil, for example—is a mixture. It is not possible to write one specific structure for corn oil, because the molecules are not all identical, yet they are very similar. They are all esters of glycerol. The carboxylic acids used to make these esters, with few exceptions, are straight-chain; they have an even number of carbon atoms, they have from 10 to 20 carbons, and they may have 0, 1, 2, or 3 double bonds. These acids are often called *fatty acids,* after their source, and Table 19.1 contains a list of the most important. A specific example of the structure of a molecule that might occur in olive oil or peanut oil is shown in Figure 19-7. Although molecules of fats and oils are very large, they are chemically very simple. Only two functional groups are present—alkene double bonds and ester groups. Fats and oils, therefore, can be hydrogenated and hydrolyzed.

FIGURE 19-5
The basic structural feature of cellulose, a linear polymer of β-glucose. It is found as a part of the cell walls of plants, and fibrous cotton is almost pure cellulose.

FIGURE 19-6
Cotton, a polymer of glucose and a product of photosynthesis. (Grant Heilman.)

TABLE 19.1
IMPORTANT FATTY ACIDS

Saturated acids

$CH_3CH_2CH_2CH_2CH_2CH_2CH_2CH_2CH_2CH_2CH_2COOH$
Lauric acid

$CH_3CH_2CH_2CH_2CH_2CH_2CH_2CH_2CH_2CH_2CH_2CH_2CH_2COOH$
Myristic acid

$CH_3CH_2CH_2CH_2CH_2CH_2CH_2CH_2CH_2CH_2CH_2CH_2CH_2CH_2CH_2COOH$
Palmitic acid

$CH_3CH_2CH_2CH_2CH_2CH_2CH_2CH_2CH_2CH_2CH_2CH_2CH_2CH_2CH_2CH_2CH_2COOH$
Stearic acid

Unsaturated acids

$CH_3CH_2CH_2CH_2CH_2CH_2CH_2CH_2CH=CHCH_2CH_2CH_2CH_2CH_2CH_2CH_2COOH$
Oleic acid

$CH_3CH_2CH_2CH_2CH_2CH=CHCH_2CH=CHCH_2CH_2CH_2CH_2CH_2CH_2CH_2COOH$
Linoleic acid

$CH_3CH_2CH=CHCH_2CH=CHCH_2CH=CHCH_2CH_2CH_2CH_2CH_2CH_2CH_2COOH$
Linolenic acid

Enzymes in the digestive tracts of human beings and other animals efficiently catalyze the hydrolysis of lipids. That is what happens when fats and oils are digested. For example, using the specific lipid of Figure 19-7:

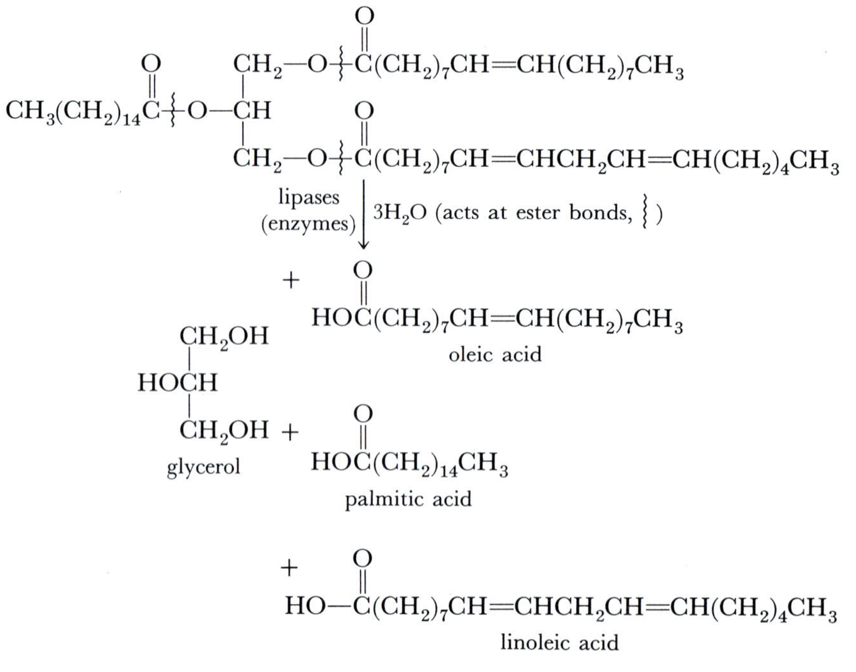

We use the lipids in our diets principally for energy. Their chemical energy is stored chiefly in carbon-hydrogen bonds, which are abundant in lipid molecules. What we do not use as energy we store—around midriffs that bulge ever larger

$$CH_3(CH_2)_{14}\overset{O}{\underset{\|}{C}}\!\!\left\{\!\!\begin{array}{c}\\\end{array}\!\!\right.\!\!O\!-\!\!\begin{array}{c}CH_2-O\!\!\left\{\!\!\begin{array}{c}\\\end{array}\!\!\right.\!\!\overset{O}{\underset{\|}{C}}(CH_2)_7CH=CH(CH_2)_7CH_3\\ |\\ CH\\ |\\ CH_2-O\!\!\left\{\!\!\begin{array}{c}\\\end{array}\!\!\right.\!\!\overset{O}{\underset{\|}{C}}(CH_2)_7CH=CHCH_2CH=CH(CH_2)_4CH_3\end{array}$$

$\xrightarrow[\text{(enzymes)}]{\text{lipases}}$ $3H_2O$ (acts at ester bonds, $\}$)

$$\begin{array}{c}CH_2OH\\ |\\ HOCH\\ |\\ CH_2OH\end{array}\quad +\quad HO\overset{O}{\underset{\|}{C}}(CH_2)_7CH=CH(CH_2)_7CH_3$$
glycerol   oleic acid

$+\quad HO\overset{O}{\underset{\|}{C}}(CH_2)_{14}CH_3$
palmitic acid

$+\quad HO-\overset{O}{\underset{\|}{C}}(CH_2)_7CH=CHCH_2CH=CH(CH_2)_4CH_3$
linoleic acid

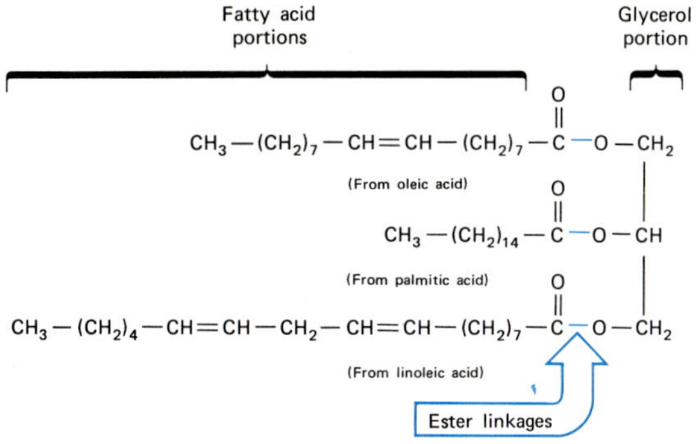

FIGURE 19-7

$CH_3-(CH_2)_7-CH=CH-(CH_2)_7-C(=O)-O-CH_2$ (From oleic acid)
$CH_3-(CH_2)_{14}-C(=O)-O-CH$ (From palmitic acid)
$CH_3-(CH_2)_4-CH=CH-CH_2-CH=CH-(CH_2)_7-C(=O)-O-CH_2$ (From linoleic acid)

Fatty acid portions — Glycerol portion — Ester linkages

This structure is typical of the molecules of mixed glycerides found in a vegetable oil. It contains three ester linkages involving oleic acid, palmitic acid, and linoleic acid. Since it has three alkene linkages, the molecule is polyunsaturated.

the more out of balance the intake of food is compared to the outgo as energy. Lipids also provide a protective cushion around internal organs.

19.12
PROTEINS

Proteins make up about half of the dry weight of the human body. They are found in all cells. They hold you together (skin); give you levers (muscles and tendons); reinforce your bones; patrol your circulatory system as antibodies; carry essential nutrients such as oxygen, fatty acids, and amino acids; catalyze all the chemical reactions in your body (enzymes); and give you a communications system (nerves). Yet even with all these diversities of function, proteins have many features of structure in common.

19.13
AMINO ACIDS

All proteins are polymers, and the building blocks or monomers are called *amino acids*. There are roughly twenty, but all have in common this system:

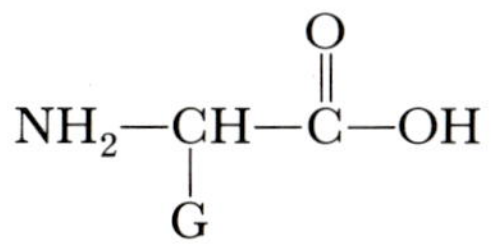

$NH_2-CH(G)-C(=O)-OH$

General structure of amino acids

19.14 HOW AMINO ACIDS ARE INCORPORATED INTO PROTEINS

The *G-* stands for an organic group attached as a side chain. Each amino acid has a different side chain (see Table 19.2).

19.14
HOW AMINO ACIDS ARE INCORPORATED INTO PROTEINS

We have learned that amines and carboxylic acids can form amides (Section 18.19), and this is what happens when amino acids are converted into proteins. The amino group of one amino acid forms an amide with the carboxyl group of another. For example:

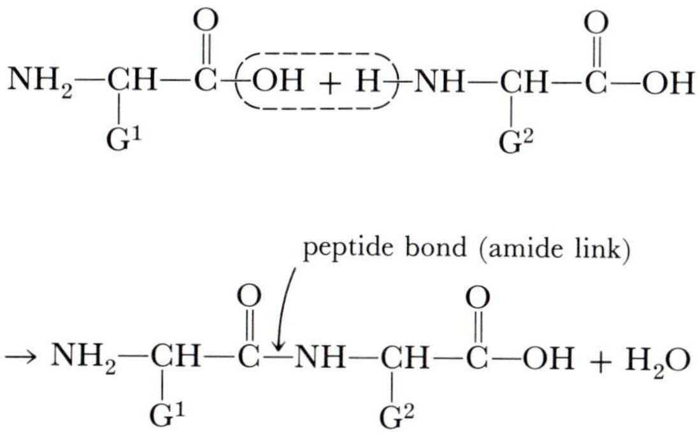

$NH_2-CH(G^1)-C(=O)-(OH + H)-NH-CH(G^2)-C(=O)-OH$

$\rightarrow NH_2-CH(G^1)-C(=O)-NH-CH(G^2)-C(=O)-OH + H_2O$

peptide bond (amide link)

Dipeptide

Specific example:

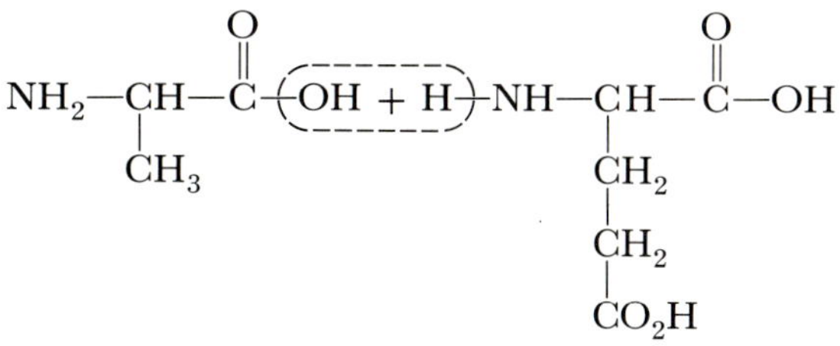

$NH_2-CH(CH_3)-C(=O)-(OH + H)-NH-CH(CH_2CH_2CO_2H)-C(=O)-OH$

Alanine (ala)   Glutamic acid (glu)

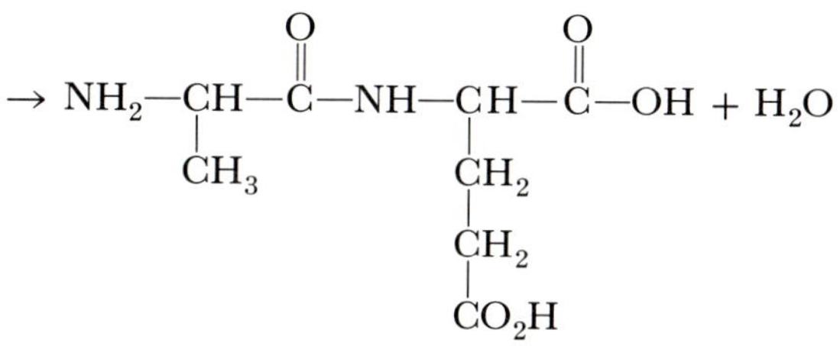

$\rightarrow NH_2-CH(CH_3)-C(=O)-NH-CH(CH_2CH_2CO_2H)-C(=O)-OH + H_2O$

Alanylglutamic acid (ala·glu)

In protein chemistry, the amide linkage is called a *peptide bond*, and a molecule made from two amino acids is called a *dipeptide*. In our example,

19 ORGANIC COMPOUNDS OF THE BIOSPHERE

TABLE 19.2
COMMON AMINO ACIDS

$$HO-\overset{\overset{O}{\|}}{C}-\underset{\underset{NH_2}{|}}{CH}-G$$

Name	Structure	Symbol
Glycine	$HO-\overset{\overset{O}{\|}}{C}-\underset{\underset{NH_2}{\|}}{CH}-H$	Gly
Alanine	$-CH_3$	Ala
Valine	$-\underset{\underset{CH_3}{\|}}{CH}-CH_3$	Val
Leucine	$-CH_2\underset{\underset{CH_3}{\|}}{CH}CH_3$	Leu
Isoleucine	$-\underset{\underset{CH_3}{\|}}{CH}CH_2CH_3$	Ile
Serine	$-CH_2OH$	Ser
Threonine	$-\underset{\underset{OH}{\|}}{CH}CH_3$	Thr
Cysteine	$-CH_2SH$	Cys
Methionine	$-CH_2CH_2SCH_3$	Met
Aspartic acid	$-CH_2\overset{\overset{O}{\|}}{C}-OH$	Asp
Asparagine	$-CH_2\overset{\overset{O}{\|}}{C}-NH_2$	Asn
Glutamic acid	$-CH_2CH_2\overset{\overset{O}{\|}}{C}-OH$	Glu
Glutamine	$-CH_2CH_2\overset{\overset{O}{\|}}{C}-NH_2$	Gln
Lysine	$-CH_2CH_2CH_2CH_2NH_2$	Lys
Arginine	$HO-\overset{\overset{O}{\|}}{C}-\underset{\underset{NH_2}{\|}}{CH}-CH_2CH_2CH_2N-\overset{\overset{NH}{\|}}{C}-NH_2$ with H on N	Arg
Histidine	$-CH_2-C=CH$ (imidazole ring with HN and N, CH)	His
Phenylalanine	$-CH_2-C_6H_5$	Phe
Tyrosine	$-CH_2-C_6H_4-OH$	Tyr
Tryptophan	$-CH_2$-(indole)	Trp

19.14 HOW AMINO ACIDS ARE INCORPORATED INTO PROTEINS

TABLE 19.2 (continued)

Name	Structure	Symbol
Proline	$HO-\overset{O}{\overset{\|\|}{C}}-CH-CH_2$ ... $H-N$ ... CH_2-CH_2 (ring)	Pro

we used alanine as the acid and glutamic acid as the amine. The other sequence could have just as well been used.

$$NH_2-\underset{\underset{\underset{CO_2H}{|}}{\underset{CH_2}{|}}}{\underset{|}{CH}}-\overset{O}{\overset{\|\|}{C}}OH + H-NH-\underset{\underset{CH_3}{|}}{CH}-\overset{O}{\overset{\|\|}{C}}OH \rightarrow$$

Glutamic acid (glu)   Alanine (ala)

$$NH_2-\underset{\underset{\underset{CO_2H}{|}}{\underset{CH_2}{|}}}{\underset{|}{CH}}-\overset{O}{\overset{\|\|}{C}}-NH-\underset{\underset{CH_3}{|}}{CH}-\overset{O}{\overset{\|\|}{C}}-OH + H_2O$$

Glutamylalanine (glu·ala)

Just as the two letters *N* and *O* can be put together to form two radically different words, *NO* and *ON*, so two different amino acids can be joined together in two different, but isomeric dipeptides. Note carefully what they have in common and how they differ. In both of them—alanylglutamic acid and glutamylalanine—the "backbones" are identical.

$$NH_2-\underset{|}{CH}-\overset{O}{\overset{\|\|}{C}}-NH-\underset{|}{CH}-\overset{O}{\overset{\|\|}{C}}-OH$$

They differ only in the sequence in which the side chains appear. In one, the side chain of alanine, a CH_3- group, is next to the $-NH_2$ group of the backbone. In the other, the side chain of glutamic acid, $-CH_2CH_2CO_2H$, is there.

As you see, the structures can get complicated, and that is why scientists use three-letter symbols for the amino acids. These are given in Table 19.2, and we used them in the examples already shown.

The dipeptides we wrote still have amino groups and carboxyl groups. This means we can operate at either one group or the other with a third amino acid to make a tripeptide. If we use three different amino acids—for example, alanine (ala), glutamic acid (glu), and phenylalanine (phe)—there are six possible tripeptides. They are all isomers of each other.

ala·glu·phe   glu·ala·phe   phe·ala·glu
ala·phe·glu   glu·phe·ala   phe·glu·ala

(By convention, the first amino acid symbolized will have a free amino group, and the last in the series will have the free carboxyl group. The dot between the three-letter symbols stands for the peptide bond.) From any of the tripeptides we could add still another amino acid to make a tetrapeptide, and you can see now the general pattern common to all *polypeptides*. They all have molecular "backbones" of repeating units of

$$-N-\overset{O}{\overset{\|\|}{C}}-C-.$$

$$NH_2-\underset{\underset{G^1}{|}}{CH}-\overset{O}{\overset{\|\|}{C}}-NH-\underset{\underset{G^2}{|}}{CH}-\overset{O}{\overset{\|\|}{C}}-NH-\underset{\underset{G^3}{|}}{CH}-\overset{O}{\overset{\|\|}{C}}-NH-\underset{\underset{G^4}{|}}{CH}-\overset{O}{\overset{\|\|}{C}}-\text{etc.}$$

Polypeptide structure

Proteins differ in the length of that backbone (how many amino acids used per protein molecule), in the identities of the amino acids used, and in the sequence of side-chain groups. These details are matters of life and death. The case of sickle cell anemia is but one illustration.

The chemical that carries oxygen in our bloodstreams is hemoglobin. Most of this molecule is made up of four polypeptide units that together comprise the globin (heme is a nonprotein pigment). In one polypeptide subunit, there are about 300 amino acids. In sickle cell anemia a valine unit with an alkanelike side chain has been replaced by a glutamic acid unit, which has a carboxyl group on its side chain. It seems like a minor change, one amino acid out of 300, but individuals having this problem in the severe

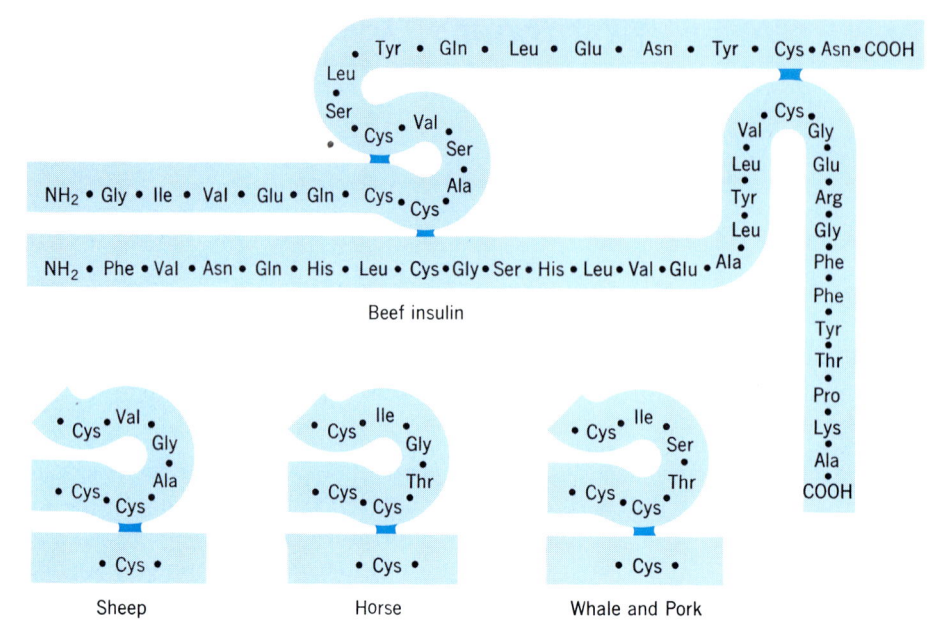

FIGURE 19-8

The primary structure of insulin. F. Sanger, British biochemist, and a team of coworkers at Cambridge worked from 1945 to 1955 on the determination of this structure. Sanger received the 1958 Nobel prize in chemistry for this work and the development of the techniques. Insulins from different species have the slight differences noted. Fortunately, these are not serious for diabetics on insulin therapy. Usually, when an alien protein gets into circulation, the body makes antibodies to combine with it. The rejection of skin grafts from one person to another and the extraordinary problems of transplanting organs are related to this formation of antibodies. Sheep insulin, however, can be given to human diabetics. Some antibodies form, but the insulin activity is reduced only slightly.

form usually die in childhood. Their red blood cells tend to adopt a sickle shape, and they are much less efficient at carrying oxygen. It is an inherited problem, and when both parents carry the trait, offspring are most likely to have the severe form.

Most native proteins, those actually found in living things, are enormous, having not just hundreds but thousands of amino acids incorporated into their molecules. Figure 19-8 shows the structure of insulin, which is actually a very small and simple protein. Made in an organ called the pancreas, insulin is essential to your use of glucose. Without it, you would have diabetes.

19.15 HANDEDNESS AND HELICES

Molecules of the naturally occurring amino acids, except those of glycine, are all "handed" in the same sense, as illustrated in Figure 19-9. Their mirror-image forms are exceedingly rare in nature. They can be synthesized, but if you were given them as your daily amino acid quota, you could not use them at all. The reason, again, is that the enzymes you have for using amino acids (or, for that matter, for digesting proteins) are mostly polypeptides and are also chiral. They work effectively only with one of the possible forms of the amino acids.

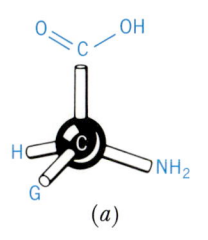

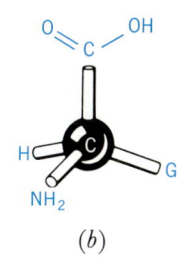

FIGURE 19-9

Molecules of all amino acids, except glycine, are chiral, and their "handedness" is that shown in part (*a*). As you look down the bond from carbon to hydrogen, your eyes must sweep counterclockwise to move from -CO_2H to -G to -NH_2, in that order. The opposite "handedness," shown in part (*b*), is only very rarely found among amino acid molecules in nature. (Glycine is excepted because its molecules are symmetric in the sense that they superimpose with their mirror images.)

Many native proteins consist of molecules that have coiled into a right-handed helix, called the *alpha-helix* (see Figure 19-10). The particular handedness of the building blocks has something to do with the helix being right- instead of left-handed. The helix is stabilized (does not "go limp") mostly because hydrogen bonds from coil to coil keep it coiled.

Should anything happen to disrupt the coil, the protein having it would lose its ability to function biologically. It would be *denatured*. Thus, one of nature's weakest recognizable bonds, the hydrogen bond, is one of its most important—precisely because it is weak! Mild heat will denature some proteins; it happens when you fry an egg. The albumin, the egg white, becomes denatured (Figure 19-11). Even the action of violent whipping or shaking is enough to denature egg white, which is what happens when meringue for lemon pies is made. Because 70 percent ethyl alcohol denatures proteins in bacteria, killing them, it is a common disinfectant for clinical thermometers. Detergents denature bacterial proteins, and that is just one reason for washing your hands before eating. Not all proteins, of course, are easily denatured. Those that have to provide protection—proteins in skin and leather, in feathers, fur, hair, fingernails, and toenails—hold up to most environmental stresses.

19.16 ENZYMES

Enzymes, the catalysts in living systems, are substances whose molecules are mostly or entirely

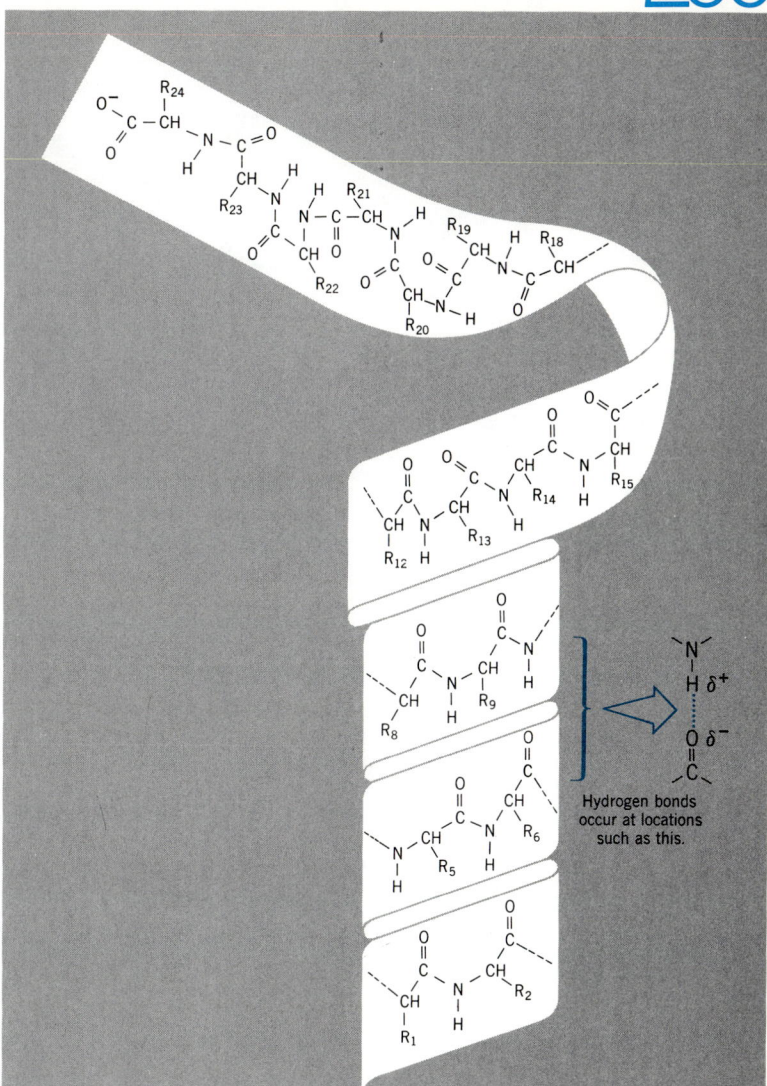

FIGURE 19-10

This representation of a section of a protein chain as a ribbon shows the right-handed coiling of an α-helix. Hydrogen bonds exist between carbonyl oxygens and hydrogens on the amide nitrogens. Although each represents a weak force of attraction between the turns of the coil, there are many of them, and the total force stabilizing the helix is more than enough to counteract the natural tendency for the chain to adopt a randomly flexing form. It is understandable that the repeating series of hydrogen bonds up and down the helix is called the "zipper" of the molecule. (From G. H. Haggis, D. Michie, A. R. Muir, K. B. Roberts, and P. M. B. Walker, *Introduction to Molecular Biology*, John Wiley and Sons, New York, 1964, page 51.)

protein. Frequently, a small, nonprotein molecule called a *cofactor* is joined to a protein portion to make the complete enzyme. Some of the vita-

19 ORGANIC COMPOUNDS OF THE BIOSPHERE

FIGURE 19-11
Denaturation. Egg "white" is not white at all in the fresh egg (left), but a clear, colorless fluid. Simply whipping this fluid, however, irreversibly changes it—denatures it. (Mimi Forsyth/Monkmeyer.)

mins, particularly the B-vitamins, are essential parts of several cofactors. Hence, without these B-vitamins, important enzymes are in short supply or nonexistent altogether. Reactions they catalyze shut dwn and disease results.

19.17
VITAMINS

By definition, a *vitamin* is any substance meeting the following criteria. It is an organic compound we cannot make ourselves, internally, at all. Its absence in the diet results in a specific disease, a vitamin deficiency disease such as scurvy or beriberi. It is found in small concentrations in various foods.

19.18
"LOCK-AND-KEY" THEORY

The polypeptide portions of enzyme molecules have particular surface shapes with various side chains exposed. The molecules to be acted upon by the enzyme have their own molecular shapes. The enzyme surface can accept only molecules to which they can fit, much as a tumbler lock will operate only with a key that fits. That is how enzymes work—by the *"lock and key theory"* (Figure 19-12). They accept reactant molecules, and while these are momentarily fitting to the enzyme, surface side chains on the enzyme or parts of the cofactor catalyze the reaction. Then the products peel away.

19.19
POISONS AND ENZYMES

Some of our most dangerous *poisons* work by binding to enzymes, changing their surface shapes and making it impossible for reactant molecules to fit anymore. That is how nerve gases work, as well as toxins in botulism (food poisoning), arsenic poisons, cyanide, and compounds of mercury or lead.

19.20
HORMONES

In higher forms of life, tissues have become so highly specialized that they cannot exist for long independent of the whole organism. The division of labor among specialized tissues resembles the division of labor found in complex societies. Brain tissue has one set of tasks; liver tissue another; the heart still another. All these tissues have to be able to send signals back and forth; they cannot "go it alone." We send signals in two ways: by the nervous system and the electrical signals shuttled back and forth through it, and by chemical signals, special molecules sent out into the bloodstream by one organ to circulate until they reach a "target" organ where they lodge in a "lock and key" type of event. These chemical messengers are called *hormones* (Greek: *hormon*, "arousing, exciting"). The tissues that make and secrete them are called the endocrine glands (Greek: *endon*, "within"; and *krinein*, "to separate," that is, to secrete internally). Hormones, unlike vitamins, are made internally. Many are proteins. The presence or absence of hormones has been related to such dramatic gross defects as cretinism, goiter, diabetes, dwarfism, and giantism. Sexual development is particularly dependent on normal hormonal action.

Important as the hormones are, science is still a long way from understanding how they accomplish their important effects at the molecular level. While hormones do not appear to *initiate* chemical events, they somehow control their rates. Increasingly, the evidence points to their acting to control how much, if any, of a particular nutrient or metabolite gets inside the cells of their target tissues and organs. Thus, insulin, the hormone whose effective absence means diabetes mellitus, apparently acts to help glucose molecules to enter cells in sufficient amounts. The human growth hormone (HGH) also appears to act by controlling what kinds of and how many metabolites get inside cells. Before cells can divide, a great deal of synthesis must be accomplished to furnish the new "daughter" cells with all the needed components. If the growth hormone is not present in sufficient amounts, amino acids do not get through the cell membranes, and growth is severely retarded. With the hormones we again have dramatic examples of our dependence on trace amounts of chemicals.

19.21 HEREDITY AND NUCLEIC ACIDS

Important as enzymes and hormones are as directors and controllers of chemical events in living cells, even more important is a family of chemicals known as the *nucleic acids*. These direct and control the synthesis of enzymes. If the cell may be likened to a chemical factory, then nucleic acids are the master blueprints.

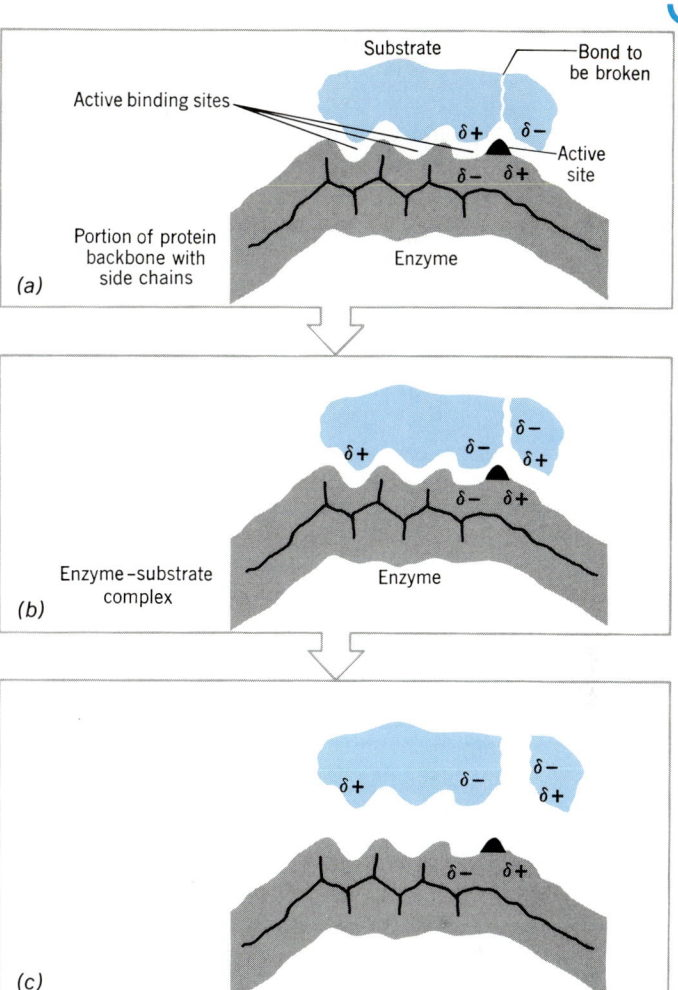

FIGURE 19-12
"Lock and key" theory of how enzymes work. (a) The substrate, the molecule to undergo some catalyzed reaction, fits to the enzyme. (b) While joined to the enzyme, the reaction proceeds. Polar sites undergo changes in electrical character, and forces that will repel the pieces of the product molecules build up. (c) The product molecules are driven away freeing the enzyme surface for another event. (From J. R. Holum, *Elements of General and Biological Chemistry*, 4th edition. 1975, John Wiley & Sons, New York.)

Nucleic acids are polymers. There are two main types, RNA or ribonucleic acid and DNA or deoxyribonucleic acid. *DNA* is the actual chemical of genes, the individual units of heredity and the chemical basis through which we inherit our characteristics from our parents.

The building blocks for nucleic acids are called nucleotides and these are made, in turn, from simpler molecules. For *RNA* the system uses a phosphate unit, $-HPO_4-$, a simple sugar known as ribose ($C_5H_{10}O_5$), and four amines. In DNA we have also a phosphate unit, a simple sugar called deoxyribose, $C_5H_{10}O_4$, and four amines.

The amines are to nucleic acids what the side chains are to proteins. The sequence in which these amines appear along a nucleic acid "backbone" makes a particular DNA or RNA molecule unique and special. In DNA, the sequence of side-chain amines is, in fact, the genetic code. Figure 19–13 shows a segment of a DNA chain. The backbone is a repeating sequence of phosphate-sugar-phosphate-sugar where the sugar is deoxyribose. The same backbone appears in RNA, except that the sugar is ribose. DNA and RNA have three of the same amines: guanine (G), cytosine (C), and adenine (A). In addition, DNA has thymine (T), and RNA, in place of thymine, has uracil (U).

19.22
CRICK-WATSON THEORY

In 1953 F. H. C. Crick and J. D. Watson proposed a structure of DNA that successfully explained how it could function as a gene. They said that the x-ray data about DNA meant that DNA occurs as a pair of intertwined strands. The now-famous *double helix* of Crick and Watson is illustrated in Figure 19–14. The functional groups and molecular geometries of adenine and

FIGURE 19–13
Segment of a DNA molecule showing the four amines that appear as side chains on the DNA backbone of phosphate-sugar-phosphate-sugar units. (a) Full structure. (b) A convenient representation of the DNA chain. The positions in (a) marked with asterisks involve a carbon as part of the five-membered ring and to which two hydrogens are attached. At these locations in RNA there is one hydrogen and one -OH group. Thymine is replaced by uracil in RNA, too, but its molecular shape is similar to that of thymine.

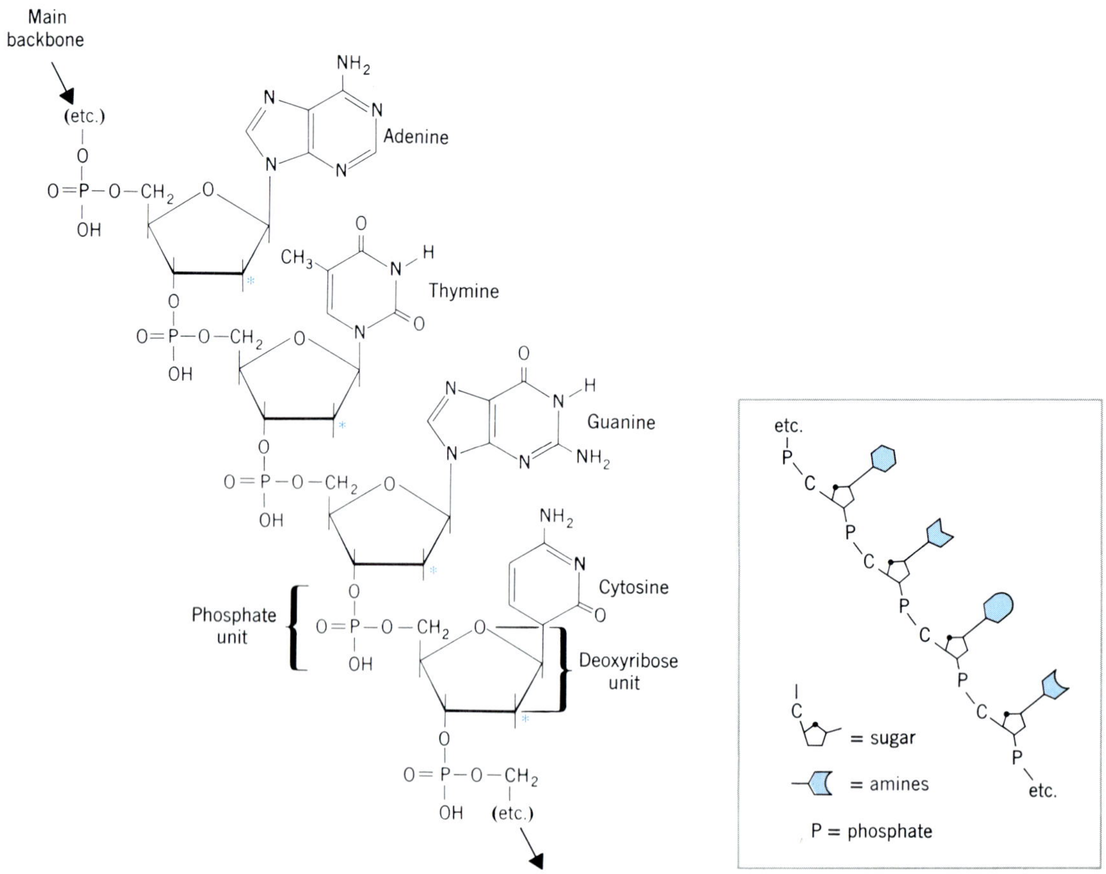

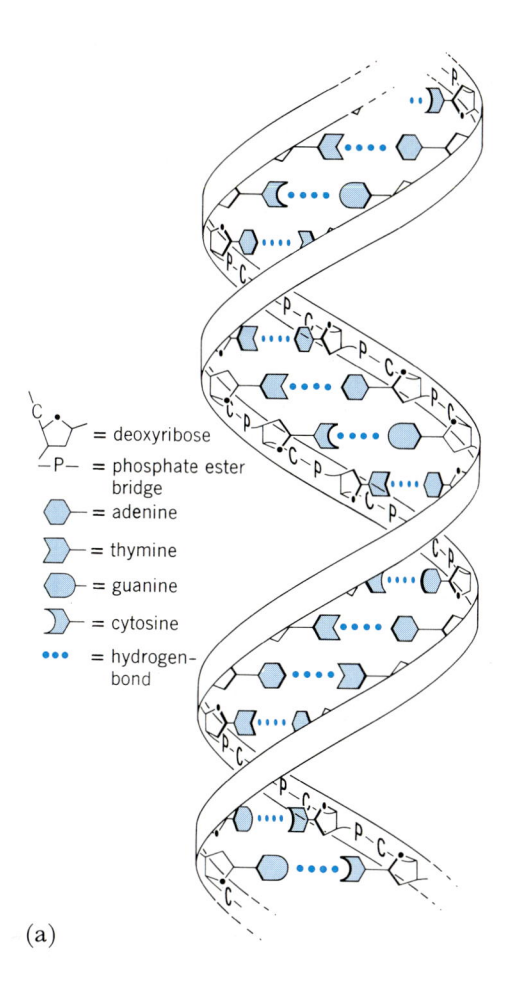

19.22 CRICK-WATSON THEORY

FIGURE 19-14
The DNA double helix according to Crick and Watson. (a) Two DNA strands form a right-handed helix. The two ribbons are the two "backbones." Geometric figures depict the side-chain amines. Dotted lines are hydrogen bonds between the amines. These forces of attraction (as in an alpha-helix of a protein) stabilize the double helix. (b) Details of the hydrogen bonds. (From J. R. Holum. *Organic Chemistry: A Brief Course.* 1975. John Wiley & Sons, Inc., New York.)

thymine match perfectly; the two can fit to each other and be held together by hydrogen bonds. Cytosine and guanine match each other, too. In DNA A pairs with T; C with G.

According to the *Crick-Watson theory* the *genetic code* is the sequence of amines on the backbone of the DNA molecule. A set of three side chain amines in a row is one code word. The code has been deciphered, and each code word is translated by the cell into a particular amino acid. The series of code words on DNA specifies an analogous series of amino acids in a polypeptide that the cell makes, a polypeptide that is or becomes part of an enzyme. Each gene, therefore, can specify one polypeptide. The sequence, in broad outline, is

$$\text{DNA} \xrightarrow[\text{(message is transferred)}]{\text{transcription}} \text{RNA} \xrightarrow{\text{translation}}$$

$$\text{protein} \xrightarrow{\text{catalysis}} \text{other compounds and reactions in cells}$$

The side-chain amine sequence in DNA determines a complementary sequence of side chains in RNA. This controls an amino acid sequence in polypeptides.

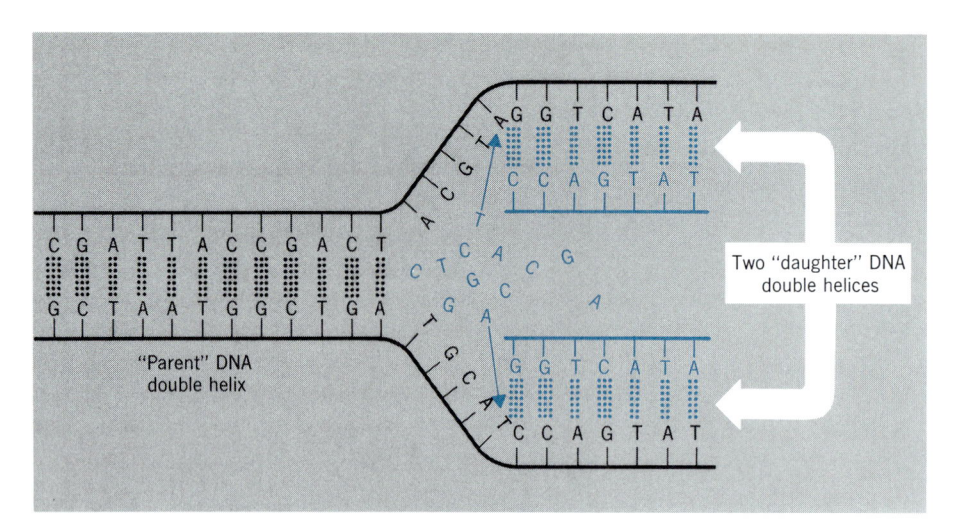

FIGURE 19-15
Replication of DNA. In this very highly schematic diagram we see the basic plan. The monomer units, shown here simply by letters, join to each other in an order determined by the complementary order on the untwining DNA double helix.

When a cell is about to divide, all its genes must be duplicated. Each new daughter cell must get exact copies of the original genes. The process whereby genes are duplicated is called replication. In broad outline, it involves an uncoiling of a double helix. Along each DNA strand a complementary strand is synthesized, as sketched in Figure 19-15. When the cell makes the RNA needed to translate a gene's message into a specific enzyme, it lines up RNA monomers along a DNA chain in a similar manner. Thus, the new RNA will be coded for assembling amino acids in just one particular sequence as the polypeptide-enzyme is put together.

SUMMARY

Photosynthesis. Complicated molecules are made in plants from simple molecules such as carbon dioxide and water with the aid of solar energy absorbed by chlorophyll. The primary products are carbohydrates, but the plant uses them partly for energy and partly to make, with the aid of minerals and ammonia, amino acids, proteins, lipids, and other things the plant needs.

Carbohydrates. Glucose typifies the simple sugars. It is a polyhydroxyaldehyde. Glucose in one form or another is involved in the three nutritionally important disaccharides—maltose, lactose, and sucrose—and in several polysaccharides—starch, glycogen, and cellulose. Glucose molecules exist in cyclic forms in which the ring easily swings open or closes shut. Depending on the relative orientation of one hydroxyl group, we can have α-glucose or β-glucose. Cellulose is a linear polymer of β-glucose and starch a polymer of α-glucose.

Lipids are hydrocarbon-like substances. Plant and animal materials that tend to dissolve in solvents such as benzene, gasoline or ether are lumped together in the broad family of lipids. The nutritionally important families are the animal fats such as lard and butterfat and the vegetable oils such as corn oil, olive oil, and peanut oil. Each of their molecules has three ester groups, and digestion simply hydrolyzes the molecules at these places to give glycerol and fatty acids.

Proteins. The body has an inventory of about 20 amino acids that are used in varying kinds and numbers to make proteins, polymers of amino acids. The amino acid units are joined to each other by peptide bonds. Protein molecules in living systems tend to be coiled into helices stabilized by hydrogen bonds. Stresses such as heat or alcohol or detergents can cause these helices to collapse, denaturing the protein.

Enzymes, the catalysts of reactions in cells, are mostly made of protein molecules, which are usually joined to a nonprotein cofactor.

Vitamins. Cofactors for enzymes are sometimes made of pieces of vitamin molecules, substances we cannot make internally, we must get in the diet, and without which we develop a deficiency disease.

Hormones. The body uses chemicals called hormones made in and secreted by very specialized tissues called the endocrine glands. Hormones circulate in the blood until they reach their target tissue where they help to regulate what can happen in cells of the tissue.

Nucleic acids. Genetic informaton in each cell is encoded in a polymeric type of molecule called deoxyribonucleic acid, or DNA. The genetic code is the sequence of amines—four possible in number—along the polymer backbone made of alternating units of deoxyribose (a special sugar) and phosphate. The gene is a particular portion of a DNA molecule, and these molecules occur as intertwined, double helices. One gene has the information for directing the synthesis of one enzyme. DNA first directs the synthesis of a complementary or matching molecule of another kind of nucleic acid called RNA, ribonucleic acid. This messenger RNA then directs the ordering of amino acids in the cell's synthesis of a protein, with the help of special enzymes and still another kind of nucleic acid. The one-gene/one-enzyme theory thus relates genetic information to enzyme synthesis.

TERMS AND CONCEPTS

photosynthesis
chlorophyll
carbohydrate
saccharide
monosaccharide
disaccharide
polysaccharide
digestion
lipid
animal fat
vegetable oil
fatty acid
protein
amino acid
peptide bond
dipeptide
polypeptide
alpha-helix
denaturation
denatured protein
enzyme
cofactor
vitamin
lock and key theory
poison
hormone
nucleic acid
DNA
RNA
double helix
genetic code
Crick-Watson theory

QUESTIONS AND EXERCISES

1. When plant materials are burned carbon dioxide is produced. What happens to it eventually in nature?

2. What are other names for (a) dextrose? (b) table sugar? (c) milk sugar?

3. What is blackstrap molasses?

4. What is the difference between starch and cellulose (a) in how the plant uses these materials? (b) in their structures (in general terms only)?

5. What is produced when each of these is digested? (a) maltose, (b) starch, (c) sucrose, (d) lactose.

6. In what form do animals store glucose units?

7. What are lipids?

8. When a corn oil product is described as "polyunsaturated," what does that mean, in structural terms?

9. What is used to make oleomargarine?

10. How do we use lipids in our diets?

11. The monomer units for proteins have which two functional groups?

12. What is the peptide bond?

13. How are all proteins alike?

14. How are proteins different?

15. What is the alpha-helix?

16. What is denaturation?

17. What is the difference between a vitamin and a hormone (a) in their origin? (b) in their function?

18. In what way do B-vitamins participate in enzyme chemistry?

19. How do certain poisons shut down key body reactions?

20. What is the chemical of a gene?

21. What is the relation between a gene and a specific polypeptide that might be an enzyme?

22. What contribution did Watson and Crick make to biology?

23. What is the difference between DNA and RNA (a) in structure? (b) in function?

24. What is meant by the genetic code, in structural terms?

25. Cite three instances where the hydrogen bond is important in stabilizing molecules found in living systems.

CHAPTER 20
SYNTHETICS AND DRUGS

20.1 THE WORLD OF SYNTHETICS

Petroleum, coal, and natural gas have long been familiar to us as fuels to heat buildings and drive machinery. Much less understood by most people is the enormous importance of these natural substances as chemical raw materials for the world of *synthetics*.

About 80 percent of the gasoline used in cars and trucks is synthetic, being made from other parts of petroleum that do not work in gasoline engines. Much of the rubber used to make tires and inner tubes is synthetic, made from petroleum.

Drip-dry, wash-and-wear clothing is made wholly or mostly from synthetics—fabrics made from the chemicals wrung from coal and petroleum and natural gas and changed to nylon, Dacron, Orlon, and the acrylates.

The entire world of plastics—polyethylenes, polypropylenes, Mylar, polystyrene—emerged from the ingenuity of scientists and engineers, the daring of administrators, managers, and investors, and the availability of raw materials.

High-yield agriculture, without which there is not the slightest hope of feeding the current world population, depends heavily, for good or ill, on considerable use of synthetic weed killers, bug killers, and fungus preventers.

High-quality medicine depends increasingly on synthetic drugs, disposable hospital items, and plastic fabrics and tubings for apparatus and for internal implantation in new surgical techniques.

Virtually all of the bright colors of modern fabrics are made possible by synthetic dyes and pigments.

Many hobbies and recreational activities depend on synthetics—strong, ultralight fabrics for tents and backpacks (Figure 20–1), sails (Figure 20–2), lines and lures for fishing, epoxy resins for fiberglass plastic boat hulls and bodies of some racing vehicles, to say nothing of surfboards and skis, and flame-resistant backing for photographic film.

In this chapter we survey in a general way the principal materials that go into the finished products of the world of synthetics. We also study certain drugs, some synthetic, some not. We begin with the nature of petroleum and coal.

20.2 PETROLEUM

Petroleum (*petra*, "rock"; *oleum*, "oil") is a complex mixture of organic compounds, mostly hydrocarbons, together with some water. Crude oil is what remains when water and organic gases in petroleum are removed. Besides the hydrocarbons, crude oil contains from 1 to 6 percent nitrogen, oxygen, and sulfur.

The petroleum deposits of the world today formed throughout hundreds of thousands of years during the Carboniferous Period (280 to 345 million years ago) while vast areas of the continents, little more than featureless plains, basked in sunlight near sea level. In the seas countless microscopic plants such as diatoms—tons of them per acre of ocean surface during early springtime—soaked up solar energy via photosynthesis. Then they died. According to one theory, death of each such plant released a microscopic droplet of oily material that eventually settled into the bottom muds. The muds grew in thickness and compacted, sometimes into shales, sometimes into limestone and sandstone deposits. The oily material changed to petroleum. In some parts of the world it managed to move through porous rock and collect into pools forming the great petroleum reserves of our planet. In other parts, this movement could not occur, and the oily substances remain to this day locked in enormous deposits of oil shales and oil sands.

20.3 COAL

Plants and animals, of course, grew and died on the lands bordering those ancient seas. In a moist and sunny setting where the land was marshy and boggy, lush vegetation flourished and died. The remains, covered by stagnant, oxygen-poor, often acidic water, decomposed, but this occurred more slowly than the plants died. The rotting mass accumulated, became fibrous, and turned into peat. Where peat layers became unusually thick, they compacted into *lignite* ("brown coal"). Lignite is over 40 percent water, but it is still an important fuel. When lignite deposits grew in thickness and became further compacted, sometimes by overlying deposits of sedimentary rock, the water was forced out. It became *bituminous coal* ("soft coal"), with less than 5 percent water (see Figures 20–3 and 20–4). Bituminous coal includes roughly 30 percent volatile, organic matter. When nature's workings forced that out, the remainder was *anthracite* ("hard coal") with over 95 percent fixed carbon.

20.4 NATURAL GAS

Both in soft coal deposits and in petroleum deposits the most volatile hydrocarbons—methane and ethane—also accumulated. *Natural gas* is largely methane.

The various coals, petroleum, and natural gas constitute the *fossil fuels,* a legacy from the past now being consumed so rapidly that for the first time in history worry about running out has become widespread. That we shall run out is quite simply just a matter of time. Almost any imaginable human use of the fossil fuels would use them up much faster than nature can replenish them (see Figures 20–5 and 20–6). We shall discuss fuel reserves further in Chapter 27.

FIGURE 20-1
Rugged backpacking tents sleeping five weigh under ten pounds when made of ultralight yet ultrastrong nylon taffeta. (Tents by Gerry; photo by J. R. Holum.)

20.5 CRUDE OIL REFINING

The object of crude oil *refining* is to separate it into products of varying uses (see Figures 20–7 and 20–8). Table 20.1 contains a list of the principal fractions. Each fraction is a mixture of hydrocarbons having an overall volatility that makes it useful in certain kinds of engines or

FIGURE 20-2
Strong, lightweight sails are made of Dacron, a polyester synthetic. (Photo by Stanley Rosenfeld; boat by Chris Craft.)

20 SYNTHETICS AND DRUGS

FIGURE 20-3
Underground coal mining. The alloy steel bits on the rotating head of this continuous mining machine rip coal loose as the head moves vertically along the face of the coal seam. The loosened coal travels back on a moving belt. (Courtesy National Coal Association.)

furnaces. Residues that do not distill yield some special products, but most are used as residual fuel oil ("resid") or are made into coke and asphalt. An estimated 500 compounds occur in the fractions that range in boiling points up to 200°C. In the lower ranges, roughly a third are alkanes, a third are cycloalkanes, and a third are aromatic hydrocarbons.

20.6
SYNTHETIC GASOLINE
The *gasoline* fraction of crude oil is not large enough by far to supply world needs. Molecules

FIGURE 20-4
Strip mining coal. Surface mining of soft coal in the foreground fuels the power plant in the background, one of several in the "four corners" region of the southwestern United States. (Mimi Forsyth/Monkmeyer.)

FIGURE 20-5
Oil under the floor of the North Sea is being pumped to Norway from this producing platform seen here passing the narrows of Stavanger Fjord on its way to the Brent field in August 1975. (An Exxon photo.)

in the gasoline fraction have from 5 to 10 or 12 carbons each. We can make such molecules by "cracking" larger molecules or by "alkylating" smaller ones. Even though alkanes are very stable thermally, if they are heated (in the absence of air) well above their boiling points, their molecules will "crack" and sometimes undergo rearrangements. The operation is called *thermal cracking*. When it is done on the mixture in gas oil, the product is naturally an exceedingly complex mixture of lower boiling alkanes and alkenes. We illustrate one of the myriad possibilities.

FIGURE 20-6
This enormous supertanker—Gulf Oil Corporation's "Universe Kuwait"—can carry 2.4 million barrels of crude oil. The 1975 United States oil needs amounted to the equivalent of 2500 loadings of a tanker this size. (Gulf Oil Corporation.)

20.6 SYNTHETIC GASOLINE

FIGURE 20-7
Fractionating towers at the crude refinery of Mobile Oil Corporation at Joliet, Illinois. (Mobil Oil Corporation.)

$$CH_3(CH_2)_{16}CH_3 \xrightarrow{500°}$$
Octadecane

$$CH_3(CH_2)_7CH=CH_2 + CH_3(CH_2)_6CH_3$$
1-Decene Octane

"Gas oil"
(a mixture) $\xrightarrow{\Delta}$

Low-octane gasoline (octane $\approx$ 70)

FIGURE 20-8
Typical modern fractionating tower used by an oil refinery to separate components of crude oil into various useful products. (American Petroleum Institute.)

TABLE 20.1
PRINCIPAL FRACTIONS FROM DISTILLING PETROLEUM

Boiling Point Range (°C) (a measure of volatility)	Number of Carbons in Molecules of the Fraction	Uses
Below 20	C_1–C_4	Natural gas; heating and cooking fuel; bunsen burner fuel; raw material for other chemicals
20–60	C_5–C_6	Petroleum "ether"; nonpolar solvent and cleaning fluid
60–100	C_6–C_7	Ligroin or light naphtha; nonpolar solvent and cleaning fluid
40–200	C_5–C_{10}	Gasoline
175–325	C_{12}–C_{18}	Kerosene; jet fuel; tractor fuel
250–400	C_{12} and higher	Gas oil; fuel oil; diesel oil
Nonvolatile liquids	C_{20} and up	Refined mineral oil; lubricating oil; grease (a dispersion of soap in oil); resid
Nonvolatile solids	C_{20} and up	Paraffin wax (purified solids that crystallize from some oils); asphalt and tar for roads and roofing

20 SYNTHETICS AND DRUGS

It is as if something like the following shuffling of bonds occurred (where R is an alkane-like group):

$$R-CH_2-CH_2-R' \xrightarrow{\Delta} R-CH=CH_2 + H-R'$$
(Alkane → Alkane)

The cracking of octadecane, illustrated above, is true to one fact about thermal cracking. Straight-chain alkanes tend to crack to straight-chain hydrocarbons. Thermal cracking does not tend to increase the population of highly branched hydrocarbons or of aromatics. Therefore the product is not of very high octane rating (Section 20.7), having a rating of about 70 octane. Three very important developments, however, help to solve this problem: catalytic cracking, catalytic reforming, and lead additives.

If cracking is done in the presence of a mixed alumina-silica powder (or granules), the proportion of branched hydrocarbons increases. The product therefore has a much higher octane rating. *Catalytic cracking* does not so much permit lower cracking temperatures as it encourages the rearrangements of carbon chains from straight to branched. The advantage is that higher-octane gasoline is obtained, gasoline of about 90 octane, without additional reforming. Again, the reactions are numerous, and we can only illustrate.

$$CH_3(CH_2)_{16}CH_3 \xrightarrow[Al_2O_3 \cdot SiO_x]{500°}$$
Octadecane

$$CH_3C(CH_3)=CH(CH_2)_5CH_3 + CH_3CH(CH_3)CH_2C(CH_3)_2CH_3$$

2-Methyl-2-Nonene   2,2,4-Trimethyl-pentane ("Isooctane")

Reforming is an operation in which lower-octane gasoline (richer in straight-chain hydrocarbons) is passed over a special catalyst such as platinum or alumina to produce a higher-octane gasoline, one richer in branched alkanes. The process also dehydrogenates some hydrocarbons. The following reactions illustrate:

$$CH_3(CH_2)_5CH_3 \xrightarrow{\text{Pt/alumina, or "chromia"/alumina, or "molybdena/alumina.}}$$
n-Heptane

$$CH_3CH(CH_3)(CH_2)_3CH_3 + \text{isomers}$$
2-Methylhexane

Methylcyclopentane $\xrightarrow[\Delta]{\text{Catalyst}}$ Cyclohexane

$\xrightarrow[\Delta]{\text{Catalyst}}$ Benzene (+H$_2$)

Methylcyclohexane $\xrightarrow[\Delta]{\text{Catalyst}}$ Toluene (+H$_2$)

The unsaturated cyclic compounds themselves are present, and some arise during reforming from open-chain compounds. Reforming, in other words, can involve increased *branching, cyclizations,* and *aromatizations.*

Benzene and toluene have high octanes, and their formation means that reformed gasoline has a higher octane than prior to the operation. Aside from that, most of the benzene and toluene that are used as industrial chemicals are now made from petroleum by reforming, cyclizations, and aromatizations. These compounds, as we shall see, are of major importance in the synthesis of plastics, drugs, dyes, and other substances of commercial value.

Alkylation is still another operation for extending the supply of high-quality gasoline. Where cracking makes smaller molecules from the larger, alkylation does the opposite. It makes larger molecules from the smaller. Using concentrated sulfuric acid or hydrofluoric acid as the catalyst, isobutane can be made to add to isobutylene.

$$CH_3C(CH_3)(CH_3)-H + CH_2=C(CH_3)-CH_3 \xrightarrow{H_2SO_4 \text{ or } HF}$$
Isobutane   Isobutylene

$$CH_3-C(CH_3)(CH_3)-CH_2CH(CH_3)CH_3$$
Isooctane

The cracking we discussed earlier is one of the

sources of these lower-formula-weight hydrocarbons. Alkylation is by no means limited to this example. Benzene can be made to add to ethylene or propylene:

$$\text{Benzene} \begin{cases} CH_2=CH_2 \text{ (ethylene)} \xrightarrow[95°, \text{pressure}]{AlCl_3-HCl} \text{Ethylbenzene (}CH_2CH_3\text{)} \\ CH_2=CH-CH_3 \text{ (propylene)} \xrightarrow[H_3PO_4, \text{pressure}]{250°} \text{Cumene (isopropylbenzene)} \end{cases}$$

Ethylbenzene is a raw material for styrene, monomer for polystyrene. Cumene, among other uses, is the raw material for both phenol and acetone. Phenol is a raw material for aspirin (Section 20.20) and acetone is an important solvent. In the early 1970s ethylbenzene production was over 6 billion pounds per year, placing it among the top 20 of *all* chemicals produced by the chemical industry.

One of the purposes of all these operations is to make a larger supply of high-quality gasoline. One of the ways of measuring quality is by the octane rating.

20.7 OCTANE NUMBER

When gasoline performs poorly the engine makes pinging sounds. We say that it "knocks." Performance is measured in carefully standardized laboratory engines that are fueled by the gasoline being rated. The compound heptane is taken as the standard of bad performance, and it has an *octane rating* of zero. An isomer of C_8H_{18}, isooctane, is taken as the standard of excellence, and it has an octane rating of 100. The special engines are run with varying combinations of heptane and isooctane. The test runs give a scale of performance from zero to 100 octane. A particular gasoline fraction is then rated by comparing its performance with the performance of standard combinations of heptane and isooctane.

Since this procedure was developed, fuels that performed even better than isooctane were found. Hence, we now can have fuels rated above 100 octane.

The most common means of improving the octane rating of a gasoline is to add tetraethyllead, $(C_2H_5)_4Pb$, about 2-3 ml per gallon of gasoline. Ethyl Fluid[1] consists of about 60 percent tetraethyllead and 38 percent lead "scavengers"—ethylene bromide and ethylene chloride that react with the lead and change it into a form volatile enough to leave the cylinder of the engine via the exhaust. The remainder is dye and kerosene. Tetramethyllead, $(CH_3)_4Pb$, is another useful antiknock compound.

20.8 COAL AND COAL BY-PRODUCTS

Most of the coal mined in the world is used to heat buildings or to generate steam for making electricity. However, some bituminous coal is used to make coke and chemicals.

If soft coal is heated in the absence of air to temperatures ranging from 900–1200°C, it breaks down to *coke* and a complex mixture of chemicals. The process is called destructive distillation or carbonization. From one ton of coal there is produced 1500 pounds of coke, 11,000 cubic feet of coal gas, 10 gallons of coal tar, and 3.5 gallons of light oil. The gas includes ammonia. If it is removed by reacting with sulfuric acid, about 20–30 pounds of ammonium sulfate, a fertilizer, is made.

Coke is impure carbon, and it is needed by the steel industry to reduce iron oxides to iron (Section 17.3). Coal oil consists mostly of simple aromatic hydrocarbons—benzene, toluene, and the xylenes. All are essential raw materials in the world of synthetics.

Coal tar is a thick, sticky, black fluid containing a huge number of commercially important substances. Three general types can be separated: hydrocarbons, tar acids, and tar bases. The hydrocarbons are mostly aromatic. There are procedures for separating and purifying all of them, because all are useful raw materials for one kind of synthetic or another.

[1] Ethyl Fluid is a registered trademark.

20.9 ETHYLENE AND PROPYLENE AS PETROCHEMICALS

Besides methane and ethane, natural gas contains varying amounts of propane (3–8 percent) and butane (2–14 percent). Because these two gases are easily changed to liquid forms, they are easily removed for other uses. Quantities of them are marketed in cylinders as Skelgas or Philgas for heating and cooking purposes in rural areas (or campers and mobile homes). Perhaps more importantly, these alkanes can be cracked. For example,

$$2CH_3CH_2CH_3 \xrightarrow{500°} CH_2{=}CH_2$$
Propane Ethylene

$$+ \; CH_2{=}CHCH_3 \; + \; CH_4 \; + \; H_2$$
Propylene Methane Hydrogen

This makes available a larger supply of ethylene, propylene, and butylenes (from C_4 alkanes). As seen in Figure 20–9, ethylene is the raw material for a large number of commercially important products. Polyethylene (Section 18.14) is just one. By one or a series of reactions, ethylene is also used to make Dacron and Mylar (Section 20.17). The two are chemically the same. Mylar is unusually resistant to tearing, making it possible to use very thin films in high-quality magnetic tapes. It is also an excellent insulator. Dacron and Mylar are made from ethylene glycol, which is important itself as a permanent antifreeze.

FIGURE 20–9
Some major uses of ethylene as a petrochemical. (From J. R. Holum, *Organic Chemistry: A Brief Course*, 1975, John Wiley & Sons, Inc., New York.)

The production of ethyl chloride from ethylene leads to two extremely important catalysts. One, tetraethyllead (Section 20.7), has been discussed. The other, triethylaluminum, is one constituent of the famous Ziegler catalyst used to polymerize alkenes and dienes.

Still another use of ethylene is in the production of styrene, a process also requiring benzene that can be obtained from petroleum or coal, via coal tar. Styrene is a raw material for synthetic rubber as well as styrofoam and other polystyrene plastics.

Ethyl alcohol is made in huge quantities from ethylene, and it is needed as an industrial solvent as well as a solvent for over-the-counter products such as pharmaceuticals.

The actual chemistry of the reactions used to make most of these products cannot be developed here. We shall in later sections illustrate just a few examples. Our purpose is simply to make the case that petrochemicals are as much a part of our daily lives as the move obvious petroleum products: gasoline and oil.

As seen in Figure 20–10, propylene also has many uses. Making polypropylene is only one (Section 20.12).

Propylene is the major source of the glycerol needed to make nitroglycerine, an important explosive. Glycerol is also used to make alkyds or glyptal resins used as coatings and varnishes. By a series of changes, propylene serves as the basis for the epoxy resins we use in fiberglass products,

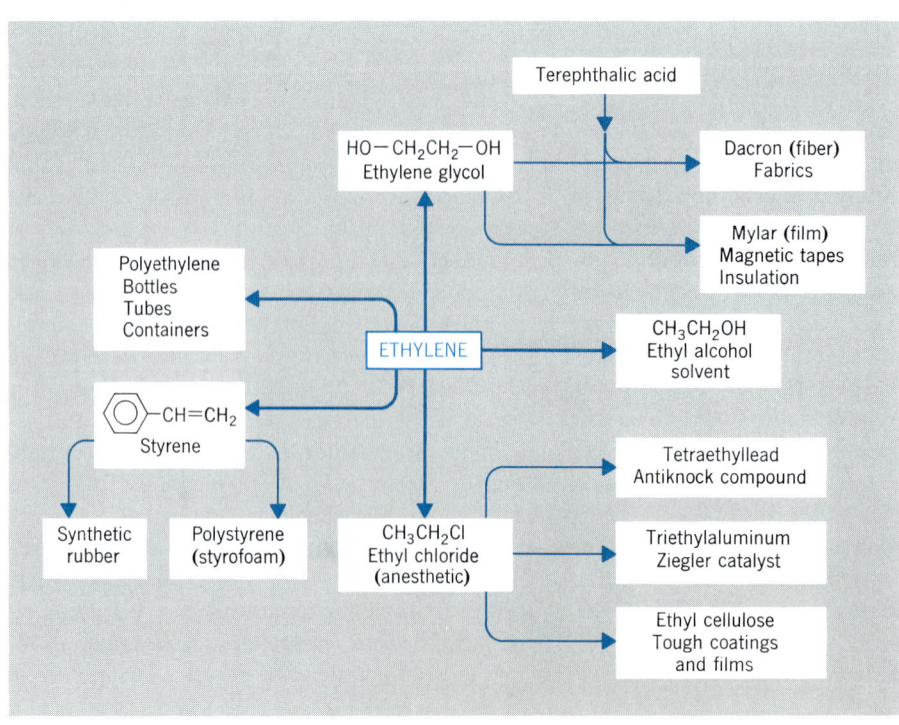

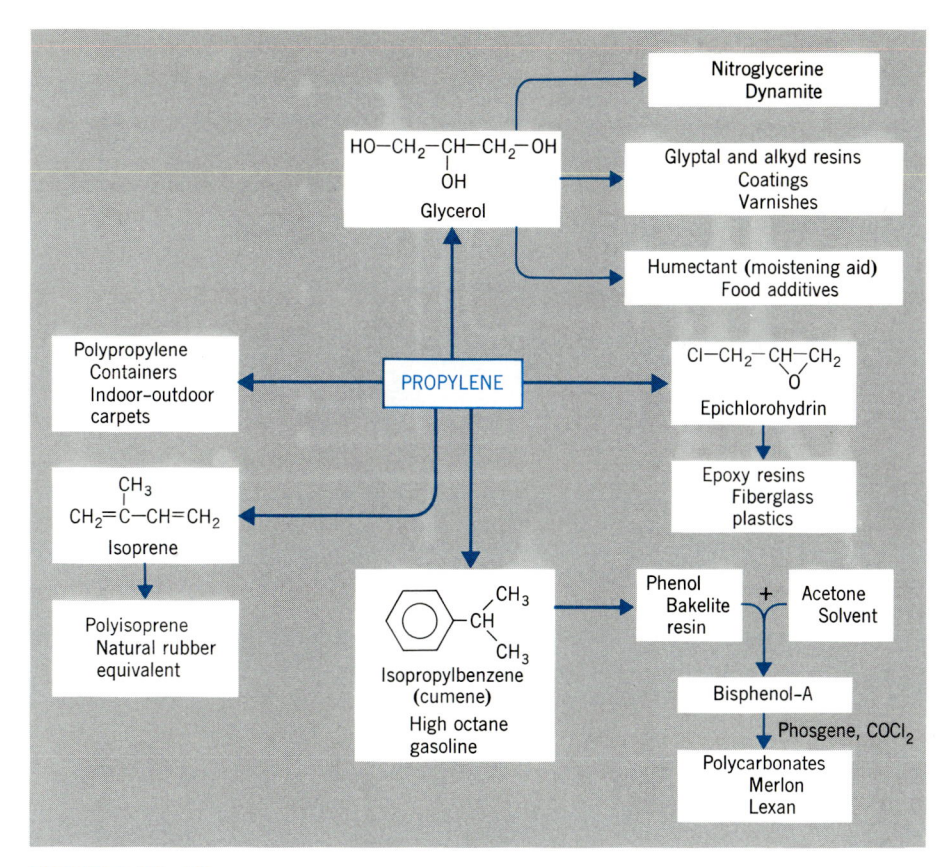

20.11 POLYMERS

FIGURE 20-10
Some major uses of propylene as a petrochemical. (From J. R. Holum, *Organic Chemistry: A Brief Course*, 1975, John Wiley & Sons, Inc., New York.)

for the polycarbonates (Merlon and Lexan) that are so tough they are used to make bulletproof shields, for Bakelite, and for high-octane fuel. After several changes, propylene yields isoprene from which polyisoprene, essentially identical with natural rubber, is made.

20.10
THE C$_4$-HYDROCARBONS AS PETROCHEMICALS

Figure 20-11 outlines several ways by which the various butanes and butylenes are used as raw materials. Two products stand out: high-octane gasolines and rubberlike polymers. Butadiene, after several changes, yields one of the important raw materials for nylon: hexamethylenediamine.

20.11
POLYMERS

We encountered the idea of a polymer in Section 18.14 where we learned how an alkene, ethylene, can react with itself to give a substance, polyethylene, whose molecules have hundreds of repeating units, -CH$_2$-CH$_2$-, furnished by ethylene molecules. (Chemists use the language of "hundreds of repeating *ethylene units*," but that is a trifle misleading. Polyethylene molecules contain no carbon-carbon double bonds as do those of ethylene. Having noted this point, we shall switch to the easier language of the organic chemist in further discussions.)

In general, the word *polymer* stands for any substance whose molecules have a high formula weight and a repeating structural unit. Not all polymers are plastics, because the family of polymers includes proteins, starches, and other natural substances fitting the definition of polymer. A *plastic* is a finished or semifinished article or substance made from a resin by molding, casting, extruding, drawing, or laminating during or after polymerization of the raw materials. A synthetic *resin* is the unfabricated polymeric material that is used to make all or most of a plastic article or to coat some surface for protection. The finished plastic often consists of something besides the resin. Sometimes when a resin produces a finished article that is too hard, stiff, and brittle, another chemical is added called a *plasticizer*. Several plasticizers are available, and when one is intimately mixed with the resin during fabrication of the plastic, the plasticizer molecules act

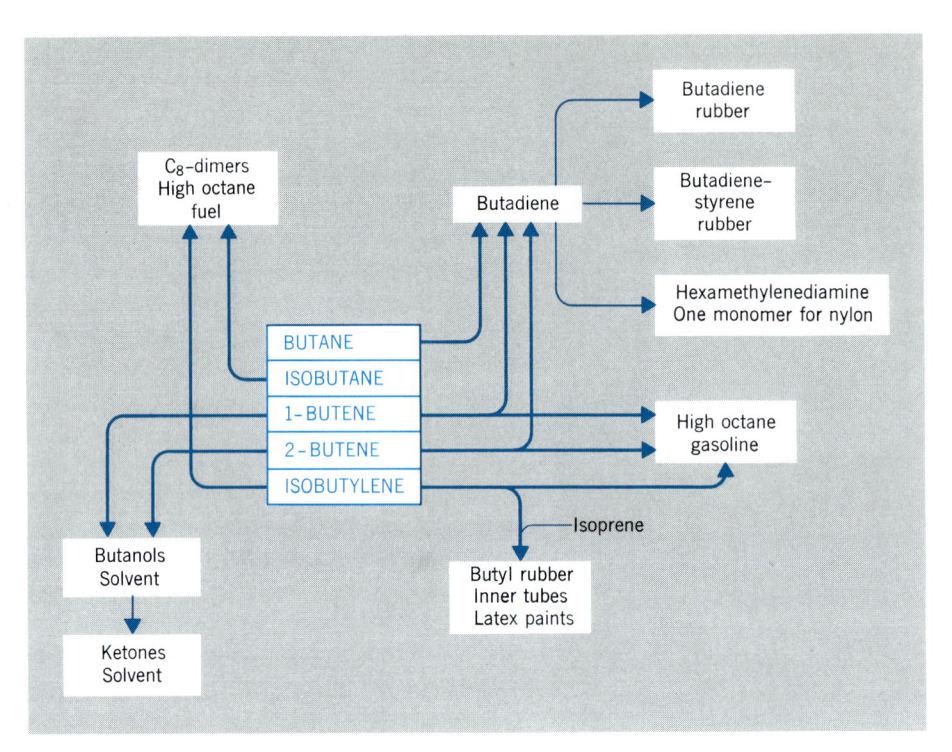

FIGURE 20-11
Uses for the C_4 petrochemicals. (From J. R. Holum, *Organic Chemistry: A Brief Course,* 1975, John Wiley & Sons, Inc., New York.)

as a lubricant, like oil. They make the article more flexible, less brittle.

20.12 POLYOLEFINS

Long before substances such as ethylene and propylene were called alkenes they were referred to as *olefins* after an old common name of ethylene. *Olefin* is a nickname for *alkene*. The polymers of olefins are therefore often called *polyolefins*. Polyethylene is one example. Polymerization of ethylene, like all polymerization reactions, does not produce one pure, homogeneous substance. When billions of ethylene molecules polymerize, some very long chains and some shorter chains form. Polyethylene therefore is really a mixture of molecules, but all have the same structural feature: repeating ethylene units. Since polyethylene is essentially a mixture of very long-chain alkanes, it has the chemical inertness of alkanes. That is why it is useful in making such a variety of plastic household articles (icebox dishes, wastebaskets, drinking glasses, funnels, bowls, and pans), laboratory apparatus, agricultural implements (see Figure 20-12), and hospital equipment (for instance, tubing for intravenous feeding and transfusions).

Polypropylene is another polyolefin.

$$n\text{CH}_2=\underset{\underset{\text{Propylene}}{\text{CH}_3}}{\text{CH}} \rightarrow -(\text{CH}_2-\underset{\underset{\text{Polypropylene}}{\text{CH}_3}}{\text{CH}})_n-$$

Made by polymerizing propylene, this polymer not only is used to manufacture most of the items that are also made of polyethylene; it has also become one of the major synthetic fibers. Indoor-outdoor carpeting is made of it (Figure 20-13). Many novel medical applications (Figure 20-14) have been made. If a polymer is to be used successfully in making fibers, its molecules must not only be long, uniform, and symmetric; they must also have some feature that makes neighbor molecules line up side by side and stick to each other. Polyethylene molecules tangle well together, but do not line up well side by side and stick. Polypropylene molecules have methyl groups on every other carbon. These help neighbor molecules catch on each other, but they do not line up well side by side unless all the methyl groups are on the same side of the chain, as illustrated in Figure 20-15. If they alternate, first one on one side then the next on the other side, and so forth, or if they are randomly oriented, good fibers cannot be made. Each of these three types of polypropylene—methyl groups all on the same side, or alternate, or random—can be made. The correct catalysts and conditions of temperature and pressure are known for each.

20.13 VINYL POLYMERS

The group $\text{CH}_2=\text{CH}-$ is called a *vinyl* group. The vinyl polymers or vinyls are a family of polymers whose molecules on alternate chain

20.13 VINYL POLYMERS

FIGURE 20–12
Polyethylene film is rolled over tomato seedlings to protect them from both heat and cold, to conserve moisture, and to inhibit the growth of weeds. The seedlings will grow through slits in the film. (Courtesy Monsanto Company.)

FIGURE 20–13
Polypropylene fiber has become very popular in the manufacture of indoor-outdoor carpeting. Shown here is a double-tufted carpet made of Herculon fiber. (Courtesy Hercules Incorporated.)

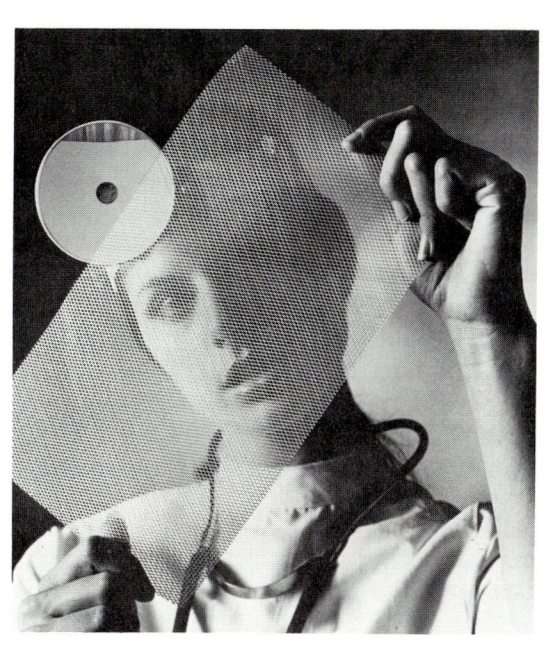

FIGURE 20–14
Polypropylene is chemically inert and has no reaction with blood or other body fluids making it useful for many medical applications as in this surgical mesh. (Exxon Chemical Co.)

carbons have different side-chain groups or atoms. The principal monomers are given in Table 20.2.

The polyolefins and the vinyls are all *addition polymers*. These are polymers resulting when one monomer molecule adds to another and nothing breaks out in the process. Very often two or more monomers are mixed together before polymerization is initiated. The various monomers copolymerize; they add to each other. The product is called a *copolymer*. Saran (Table 20.2) is one example.

FIGURE 20–15
Propylene, one geometric form. All the methyl groups (CH_3) project on the same side of the main chain. (From J. R. Holum, *Organic Chemistry: A Brief Course*, 1975, John Wiley & Sons, Inc., New York.)

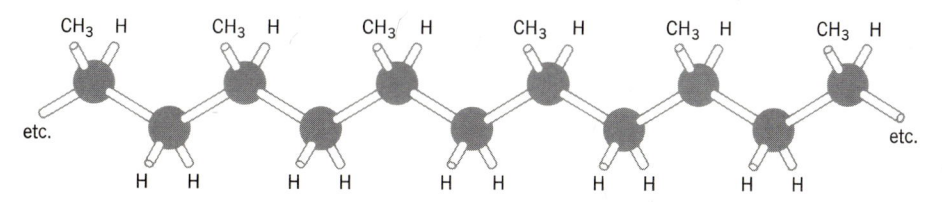

TABLE 20.2
VINYL MONOMERS AND POLYMERS

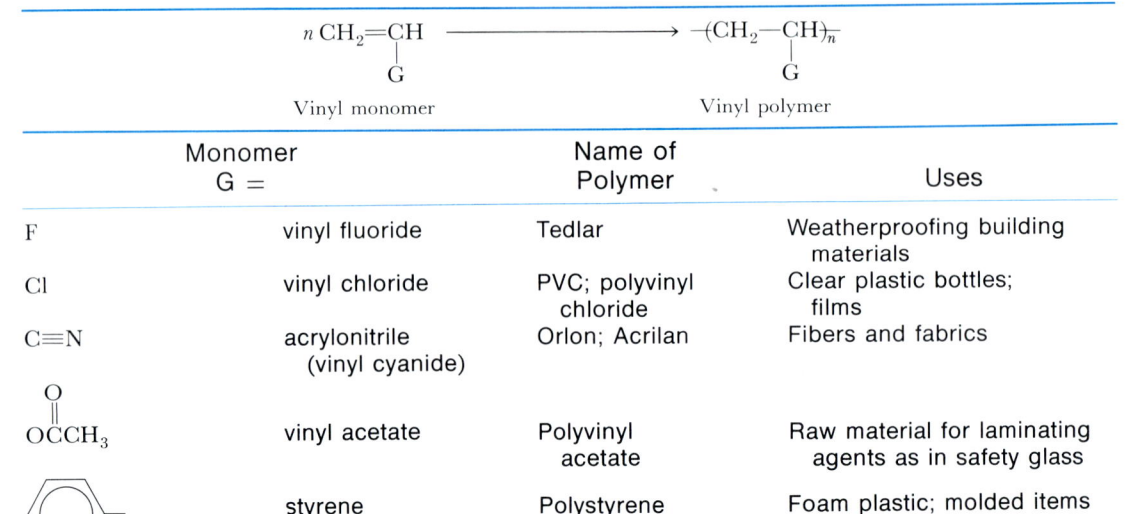

$$n\ CH_2=CH(G) \longrightarrow -(CH_2-CH(G))_n-$$

Vinyl monomer → Vinyl polymer

Monomer G =	Name of Polymer	Uses
F	vinyl fluoride / Tedlar	Weatherproofing building materials
Cl	vinyl chloride / PVC; polyvinyl chloride	Clear plastic bottles; films
C≡N	acrylonitrile (vinyl cyanide) / Orlon; Acrilan	Fibers and fabrics
$OCCH_3$ (O=)	vinyl acetate / Polyvinyl acetate	Raw material for laminating agents as in safety glass
(phenyl)	styrene (vinylbenzene) / Polystyrene	Foam plastic; molded items

Other Substituted Olefin Monomers

$CH_2=CCl_2$ (Cl,Cl)	dichloroethylene / Saran[a]	Packaging film; fibers; tubing and pipe
$F_2C=CF_2$	tetrafluoroethylene / Teflon	Circuit boards in computers; nonstick surfaces in kitchen utensils; insulators
$CH_2=C(CH_3)-C(=O)-OCH_3$	methyl methacrylate / Lucite; Plexiglas	Windows; molded items, coatings

[a] A copolymer of dichloroethylene and vinyl chloride.

20.14 DIENE POLYMERS

Natural rubber is a polymer of isoprene. Synthetic rubber with all the properties of natural rubber can be made from isoprene using a special catalyst. Other polymers having elastic

$CH_2=C(CH_3)-CH=CH_2 + CH_2=C(CH_3)-CH=CH_2 + CH_2=C(CH_3)-CH=CH_2 + CH_2=C(CH_3)-CH=CH_2 +$ etc.

Isoprene

↓ catalyst

Natural rubber
(Note the isoprene units between the dashed lines)

properties can be made from a variety of monomers, and they form a family called the *elastomers* (*elast*ic poly*mers*). Butyl rubber, for example, is a copolymer between isoprene and isobutylene.

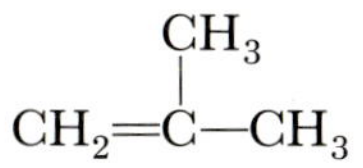

$$CH_2=\underset{\underset{CH_3}{|}}{\overset{\overset{CH_3}{|}}{C}}-CH_3$$
Isobutylene

It is not really possible to describe the precise structure of the polymer. Mixtures of molecular types are produced by copolymerization. When two or more different kinds of monomer molecules are involved, they may end up in the polymer alternated with each other, or grouped as blocks within the chain, or grafted on each other, or in a fully random sequence. If molecules of A and B copolymerize we have these possibilities:

Polymer scientists and engineers can control copolymerization by changing conditions of temperature, pressure, catalyst, time, and order of mixing. They tailor-make molecules that will give the desired properties to the final plastic or elastomer.

Elastomers with carbon-carbon double bonds have functional groups that make it possible to vulcanize them. *Vulcanized rubber* is made by heating a mixture of rubber latex, a vulcanizing agent (usually sulfur), and other chemicals (Figure 20-16). The long molecules in rubber become joined by units of the vulcanizing agent to create interlacing networks of bonds that improve the material's resistance to abrasion, increase its resiliance, and improve its ability to withstand high temperatures.

```
Alternate  —A—B—A—B—A—B—A—B—A—B—A—B—A—B—
Block      —A—A—A—B—B—A—A—A—B—B—A—A—A—B—B—
Graft      —A—A—A—A—A—A—A—A—A—A—A—A—A—A—
                 |         |         |         |         |
                 B         B         B         B         B
                 |         |         |         |         |
                 B         B         B         B         B
Random     —A—B—B—A—A—A—B—A—B—B—A—A—A—B—A—B—
```

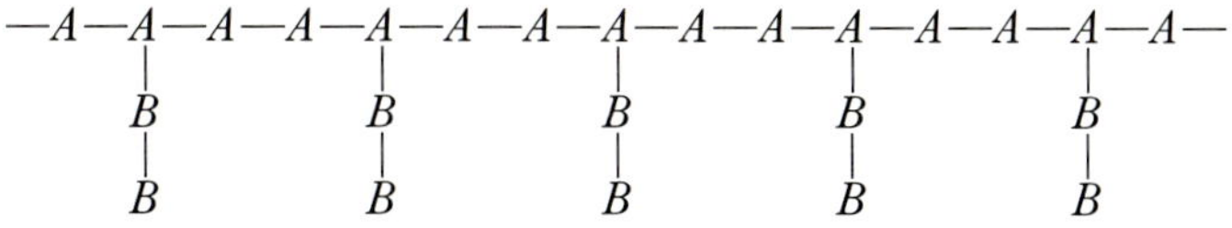

20.15 CONDENSATION POLYMERS

Addition polymers form one large family. Another family includes *condensation polymers*—those made in a polymerization reaction that splits out a small molecule. Polyamides and polyesters are two common types.

20.16 POLYAMIDES—NYLON

The term *nylon* is a coined name that applies to any synthetic, long-chain, fiber-forming polymer with repeating amide linkages. One of the most common members of the nylon family is so-called nylon-66. It is made from hexamethylenediamine and adipic acid according to the following reaction in which water molecules split out:

(The "66" comes from the fact that each monomer contains six carbons.) To be useful as a fiber-forming polymer, each nylon-66 molecule should contain from 50 to 90 of each of the monomer units. Shorter nylon molecules form weak or brittle fibers.

20.17 POLYESTERS—DACRON

Dacron shares most of the desirable properties of nylon. Most distinctive is its ability to be set into permanent creases and pleats. Dacron is a polyester. It is made by the copolymerization of

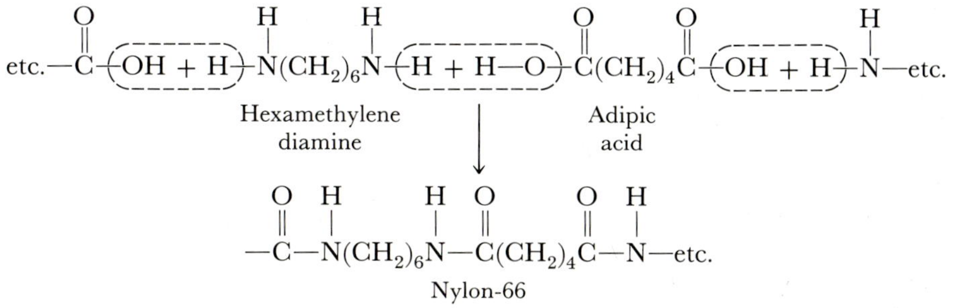

$$\text{etc.}-\overset{O}{\overset{\|}{C}}{-}(\text{OH} + \text{H}){-}\overset{H}{\overset{|}{N}}(CH_2)_6\overset{H}{\overset{|}{N}}{-}(H + H{-}O){-}\overset{O}{\overset{\|}{C}}(CH_2)_4\overset{O}{\overset{\|}{C}}{-}(\text{OH} + \text{H}){-}\overset{H}{\overset{|}{N}}-\text{etc.}$$

Hexamethylene   Adipic
diamine   acid
↓

$$-\overset{O}{\overset{\|}{C}}-\overset{H}{\overset{|}{N}}(CH_2)_6\overset{H}{\overset{|}{N}}-\overset{O}{\overset{\|}{C}}(CH_2)_4\overset{O}{\overset{\|}{C}}-\overset{H}{\overset{|}{N}}-\text{etc.}$$
Nylon-66

ethylene glycol with terephthalic acid. Water molecules split out:

When Dacron is cast as a thin film, it is called Mylar (Figure 20–17).

$$\text{etc.} \rangle\!-\!\overset{\overset{\displaystyle O}{\|}}{C}\!-\!(OH + H)\!-\!OCH_2CH_2O\!-\!(H + HO)\!-\!\overset{\overset{\displaystyle O}{\|}}{C}\!-\!\bigcirc\!-\!\overset{\overset{\displaystyle O}{\|}}{C}\!-\!(OH + H)\!-\!O\!-\!\text{etc.}$$

Ethylene glycol   Terephthalic acid

$$\downarrow$$

$$\text{etc.} \rangle\!-\!\overset{\overset{\displaystyle O}{\|}}{C}\!\left(\!-\!O\!-\!CH_2\!-\!CH_2\!-\!O\!-\!\overset{\overset{\displaystyle O}{\|}}{C}\!-\!\bigcirc\!-\!\overset{\overset{\displaystyle O}{\|}}{C}\!\right)_n\!-\!O\!-\!\text{etc.}$$

Dacron

20.18
SOME SYNTHETIC DRUGS

A *drug* is here defined as a substance, other than a food or a vitamin, intended for the diagnosis, treatment, abatement, prevention, or cure of a disease in humans and other animals. Many drugs still must be obtained from plants or animals; they are too difficult or too costly to make on a commercial scale synthetically from petroleum or coal chemicals. These include *antibiotics* that, by definition, are chemical agents extracted from living organisms such as fungi, bacteria, and other microorganisms and that can inhibit the growth of or even destroy other microorganisms. Antibiotics work against bacterial infections. Penicillin is one of the most common examples. It interferes with the chemical work bacteria must do to make their cell membranes. Other examples of antibiotics are Terramycin, Aureomycin, streptomycin, and chloramphenicol. Antibiotics are part of a larger family of *antimetabolites*—drugs that in general work specifically by inhibiting enzymes that bacteria need for their own lives. An example is the family of sulfa drugs.

FIGURE 20–16
To vulcanize the world's largest tires, three at a time, a five-story "pot heater" is needed, seen here with its hemispherical top removed. (Courtesy Goodyear.)

20.19
SULFA DRUGS

The sulfa drugs function by interfering with an important enzyme system in several disease-causing germs. These germs, unlike human beings, require *para*-aminobenzoic acid, a B-vitamin, to complete the formation of folic acid, an important vitamin itself. (Human beings must apparently obtain folic acid intact in the diet.) Structures of several important sulfa drugs are indicated in Figure 20–18. To affected bacteria these molecules apparently look enough like *para*-aminobenzoic acid to "fool" them into using sulfa drug molecules to make a needed piece for an enzyme. However, the altered enzyme does not work. This means that an important metabolic process in the microbe is inhibited.

20.20
SALICYLATES

The sulfa drugs work to combat causes of disease. The salicylates are a family of drugs that work to ease painful symptoms. They are all derivatives of salicylic acid, which can be made from carbon dioxide and phenol, a coal-tar acid.

Salicylic acid, its esters, and its salts, taken internally, have both an analgesic effect (depressing sensitivity to pain) and an antipyretic action (reducing fever). As analgesics, they act to raise the threshold of pain by depressing pain centers in the brain. Salicylic acid by itself is too irritating to be used internally, but sodium salicylate and especially acetylsalicylic acid (aspirin) are widely used.

Sodium salicylate

Acetylsalicylic acid (aspirin)

Methyl salicylate (oil of wintergreen)

Phenyl salicylate (salol)

Methyl salicylate is used in liniments. Phenyl salicylate has been used in ointments that protect the skin against ultraviolet rays.

20.21 ALKALOIDS

The *alkaloids* are physiologically active compounds, obtained from plants, whose molecules are amines or derivatives of amines. Because they are amines they are basic compounds. (*Alkaloid* means "alkali-like.") They exert their physiological action in one or more ways on the nervous system. Long-term effects may involve other systems in the body. They include drugs of enormous value in relieving human suffering. Morphine (for severe pain) and quinine (antimalaria) are two examples. They also include drugs whose abuse is the cause of enormous human misery, such as cocaine. Heroin is another example, although it actually is a synthetic derivative of a natural alkaloid.

In these sections we shall survey some of the principal alkaloids and, more importantly, see examples of an important fact in drug chemistry and drug application. That fact is that very small variations in molecular structure can give huge changes in drug action. The main strategy for rational drug research (as opposed to the accidental discovery of new drugs) is to use the occasional discovery of a new kind of drug to suggest the new lines of research. The discovered drug may (and usually does) have unfortunate

FIGURE 20-17
Mylar, the film form of the polyester Dacron, has exceptional clarity. In this unretouched photo, the difference in clarity between new and old film is evident. (DuPont Company.)

FIGURE 20-18
Sulfa drugs as antimetabolites. When a microbe selects a sulfa drug to synthesize folic acid, it fashions an altered folic acid that is ineffective. Reactions requiring it do not occur, and the growth of the microbe is inhibited. It suffers, in effect, from a vitamin-deficiency disease. (From J. R. Holum, *Organic Chemistry: A Brief Course,* 1975, John Wiley & Sons, Inc., New York.)

side effects. The research team then asks for substances whose molecules differ from the "parent" in small ways. These are tested by other members of the team. At least two animals have to be used in the testing, and various ways to administer the drug must be tried. Finally human volunteers are sought. Millions of dollars of research go into the development of one new drug, and during the effort scores of compounds may be tried, tested, and abandoned. The team effort brings together not only pharmacologists, but also synthetic organic chemists, toxicologists, zoologists, microbiologists, physicians, medical statisticians, computer programmers, and the like, to say nothing of investment teams.

20.22
β-PHENYLETHYLAMINES

β-Phenylethylamines are borderline members of the alkaloid family; not all are from plants, and the basic amine function is not part of a ring. Nonetheless, they are usually treated in the alkaloid group. Moreover, they illustrate how drug activity can vary dramatically with structure. Important examples are in Table 20.3.

Epinephrine and norepinephrine are two hormones produced in the adrenal gland. They act on the sympathetic nerve fibers to strengthen the heartbeat, raise the blood pressure, and dilate the pupils of the eyes.

TABLE 20.3
SOME β-PHENYLETHYLAMINE DRUGS

The parent molecular framework[a]

Two hormones

(−)-Epinephrine[b]
(adrenaline)

(−)-Norepinephrine
(arterenol)

Two "pep pills"—amphetamines

(±)-Benzedrine
(+)-Dexedrine
("speed")

(+)-Methedrine
("crystal," "meth")

Two bronchial aids

(−)-Ephedrine

(−)-Neo-Synephrine

An antibiotic

Chloramphenicol
(chloromycetin)

A hallucinogen

Mescaline

[a] The drugs shown here are usually obtained as hydrogen sulfate or hydrogen chloride salts.
[b] The plus and minus signs appearing in parentheses before some of the names designate that the molecules of the named substance have a particular kind of "handedness," as defined in Section 18.9. One form can sometimes be as much as 20 times as effective as the other.

The natural actions of these two hormones, which are released in times of stress, fright, or awareness of danger, are to increase your alertness and make you feel less fatigued for a while. In short, they help get you ready to handle emergencies or other kinds of stress, such as depression. Some drugs mimic these effects. Unfortunately, especially when abused, they also bring on sleeplessness and increased irritability followed, ironically, by mental depression and fatigue. The *amphetamines* are notorious examples. Medically, they are classified (and prescribed) as stimulants and antidepressants. Sometimes they are prescribed in weight-control programs or to treat certain mild mental disorders. While they are legally available only by prescription, they are available illegally in enormous amounts. Billions of amphetamine pills are made annually. Known generally as "pep pills" or "uppers," individual compounds are called "bennies" (Benzedrine), "dexies" (Dexedrine), or "speed" (methedrine) (see Table 20.3). The dangers of abuse include suicide, hostility and belligerence, paranoia and hallucinations, and hepatitis. Some scientists consider them vastly more dangerous than heroin.

Also included in Table 20.3 are substances long used to treat bronchial asthma. Ephedrine, obtained from a Chinese herb, is one. Neo-Synephrine, used in nose drops, is synthetic.

Mescaline is isolated from the mescal button, a growth on the top of the peyote cactus. Indians of northern Mexico and the southwestern United States have long used it in religious ceremonies. Medically, it is one of the family of *hallucinogens,* substances that cause illusions of time and place, make unreal experiences or things seem real, or distort the qualities of objects.

Chloromycetin (chloroamphenicol), also in Table 20.3, is unusual in that it contains both a nitro group and chlorine, neither of which had ever been observed before in naturally occurring organic compounds. It is an antibiotic effective against Rocky Mountain fever and typhus.

20.23
OPIUM ALKALOIDS

When unripe seed capsules of the opium poppy (*papaver somniferum*) are scored with a knife, a milky sap oozes out and dries. Roughly 10 percent is morphine. Medically, morphine is used as a painkiller, and for severe pain there is essentially nothing as effective. Overuse unfortunately leads to addiction. Codeine is another (but trace) alkaloid in poppy sap. It is used as a cough suppressant in cold medicines.

Morphine

Codeine

Heroin

Heroin is made from morphine. Known by a variety of slang names—"horse," "H," "smack," "junk," and "dope," to name just a few—it has no legal use in the United States. While it produces a brief period of euphoria, it also causes lethargy, confusion, nausea, constricted pupils, and respiratory depression. Those who become addicts face great difficulties in stopping. The controversial methadone treatment program substitutes a legal drug for an illegal one, a drug, so it is claimed, that blocks out the heroin addict's hunger for heroin and a drug that can be taken orally.

Methadone

20.24 COCA ALKALOIDS

The leaves of the coca bush found in the Andes Mountains of Peru and Bolivia contain cocaine ("snow," "coke," "C"). It has apparently been a native practice for centuries to chew these leaves for their local anesthetic action as well as to promote physical endurance. Overconfidence in one's abilities often results, and large doses cause hallucinations, long periods of depression, and

Cocaine

sometimes convulsions and death. In some addicts, antisocial acts become violent and murderous. The image some people have of the drug addict as a dangerous maniac probably originated when cocaine became one of the first drugs to be widely abused in nineteenth-century Europe. Abusers of cocaine sometimes went on maniacal binges of killing.

20.25 ERGOT ALKALOIDS

Various amides of lysergic acid are obtained by extracting the ergot fungus that grows on diseased rye grain. Because they possess some medicinal properties, these derivatives of lysergic acid have been extensively studied by drug companies. A. Stoll and A. Hofmann of Switzerland prepared a large number of derivatives of lysergic acid to test as therapeutic agents. Among them was the diethylamide, LSD (from the German, "*l*ysergs*a*ure*d*iathylamid"). In 1943 Hofmann accidentally discovered that LSD is a powerful hallucinogen. A speck of a few micrograms in size can produce the trancelike state reported by users. Just a trace more, still less than a grain of salt, can cause severe psychotic reactions that lead the user to self-destructive acts. How it works, biochemically, is not known.

Lysergic acid diethylamide
LSD

20.26 MARIJUANA

The active principle in marijuana is not an alkaloid nor does it have an amine group, but because it is a mild hallucinogen, we include a brief mention of it here. Marijuana is made by chopping the dried leaves and flowers of the Indian hemp plant, *cannabis sativa*. The active principle is THC or tetrahydrocannabinol.

Tetrahydrocannabinol
THC

Marijuana or its related preparations—hashish, dagg, bhang, ganga, charas, etc.—is the most widely used illicit drug in the world. Widespread disagreement exists over the wisdom of making it legal. Some facts are well established. THC or its metabolites persist in circulation far longer than those of alcohol. Marijuana is not addictive; it is not a narcotic (except that it is defined that way in federal law); it does not lead to an increase in sexual activity. Physiologically, its biochemical workings do not create a craving for heroin. It creates fantasies of increased creativity but not actual increases. Whether it is no worse than alcohol (which is bad enough when abused) has not been established to the point of widespread acceptance among physical and medical scientists. Low doses of either alcohol or marijuana have a sedative-depressant action on the central nervous system, including muscular weakness, poor coordination, and loss of balance. Overdoses of marijuana have been reported to cause depression, paranoid reactions, catatonic excitement, and various hallucinations.

SUMMARY

Synthetics. Low-formula-weight petrochemicals—compounds wrung from petroleum, natural gas, or coal—are the raw materials for nearly all synthetics: fibers, plastics, dyes, drugs, and even special fuels. The chief petrochemicals are alkanes and alkenes of one to four carbons.

Fossil fuels. Petroleum is mostly a mixture of hydrocarbons together with some water and natural gas. Crude oil is what is left when water and natural gas are removed from petroleum. Natural gas is mostly methane. Coal consists mostly of carbon but includes many compounds.

Petroleum refining. At refineries crude oil is distilled to separate it into various fractions that are commercially valuable in various engines or as petrochemicals. Some fractions are "cracked" to yield synthetic gasoline. Other fractions can be reformed into better gasoline.

Gasoline. A mixture of hydrocarbons whose molecules have from 5 to 10 or 12 carbons has the volatility most appropriate for gasoline engines. To prevent or minimize engine pinging or knocking, their molecules should be highly branched. Otherwise, tetraethyllead or aromatic hydrocarbons can be added to improve the octane rating. Straight-chain heptane causes the most knocking in the test engines used and has an octane rating of zero. Isooctane is the standard for good performance and has an octane rating of 100. Mixtures of these two give standards for intermediate octane ratings.

Polymers. Molecules that can react with each other hundreds of times to produce molecules with repeating structural units, polymers, are themselves called monomers. If monomers simply add to each other, the reaction is called addition polymerization. All the polyolefins, the "vinyls," and the polydienes (such as rubber) are addition polymers. Condensation polymers such as nylon and Dacron are formed when a small molecule such as water splits out as the monomer units react. Nylon is a polyamide; Dacron is a polyester.

Copolymers. When two or more monomers are mixed and allowed to react, the copolymer that forms often has properties superior to other polymers. Elastomers are an example; they are synthetic copolymers with the elastic properties of rubber. Rubber and similar polymers lose some of their elasticity but gain some resistance to wear and tear when they are vulcanized. This process places cross links between polymer chains.

Drugs. Some drugs are synthetics and some are obtained from nature directly. Antibiotics are specifically obtained from living sources and are examples of antimetabolites—drugs that act specifically on enzyme systems in bacteria. The sulfa drugs are synthetics that combat bacteria. The salicylates (e.g., aspirin) are also synthetics, but they act to relieve symptoms not cause cures.

Alkaloids. Many nitrogenous, alkali-like compounds called alkaloids are obtained from natural plant sources and have unusual physiological activities. Morphine is still the best single agent to use against severe pain. Not all alkaloids are useful drugs, and many are very dangerous.

TERMS AND CONCEPTS

synthetic
petroleum
coal
lignite
bituminous coal
anthracite
natural gas
fossil fuel
oil
refining
gasoline
cracking
octane number
coke
plastic
resin
plasticizer
polyolefin
vinyls
addition reaction
addition polymer
copolymer
elastomers
vulcanization
condensation polymer
drug
antibiotic
antimetabolite
alkaloid
amphetamines
hallucinogen

QUESTIONS AND EXERCISES

1. What is the difference between petroleum and crude oil?

2. What are the principal differences between anthracite and bituminous coal?

3. What is "cracking" as applied to petroleum refining and processing?

4. What is the difference between thermal and catalytic cracking?

5. Why, in general, is it desirable to have the more highly branched alkane molecules in gasoline instead of the less branched molecules?

6. What is "octane number"?

7. Name three important polymeric materials that are dependent on ethylene as a raw material.

8. Name six polymeric materials dependent on propylene as a raw material.

9. Write an equation that illustrates each reaction: (a) thermal cracking of hexadecane, (b) catalytic cracking of hexadecane, (c) reforming of *n*-octane.

10. If a polymer is to be a successful fiber, what structural characteristics should its molecules have?

11. What is the difference between a resin and a plastic?

12. What is the difference between a plastic and a polymer?

13. What is an elastomer?

14. What gives polyethylene or polypropylene its exceptional resistance to chemical attack by substances in the environment?

15. What is the difference between a polymer and a copolymer?

16. What is the difference between an antimetabolite and an antibiotic?

17. In general terms, how does penicillin work?

18. In general terms, how do the sulfa drugs work?

19. What is the name of the principal, active substance in each of the following: (a) coca alkaloids? (b) ergot alkaloids? (c) opium alkaloids?

GLOSSARY

Absorptivity is the ratio of the light energy that will be absorbed by an object per unit area in unit time to the light energy that a black body will absorb under the same conditions.

Acid, in the Arrhenius view is any substance that can furnish hydrogen ions in water. Any substance that can neutralize an alkali (base).

Actinide is any element of atomic number 89 through 103.

Addition reaction, in organic chemistry, is a reaction in which one reagent adds to a double or triple bond of another reagent. Addition polymerization occurs when monomer molecules add to each other and no molecular fragments break out (in contrast to so-called condensation polymerization).

Adiabatic change occurs in a system without any transfer of heat or other energy into or out of the system.

Aerosol is any suspension of finely divided liquid droplets in air. In commerce, "aerosol" often refers to the propellant in an "aerosol spray can" and is frequently (but not always) Freon.

Alcohol is any organic compound possessing the -OH group attached to a saturated carbon atom.

Aldehyde is any organic compound in which the $\overset{O}{\underset{\parallel}{-C}}-H$ group is attached either to hydrogen or to a carbon atom of the same group.

Aliphatic compound is any organic compound that either is or may be regarded as a derivative of an open-chain or cyclic alkane, alkene, or alkyne; any organic compound without the benzene ring system or one like it.

Alkaloid is a physiologically active compound obtained from plants whose molecules include the amine functional group and are therefore basic.

Alkane is any hydrocarbon with only single bonds, cyclic or open-chain.

Alkene is any hydrocarbon with at least one double bond, open-chain or cyclic.

Alkylation is an operation in petroleum refining whereby molecules too small to be of much use in gasoline (because their compounds are too volatile) are converted into larger molecules.

Alkyne is any hydrocarbon with at least one triple bond.

Allotropy is the phenomenon of the existence of an element in two or more crystalline forms. Each form is called an allotrope of the other. Diamond, for example, is one allotrope of carbon, and graphite is another.

Alloy is a metallic substance consisting of an intimate mixture or solution of two or more elements (at least one being a metal) prepared by stirring the molten components together.

Alpha helix is an important kind of gross shape or form of a protein; a twisting of a long protein molecule into a helix.

Amide is any organic compound with a carbonyl-nitrogen system of $-\overset{O}{\underset{\parallel}{C}}-\overset{|}{N}-$

Amine is any organic compound in which a trivalent nitrogen atom is attached to one, two, or three carbons either of the alkyl group type or of a benzene ring.

Amino acid is an organic compound whose molecules have both an amino group and a carboxylic acid group. In all of the amino acids involved in protein structure, these two groups are attached to the *same carbon;* they are, in other words, alpha amino acids.

Amorphous is having no regular crystalline shape and therefore not giving a definite x-ray diffraction pattern.

Analgesic describes a drug that deadens pain without producing loss of consciousness, such as aspirin.

Animal fat is a mixture of triglycerides obtained from an animal source. Examples are lard, tallow, and butter.

Anode is an electrode to which negatively charged ions will be attracted because it bears a deficiency of electrons.

Anthracite is a type of coal without water and over 95 percent fixed carbon. It is hard coal—a black, lustrous solid with little or no loose soot.

Antibiotic is an antimetabolite made by and isolated from microorganisms.

Antimetabolite is an agent in chemotherapy that inhibits the growth of bacteria.

Antipyretic describes a drug such as aspirin that allays, reduces, or prevents fever.

Archimedes principle. The buoyant force exerted by a fluid on a submerged object equals the weight of the fluid that the object has displaced. For objects that do not become completely submerged, this means that the floating object displaces a volume of the liquid equal to its own weight.

Aromatic compound is any organic compound that contains the benzene ring system or something like it, such as naphthalene.

Aromatization is an operation in petroleum refining in which saturated cycloalkanes are made to lose hydrogen to become aromatic hydrocarbons; these improve octane rating.

Arrhenius theory explains how acids in water have so many common properties because they all release hydrogen ions and how bases in water also have many common properties because they all release hydroxide ions.

Atmosphere is the gaseous envelope of the earth; also, wherever dry component gases are found.

Atom is an exceedingly small particle having only one nucleus and no electric charge. It is the smallest particle of any element that can be said to be unique for that element.

Atomic number is the electric charge on an atom's nucleus and is therefore the number of protons possessed by the atom, as well as the number of electrons.

Atomic weight is, in atomic mass units, the sum of the number of protons and neutrons in an atom.

Aufbau rules for reconstructing an electronic configuration of an atom of an element given only its atomic number: (1) electrons go into the lowest energy level available, (2) electrons tend to become distributed among different orbitals of the same general energy level insofar as room is available (Hund's rule), (3) no more than two electrons may occupy the same space orbital at the same time, and two may be present only if they have opposite spins (Pauli exclusion principle).

Avogadro's number is the number of formula units (whether those units be atoms, molecules, ion groups, etc.) in one mole of a substance and is equal to 6.023×10^{23}.

Base is any substance, in the Arrhenius view, that releases hydroxide ions in water; also, any substance that can combine with an acid to form a salt or that can neutralize an acid.

Basic oxygen process is a modern method of making steel from pig iron in which nearly pure oxygen is blasted through molten pig iron to change impurities to forms that can be removed.

Biodegradable describes the property (often used in connection with detergents) of being broken down into small molecules by microbial action in soil, septic tanks, and sewage treatment plants.

Biosphere is the sum total of all regions of the planet where living things are found.

Bituminous coal is less than 5 percent water but roughly 30 percent volatile organic material; soft coal.

Black body is an object that cannot reflect or transmit light. It can only absorb and emit electromagnetic radiation. The cavity within a hollowed-out, opaque object having only a small pinhole to the outside behaves like a black body.

Blast furnace is a tall, somewhat cylindrical structure used in converting iron ore to pig iron. Coke is burned to hot carbon monoxide, which reduces iron oxides to molten iron. Limestone is added to combine with impurities such as sand and similar silicaceous matter to form a molten slag that can be separated from molten iron.

Bohr model of the atom is a visual picture based on the assumptions of Niels Bohr: that electrons in an atom are confined to definite, particular energy states or levels and that if an electron moves from one level to a higher one, this means that the atom has absorbed an amount of energy exactly equal to the difference of the two states.

If an electron drops to a lower level, it will emit (sometimes as visible light) the energy corresponding to the difference. Bohr's model is the solar system model—the nucleus is like the sun; the electrons are like the planets, and they move in definite orbits.

Boiling is a turbulent phenomenon in a liquid when its vapor pressure equals the atmospheric pressure and the liquid absorbs heat without undergoing a change in temperature.

Bond angle is the number of degrees of the angle formed by the lines connecting two atoms to a third atom in a molecule.

Bond, chemical, is a net force that prevents the atomic nuclei within compounds from repelling each other to infinity. We recognize three different kinds of organization of atomic pieces: organizations in molecular compounds, in ionic compounds, and in metals.

Branched chain, in organic chemistry, is a feature of molecular structure in which one or more carbons is attached at points other than the ends of a continuous sequence of carbon atoms covalently bound together.

Brass is an alloy of copper and zinc, with zinc about 20–45 percent.

Bronze is an alloy of copper and tin, with tin about 5–10 percent. Coinage bronze usually contains a trace of zinc.

Brownian movement is the random, chaotic movement of particles in a colloidal dispersion seen with a microscope.

Buoyancy is the property of floating or tending to be supported by a fluid, either liquid or gas.

Buoyant force is the upward force exerted by a fluid (liquid or gas) on an object suspended or floating in the fluid.

Carbohydrate is any naturally occurring substance with the properties of polyhydroxyaldehydes or polyhydroxyketones or substances that by simple hydrolysis will yield these.

Carbon cycle is the cycling of carbon nuclei in nature from carbon dioxide, carbonate sediments, and fossil fuels through bioorganic chemicals in living systems and back again to carbon dioxide, carbon sediments, and (exceedingly slowly) fossil fuels. The carbon cycle is intimately involved via photosynthesis with the oxygen cycle.

Carbon family is Group 4 in the periodic table—carbon, silicon, germanium, tin and lead.

Carbon ring, in organic chemistry, is a feature of molecular structure in which three or more carbon atoms are joined in a closed loop.

Carbonization is an operation converting coal to coke and extracting from the coal various mixtures of chemicals in coal oil, coal tar, and coal gas.

Carbonyl group is the carbon-oxygen double-bond system, $>\!\!C\!\!=\!\!O$.

Carboxylic acid is any organic compound containing the carboxyl group -CO_2H.

Catalysis is the phenomenon of the increase in the rate of a chemical reaction that is brought about by the presence of some usually small amount of chemical—the catalyst—that can be recovered at the end of the reaction.

Catalyst is a substance that in relatively small amounts can accelerate a chemical reaction without being permanently changed in a chemical sense itself. The catalysts present in all living things are called enzymes.

Cathode is an electrode to which positively charged ions will be attracted because it bears an excess of electrons.

Cellulose, the constituent of cotton fibers, is a linear polymer of beta-glucose; it also occurs in woody plants.

Chain (carbon), in chemistry, is a feature of molecular structure characterized by two or more carbon atoms joined together by covalent bonds, one following another like links in a chain.

Chemical energy is the energy potentially available from chemicals that can undergo chemical reactions.

Chemical family is a main group (vertical column) of elements in the periodic table whose atoms are characterized by having identical outside-level configurations. Examples are the alkali metals (Group 1), alkaline earth metals (Group 2), the halogens (Group 7), and the noble gases (Group 0).

Chemical property is any one of the chemical reactions that a particular substance is known to undergo.

Chemical reaction is one event that can happen in nature and whose unique characteristic is that a rearrangement or redistribution of electrons relative to nuclei takes place and new chemical substances are formed.

Chemistry is the study of the nature of matter, its properties, structure, and changes.

Chirality, in molecular or crystal structure, is the possession of handedness such that the mirror image of the structure does not superimpose with it.

Chlorophyll is green pigment in plants that absorbs sunlight and changes it into chemical energy for use by the plant in photosynthesis.

Coal is a black or brownish-black or brown combustible solid dug from the ground consisting of the altered remains of vegetable material transformed over long periods of time in oxygen-poor conditions. Coal is a fossil fuel.

Coal oil is a liquid mixture of chemicals obtained when soft coal is carbonized; it provides several compounds for making synthetics.

Coal smog is the type of smog that forms in an urban industrialized area from the burning of sulfur-laden soft coal or petroleum. The primary pollutants are sulfur oxides and particulates with some carbon monoxide.

Coal tar is a thick, viscous mixture of chemicals obtained when soft coal is carbonized; it is used in the chemical industry.

Coefficient, in chemistry, is a small whole number placed in front of the formula of a substance in a chemical equation to indicate, together with other coefficients, the relative amount of that substance (by formula units or by moles) that interacts in the reaction being symbolized by the equation.

Cofactor is a nonprotein substance essential to the completion of the synthesis of an enzyme or to its ability to function; not all enzymes need cofactors.

Coke is the porous black residue from carbonizing coal. Almost entirely carbon, it is used as a fuel and as a reducing agent in metal refining.

Colloidal dispersion is a reasonably stable, uniform distribution in some dispersing medium of particles of colloidal dimension; that is, having at least one dimension 1 to 100 nanometers—next up in size from an ion, an atom, or a small molecule.

Colloidal state is a state of subdivision of any matter into particles having at least one dimension on the order of 1 to 100 nanometers. Smoke particles are in the colloidal state. Some substances consist of molecules large enough to be in the colloidal state—e.g., gelatin.

Combustion is a rapid chemical change involving oxygen producing heat, light, and usually sound in addition to gaseous products.

Compound is a substance made up of the pieces of the atoms of two or more elements; in a fixed, definite ratio, the pieces of atoms (electrons and nuclei) are organized either into molecules or into oppositely charged ions.

Concentration is the amount of some substance present in a unit of volume of a solution or colloidal dispersion.

Condensation is the change of vapor to its corresponding liquid or a liquid that has freshly formed from a vapor. In organic chemistry, condensation is a reaction between two or more large molecules accompanied by the splitting out of a small molecule such as water.

Condensation polymerization occurs when molecules of different monomers react and in the process expel a small molecular fragment, such as a water molecule. Nylon and Dacron are formed by condensation polymerization.

Conduction, in the physics of heat, is the transfer of heat by the transfer of kinetic energy from ions, atoms, or molecules to neighbors.

Convection is the transfer of heat by the circulation of a warmer fluid stream throughout the rest of the fluid.

Coordination compound is the product of the interaction between a metallic ion and a ligand.

Coordinate covalent bond is a bond in which both electrons in the shared pair were donated by one of the atoms involved in the bond.

Copolymer is a polymer formed by additional polymerization of at least two monomers.

Covalence is the number of electron-pair (covalent) bonds an atom can have when used to make a molecule.

Covalent bond is the net force of attraction that arises when two (or sometimes more) atomic nuclei share a pair of electrons. In a single bond, one pair of electrons is shared; two are shared in a double bond; three in a triple bond.

Cracking is an operation in refining petroleum where large hydrocarbon molecules are broken up into smaller ones. High temperature alone is used in thermal cracking; high temperature and a catalyst are used in catalytic cracking, which produces higher-octane gasoline.

Crick-Watson theory uses the double-helix structure of DNA in genes to explain how genes store genetic information and how this information is translated into sequences of amino acids in proteins.

Crude oil is what is left after water and natural gas are removed from petroleum.

Crystalline is having a regular arrangement of atomic nuclei in a characteristic latticelike array and therefore giving a regular x-ray diffraction pattern.

Cycloalkane is a hydrocarbon whose molecules include at least one ring of carbon atoms but has only single bonds.

Dacron is a polymer made of units from terephthalic acid and ethylene glycol.

Dalton's theory is an atomic theory of matter proposed by John Dalton to account for the laws of chemical combination. It postulates that matter consists of atoms that are indestructible and that all atoms of one element are identical in weight and other properties. When a compound forms, a definite but small number of atoms of each of its constituent elements join to make up the fundamental particles of the compound.

Decay, in the biosphere, consists of slow chemical changes in dead matter that return organic and inorganic material to simple substances. In radioactivity, decay is a phenomenon whereby radioactive elements lose small particles and energy and change to other elements. In geology, decay means chemical changes in rocks producing smaller particles sometimes of different chemical composition from the original rock.

Denaturation, in protein chemistry, is an event whereby some agent—physical or chemical—causes a protein molecule to lose its natural shape and form without actually being hydrolyzed. Denaturation is usually accompanied by an irreversible loss of solubility and of biological function. An agent that can do this is a denaturing agent.

Denitrification is the reduction of nitrate or nitrite ions to ammonia or to molecular nitrogen by soil bacteria.

Density is the ratio of the mass of an object to its volume or the mass per unit volume with the usual units being grams per cubic centimeter.

Deoxyribonucleic acid (DNA) is the chemical of a gene; a polymer in which the repeating units are contributed by four nucleotides, each carrying one of four heterocyclic amines—adenine (A), thymine (T), guanine (G), and cytosine (C).

Dessicant is a substance that reduces the concentration of water vapor in an enclosed space by combining with the water to form a hydrate.

Detergent is anything used to make water a better cleansing agent.

Digestion is the set of chemical reactions in the body whereby molecules in foods are broken down into small molecules that are then absorbed into the bloodstream.

Disaccharide is a carbohydrate that can be hydrolyzed to two monosaccharide units.

DNA. See deoxyribonucleic acid.

Double helix, in the chemistry of heredity, is the structural arrangement of two DNA strands directed in opposite ways and intertwined in a right-handed corkscrewlike helix with the two strands held together by hydrogen bonds between appropriate pairs of heterocyclic amines.

Drug is a substance, other than a food or vitamin, intended for the diagnosis, treatment, abatement, prevention, or cure of a disease in humans and other animals.

Ductility is the physical property of being capable of being drawn out permanently into a longer form, as into a wire.

Elastomer is a polymer with elastic properties.

Electrochemistry is a branch of chemistry that studies how electricity can be produced by chemical reaction or how electricity causes chemical reactions.

Electrode is a solid object, usually a wire, suspended in an electric conducting medium through which electricity passes to or from the external circuit.

Electrolyte is any substance whose solution in water will conduct an electric current.

Electrolytic refining is a method of purifying metals by using the impure metal as one electrode and transferring it to the other electrode by electrolysis.

Electron gas model (electron sea model) is the model of the structure of a metal that postulates that atomic nuclei in a metal are fixed in a sea (or a gas) of outer-level electrons moving in multicentered molecular orbitals formed when partially filled d- (or f-) orbitals of the metallic atoms overlap.

Electronegativity is the ability of an atom joined to another by a covalent bond to attract toward itself the electrons of that bond.

Electronic configuration is the most stable (i.e., the lowest-energy) arrangement of the electrons of an atom, ion, or molecule. There are only 105 fundamentally different kinds of electronic configurations among nature's atoms—one for each of the 105 elements.

Electroplate is to cause a metal to be deposited onto an electrode by passing an electric current through a solution containing ions of that metal.

Electropositivity is a tendency of an atom in a molecule to become positive, an ability of the atom to release electrons or electron density from itself.

Electrovalence is the electric charge (including its sign) that an atom can have by changing into an ion.

Element is one of the three broad classes of matter (besides compound and mixture). All the atomic nuclei within a sample of matter classified as an element have identical electric charges. (Because of the existence of isotopes, the atoms in an element may have small differences in their masses.)

Emissivity is the amount of energy emitted by an object per square centimeter of its surface area per second—the emitting power of the object.

Emulsion is a colloidal dispersion of tiny microdroplets of one liquid in another liquid.

Endergonic describes requiring energy.

Endocrine gland is one of a small number of highly specialized organs that make hormones and participate in controlling their secretion into the bloodstream.

Endothermic describes requiring heat.

Energy of activation is the minimum amount of energy that must be received by particles (ions, atoms, or molecules) engaged in a potential chemical reaction before they will actually undergo that reaction.

Energy level is a general region of space near an atomic nucleus where an atom's electrons exist. A low energy level is one close to the nucleus, and a high energy level is one farther away. After the lowest level, the first level (principal quantum number, $n = 1$), succeeding energy levels possess sublevels that are lettered *s, p, d,* and *f*.

Enzyme is a substance in living organisms that acts as a catalyst. These materials are wholly or mostly proteins whose syntheses are under the guidance of genes.

Equation (chemical) is the shorthand representation of a chemical reaction employing the correct formulas for the reactants and products and using the smallest whole numbers as coefficients, put in front of the formulas, to reveal the proportions of the chemicals involved. An equation is balanced when all atomic nuclei among the reactants are accounted for among the products.

Equilibrium is the situation when two opposing events are occurring at the same rates, resulting in no further net change.

Ester is any organic compound with a carbonyl-oxygen-carbon system:
$$-\overset{\overset{\displaystyle O}{\|}}{C}-O-\overset{\displaystyle |}{\underset{\displaystyle |}{C}}-$$

Ether is any organic compound possessing an oxygen attached to two saturated carbon atoms.

Excited state is any electronic configuration, in atomic or molecular structure, other than the ground state.

Exergonic describes liberating energy.

Exothermic describes evolving heat.

Fatty acid is any monocarboxylic acid usually but not necessarily with a long carbon chain and obtained from a natural product (e.g., by hydrolysis of a fat or oil); it may have one or more carbon-carbon double bonds.

Flotation (froth flotation) is a method in the refining of ores whereby the ore is slurried with water containing special chemicals while air or some other gas is bubbled through. The foam and froth attract either the desired mineral or the waste material and separate it.

Formula (chemical) is the shorthand symbol of a substance written by combining the abbreviations of the constituent elements in the correct number and proportion to show the substance's elementary composition.

Formula unit is the smallest piece of any pure substance (i.e., of any compound or element) that can be imagined to have at least the chemical properties of the compound or element or that, at least in principle, could be capable of an isolated existence. A formula unit has the composition given by the formula for the pure substance. The molecule is one kind of formula unit, and the pair or the small cluster of ions represented in the formula of an ionic compound is another. The atom is a formula unit of an element.

Formula weight is the sum of the atomic weights of all the atoms represented in the chemical formula of a compound or element.

Fossil fuel is petroleum, coal, natural gas, or the petroleum-like materials in oil or tar sands and oil or tar shales. Fossil fuels were formed in an age and under conditions similar to those producing fossils.

Free rotation is the absence of any insurmountable barrier to the rotation of two groups with respect to each other when they are joined by a single covalent bond in an open-chain molecular environment.

Freon is one of a family of easily vaporized liquids whose molecules have one, two, or just a few carbon atoms combined with varying numbers of fluorine and chlorine atoms. (Freon is not to be confused with fluorocarbon, a substance that by definition consists only of fluorine and carbon. Most Freons contain chlorine.) The two most commonly used propellant Freons are Freon-11 ($CFCl_3$) and Freon-12 (CF_2Cl_2).

Functional group is a feature of molecular structure—that part of a molecule that displays a characteristic set of reactions displayed by all substances whose molecules possess it. Examples are the alcohol group, -OH, and the carbon-carbon double bond.

Fusion is the transition of a solid to a liquid. In atomic physics, fusion is the merging of two atomic nuclei into a larger one with the liberation of energy.

GLOSSARY

Galvanic action is the corrosion that occurs when two different metals are in contact with each other and with a conducting medium.

Gas is a state of matter in which a substance has no definite shape or volume unless confined; where the individual particles move about most freely, collide often, but are not generally in constant physical contact with their neighbors.

Gasoline is a liquid mixture of hydrocarbons, mostly branched-chain alkanes, that boils in the range 40–200°C. When sold as fuel for engines, gasoline often contains tetraethyllead ("lead") to improve its octane.

Gel is a colloidal dispersion of a solid in a liquid that has assumed a semisolid, semirigid form, such as gelatin desserts.

Gene is a hereditary unit carried within a cell on its chromosomes and consisting of DNA.

Genetic code is the sequence of triads (groups of three in a row) of heterocyclic amines arrayed along the "backbone" of a DNA strand in a gene.

Geometric isomerism is isomerism caused by differences in the geometries of the isomers, differences that cannot be erased because of the lack of free rotation about either a double bond or atoms joined as part of a cyclic portion or ring in a molecule.

Glycogen is the starchlike polymer of alpha-glucose that is an animal's means of storing glucose units.

Gram-formula weight is the practical reacting unit of any pure substance, whether an element or compound; the quantity of the substance equal to its formula weight taken in units of grams.

Ground state, in atomic or molecular structure, is the electronic configuration corresponding to the lowest total electronic energy.

Group, in chemistry, is a vertical column in the periodic table or one family of elements.

Hallucinogen is a substance that causes illusions of time and place, makes unreal experiences or things seem real, or distorts the qualities of objects. LSD and mescaline are examples.

Heat capacity (specific heat) is the ability of a substance to absorb heat; specifically, the number of calories of heat that one gram of a substance can absorb for every degree change in Celsius temperature. Heat capacity = $\frac{\text{calories}}{\text{gram} \times °C}$

Heat of fusion is the number of calories that one gram of a solid will absorb in going from the solid to the liquid state without changing its temperature. (These calories are released in the reverse change, from liquid to solid.)

Heat of vaporization is the number of calories that one gram of a liquid will absorb in going from the liquid to the gaseous state without changing its temperature. (These calories are released in the reverse change, from gas to liquid.)

Heredity is the total of the qualities, capabilities, and potentials of an individual organism that were received from its ancestors by means of the chemical carriers called genes.

Homogeneous means everywhere and in every way alike.

Hormones are chemical messengers made in specialized tissues—the endocrine glands—and secreted by those tissues in response to some stimulus; they are carried by the bloodstream to their respective "target organ," where they trigger particular responses.

Hund's rule is that electrons tend to become distributed among different orbitals of the same general energy level insofar as room is available.

Hydrate is a chemical compound in which intact molecules of water are entrapped within the crystalline lattice in a definite ratio with the other components.

Hydration, in chemistry, is the association of molecules of water with dissolved ions or polar molecules helping them to remain in solution.

Hydrocarbon is any compound consisting entirely of hydrogen and carbon.

Hydrogen bond is a force of attraction between opposite but only partial charges—$\delta+$ on a hydrogen bound to oxygen or nitrogen, $\delta-$ on a nearby oxygen or nitrogen. A secondary structural feature in proteins that helps to stabilize the gross shape and form of a protein molecule; it is easily disrupted by denaturing agents.

Hydrogen ion is, strictly speaking, nothing but the bare proton, H^+, but in common use the term is often used when *hydronium ion* is meant.

Hydronium ion is the ion, H_3O^+, that forms when a water molecule accepts a bare proton.

Hydrosphere is the sum total of all parts of the earth where water is found in any of its states—solid, liquid, or vapor.

Hydrostatic pressure is the pressure exerted by a liquid or within a liquid by virtue of the force per unit area generated by the weight of the liquid. It is directly proportional to depth.

Indicator is a dye that has one color in an acidic medium and a different color in a basic medium. Litmus is an example.

Inorganic describes any substance of mineral origin, as distinguished from plants or animals—common acids, bases, salts (including carbonates and cyanides), metals, nonmetals, rocks, ores, and the like; anything that is not organic.

Inversion (thermal) is a condition of the atmosphere in which the air temperature increases with increasing altitude, at least for a distance, until the inversion layer is reached; this is the narrow band of air in the atmosphere in which the temperature switches back to its normal decrease with increasing altitude.

Ion is an electrically charged particle having one or a few atomic nuclei and either one or two (seldom three) too many electrons or too few electrons to render the particle electrically neutral. Ions of opposite electric charge are the building blocks of ionic compounds. The ion group corresponding to the formula of the ionic compound is the smallest representative sample of that substance. The polyatomic ion contains two or more atomic nuclei that are held together by covalent bonds, but the cluster as a whole bears a small net electric charge.

Ionic bond is the force of attraction between oppositely charged ions in an ionic compound.

Ionic compound is a substance made up of a more or less orderly aggregation of oppositely charged ions occurring in a definite ratio that ensures that all of the electric charges on all of the assembled ions "cancel out" and the substance is electrically neutral.

Ionization is a reaction, usually with solvent molecules, whereby molecules break apart into ions.

Isomer is one of a set of compounds having identical molecular formulas but different structures.

Isomerism is the phenomenon of the existence of two or more compounds having identical molecular formulas but different structures.

Isothermal change is a constant temperature change in a system. Heat flows in or out of the system in whatever amounts are needed to keep the temperature constant while the system changes in some other way.

Isotope. Two (or more) atoms are said to be isotopes if their nuclei have identical nuclear (positive) charges but they differ slightly in their numbers of neutrons (and therefore their atomic masses). Isotopes have identical electronic configurations and therefore essentially identical chemical properties.

Kernel (atomic) is an atom minus its outside shell electrons.

Kerosene is a mixture of compounds, mostly hydrocarbons, obtained from petroleum; it boils in the range 175–325°C.

Ketone is any organic compound in which the carbonyl group is attached directly to two carbon atoms:

$$-\overset{|}{\underset{|}{C}}-\overset{O}{\underset{\|}{C}}-\overset{|}{\underset{|}{C}}-$$

Lanthanide (rare earth) is any element of atomic number 57 through 71.

Latent heat is heat that appears to be hidden in some way in a substance and that will be released by the substance when it goes through a change of state on cooling. See also *heat of fusion* and *heat of vaporization*

Law of conservation of mass. Matter is neither created nor destroyed in chemical reactions; the mass of all products equals the mass of all reactants.

Law of definite proportions. The elements that make up a compound occur in it in definite proportions by weight and therefore definite proportions by atoms.

Law of multiple proportions. When two elements combine to form more than one compound, the different weights of one that combine with the same weights of the second are in the ratio of small whole numbers.

Le Châtelier's principle is that if a system is in stable equilibrium and something happens to change the conditions, the system will shift in whatever way will most restore the equilibrium.

Lewis octet theory, named after Gilbert Norton Lewis, exploited an octet rule to explain the combining abilities (valences) of atoms of elements and postulated that outside shells of different atoms could interpenetrate and thereby achieve outer octets by a sharing process that resulted in covalent bonds.

Ligand is a molecule or a polyatomic ion that can tie up a metallic ion by means of a coordinate covalent bond. The tripolyphosphate ion is the metal ion binder in detergents containing "phosphates."

Lignite is a type of coal containing over 40 percent water and therefore having lower heating value than most other forms of coal.

Lipid is a plant or animal product that is soluble in a nonpolar solvent such as ether, carbon tetrachloride, or benzene.

Liquid is a state of matter in which the substance has no definite shape (except that of the container when it is contained) but does have a definite volume. Individual particles move about freely, collide often, and generally are always in physical contact with neighboring molecules, but not for long with the same molecules.

Lithosphere is the sum total of all solid mineral portions of the earth.

Lock and key theory explains the specificity of an enzyme for a particular substrate (or family of closely related substrates) in terms of fitting a substrate molecule to an enzyme that is as sensitive to molecular geometry as a key is to a tumble lock.

Malleability is the physical property of being capable of being shaped and formed by hammering or by action of rollers.

Metabolism is any and all reactions within living systems.

Metal is any element whose atoms have one, two, or three electrons in their outside shells (hydrogen and helium excepted). Metals are generally good conductors of electricity, but this decreases with temperature.

Metallic bond is the force of attraction within a metal that holds the atoms together; the bond formed by overlapping of partially filled *d*-orbitals (or *f*-orbitals) in metallic atoms.

Metalloid (semimetal) is an element with some metallic properties (electric conductivity, luster) and some nonmetallic properties (not malleable, not ductile). Carbon and silicon are examples.

Mineral is a solid, homogeneous chemical element or compound produced by processes in nature, possessed of a definite chemical composition (or range of compositions), and characterized by a definite crystalline structure and a set of particular chemical properties.

Mixture is one of three types of matter (the others are elements and compounds). Mixtures consist of two or more of the other types physically combined in an indefinite ratio and separable by physical means.

Molar (M) describes any solution's concentration in units of moles of solute per liters of solution. A one-mole per liter concentration is noted as 1 M.

Mole is identical with gram-formula weight.

Molecular compound is a compound whose smallest representative particles are molecules; sometimes called a covalent compound.

Molecular formula is a formula of a covalent compound that states the composition of an individual molecule without (in most examples) revealing its structure.

Molecular orbital is a region in space enveloping two (sometimes more) atomic nuclei within which a shared pair of electrons of a covalent bond resides.

Molecule is an electrically neutral (but frequently polar) particle made up of the nuclei and electrons of two or more elements and held together by covalent (or coordinate covalent) bonds. The molecule is the smallest representative sample of a covalent compound.

Monomer is a molecule that can be made to join its own kind many times to produce a polymer.

Monosaccharide is a carbohydrate that cannot be hydrolyzed.

Natural gas is a mixture of low-boiling hydrocarbons, mostly methane, that occurs naturally in underground deposits and usually occurs as part of petroleum fields; it is one of the fossil fuels.

Neutral, in the electric sense, is the condition of being neither positively nor negatively charged. In acid-base chemistry involving water as the solvent, neutral is the condition of a solution in which the molar concentration of hydronium ions exactly equals the molar concentration of hydroxide ions.

Neutralization, in chemistry, is the conversion of an acid or a base by any means into a salt; the destruction of acidic or basic properties.

Nitrification is the oxidation of ammonia to nitrite ion and then to nitrate ion by soil bacteria.

Nitrogen cycle is the series of events in nature by which nitrogen moves from the atmosphere into nitrogen compounds used by plants and then animals and is then returned by decay to nitrogen.

Nitrogen family is Group 5 in the periodic table—nitrogen, phosphorus, arsenic, antimony, and bismuth.

Nitrogen fixation is the conversion of molecular nitrogen from air into ammonia or nitrite ions or nitrate ions by soil microorganisms.

Nitrogen oxides (NO_x) are primary air pollutants, principally nitric oxide (NO) from which nitrogen dioxide (NO_2) forms. They arise when any fuel is burned in an internal combustion engine or furnace because at high temperatures nitrogen and oxygen—the constituents of air—will undergo some reaction to form nitric oxide.

Noble gas is one of six elements making up Group O in the periodic table—helium, neon, argon, krypton, xenon, and radon; so named because they enter into virtually no chemical reactions.

Nonfunctional group is that part of the structure of a molecule that ordinarily does not enter into chemical reactions; usually an entirely alkane-like group consisting only of carbon and hydrogen and single bonds.

Nonmetal is any element whose atoms have four, five, six, seven, or eight electrons in their outer shells, as well as hydrogen and helium. They are generally poor conductors of electricity but get better at conducting at higher temperatures.

Nucleic acid is a polymer of nucleotides where the repeating unit is a phosphate-pentose system and the side chains, attached to the pentose units, are heterocyclic amines.

Nucleotide is a monomer for a nucleic acid consisting of a phosphate ester of either ribose or deoxyribose attached, in turn, to one of a small set of heterocyclic amines.

Nylon is one of a family of fiber-forming polymers characterized by being polyamides.

Octane number ("octane") is a rating of gasoline according to its ability to function in a standard test engine without knocking. A rating of zero octane is produced by *n*-heptane; a rating of 100 octane is produced by isooctane. Blends of known composition of these two provide standard mixtures of intermediate values of octane.

Octet, in chemistry, is a condition in which the atom has eight electrons in its highest occupied energy level.

Octet rule. When elements react with each other, their atoms tend to change in whichever way most directly leads to their having outer octets, either through transferring electrons and forming ions or by sharing electrons and forming molecules. If the outer energy level is the first level, then instead of tending toward an outer octet, the tendency is toward a filled first level of two electrons. The octet rule can predict how many electrons an atom most likely will donate or accept or how many electrons it must obtain by a sharing process.

Oil is a mixture of compounds, mostly hydrocarbons, obtained from petroleum (sometimes coal) that boils in the range 250–400 °C. Specific types are gas oil, fuel oil, and diesel oil. Mineral oil and lubricating oil have higher boiling points but are similarly defined.

Optical isomer is one of a set of compounds whose molecules differ only in their chirality (handedness).

Optical isomerism is the phenomenon of the existence of compounds having identical molecular formulas and structures, except that structures of molecules with this form of isomerism have different chiralities.

Orbital is a small, three-dimensional portion of the space near an atom's nucleus in which an electron having a particular energy can be.

Ore is a mixture of minerals from which a metal can be obtained at a profit.

Organic describes any substance containing carbon other than pure carbon, or carbonic acid and its salts, or hydrocyanic acid (HCN) and its salts (the cyanides).

Oxidation is an event accompanied by the loss of one or more electrons from an atom, molecule, or ion. In organic chemistry, oxidation may be regarded as either the loss of hydrogen or the gain of oxygen by a molecule or organic ion.

Oxidizing agent is any substance that can cause something else to be oxidized.

Oxygen cycle is the cycle of consumption and regeneration of oxygen in the environment involving respiration, decay, and combustion—all processes of consumption—and photosynthesis—a process of regeneration. Because of the nature of photosynthesis, the oxygen cycle is intimately involved with the carbon cycle.

Oxygen family is Group 6 in the periodic table—oxygen, sulfur, selenium, tellerium, and polonium.

Ozone is triatomic oxygen, O_3. A very powerful oxidizing agent, it is one of the eye, nose, and lung irritants in certain kinds of smog; a substance in the stratosphere that enables the conversion of high-energy ultraviolet rays into heat.

Ozone cycle is a series of chemical reactions powered by high-energy ultraviolet rays from the sun that are converted to heat as a small, steady-state concentration of ozone is maintained. The initial event is the breaking of an oxygen molecule into two oxygen atoms by the absorbtion of the energy of the ultraviolet ray.

PAN is a powerful eye irritant in photochemical smog; it is a member of a family of organic compounds called *p*eroxy*a*cyl *n*itrates.

Parffin is a solid mixture of hydrocarbons of too high formula weight to be collected by boiling; paraffin wax.

Particulate is any finely divided particle suspended in polluted air. Soot is an example, but any dust is a particulate.

Parts per billion (ppb) is the number of parts in a billion parts. Two drops of water in a railway tank car of 34,000 gallons corresponds roughly to 1 ppb.

Parts per million (ppm) is the number of parts in a million parts. Two drops of water in a 32-gallon garbage can corresponds roughly to 1 ppm.

Pauli exclusion principle. No more than two electrons may occupy the same space orbital at the same time, and two may be present only if they have opposite spin.

Peptide bond is the amide linkage in a protein; a carbonyl-to-nitrogen bond.

Percent, if volume/volume percent, is the number of volumes of solute in 100 volumes of solution (often used for mixtures in the gaseous state and sometimes when a liquid is dissolved in a liquid). If weight/weight percent, the number of grams of solute in 100 grams of solution (seldom used). If weight/volume percent, the number of grams of solute in 100 cm³ of solution (very commonly used when a liquid or a solid is dissolved in a liquid).

Period, in chemistry, is a horizontal row in the periodic table.

Periodic law. Many properties of the elements are periodic functions of their atomic numbers.

Periodic property is any property that reoccurs with a succession of changes in time or distance or some other quantity of measure. Full phases of the moon are a periodic function of time. Street intersections are a periodic function of distance. Physical and chemical properties of the elements are a periodic function of atomic number (periodic law).

Periodic table is a display of the elements that emphasizes family relationships. All elements in the same family have identical outside shell electronic conditions—the number of electrons.

Petrochemical is one group of chemicals obtained from petroleum and used to make synthetics. The most common petrochemicals are saturated and unsaturated hydrocarbons of one to four carbons—e.g., ethylene and propylene.

Petroleum is a mixture of hundreds of organic compounds occurring naturally in underground pools, often under pressure created by its gaseous components (natural gas). Most of the compounds are hydrocarbons. Water is often present.

pH is the negative power to which the base 10 is to be raised to express in moles per liter the concentration of hydrogen ions in an aqueous solution. The best definition is its defining equation $[H^+] = 1 \times 10^{-pH}$.

Phenol is any organic compound posessing the -OH group attached to a carbon of a benzene ring.

Photochemical describes any event for which the energy comes directly from sunlight or any artificial light.

Photosynthesis is the reaction in green plants that uses solar energy trapped by chlorophyll to convert carbon dioxide and water into oxygen and molecules that the plant needs.

Physical change is one of the kinds of events where the observed change, when explored at the atomic or molecular level, involves no deep-seated rearrangement or redistribution of electrons relative to atomic nuclei.

Physical property is any observable characteristic of a substance other than a chemical property—color, density, melting point, boiling point, temperature, ability to reflect, transmit, or block light, ability to conduct electricity (although some substances undergo chemical changes if they are made to conduct an electric current), and a number of other "observables."

Pig iron is the cooled solid iron obtained in a blast furnace before being converted into steel.

Plastic is a finished or semifinished article or substance made from a resin by molding, casting, extruding, drawing, or laminating during or after polymerization.

Plasticizer is one of a group of chemicals added to a resin when it is converted into a plastic article to affect the softness of the finished article.

Poison is a substance that in very small amounts will injure or kill. The most potent poisons inactivate enzymes in respiration or the nervous system.

Polar molecule is one possessing a permanent electrical dipole, with sites of net partial positive and partial negative charges.

Polarity is a condition of a molecule wherein the center of density of all its positive charges (contributed by its atomic nuclei) does not coincide with the center of density of all its negative charges (furnished by all its electrons).

Pollutant is anything added to the environment by human activities that creates hazards to the health or life of living things or adversely affects materials. In common usage, "pollutants" are not limited to chemicals; heat and light are also capable of polluting.

Polyatomic ion is any ion made of two or more atoms (e.g., OH^-, CO_3^{2-}).

Polymer is the molecule whose structure arises from the joining together many successive times of small units—the monomers—in a reaction called polymerization.

Polymerization is a chemical reaction in which the molecules of one substance, called the monomer, react with each other to form another substance, called the polymer.

Polyolefin is a polymer whose monomer units came from olefins (e.g., polyethylene, polypropylene).

Polypeptide is a term often used synonomously with protein; it is usually (but not always) used when discussing proteins with relatively short molecules.

GLOSSARY

Polysaccharide is a carbohydrate that can be hydrolyzed to many monosaccharide units.

Polyunsaturated describes a vegetable oil whose molecules usually have three or more carbon-carbon double bonds.

***p*-orbital** is one kind of atomic orbital found at all main energy levels except the lowest one (the so-called K-shell). There are three *p*-orbitals at each of the higher levels, and each has the symmetry possessed by an hourglass (or a figure eight if it is rotated about its long axis). The long axes of the three at each level are directed at right angles to each other.

Potentiation is an interaction between two harmful substances occurring within an organism that makes their combined harm actually greater than the sum of the two acting individually.

Product, in chemistry, is a substance that forms in a chemical reaction.

Property is a quality or trait of something, e.g., a chemical or material object. The two kinds of properties in science are chemical and physical.

Protein is a naturally occurring polymer of alpha-amino acids.

Proton is a subatomic particle bearing one unit of positive charge. On the atomic mass (or weight) scale, each proton has a mass of 1 amu (atomic mass unit). In atoms, all protons are found in atomic nuclei. The proton is also the nucleus of a hydrogen atom, H. If this atom's lone electron is lost, a hydrogen ion, H^+, remains. Thus, proton also means hydrogen ion, H^+. In common usage, *proton* is often used when *hydronium ion* is meant.

Rare earth. See lanthanide.

Rayleigh scattering law. The intensity of the light scattered from its initial direction by particles in the atmosphere varies directly as the fourth power of the frequency. The blue part of sunlight is much more scattered than the red or orange parts.

Reactant is any substance that reacts in a chemical change.

Reducing agent is any substance that can cause something else to be reduced.

Reduction is an event accompanied by the gain of one or more electrons by an atom, molecule, or ion. In organic chemistry, reduction may be regarded as the gain of hydrogen or the loss of oxygen by an organic molecule or ion.

Refining is a procedure for separating crude or impure mixtures into components that have different uses, as in petroleum refining, or into pure materials, as in the electrolytic refining of metals.

Reforming is an operation in petroleum refining in which straight-chain alkane molecules are rearranged into more branched-chain molecules to improve octane rating.

Resin is an unfabricated polymer used to make all or most of a plastic article or to coat some surface for protection.

Respiration is the process whereby an organism exchanges gases between itself and its environment; the two gases of importance are oxygen and carbon dioxide.

Ribonucleic acid (RNA) is one of a family of polymers in which all members have the same kind of basic structural features but have different functions. All RNAs are polymers of ribonucleotides, and the principal heterocyclic amines are adenine (A), uracil (U), guanine (G), and cytosine (C). (Uracil replaces thymine for informational purposes in moving from DNA to RNA.) A few other rarer amines are also used in various RNAs.

RNA. See ribonucleic acid.

Saccharide is another name for carbohydrate.

Salt is any crystalline substance consisting of a reasonably orderly aggregation of oppositely charged ions (other than H^+ or OH^-) occurring in a definite ratio.

Saturated is the condition in which a solution cannot dissolve any more of a specified solute at that temperature and pressure. In organic chemistry, *saturated* describes any molecule having only single bonds.

Saturated compound, in chemistry, is a compound whose molecules are joined together by single bonds only. When an atom is holding a maximum number of groups, we say its valence or combining ability is saturated.

Slag is the waste product from a blast furnace resulting from limestone combining with impurities in the ore used.

Smog is air so contaminated by pollutants and irritants that visibility is reduced (from "*sm*oke" and "*fog*").

Smoke is a colloidal dispersion of tiny dust particles in a gas.

Sol is a colloidal dispersion of tiny dust particles in a liquid.

Solid is a state of matter in which a substance has definite shape and volume, where individual particles can only vibrate about fixed positions and cannot move from point to point; they retain the same neighbors.

Solubility is the extent to which a given substance will dissolve in a fixed volume (or weight) of a solvent.

Solute is the component of a solution that is taken to be dissolved or dispersed in a continuous solvent.

Solution is a reasonably stable, uniformly distributed mixture at the smallest particle level (ions, atoms, and/or molecules) of two or more substances.

Solvent is the component of a solution present as a continuous phase—the medium in which the solute is dispersed (the term is commonly applied to liquids).

s-Orbital is one kind of atomic orbital found at all main energy levels and having the same kind of symmetry as a sphere.

Specific heat. See heat capacity.

SST is a supersonic transport designed to fly in the stratosphere at a speed greater than that of sound. The British-French Concorde and the Soviet Tupelov are SSTs designed for civilian air travel.

Starch is the nutritionally important polysaccharide, a polymer of alpha-glucose, that consists of a mixture of amylose and amylopectin.

Steel is a commercial form of iron containing up to 1.7 percent carbon and, in several special alloy steels, small percentages of other metals.

Stefan-Boltzmann law. The emissivity of a perfect black body is proportional to the fourth power of its Kelvin temperature: emissivity $\propto T^4$.

Structural formula is the nucleus-to-nucleus sequence within an individual molecule.

Structure, in chemistry, is a synonym for structural formula, a symbol using atomic symbols and lines (bonds) to show how the atoms in a molecule are organized.

Subatomic particle is any of the particles common to all atoms—the electron, proton, and neutron. (The hydrogen atom is an exception; it has no neutron.) The nucleus may also be considered a subatomic particle.

Substrate is the molecule whose reaction an enzyme catalyzes.

Sulfur oxides (SO_x) are primary air pollutants, principally sulfur dioxide (SO_2), with small amounts of sulfur trioxide (SO_3) or sulfuric acid (H_2SO_4) or salts of sulfuric acid as suspended particulates. They arise when sulfur-containing fuels are burned.

Superimposability is the property of an object's being able to blend exactly at every point simultaneously with its mirror image.

Suspension is a mixture in which one or more substances are dispersed in some fluid medium, the dispersed particles being large enough to be influenced by gravity to settle. Muddy water is an example.

Synthetic is one of a large, diverse family of materials made from the chemicals obtained from fossil fuels; often refers to plastics, but not limited to them.

Taconite is a low-grade iron ore consisting of fine grains of magnetite embedded in silica-like minerals.

Thermal contact is a condition of relative location of two objects in which heat can be exchanged between them by any means—radiation, convection, or conduction.

Thermal property is an ability of an object to respond in some physical way to being in thermal contact with another object having a different temperature. Examples of thermal properties are heat capacity, latent heat of fusion, and latent heat of vaporization.

Tinplate is steel with a thin coating of tin to protect it from corrosion.

Transition element is one occurring between Group 2 and Group 3 in the long periods of the periodic table; a metallic element other than one in Group I or II and other than an actinide or a lanthanide.

Triglyceride is a lipid whose hydrolysis products are glycerol and a mixture of fatty acids; an ester between glycerol and three fatty acids.

Tyndall effect is the scattering of light by particles in a colloidal dispersion.

Unsaturated is a condition where a solution could dissolve more of a specified solute at that temperature. In organic chemistry, *unsaturated* describes any molecule having one or more double or triple bonds.

Unsaturated compound, in chemistry, is any compound whose molecules contain one or more double or triple bonds.

Unsaturated fat is a triglyceride normally of vegetable origin, whose molecules usually have three or more carbon-carbon double bonds.

Valence is chemical binding ability.

Vapor pressure is the equilibrium pressure exerted by the vapor over a liquid at a given temperature.

Vegetable oil is a mixture of triglycerides obtained from plants—e.g., corn oil, olive oil, and cottonseed oil.

Vinyls are a family of polymers used to make a variety of useful plastic articles. Their monomers all contain a unit of molecular structure consisting of two carbons joined by a double bond (a vinyl group).

Vitamin is an organic compound that cannot be synthesized by the organism; its absence leads to a "deficiency disease," and its presence is essential to normal growth. It is usually present in trace amounts in various foods.

Vulcanization is an operation performed on a polymer during fabrication to make it harder and more resistant to abrasion. It forms crosslinks between polymer molecules; the crosslinks are usually provided by the molecules or atoms of an added chemical.

Wien's displacement law. The wavelength at which the emission of radiation from a black body radiator is a maximum, λ_{max}, is inversely proportional to the Kelvin temperature: $\lambda_{max} \propto \frac{1}{T}$.

PART THREE
EARTH SCIENCE
Arthur N. Strahler

CHAPTER 21
ENERGY FLOW IN ATMOSPHERE AND OCEANS

21.1 THE ATMOSPHERE AND OCEANS

The *atmosphere* is the earth's envelope of gases in which are suspended countless minute liquid and solid particles. The *oceans* form the vast bulk of the earth's *hydrosphere,* or liquid realm. At the same time, the oceans hold a large quantity of mineral matter, most of it dissolved as sea salts. Both air and water are fluids, substances that flow easily when subjected to unbalanced forces. Both the atmosphere and the oceans are in ceaseless motion. Both have a common role in the physical processes of our planet. That role is to transfer heat, received by electromagnetic radiation from the sun, from one part of the globe to another. Inequalities in the heating of the atmosphere and oceans are basically responsible for generating and sustaining enormous flow systems. These currents of air and water serve to regulate the earth's heat budget.

So far as we know, no other planet has both a dense atmosphere and extensive oceans of liquid matter. Earth possesses a unique combination of surface fluids that together have provided a favorable environment for the development of life, sustained with remarkable constancy for hundreds of millions of years.

The science of *meteorology* deals with the physics of the atmosphere; the science of *physical oceanography* deals with the physics of the oceans.

21.2 COMPONENTS OF THE ATMOSPHERE

We learned in Chapter 16 that the earth's atmosphere consists of a mixture of various gases surrounding the earth to a height of many kilometers. Held to the earth by gravitational attraction, this spherical shell of gases is densest at sea level and thins rapidly upward.

The uniform atmospheric layer consists of (1) a mixture of gases referred to collectively in Chapter 16 as the *pure dry air,* (2) water vapor, (3) dust particles, and (4) water droplets and ice particles. The first two components are true gases composed of individual molecules. Figure 21–1 shows the proportions of gases most important in the composition of pure dry air. Dust consists of solid particles much larger than molecules, but so tiny that they mix freely with the gases and stay aloft almost indefinitely. Clouds and fog, composed of minute water droplets or ice crystals, are also present much of the time in the lower atmosphere the world over.

The second major component of the atmosphere, *water vapor,* is the gaseous state of water in which individual water molecules have the same freedom of movement as molecules of nitrogen or oxygen gas. The water molecules diffuse completely among gases of the pure dry air. Water vapor supplies the water for all clouds and rain, and during condensation it releases latent heat, which supplies the energy for storms.

21 ENERGY FLOW IN ATMOSPHERE AND OCEANS

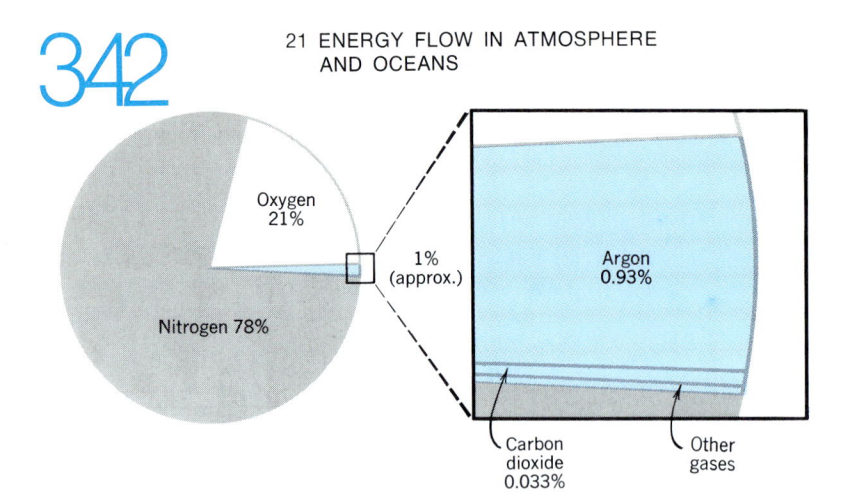

FIGURE 21-1
Component gases of the lower atmosphere. Figures tell percentage by volume.

Dust in the lower atmosphere, the third component, consists of particles so small that, for example, 100,000 of them placed side by side would be needed to make a line 1 cm long. Most atmospheric dust comes from the earth's surface.

Both water vapor and dust originate mainly from the earth's surface and depend on air motions to be lifted vertically. As a result, these components tend to be most heavily concentrated in the lowermost air layers and to diminish to nearly zero values in the upper atmosphere, above 100 km.

21.3 ATMOSPHERIC PRESSURE

Molecules of the atmospheric gases are attracted earthward and tend to crowd together. The crowding becomes closer from the outer limits of the atmosphere to sea level, because any one layer of the atmosphere is being compressed by the weight of all the layers above it.

Any surface exposed to the atmosphere is under a force represented by the weight of the column of gases lying above the surface. This force, when measured for a standard unit of area, is known as *atmospheric pressure*, or simply *air pressure*.

Atmospheric pressure has an average value at sea level of about 1 kg/cm^2. The pressure graph in Figure 21-2 shows how rapidly air pressure decreases with increase in altitude above sea level. Taking the sea-level pressure as 100 per-

FIGURE 21-2
Structure of the atmosphere.

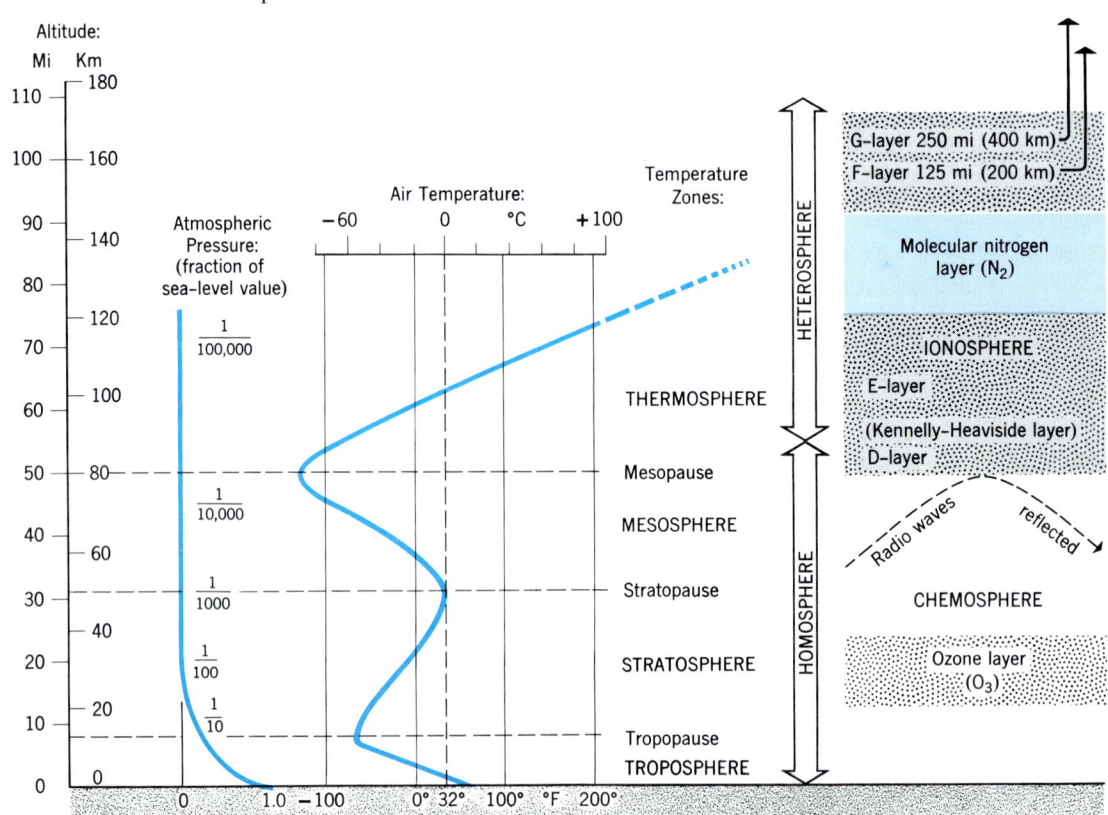

cent, we see that air pressure has fallen to 10 percent at 18 km and to a mere 1 percent at 32 km. At 110 km air pressure is only 1/100,000 of the sea-level value.

Air pressure can be demonstrated and measured by a very simple device, the *mercurial barometer* (Figure 21-3). The demonstration is often called Torricelli's experiment, after the man who first performed it in 1643. A glass tube of very narrow bore and about 1 m long is sealed at one end, filled with mercury, and inserted open end down into a dish of mercury. Instead of pouring out of the tube, the mercury column stands at rest with a height of about 76 cm. A vacuum occupies the section of empty tube above it. The column of mercury is counterbalanced by pressure of the entire column of atmosphere extending from the solid earth to interplanetary space. The average height of the mercury column at sea level is 76 cm; this value is taken as the *standard sea-level pressure* of the atmosphere. In modern weather science the unit of air pressure is the *millibar*. Standard sea-level pressure is 1013.2 mb.

21.4 TEMPERATURE LAYERS OF THE ATMOSPHERE

Figure 21-2 includes a graph of temperature from ground level to 160 km altitude. Starting at the ground, let us examine the great range of temperatures we would encounter in a vertical ascent. Most of us know from experience that as we climb a mountain or rise in an airplane, the temperature falls steadily. This rate of temperature drop, averaging about 6.4 C° per km, is called the *environmental temperature lapse rate*.

Air temperature falls steadily upward to an altitude of about 11–12 km in middle latitudes. At this level the curve breaks sharply and reverses. Here, at the curve break, we have reached the *tropopause*. It marks the top of the basal atmospheric layer known as the *troposphere*. Above lies the *stratosphere*.

The troposphere contains almost all the water vapor of the atmosphere. Consequently, this basal layer contains nearly all clouds, precipitation, and storms. Today, jet aircraft can maintain flight in the stratosphere in middle and high latitudes and thus avoid many weather hazards.

Upward through the stratosphere, temperatures rise gradually up to a level of about 50 km, where the *stratopause* is encountered (see Figure 21-2). Here the temperature reaches a maximum of about 0°C. Above the stratopause lies a zone of diminishing temperature, the *mesosphere*. At the *mesopause*, about 85 km, temperatures reach a minimum value, averaging about −83°C.

Above the mesopause lies the *thermosphere*, a zone of rapid temperature increase to extremely high values of over 700°C at an altitude of 200 km. Keep in mind that at these great heights the rarefied air is a near vacuum, so that little heat is actually held in the air. We must now explain the heating of both stratosphere and thermosphere.

21.5 THE IONOSPHERE

Electromagnetic radiation was explained in Chapter 6 (see Figure 6-8). The sun emits this radiant energy into space in all directions from its spherical surface. The solar energy spectrum includes gamma rays, x-rays, ultraviolet rays, visible light rays, and infrared rays. (Energy distribution within this solar spectrum is shown in Figure 21-10.)

As solar radiation penetrates our upper atmosphere, its rays strike atoms and molecules of the various gases. First, molecules and atoms of nitrogen and oxygen of the upper atmosphere absorb the highly energetic gamma rays, x-rays, and ultraviolet rays. In so doing, each affected molecule or atom loses an electron and becomes a positively charged ion. This process of ionization begins at a height of about 1000 km. Within a region from about 80 to 400 km above the earth, the concentration of positive ions is most dense; this region is known as the *ionosphere* (see Figure 21-2). Heating of the thermosphere is

FIGURE 21-3
Principle of the mercurial barometer.

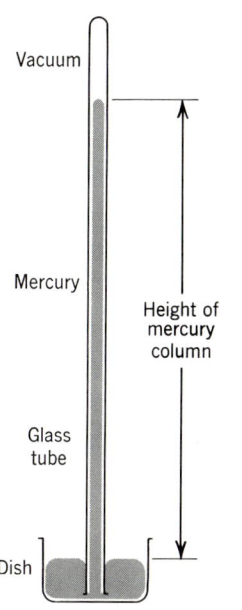

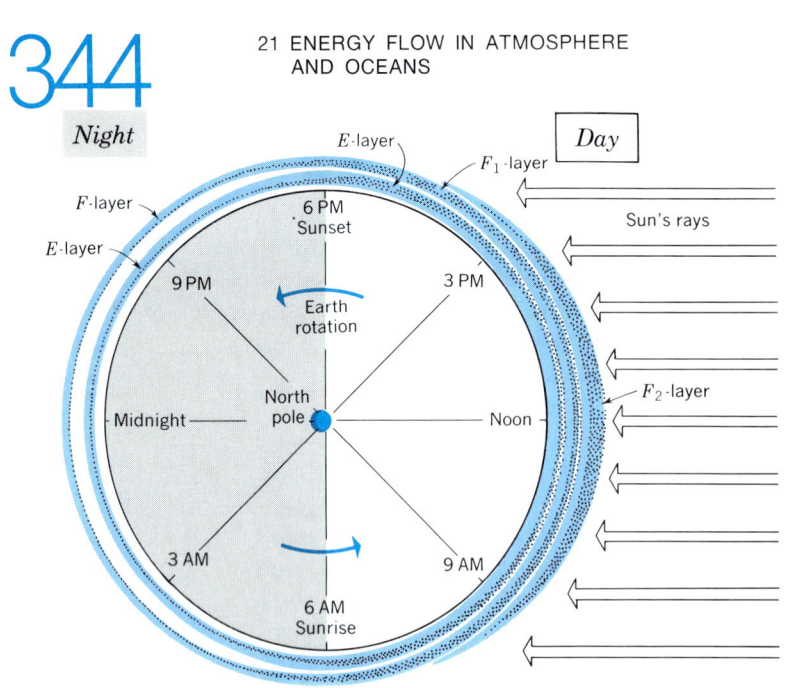

FIGURE 21–4

This diagram of the ionospheric layers is a cross section through the earth's equator. The observer looks down upon the cross section from a point over the north pole. (Based on a figure by B. F. Howell, Jr., 1959, *Introduction to Geophysics,* McGraw-Hill, New York. From THE EARTH SCIENCES, 2nd Edition by Arthur N. Strahler. Copyright © 1963, 1971 by Arthur N. Strahler. By permission of Harper & Row, Publishers, Inc.)

FIGURE 21–5

Radio waves are reflected alternately from the lower ionosphere and the earth's surface, making long-distance communication possible. (From PRINCIPLES OF EARTH SCIENCE by Arthur N. Strahler. Copyright © 1976 by Arthur N. Strahler. By permission of Harper & Row, Publishers, Inc.)

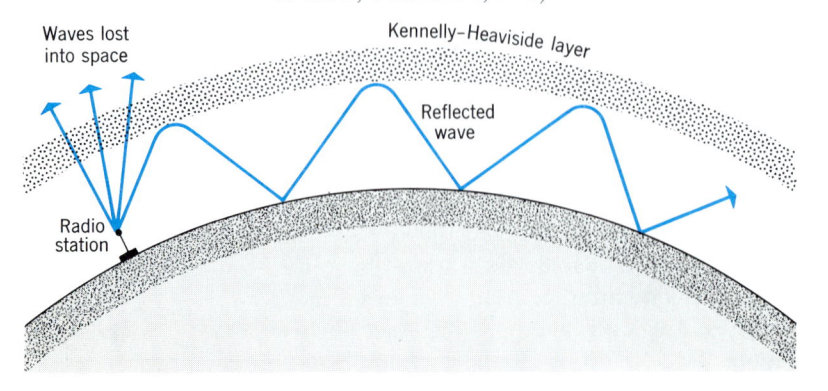

explained by the absorption of solar energy during the ionization process.

Electrons ejected from molecules of oxygen and nitrogen during the ionization process are free to travel as an electric current. We can think of the ionosphere as an electrically conducting layer. The earth itself is an electrical conductor, but the intervening atmospheric layer from the ground to the 80-km level is a poor conductor and can be imagined as an insulator. In a way, then, the earth and ionosphere are like the two wire conductors of an electric cable with protective insulation between.

Because the formation of ions in the ionosphere requires solar radiation, we should expect the greatest density of ions to be on the illuminated side of the earth and the least on the dark side. Figure 21–4 shows how the ions form into a number of dense layers on the sunlit hemisphere. In hours of darkness, the ionized molecules recapture the free electrons and ionization rapidly disappears. Only one or two thin layers remain on the darkened hemisphere.

The ionosphere is extremely important in radio-wave transmission. Figure 21–5 shows how the paths of radio waves sent out from a transmitter on the earth are reflected back to earth by the ionosphere. Before the nature of the ionosphere was understood, two early radio experimenters independently discovered the radio-wave reflection principle. The *Kennelly-Heaviside layer,* named in their honor, was first applied to the lower part of the ionosphere, about at the 80–100 km level, where the principal reflection of radio waves makes possible long-distance radio-signal transmission. Extremely long radio waves, with lengths over 300 m, travel by reflection with very little loss and are used in the transoceanic radiotelephone transmission.

21.6
THE OZONE LAYER

Let us continue to trace the effects of solar radiation as it penetrates deeper into the earth's atmosphere. Solar x-rays and the shorter ultraviolet rays are almost completely absorbed within the ionospheric region. However, certain ultraviolet rays pass readily into lower levels, reaching the stratosphere.

In a stratospheric region known as the *ozone layer,* largely concentrated in the altitude range of 20–35 km, the absorption of the longer ultraviolet rays produces ozone from atmospheric oxygen. The process was described in Chapter 16. The ozone layer also absorbs much of the longer ultraviolet radiation and some of the visible and infrared wavelengths as well. This absorption heats the ozone layer, causing the temperature maximum at the stratopause, reached at about 50 km.

21.7 THE MAGNETOSPHERE

The earth also has a magnetic atmosphere, but it is entirely different from the atmosphere of gases we have examined so far in this chapter. The magnetic atmosphere is called the *magnetosphere*, for short. The earth acts as a great magnet (see Chapter 24). Lines of force extend far out into space, making up the earth's *external magnetic field*. This field extends many times farther from earth than the gaseous atmosphere (Figure 21–6).

A magnetic field is a form of energy and has no matter, but it can trap and hold matter in the form of particles. These particles are electrons and protons that stream from the sun and pass by the earth and other planets in the form of a *solar wind*. When a particle from the solar wind reaches the earth's magnetic field, it can become entrapped between lines of force. Countless billions of these particles are continually entrapped in the magnetic field, and they make up the magnetosphere.

The earth's magnetic field would be symmetric and doughnut-shaped were it not for the pressure exerted by the solar wind. As Figure 21–6 shows, the magnetosphere is deformed into a teardrop shape, with a long, tapering tail pointed away from the sun. Streamlines of the solar wind are sent into wavy ripples along a *shock front*, where the wind first encounters the magnetic field. The streamlines of the solar wind are then forced to flow around the magnetosphere, enclosing it in a sharp boundary called the *magnetopause*.

Trapped particles in the magnetosphere are said to be energetic, meaning that they produce *ionizing radiation*, much like that generated by an x-ray machine or a radioactive isotope, like radium. Early space probes carried geiger counters, which are used to detect radiation coming from radioactive isotopes. The geiger counters in the spacecraft sent back information that they were passing through belts of heavy radiation found several thousand kilometers out. Within a doughnut-shaped belt surrounding the earth, the radiation is intense. This region was named the *Van Allen radiation belt*, after the scientist who first reported its existence. The Van Allen belt is shown to scale in Figure 21–6, by small black patches beside the earth.

FIGURE 21–6
Magnetosphere and magnetopause. The Van Allen radiation belt is shown by black areas on either side of earth. (After C. O. Hines, *Science*, 1963, and B. J. O'Brien, *Science*, 1965.)

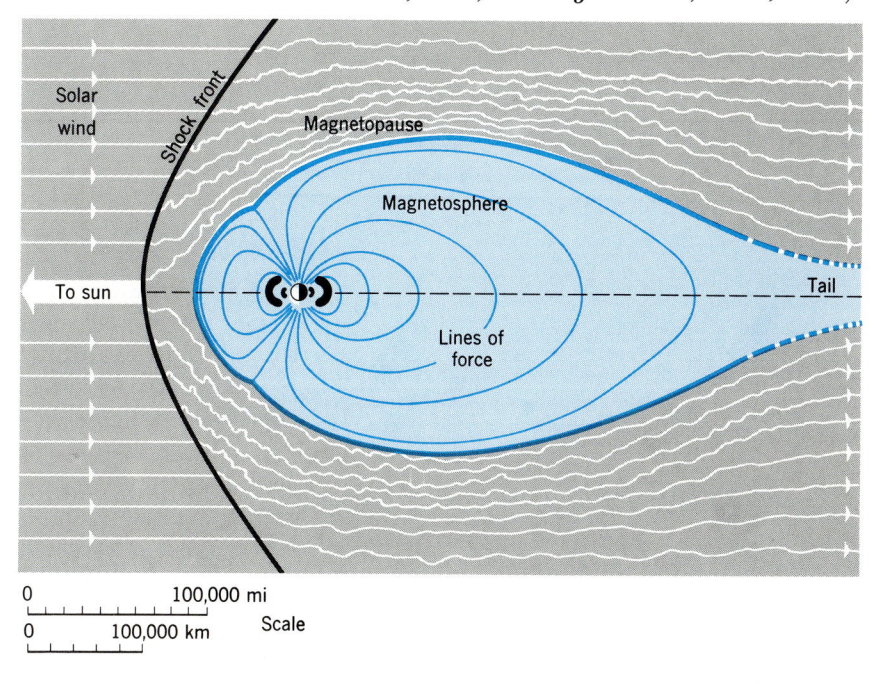

21.8
THE OCEANS

The term *world ocean* designates the combined oceans of the earth. The world ocean occupies about 71 percent of the earth's surface and has a mean depth of about 4 km. Volume of the world ocean is estimated as 1.4×10^9 km^3 and constitutes 97.2 percent of the world's free water. Most of the remaining 2.8 percent is locked up in ice sheets.

The atmosphere, being a gas, has no distinct upper boundary. In contrast, the fluid of the oceans has a clearly defined and visible upper surface where the water is in sharp contact with the air. The direct absorption and radiation of heat occur through the upper ocean boundary; the loss of moisture by evaporation and the gain by precipitation take place here; the dragging force of the wind to produce waves and surface currents is exerted on this surface.

The oceans occupy basins, in which the continental margins act as sidewalls to restrict the motion of the ocean water. On the other hand, the atmosphere has no such compartmentation by high physical barriers and is free to circulate on a global scale. Air, a substance of low density and viscosity, moves freely and quickly attains high velocities; it also quickly comes to rest. In contrast, ocean water, with its much greater density and viscosity, is comparatively sluggish in motion.

21.9
COMPOSITION OF SEAWATER

Seawater is *brine,* a solution of salts. Table 21.1 lists the five most important constituent salts. Notice that in these five salts the first element is always a metal: sodium, magnesium, calcium, or potassium. Notice also that in four of the five salts chlorine (as chloride) is the second element.

The element chlorine makes up 55 percent of the total weight of all matter dissolved in seawater; sodium is next with 31 percent.

Less abundant but important elements not appearing in the table are bromine, carbon, strontium, boron, silicon, and fluorine. A complete list of elements known to be present in seawater would include at least half of all the naturally occurring elements.

21.10
TEMPERATURE LAYERS OF THE OCEANS

The temperature structure of the oceans over middle and low latitudes can be described as a three-layer system (Figure 21-7). The surface water is subjected to intense solar radiation—all year in low latitudes and in summer in middle latitudes. The heated water takes the form of an upper layer of quite uniform temperature, a result of mixing within the layer. This warm surface layer attains a thickness of 500 m and a temperature of 20–25°C or higher in equatorial latitudes.

Immediately below the warm layer, water temperatures drop sharply. This layer of rapid temperature change is called the *thermocline* (Figure 21-8). It is 500–1000 m thick. Below the thermocline temperatures decline gradually from about 5°C immediately below the thermocline to about 1°C close to the bottom. This statement applies to low latitudes. In arctic and antarctic latitudes, surface water temperatures are close to 0°C. Here the temperature changes with increasing depth are very slight. The vast bulk of the ocean waters lies cold and dark upon a solid floor. Only a very small flow of heat takes place from within the earth and is of practially no importance in warming the oceans.

21.11
SALINITY AND DENSITY OF SEAWATER

Salinity of seawater is the ratio of the weight of dissolved salts to the weight of water. In Table 21.1 we showed a salinity of 34.5 grams of salt for 1000 grams of water (34.5 g/kg). This is a good average figure, but salinity shows important differences from place to place over the oceans. The full range in salinity is from as low as 33 g/kg to as high as 40 g/kg. High values, because of intense evaporation, occur in certain bays or gulfs largely shut off from the ocean in tropical desert areas. The Red Sea is an example of a gulf with very high salinity. Low salinity is

TABLE 21.1
PRINCIPAL CONSTITUENTS OF SEAWATER

Name of Salt	Formula	Grams of Salt per 1000 Grams of Water
Sodium chloride	NaCl	23
Magnesium chloride	MgCl$_2$	5
Sodium sulfate	Na$_2$SO$_4$	4
Calcium chloride	CaCl$_2$	1
Potassium chloride	KCl	0.7
With other minor ingredients to total		34.5

found over the equatorial oceans where rainfall is very heavy and the salt water tends to be diluted.

Sea water, because of the presence of dissolved salts, is slightly denser than pure fresh water. Compared with a density of 1.000 g/cm³ for pure fresh water, seawater has a density of about 1.026–1.028 g/cm³. Both temperature and salinity affect the density.

Seawater becomes increasingly dense as it becomes colder until the freezing point is reached, at about −2°C. This is an important principle because it means that seawater chilled near the surface will tend to sink, displacing water of less density.

Density also becomes greater as salinity increases, so that where surface evaporation is great, the water near the surface may become slightly denser than that below it and will sink to a lower level. Because temperature is the stronger of the two controls of density, the densest seawater is formed in the cold arctic and polar seas. This very cold water sinks to the bottom and tends to remain close to the floor of the deep ocean basins.

21.12 PRESSURE IN THE OCEAN

Downward into the ocean, the pressure that water exerts equally in all directions on any exposed surface (hydrostatic pressure) increases in direct proportion to depth. This rule applies because water is practially incompressible, as compared to air. The pressure increase will be about 1 g/cm² for each cm of depth and will amount to 1 kg in 10 m, or 100 kg in 1 km. Of course, we must add the pressure of the overlying atmosphere, which is equal to the weight of a water layer about 10 m thick, and is therefore 1 kg/cm². At a depth of 4000 m, on the deep ocean floor, the pressure will be about 400 kg/cm², or 4 million (4×10^6) kg/m². Constructing a deep-diving submarine to reach these depths under such enormous pressures is obviously a major engineering problem.

21.13 THE SUN-EARTH-SPACE RADIATION SYSTEM

Planet earth intercepts radiant energy from the sun. This energy powers the motions of the atmosphere and oceans and is the source of energy for all precipitation and storms. But the earth is also a radiator of energy, emitting energy into outer space. As Figure 21–9 shows, the input of energy from the sun and the output of earth's

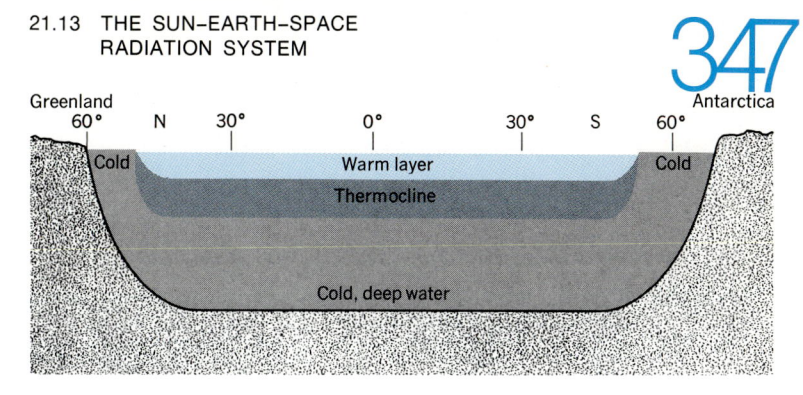

FIGURE 21–7
A schematic north-south cross section of the world ocean shows that the warm surface water layer disappears in arctic latitudes, where very cold water lies at the surface.

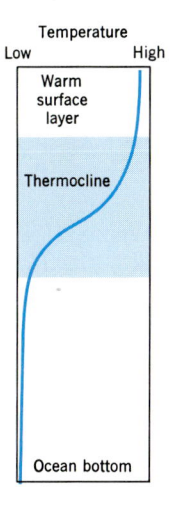

FIGURE 21–8
Schematic diagram of temperature change with depth in the ocean, showing the thermocline.

FIGURE 21–9
Schematic diagram of the sun-earth-space radiation system.

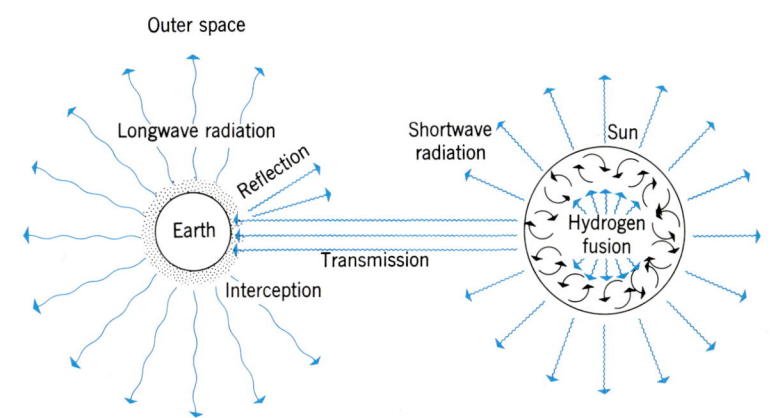

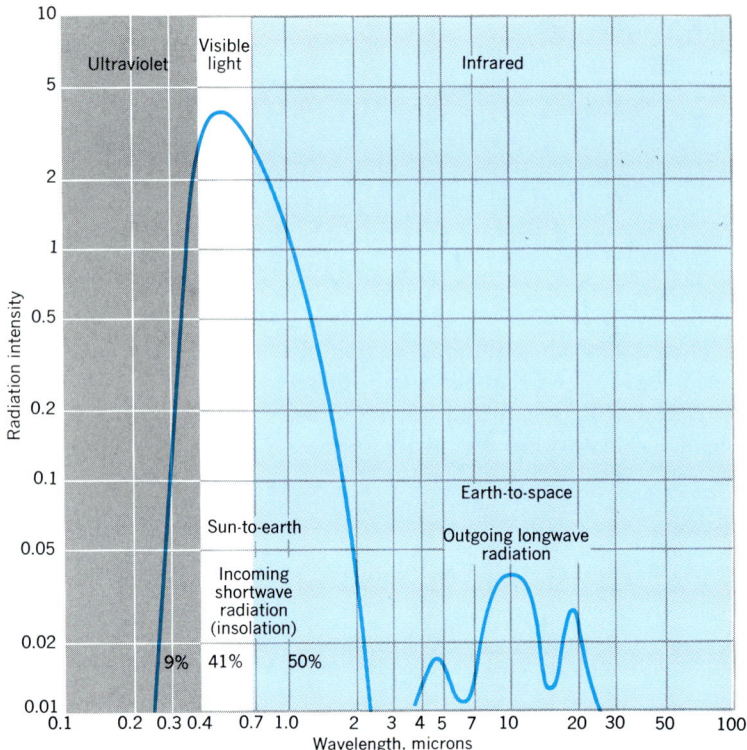

FIGURE 21-10

Intensity of incoming shortwave radiation at the top of the atmosphere (*left*); longwave radiation from earth to outer space (*right*).

radiation of energy into space go on simultaneously and constantly. As a result there is a *radiation balance* between incoming energy and outgoing energy.

Of the solar energy we receive, about half is within the shorter wavelengths of the electromagnetic spectrum—x-rays, ultraviolet rays, and visible light. On the other hand, energy radiated by the earth is entirely of the longer waves, or infrared radiation.

Figure 21-10 shows the curve of radiation intensity for our sun. We are approximating its activity by a perfect radiator with a temperature of 6000°K. The meaning of the units of radiation intensity on the vertical scale does not concern us here; we are only interested in comparing sun with earth. The scale is a logarithmic scale, based on powers of ten. On the horizontal scale of the graph is wavelength. We have used *microns* as our unit; one micron is equal to 0.0001 cm. The scale of microns is also a logarithmic scale. The peak intensity of our model sun is right in the middle of the visible light band. About 41 percent of the solar energy lies in the visible light portion, about 9 percent in the ultraviolet portion, and about 50 percent in the infrared portion. Very little energy is carried by infrared rays longer than 50 microns.

Now let us consider the earth as a radiator. The ideal model for our earth is a black body with a temperature of 300°K. This is equal to 27°C, which would be a pleasant outdoor temperature on a summer day. The radiation intensity curve for the earth model is shown at the right side of Figure 21-11. Most of the energy is emitted between 5 and 50 microns; practically all of it is in longer wavelengths than the solar spectrum. The peak is about at 10 microns. However, the peak for earth radiation has a value of about 5/100 of a unit, whereas the peak for the sun is about 4 units. This means that the peak intensity of the sun's radiation is about 80 times that of earth for each square centimeter of surface area.

Scientists refer to solar radiation as *shortwave radiation* and to the earth's radiation as *longwave radiation*. Figure 21-10 makes clear the reason for using these terms.

Because the sun's rays diverge radially outward from its surface, the intensity of the sun's energy diminishes inversely as the square of the distance. At the average distance of the earth, solar radiation beyond the atmosphere has an intensity of about 2 cal/cm^2/min on a surface held at right angles to the sun's rays. This quantity is called the *solar constant*. Only minor fluctuations of one percent or so can be observed in the solar constant, and there is no reason to believe that it has changed significantly in many decades of observation.

21.14 SOLAR RADIATION OVER A SPHERICAL EARTH

Because the earth is almost a true sphere, only one point on earth presents a surface at right angles to the sun's rays. This *subsolar point* coincides with the occurrence of solar noon at the latitude where the sun reaches the zenith position for an observer (see Figures 21-12 and 21-13). In all directions away from the subsolar point, the earth's curvature causes the receiving surface to be turned away from the sun at an increasing angle with respect to the rays. At the *circle of illumination*, a horizontal surface parallels the rays. The hemisphere lying beyond this great circle is, of course, in shadow.

Assuming for the moment that the earth has no atmosphere, the total quantity of solar energy received by one square centimeter of horizontal surface in one day will depend on two factors: the angle at which the sun's rays strike the earth and the length of time of exposure to rays. These factors are varied by latitude and by the seasonal changes in the path of the sun in the sky.

21.14 SOLAR RADIATION OVER A SPHERICAL EARTH

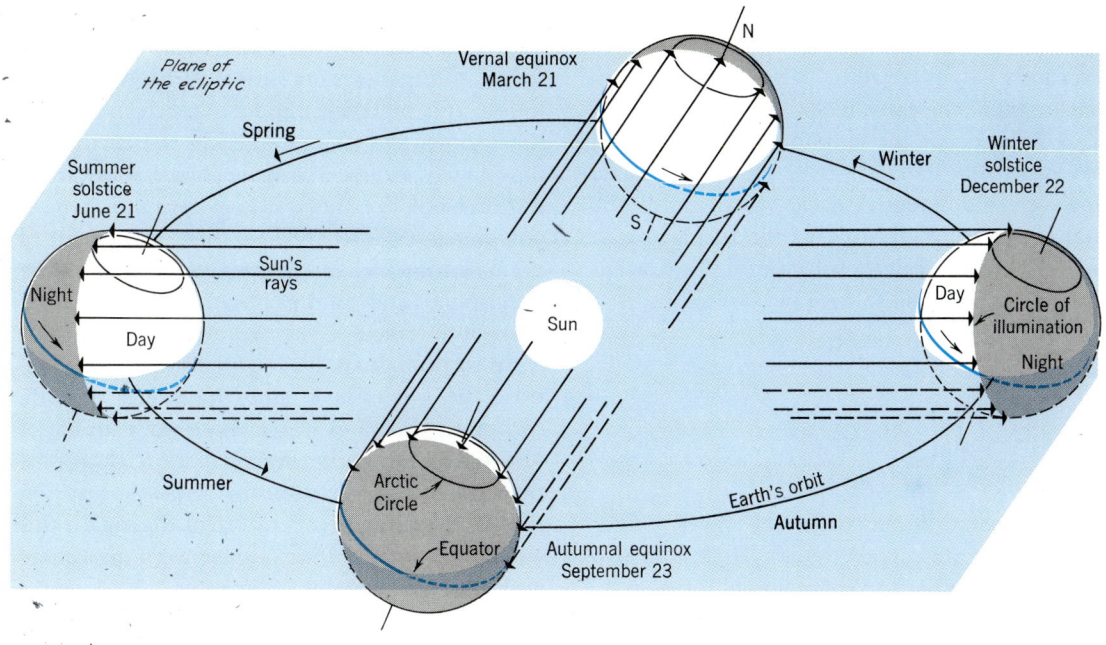

FIGURE 21-11

The seasons result because the tilted earth's axis keeps a constant orientation in space as the earth revolves about the sun.

It is important to understand that the earth's axis of rotation is not oriented perpendicular to the plane of the earth's orbit. Instead, as shown in a perspective drawing in Figure 21-11, the earth's axis is inclined at an angle of $23\frac{1}{2}$ degrees away from the perpendicular. The earth's axis constantly maintains this angle while always aimed at the same point in space. As a result, there are two points in the orbit at which the axis is inclined neither toward nor away from the sun—the *equinoxes* (Figure 21-12). There are also two other points in the orbit in which the full value of axis inclination is directed toward the sun—the *solstices* (Figure 21-13). Names and dates of equinoxes and solstices are given in Figure 21-11.

If the earth's axis were perpendicular to the plane of the ecliptic (that is, if there were no axial tilt), the conditions of equinox would prevail throughout the entire year. At equinox,

FIGURE 21-12

Equinox conditions. The sun's rays are tangent to the globe at both north and south poles. (The observer's eyepoint is in the plane of the earth's equator, so that the axis does not seem to be inclined.)

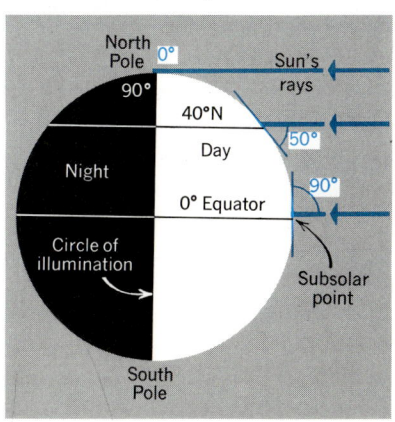

FIGURE 21-13

Solstice conditions. At summer solstice all of the region lying poleward of the arctic circle experiences solar radiation throughout 24 hours of the day, whereas at winter solstice the region south of the antarctic circle receives 24-hour solar radiation.

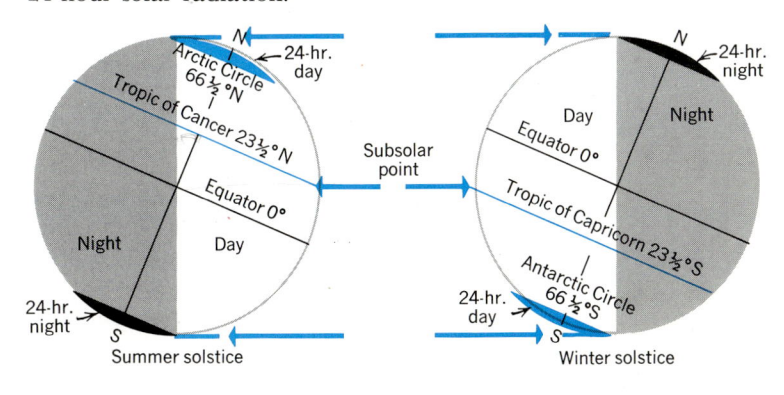

radiation at the equator is 100 percent, with a value of 2.0 cal/cm² /min; at 90 degrees north and south, values are zero. These facts lead us to conclude that the earth receives its greatest total solar radiation at the equator and the least at the poles, considered on a yearly average basis. However, the seasonal cycle redistributes the total global radiation toward the poles, so that the polar zones receive a substantial share, accumulated during one solstice period.

21.15
DEPLETION OF INCOMING SOLAR RADIATION

As solar radiation penetrates the lower, denser layers of the atmosphere, energy is lost in two ways: reflection back to space and absorption by the atmosphere.

Gas molecules of the atmosphere cause the visible light rays to be turned aside in all possible directions. In this process of *scattering,* some of the energy is turned back into outer space and lost. Dust particles and cloud particles reflect part of the incoming shortwave energy back into space.

Part of the shortwave energy reaching the earth's land and ocean surfaces is reflected back into space. Altogether, combining all forms of scattering and reflection, about 32 percent of the solar shortwave energy entering the earth's atmosphere is returned to space. This is an average for the whole earth and is known as the *albedo* of our planet. On the average, then, 68 percent of the incoming radiant energy is absorbed by the atmosphere and solid earth and thus converted into sensible heat.

Absorption takes place as the sun's rays penetrate the atmosphere, since molecules of both carbon dioxide and water vapor are capable of directly absorbing infrared radiation. Dust particles and clouds also absorb energy. Absorption results in a rise of temperature of the air. In this way some direct heating of the lower atmosphere is caused by incoming solar radiation. A world average figure for absorption is 18 percent. Adding this 18 percent to the 32 percent for average losses to space by scattering and reflection, we have a total of 50 percent, representing the portion of incoming solar (shortwave) energy actually absorbed by the earth's surface, including both land and ocean surfaces. Figure 21–14 summarizes the figures we have given.

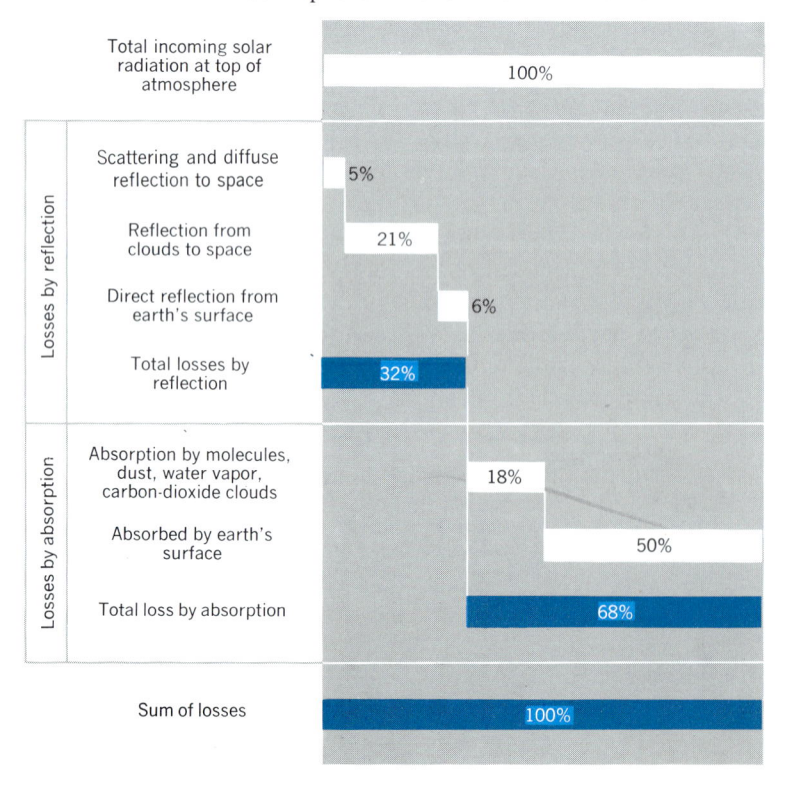

21.16
LONGWAVE RADIATION

The surface layer of the continents and oceans holds sensible heat derived originally from absorption of the sun's rays. Land and water surfaces continually radiate this energy back into the atmosphere. This infrared radiation occurs at wavelengths longer than 4 microns, as shown in Figure 21–10. All of this energy is what we have called longwave radiation.

The atmosphere also radiates longwave energy both toward the earth and outward into space, where it is lost. Longwave radiation from both ground and atmosphere continues during the night, when no solar radiation is being received.

Because longwave radiation goes on through the night, it is possible to obtain *infrared imagery* of ground features using special cameras mounted on aircraft. Figure 21–15 is an example. This image was obtained in the early morning hours in total darkness. Pavements and streams appear bright on the print because they are warm and radiate more intensely. In contrast, moist soil surfaces of fields are cooler and appear dark.

Energy radiated from the ground is easily absorbed by water vapor and carbon dioxide in the atmosphere. However, some longwave radi-

FIGURE 21–14
A summary of the energy losses by reflection and absorption as solar energy penetrates the atmosphere to reach the earth's surface.

	Total incoming solar radiation at top of atmosphere	100%
Losses by reflection	Scattering and diffuse reflection to space	5%
	Reflection from clouds to space	21%
	Direct reflection from earth's surface	6%
	Total losses by reflection	32%
Losses by absorption	Absorption by molecules, dust, water vapor, carbon-dioxide clouds	18%
	Absorbed by earth's surface	50%
	Total loss by absorption	68%
	Sum of losses	100%

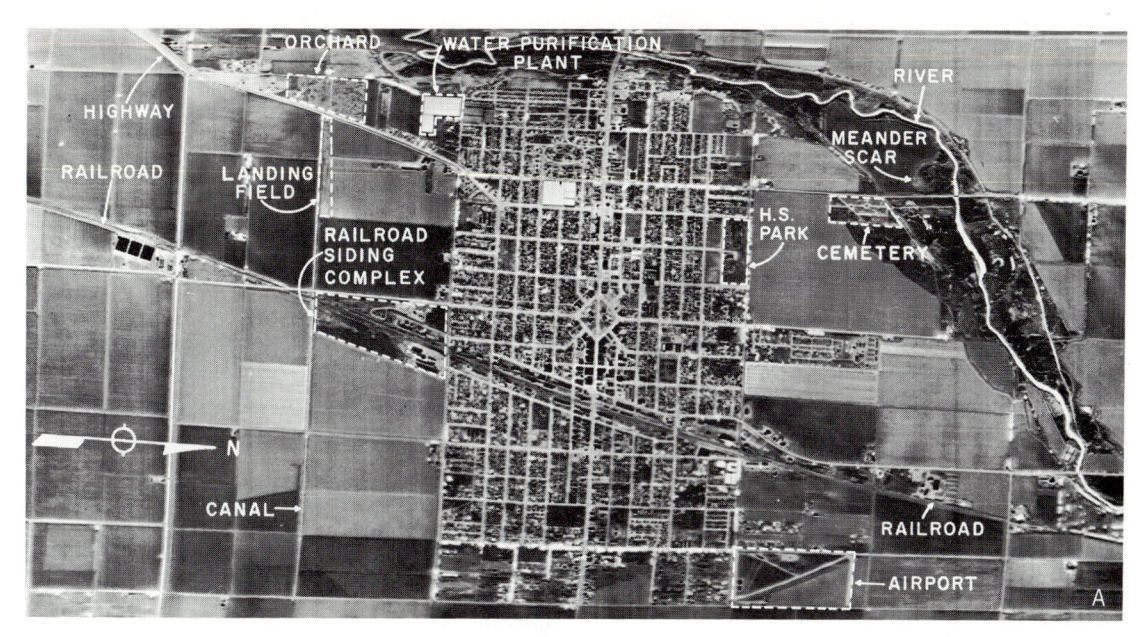

FIGURE 21-15
This infrared imagery of Brawley, a small town in the Imperial Valley of California, looks like an air photograph. It was taken in darkness between 2 and 4 A.M. Special lenses and films can record the infrared rays coming up from the surface. (Environmental Analysis Department, HRB—Singer, Inc.)

ation passes freely through the earth's atmosphere and into outer space. About 8 percent of the longwave radiation directed outward leaves the atmosphere in this manner.

So we see that the atmosphere receives much of its heat by an indirect process; much of the incoming energy in shortwave form is allowed to pass through, whereas outgoing longwave energy cannot easily escape and is absorbed. For this reason the lower atmosphere with its water vapor, carbon dioxide, and clouds acts as a warm blanket. Longwave radiation from atmosphere back to the surface is called *counterradiation*. It returns heat to the earth and helps to keep surface temperatures from dropping excessively during the night or in winter at middle and high latitudes.

Somewhat the same principle is used in greenhouses and in homes using large glass walls as a solar heating method. Here the glass permits entry of shortwave energy. Accumulated heat cannot escape by mixing with cooler air outside. The expression *greenhouse effect* is used by meteorologists to describe this atmospheric heating principle.

The atmosphere radiates longwave energy into outer space, balancing the total global energy budget.

21.17 NET RADIATION AND LATITUDE

When all forms of radiation, both incoming and outgoing, are summed for a given unit of area, we obtain the *net radiation*. We have already concluded that the earth's net radiation is equal to zero, but this is the total planetary condition, year-in and year-out. Net radiation is not zero at a particular time and place on the earth. Some places show a positive yearly value or *annual surplus;* others show a negative yearly value, or *annual deficit*. Where a surplus exists, either the temperature of the air and surface will be rising or the surplus energy is being exported by some process other than radiation as fast as it accumulates. Where a deficit exists, the temperatures will be falling, or heat is being imported to counteract the deficit.

Figure 21-16 shows a pole-to-pole profile of the net radiation of the earth, including both the earth's surface and atmosphere. An energy surplus exists from 40 degrees north to 40 degrees south latitude; a deficit exists poleward of the 40th parallel in both hemispheres. The surplus equals the combined deficits of the two hemispheres. Temperature records show that the

21 ENERGY FLOW IN ATMOSPHERE AND OCEANS

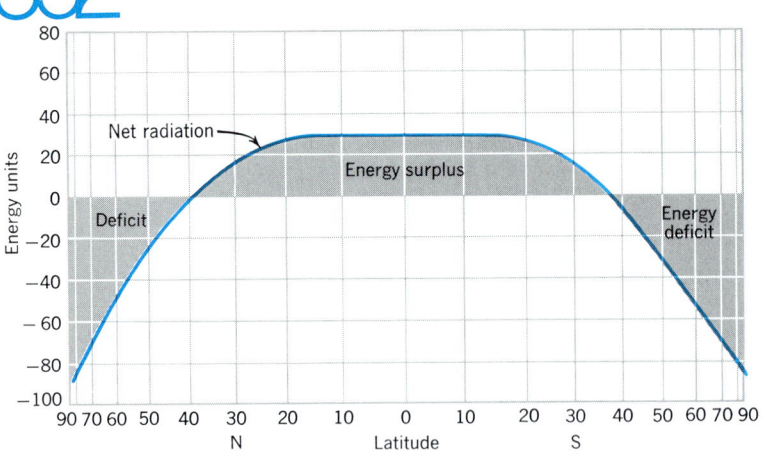

FIGURE 21-16

Net radiation from pole to pole shows two polar regions of energy deficit, matching a large low-latitude region of energy surplus.

yearly averages stay more or less constant at any given observing station no matter where it is located. This constancy requires that heat be transported from the belt of surplus to the regions of deficit. We shall find that heat transfer mechanisms exist in the circulation of the atmosphere and oceans and that these mechanisms effectively keep temperatures under control in all latitude belts.

FIGURE 21-17

Isobaric surfaces and the pressure gradient.

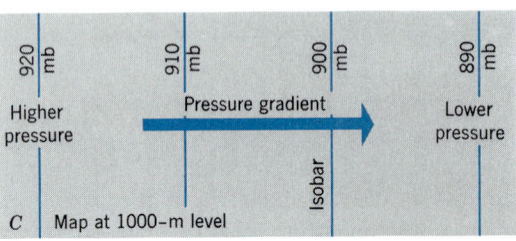

21.18 WINDS AND THE PRESSURE GRADIENT

Recall that atmospheric pressure, or *barometric pressure*, decreases from the ground up. For an ideal atmosphere at rest, the barometric pressure will be the same throughout a given horizontal surface at any height. This condition is shown in the upper part of Figure 21–17. Surfaces of equal barometric pressure, called *isobaric surfaces*, appear in cross section in this diagram as horizontal, parallel lines. Pressures in millibars have been assigned to these surfaces.

Most of the time, over the earth generally, isobaric surfaces are not horizontal. Instead, as shown in the middle diagram of Figure 21–17, the isobaric surfaces slope down or up from one area to another. Consequently, at some given height, such as 1000 meters, pressure is higher in one place and lower in another. In the diagram, pressure at 1000 meters is higher at the left than at the right. Specifically, pressure at the left is 920 mb; at the right it is 890 mb. We describe this situation as a *pressure gradient*, in this case sloping down from left to right.

Next, we make a map of the situation. As shown in the lower diagram of Figure 21–17, a map of the 1000-meter level consists of lines of equal pressure, called *isobars*. These lines cut across the map from top to bottom and are labeled with their pressure values. A broad arrow shows the direction of the pressure gradient.

Where a pressure gradient exists, air tends to move in the same direction as the gradient—that is, from higher to lower pressure. For simplicity, we say that a *pressure-gradient force* acts on the air, urging it to move. The stronger (steeper) the pressure gradient, the stronger the force, and so the stronger will be the wind.

21.19 THE CORIOLIS EFFECT

If the earth did not rotate on its axis, wind would move exactly in the direction of the pressure gradient. Wind arrows would cut across the isobars at right angles. However, because of earth rotation, another factor comes into play, changing the wind direction. This factor is the *Coriolis effect*. In the northern hemisphere, the Coriolis effect acts like a force pulling toward the right on any substance in horizontal motion, whether that substance be a solid object, a liquid, or a gas. In response to the Coriolis effect, air motion is deflected toward the right. As shown in Figure 21–18, it does not matter in what

21.20 THE GEOSTROPHIC WIND

compass direction the air is moving—east, west, north, or south—the deflection is always toward the right. This tendency for deflection is named after a French mathematician, G. G. Coriolis, who presented the first analysis of the phenomenon in 1835. Physicists are careful to point out that the Coriolis effect is not a true force, although it can be visualized as a force in explaining the relationship between wind and the pressure gradient.

The Coriolis effect is nonexistent precisely on the equator; it increases in strength to a maximum at each pole, as the widening arrows in Figure 21-18 suggest. At the 30th parallel, the deflecting effect is 50 percent of the polar value; at the 60th parallel, it is about 87 percent. The Coriolis effect also becomes stronger as the velocity of motion increases.

A full explanation of the Coriolis effect would be too lengthy and complex to present here. However, the principle is easy to illustrate with a rotating disk, such as a record turntable (Figure 21-19). Imagine this disk to represent a small region centered on the north pole. Mount a straightedge above the table. Using this straightedge as a guide, move a pencil poleward over the turning disk. The mark you have made will be a curving path, deflected toward the right of the direction of motion when the disk is turned counterclockwise (northern hemisphere). Your pencil point followed a straight path in space, but its track was curved on the turning surface. If you reverse direction and draw the line from the pole toward the outer edge of the disk, the track will again be curved toward the right.

The principle involved here is Newton's first law of motion, stating that any body in motion will follow a straight line unless compelled to change its path by some external force. The straight line mentioned in the law is a line fixed in space with respect to the stars. Of course, objects moving on the earth's surface are held to that surface by gravity and are also subjected to friction, so that they cannot move freely in a straight space path. The deflecting effect is nevertheless acting on them. Air and water respond very readily to the Coriolis effect.

21.20
THE GEOSTROPHIC WIND

Let us now apply the Coriolis effect to winds. Visualize the motion of a small parcel of air starting at rest from point *A* of Figure 21-20. Isobars show the pressure gradient to be uniformly lower toward the left. The pressure-gradient force acts toward the left with a con-

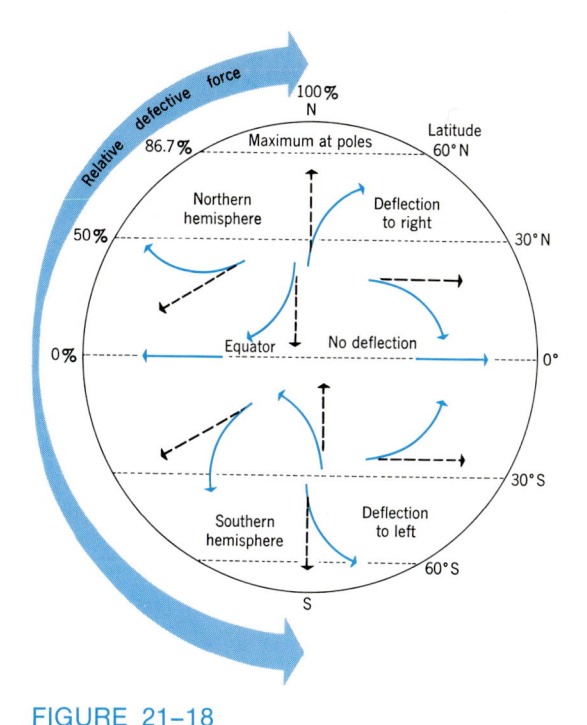

FIGURE 21-18
Deflective force of the earth's rotation.

stant value at all times. The pressure-gradient force tends to set the air parcel in motion at right angles across the isobars. However, as soon as motion starts, the Coriolis effect acts at right angles to the path of motion, as shown by very

FIGURE 21-19
The Coriolis effect can be demonstrated by drawing a pencil line on a rotating turntable. (From PRINCIPLES OF EARTH SCIENCE by Arthur N. Strahler. Copyright © 1976 by Arthur N. Strahler. By permission of Harper & Row, Publishers, Inc.)

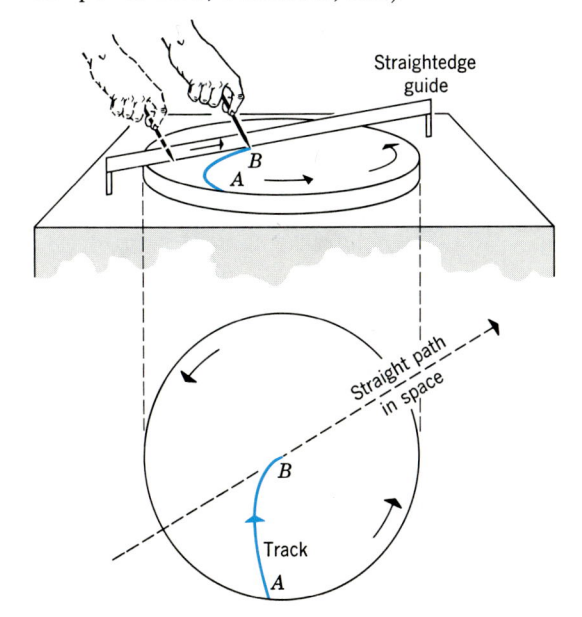

FIGURE 21-20
Deflection of a parcel of air by the Coriolis effect, leading to development of the geostrophic wind. (From THE EARTH SCIENCES, 2nd Edition by Arthur N. Strahler. Copyright © 1963, 1971 by Arthur N. Strahler. By permission of Harper & Row, Publishers, Inc.)

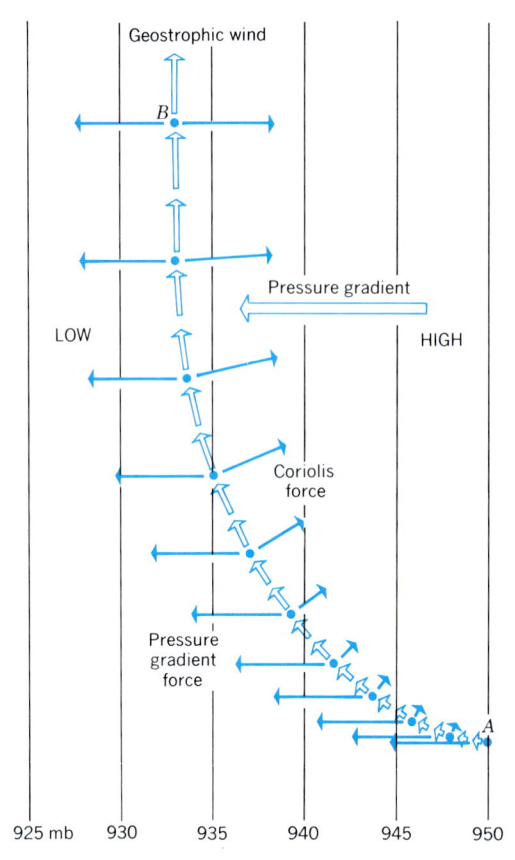

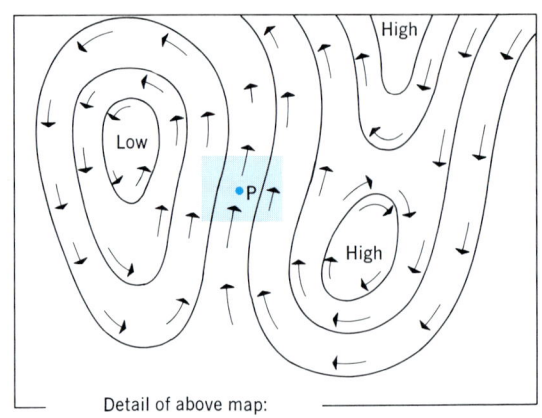

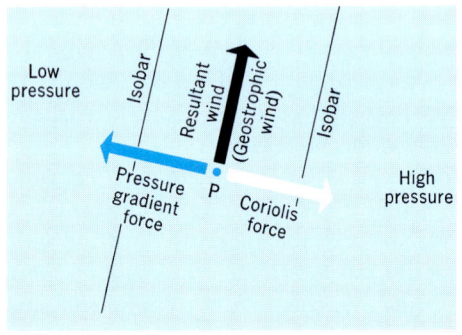

FIGURE 21-21
Wind follows the isobars at high levels.

small arrows. The air parcel responds by turning toward the right. As speed of motion increases, the Coriolis effect also increases. Finally, the path is turned to achieve a direction at right angles to the pressure gradient but parallel with the isobars (point B of Figure 21-20). Here the Coriolis effect exactly balances the pressure-gradient force and no further turning ensues. In the case of straight, parallel isobars, the flow of air at point B is termed the *geostrophic wind*.

21.21 CYCLONES AND ANTICYCLONES

Where isobars are curved, air also moves in a curved path. Motion in a curved path brings into play centrifugal force. We shall not go into details except to note that centrifugal force acts either to increase or to decrease the speed of the air motion, depending on the geometry of the given situation. So far as we are concerned here, the wind will follow closely the curving isobars along whatever configurations they happen to take.

Using the Coriolis principle, analyze a simple map showing isobars and winds (Figure 21-21). This weather map shows conditions several thousand meters above the surface. A low-pressure center lies at the left; a high-pressure center at the right. The lower diagram is an enlarged portion of the map between the high and the low. At this point the geostrophic wind blows northward, paralleling the isobars. Arrows on the map show the airflow paralleling the isobars and circling the pressure centers.

Winds circling a low-pressure center constitute a *cyclone*; winds circling a high-pressure center constitute an *anticyclone*. Figure 21-22 illustrates cyclones and anticyclones. In the northern hemisphere, winds move counterclockwise in the cyclone, clockwise in the anticyclone. Reverse directions apply to the southern hemisphere, as you see in the lower half of the figure.

If you were an airline pilot, plotting your course so as to have a tailwind at all times, your rule would be "keep the lows on your left and the highs on your right" (northern hemisphere).

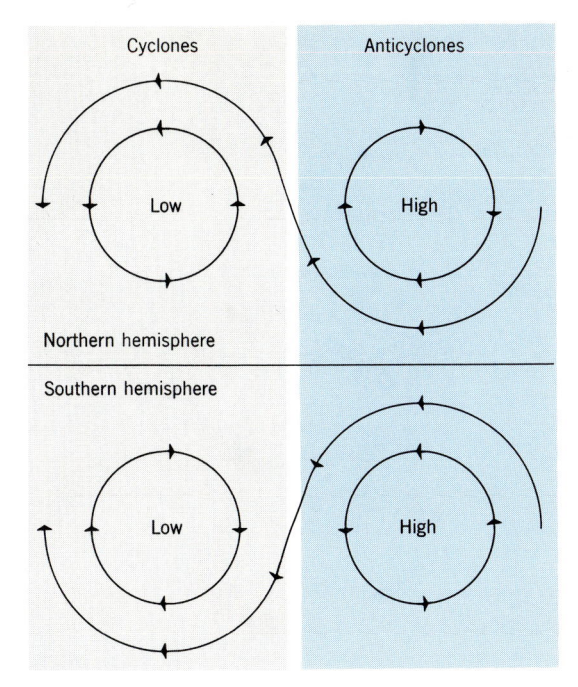

FIGURE 21-22

Winds at high levels around cyclones and anticyclones.

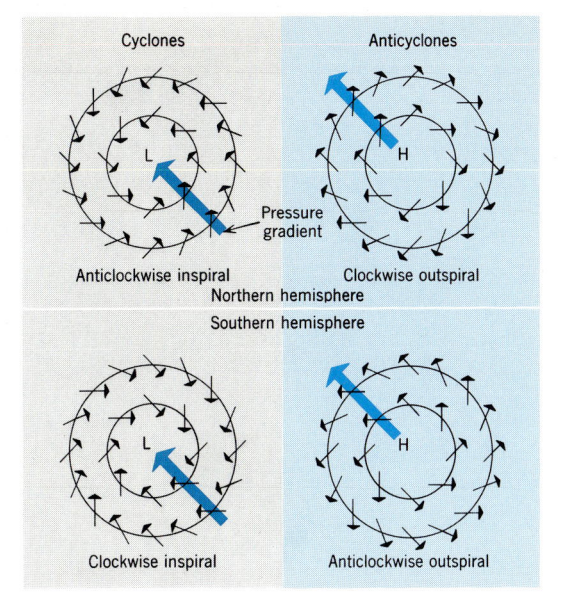

FIGURE 21-23

Surface winds within cyclones and anticyclones.

21.22
SURFACE WINDS

Air moving close to the surface encounters friction with the land or water surface beneath it. This drag not only retards the airflow, reducing the wind speed, but also changes the angle between wind and isobars.

Surface winds within cyclones and anticyclones are illustrated in Figure 21-23. The broad arrows show the pressure gradient—inward toward the center of the cyclone but outward from the center of the anticyclone. Cyclones have in-spiraling winds; anticyclones have out-spiraling winds. Directions of spiraling are reversed in the southern hemisphere. Now it must be obvious that where air is spiraling into the center of a cyclone it is converging and must be disposed of by rising to higher levels. Air spiraling out from an anticyclone is diverging and must be replaced by air subsiding from higher levels. Therefore, *convergence* characterizes a cyclone; *divergence* an anticyclone.

21.23
THE GLOBAL CIRCULATION

The atmosphere is heated most intensely in the equatorial zone. Here the heated air tends to expand, producing a low-pressure belt at the surface. There is a general rise of the warm air over the equatorial zone, and at high levels this air begins to move poleward. As this happens the air motion is deflected to the right (northern hemisphere) to become a system of upper-air westerly winds paralleling the isobars at about the 30th parallel (Figure 21-24). Correspondingly, in the southern hemisphere, deflection to the left also turns the poleward flow at high level into a westerly wind system. Westerly winds also extend down to the earth's surface in middle and high latitudes, and are called the *prevailing westerlies*.

Because air moving at high levels poleward from the equatorial belt has been turned into westerly flow, following the earth's parallels of latitude, the air here tends to accumulate more rapidly than it can escape poleward. This accumulation, or banking up, of air takes place in zones between 20 and 30 degrees north latitude and between 20 and 30 degrees south latitude. Here accumulation aloft produces at the surface two belts of high barometric pressure known as the *subtropical high-pressure belts,* one in each hemisphere.

Part of the air subsiding within the subtropical high-pressure belt spreads toward the equator. As this air follows the barometric pressure gradient from subtropical high to equatorial low, it is deflected westward to create a system of prevailing winds known as the *tropical easterlies* (Figure 21-24, right). The tropical easterlies form a broad, steady, deep airstream moving around the earth over the equatorial regions and extending to high altitudes. At low levels, these winds are the *trade winds,* or *trades*.

The atmospheric circulation system of equatorial and tropical latitudes thus consist of two

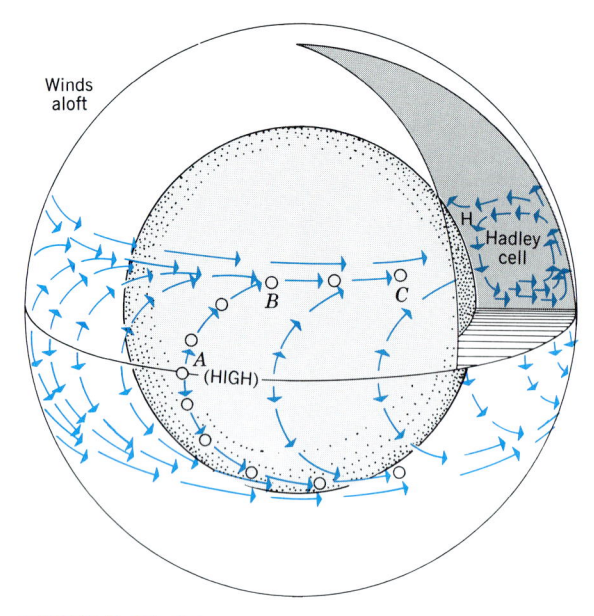

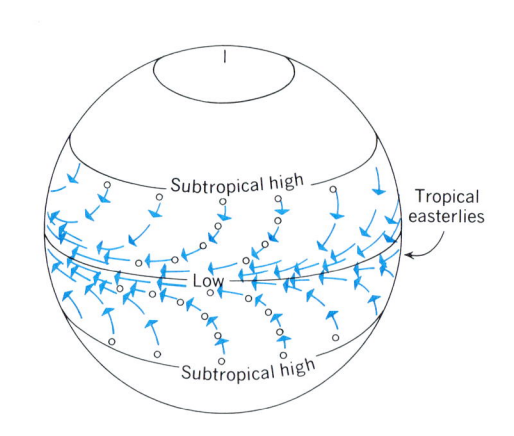

21 ENERGY FLOW IN ATMOSPHERE AND OCEANS

FIGURE 21-24
Idealized diagram of formation of the Hadley cell circulation and the tropical easterlies. (From THE EARTH SCIENCES, 2nd Edition by Arthur N. Strahler. Copyright © 1963, 1971 by Arthur N. Strahler. By permission of Harper & Row, Publishers, Inc.)

cells, one in each hemisphere. Seen in cross section, and neglecting east-west components of motion, the meridional circulation within each cell consists of horizontal and vertical motions, together forming a complete circuit. The existence of such a circulation system was first postulated by George Hadley in 1735 and is now called the *Hadley cell* by meteorologists.

21.24 THE UPPER-AIR WESTERLIES

Poleward of the subtropical high-pressure belts, circulation in the troposphere takes the form of a prevailing system of *upper-air westerlies*. These winds are shown schematically in Figure 21-25. Air moving northward is deflected by the Coriolis effect to the right (toward the east) and becomes a west wind. The flow constitutes a great vortex moving counterclockwise (northern hemisphere) around a prevailing center of low barometric pressure, the *polar low*. A corresponding system of upper-air westerlies exists in the southern hemisphere.

The simple west-to-east flow of the westerlies is disturbed by a ceaseless succession of wavelike undulations. These undulations are *upper-air waves,* or *Rossby waves,* named for C.-G. Rossby, a meteorologist who developed the mathematical equations governing the waves.

The upper-air waves may grow, change in form, and dissolve. They may remain essentially stationary for many days and may also drift slowly in the east-west direction.

21.25 HOW UPPER-AIR WAVES DEVELOP

Let us analyze the patterns of development of upper-air waves and their relation to atmospheric temperatures. Figure 21-26 is a schematic diagram showing wave evolution in four stages. Long, heavy arrows show the location of a high-speed airstream called the *jet stream*. The jet stream defines the position of the waves. Conditions shown are those existing near the top of the troposphere, which ranges in height from about 9 km over the poles to about 17 km over the equator.

The troposphere lying poleward of the jet stream consists of cold polar air, whereas that on the equatorward side consists of warm tropical air. Such large bodies of the atmosphere are referred to as *air masses*. Air masses are identified on the basis of both temperature and water-vapor content. We will explain in greater detail in Chapter 22 that polar air masses, because they are cold, can hold little moisture. In contrast, warm tropical air masses can hold comparatively large quantities of moisture.

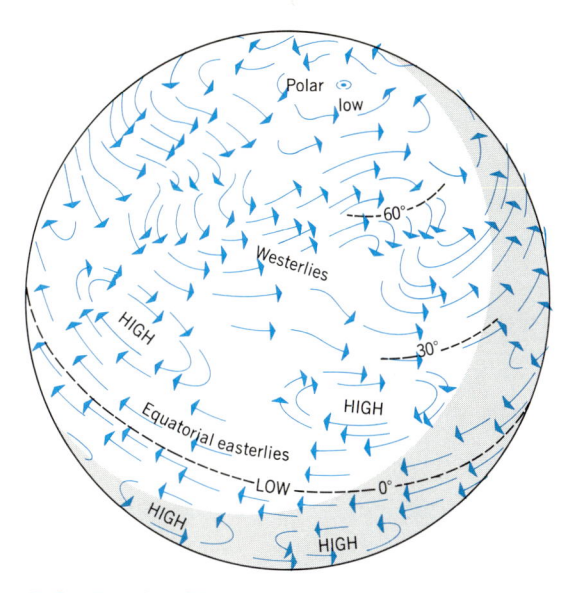

FIGURE 21-25
Schematic representation of circulation in the upper part of the troposphere, 6 to 12 km.

The jet stream in middle latitudes occupies a position at the contact between the *polar air mass* and the *tropical air mass*. A contact surface between adjacent air masses is known as a *front*. In the case we are examining, the front lying beneath the jet stream is known as the *polar front*.

Figure 21-26A shows the jet stream lying over the high latitudes and with only small undulations. As waves form (Figure 21-26B), the polar air pushes south at one place and the tropical air moves north at another. Soon great tongues of air form an interlocking pattern, with the jet stream taking a sinuous path between them (21-26C). Finally, a wave constricts at the base, and a mass of cold or warm air is detached, forming an isolated *pressure cell* (21-26D).

A cell of stranded cold air aloft at subtropical latitudes forms a low-pressure center with counterclockwise circulation. An isolated cell of warm air aloft at the higher latitude becomes a high-pressure center with clockwise airflow. At the close of the wave-development cycle, which takes four to six weeks to complete, the isolated cells dissolve. The jet stream then resumes its simple course over the high latitudes.

The cycle of upper-air wave development explains how great quantities of heat are transferred from equatorial regions to polar regions. North-moving tropical air carries heat to the high latitudes where the heat is lost. South-moving tongues bring cold air to the low latitudes. Here the cold air mass absorbs part of the excess heat. Although this form of heat and moisture transfer is fluctuating in intensity and location, the average effect, year in and year out, is to maintain in balance the earth's heat budget.

You can think of the jet stream as resembling the high-pressure flow of water from a hose nozzle held submerged and pointed horizontally in the direction of flow of a slowly moving stream (Figure 21-27). The jet stream is shaped like a tube that lies roughly horizontally. The tube may curve from side to side to change direction from, say, northwest to west to southwest. Its core lies at the top of the troposphere.

The jet stream derives its high wind speed from a very steep pressure gradient found at the polar front, along the contact of the polar and tropical air masses. The steep pressure gradient is explained by the great contrast in air temperatures on the two sides of the polar front.

21.26
WINDS AND OCEAN CURRENTS

Winds blowing over the ocean surface transfer vast quantities of energy from the atmosphere to the oceans. One can think of the atmospheric circulation systems, such as the Hadley cell and the prevailing westerlies, as great gear wheels. Meshed with the sea, these wheels turn related systems of surface water motion.

As wind blows over a smooth water surface a frictional drag is exerted by the air on the water. Frictional drag sets the surface water in motion. As the uppermost water layer is dragged in the downwind direction, it in turn drags along the next lower layer. In this way the motion is propagated toward increasing depth. The moving water is subjected to the Coriolis effect. In the northern hemisphere Coriolis acts as a force pulling toward the right in a direction perpendicular to the direction of motion. The action is the same as for air in motion.

All horizontal motions of the surface ocean layer are included in the general term *ocean currents*.

Within each ocean a characteristic pattern of surface currents and drifts is repeated. Figure 21-28 shows schematically the major elements of the flow system. The major features are two great *gyres*, or circular flow systems. One gyre is located in each hemisphere, centered approximately on the subtropical cell of high barometric pressure. Water motion is clockwise about the gyres.

The trade winds set in motion a west-moving *equatorial current*, paralleling the equator. The prevailing westerlies set in motion the east-moving *west-wind drift* in middle latitudes. The

21 ENERGY FLOW IN ATMOSPHERE AND OCEANS

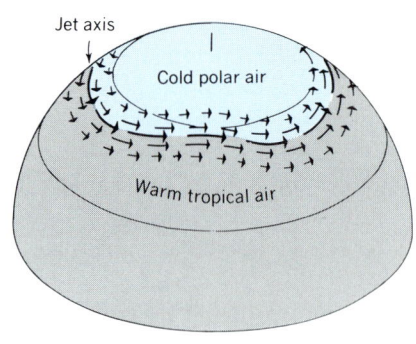

A. Jet stream begins to undulate

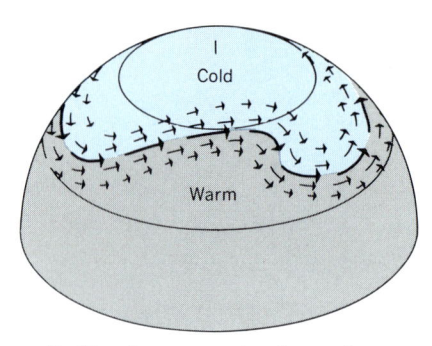

B. Rossby waves begin to form

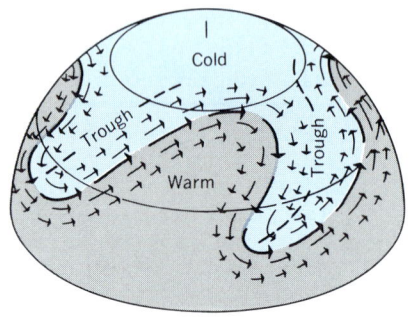

C. Waves strongly developed

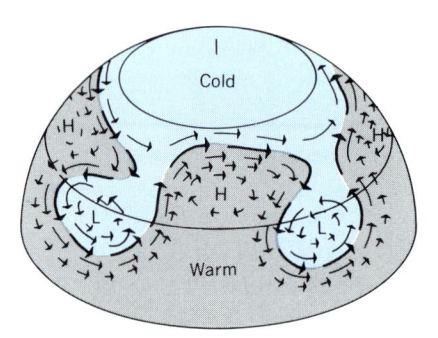

D. Cells of cold and warm air bodies are formed

FIGURE 21-26
Development of upper-air waves in the westerlies. Modified from diagrams by J. Namias, NOAA National Weather Service. (From THE EARTH SCIENCES, 2nd Edition by Arthur N. Strahler. Copyright © 1963, 1971 by Arthur N. Strahler. By permission of Harper & Row, Publishers, Inc.)

gyre is completed by a strong poleward flow at the western side of the gyre.

Because of the Coriolis effect, the gyres are pushed toward the west side of the ocean. Consequently, the poleward currents on the west sides are intensified. Examples are the Gulf Stream off eastern North America, and the Kuroshio Current off Japan. These currents are relatively warm and serve to transport heat from low to high latitudes.

On the eastern sides of the gyres, the drift is turned equatorward, bringing cool water of arctic and antarctic origin into low latitudes. These equatorward currents bring unusually cool air temperatures to continental west coasts at low latitudes. The effect of the great oceanic gyres on

FIGURE 21-27
The jet stream. Figures give typical wind speeds in km/h. (From National Weather Service.)

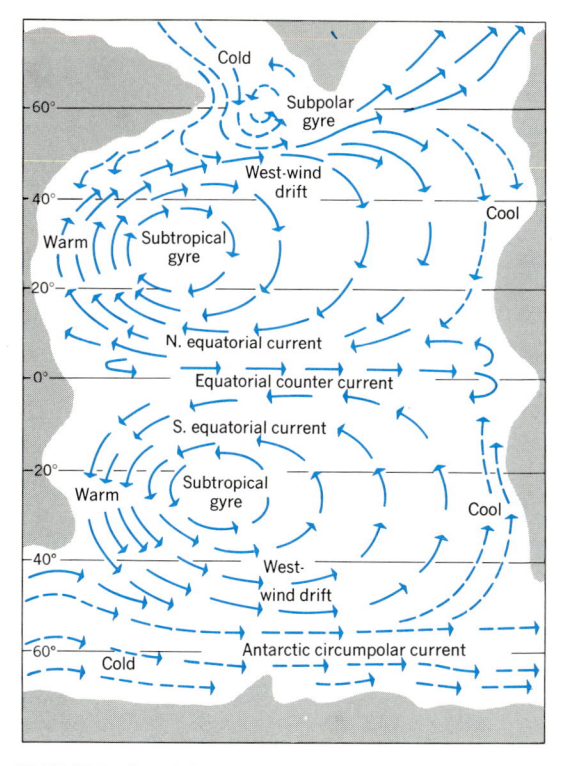

TERMS AND CONCEPTS

FIGURE 21-28
Schematic map of system of ocean currents.

the earth's heat budget is most important, for enormous quantities of water are exchanged by the flow.

Figure 21-29 summarizes the roles of atmospheric circulation and ocean currents in transporting surplus heat from low latitudes to high latitudes.

FIGURE 21-29
The global heat balance.

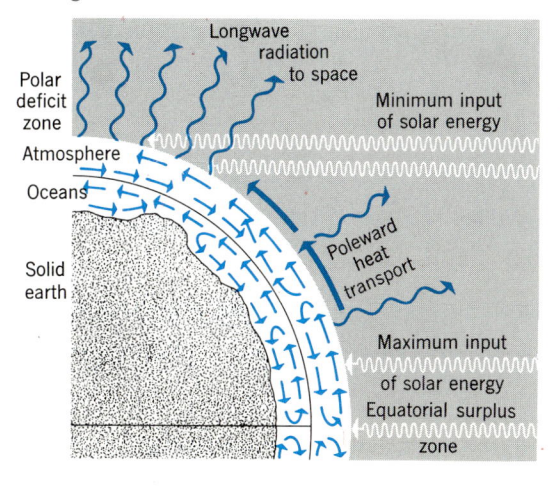

SUMMARY

The atmosphere. Besides the nonvarying gases, our atmosphere contains important quantities of water vapor, dust, water droplets, and ice particles. Drawn to the earth by gravity, the atmosphere is densest at the surface, where barometric pressure is highest. Temperature layers and layer boundaries of the atmosphere from base to top are: troposphere, tropopause, stratosphere, stratopause, mesophere, mesopause, thermosphere.

Ionosphere, ozone layer, magnetosphere. Shortwave solar radiation ionizes atmospheric gases in the thermosphere, generating the ionosphere. Lower down, in the stratosphere, ultraviolet radiation absorption forms the ozone layer. Surrounding the earth is a vast external magnetic field, the magnetosphere, in which energetic particles are trapped, producing intense ionizing radiation in the Van Allen radiation belt.

The world ocean. Containing over 97 percent of the world's free water, and averaging 4 km in depth, the world ocean is a brine containing salts, principally sodium chloride. A three-layer temperature structure is found in all but arctic regions, with warm water forming a surface layer above a strong thermocline.

The radiation balance. Incoming shortwave solar radiation is exactly balanced by outgoing longwave radiation for the globe as a whole. The atmosphere reflects and absorbs about half of the incoming shortwave radiation. The atmosphere is heated largely by longwave radiation, retaining heat through the greenhouse effect.

Winds. The pressure gradient force sets air in motion, while the Coriolis effect turns the flow to a trend parallel with the isobars. Airflow forms spiral patterns in cyclones and anticyclones.

Global circulation. A great flow of westerly winds encircles the globe in middle and high latitudes. Here Rossby waves develop along the jet stream, where cold and warm air masses are in contact. In low latitudes the general flow is opposite, within the system of tropical easterlies.

Ocean currents. Driven by surface winds, the surface layer of the oceans is set into motion in a system of great gyres. Within each gyre, an equatorial current lies beneath the tropical easterlies, a west-wind drift beneath the prevailing westerlies. Both winds and ocean currents transport excess heat from low to high latitudes, maintaining the global energy balance.

TERMS AND CONCEPTS

atmosphere
oceans
hydrosphere
meteorology

physical oceanography
pure dry air
water vapor
atmospheric pressure (air pressure)
mercurial barometer
standard sea-level pressure
millibar
environmental temperature lapse rate
troposphere
tropopause
stratosphere
stratopause
mesosphere
mesopause
thermosphere
ionosphere
Kennelly-Heaviside layer
ozone layer
magnetosphere
external magnetic field
solar wind
shock front
magnetopause
ionizing radiation
Van Allen radiation belt
world ocean
brine
thermocline
salinity
radiation balance
micron
shortwave radiation
longwave radiation
solar constant
subsolar point
circle of illumination
equinoxes
solstices
scattering
albedo
infrared imagery
counterradiation
greenhouse effect
net radiation
annual surplus, deficit
barometric pressure
isobaric surfaces
pressure gradient
isobars
pressure-gradient force
Coriolis effect
geostrophic wind
cyclone

anticyclone
convergence
divergence
westerly winds
prevailing westerlies
subtropical high-pressure belts
tropical easterlies
trade winds (trades)
Hadley cell
upper-air westerlies
polar low
upper-air waves (Rossby waves)
jet stream
air masses, polar, tropical
front
polar front
pressure cell
ocean currents
gyres
equatorial current
west-wind drift

QUESTIONS

1. Name and describe the component substances of the earth's atmosphere. What important function has each component?

2. What causes atmospheric pressure? How does atmospheric pressure change with increasing altitude? Describe the Torricelli experiment and explain how it makes possible the construction of a simple barometer.

3. Describe the temperature layers and layer boundaries of the atmosphere from the surface upward. Give the value of the environmental temperature lapse rate in the troposphere.

4. What causes the ionosphere to form? Where is the ionosphere located? How does ionization explain heating of the thermosphere? How are radio waves affected by the ionosphere?

5. Where is the ozone layer located? What causes this layer to form? What importance has the ozone layer to humans?

6. Describe the magnetosphere and explain its existence. How does the solar wind play a part in shaping the magnetosphere? How is ionizing radiation produced within the magnetosphere, and where is it concentrated?

7. Describe the world ocean in terms of surface area and volume. Compare the oceans with the atmosphere in terms of fluid motion. What is the composition of seawater?

8. Describe the temperature structure of the oceans. How is density of seawater related to salinity? How does pressure change with depth in the ocean?

9. Describe the earth-sun-space radiation system and explain the concept of a radiation balance. Describe the radiation spectrum of the sun and compare it with the earth's radiation in terms of wavelength and intensity.

10. What factors determine the intensity of solar radiation received by a unit area of horizontal surface? How do the seasons cause variation in the intensity of solar radiation at various latitudes?

11. What energy losses does incoming solar radiation experience in passing through the earth's atmosphere? Give a value for the earth's albedo. What proportion of energy reaches the earth's surface?

12. Explain the concept of longwave radiation from the earth and its atmosphere. How is infrared imagery obtained? Explain the greenhouse effect.

13. How does net radiation vary from low latitudes to the poles? Explain. Where is there a radiation surplus? Where is there a deficit?

14. Explain the relationship of the pressure-gradient force to isobaric surfaces and isobars. How does the Coriolis effect influence air motion? Can you explain the Coriolis effect?

15. Explain why air motion in the geostrophic wind parallels the isobars. What patterns of airflow are found in cyclones and anticyclones? Is the motion the same in the northern hemisphere as in the southern hemisphere? How are winds near the earth's surface influenced by the frictional effect? Describe airflow near the surface in cyclones and anticyclones.

16. Describe the general patterns of global atmospheric circulation. Explain both the upper-air westerlies and the tropical easterlies. Why does high pressure exist in the subtropical zone?

17. What changes do upper-air waves (Rossby waves) experience? How are the waves related to the jet stream? What air masses are involved in wave development?

18. What causes ocean currents? Describe the general pattern of oceanic gyres and relate the flow components to prevailing winds.

PROBLEMS

1. At Chicago, Illinois, on a day in mid-April, the surface air temperature measures 10°C. A vertical sounding of the atmosphere by balloon reveals a nearly constant environmental temperature lapse rate of 6.4C°/km. At the tropopause a temperature of −50°C is reported. Assuming the altitude of Chicago to be sea level, calculate the altitude of the tropopause.

2. On Nauru Island, located in mid-Pacific close to the equator, a temperature sounding is made in a manner similar to that described in problem 1. Here the sea-level air temperature is 30°C and the lapse rate is constant at 6.6C°/km. The tropopause is encountered at an altitude of 16.8 km. Calculate the air temperature at the tropopause.

3. The table below gives standard values for the barometric pressure in millibars at various altitudes above sea level. Using a sheet of graph paper, plot these data as a series of points and connect them with a smooth curve.

Pressure (mb)	Altitude (m)
400	7425
500	5643
600	4186
700	2955
800	1889
900	984
1000	106
1013	0

(a) Using your graph, estimate the standard barometric pressure for each of the following places:

Area	Altitude (m)
Canton, Ohio	30
Las Vegas, Nevada	620
Cheyenne, Wyoming	1860
Mount Hood, Oregon	3400
Mount Whitney, California	4420

(b) Estimate the rate of decrease of pressure with altitude, in millibars per km, between the following levels:

Between 1 km and 2 km
Between 3 km and 4 km
Between 6 km and 7 km

4. Calculate the hydrostatic pressure in kg/cm^2 at a depth of 300 m in the ocean. Be sure to include the atmospheric pressure in your calculation.

5. Using a value of 2.0 cal/cm^2/min for incoming solar radiation intensity at the equator at equinox (assuming no atmosphere), what will be the radiation intensity (a) at 30 degrees latitude? (b) at 60 degrees latitude? (Hint: Radiation intensity diminishes as the sine of the angle of latitude.)

CHAPTER 22
DYNAMICS OF THE WEATHER—PRECIPITATION AND STORMS

22.1 WATER VAPOR IN THE ATMOSPHERE

The role of water in the atmosphere is the subject of this second of two chapters on meteorology, the physical science of the atmosphere. Using principles developed in earlier chapters, we first examine atmospheric water in the vapor state, then follow with condensation of that water vapor and the liberation of latent heat. This pathway leads us to investigate clouds and precipitation, and finally major atmospheric disturbances—storms—powered by the liberation of latent heat during the condensation process.

The amount of water vapor that may be present in the air at a given time varies widely from place to place. It ranges from almost nothing in the cold, dry air of arctic regions in winter to as much as 4 or 5 percent of the volume of the atmosphere in the warm equatorial belt.

Water vapor enters the atmosphere by evaporation from exposed water surfaces such as oceans, lakes, rivers, or moist ground. Some is supplied by plants that transpire water (a form of evaporation). With large expanses of ocean and densely forested lands over the globe, there is no lack of surface for evaporation.

We use the term *humidity* to refer generally to the degree to which water vapor is present in the air. For any specified temperature there is a definite limit to the quantity of moisture that can be held by the air. When this limit is reached, the air is said to be *saturated*.

The proportion of water vapor present relative to the maximum quantity is the *relative humidity*, expressed as a percentage. For saturated air, relative humidity is 100 percent. When half of the total possible quantity of vapor is present, relative humidity is 50 percent, and so on.

22.2 RELATIVE HUMIDITY AND AIR TEMPERATURE

A change in relative humidity of the atmosphere can be caused in one of two ways. If an exposed water surface is present, the humidity can be increased by evaporation. This is a slow process, requiring that the water vapor diffuse upward through the air.

The other way relative humidity can change is through a change of air temperature. Even though no water vapor is added, a lowering of temperature results in a rise of relative humidity. This is an automatic change. It is a consequence of the fact that the capacity of the air to hold water vapor has been reduced by cooling. For cooler air, the existing amount of vapor represents a higher percentage of the total capacity of the air. Going the opposite way, a rise of air temperature results in decreased relative humidity, even though no water vapor has been taken away.

A simple example illustrates these principles (Figure 22–1). At a certain place the midmorning temperature of the air is 16°C; the relative humidity (RH) is 50 percent. In midafternoon the

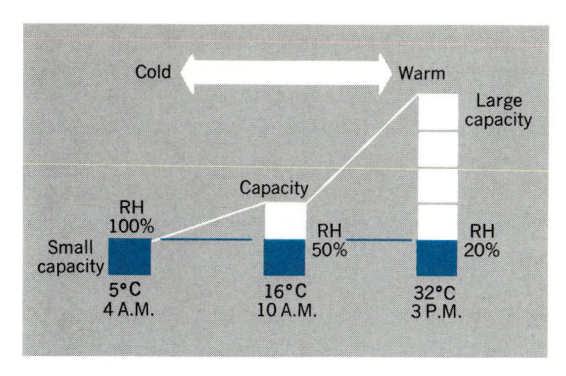

FIGURE 22-1
Relative humidity decreases as air temperature increases because the capacity of warm air is greater than the capacity of cold air.

air becomes warmed by longwave radiation from the sun and ground surface to reach 32°C. The relative humidity has now dropped to 20 percent, which is very dry air. By outgoing longwave radiation, the air becomes chilled during the night; by early morning, its temperature falls to 5°C. Now the relative humidity has automatically risen to 100 percent, the saturation value.

Any cooling of saturated air will usually cause condensation of the excess vapor into liquid form. As the air temperature continues to fall, the humidity remains at 100 percent, but condensation continues, taking the form of minute droplets of dew or fog. If the temperature falls below freezing, condensation usually occurs as frost upon exposed surfaces.

22.3 DEW-POINT TEMPERATURE

The *dew point* is that critical temperature at which the air is fully saturated. Below this temperature, condensation normally occurs. We have an excellent illustration of condensation due to cooling in summer when beads of moisture form on the outside surface of a pitcher filled with ice water. Air immediately adjacent to the cold glass surface is chilled enough to fall below the dew-point temperature. Moisture then condenses on the surface of the glass.

22.4 SPECIFIC HUMIDITY

Relative humidity has only limited use in weather science, because of the fact that an air temperature change causes the relative humidity to change. Besides, relative humidity, being a percentage, does not tell how much moisture is actually held in the air. Meteorologists use a more definite measure—specific humidity—to describe the moisture content of an air mass.

Specific humidity is the ratio of weight of water vapor to weight of moist air (including the water vapor). Units are grams of water vapor per kilogram of moist air. When a given mass of air is lifted to higher altitudes without gain or loss of moisture, the specific humidity remains constant even though the air expands to occupy a larger volume.

Specific humidity is used to describe the moisture characteristics of a large mass of air. For example, extremely cold, dry air over arctic regions in winter often has a specific humidity as low as 0.2 g/kg, whereas extremely warm, moist air of tropical regions may hold as much as 18 g/kg. The total natural range on a worldwide basis is such that the largest values of specific humidity are 100–200 times as great as the least.

In a sense, specific humidity is a yardstick of a basic natural resource—fresh water—to be applied from equatorial to polar regions. It is a measure of the quantity of water that can be extracted from the atmosphere as precipitation. Cold air can supply only a small quantity of rain or snow; warm air is capable of supplying large quantities. This is a very important concept to keep in mind, and we shall apply it in our study of storms.

22.5 CONDENSATION AND THE ADIABATIC PROCESS

Precipitation is the general term applying collectively to actively falling rain, snow, sleet, or hail. Precipitation can result only where large masses of air are experiencing continued drop in temperature below the dew point. This required drop in temperature cannot be brought about by the simple process of chilling of the air through loss of heat by radiation during the night. Instead, it is essential that a large mass of air be rising to higher elevations. To understand why this must be so requires us to apply the concept of the adiabatic process.

One of the most important laws of meteorology is that rising air experiences a drop in temperature, even though no heat energy is lost to the outside (Figure 22-2). The drop of temperature is a result of the decrease in air pressure at higher altitudes, permitting the rising air to expand. The opposite relationship also holds true: air sinking to lower levels is compressed into a smaller volume and experiences a rise in temperature.

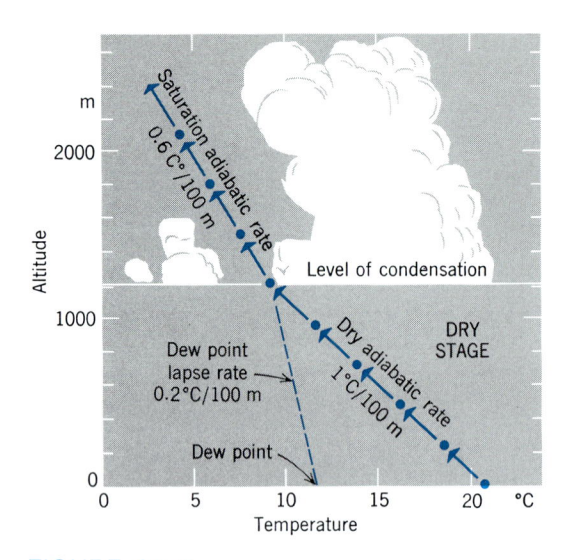

FIGURE 22–2
Adiabatic changes of temperature in a rising air mass.

When no condensation is occurring, the rate of drop of temperature of rising air is the *dry adiabatic lapse rate*; it has a value of about 1 C°/100 m of vertical rise. The dew point also declines as air rises; the rate is 0.2 C° per 100 m. In Figure 22–2, the drop in dew-point temperature is labeled *dew-point lapse rate*.

When water vapor in the air is condensing, the adiabatic rate is less; a typical value is 0.6 C° per 100 m. The reduced rate results from the liberation of latent heat during the condensation process. The production of latent heat partially compensates for the adiabatic cooling. This modified rate is referred to as the *wet* (or *saturation*) *adiabatic lapse rate* (Figure 22–2).

Do not confuse the adiabatic lapse rate with the environmental temperature lapse rate, explained in Chapter 21. The environmental lapse rate applies to still air whose temperature is measured at successively higher levels by a thermometer carried upward. As Figure 22–2 shows, the rise of air according to the dry adiabatic rate falls within the *dry stage*. When the converging dew-point lapse rate and dry adiabatic lapse rate meet, the air is saturated and the *level of condensation* is reached. With further rise, condensation sets in and a cloud forms; this is within the *rain stage*.

22.6
CLOUDS

Clouds are dense concentrations of suspended minute water droplets or ice crystals. These particles have diameters in the range 0.02–0.06 mm. Under a microscope cloud droplets look like tiny spheres. Each droplet forms by condensation of water on a tiny nucleus. Usually the nucleus is a bit of *hygroscopic* salt—that is, salt with an affinity for water. Common table salt is hygroscopic, unless specially treated, and it turns damp and sticky in moist weather.

Salt particles are abundant in the atmosphere because the turbulent winds blowing over ocean waves lift bits of salt spray into the air. Evaporation of these spray droplets leaves salt particles that travel easily into all parts of the troposphere and make excellent cores upon which cloud particles can be formed. Growth of cloud droplets begins while air is still not fully saturated, and the droplets become rapidly larger when the saturation point is reached.

Minute particles of water can remain in the liquid state at temperatures far below the normal freezing point of 0°C. Such liquid water is said to be *supercooled*. At temperatures down to about −10°C, cloud particles are almost entirely in the liquid state. In colder air the cloud is a mixture of liquid and ice particles. Below −40°C the cloud consists entirely of ice crystals. Generally speaking, the highest forms of clouds—above about 6 km—are composed entirely of ice crystals because air temperatures here are very low.

Clouds have such a high capacity to reflect sunlight that reflection of the entire visible spectrum occurs, accounting for their brilliant snowy appearance when lighted by the sun. Dense cloud masses appear gray or black on the underside because sunlight is unable to pass through. Thin layers transmit enough sunlight to appear gray, whereas some of the thinnest veillike clouds seem scarcely to weaken the intensity of direct sunlight.

Apart from their major role as producers of rain, snow, sleet, and hail, clouds are excellent indicators of the general weather situation, the direction and speed of air movement, and the moisture state of the air.

22.7
CLOUD FORMS

Clouds formed into blanketlike layers are described as *stratiform*. Obviously, such layers could not exist with large-scale vertical motions of the air. However, stratiform clouds may indicate that air is moving in layers, one sliding over the other and gradually rising.

Flat-based clouds of massive globular shape, often higher than wide, are described as *cumuliform*. These forms generally indicate strong rising air currents, or updrafts, carrying moist air rapidly to higher levels and causing continued adiabatic cooling and condensation.

22.7 CLOUD FORMS

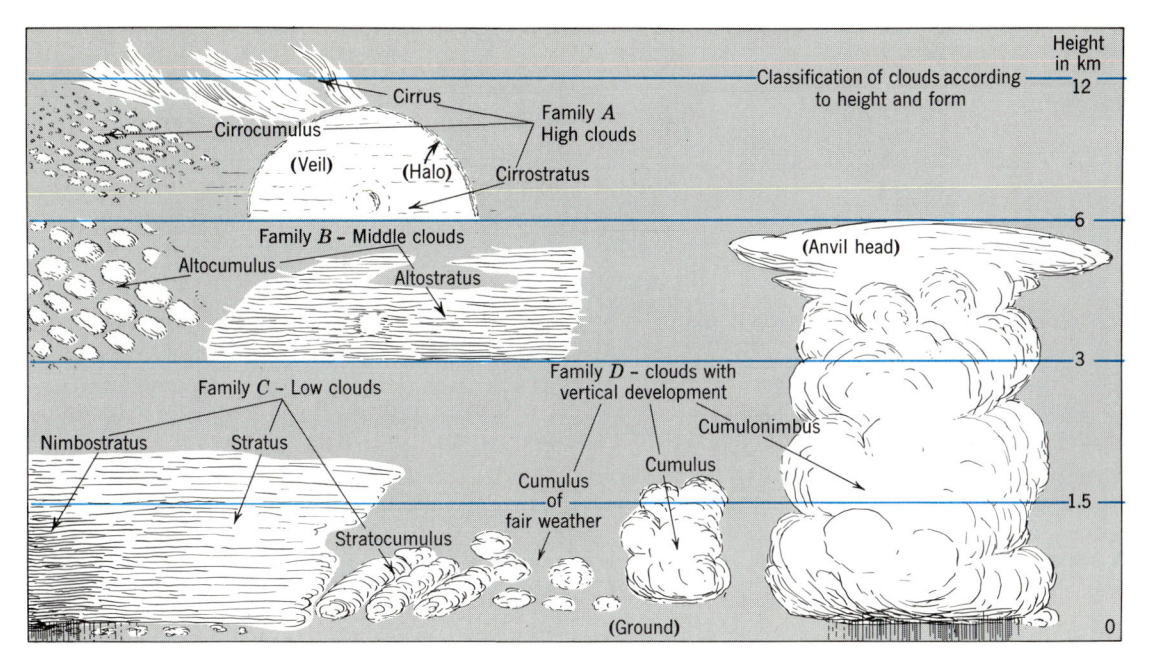

FIGURE 22-3
Cloud types are grouped into families according to height range and form.

FIGURE 22-4
Cirrus clouds formed of ice crystals. (National Weather Service.)

Clouds are grouped into classes and named according to height and general form, whether stratiform or cumuliform. An international system of classification recognizes four *cloud families:* high clouds, middle clouds, low clouds, and clouds of vertical development (Figure 22-3).

The high-cloud family above 7 km includes individual types named cirrus, cirrocumulus, and cirrostratus. All are composed of ice crystals. *Cirrus* is a wispy, featherlike cloud, commonly forming streaks or plumes named "mares' tails" (Figure 22-4). Cirrus clouds are so thin as to make no barrier to sunlight. The streaked cirrus bands usually indicate the presence of a high-altitude jet stream, with the wind direction paralleling the long lines of the cloud.

The middle-cloud family, extending from 2 to 7 km in height, includes two cloud types—altocumulus and altostratus.

The low-cloud family, found from ground level to a height of 2 km above the earth's surface includes three types: stratus, nimbostratus, and stratocumulus. *Stratus* is a uniform cloud sheet at low height; it usually completely covers the sky. The gray undersurface is foglike in appearance. Where stratus thickens to the point that rain or snow begins to fall from it, the cloud becomes *nimbostratus*, the prefix *nimbo* meaning rain. Nimbostratus is usually dense and dark gray, shutting out much daylight.

Clouds of the fourth family, those of vertical or upright development, are all the cumuliform

22 DYNAMICS OF THE WEATHER—PRECIPITATION AND STORMS

FIGURE 22-5
Cumulus clouds of fair weather. (National Weather Service.)

type. The smallest and most pleasant are the simple *cumulus* of fair weather. These are snow-white cottonlike clouds, generally with rounded tops and rather flattened bases (Figure 22-5); their shaded undersides are gray. The accompanying weather is fair with much sunshine.

Small cumulus can grow larger and denser to form congested cumulus with rounded tops resembling heads of cauliflower and flat, dark gray bases. These larger cumulus in turn sometimes grow into gigantic *cumulonimbus* clouds, or thunderheads. From these clouds we get heavy rain, hail, wind gusts, and thunder and lighning (Figure 22-6). Cumulonimbus clouds on occasion extend upward to heights of 18 km in the tropical zone.

FIGURE 22-6
A large cumulonimbus cloud producing heavy rainfall (dark area below flat cloud base). (U.S. Navy and National Weather Service.)

22.8 FOG

Fog is simply a cloud at the earth's surface. Dense fog is an indication that the air is at or close to the dew-point temperature and that sufficient moisture has condensed to produce abundant cloud droplets or ice particles. Perhaps the simplest type of fog is *radiation fog*, produced at night when a cold land surface conducts heat away from the lowermost layer of the atmosphere.

Another type of fog is *advection fog*. Advection simply means horizontal transfer of air. One type of advection fog is formed when moist warm air blows over a colder surface, whether land or water. Air passing close to the surface loses heat by conduction to the colder surface beneath, and its temperature is brought to the dew point. Among the most famous of advection fogs is that over the Grand Banks off Newfoundland, where warm, moist air overlying the Gulf Stream is close to the cold Labrador Current.

22.9 FORMS OF PRECIPITATION

Precipitation includes all forms of water particles that fall from the atmosphere and reach the ground. Excluded from precipitation are dew and hoarfrost, which are produced when moisture condenses directly on soil or plant surfaces.

Commonly recognized forms of precipitation are rain, snow, hail, and sleet. *Rain* consists of water droplets larger than 0.5 mm in diameter. The droplets form by rapid condensation and grow by joining with other droplets in frequent collisions. Raindrops are also produced by the melting of snowflakes falling to lower, warmer levels. The average raindrop contains roughly one million times the quantity of water found in a single cloud particle and may grow as large as 5 mm in diameter. Above this size the drop is unstable and will break apart as it falls. Drizzle is simply precipitation composed of tiny droplets, each less than 0.5 mm in diameter. Drizzle falls from low-lying nimbostratus clouds.

Snow is a form of ice in tabular or branched hexagonal (six-sided) crystals. These crystals mat together to form snowflakes. *Sleet* consists of small grains or pellets of ice formed by the freezing of raindrops falling through a cold air layer.

Hail consists of rounded pieces of ice, often made up of concentric ice layers much like the layers of an onion (Figure 22-7). They are formed only in cumulonimbus clouds and indicate powerful updrafts within the clouds. The ice layers are formed by repeated lifting and delayed fall of hailstones within a moist air layer where subfreezing temperatures exist.

22.10 CONDITIONS THAT PRODUCE PRECIPITATION

Precipitation can occur only when large masses of moist air are cooled rapidly below the dew-point temperature. Condensation must continue until large droplets or ice particles are formed. Only through the vertical rise of large air masses can such continued cooling take place. So we must investigate the ways in which large masses of air are made to rise through several kilometers of altitude.

The rise of large masses of air may be either spontaneous or forced. A forced rise often triggers spontaneous rise. Precipitation resulting from spontaneous rise of air is described as *convectional precipitation*. Forced ascent of air produces precipitation of two quite different types. *Orographic precipitation* is caused by the forced ascent of air in crossing a mountain barrier. *Frontal precipitation* is caused by the forced rise of air occurring when unlike air masses meet.

22.11 CONVECTION AND THUNDERSTORMS

Convectional rainfall forms from rising columns of moist, warm air. The rainfall usually takes the form of the torrential downpour of a *thunderstorm*, with its massive cumulonimbus cloud and associated lightning, thunder, and occasionally hail. You can think of the rising convection column as similar to the updraft in a chimney caused by the rise of the less dense heated air produced in a fireplace.

How can convection, once begun, intensify itself into a thunderstorm? Local heating of the ground layer is one of several triggering mechanisms that may set off spontaneous growth of powerful convection columns. Another, much larger source of energy is necessary to keep the air rising for thousands of meters. This energy source is the latent heat of condensation present in the moist air and spontaneously released. You can think of the energy as the heat of a fire obtained by release of energy stored in another form as fuel. A tiny match can start a great conflagration; similarly a small rising air current can grow into a violent storm.

When condensation begins, the wet adiabatic rate takes over and liberated heat is added to the rising air. Because of this added heat, the rising air continues to be warmer and less dense than the surrounding air. Consequently, the convection grows stronger and the upward currents intensify. The result is a violent storm with heavy bursts of rain. Finally, at a high altitude, condensation in the rising air becomes greatly reduced, and its temperature falls to the same level as the surrounding air. The rising air is now of the same density as the surrounding air. The upward movement of the convection system ceases, and the air spreads sideways.

Thunderstorms consist of individual parts, called cells. Within each cell air rises in a succession of bubblelike masses, instead of in a single continuous column (Figure 22-8). At all levels air is brought into the cell from the sides in the wake of the rising bubble by a process called entrainment.

FIGURE 22-7
These hailstones, larger than hens' eggs (arrow), fell during an August thunderstorm in Illinois. (National Weather Service.)

FIGURE 22-8
Schematic diagram of the interior of a thunderstorm cell.

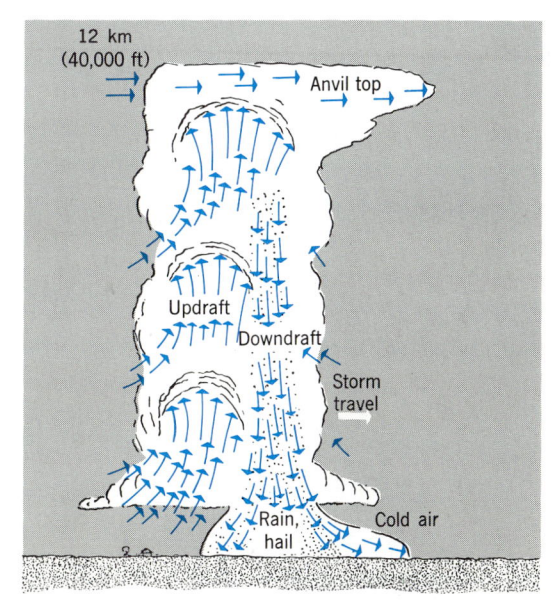

As the rising air bubble travels upward to high levels, heavy precipitation occurs. The top of the cloud, above the freezing level, spreads laterally to form an *anvil top*. Falling ice particles cool the cloud. These particles also serve as nuclei for condensation. Their action is described as *cloud seeding*, a process that causes rapid condensation. Falling drops or ice particles actually drag the air downward to produce a strong downdraft of cold air that strikes the ground at the time of the heavy initial burst of rain (Figure 22-8). This gusty squall wind spreads out horizontally along the ground.

Air turbulence within a thunderstorm is violent and will seriously damage or destroy light aircraft that venture into the cell. Raindrops caught in the updraft and carried above the freezing level become frozen pellets. Coated by supercooled water, these pellets grow into hailstones.

22.12
LIGHTNING

A familiar effect of the thunderstorm is *lightning*, an electrical discharge. It is simply a great spark, or arc, from one part of a cloud to another or from the cloud to the ground. Regions of positive and negative electrical charges accumulate in the cumulonimbus cloud.

As shown in Figure 22-9, the ground beneath the negatively charged part of a cloud in turn develops a positive charge. When the electric potential has reached sufficient magnitude (some 20–30 million volts), a lightning stroke occurs, traveling first from cloud to ground, then returning to the cloud. Several alternations between cloud and ground eliminate the difference in electric pressure. The whole process takes less than one-tenth of a second and appears as a single flash.

An electric current of perhaps 60,000–100,000 amperes may flow during a lightning discharge. *Thunder* is the sound produced by lightning. What we hear is the shock wave of sound sent out by the lightning stroke. Intense heating by the arc causes a sudden expansion of the air along the path of the stroke. Because sound travels at a rate of roughly 330 m/s, thunder follows the lightning flash by a time interval depending on distance. If both seem to occur at the same instant, the strike is very close by, whereas a delay of 3 seconds in the sound would indicate a strike roughly 1 km away.

FIGURE 22-9
Distribution of electrical charges inside a typical thunderstorm cell. (After U.S. Dept. of Commerce, *C. A. A. Technical Manual 104*.)

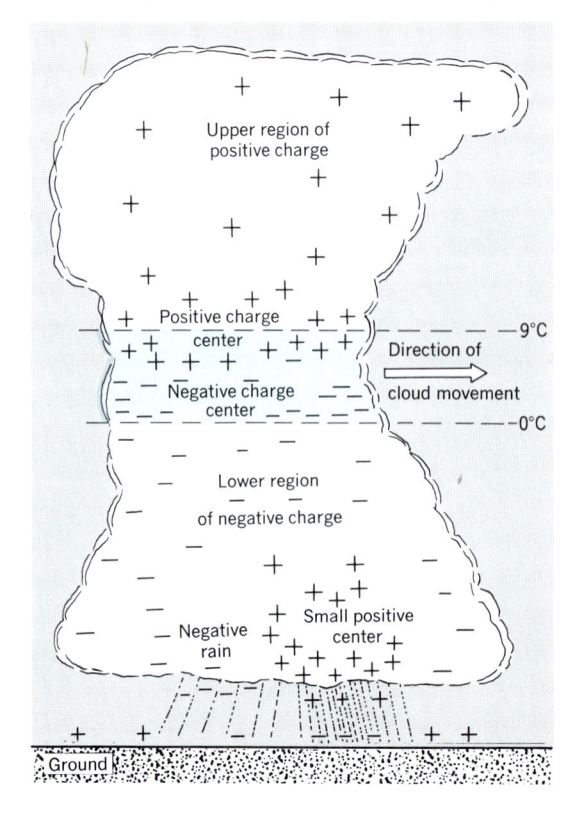

22.13
OROGRAPHIC PRECIPITATION

The word "orographic" is an adjective meaning "relating to mountains." Orographic precipitation is therefore related to the existence of a mountainous terrain. The principle of orographic rainfall is explained in Figure 22-10. Where prevailing winds blow from an ocean across a mountainous coast, air is forced to rise, often through many thousands of meters. If the moisture content is high, as it normally is in an air mass passing over an ocean, the dew-point temperature is quickly reached. Further ascent of the air produces rain or snow, which falls on the windward slopes and crest of the range.

Orographic precipitation can take two forms: a persistent rain or snow resulting from the steady lift of air, or heavy convectional showers or thunderstorms set off by the forced rise.

Orographic rainfall is heavy along the western coasts of both North America and Europe wherever mountains lie close to the sea and are exposed to prevailing westerly winds bringing a moist air mass. For example, the Klamath Mountains of northern California receive over 250 cm of precipitation annually.

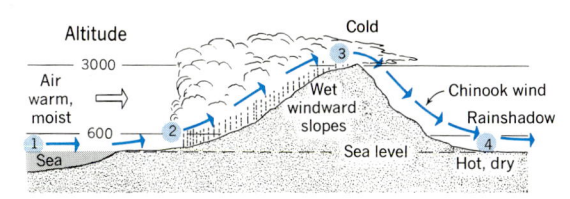

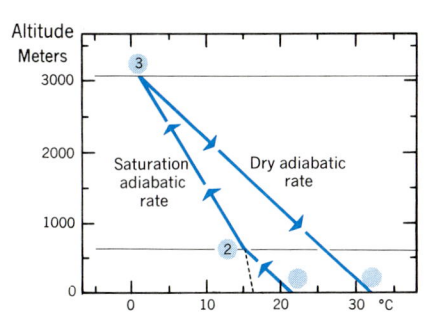

22.15 AIR MASSES

FIGURE 22-10
Forced ascent of oceanic air masses, producing precipitation and a rainshadow desert. (From THE EARTH SCIENCES, 2nd Edition by Arthur N. Strahler. Copyright © 1963, 1971 by Arthur N. Strahler. By permission of Harper & Row, Publishers, Inc.)

22.14 RAIN-SHADOW DESERTS

Subsiding air is warmed at the dry adiabatic rate. As Figure 22-10 shows, air descending the lee slope of a range warms rapidly and the relative humidity also declines rapidly. The remaining water droplets of the clouds quickly evaporate, clearing the air. Since no more moisture is available, the air temperature continues to increase at the dry adiabatic lapse rate of 1 C° per 100 m. Notice that the graph informs us that at any given level, say 600 m, the descending air is considerably warmer than the rising air at the same level on the windward slope of the mountain range.

By the time the air has reached the valley floor at sea level on the lee side of the range, it is hot and very dry. A desert, called a *rain-shadow desert*, will exist here if winds blow prevailingly from the ocean toward the continent throughout the year. An example is the rain-shadow desert in Owens Valley and Death Valley, east of the Sierra Nevada in California.

Another important example of dryness associated with sinking air is seen in the high-pressure center, or anticyclone, explained in Chapter 21. Air in a high-pressure center is slowly sinking and spreading outward toward areas of lower barometric pressure. The air is warmed adiabatically and becomes drier as it descends. This process tends to produce fair skies and sunny weather. Great tropical deserts, such as the Sahara Desert of North Africa, lie beneath the subtropical high-pressure belt, described in Chapter 21.

22.15 AIR MASSES

Important in your understanding of weather phenomena such as rain, snow, thunderstorms, and hurricanes is the concept of the air mass. We referred briefly to air masses in the previous chapter. Recall that a cold polar air mass lies poleward of the polar jet stream, while a warm tropical air mass lies on the equatorward side. We now go into further detail about air masses.

In its dimensions, a single *air mass* is a body of air extending horizontally over a substantial part of a continent or ocean. The air mass extends vertically through a major fraction of the troposphere. A given air mass possesses nearly uniform temperature and water-vapor content in all horizontal directions at any given altitude. An important descriptive property of an air mass is the rate in which its temperature changes with altitude. Typically, a given air mass has a sharply defined boundary in contact with a different air mass adjacent to it.

An air mass is a product of its *source region*—the ocean or land surface from which it derives its physical properties. For example, over a warm ocean, the air mass derives a large water-vapor content in combination with high temperature. Over a cold continent in winter, an air mass is not only intensely cold, especially in the lower layer, but its water-vapor content is also extremely small.

Although an air mass may be stagnant for long periods over the source region, it can travel into other regions. It follows the movement of regional winds in response to the pressure gradient. Modification of the air mass gradually takes place during migration, depending on whether heat is gained or lost to the ground surface through longwave radiation. An air mass can also change by mixing of warm lower air layers with colder air aloft. Water vapor may be added by evaporation from a sea surface below and find its way upward through such mixing.

The meteorologist's classification of air masses is based on the global position of the source area and the nature of the underlying surface, whether continent or ocean. Latitude relates closely to air temperature. Nature of the surface below strongly influences the water-vapor content. For the sake of simplicity, we introduce a two-by-two classification of air masses: two classes based on latitude; two based on source region.

22 DYNAMICS OF THE WEATHER— PRECIPITATION AND STORMS

Polar air masses (P) have source regions in the latitude zone from 50 to 65 degrees and are cold in winter and cool in summer. *Tropical air masses* (T) have source regions in the latitude zone from 20 to 35 degrees and range from warm to hot, depending on season. *Continental air masses* (c) originate over large landmasses and are typically dry. *Maritime air masses* (m) originate over large ocean bodies and are typically moist.

The four possible combinations in the two-by-two classification are as follows:

	continental (c)	maritime (m)
Polar (P)	cP	mP
Tropical (T)	cT	mT

Table 22.1 gives examples of the qualities of each of these air mass types.

Continental polar (cP) air masses originate over the large landmasses of North America and Eurasia. They are very cold in winter and have a low moisture content. Maritime polar (mP) air masses originate over the cold waters of the North Atlantic and North Pacific. They are cool to moderately cold and hold substantial amounts of water vapor. Tropical continental (cT) air masses originate over land areas in the subtropical high-pressure belt. They are warm to hot and carry substantial amounts of water vapor. The maritime continental (mT) air masses originate over tropical oceans. They are warm and very moist, capable of yielding very large amounts of rainfall.

22.16
WEATHER FRONTS

In Chapter 21, we noted that polar and tropical air masses lie in contact along the polar front. To understand weather phenomena of middle and high latitudes, we must go into further detail about the various kinds of fronts.

A *weather front* is the boundary separating two air masses of unlike properties. Air masses do not mix easily. Instead, because of their different properties, unlike air masses tend to have distinct boundaries between them, just as oil and water tend to remain in separate layers or drops without mixing.

Along most weather fronts one mass of air is invading a region occupied by an unlike air mass. Not only is the air on both sides of the front in motion, but the front itself is also moving over the earth's surface beneath it. The primary rule of weather fronts is that the air of the colder mass, being the denser, stays close to the ground, forcing the warm (or less cold) air to slide over it and to rise upward. In other words, the interaction of fronts leads to forced rise of air masses.

Putting these principles to use, consider the three basic types of fronts that may develop. The *cold front,* shown in Figure 22–11, is formed by a cold air mass invading the region occupied by a warm (or less cold) air mass. Staying close to the ground, the cold air forms a wedge, pushing the warm air upward from its advancing edge. Ground friction slows the advancing cold air close to the ground, so that it may develop a steep or blunt leading edge. The lifting of the warm air is therefore abrupt and violent on occasions.

If the warm air is of maritime tropical (mT) type, as is most often the case, it may break into spontaneous convection. Warm, moist air masses are said to be *unstable* because a small amount of forced lift sets off spontaneous rise. Convection produces dense cumulus and cumulonimbus clouds (thunderstorms) extending to extreme heights. Often a cold front produces a line of thunderstorms 300–800 km long. Such fronts are important to airline pilots, for the storms often rise too high to be surmounted and extend too low to be flown beneath. The pilot must try to pass between individual convection cells.

FIGURE 22–11
A cold front.

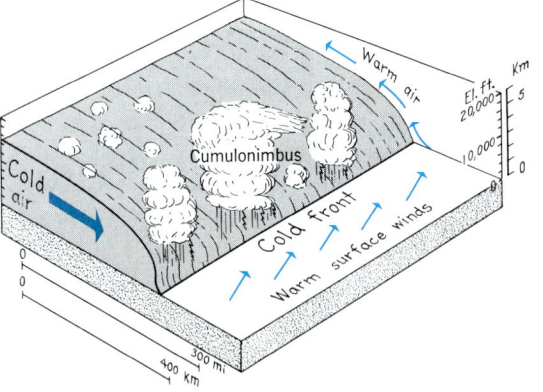

TABLE 22.1
TYPICAL EXAMPLES OF AIR MASSES

Air Mass Name	Symbol	Properties	Temperature (°C)	Specific Humidity (g/kg)
Continental polar	cP	Cold, dry (winter)	−11	1.4
Maritime polar	mP	Cool, moist (winter)	4	4.4
Continental tropical	cT	Warm, dry	24	11
Maritime tropical	mT	Warm, moist	27	17

The *warm front,* shown in Figure 22-12, is formed by a relatively warm air mass moving into a region occupied by colder air. The cold air remains close to the ground, while the warm air slides up over it on a broad, gently sloping frontal surface. If the warm air mass is stable, stratiform clouds mark the overriding air layer, because the forced ascent of air causes steady adiabatic cooling and condensation. The highest fringe of advancing warm air is marked by cirrus clouds. As the front comes nearer, these clouds are replaced by altostratus, then by a dense stratus, and finally by a nimbostratus, with a broad zone of light, steady, and prolonged precipitation.

A third type of front is the *occluded front,* diagrammed in Figure 22-13. Here a cold front has caught up with and pushed into a warm front, completely lifting the warm air off the ground. The invading cold air remains close to the ground and comes in contact with less cold air under the warm front. The warm front is now said to be occluded—that is, cut off from contact with the ground.

Cold, warm, and occluded fronts form in the middle and high latitudes wherever unlike air masses meet in conflict. Fronts will be most frequent in the region of conflict between polar and maritime air masses along the polar front beneath the jet stream (see Figure 21-26).

22.17 WAVE CYCLONES OF MIDDLE AND HIGH LATITUDES

We have defined cyclone as a center of low barometric pressure. In the northern hemisphere a cyclone has a counterclockwise flow of winds about its center. Near the ground the air spirals inward toward the center of the cyclone (see Figure 21-23).

FIGURE 22-12
A warm front.

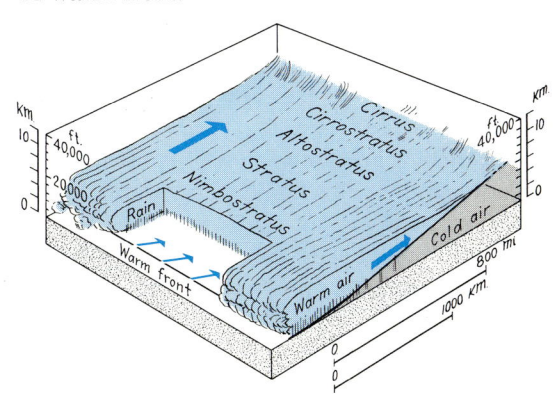

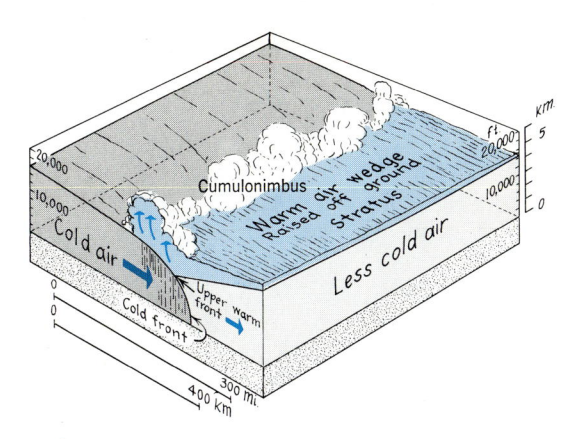

FIGURE 22-13
An occluded front.

Cyclones occur at all latitudes and vary greatly in size and intensity. Some cyclones of the tropical zone are of a violent type, such as the hurricane or typhoon. Our first concern here is with cyclones in the middle and high latitudes. They are produced by wavelike kinks developing on the fronts separating cold from warm air masses. For this reason such disturbances are called *wave cyclones.*

Stages in the development of a wave cyclone are shown in Figure 22-14. Block *A* shows a portion of a front between a polar and a tropical air mass. Because the airflow on the two sides of this front is in opposite directions, a shearing or dragging action is set up between the air masses. Under such conditions the front cannot remain smooth, but will tend to develop a bend or kink, such as that shown beginning to form in Block *A*.

As the frontal wave in Block *A* develops, the flow of air is modified so that cold air begins to push into the region of the warm, and the warm begins to move into the region occupied by the cold. This motion resembles a revolving door, with one person going out of a building while another enters.

Block *B* shows the open stage of the wave cyclone. The frontal wave is well developed and forms a sharp point toward the region of cold air. To the left of this point, or crest, lies a cold front with cold air actively pushing south. To the right lies a warm front curved in a great arc bowed to the east.

Block *C* shows that the wave has now steepened to the point that the crest has been cut off. Here the cold front has caught up with the warm front to produce an occluded front separating a layer of warm air from contact with the ground. The cyclone has reached the occluded stage.

After the occluded stage the cyclone enters the dissipating stage shown in Block *D*. The cyclone

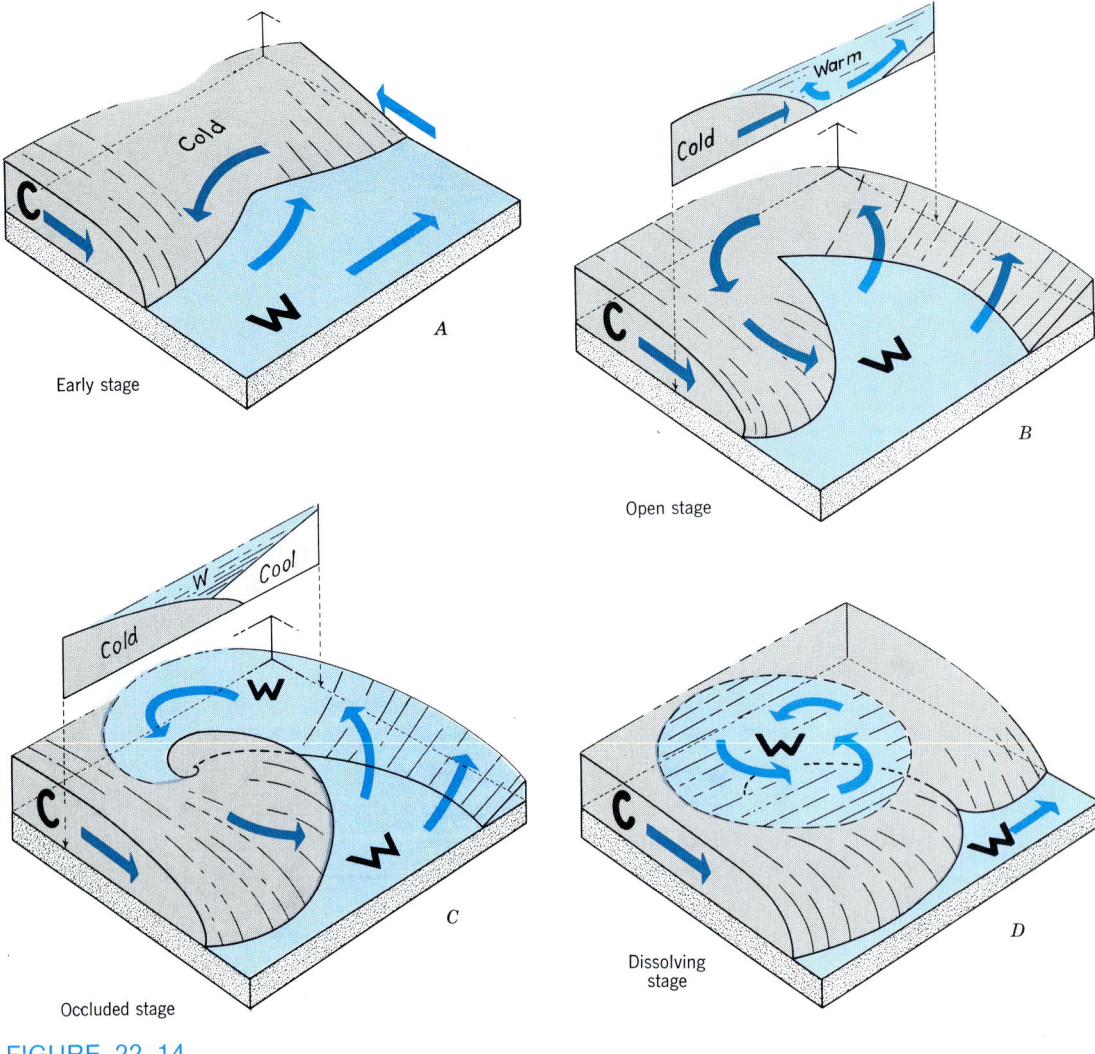

FIGURE 22-14
Development of a wave cyclone in four stages.

is no longer supplied with moist warm air. Lacking condensation to supply latent heat energy, the cyclone dies out. Now the front is reformed as a smooth line between polar and tropical air masses.

22.18
WAVE CYCLONES ON THE DAILY WEATHER MAP

The anatomy of a wave cyclone is shown on the surface weather map. Figure 22-15 is a pair of such maps, showing conditions on two successive days in the early spring. Areas experiencing precipitation are shaded. Pressure is shown by isobars at intervals of 3 mb; centers of cyclones are indicated by "L" for low; of anticyclones by "H" for high. Fronts are shown by standard conventions, explained on the first map. A dashed line crosses the map; it is an *isotherm*, connecting all points at which the air temperature is at the freezing point, 0°C.

Map *A* shows a cyclone in the open stage, approximately equivalent to Figure 22-14*B*. The storm is moving northeastward. Notice the following points: (1) Isobars of the low are closed to form an oval-shaped pattern. (2) Isobars make a sharp V where crossing the cold front. (3) Wind directions, indicated by arrows, are at an angle to the trend of the isobars and form a pattern of counterclockwise inspiraling. (4) In the warm air sector, there is northward flow of moist tropical air toward the direction of the warm front. (5) There is a sudden shift of wind direction accompanying the passage of the cold front. (6) There is a severe drop in temperature accompanying the passage of the cold front. (7) Precipitation is occurring over a broad zone near the warm front and in the central area of the

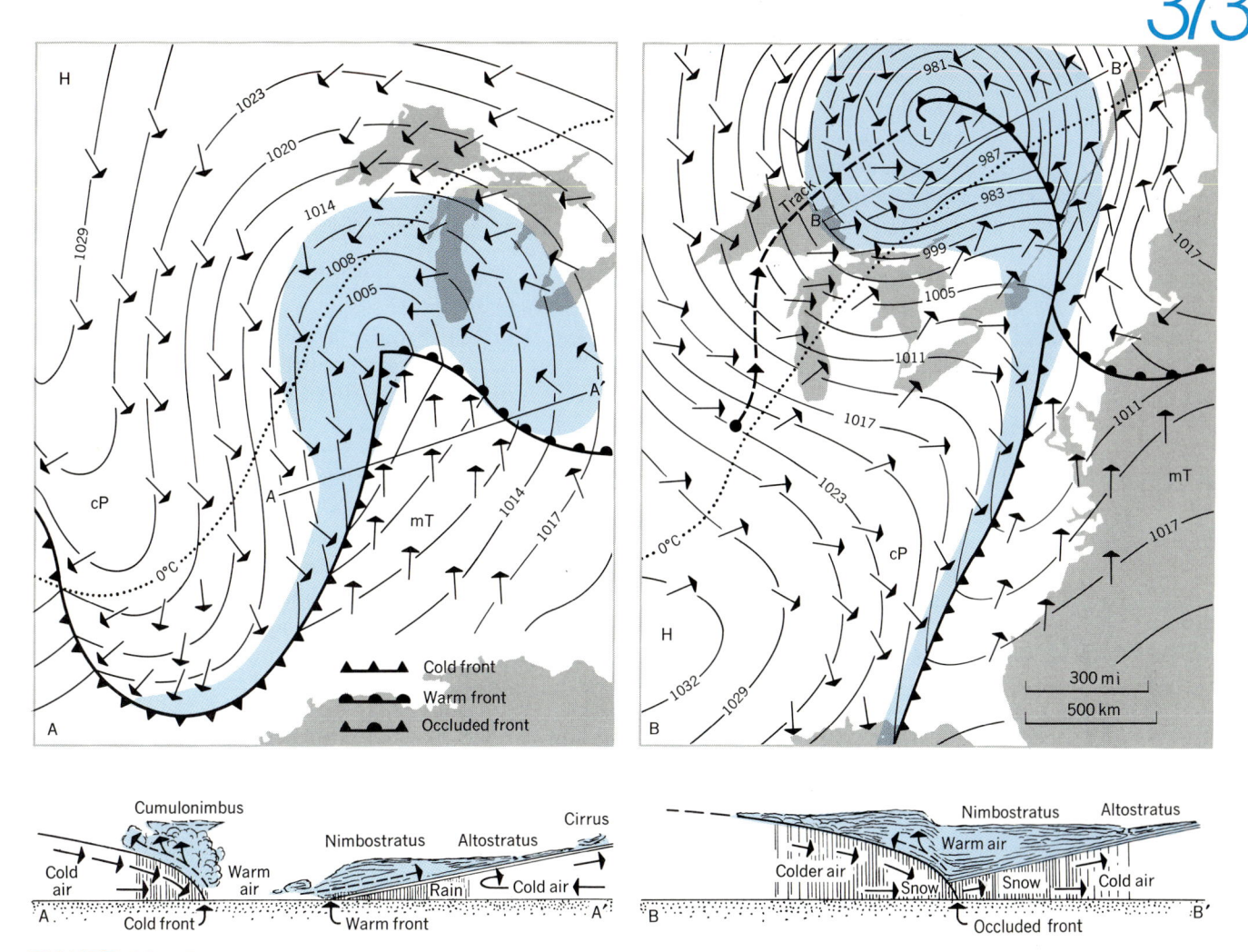

22.19 CYCLONE TRACKS AND FAMILIES

FIGURE 22-15
Simplified weather maps and cross sections through a wave cyclone. (*A*) Open stage. (*B*) Occluded stage.

cyclone, but extends as a thin band down the length of the cold front. (8) The low is followed on the west by a high (anticyclone) in which low temperatures and clear skies prevail. (9) The 0°C isotherm crosses the cyclone diagonally from northeast to southwest, showing that the southeastern part is warmer than the northwestern part. A cross section through map *A* along the line *AA'* shows how the fronts and clouds are related.

The second weather map (Figure 22-15*B*) shows conditions 24 hours later. The cyclone has moved rapidly northeastward into Canada, its path shown by the line labeled *storm track*. The center has moved about 1600 km in 24 hours, a speed of just over 65 km/hr. The cyclone has occluded. An occluded front replaces the separate warm and cold fronts in the central part of the disturbance. The high-pressure area, or tongue of cold polar air, has moved in to the west and south of the cyclone. A cross section below the map shows conditions along the line *BB'*, cutting through the occluded part of the storm. Note that the warm air mass is being lifted higher off the ground and is giving heavy precipitation.

22.19 CYCLONE TRACKS AND FAMILIES

Cyclones travel in a generally easterly direction in the northern hemisphere at speeds ranging from 30 to 65 km/hr. Some travel a distance of one-third to one-half of the way around the earth during their life cycle. The intensity of these cyclones is extremely varied. Considering that a cyclone is experienced every few days throughout the year by persons living in North America and Europe, it is obvious that most cyclones pass almost unnoticed as spells of cloudy or rainy weather.

On the other hand, a large intense wave cyclone can be a powerful and devastating storm. The storm can have winds up to 110 km/hr or more and can bring flooding rains or deep snows. At sea, over the North Atlantic and Pacific oceans and over the Southern Ocean in 40 to 70 degrees latitude, cyclonic storms tend to intensify. These deep cyclones become extremely severe in the winter season, causing high seas and great peril to shipping.

Figure 22-16 shows the principal tracks of cyclones of the middle latitudes. In the southern hemisphere the storms are rather uniformly distributed around the Southern Ocean. In the northern hemisphere there is a definite concentration into two characteristic paths. One originates in eastern North America and crosses the North Atlantic to northern Europe. Another originates in easternmost Asia and the Japanese Islands and extends across the North Pacific to Alaska and the northwestern coast of North America. Along these prevailing tracks wave cyclones tend to form in succession as *cyclone families*, shown in Figure 22-17. Within a single family the stage of development ranges from open wave to occlusion, from southwest to northeast.

Cyclones are closely tied in with the upper-air, or Rossby, waves described in Chapter 21. Cyclones typically form near the southerly base of a well-developed wave and are dragged along beneath the jet stream. This is the reason that accurate forecasting of the direction of storm travel depends on a knowledge of airflow at high levels.

22.20
TORNADOES

The *tornado* is a small but very intense wind vortex. It extends down from a cumulonimbus cloud, taking the form of a tapering funnel (Figure 22-18). Although only a few hundred meters in diameter, the funnel cloud contains winds with speeds up to 400 km/hr. Within the center of the funnel cloud is a vortex in which air pressure is but a fraction of normal pressure.

Tornadoes commonly move northeastward across country at 40-65 km/hr, following the general motion of thunderstorms and associated cold fronts. Where the funnel cloud touches ground, there is complete destruction along a swath on the order of 300 m wide and often many kilometers long. The great wind speed is disastrous, and, even worse, the sudden lowering of air pressure as the vortex passes may cause buildings to explode from the expansion of air entrapped within.

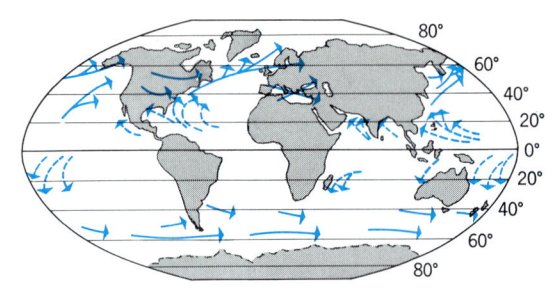

FIGURE 22-16
Principal tracks of wave cyclones are shown by solid lines; those of tropical cyclones by dashed lines. (After Petterssen, *Introduction to Meteorology*.)

Tornadoes are eddies of intense turbulence generated by the mixing of dry polar air masses (*mP*) with unstable moist maritime tropical air (*mT*). The region most favorable for such tornado-producing interactions between unlike air masses is the Great Plains region and the Mississippi Valley.

Tornadoes are most likely to occur in months in which the greatest contrast exists between polar and tropical air masses. These months are from early spring to late summer, with May leading, followed by June and April.

22.21
TROPICAL CYCLONES

The belt of tropical easterlies breeds the most violent of all large cyclonic storms, the *tropical cyclone*. It is a nearly circular storm with extremely low pressure at the center, accompanied by high winds, dense clouds, and heavy precipitation (Figure 22-19). The names *hurricane* (West Indies) and *typhoon* (western Pacific) are acceptable equivalent names for severe tropical cyclones.

Tropical cyclones originate only over oceans. At first a weak center of low pressure forms. If conditions are favorable, the low deepens rapidly, with the isobars taking on the form of nearly concentric circles. Not all such tropical cyclones deepen into severe storms—some die out quickly, others travel long distances as mild disturbances. However, if the pressure becomes extremely low in the storm center, winds will increase to speeds of 120 km/hr or much higher, with the accompaniment of dense clouds, extreme air turbulence, and heavy rain. The storm is then designated as a hurricane or typhoon and is a serious menace to ships at sea and to islands or continental coasts over which it may pass. The storm normally travels westward at a rate of 10-20 km/hr in the belt of tropical easterlies.

22.21 TROPICAL CYCLONES

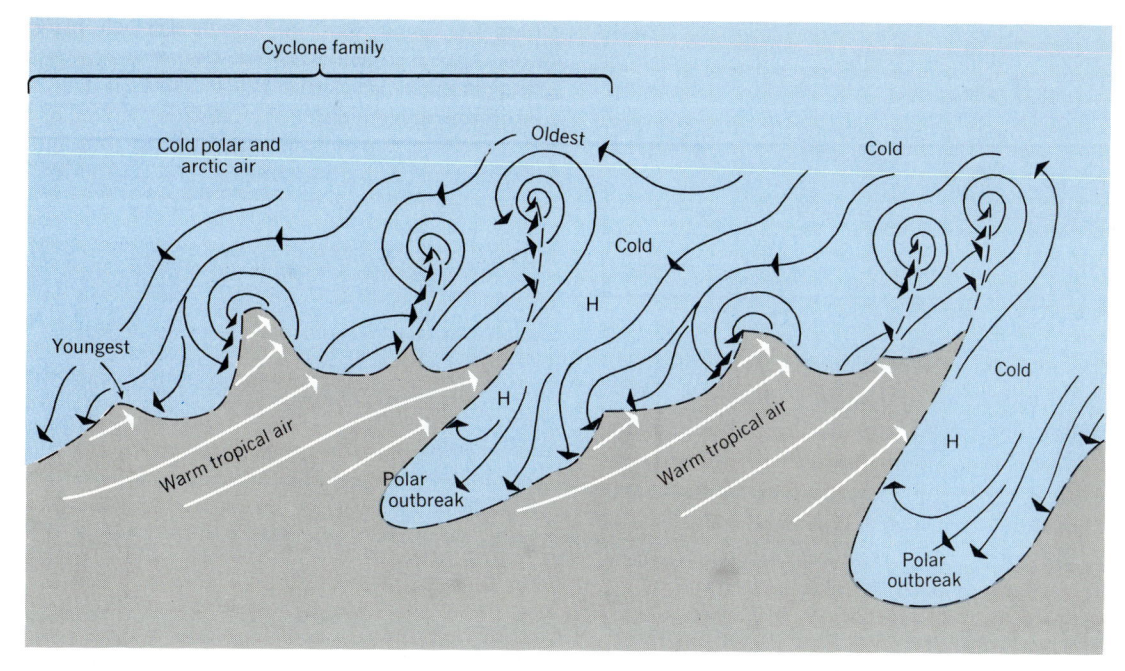

FIGURE 22-17
Two families of wave cyclones in the northern hemisphere, as seen on a schematic weather map. After Bjerknes and Solberg; Petterssen. (From THE EARTH SCIENCES, 2nd Edition by Arthur N. Strahler. Copyright © 1963, 1971 by Arthur N. Strahler. By permission of Harper & Row, Publishers, Inc.)

FIGURE 22-18
A tornado funnel cloud seen at Hardtner, Kansas. (National Weather Service.)

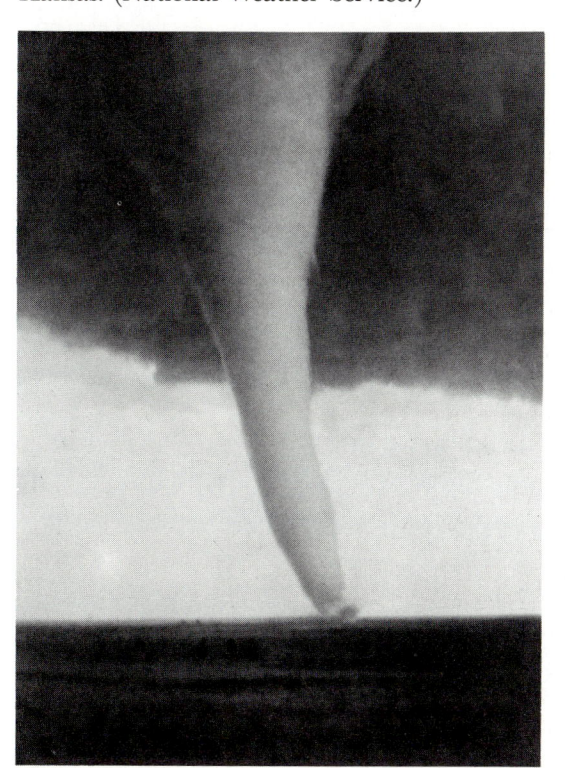

FIGURE 22-19
Hurricane Gladys, photographed on October 8, 1968, from Apollo 7 spacecraft at an altitude of about 180 km. The storm center was about 240 km southwest of Tampa, Florida. (NASA)

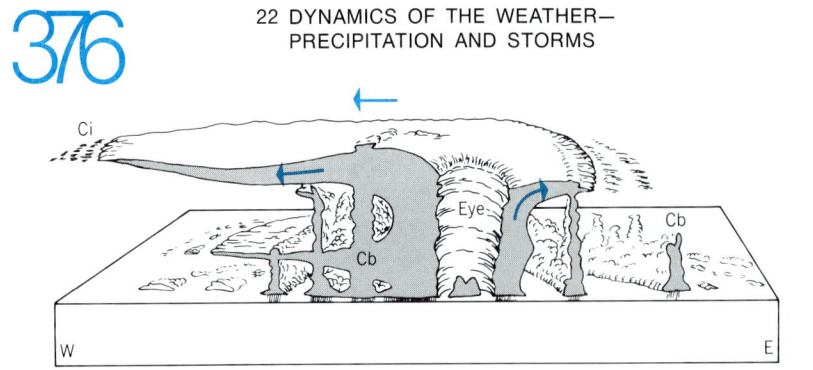

22 DYNAMICS OF THE WEATHER— PRECIPITATION AND STORMS

FIGURE 22-20
Schematic three-dimensional drawing of a typical hurricane. The front section cuts through the eye. Cumulonimbus clouds rise stalklike through dense stratiform cloud layers. Width is about 1000 km; highest clouds are at an altitude often over 9 km. (Redrawn from R. C. Gentry, 1964, *Weatherwise*, Vol. 17, p. 182. Data of NOAA, National Weather Service.)

Figure 22-20 is a schematic cross section through a tropical cyclone, showing cloud formations and rain bands. Upon reaching high levels, air flows outward, producing a cirrus cloud cap. The cyclone has a *central eye*, a strange hollow vortex several kilometers wide surrounded by a dense cloud wall.

Severe tropical cyclones occur in all oceans of the world except the South Atlantic (see Figure 22-16). Tropical cyclones occur in that part of the year including and immediately following the period of high sun, or summer season, in the hemisphere in question. For example, the hurricanes and typhoons of the Caribbean Sea occur from June through November.

The vast destruction and loss of life brought about by tropical storms rank them among the great catastrophes nature inflicts on human beings. Although of prime importance as a hazard to ships at sea, these storms do their greatest damage when passing over densely inhabited islands and low-lying coasts.

SUMMARY

Water vapor, humidity. Water vapor in the atmosphere is the source of all clouds, precipitation, and storms. Relative humidity changes with air temperature, relating vapor content to the saturation quantity and to dew point. However, specific humidity tells water vapor content in absolute terms.

Adiabatic cooling, condensation. Rising air expands and is automatically cooled, eventually reaching the dew-point temperature at which condensation sets in and clouds are formed. Latent heat is released during condensation and partially counteracts the rate of adiabatic cooling.

Clouds, fog. Clouds consist of minute water droplets or ice crystals, depending on air temperature. Hygroscopic nuclei are needed for clouds to form. Stratiform clouds indicate slow rise and cooling of an extensive air layer. Cumuliform clouds indicate rapid updrafts in narrow columns. Fog is a cloud layer close to the ground.

Precipitation. Precipitation in the form of rain, snow, or hail requires rapid condensation within clouds, where large masses of air are rising. Precipitation can be convectional, orographic, or frontal in origin. Convectional precipitation takes place in thunderstorms, powered by release of latent heat during condensation. Orographic precipitation occurs when moist air is forced to rise over a mountain range. Air descending on the lee side is warmed and becomes dry.

Air masses, fronts. Large sections of the troposphere having uniform conditions of temperature and water vapor content comprise air masses. Air masses are classified as polar (cold) or tropical (warm), and as continental (dry) or maritime (moist). Air masses meet in conflict along fronts, classified as cold, warm, and occluded. Precipitation occurs as moist warm air masses are lifted along moving fronts.

Wave cyclones. Kinks along the polar front, beneath the jet stream, deepen into wave cyclones in which frontal precipitation occurs and winds may attain high speeds. The wave cyclone occludes as it travels eastward in the belt of westerly winds. Tornadoes may occur locally along cold fronts within wave cyclones.

Tropical cyclones. Within the belt of tropical easterlies, intense centers of low pressure develop into violent tropical cyclones. These storms travel westward across the oceans and often impinge upon the margins of continents, bringing severe destruction.

TERMS AND CONCEPTS

humidity
saturated air
relative humidity
dew point
specific humidity
precipitation
dry adiabatic lapse rate
dew-point lapse rate
wet (saturation) adiabatic lapse rate
dry stage
level of condensation
rain stage
clouds
hygroscopic nuclei
supercooled water
stratiform, cumuliform clouds

cloud families
cirrus
stratus
nimbostratus
cumulus
cumulonimbus
fog
radiation fog
advection fog
advection
rain
snow
sleet
hail
convectional precipitation
orographic precipitation
frontal precipitation
thunderstorm
anvil top
cloud seeding
lightning
thunder
rain-shadow desert
air mass
source region
polar air masses
tropical air masses
continental air masses
maritime air masses
weather front
cold front
unstable air mass
warm front
occluded front
wave cyclones
isotherm
storm track
cyclone families
tornado
tropical cyclone
hurricane
typhoon
central eye

QUESTIONS

1. What role does water vapor play in atmospheric processes? Explain how relative humidity changes as air temperature changes. What is the significance of the dew-point temperature? For what uses is specific humidity a necessary measure of atmospheric water vapor content?

2. Explain fully the adiabatic process and how it leads to formation of clouds and precipitation. What is the value of the dry adiabatic lapse rate? The dew-point lapse rate? The wet (or saturation) adiabatic lapse rate? Why is the wet rate less than the dry rate?

3. In what state (liquid, solid) are clouds particles found? How is a cloud droplet formed? What are the basic forms of clouds? What does each form tell us about the state of the atmosphere? Name and describe four or five common cloud types. Under what conditions does fog form?

4. Name the forms of precipitation. What atmospheric conditions and processes are essential to the production of precipitation? Name three basic types of precipitation.

5. Describe the processes acting within a thunderstorm. What is the energy source for the storm? What is the significance of the anvil top of a cumulonimbus cloud? How are lightning and thunder produced?

6. Describe the stages in development of orographic precipitation and a rain-shadow desert. Make full use of the adiabatic principle in your explanation. Why is the desert air warmer than the same air prior to its ascent from sea level?

7. On what basis is one air mass recognized as different from another air mass? How does the source region determine air mass properties? Describe a simple air-mass classification system. Which air-mass types are capable of producing large amounts of precipitation? Which are not?

8. Describe the common types of weather fronts. What types of air masses interact within each of these fronts? What cloud forms are characteristic of each?

9. Describe the wave cyclone of middle and high latitudes. What stages of evolution are typical of a wave cyclone? Describe the important features of a typical wave cyclone as it appears on the daily weather map. What locations and tracks are typical for wave cyclones and cyclone families in the northern hemisphere? What position does a tornado occupy with respect to a wave cyclone and its fronts?

10. Describe a tropical cyclone in terms of its size, internal structure, precipitation, and wind speeds. What causes the central eye? Where do tropical cyclones originate? What paths do they usually follow?

PROBLEMS

1. Use the typical values for dry and wet adiabatic lapse rates and dew-point lapse rate given in the text. (a) A mass of rising air has a temperature of 20°C at an altitude of 200 m. Using the dry adiabatic lapse rate, what will be the temperature of this mass of air when it has reached an altitude of 800 m?

(b) Assuming that this same air mass has a dew-point temperature of 14°C at an altitude of 200 m, at what altitude will the saturation point be reached by the rising air?

22 DYNAMICS OF THE WEATHER— PRECIPITATION AND STORMS

2. Refer to Figure 22-10, showing orographic precipitation and the rain-shadow effect. We find that a mass of air at an altitude of 200 m over the ocean has a temperature of 23°C. The dew-point temperature of this same air mass is 15°C. The air mass is forced to rise over the mountain barrier.

(a) At what altitude will condensation set in?

(b) What will be the air temperature at the saturation point?

(c) At an altitude of 2600 m the air mass reaches its highest point of lift. What is the air temperature at this level?

(d) The air mass now descends from 2600 m to a point at 200 m altitude in a deep valley on the lee side of the mountain range. What is the air temperature at the 100-m level?

(Hint: It will be helpful to plot the data on a sheet of graph paper, similar to that shown in Figure 22-10.)

CHAPTER 23
EARTH MATERIALS— MINERALS AND ROCKS

23.1 ROCKS AND MINERALS

What is rock? In a broad sense, *rock* is the solid substance comprising the earth's outer shell, or *lithosphere*. Rock is composed of mineral matter in the solid state. Most rock consists of several minerals in combination. (Mineral matter has been defined in Chapter 17.) The minerals commonly occur as individual grains, so that the rock is a physical mixture of such mineral grains.

The various rock groups and individual rock types are distinguished in terms of the mineral varieties present and the proportions in which they occur. Then there are further distinctions among rock types based on the size of the individual crystal grains, or the lack of any observable crystalline structure.

23.2 ELEMENTS OF THE EARTH'S CRUST

Before beginning a study of the common minerals and rocks, it will be helpful to examine figures on the abundance of elements in the earth's outermost layer, or crust, with an average thickness of about 17 km.

Table 23.1 lists the eight most abundant chemical elements in the earth's crust. The order of listing is according to percentage by weight. Several points are of interest in this table. Notice, first, that the eight elements constitute 98–99 percent of the crust by weight and that almost half of this weight is oxygen. Measured in other ways, the importance of oxygen is even greater—in numbers of atoms it makes up over 60 percent of the total. Because oxygen is an atom of comparatively large radius, it represents almost 94 percent by volume. Notice that silicon is in second place with about 28 percent, or roughly half the value for oxygen. Aluminum and iron occupy intermediate positions, while the last four elements—calcium, sodium, potassium, and magnesium—are subequal in the range of 2–4 percent.

If the table were extended, you would find that the ninth most abundant element is titanium, followed in order by hydrogen, phosphorus, barium, and strontium.

TABLE 23.1
THE MOST ABUNDANT ELEMENTS IN THE EARTH'S CRUST

Element	Chemical Symbol	Percentage by Weight
Oxygen	O	46.6
Silicon	Si	27.7
Aluminum	Al	8.1
Iron	Fe	5.0
Calcium	Ca	3.6
Sodium	Na	2.8
Potassium	K	2.6
Magnesium	Mg	2.1
Total		98.5

23.3 THE SILICATE MINERALS

Our first concern is with a class of rocks designated as *igneous rocks,* meaning that the rock has solidified from a high-temperature molten condition. In other words, the molten mineral matter, or *magma,* has undergone a transformation from liquid to solid state.

The vast bulk of all igneous rock consists of mineral compounds containing the elements silicon and oxygen. Collectively, these minerals are known as *silicates.* In a *silicate mineral* both silicon and oxygen are combined with one or more of the metallic elements listed in Table 23.1.

We can gain a good appreciation of the nature of igneous rocks as a class by noting the proportions of only seven silicate minerals, or mineral groups. These are shown in Figure 23–1. The mineral list begins with *quartz,* containing only silicon and oxygen, as silicon dioxide, SiO_2. The next five items are mineral groups; they are silicate compounds, all containing aluminum, and can be designated *aluminosilicates.*

The first aluminosilicate is *potash feldspar* $(K,Na)Si_3O_8$. Potassium is the dominant metallic element in potash feldspar. The mineral name for a common kind of potash feldspar is *orthoclase.*

Next come the *plagioclase feldspars*. They span a continuous range from *sodic plagioclase* $(NaAlSi_3O_8)$, at one end of the series, with sodium making up 100 percent of the variable metallic element, to *calcic plagioclase* $(CaAl_2Si_2O_8)$ at the other end, with calcium

FIGURE 23–1

A simplified chart of common silicate minerals and abundant igneous rocks. (From PLANET EARTH: ITS PHYSICAL SYSTEMS THROUGH GEOLOGIC TIME by Arthur N. Strahler. Copyright © 1972 by Arthur N. Strahler. By permission of Harper & Row, Publishers, Inc.)

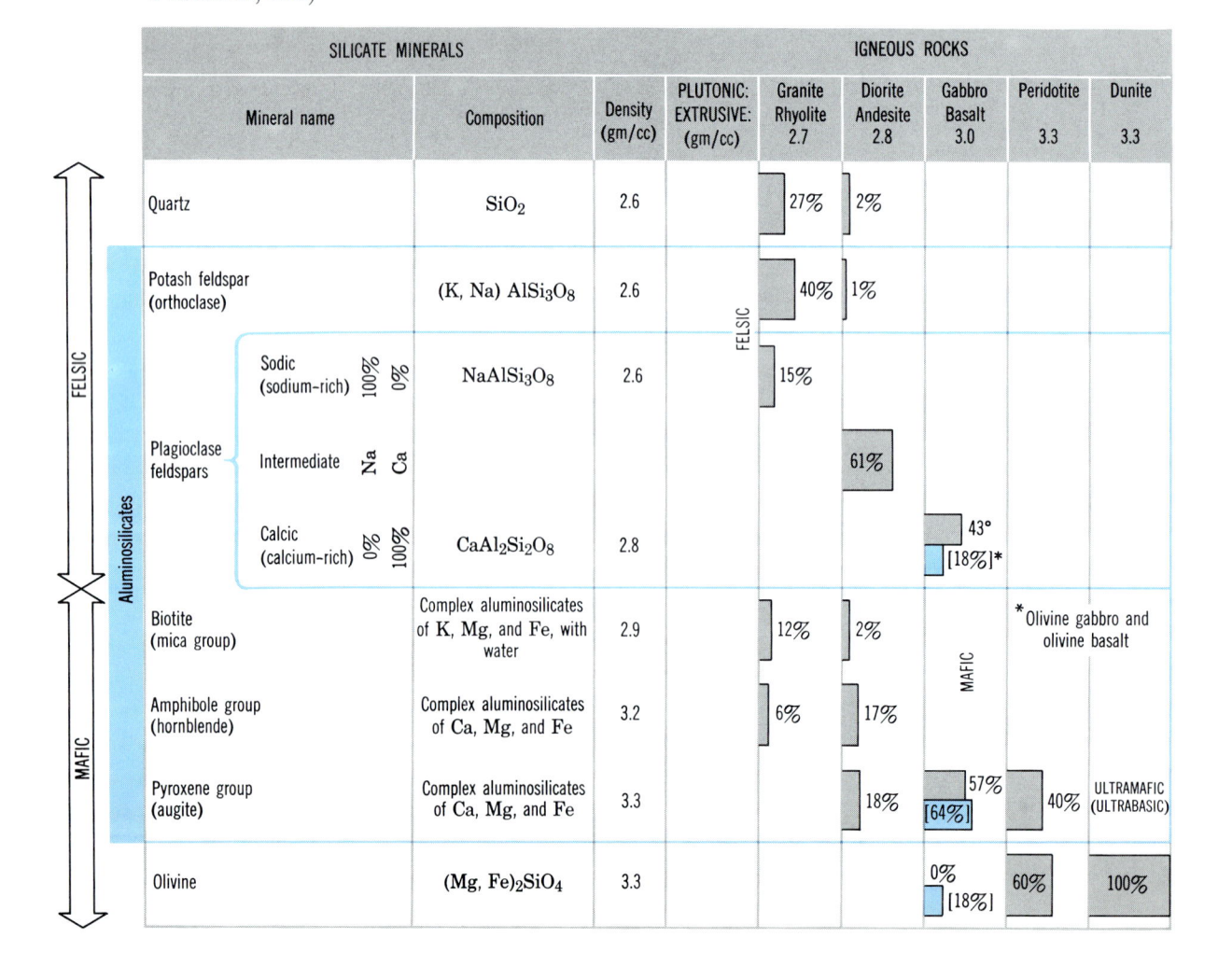

Silicate Minerals			Igneous Rocks				
Mineral name	Composition	Density (gm/cc)	Plutonic: Granite / Extrusive: Rhyolite (2.7)	Diorite / Andesite (2.8)	Gabbro / Basalt (3.0)	Peridotite (3.3)	Dunite (3.3)
Quartz	SiO_2	2.6	27%	2%			
Potash feldspar (orthoclase)	$(K, Na) AlSi_3O_8$	2.6	40%	1%			
Plagioclase feldspars — Sodic (sodium-rich) 100%/0%	$NaAlSi_3O_8$	2.6	15%				
Plagioclase feldspars — Intermediate (Na/Ca)				61%			
Plagioclase feldspars — Calcic (calcium-rich) 0%/100%	$CaAl_2Si_2O_8$	2.8			43% / [18%]*		
Biotite (mica group)	Complex aluminosilicates of K, Mg, and Fe, with water	2.9	12%	2%			
Amphibole group (hornblende)	Complex aluminosilicates of Ca, Mg, and Fe	3.2	6%	17%			
Pyroxene group (augite)	Complex aluminosilicates of Ca, Mg, and Fe	3.3		18%	57% / [64%]	40%	
Olivine	$(Mg, Fe)_2SiO_4$	3.3			0% / [18%]	60%	100%

*Olivine gabbro and olivine basalt

FELSIC / MAFIC; Aluminosilicates; FELSIC / MAFIC / ULTRAMAFIC (ULTRABASIC)

making up 100 percent of the variable metallic element. Plagioclase of *intermediate* composition contains about equal proportions of sodium and calcium. Quartz and the feldspars are light in color.

Biotite is the dark-colored representative of the *mica group* of silicate minerals. Biotite is a complex aluminosilicate of potassium, magnesium, and iron, with some water. Continuing down the list, we come to two more mineral groups. In each group are several closely related minerals, each with its own name and distinctive chemical composition. The *amphibole group* is represented by the mineral *hornblende;* the *pyroxene group* by *augite*. Both of these groups are complex aluminosilicates of calcium, magnesium, and iron. Minerals of this group are usually very dark in color. Finally, there is *olivine* $(Mg,Fe)_2SiO_4$, a dense greenish mineral that is a silicate of magnesium and iron but without aluminum.

Figure 23–2 shows specimens of some of the minerals we have listed. These are showpiece specimens of the pure minerals. Notice that pure quartz is glasslike; the feldspars are white and porcelain-like; the remaining minerals are dark in color and opaque to light.

23.4
FELSIC AND MAFIC MINERAL GROUPS

A very important concept in understanding how minerals and rocks are arranged in the earth's crust is that of mineral density. In Figure 23–1, densities are listed opposite each mineral. Notice that density increases from top to bottom of the list. Quartz and the feldspars range from 2.6 to 2.8 g/cm^3; they comprise the *felsic group*. "Felsic" is a coined word; it is derived from "fel" in "feldspar" and "si" in "silica." The felsic minerals have comparatively low density, and they are typically light in color. The remaining minerals range in density from 2.9 g/cm^3 for biotite to 3.3 g/cm^3 for augite and olivine. These denser minerals comprise the *mafic group*. "Mafic" is a word coined from the syllable "ma" in "magnesium" and "fic," a contraction of "ferric" (an adjective describing iron). The presence of iron, a high-density element, largely accounts for the greater densities of the mafic minerals.

23.5
MINERAL CLEAVAGE

Certain silicate minerals—and a wide range of other minerals as well—exhibit a physical property known as *cleavage*. Such minerals, when struck a sharp blow or crushed, break apart along planelike partings. These planes tend to be oriented parallel with one another to form a set. A mineral of this type separates (cleaves) into sheets. There may also be another set of parting planes at a right angle to the first, or at some intermediate angle. In such a case the mineral breaks apart into prisms. Even a third set of planes may exist, yielding cubes or rhombohedrons when the mineral is broken. Calcite is a mineral illustrating perfect rhombohedral cleavage (Figure 23–3).

What causes mineral cleavage? To answer this question, we must move into the dimension of atoms within the mineral crystal, applying principles of crystal structure developed in Chapter 14.

The essential building block in the crystal lattice structure of all of the silicate minerals is the *silicon-oxygen tetrahedron*, pictured in unexpanded form in Figure 23–4. It consists of four oxygen ions surrounding a single silicon ion. The expanded form of this tetrahedron (Figure 23–5) shows the positions of the bonds. In its central position, the small silicon ion fits neatly into the space between the four surrounding large oxygen ions.

The lattice structure of quartz is illustrated in Figure 23–6. The tetrahedra are arranged so that each oxygen ion is common to two tetrahedra—in other words, the oxygen ions are shared. The electric charges between ions are balanced and a stable compound is produced. Because the ionic bonds are equally strong in all directions, no planes of weakness exist in the structure of quartz, and it exhibits no cleavage. The mineral olivine, like quartz, is structured so as to have strong bonds between ions; it too has no cleavage.

Lattice structures of the other silicate minerals are more complex, and we can treat them only by more generalized descriptions. The structures can consist of either chains or sheets. A single chain of tetrahedra, linked by their shared oxygen ions, is the basic arrangement of the pyroxene group (for example, augite), as shown in Figure 23–7. Adjacent chains are linked by ions of magnesium, iron, calcium, or aluminum. The arrangement is such that two planes of weaker bonding exist at about right angles, giving the prismatic cleavage seen in augite.

A double chain of silicon-oxygen tetrahedra characterizes the amphibole group (for example, hornblende), as shown in Figure 23–8. In this arrangement the tetrahedra alternately share two and three oxygen ions. Ions of magnesium, iron, calcium, sodium, or potassium occupy positions between chains. This arrangement results in two directions of weaker bonding and yields

23 EARTH MATERIALS—MINERALS AND ROCKS

(A) Crystals of pure quartz, about one-half natural size. The crystals are hexagonal (six-sided) with pyramidal free ends. (Ward's Natural Science Establishment, Inc., Rochester, N.Y.)

(B) A mass of clear quartz of a variety often called "rock crystal." It is enclosed by fracture surfaces and resembles a chunk of clear glass. The needlelike objects inside are crystals of tourmaline. (Ward's Natural Science Establishment, Inc., Rochester, N.Y.)

(C) Microcline, a potash feldspar, with well-developed cleavage surfaces. (Ward's Natural Science Establishment, Inc., Rochester, N.Y.)

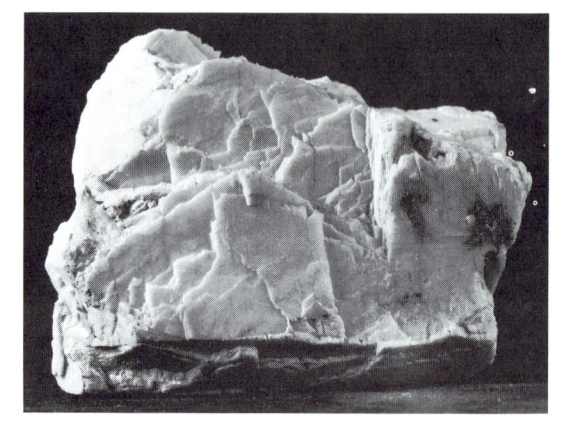

(D) Albite, a sodic plagioclase feldspar, with well-developed cleavage surfaces. (American Museum of National History.)

(E) A cleavage piece of biotite mica. (Ward's Natural Science Establishment, Inc., Rochester, N.Y.)

(F) Olivine, showing a glassy fracture surface. (Ward's Natural Science Establishment, Inc., Rochester, N.Y.)

FIGURE 23-2
Silicate mineral specimens.

FIGURE 23-3
These cleavage rhombohedrons of pure calcite are of a variety known as Iceland spar. (Ward's Natural Science Establishment, Inc., Rochester, N.Y.)

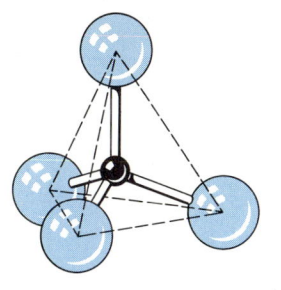

FIGURE 23-5
An expanded model of the silicon-oxygen tetrahedron. (From THE EARTH SCIENCES, 2nd Edition by Arthur N. Strahler. Copyright © 1963, 1971 by Arthur N. Strahler. By permission of Harper & Row, Publishers, Inc.)

two sets of cleavage planes. One of the angles between planes is acute; the other is obtuse.

The micas possess an interesting sheet structure, pictured in Figure 23-9. The silicon-oxygen tetrahedra are arranged in sheets in which each of the three basal oxygen ions is shared with an adjacent tetrahedron. A single layer, seen from above the plane of the sheet, forms a pattern of hexagonal cells. Alternate sheets of tetrahedra are inverted, as seen in the side view of Figure 23-9. Sheets are alternately separated by layers of positively charged ions, but one of these intervening layers has very weak bonding. As a result, the sheets separate readily, giving perfect cleavage in parallel planes.

23.6 SILICATE MAGMA

Igneous rocks are composed almost entirely of the seven silicate minerals or mineral groups we have studied. Other minerals are of secondary importance. In fact, it is estimated that the seven silicate minerals or groups make up as much as 99 percent of all igneous rocks.

A *silicate magma* is a mass of molten mineral matter capable of yielding silicate minerals as it solidifies. In some cases the magma solidifies within the surrounding rock; in other cases it is able to reach the surface, emerging as lava.

Crystallization begins to take place in a gradually cooling silicate magma at a certain critical

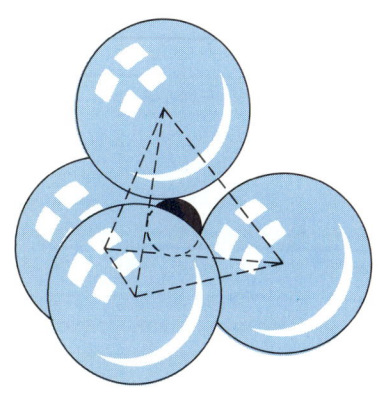

FIGURE 23-4
A silicon-oxygen tetrahedron in which one silicon atom fits between four oxygen atoms. (From THE EARTH SCIENCES, 2nd Edition by Arthur N. Strahler. Copyright © 1963, 1971 by Arthur N. Strahler. By permission of Harper & Row, Publishers, Inc.)

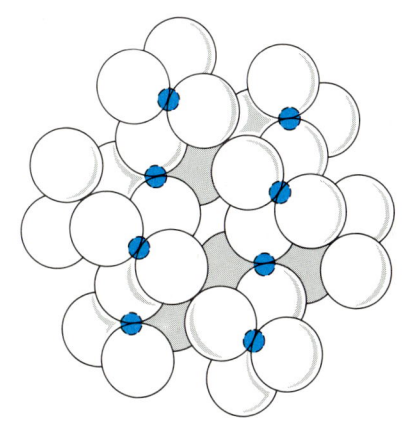

FIGURE 23-6
Arrangement of silicon-oxygen tetrahedra in quartz. (From THE EARTH SCIENCES, 2nd Edition by Arthur N. Strahler. Copyright © 1963, 1971 by Arthur N. Strahler. By permission of Harper & Row, Publishers, Inc.)

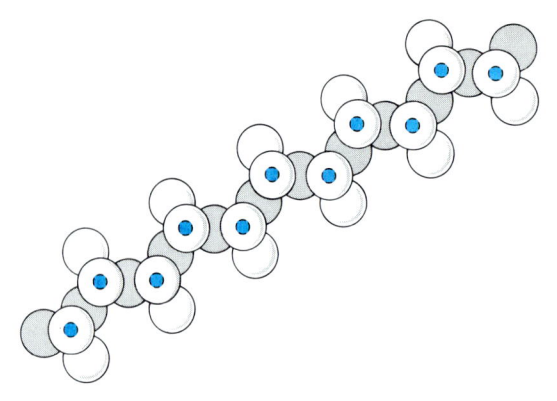

FIGURE 23-7
A single chain of silicon-oxygen tetrahedra, the basic arrangement in the pyroxene group. (From THE EARTH SCIENCES, 2nd Edition by Arthur N. Strahler. Copyright © 1963, 1971 by Arthur N. Strahler. By permission of Harper & Row, Publishers, Inc.)

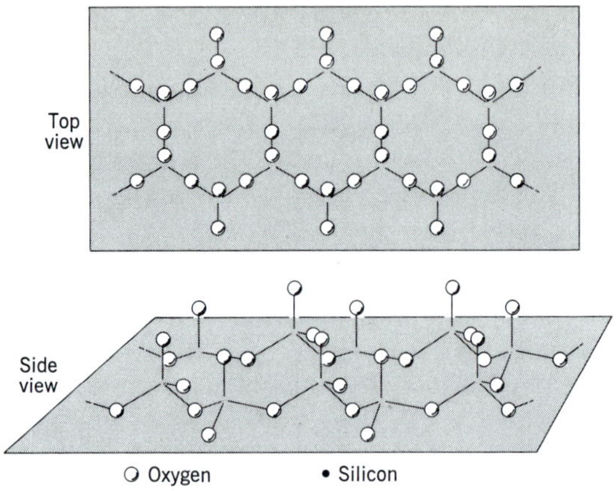

FIGURE 23-8
Perspective drawings of a double chain of tetrahedra in the structure of amphibole. (From *General Crystallography: A Brief Compendium* by W. F. de Jong, W. H. Freeman and Company, © 1959.)

combination of temperature and pressure. However, all minerals do not begin to crystallize at the same time. Moreover, a mineral, once formed, does not necessarily remain intact and unchanged from that point on. Instead, the early-formed minerals may subsequently be changed gradually in composition. Also, certain silicate minerals are dissolved and reformed as temperatures continue to fall.

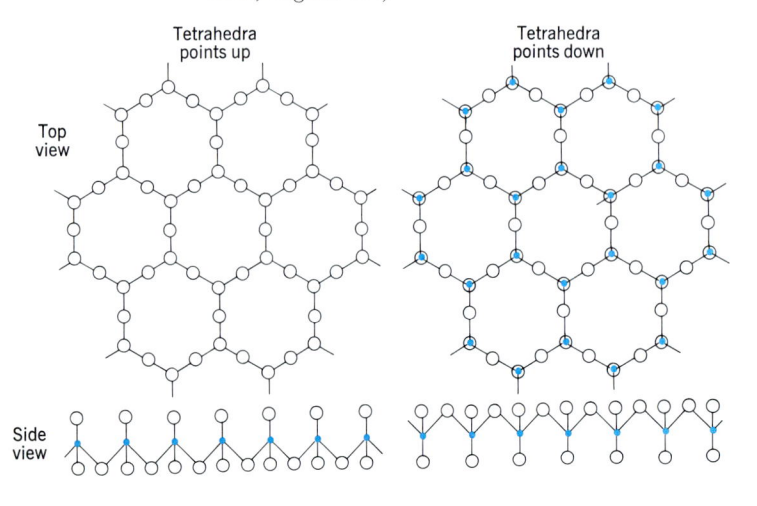

Generally speaking, the first minerals to crystallize in a cooling silicate magma are those of the mafic group, beginning with olvine. Pyroxene, amphibole, and calcic plagioclase feldspar follow olivine; then come sodic plagioclase feldspar and biotite mica. When these minerals have crystallized, potash feldspar appears, and finally quartz—both felsic minerals.

23.7 INTRUSIVE AND EXTRUSIVE IGNEOUS ROCKS

Igneous rocks are classified not only by mineral composition, but also in terms of the sizes of individual crystals that make up the rock. The word *texture* covers crystal sizes and arrangements.

Crystal size is largely dependent on the rate of cooling of the magma through the stages of crystallization. As a general rule, rapid cooling results in very small crystals, while extremely sudden cooling produces a natural glass. Very slow cooling, on the other hand, tends to produce large crystals.

From this principle we can deduce that igneous rocks cooling in huge masses at great depths where escape of heat is extremely low, will tend to develop a texture consisting of quite large crystals. The specimen of granite shown in Figure 23-10 illustrates such texture. Individual mineral grains can easily be distinguished with the unaided eye. We say that this granite is *coarse-grained* in texture. Large bodies of coarse-grained igneous rock, crystallized slowly at great

FIGURE 23-9
Crystal lattice structure of mica. The diagram shows alternate sheets of silicon-oxygen tetrahedra. Not shown are intervening layers of potassium and magnesium ions. (Modified from W. A. Deer, R. A. Howie, and J. Zussman, 1966, *An Introduction to the Rock-Forming Minerals,* London, Longman Group Ltd., Figure 69.)

FIGURE 23-10
Granite, a coarse-grained intrusive rock, is made up of tightly interlocking crystals of a few kinds of minerals. (A. N. Strahler.)

FIGURE 23-11
A frothy, gaseous lava solidifies into a light, porous scoria (*left*). Rapidly cooled lava may form a dark volcanic glass (*right*). (A. N. Strahler.)

depth below the surface, are described as *plutonic* igneous masses, or simply as *plutons*. Plutonic igneous rocks belong to a general class of igneous rocks called *intrusive igneous rocks*. Rocks of this class solidify beneath the surface, completely enclosed in preexisting solid rock. We give the name *country rock* to this older, surrounding rock.

The intrusion and solidification of an enormous mass of plutonic rock results in a *batholith*. Most commonly batholiths are of felsic igneous rock, typically granite or a closely related rock type. The horizontal extent of a single batholith may amount to many thousands of square kilometers; the depth may be on the order of 10 km.

Igneous rock that emerges at the earth's surface, cooling rapidly in contact with the atmosphere or ocean, is referred to as *extrusive igneous rock*. Lava, which is fluid magma pouring out from a vent and spreading over the surface, is one expression of extrusive action. Because the lava cools rapidly, mineral crystals are extremely small—most cannot be distinguished with the unaided eye, or even with a good magnifying lens. These rocks are said to be *fine-grained* in texture. Then, in some instances, cooling is so rapid that the magma solidifies into a *volcanic glass*, or *obsidian* (Figure 23-11). Notice the typical glassy fracture of the dark obsidian fragment. Next to the obsidian is a specimen of a rock full of spherical cavities; this is *scoria*. It is a type of lava in which expanding gases produced countless bubble-holes in the rock. No known igneous rock on earth exceeds an age of about $3\frac{1}{2}$ billion years. It is generally agreed that the earth achieved its identity as a planet from $4\frac{1}{2}$ to 5 billion years ago. Note that this leaves about one billion years or more of earth history unaccounted for by rock records. We are led to conclude that all known igneous rocks are formed of magma that invaded and completely replaced whatever older crustal rock may have previously existed.

23.8
THE GRANITE-GABBRO ROCK SERIES

In terms of bulk composition, most igneous rock of the earth's crust belongs to members of the *granite-gabbro series*. These rocks are presented in sequence from the felsic, or less dense, end toward the mafic, or more dense, end.

The right-hand part of Figure 23-1 lists rocks of the granite-gabbro series. In each column, under the rock name, bars show the proportions of minerals making up the rock in a typical example. Notice that each plutonic (intrusive) rock has an equivalent extrusive (lava) type. Although the name of the extrusive rock differs from that of the equivalent intrusive rock, their compositions are alike.

Granite is dominated in composition by the feldspars and quartz. Potash feldspar of the orthoclase variety is the most important mineral, while sodic plagioclase may be present in moderate amounts or absent. Quartz, which accounts for perhaps a quarter of the rock, reaches its most abundant proportions in granite. Biotite and hornblende are common accessory minerals.

Granite is described as a light-colored igneous rock and is grayish to pinkish, depending on the variety of potash feldspar present. Its density, about 2.7 g/cm^3, is comparatively low among the igneous rocks. Most granites are sufficiently coarse in texture for the component minerals to be identified with the unaided eye (Figure 23-10). The grayish cast of the quartz grains, with their glassy luster, sets them apart from the milky white or pink feldspars. Black grains of biotite or hornblende contrast with the light minerals. The extrusive equivalent of granite is *rhyolite,* a light gray to pink form of lava.

Diorite is the next important plutonic rock on our list. Its extrusive equivalent, *andesite,* occurs very widely in lavas associated with volcanoes. Looking at the bars in Figure 23-1, we see that diorite is dominated by plagioclase feldspar of intermediate composition, while quartz is a very minor constituent. At this point in the granite-gabbro series, pyroxene makes its appearance and is of the augite variety. Amphibole, largely hornblende, is also important, and some biotite is present.

Gabbro is an important although not abundant plutonic rock, but it is greatly overshadowed in importance by its extrusive equivalent, *basalt.* We find that basalt makes up vast areas of lava flows and is the predominant igneous rock underlying the floors of the ocean basins. Basalt is also a major rock type at the surface of the moon. Gabbro and basalt are composed largely of pyroxene and calcic plagioclase feldspar with varying amounts of olivine (some types lack olivine). Gabbro and basalt are dark-colored rocks—dark gray, dark green, to almost black—of relatively high density.

We can now apply the adjectives *felsic* and *mafic* to igneous rocks as well as to silicate minerals. *Felsic igneous rocks* are those rocks dominantly composed of felsic minerals; they include granite and diorite. *Mafic igneous rocks* are those rocks composed dominantly of mafic minerals. As shown in Figure 23-1 by numbers beneath the rock names, the felsic rocks have densities less than 3.0 g/cm^3. The mafic rocks have densities of 3.0 g/cm^3 or higher.

Continuing the igneous-rock series depicted in Figure 23-1, we arrive at *peridotite,* a rock composed almost entirely of two mineral constituents: olivine and pyroxene. Although widespread in occurrence, peridotite occurs in relatively small plutonic bodies. Peridotite is a dark-colored rock of high density (3.3 g/cm^3) and belongs to a group designated as *ultramafic rocks.*

Finally, we include a variety of igneous rock consisting almost entirely of the mineral olivine. This rock is called *dunite.* Like peridotite, dunite is an ultramafic rock and has a density of 3.3 g/cm^3. Although dunite occurs rarely at the earth's surface, its geologic importance is very great; there is evidence that dunite comprises much of the earth beneath the outer crust.

23.9 VOLCANOES

Volcanism is a term applied generally to the formation of extrusive igneous rocks. Volcanism includes both the outpourings of lavas in broad sheets and the more localized accumulations of magma in the form of individual volcanoes and groups of volcanoes. By *volcano,* we mean a massive structure built by emission of magma and its contained gases from a pipelike conduit. As eruption continues through time, the accumulated igneous rock must form a more or less conical mountain mass, a *volcanic cone.* The cone surrounds the vent, or point of emergence, of the conduit.

Eruption of rhyolite and andesite lava typically produces a tall, steep-sided cone. The cone characteristically steepens to the summit. At the summit is a depression, the *crater,* marking the position of the vent. Familiar to all are the graceful conical profiles of such volcanoes as Mt. Fuji in Japan, Mt. Mayon in the Philippines, Mt. Hood in the Cascade range, and Shishaldin in the Aleutian Islands (Figure 23-12). Cones of this type are called *composite volcanoes,* because they consist in part of lava flows and in part of volcanic ash. The internal structure of such a volcano is shown in Figure 23-13. By *ash* we mean finely divided igneous rock that results from the explosive emission of magma heavily charged with gases under high pressure.

The continued outpouring of great quantities of highly fluid basalt lavas from a radiating series of fissures (cracks) produces the *shield volcano.* Unquestionably the greatest assemblage of shield volcanoes is the Hawaiian Islands. Each island is formed of one or more such volcanoes (Figure 23-14). Another locality famous for its shield volcanoes is Iceland.

Most of the world's active and recently active volcanoes tend to be concentrated in chainlike fashion in long, narrow belts. Others are clustered, and some are geographically quite isolated, seemingly not part of any recognizable belt. The greatest of all volcano chains is the *circum-Pacific ring,* sometimes called "The Ring of Fire" (see Figure 24-5).

Interpretation of great volcano chains and their origin is closely tied in with the breaking and bending of the lithosphere. We shall develop this topic further in Chapter 24.

FIGURE 23-12
Mt. Shishaldin, an active composite volcano on Unimak Island in the Aleutian Islands, rises to an elevation just over 2800 m. A plume of condensed stream marks the summit crater. (U.S. Navy Department, The National Archives.)

FIGURE 23-13
Idealized cross section of a composite volcanic cone with feeders from magma chamber beneath. (From PLANET EARTH: ITS PHYSICAL SYSTEMS THROUGH GEOLOGIC TIME by Arthur N. Strahler. Copyright © 1972 by Arthur N. Strahler. By permission of Harper & Row, Publishers, Inc.)

FIGURE 23-14
This air view of Mauna Loa, a great shield volcano built of basalt, shows a row of pit craters leading up to the large central depression at the summit. (U.S. Army Air Force.)

23.10 SEDIMENTS AND SEDIMENTARY ROCKS

Exposed at the earth's surface, igneous rock comes under attack from atmospheric forces, both physically and chemically. Physical processes that disintegrate exposed rock are a subject of Chapter 26. Chemical processes include the breakdown of silicate minerals in the presence of heat and water and with abundant atmospheric free oxygen present. All these forms of rock disintegration and decomposition are included in the term *rock weathering*. Weathering produces *sediment*, broadly defined as any finely divided mineral or organic matter derived directly or indirectly from the disintegration, decomposition, and reprocessing of preexisting rock, and from life processes. Other earth processes such as streams, waves, and wind, transport and redistribute sediment, producing accumulations that constitute the second major rock class—the *sedimentary rocks*.

We can classify sediments and sedimentary rocks, according to their origin, into three major groups:

1. Clastic sediments.
2. Chemically precipitated sediments
3. Organic sediments.

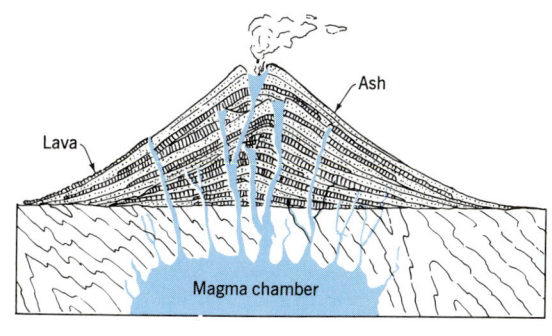

The clastic sediments consist of physically broken particles derived from preexisting rock of any variety. The chemically precipitated sediments consist of new inorganic compounds formed by chemical precipitation from a water solution or secretion within organisms in oceans or freshwater lakes. The organic sediments consist of organic hydrocarbon compounds pro-

23 EARTH MATERIALS—MINERALS AND ROCKS

TABLE 23.2
SIZE GRADES OF SEDIMENT PARTICLES

Grade Name	Diameter (mm)
Boulders	Over 256
Cobbles	64–256
Pebbles	2–64
Sand	0.64–2
Silt	0.004–0.06
Clay	Under 0.004

duced by life processes, usually in water bodies or in a bog environment.

Sedimentary rocks are easily recognizable through the presence of distinct layers resulting from changes in particle size and composition during the period of deposition. These layers are termed *strata*, or beds. The planes of separation between layers are *planes of stratification*, or bedding planes. The rock is described as being *stratified*, or bedded (see Figure 25–1). Bedding planes in their original condition are nearly horizontal, but they may have become steeply tilted (see Figure 24–4) or otherwise distorted into wavelike folds by subsequent movements of the earth's crust.

23.11 CLASTIC SEDIMENTARY ROCKS

Clastic rocks are named and classified according to the grade sizes of the mineral particles of which they are composed. Table 23.2 gives the names and diameter ranges of six classes of sediment particles ranging from boulders to clay.

Conglomerate consists of pebbles or cobbles, usually quite well rounded in shape, embedded in a fine-grained matrix of sand or silt (Figure 23–15). Rounding of the conglomerate pebbles is a result of abrasion (wearing action) during transportation in stream beds or along beaches. Essentially, then, conglomerates represent solidified stream gravel bars and gravel beaches.

Sandstone is composed of grains in the range from 2—0.06 mm. Perhaps the most abundant and familiar form is quartz sandstone, in which quartz is the predominant constituent. Beautifully rounded quartz grains, many of spherical form, extracted from a sandstone, are pictured in Figure 23–16. In this example rounding was perfected by wind transport in ancient sand dunes.

Loose sand or gravel can become hardened into solid rock, a process called *lithification*. The pore spaces between grains are partially or entirely filled with new mineral matter, acting as a cement. This cementation is accomplished by slowly moving underground water importing the mineral matter as ions in solution. The cementing mineral of sandstone may be silica (SiO_2); in this case the sandstone is extremely hard. If the cementing material consists of calcium carbonate ($CaCO_3$), a less durable rock results.

The compaction and cementation of layers of silt gives a dense, fine-grained rock known as *siltstone* when largely free of clay particles. Siltstone has the feel of very fine sandpaper and is closely related to fine-grained sandstone, with which there is a complete intergradation.

The compaction of clay-sized particles—finer than 0.004 mm in diameter—yields *shale*. The bulk of most shale consists of *clay minerals*, a class

FIGURE 23–15
Conglomerate is a mixture of pebbles and sand cemented into a hard rock. (A. N. Strahler)

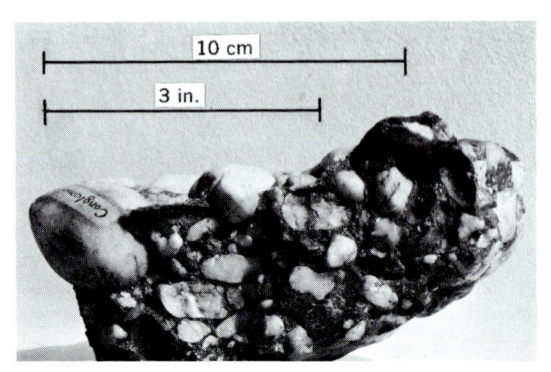

FIGURE 23–16
Rounded quartz grains from an ancient sandstone. The grains average about 1 mm in diameter. (Andrew McIntyre, Columbia University.)

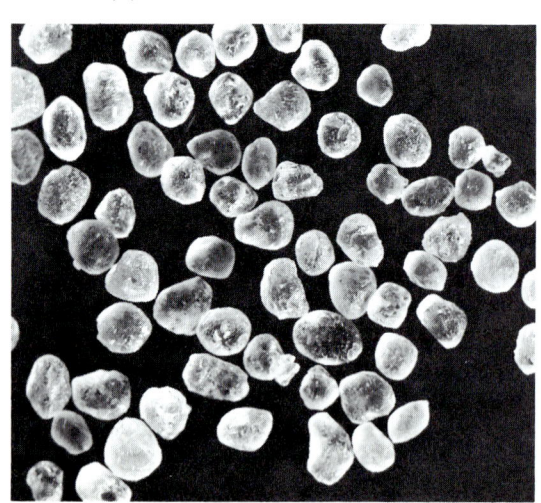

of minerals produced by chemical alteration of silicate minerals. Minerals such as feldspar, pyroxene, and amphibole are altered in the presence of water at the earth's surface. A chemical change known as *hydrolysis* takes place, in which molecules of water are combined with the silicate compounds. The resulting clay minerals are soft and typically take the form of microscopic scales and platelets, as shown in Figure 23–17. Clay minerals are typically plastic when soft. An example is *kaolinite,* formed by hydrolysis of potash feldspar. In pure form, kaolinite is used in the manufacture of ceramics. Another common clay mineral is *illite* (Figure 23–17), formed by hydrolysis of feldspars or mica and one of the commonest constituents of shale. Shale becomes lithified by the process of compaction under the load of overlying sediment, driving out the free water and forming a dense, fine-grained rock. Most shales have a structure consisting of very thin laminations or plates, splitting apart readily on these planes.

23.12
CHEMICALLY PRECIPITATED SEDIMENTARY ROCKS

Most of the chemically precipitated sediments and sedimentary rocks fall into two major classes: carbonates and evaporites.

FIGURE 23–17
Seen here enlarged about 20,000 times are tiny flakes of the clay minerals *illite* (sharp outlines) and *montmorillonite* (fuzzy outlines). These particles have settled from suspension in San Francisco Bay. (Harry Gold, San Francisco District Corps of Engineers, U.S. Army.)

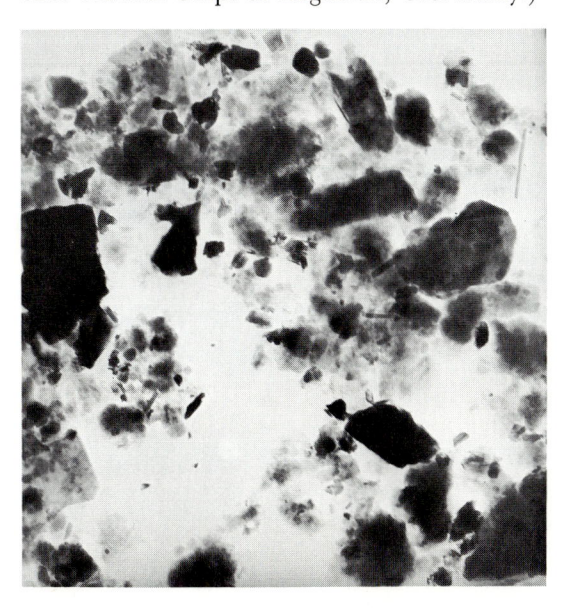

Many carbonate sediments and rocks consist largely of calcium carbonate ($CaCO_3$) represented, in the pure form, by the mineral *calcite*. Rock of this composition is called *limestone*. Also common is the mineral *dolomite,* a calcium-magnesium carbonate with the formula $CaMg(CO_3)_2$. A rock composed largely of this mineral also has the name dolomite.

Limestone comes in a wide variety of forms and has many and varied conditions of origin. Most limestones were formed as accumulations of carbonate mineral matter on the floors of shallow seas. The material comprising some limestones is made up largely of hard parts or shells of lime-secreting organisms.

Evaporites are salts crystallized from solutions undergoing intense evaporation in shallow lakes or in coastal estuaries in a dry, warm climate. One of the commonest of the evaporite minerals is *gypsum,* a hydrous form of calcium sulfate ($CaSO_4 \cdot 2H_2O$). Another is *halite,* often called "rock salt," which is largely sodium chloride (NaCl).

23.13
ORGANIC SEDIMENTS

Organic sediments, consisting of hydrocarbon compounds, were described in Chapter 20. Only lignite and coal qualify for designation as rock. These solid forms of hydrocarbon compounds remain in the place of original accumulation. On the other hand, the liquid and gaseous forms—petroleum and natural gas—can migrate far from the places of origin to become concentrated in distant rock reservoirs.

Coal occurs in layers, known as *seams,* interbedded with sedimentary strata, usually thinly bedded shales, sandstones, and limestones (Figure 23–18). Collectively, these accumulations are known as *coal measures*. Individual coal seams range in thickness from a fraction of a centimeter to ten meters or more.

23.14
METAMORPHIC ROCKS

Metamorphic rocks comprise the third major class of rocks. The process by which they are formed is called *rock metamorphism;* it consists of changes in mineral composition and mineral structure through application of heat, pressure, and kneading action. Metamorphism of this type usually takes place at great depths beneath the surface in zones of active mountain-building. The basic geologic circumstances under which metamorphism occurs will become clear in the next chapter.

FIGURE 23-18
A coal seam, 2.4 m thick, exposed in a river bank, Dawson County, Montana. (M. R. Campbell, U.S. Geological Survey.)

The total process of rock metamorphism is felt in two ways. First, original minerals recrystallize and new minerals are formed. Second, a new set of structures is imposed on the rock and may replace or obliterate original bedding structures. This type of metamorphism has affected enormous bodies of rock within the root zones of mountain chains. Consequently, the effects are seen today in surface rocks over large areas.

Application of high temperatures alone can also cause metamorphism. Such effects are often conspicuous in country rock close to an igneous intrusion. The changes are essentially those of baking in a high-temperature oven. A shale rock close to an igneous contact may experience a hardening and color change not unlike that caused by baking of brick or tile. However, most large igneous intrusions cause metamorphism through emanations of hot watery solutions containing many different kinds of ions. These are highly active solutions and cause mineral alteration of the country rock.

Metamorphism is a change of mineral state in response to a change in environment. Many common igneous and sedimentary minerals are unsuited to the environment of deformation under stress at high pressures and high temperatures. They will be altered to form minerals capable of attaining equilibrium under those environments. There will often be changes in grain size and shape as well.

Many of the most common and abundant minerals of metamorphic rocks are also abundant in igneous and sedimentary rocks. For example, quartz, one of the most abundant of minerals in rocks of all classes, persists unaltered in metamorphic rocks.

Among the more important new minerals distinctive of metamorphism are several aluminosilicates. As a group they are hard minerals with densities comparable to those of the mafic minerals. Crystals of distinctive form grow in the metamorphic rock and are often easily recognized. An example of a metamorphic mineral of aluminosilicate composition is garnet. Most of us know garnet as a semiprecious gemstone of red color. Figure 23–19 shows beautifully formed crystals of garnet embedded in a metamorphic rock rich in mica.

Some metamorphic rocks are characterized by *foliation*, a crude layering along which the rock easily separates. A second structure is *banding*, a layered arrangement of strongly knit crystals forming a massive rock.

Slate is a very fine-grained rock that splits readily into smooth-surfaced sheets along cleavage surfaces. Slate is largely derived from shale. The cleavage of slate is a new structure imposed by metamorphism and usually cuts across the original bedding.

Schist is a foliated rock that comes in many varieties. Foliation results from the parallel alignment of easily cleavable minerals such as mica. The reflecting surfaces of these minerals give a charateristic glistening sheen to the foliation surfaces. Schists have undergone a high degree of metamorphism, and their origin is not always clear. Most schists are interpreted as altered clastic sedimentary strata rich in aluminosilicate minerals. It is commonly inferred that slates represent an intermediate grade of metamorphism between shale and schist. This sequence is actually demonstrated in some localities by tracing the changes continuously from shale, through slate, to schist.

FIGURE 23-19
Garnet crystals in schist. The larger crystal is about 2 cm across. (A. N. Strahler.)

FIGURE 23-20
Banded gneiss of Precambrian age, along the east coast of Hudson Bay, Canada. (Photograph G.S.C. No. 125221 by F. C. Taylor, Geological Survey of Canada, Ottawa.)

Gneiss is a general term for a metamorphic rock showing banding or lineation. The strongly banded forms of gneiss, such as that shown in Figure 23-20, are often interpreted as highly altered sedimentary strata. It is thought that highly heated, chemically active solutions from nearby magma permeated the sedimentary layers, bringing in and precipitating new minerals, such as quartz, feldspar, and hornblende. In this way, layers resembling intrusive igneous rock have been formed. Other kinds of gneiss may simply represent plutonic rock, such as granite, that has been severely deformed, so that the mineral crystals have been drawn out into long, thin, pencillike shapes.

Limestone and dolomite are metamorphosed into *marble,* which is typically a light-colored granular rock exhibiting a sugary texture on a freshly broken surface. Although white when pure, marbles come in many colors, depending on the presence of impurities.

23.15
THE ROCK CYCLE

This is a good place to summarize the relationships among the three major rock classes in terms of a *cycle of rock transformation,* or, simply, the *rock cycle.*

Figure 23-21 is a triangular diagram showing that any one of the three major rock classes—igneous, sedimentary, and metamorphic—can be derived from either of the other two classes. Sequences of changes examined in this chapter are labeled on the sides of the triangle; they need no further explanation here.

Let us relate the rock cycle to the contrasting physical-chemical environments found at depth within the earth and at the earth's surface (Figure 23-22). The completed diagram now represents a schematic vertical cross section of the crust, say to a depth of about 30 km. Throughout the rock cycle, large masses must be moved from the *deep environment* of high temperatures and pressures to the *surface environment* of low temperatures and pressures. We see that this change of environment can be accomplished by two processes: first, rise of magma may bring igneous rock to various intermediate positions, where it solidifies into plutonic rock bodies, or by extrusion it may reach the surface to form volcanic rocks; second, rock formed at depth can appear at the surface of the earth by uncovering as a result of the combined processes of crustal uplift and erosion. Thus, large bodies of igneous, metamorphic, or sedimentary rock can slowly migrate upward from the deep environment to the surface environment.

Transition from the surface environment to the deep environment can be accomplished by burial and down-sinking of the earth's crust.

FIGURE 23-21
Three major rock classes as corners of a triangle. (From THE EARTH SCIENCES, 2nd Edition by Arthur N. Strahler. Copyright © 1963, 1971 by Arthur N. Strahler. By permission of Harper & Row, Publishers, Inc.)

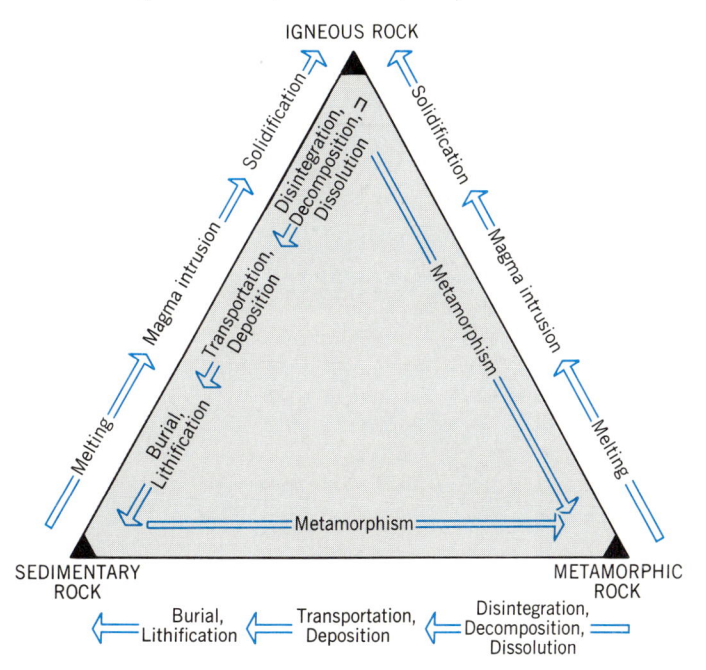

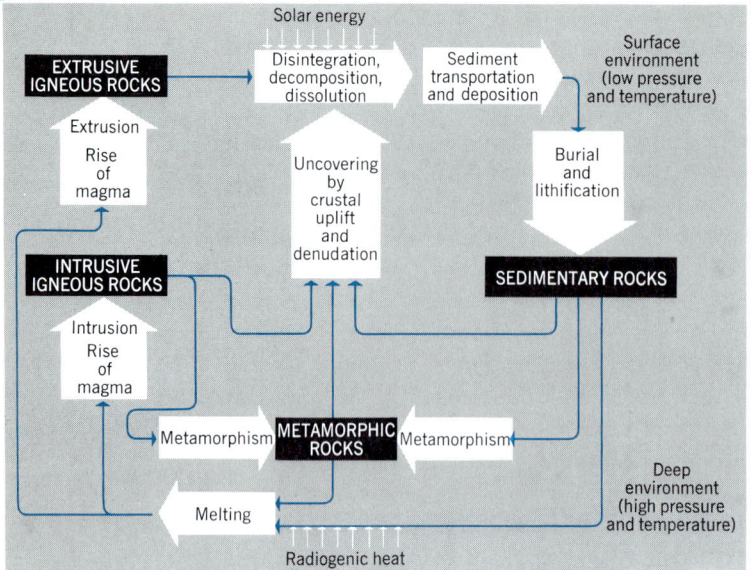

FIGURE 23-22
Schematic diagram of the cycle of rock transformations. (From THE EARTH SCIENCES, 2nd Edition by Arthur N. Strahler. Copyright © 1963, 1971 by Arthur N. Strahler. By permission of Harper & Row, Publishers, Inc.)

Both sedimentary strata and extrusive volcanic rocks can eventually reach the deep environment where either metamorphism or remelting can take place.

23.16 EARTH RESOURCES

Human beings are rapidly consuming mineral resources that required geologic spans of time to accumulate. Geological processes operate with extreme slowness. Natural rates of replenishment are infinitesimally small in comparison with the present rates of consumption; therefore, geologic resources are finite. These nonrenewable mineral resources can be grouped as follows:

Metalliferous deposits (examples: ores of iron, copper, tin).
Nonmetallic deposits
    Structural materials (examples: building stone, gravel, and sand).
    Materials used chemically (examples: sulfur, salts).
Fossil fuels (coal, petroleum, and natural gas).
Nuclear fuels (uranium, thorium).

Notice that the last two groups represent sources of energy, whereas the first two groups are sources of matter (materials).

23.17 METALS IN THE EARTH'S CRUST

Metals occur in useful concentrations as ores. Recall from Chapter 17 that an ore is a mineral accumulation that can be extracted at a profit for refinement and industrial use. A number of important metallic elements are listed in Table 23.3 with their abundances, as percentage by weight, in the average crustal rock. Whereas aluminum and iron are relatively abundant, most of the essential metals of our industrial civilization are present in extremely small proportions—witness mercury and silver with abundances of only 0.000008 and 0.000007 percent, respectively.

Obviously, most metals must occur greatly concentrated by geologic processes to be ores extractable at a profit. For example, chromium has an average crustal abundance of only 0.01 percent; it must be found concentrated by a factor of about 1500 times to be an ore sufficiently rich to be extracted. For lead, the crustal abundance must have been concentrated by a factor of 2500 times to constitute an ore.

TABLE 23.3
METALLIC ABUNDANCES IN AVERAGE CRUSTAL ROCK

Symbol	Element Name	Abundance (percentage by weight)
Al	Aluminum	8.1
Fe	Iron	5.0
Mg	Magnesium	2.1
Ti	Titanium	0.44
Mn	Manganese	0.10
V	Vanadium	0.014
Cr	Chromium	0.010
Ni	Nickel	0.0075
Zn	Zinc	0.0070
Cu	Copper	0.0055
Co	Cobalt	0.0025
Pb	Lead	0.0013
Sn	Tin	0.00020
U	Uranium	0.00018
Mo	Molybdenum	0.00015
W	Tungsten	0.00015
Sb	Antimony	0.00002
Hg	Mercury	0.000008
Ag	Silver	0.000007
Pt	Platinum	0.000001
Au	Gold	0.0000004

Source: (Data from B. Mason, 1966, *Principles of Geochemistry*, 3rd ed., New York, Wiley, pp. 45–46, Table 3.3 and Appendix III)

In the next 25 years, demands by the United States for essential industrial metals will increase greatly. A doubling of need will be felt for many metals, among them iron, manganese, zinc, cobalt, lead, antimony, mercury, and silver. For others, the factor of increased demand will be much greater. For example, our need for aluminum will increase about sixfold, for titanium twelvefold, and uranium twentyfold. Can these increased demands be met from reserves within the United States? The answer is an emphatic "no," because even at the present moment, the United States depends on foreign imports for a large proportion of nearly all industrial metals consumed.

23.18 MINERAL RESOURCES FROM THE SEA

If the prospect of eventually running out of various mineral resources from the lands seems all too real, we may want to turn to consider possible substitutions of mineral resources from the sea. Seawater has long provided the bulk of the world's supply of magnesium and bromium, as well as much of the sodium chloride.

The list of elements present in seawater includes most of the known elements and, despite their small concentrations, these are potential supplies for future development. It is thought that sodium, sulfur, potassium, and iodine lie in the category of recoverable elements. But it seems beyond reason to hope for extraction of ferrous metals (principally iron) and the ferro-alloy metals in significant quantities to provide substitutes for ore deposits of the continents.

The continental margins, with their shallow continental shelves and shallow inland seas, are already being exploited for mineral fuel production. The petroleum resources of the North American continental shelf are under development along the Gulf Coast; zones of potential development are believed to exist on the shelf off the Atlantic Coast as well.

Exploration of the deep ocean floor as a source of minerals is still in an early stage. Already the surface layer of manganese nodules found in parts of all of the oceans is regarded by some as a major future source of manganese, along with a number of metals in lesser quantities.

23.19 SUSTAINED-YIELD SOURCES OF ENERGY

Before looking into the sources of energy that are derived from the solid earth, let us review the full picture of world energy resources to gain a better perspective. Sources of energy are found in both sustained-yield and exhaustible categories. A *sustained-yield* source is one that undergoes no appreciable diminution of energy supply during the period of projected use. Consider first the sutained-yield sources.

Solar energy is a sustained-yield source of extreme constancy and reliability. Stated in terms of power, solar radiation intercepted by one hemisphere is calculated to be about 100,000 times as great as the total existing electric power-generating capacity. The problem is, of course, that solar radiation derived from a large receiving area must be concentrated into a very small distribution center. To produce power equivalent to that of a large generating plant (about 1000 megawatts capacity) would require at an average location a collecting surface represented by a square measuring 6.5 km (4 miles) on a side. While there seems to be no technological barrier to building such a plant, the cost at present is too high to make this energy source practical.

Solar energy has now been recognized as the most promising long-range energy source capable of substituting for mineral fuels on a large scale. We can look forward to intensified research on practical and economical ways to convert direct solar energy to electric energy. Other forms of solar energy are also being explored. For example, the temperature difference between the warm surface ocean layer and the cold mass of deep water can, in theory, be utilized through heat exchange systems to derive enormous supplies of energy. Another avenue of research lies in the use of solar energy to promote intense growth of plants, such as the algae, thereby converting solar energy into stored chemical energy within carbohydrate molecules.

Wind power is another form of solar energy, capable of exploitation as a substitute for mineral fuels. Recall from Chapter 21 that winds are driven by differences in intensity of atmospheric heating from place to place, setting up pressure gradients. Windmills can convert the kinetic energy of air motion into electric energy and thereby povide valuable supplements to the total energy supply.

Hydropower, water power of streams moving by gravity flow, is a valuable source of sustained energy and has been developed to a point just over one-quarter of its estimated ultimate maximum capacity in the United States. Water power now supplies about 4 percent of the total energy production of the United States. For the world as a whole, present development is estimated to be about 5 percent of the ultimate maximum capacity.

Tidal power is another sustained-yield energy source. To utilize this power, a bay is chosen along a coast subject to a large range of tide. Narrowing of the connection between bay and open ocean intensifies the differences of water level that are developed during the rise and fall of tide. A strong hydraulic current is produced and alternates in direction of flow. The flow is used to drive turbines and electrical generators. The world total of annual energy potentially available by exploitation of all suitable sites comes to only 1 percent of the energy potentially available through hydropower development.

23.20
GEOTHERMAL ENERGY

Another form of sustained-yield energy is *geothermal energy*, drawn from heat within the earth at points of locally high concentration. Usually the surface manifestations of such heat are hot springs, fumaroles (holes from which steam issues), and active volcanoes. Wells drilled at these places yield superheated steam, which can be used to power turbines. Electric power is presently being generated from a number of geothermal fields. The most important fields at present are in Italy, New Zealand, and the United States. The total output of these fields is on the order of 1 million kw (10^6 kw). New developments now under construction or planned will about double that total. It is estimated that the ultimate development of known sources would give a world geothermal energy output on the order of ten times that of today. This total is about the same as for the ultimate yield of all potential tidal power projects. Both of these energy sources represent only a small fraction of existing energy requirements.

23.21
COAL AND LIGNITE

We turn next to our major source of energy for the past century, the *fossil fuels:* coal and lignite, petroleum, and natural gas. The chemical composition of these hydrocarbon substances was discussed in Chapter 20.

World coal reserves are most unevenly distributed among the continents. Figure 23–23 shows a breakdown into eight world regions; figures are in billions of metric tons. This estimate shows the Soviet Union and the United States to be in very strong positions, while Canada and Western Europe are also very favorably endowed with coal. In contrast, Africa, Australia, Japan, and the Latin American countries are poorly endowed.

4300	Soviet Union in Asia and Europe
1500	United States
680	Asia exclusive of Soviet Union
600	North America exclusive of United States
375	Western Europe
110	Africa
60	Oceania, including Australia
15	Central and South America

World total: 7640 billion metric tons

FIGURE 23–23
Estimated coal reserves of the world, in billions of metric tons, based on data of the United States Geological Survey.

In terms of present mining technologies, only about one-tenth of this amount of coal can be mined on an economically successful basis. So the figure of 1500 billion metric tons for the United States reduces to 150 billion tons, in practical terms. Converted into crude oil energy equivalent, 150 billion tons represents about 90 billion metric tons. As we shall find in assessing oil reserves, this quantity is roughly 18 times larger than the proved crude oil reserves of the United States.

23.22
WORLD PETROLEUM RESERVES

Proved petroleum reserves are mostly concentrated in a few world regions. As Figure 23–24 shows, the Middle East holds more than half the known world reserves of crude oil. Of this enormous accumulation—some 50 billion metric tons—Saudi Arabia has about 40 percent (20 billion tons), which is almost four times as much as United States reserves (5 billion tons). Kuwait, Iran, and Iraq hold most of the remainder of the Middle East oil. Another important center of oil accumulation is in lands surrounding the Gulf of Mexico and Caribbean Sea, with major reserves in the United States Gulf Coast region and Venezuela.

23.23
OIL SHALE

Oil shale constitutes a tremendous reserve of hydrocarbon fuel. Important oil shale beds occur over a large area in northeastern Utah, northwestern Colorado, and southwestern Wyoming.

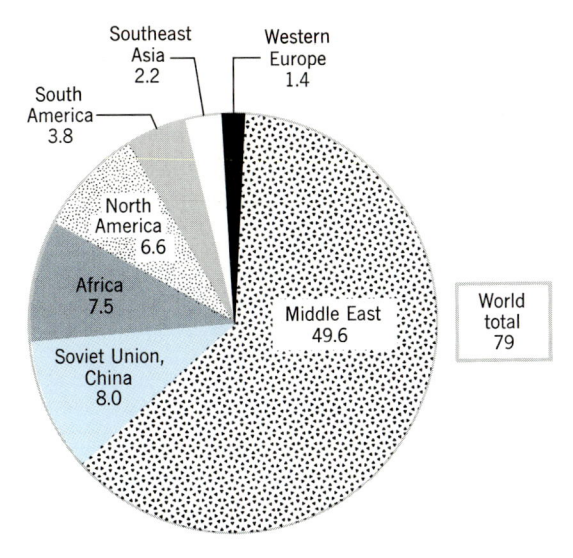

FIGURE 23-24
The world petroleum pie. Figures show proven reserves of crude oil in billions of metric tons. (Data of National Petroleum Council.)

The hydrocarbon material of oil shale is a waxy substance, called *kerogen*, which adheres to the tiny grains of carbonate material constituting the shale. When oil shale is crushed and heated to a temperature of 480°C, the kerogen is altered to petroleum and driven off as a liquid. The shale may be mined and processed in surface plants or burned in underground mines, from which the oil is pumped to the surface.

It is estimated that the equivalent of some 17 billion metric tons of petroleum lie in reserve in prime beds of the western United States shale formation. When we compare this figure with proved world petroleum reserves of about 80 billion tons and a proved United States petroleum reserve of about 5 billion tons, we realize that the prime oil shale deposits of the United States are indeed an enormous energy resource.

Unhappily, it takes great quantities of water to process oil shale. The Federal Energy Administration calculated that if an oil shale industry were to produce 200,000 metric tons (1.5 million barrels) of oil per day, it would require 80 percent of the undeveloped water in Colorado and Utah and 45 percent of the undeveloped water in Wyoming. The smallest commercial operation would be in the neighborhood of 7000 metric tons of oil per day, and it would generate 100,000–150,000 tons of used shale daily. For each cubic meter of oil produced, from three to six cubic meters of water would be required—largely to reduce dust. The spent water, if allowed to get into the network of streams and rivers, would carry dissolved and leached salts downstream. The Federal Energy Administration calculates that the increase in salinity in the Colorado River basin, which furnishes water to large agricultural areas in southern California and Mexico, could create $3–6 billion a year in increased costs of agriculture.

Plans to change lignite and soft coal in the Dakotas and Montana to liquid or gaseous fuels that are easier to transport likewise depend on using considerable portions of the flowage of the Yellowstone and Missouri Rivers—waters also needed for irrigation.

23.24 NUCLEAR ENERGY AS A RESOURCE

As explained in Chapter 9, the controlled release of energy from concentrated radioactive isotopes can be achieved through one of two processes: fission and fusion. Atomic fission power plants use uranium-235. The fission of 1 gram of this substance yields an amount of heat equivalent to the combustion of about 3 metric tons of coal or about 2 metric tons of crude oil. Uranium-235 is a rare isotope of a very rare element (see Table 23.3).

The United States reserve of uranium ore profitably recoverable at today's prices of uranium oxide and with existing technology is estimated to have the equivalent energy value of 12 billion metric tons of crude oil. Compare this figure with a United States proven crude oil reserve of 5 billion tons. Estimated on the basis of a price about triple that being paid today for uranium oxide, the United States reserve of uranium is equivalent to 35 billion tons of oil. Compare this figure with our estimate of United States coal reserves, recoverable with existing mining technology, equivalent to about 90 billion tons of oil. Coal appears to be much the larger United States energy resource.

The world supply of uranium ore would be rapidly exhausted if no other fissionable material were used. It is, however, possible to induce fission in other isotopes—notably, other isotopes of uranium and isotopes of plutonium and thorium. This induced fission, known as *breeding*, can greatly reduce the expenditure of uranium-235. Some scientists have expressed great concern over the necessity to develop breeder reactors to conserve uranium. If breeder development were successful, low-grade deposits of uranium could be exploited, making available a source of energy judged to range from hundreds to thousands of times greater than all reserves of fossil fuels. So far, controlled release of energy through hydrogen fusion has not been achieved, although research is in progress. In theory, the quantities of energy available through fusion could exceed that of all fossil fuels by a factor ranging up into the hundreds of thousands.

SUMMARY

Rocks, minerals. Rocks are composed of minerals, which are homogeneous, naturally occurring, inorganic substances having definite chemical compositions and characteristic atomic structures. Dominant crustal elements are oxygen (47 percent), and silicon (28 percent).

Silicate minerals. Igneous rocks, solidified from magma, are composed largely of silicate minerals: quartz, potash and plagiclase feldspars, mica, amphibole, pyroxene, and olivine. Quartz and feldspars are felsic minerals, low in density; the others are mafic minerals, higher in density.

Mineral cleavage. Crystal lattice structure determines cleavage properties. The silicon-oxygen tetrahedron is basic to all silicate minerals; chains and sheets of tetrahedra result in good cleavage.

Igneous rocks. Igneous rocks occur in intrusive and extrusive forms, the former with coarse texture in batholiths, the latter with fine-grained or glassy texture in lavas. The granite-gabbro series arranges igneous rocks in order from felsic, through mafic, to ultramafic. Volcanoes are built of extrusive igneous rock—lava and ash.

Sedimentary rocks. Sediment, produced by rock weathering and by organic activity, accumulates in strata and becomes hardened into sedimentary rock of clastic, chemically precipitated, or organic types.

Metamorphic rocks. Under conditions of high temperatures and stresses, rock of either igneous or sedimentary origin is altered into metamorphic rock, with new structures and new minerals commonly present.

The rock cycle. The three rock classes can be arranged into a rock cycle in which any one of the three can originate from or lead to the production of either of the other two classes, moving from the deep environment to the surface environment, or vice versa.

Mineral resources. Metalliferous ores represent unusual concentrations of metals rare in the earth's crust. United States demands for ores greatly exceed production. Supplies of nonrenewable energy resources, the fossil fuels and uranium, are limited and face eventual depletion. Sustained energy sources must be developed as substitutes.

TERMS AND CONCEPTS

rock
lithosphere
crystalline solid
mineral
igneous rocks
magma
silicates (silicate minerals)
quartz
aluminosilicates
potash feldspar
orthoclase
plagioclase—sodic, calcic, intermediate
biotite
mica group
amphibole group
hornblende
pyroxene group
augite
olivine
felsic group
mafic group
cleavage
silicon-oxygen tetrahedron
silicate magma
texture—coarse-grained, fine-grained
plutonic, pluton
intrusive igneous rocks
country rock
batholith
extrusive igneous rock
lava
volcanic glass (obsidian)
scoria
granite-gabbro series
granite
rhyolite
diorite
andesite
gabbro
basalt
felsic igneous rocks
mafic igneous rocks
peridotite
ultramafic rocks
dunite
volcanism
volcano
volcanic cone
crater
composite volcano
shield volcano
circum-Pacific ring
rock weathering
sediment
sedimentary rocks
clastic sediments
chemically precipitated sediments
organic sediments

strata, stratified
planes of stratification
conglomerate
sandstone
lithification
siltstone
shale
clay minerals
hydrolysis
kaolinite
illite
calcite
limestone
dolomite
evaporites
gypsum
halite
coal seams
coal measures
metamorphic rocks
rock metamorphism
foliation
banding
slate
schist
gneiss
marble
cycle of rock transformation (rock cycle)
nonrenewable earth resources
metalliferous deposits
nonmetallic deposits
fossil fuels
nuclear fuels
ores
sustained yield energy source
solar energy
hydropower
tidal power
wind power
geothermal energy
oil shale
kerogen
nuclear fission
nuclear fusion
breeding

QUESTIONS

1. What physical and chemical properties enter into the definition of a mineral? Is water a mineral? Is coal a mineral? Can a rock consist of only one mineral variety?

2. List in order of abundance the eight most abundant elements in the earth's crust. Give an approximate percentage figure for each. How does this list compare with the most common elements in silicate magma and silicate minerals?

3. Name and give the chemical composition of the important silicate minerals and mineral groups comprising the bulk of igneous rocks. Which of these minerals fall in the felsic group? Which in the mafic group? How do densities of felsic minerals compare with densities of mafic minerals? In what way is this density difference significant?

4. What causes cleavage in minerals? Explain how crystal lattice structure, based on the silicon-oxygen tetrahedron, is responsible for cleavage or lack of cleavage in the silicate minerals. Why do quartz and olivine lack cleavage? Why does the mica group show such excellent cleavage into thin layers?

5. In what order are the silicate minerals formed as crystals in a gradually cooling magma? How does rate of cooling affect crystal size and rock texture? Compare the texture of intrusive igneous rocks with that of the extrusive types.

6. Name several important members of the granite-gabbro igneous rock series. Give names of both intrusive and extrusive types. Which are felsic, which are mafic, and which are ultramafic?

7. Describe a volcano in terms of its internal structure and mode of growth. In what respects are composite volcanoes different from shield volcanoes? Where are most of the world's composite volcanoes found?

8. How does rock weathering lead to the production of sediment? What are the three major groups of sediments and sedimentary rocks? What causes stratification in sediments?

9. Describe each of the common clastic sedimentary rocks. Name the various grade sizes. How do clastic sediments become lithified? How are the clay minerals formed? Name two. What are the two major classes of chemically precipitated sedimentary rocks? Give the names and compositions of two common carbonate minerals and two evaporite minerals. What forms do the organic sediments take within sedimentary rocks?

10. Describe the process of rock metamorphism. Where does it take place? What changes do rocks and minerals experience during metamorphism? What new structures are formed? Name and describe four common metamorphic rocks.

11. Explain the concept of a cycle of rock transformation. What part of the rock cycle operates in the deep environment? In the surface environment?

12. Name the major groups of nonrenewable earth resources. Which are sources of materials and which are energy sources? Are most of the metals essential to industry abundant in the earth's crust? Is the United States self-sufficient in supplies of essential metals?

13. Review the various sources of energy. Which are sustained-yield types? Which are nonrenewable? Evaluate solar energy, hydropower, tidal energy, geothermal energy, and wind power in terms of adequacy to fulfill our energy needs.

14. What are the future prospects for supplies of fossil fuels and uranium in terms of world needs? Compare coal reserves with petroleum reserves. Where does oil shale stand as a resource with respect to coal and petroleum? How can breeding reduce uranium consumption? What is the future prospect for fusion as an energy resource?

CHAPTER 24
THE DYNAMIC EARTH– PLATE TECTONICS

24.1 THE EARTH'S CORE AND MANTLE

Figure 24–1 is a cutaway diagram of the earth showing its principal interior subdivisions. The *crust* is a silicate rock skin too thin to show to correct scale on the diagram. Beneath the crust lies the *mantle*, a solid rock shell 2895 km thick. It is generally believed that the deeper mantle consists largely of an ultramafic silicate rock of olivine composition resembling dunite. The upper mantle is probably rock resembling peridotite.

Beneath the mantle lies the earth's *core*, a sphere 3475 km in diameter. There is good evidence to show that the core is composed of iron, mixed with some nickel. In other words, the core is believed to be metallic in contrast to the silicate rock mantle surrounding it. From data of earthquake waves it is known that the outer portion of the core is in a liquid state. However, the inner portion of the core, with a radius of 1255 km, is in a solid state.

We can make our own inference that the earth must have a very dense core, even without evidence from earthquake waves. Consider, first, that the physicist can measure the earth's mass with considerable accuracy and then measure the earth's radius, thus calculating its volume. Dividing the earth's volume by its mass, we arrive at an average earth density of about 5.5 g/cm^3. Now, the density of the ultramafic rocks at the earth's surface is only about 3.3 g/cm^3. If the earth has a thick, rocky mantle of that density, it must have a very dense core to bring the average density up to 5.5 g/cm^3.

This line of reasoning is supported by studies of the composition of meteorites, solid fragments of matter from outer space reaching the earth's surface. As we shall find in Chapter 28, many meteorites are of nickel-iron composition; they are interpreted as the fragmented core material of an early planet formed at about the same time as the earth.

FIGURE 24–1
The earth's core and mantle.

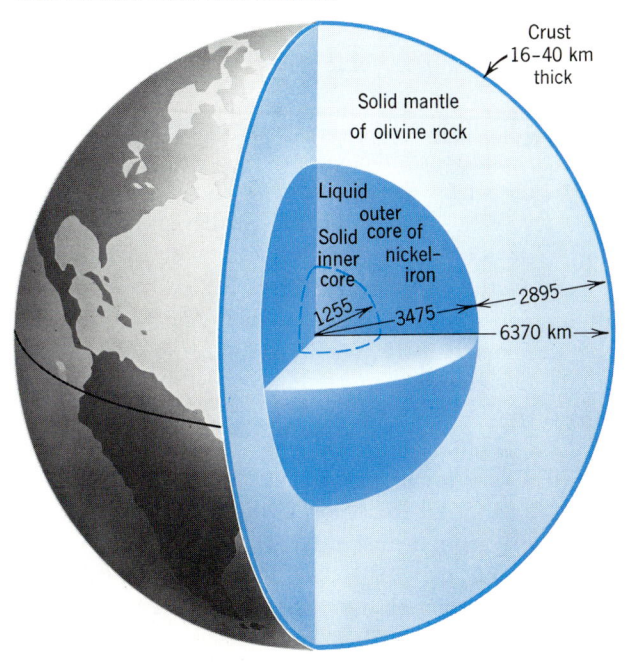

24.2
HEAT WITHIN THE EARTH

Temperaures deep within the earth are extremely high. At a depth of about 1000 km, which is within the upper mantle, the temperature is estimated to be about 2000 K. This value would be well over the melting point of silicate rock under conditions prevailing at the earth's surface, but here, under great confining pressure, it is somewhat below the melting point. At the core boundary, temperature is about 2700 K, higher than the melting point of iron. At the earth's center the temperature is perhaps 3000 K.

The earth's internal heat is best explained as the result of radioactivity. Radioactive decay of certain isotopes of uranium and thorium generates most of this heat. The heat-producing isotopes are most heavily concentrated in the crust and upper mantle—in rocks of granite and basalt composition.

Heat flows very slowly upward to the earth's surface. The amount reaching the surface in one year is only about enough to melt a layer of ice 6 mm thick; it has no appreciable effect in warming the atmosphere or causing circulation of the atmosphere and oceans.

FIGURE 24-2
Schematic cross section of crust and mantle beneath continent and oceans.

24.3
STRUCTURE OF THE EARTH'S CRUST

Simple reasoning would lead us to conclude that if the silicate rock-forming minerals were permitted to assemble freely under the attractive force of the earth's gravity, the three rock groups would be found in a layered sequence. We should expect felsic rocks (least dense) in a surface layer, mafic (more dense) rocks next below, and ultramafic (most dense) at the bottom. This general arrangement is accepted as the basic model for the earth's crust and mantle.

When describing the earth's crust we must start with the fact that the crust beneath the continents is very different in both composition and thickness from the crust beneath the ocean floors. Roughly one-third of the globe is covered by *continental crust*, about two-thirds by *oceanic crust*.

Figure 24-2 shows this crustal arrangement in a very rough way for continents and ocean basins. Felsic rock forms the upper part of the continental crust in a layer with an average thickness of perhaps 16 km. This felsic rock is largely of the composition of granite and can also be described as *granitic rock*.

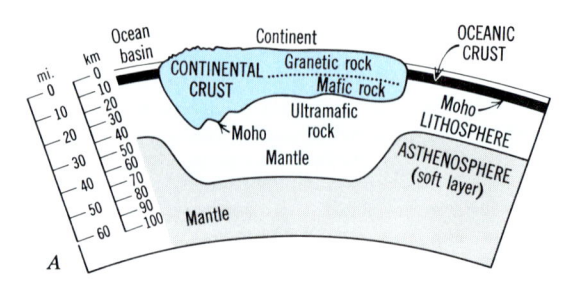

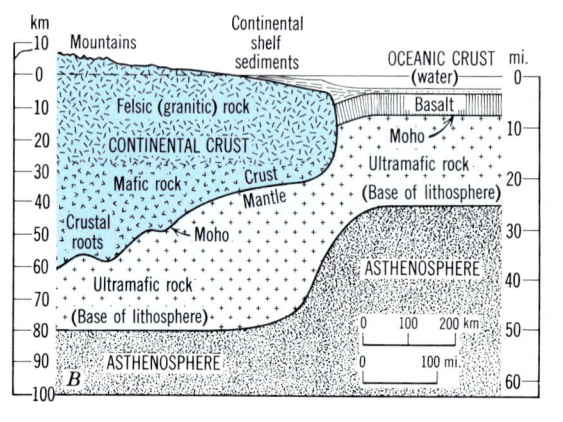

The lower part of the continental crust is probably largely of mafic rock, down to an average depth of 40 km. This mafic rock is largely of the composition of basalt and can also be described as *basaltic rock*. At the base of the basaltic layer there is an abrupt change to a denser mantle rock, which we interpret to be ultramafic rock with a composition resembling peridotite.

The surface of abrupt change from mafic to ultramafic rock is known as the *Moho*. This word is the first part of the name of a Yugoslav scientist, A. Mohorovičić, who discovered the discontinuity on the basis of earthquake studies.

Perhaps the most striking point driven home by the crustal diagram in Figure 24-2 is the difference between continental crust and oceanic crust. First, the continental crust is much the thicker—ranging mostly 30-60 km in thickness. The thickest portions lie beneath the highest mountain ranges, as the diagram shows. In other words, the mountains have *crustal roots*. In contrast, oceanic crust averages only 7-8 km thick. As a result, the Moho is encountered at much shallower depths under the ocean basins—about 11-13 km below the ocean surface—than under the continents. A second striking point of difference between the two types of crust is that the felsic, or granitic, upper layer of the continental crust is missing from the oceanic crust. Instead, basaltic rock comprises the entire igneous portion of the oceanic crust.

24.4 A SOFT LAYER IN THE MANTLE

A layer within the upper mantle is of particular interest in connection with mountain-making activity. This layer is found in the depth range of 60–200 km; it has very indefinite upper and lower boundaries. Here the mantle rock is at a temperature very close to its melting point and is therefore in a condition of reduced strength. In other words, the rock is soft, just as white-hot iron is soft compared with cold iron. The term *plastic layer,* or *soft layer,* has been applied to this part of the mantle. While remaining solid and in a crystalline form, the mantle rock at this depth develops a plastic quality because here it is close to its melting point. The plastic quality of rock in the soft layer makes possible very slow flowage movements when this rock is subjected to unequal stresses. As we shall see, flowage of mantle rock plays a vital part in modern concepts of crustal changes and mountain-making.

Geologists have given the name *asthenosphere* to the soft layer of the mantle and have restricted the term *lithosphere* to the stong, rigid overlying zone (including both the crust and part of the upper mantle). As shown in Figure 24–2, the base of the lithosphere is placed at a depth of about 60 km as a rough average figure.

24.5 THE LITHOSPHERE IN MOTION

The strong, rigid lithospheric shell is capable of rotating independently of the weaker mantle below; motion between the two bodies takes place by yielding within the soft asthenosphere. In this way the crust of the continents and ocean basins can shift widely in latitude and longitude.

Yielding in the soft layer takes place much like the gliding of playing cards over one another when a card deck is pushed from left to right, as in Figure 24–3. Such gliding in parallel layers is referred to as *shearing*. Shearing will be most rapid in the soft layer, but it will be entirely missing in the rigid lithosphere. (Imagine that the uppermost cards in the deck are glued together to make a solid block.) Of course, the rock layers we refer to as gliding over one another are extremely thin, approaching the thickness of layers of atoms in the crystal lattices of the minerals. Ice deep within glaciers experiences such shearing motion.

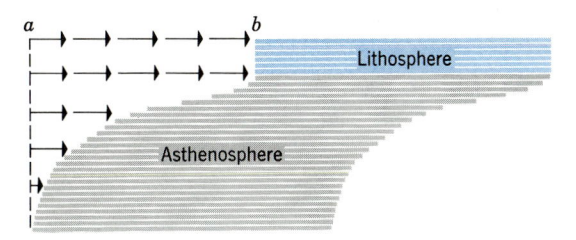

FIGURE 24–3
A deck of cards illustrates the concept of slow shearing within the soft asthenosphere. (From PRINCIPLES OF EARTH SCIENCE by Arthur N. Strahler. Copyright © 1976 by Arthur N. Strahler. By permission of Harper & Row, Publishers, Inc.)

Because the lithosphere is a brittle layer, it can break up into numerous *lithospheric plates,* and these plates can move over the asthenosphere independently. For example, two plates might move away from each other. Or, in another place, two plates might slide past each other, going in opposite directions. This concept is basic to an understanding of geologic processes associated with breaking and bending of the crust.

24.6 MOUNTAIN-MAKING BELTS

A geologist examining exposed rocks will find unmistakable evidence that certain parts of the earth's crust have been crumpled and fractured on an enormous scale. In certain localities rock is of sedimentary composition—sandstone, shale,

FIGURE 24–4
These limestone strata, originally horizontal, have been folded during tectonic activity. (Courtesy of EXXON Corporation.)

24 THE DYNAMIC EARTH—PLATE TECTONICS

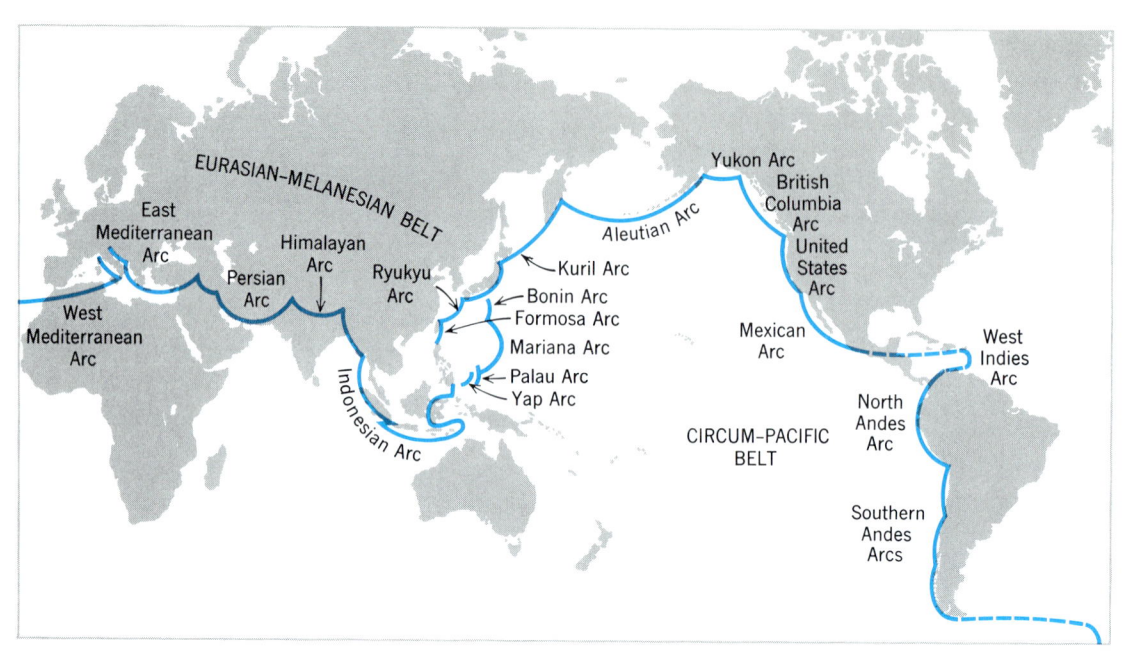

FIGURE 24-5
World map of the primary arcs. (From PRINCIPLES OF EARTH SCIENCE by Arthur N. Strahler. Copyright © 1976 by Arthur N. Strahler. By permission of Harper & Row, Publishers, Inc.)

or limestone beds—but instead of lying horizontally the strata are wrinkled into a series of folds (Figure 24-4). It might seem as if the rock had been plastic, like clay, at the time crumpling took place.

All forms of breaking and bending of crustal rock are covered by the term *tectonic activity*. The field of study of these processes is called *tectonics*. Only in the past two decades have geologists developed a complete and unified concept of the way in which tectonic activity takes place on a global scale. This new scientific revolution in geology is the main theme of later pages of this chapter.

Present-day tectonic and volcanic activity, resulting in uplift of high mountain chains and downsinking in deep trenches, is principally concentrated in long, narrow, broadly curving zones known as the *primary arcs*. Examine the location of these arcs on a world map (Figure 24-5). The arcs fall into two chains. First is the *circum-Pacific belt*, which we referred to in Chapter 22 as the Ring of Fire. Throughout North and South America, arcs of the circum-Pacific belt lie along the western continental margins, except for the West Indies Arc, which is oceanic in location.

Starting with the Aleutian Arc, and continuing south along the western side of the Pacific, the primary arcs are in oceanic positions at some distance from the Asiatic shoreline. In large part, these *island arcs* are represented by chains of volcanic islands (Figure 24-6).

Second of the chains of primary arcs is the *Eurasian-Melanesian belt*, extending from the Mediterranean region, eastward through southern Asia. This belt terminates in the Indonesian Arc, which appears to intersect the circum-Pacific belt in a T-junction. Notice that the primary arcs are convexly bowed outward from the continental centers.

The primary arcs contain the world's great mountain ranges; they are commonly described as belonging to the *alpine system*. Fine examples are the Alps of Europe, the Himalayas of southern Asia, the Andes of South America, and the Cordilleran ranges of western North America. Comparative recency of the uplift of these ranges is well established by identification and dating of fossils of marine origin among the summit rocks.

Of equal interest to the alpine ranges are deep *oceanic trenches*, typically located adjacent to island arcs. Trenches of the western Pacific Ocean are particularly striking (Figure 24-6). These oceanic trenches represent narrow zones of extreme crustal sinking. They have bottom depths on the order of 7–10 km below sea level. Equally impressive is the Peru-Chili Trench (Figure 24-7). Because sources of sediment are quite limited in the vicinity of oceanic trenches, these depressions are only partly filled with sediment.

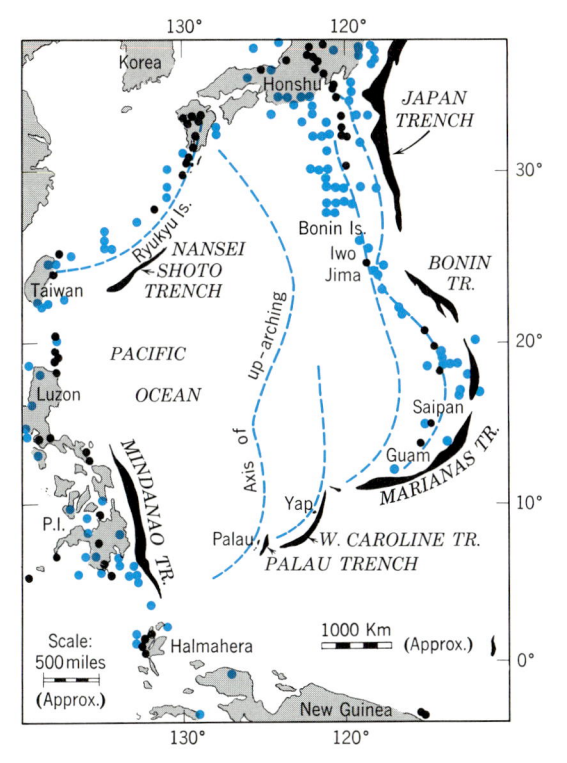

24.7 EARTHQUAKES AND FAULTS

FIGURE 24-6
Map of the western Pacific showing trenches (solid black), island arcs (dashed lines), active volcanoes (black dots), and deep-focus earthquake centers (color dots). (After H. H. Hess.)

FIGURE 24-7
The Peru-Chile Trench, off the west coast of South America. (Portion of *Physiographic Diagram of the South Atlantic Ocean*, 1961, by B. C. Heezen and M. Tharp, Geol. Soc. of Amer., Boulder, reproduced by permission.)

In summary, each primary arc consists basically of an elevated mountain chain or island volcano chain bordered by a deeply depressed trench located on the oceanic side of the arc. Such great features must involve deep-seated crustal processes.

24.7 EARTHQUAKES AND FAULTS

Tectonic activity is always accompanied by numerous, intense earthquakes. The active primary arcs are the source of many of the world's earthquakes. The *earthquake* is known to humans directly as a trembling or shaking of the ground. Commonly the motion is barely perceptible to the senses. On occasion the motion is so violent as to crack or collapse strong buildings, break water and gas mains, cause gaping cracks in the ground, and so bring great loss of life and property.

The vast majority of all important earthquakes are produced by *faulting* in the earth's crust. This phenomenon is simply a sudden slippage between two rock masses separated by a fracture surface.

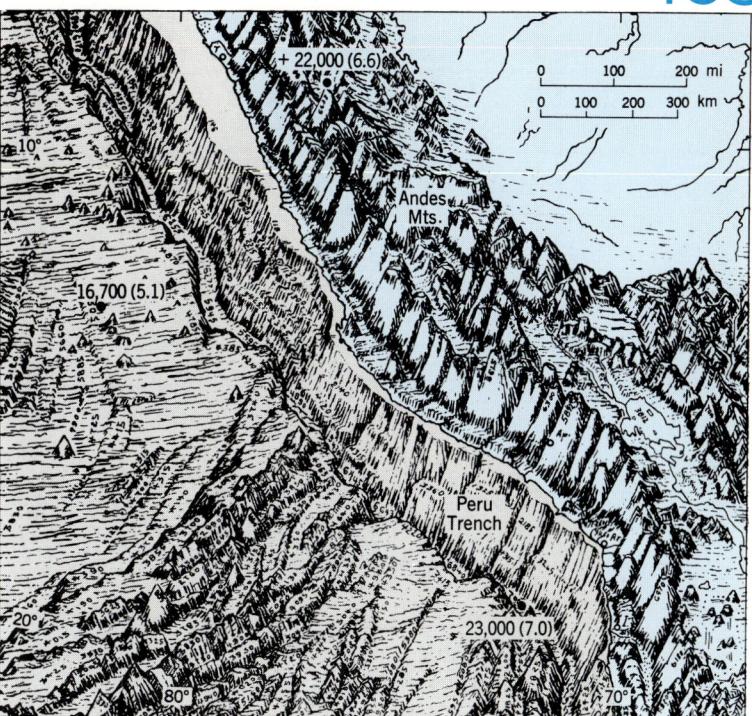

The fracture surface upon which slippage occurs is called the *fault plane;* it may take any orientation with respect to the horizontal. The accumulated displacement of corresponding points on two sides of the fault may range from a few centimeters in very small faults to several tens of kilometers in certain great faults. Movement occurs by a series of slips, each involving but a few centimeters to a few meters of displacement and occurring almost instantaneously. Each slip generates an earthquake shock.

Over spans of thousands of years, the slippage along a large fault may total tens of kilometers. It is by such long-contined faulting that small and large blocks of the earth's crust are displaced with respect to one another, bringing major landscape features into existence.

Figure 24-8 illustrates two common kinds of faults. In the *normal fault* the fault plane is steeply inclined in the direction of the downthrown block. Normal faulting produces a clifflike *fault scarp.*

The *transcurrent fault* is characterized by displacement only in the horizontal direction, along a near-vertical fault plane. Thus, one block slides past the other. Where transcurrent faulting occurs on a flat plain, as the diagram

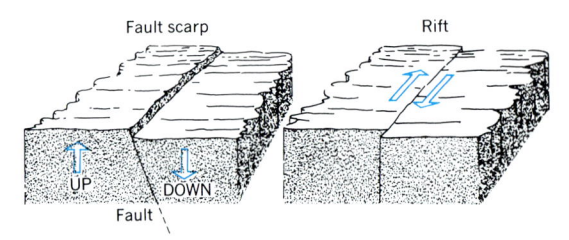

24 THE DYNAMIC EARTH—PLATE TECTONICS

FIGURE 24-8
Normal fault (*left*) and transcurrent fault (*right*). (From THE EARTH SCIENCES, 2nd Edition by Arthur N. Strahler. Copyright © 1963, 1971 by Arthur N. Strahler. By permission of Harper & Row, Publishers, Inc.)

suggests, no topographic scarp will result. In hilly terrain a discontinuous narrow trench, or *rift*, marks the line of the fault.

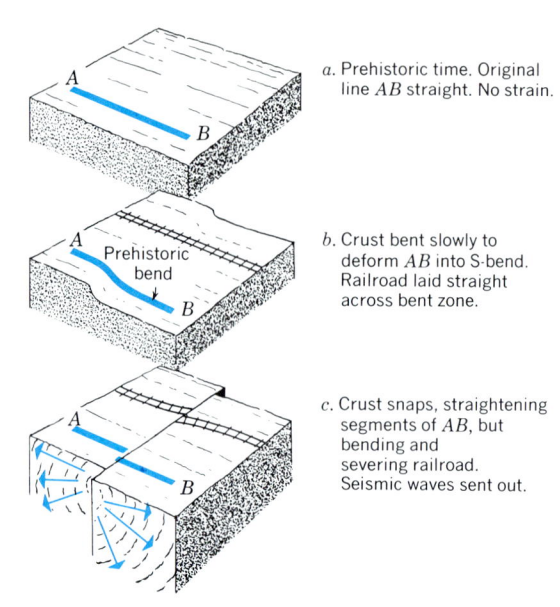

24.8
HOW EARTHQUAKES ARE GENERATED

Hard rocks of the earth's crust and upper mantle—the lithosphere—are both strong and brittle. However, this rock is subjected to enormous deforming stresses when one lithospheric plate is forced down beneath another during subduction. Under such stress rock actually bends elastically like steel. The amount of bending is scarcely detectable in small masses. Despite its ability to withstand great stress with only slight bending, or *strain*, a given rock has an *elastic limit*. If it is strained beyond this limit, a fault is formed and the bent rock snaps suddenly back to its normal shape.

An earthquake is the disturbance set off by the sudden release of elastic strain. As in a bow slowly bent, the rocks have gradually accumulated energy, only to release it with great suddenness. Figure 24–9 shows steps in production of an earthquake along a transcurrent fault.

Earthquakes are generated in the earth's crust and outer mantle down to depths as great as 650 km. The point of slippage and energy release is known as the *focus*, while the ground-surface point directly above the focus is the *epicenter*. Energy is dispersed from the focus in the form of *seismic waves*.

24.9
SEISMIC WAVES

There are three basic seismic waveforms: primary waves, secondary waves, and surface waves. *Primary waves* show the same kind of motion as observed in sound waves. As illustrated in Figure 24–10, particles transmitting the primary wave form move only forward and backward in

FIGURE 24-9
Sudden release of elastic strain accumulated by slow bending of rock along a transcurrent fault. (From THE EARTH SCIENCES, 2nd Edition by Arthur N. Strahler. Copyright © 1963, 1971 by Arthur N. Strahler. By permission of Harper & Row, Publishers, Inc.)

a. Prehistoric time. Original line AB straight. No strain.

b. Crust bent slowly to deform AB into S-bend. Railroad laid straight across bent zone.

c. Crust snaps, straightening segments of AB, but bending and severing railroad. Seismic waves sent out.

the direction of wave travel. This motion is described as alternate compression and rarefaction and constitutes a *longitudinal wave*. The primary wave is commonly called the *P-wave*. We can remember this relation by thinking of the P-wave as a "push" wave. This is easy to keep in

FIGURE 24-10
Particle motions in longitudinal and transverse seismic waves. (From THE EARTH SCIENCES, 2nd Edition by Arthur N. Strahler. Copyright © 1963, 1971 by Arthur N. Strahler. By permission of Harper & Row, Publishers, Inc.)

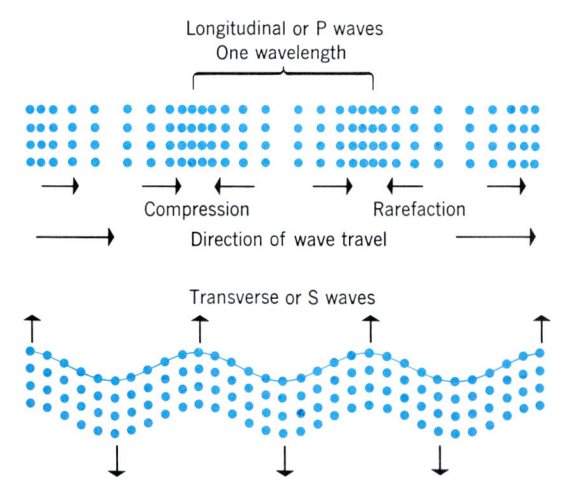

mind because the words "primary" and "push" both begin with "P."

In *secondary waves* particles transmitting the waves move back and forth at right angles to the direction of wave travel (Figure 24–10). Consequently, the secondary wave motion is termed a *transverse wave*. It is commonly designated the *S-wave*. We may think of these waves as "shake" waves; like "secondary," the key word begins with "S." In the earth, P-waves travel approximately 1.7 times more rapidly than S-waves.

Surface waves are the third major type of seismic wave. They are much like waves sent out by a stone thrown into a calm pond. As each wave crest passes, the particles move in a complete circle, oriented vertically.

24.10 SEISMOGRAPHS

Seismologists—scientists who study seismic waves—must record an earthquake and analyze the directions and amounts of the earth motions involved, using the *seismograph*. The mechanical problem facing the seismologist is that the instrument itself must be resting on the ground and will therefore also move with the ground. Because the instrument cannot be physically separated from the earth, the seismograph designer must make use of the principle of inertia to overcome the effect of the attachment.

To record an earthquake, then, a very heavy mass, such as an iron ball, might be suspended from a very thin wire or from a flexible coil spring, as shown in Figure 24–11. When the earth moves back and forth or up and down in earthquake wave motion, the large mass will stay almost motionless because the supporting wire or spring flexes easily and does not transmit the motion through to the weight. If a pen is now attached to the mass, so that the point is just touching a sheet of paper wrapped around a moving drum, the pen will produce a wavy line, or *seismogram*, on the paper. Strong shocks will give waves of high amplitude (distance of side-to-side or up-and-down swing). Weak shocks will give waves of low amplitude. The number of back-and-forth movements per second is the frequency. When frequency is greater, the undulations of the line will be more closely crowded.

The seismograph as thus far described is too simple to be actually workable. The movement of the ground is so very small that the motion must be greatly magnified if it is to produce a record suitable for study. Modern seismographs make use of magnetic and electronic devices to pick up, amplify, filter, and record the motions of the earth (Figure 24–12).

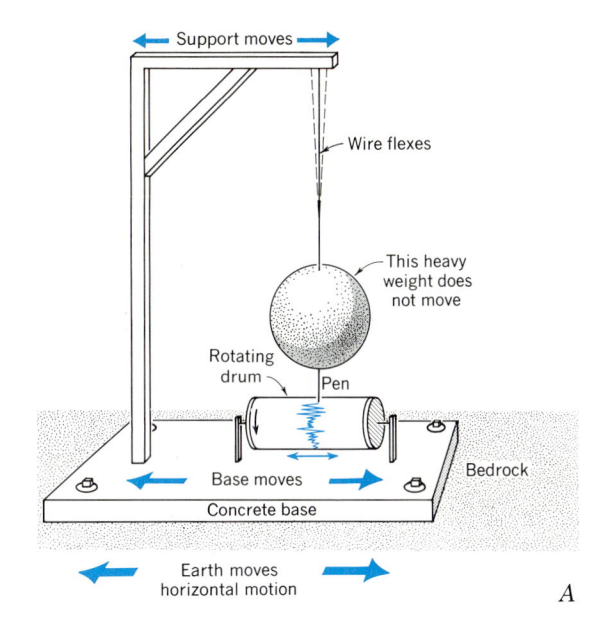

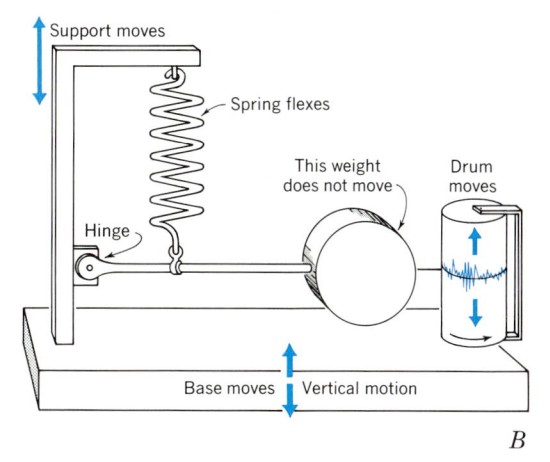

FIGURE 24–11
Basic principle of the seismograph. (From THE EARTH SCIENCES, 2nd Edition by Arthur N. Strahler. Copyright © 1963, 1971 by Arthur N. Strahler. By permission of Harper & Row, Publishers, Inc.)

24.11 INTERPRETING THE SEISMOGRAM

Figure 24–13 shows a seismogram produced by a typical distant earthquake. The epicenter was located at a surface distance of 8460 km from the observing station. Figure 24–14 is a cross section of the earth showing how the waves traveled through the mantle and core.

The first indication that a severe earthquake had occurred at a distant point was the sudden beginning of a series of larger-than-average waves; they were the primary waves (P-waves). These waves died down somewhat, then a few minutes later a second burst of activity set in

24 THE DYNAMIC EARTH—PLATE TECTONICS

FIGURE 24-12
This vertical seismograph uses the principle shown in the lower diagram of Figure 24-13. The heavy weight is at the lower left, suspended by a diagonal coil spring from the vertical support at the right. (Lamont-Doherty Geological Observatory of Columbia University.)

with the beginning of the secondary waves (S-waves). These waves were at first considerably greater in amplitude than the primary waves. There followed smooth waves that increased greatly in amplitude to a maximum and then slowly died down. These last, very high-amplitude waves were the surface waves. While the primary and secondary waves had traveled through the earth, the surface waves had traveled along the ground surface, much as storm swells travel over the sea surface.

For an earthquake occurring one-quarter of the globe's circumference away—that is, about 10,000 km—the primary waves will take about 13 minutes to reach the receiving station, and the secondary waves will begin to arrive about 11 minutes later.

The farther away the earthquake focus, the longer the spread of time between the arrival of the P- and S-wave groups. Based on many years of seismogram analysis, tables have been prepared to show the relationship of distance to timespread of P- and S-wave arrivals. These tables allow the seismologist to calculate at once the distance from the observatory to the earthquake epicenter. As shown on the globe (Figure 24-15), a circle can be drawn from the observatory to all surface points at the known distance. The epicenter lies somewhere on this circle. When three such circles are drawn from three widely separated observing stations, the earthquake epicenter can be located within the limits of a small triangle of error.

FIGURE 24-13
Seismogram shows the record of an earthquake whose epicenter was located at a surface distance of 8460 km from the receiving station, equivalent to 76.4 degrees of arc of the earth's circumference. Figure 24-14 shows the ray paths for this earthquake. (After L. Don Leet, 1950, *Earth Waves,* Cambridge, Harvard Univ. Press. © 1950 by the President and Fellows of Harvard College.)

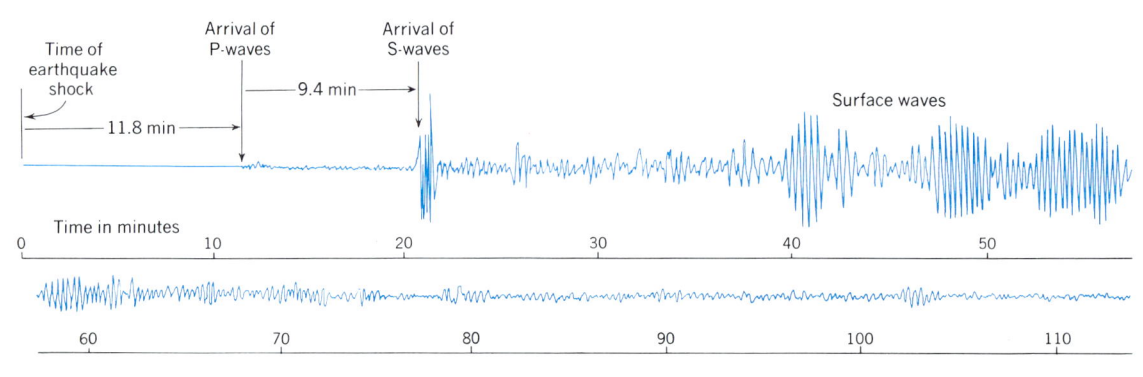

24.13 EARTHQUAKE ENERGY

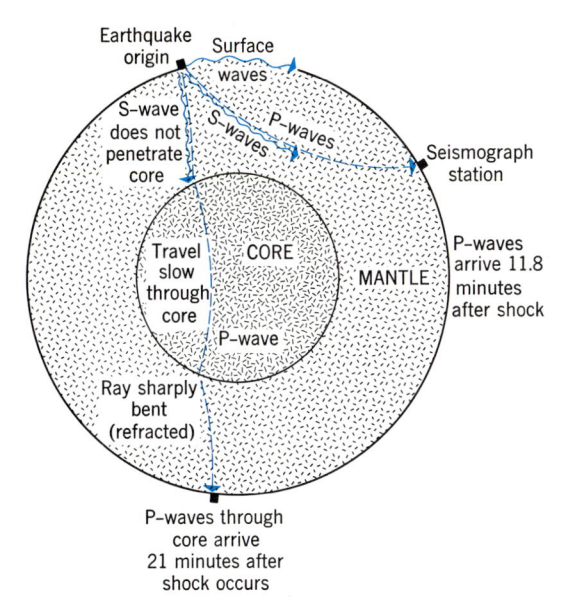

FIGURE 24–14
Cross section of the earth showing diagrammatically the paths of P-waves, S-waves, and surface waves. (Based on same data source as Figure 24–15.)

FIGURE 24–15
Circles drawn from three seismological observatories yield the location of an earthquake epicenter. (From THE EARTH SCIENCES, 2nd Edition by Arthur N. Strahler. Copyright © 1963, 1971 by Arthur N. Strahler. By permission of Harper & Row, Publishers, Inc.)

FIGURE 24–16
Diagrammatic representation of many possible ray paths from a single earthquake source. (After B. Gutenberg, 1951, *Internal Constitution of the Earth*, Dover, New York.)

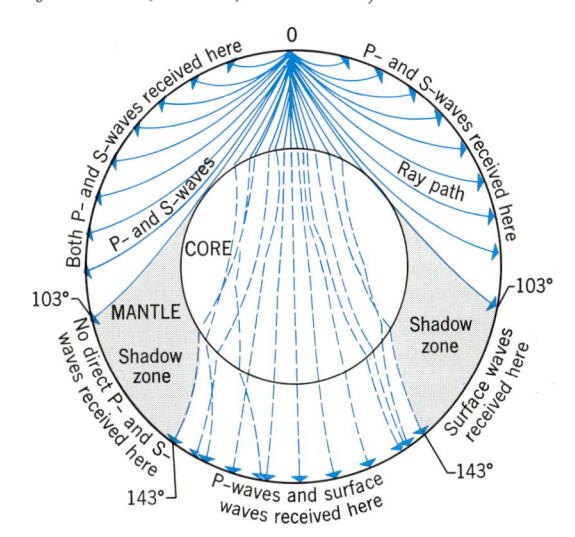

24.12 SEISMIC WAVES AND THE EARTH'S CORE

Study of earthquake waves has confirmed the existence of the spherical core at the earth's center and has added insight into its physical nature. If the earth were in a solid state entirely throughout, the P-waves and S-waves would travel through the center in all possible directions. With a completely solid earth the various seismic waves of any large earthquake could be recorded by a seismograph located directly opposite on the globe.

It was soon found, however, that there is a large region on the side of the globe opposite the earthquake focus where simple S-waves are not received. Evidently they are prevented from passing through a central region in the earth. Figure 24–16 shows many ray paths to illustrate the evidence. Physicists know that transverse waves, or S-waves, cannot be sent through a liquid; so they are agreed that the outer earth's core is in a liquid state in contrast to the surrounding mantle, which is in a solid state.

24.13 EARTHQUAKE ENERGY

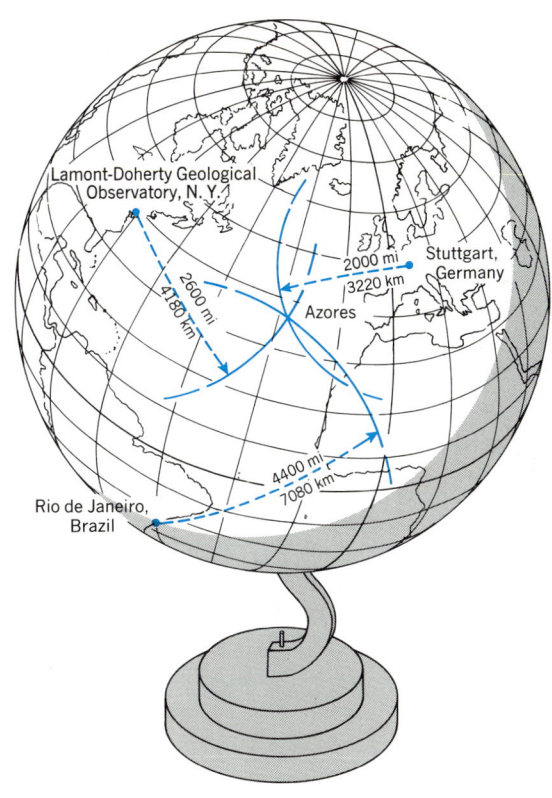

Interpretation of seismograms has made possible a calculation of the quantities of energy released as wave motion by earthquakes of various magnitudes. In 1935 a leading seismologist, Charles F. Richter, brought forth a scale of earthquake magnitudes describing the quantity of energy released at the earthquake focus. The *Richter scale* consists of numbers ranging from 0 to 8.6. The scale is logarithmic, which is to say that the energy of the shock increases by powers of 10 in relation to Richter magnitude numbers. Some facts concerning various magnitudes are given below:

Magnitude (Richter scale)	
0	Smallest detectable quake.
2.5–3	Quake can be felt if it is nearby. About 100,000 shallow quakes of this magnitude per year.
4.5	Can cause local damage.
5	Energy release about equal to first atomic bomb, Alamagordo, New Mexico, 1945.
6	Destructive in a limited area. About 100 shallow quakes per year of this magnitude.
7	Rated a major earthquake above this magnitude. Quake can be recorded over whole earth. About 14 per year this great or greater.
7.8	San Francisco earthquake of 1906.
8.4	Close to maximum known. Examples: Honshu, 1933; Assam, 1950; Alaska, 1964.
8.6	Maximum observed between 1900 and 1950. Three million times as much energy released as in first atomic bomb.

Total annual energy release by earthquakes is roughly fifty times the quantity released by a single quake of magnitude 8.4. Most of this annual total is from a few quakes of magnitude greater than 7.

The actual destructiveness of an earthquake also depends on factors other than the energy release given by Richter magnitude—for example, closeness to the epicenter is a major factor. Nature of the subsurface earth materials is another.

24.14
WORLD DISTRIBUTION OF EARTHQUAKES

A world map shows the plotted epicenters of many large shallow earthquakes (Figure 24–17). The patterns on this map reveal much about crustal activity. The picture is about the same for earthquakes of all lower magnitudes and for different spans of time.

Notice particularly the earthquake concentration in the circum-Pacific belt of primary mountain and island arcs and their related trenches. It is estimated that earthquakes of this belt account for about 80 percent of the total world earthquake energy release. Large earthquakes are shown by black dots; they are abundant around the circum-Pacific belt. Earthquakes also correspond with the Eurasian-Melanesian belt of primary arcs but are somewhat fewer and more dispersed than in the circum-Pacific belt. This belt accounts for about 15 percent of the total energy release.

Another location of frequent earthquakes is in midocean in the Atlantic, Indian, and southern Pacific Oceans. This seismic zone coincides with a great submarine mountain chain.

24.15
THE OCEAN BASINS

We turn next to the ocean basins, in search of further information on the global tectonic patterns and a general theory of tectonic activity that will link the ocean basins with the continents.

The topographic features of the *ocean basins* fall into three major divisions: (1) the *continental margins*, (2) the *ocean-basin floors*, and (3) the *mid-oceanic ridge*.

The continental margins lie in belts directly adjacent to the continent, while the mid-oceanic ridge divides the basin roughly in half. Thus, the deep floor of an ocean basin lies in two parts, one on either side of that ridge. Figure 24–18 shows these major topographic divisions as they apply to the North Atlantic basin.

Perhaps the best known and most easily studied of the units within the continental margins are the continental shelves belonging to the continents. These shelves fringe the continents in widths from a few kilometers to more than 320 km. Generally having very smooth and gently sloping floors, the continental shelves are for the most part less than 180 m deep. A particularly fine example is the continental shelf of the eastern coast of the United States (Figure 24–19). Great thicknesses of sedimentary strata lie beneath the broader continental shelves.

24.15 THE OCEAN BASINS

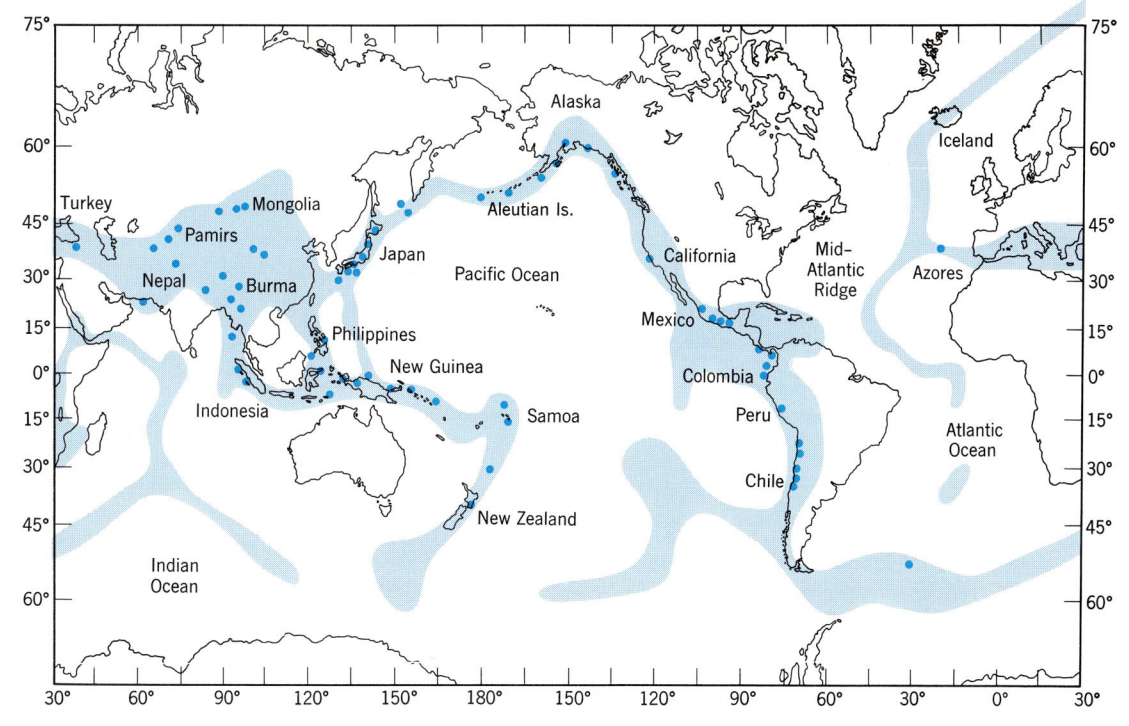

FIGURE 24-17
World distribution of most earthquakes. The black dots show epicenters of major earthquakes, measuring 7.9 or higher on the Richter scale. The shaded zones are the principal areas of abundant shallow-focus earthquakes. (From PRINCIPLES OF EARTH SCIENCE by Arthur N. Strahler. Copyright © 1976 by Arthur N. Strahler. By permission of Harper & Row, Publishers, Inc.)

FIGURE 24-18
Schematic block diagram showing the ocean basins as symmetrical elements on a central axis. The model applies particularly well to the Atlantic Ocean.

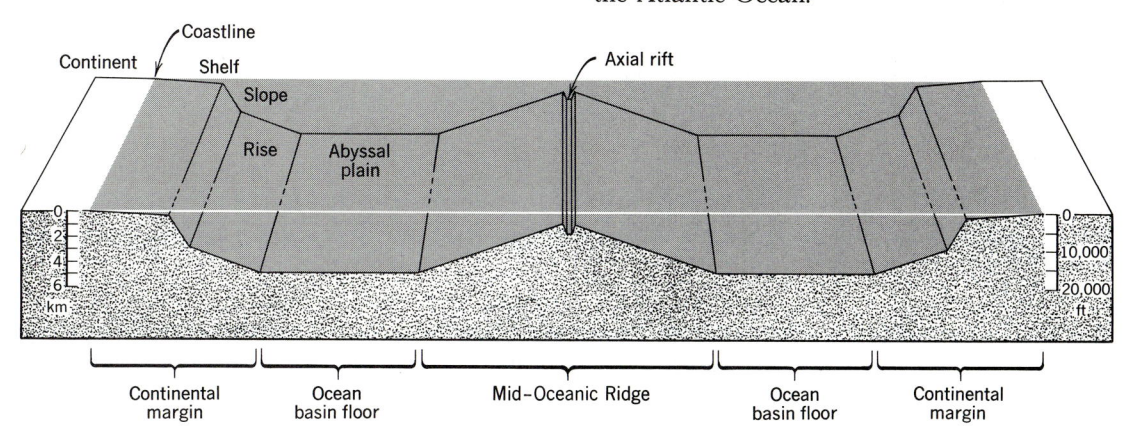

Along its seaward margins, a continental shelf gives way to the *continental slope*. The slope appears quite precipitous on the highly exaggerated diagram in Figure 24-19. We see that the continental slope drops from the sharply defined brink of the shelf to depths of 1400–3200 m. Here the slope lessens rapidly, although not abruptly, and is replaced by the *continental rise*, a surface of much gentler slope decreasing in steepness toward the ocean-basin floor. At its outer margin the continental rise reaches depths of 5000 m where it merges with the deep floor of the ocean basin.

Second of the major topographic divisions is the ocean-basin floor, generally lying in the depth range of 4500–5500 m. The ocean-basin

FIGURE 24-19

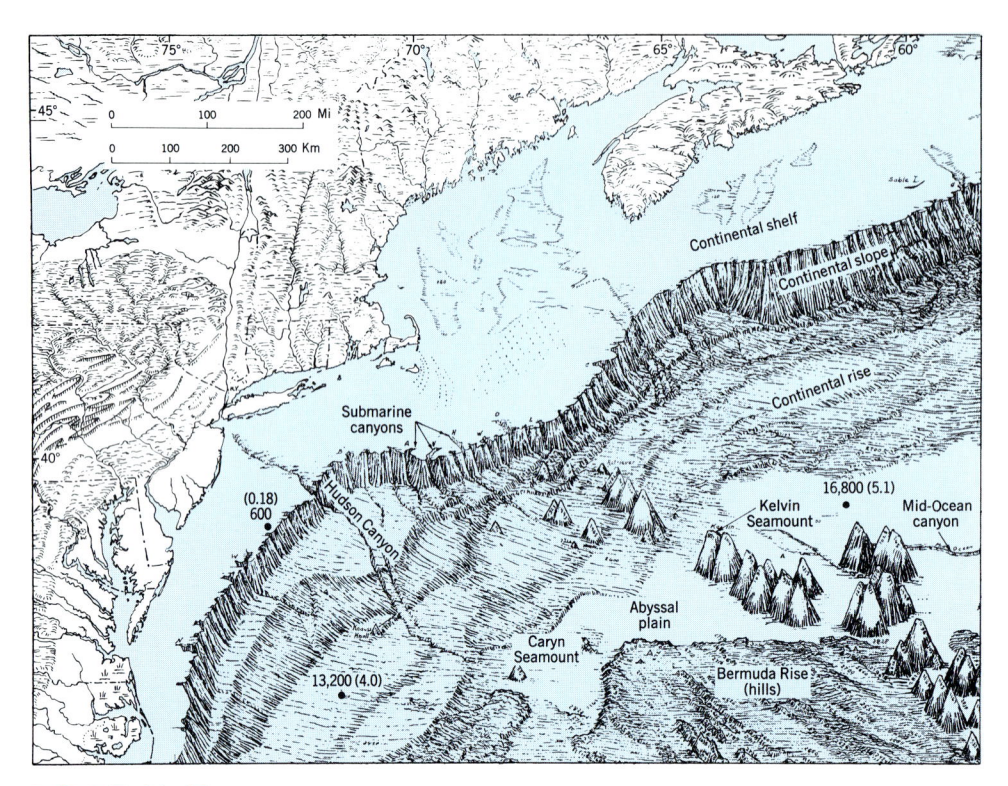

Features of the continental margin and ocean-basin floor off the coast of the northeastern United States. Depth in feet (km in parentheses). (Portion of *Physiographic Diagram of the North Atlantic Ocean*, 1968, revised, by B. C. Heezen and M. Tharp, Geol. Soc. of Amer., Boulder, reproduced by permission.)

floor includes abyssal plains and seamounts.

An *abyssal plain* is an area of the deep ocean floor having a flat bottom with a very faint slope. The only reasonable explanation for such nearly perfect flatness is that the abyssal plains are surfaces formed by long-continued sediment deposition. Previously existing irregularities of the ocean floor have thus been almost entirely buried over large areas.

Perhaps the most fascinating of the strange features of the ocean basins are the *seamounts*, isolated peaks rising 1000 m or more above the sea floor. They are most conspicuous on the ocean-basin floors. Most seamounts are interpreted as volcanoes of basaltic lava, built on the sea floor.

The central unit of the ocean basins is the mid-oceanic ridge (Figure 24-18). One of the most remarkable of the major discoveries coming out of oceanographic explorations of the mid-twentieth century has been the charting of a great submarine mountain chain extending for a total length of some 64,000 km. A world map (Figure 24-20) shows the location and extent of this feature.

24.16 SEA-FLOOR SPREADING

A distinctive feature of the mid-oceanic ridge is the lack of a single high crest line, which many narrow mountain chains of the continents have. Instead, there is a characteristic trenchlike depression, or *axial rift*, running precisely down the midline of the highest part of the ridge. Along with other parallel scarps and steplike rises on both sides, the rift strongly suggests that the crust has been pulled apart along the central zone.

A particularly significant feature of the axial rift is that it is broken into many segments, the ends of which appear to be offset along transverse fractures, shown in Figure 24-20. This arrangement of offset segments is particularly striking in the equatorial zone of the Atlantic Ocean, where a single offset displaces the main axial rift by as much as 600 km. It might seem obvious by inspection that the transverse fracture zones are transcurrent faults.

The form of the rift zone itself strongly sug-

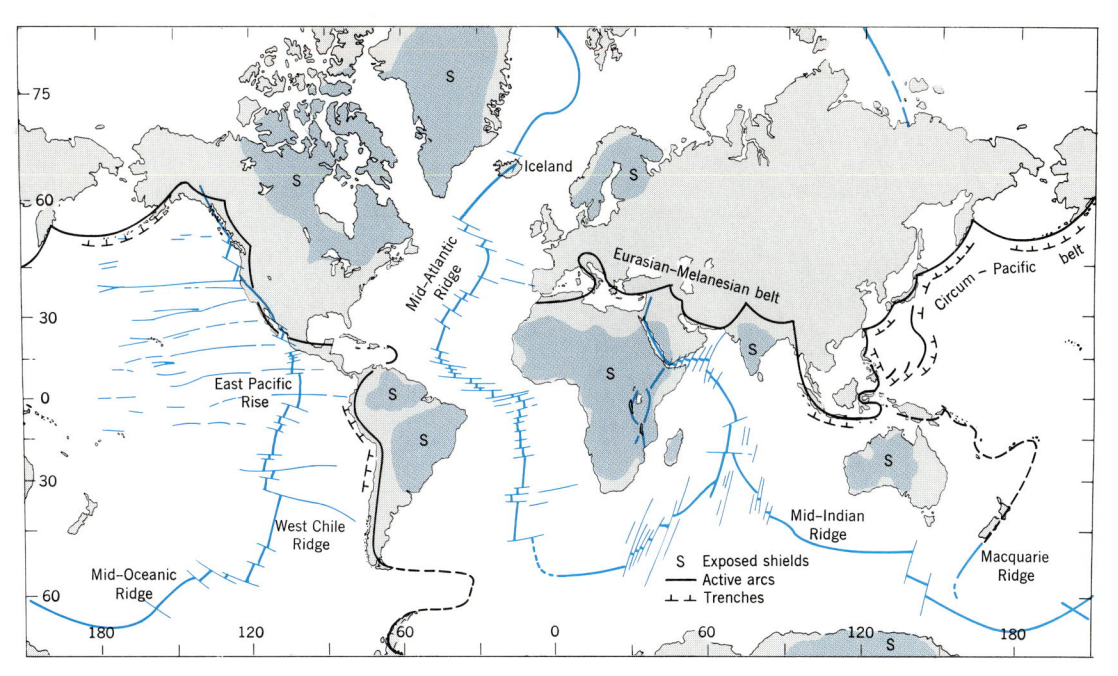

FIGURE 24–20
Mid-oceanic ridge system. The central rift zone is shown by a bold line; related fracture zones by light lines. (Based on data of L. R. Sykes, B. C. Heezen, M. Tharp, H. W. Menard, and others.)

gests that the crust is being pulled apart along the line of the rift. A picture emerges of rising basalt magma filling the gap that would otherwise be created by crustal pulling apart. This concept of *sea-floor spreading,* as it is now called, must be backed up by direct evidence, available through the phenomenon of natural magnetism in rocks.

24.17
THE EARTH AS A MAGNET

In its most simple aspect, the earth's magnetic field resembles that of a bar magnet located at the earth's center (Figure 24–21). The axis of the imaginary bar magnet is situated approximately coincident with the earth's geographic axis. At points where the projected line of the magnetic axis, or *geomagnetic axis,* emerges from the earth's surface are the *north magnetic pole* and *south magnetic pole.*

Figure 24–21 shows lines of force of the earth's magnetic field in relation to the earth's core. The force lines pass through a common point close to the earth's center. Visualized in three dimensions, the lines of force of the earth's magnetic field form a succession of doughnutlike rings. The small arrows show the attitude that would be assumed by a small compass needle, free to orient itself parallel with the force lines close to the earth's surface. As we explained in Chapter 21, force lines extend out into space surrounding the earth to great distances, forming the magnetosphere.

Earth magnetism is explained by the *dynamo theory,* which postulates that the liquid iron of the core is in slow rotary motion with respect to the solid mantle surrounding it. It can be shown that such motion will cause the core to act as a great dynamo, generating electric currents. These currents at the same time set up a magnetic field. A single symmetrical current system can thus explain the magnetic field as essentially resembling a single bar magnet.

24.18
PALEOMAGNETISM

One of the most remarkable scientific discoveries of recent decades has been sound evidence that the earth's magnetic field has undergone repeated changes in polarity. In other words, the magnetic north pole and south pole have switched places, but with the axis unchanged in position. The record of such changes can be read from basalt rock, which preserves a record of magnetic polarity at the time it solidified.

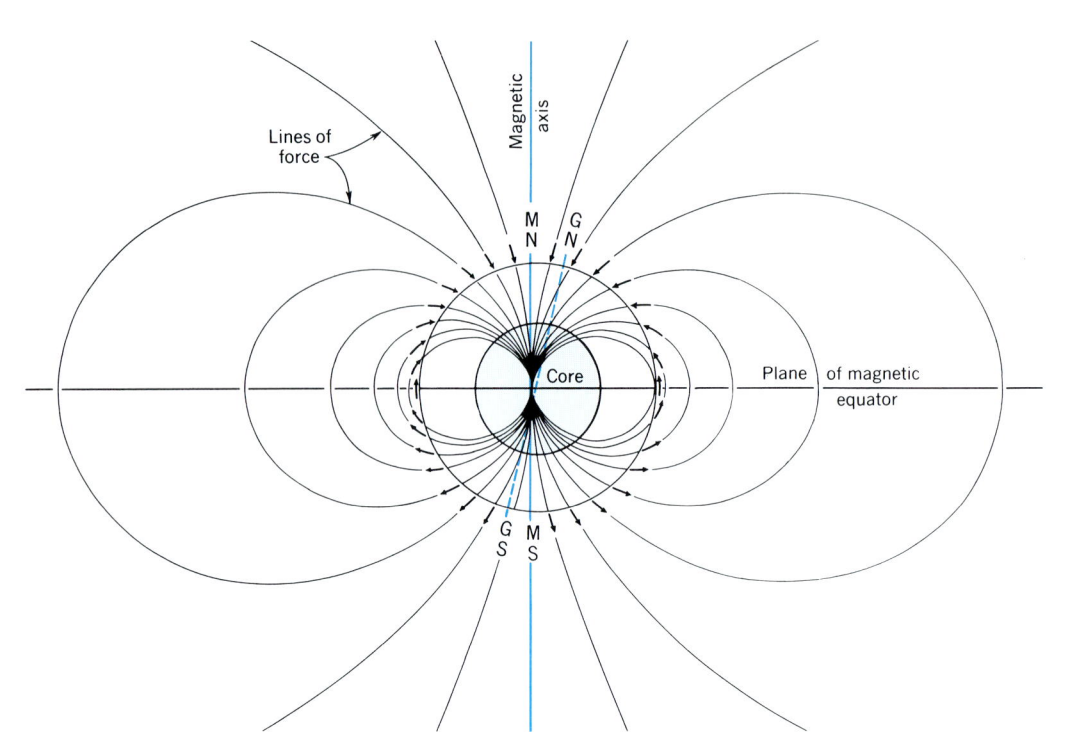

FIGURE 24-21
Lines of force of the earth's magnetic field are shown diagrammatically in a cross section drawn through the magnetic and geographic poles. The geographic axis *GN* is shown by a dashed line. The small arrows show the inclination of lines of force at surface points over the globe. (From THE EARTH SCIENCES, 2nd Edition by Arthur N. Strahler. Copyright © 1963, 1971 by Arthur N. Strahler. By permission of Harper & Row, Publishers, Inc.)

Basaltic lavas contain minor amounts of oxides of iron and titanium. At the high temperatures in a basalt magma, these iron and titanium minerals have no natural magnetism. However, when cooling sets in, the mineral is magnetized by lines of force of the earth's field. In a solid rock, magnetism is permanent. The term *paleomagnetism* is used for such locked-in magnetism dating far back into the geologic past. Paleomagnetism can be compared with present conditions and with the magnetic field at other locations and in different times in the geologic past.

As early as 1906, Bernard Brunhes, a French physicist, observed that the magnetic polarity of some samples of lavas is exactly the reverse of present conditions. He concluded that the earth's magnetic polarity must have been in a reversed condition at the time the lava solidified.

In addition to the magnetic data of the lava specimen, we need a determination of the age of the rock, using methods described in the next chapter. This information gives the date of solidification of the magma. Extensive determinations of both magnetic polarity and rock age have revealed that there have been at least nine *magnetic reversals* of the earth's magnetic field in the last $3\frac{1}{2}$ million years of geologic time.

Figure 24–22 shows the timetable of magnetic polarity changes. Polarity such as that existing today is referred to as a *normal epoch;* opposite polarity as a *reversed epoch*.

24.19
PALEOMAGNETISM AND SEA-FLOOR SPREADING

We are now prepared to return to the subject of sea-floor spreading along the mid-oceanic ridge. If the axial rift valley is a line of upwelling of basaltic lavas, and if crustal spreading is a continuing process, the lava flows that have poured out in the vicinity of the axial rift will be slowly

24.19 PALEOMAGNETISM AND SEA-FLOOR SPREADING

FIGURE 24-22

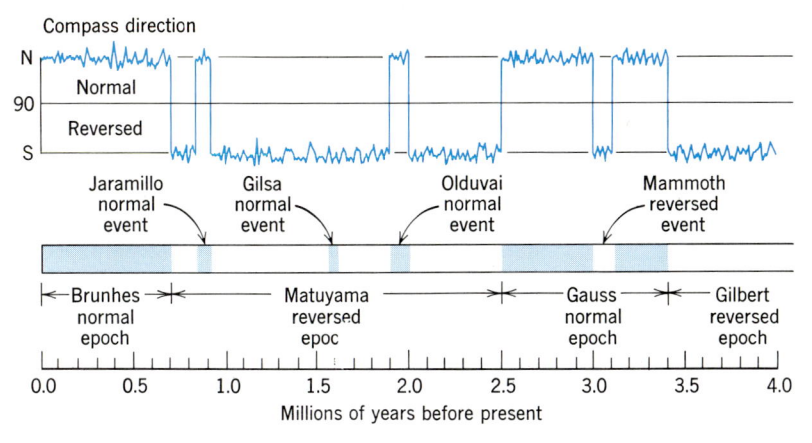

Magnetic polarity epochs and events with time scale. The upper graph suggests fluctuations in compass direction as recorded by a magnetometer towed over the ocean floor. (Based on data of Cox, Doell, and Dalrymple. From PRINCIPLES OF EARTH SCIENCE by Arthur N. Strahler. Copyright © 1976 by Arthur N. Strahler. By permission of Harper & Row, Publishers, Inc.)

moved away from the rift. Lavas of a given geologic age will thus become split into two narrow stripes, one on each side of the rift. As time passes, these stripes will increase in distance of separation, as shown in Figure 24-23. The lavas are identified and classified in terms of the epochs of normal and reversed magnetic field; these epochs will be represented by symmetrical striped patterns.

We cannot take oriented core samples of lavas from the ocean floors. However, it is possible to operate a sensitive magnetometer during a ship's traverse across the mid-oceanic ridge. When this is done, it is found that there are minute variations in the magnetic field. These departures from a constant normal value are referred to as *magnetic anomalies*. When several parallel crosslines of magnetometer surveys have been run across the mid-oceanic ridge, the magnetic anomalies can be resolved into a pattern. Figure 24-24 is a map showing the striped pattern of magnetic anomalies near Iceland. Notice the mirror symmetry of the striped pattern with respect to the axis of the ridge. From a study of the anomaly pattern, it is possible to identify the normal and reversed epochs.

Finding the magnetic stripes on the ocean floor proved to be the key evidence leading to culmination of the new revolution in geology. A series of magnetic surveys along many sections of the mid-oceanic ridge followed rapidly, all revealing similar striped patterns in mirror image.

Magnetic evidence not only made a virtual certainty of sea-floor spreading, but also allowed the rates and total distances to be estimated.

Take, for example, the case of the anomaly pattern shown in Figure 24-24, which is part of

FIGURE 24-23

Symmetrical stripes of polarity epochs are generated by sea-floor spreading and the rise of basalt magma in the axial rift. (From THE EARTH SCIENCES, 2nd Edition by Arthur N. Strahler. Copyright © 1963, 1971 by Arthur N. Strahler. By permission of Harper & Row, Publishers, Inc.)

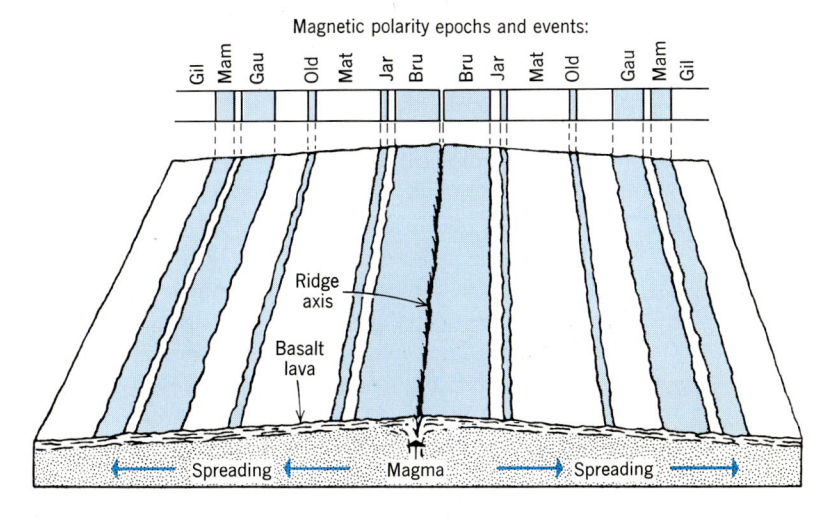

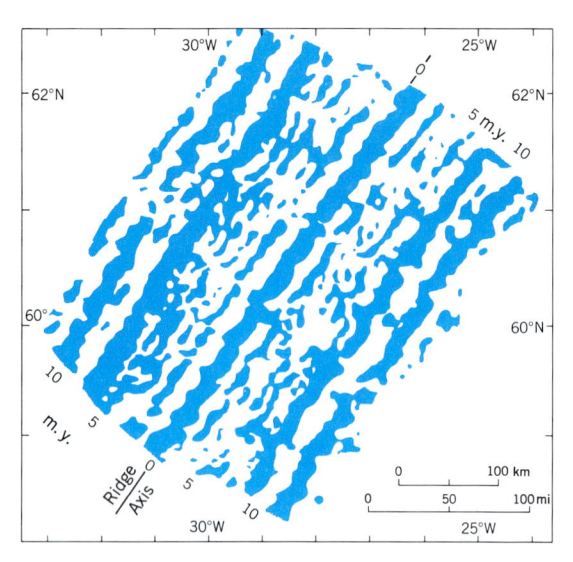

FIGURE 24-24
Magnetic anomaly pattern for Reykjanes Ridge, located on the Mid-Atlantic Ridge southwest of Iceland, with approximate rock ages in millions of years. (After J. R. Heirtzler, X. Le Pichon, and J. G. Baron, 1966, *Deep-Sea Research,* Vol. 13, p. 427.)

the Mid-Atlantic Ridge south of Iceland in the North Atlantic. Here the width of the anomaly zone is 200 km, which represents the total distance of crustal separation in about 10 million years. The average rate of horizontal motion of each lithospheric plate during this time has been about 1 cm/yr, which means that the rate of separation is double this value, or 2 cm/yr. Elsewhere some rates of spreading are found to be higher, up to 5 cm/yr.

24.20 PLATE THEORY OF GLOBAL TECTONICS

Rapid advances in our knowledge of the oceanic crust, and in particular of the wide extent of sea-floor spreading along the mid-oceanic ridge, led in the late 1960s to a general hypothesis of global tectonics. Wide acceptance of this model of earth behavior was the climactic phase in the new geological revolution. The term *plate tectonics* was applied by geologists to the new hypothesis, which features the horizontal movements of platelike masses of the strong, brittle lithosphere over a readily yielding, soft asthenosphere.

The earth's lithosphere consists of several major plates. Each plate moves horizontally as a unit and may also rotate as it moves over the asthenosphere. Obviously, two major possibilities are that adjacent plates may move apart, creating a widening gap between them, or they may move together, causing severe rupture of the edges of the plates. A third possibility is that they may slide alongside each other on transcurrent faults.

Figure 24-25 is a three-dimensional schematic diagram showing relationships among lithospheric plates. Plates spreading apart beneath the oceans produce the mid-oceanic ridge system with its axial rift and many transcurrent faults. Where plates converge, the edge of one plate is bent down and forced to descend into the mantle by a process called *subduction*. Deep in the asthenosphere, the plate margin is heated, melted, and absorbed into the mantle rock.

Figure 24-26 shows details of lithospheric plates in a hypothetical cross section. Notice the rising of the mantle rock under the mid-oceanic ridge to provide new oceanic crust as spreading occurs. Rising motions tend to elevate the axis of the mid-oceanic ridge. Ultramafic rock of the mantle provides the material for basaltic magma, and we must infer that the denser magma fractions sink back into the mantle. The opposite edge of the moving plate descends by subduction into the mantle beneath the margin of a continent. Two kinds of lithosphere are shown in Figure 24-26: *oceanic lithosphere,* composed of basalt and gabbro, is comparatively thin and has a high density; *continental lithosphere,* composed of felsic rock, is comparatively thick and has a low density. Consequently, when oceanic lithosphere is forced against continental lithosphere, the former always dives under the latter.

24.21 THE GLOBAL SYSTEM OF PLATES

Figure 24-27 shows the major plates of the globe. The American plate includes the North and South American continental crust and all of the oceanic crust as far east as the Mid-Atlantic Ridge. The Pacific plate is the only unit bearing only oceanic crust. It occupies all of the Pacific region west and north of the mid-oceanic ridge. It undergoes subduction beneath the American plate along the Alaskan-British Columbia coastal zone.

24.22 CONVECTION CURRENTS IN THE MANTLE

Lithospheric plate motions require a driving mechanism. Little is known of actual rock movements within the mantle, but most models of global tectonics have been referred to systems of very slow mantle circulation under the general heading of *convection currents.*

24.23 THE ORIGIN OF VOLCANIC ARCS

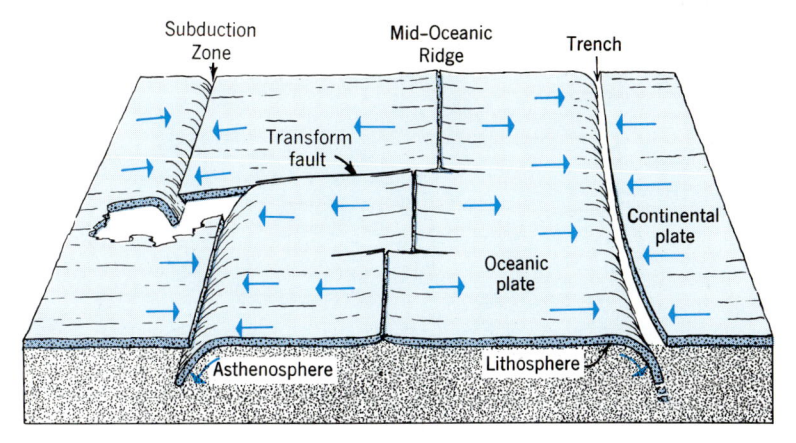

FIGURE 24-25
Schematic block diagram showing major features of plate tectonics. Earth curvature has been removed. (From THE EARTH SCIENCES, 2nd Edition by Arthur N. Strahler. Copyright © 1963, 1971 by Arthur N. Strahler. By permission of Harper & Row, Publishers, Inc.)

Figure 24-28 shows a model of convection involving the entire thickness of the mantle. Rising of less-dense mantle rock under the mid-oceanic ridge and corresponding sinking beneath the subduction zones of the trenches and island arcs are key activities within the convection system. Dominantly horizontal motion under the lithospheric plate would exert a drag, causing the plate to move away from the mid-oceanic ridge and toward the subduction zone.

24.23 THE ORIGIN OF VOLCANIC ARCS

Applying concepts of plate tectonics, we can now explain chains of volcanic mountain ranges adjacent to deep oceanic trenches. Consider first the case of a volcanic chain, such as the Andes Mountains of South America, lying on the edge of a continent and a bordering trench a short

FIGURE 24-26
The lithosphere, a rigid upper plate, glides over the soft asthenosphere. As new lithospheric plate is being formed during sea-floor spreading on an axial rift, the opposite margin is disappearing into the mantle by subduction.

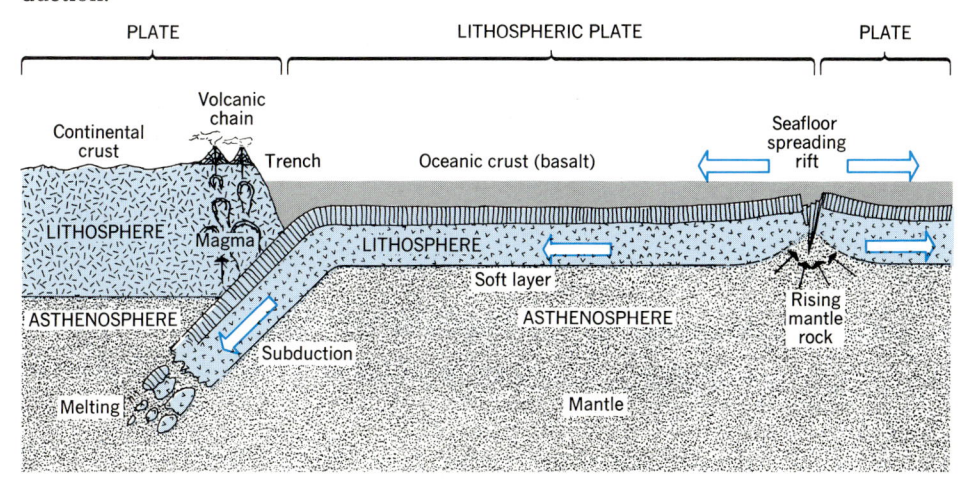

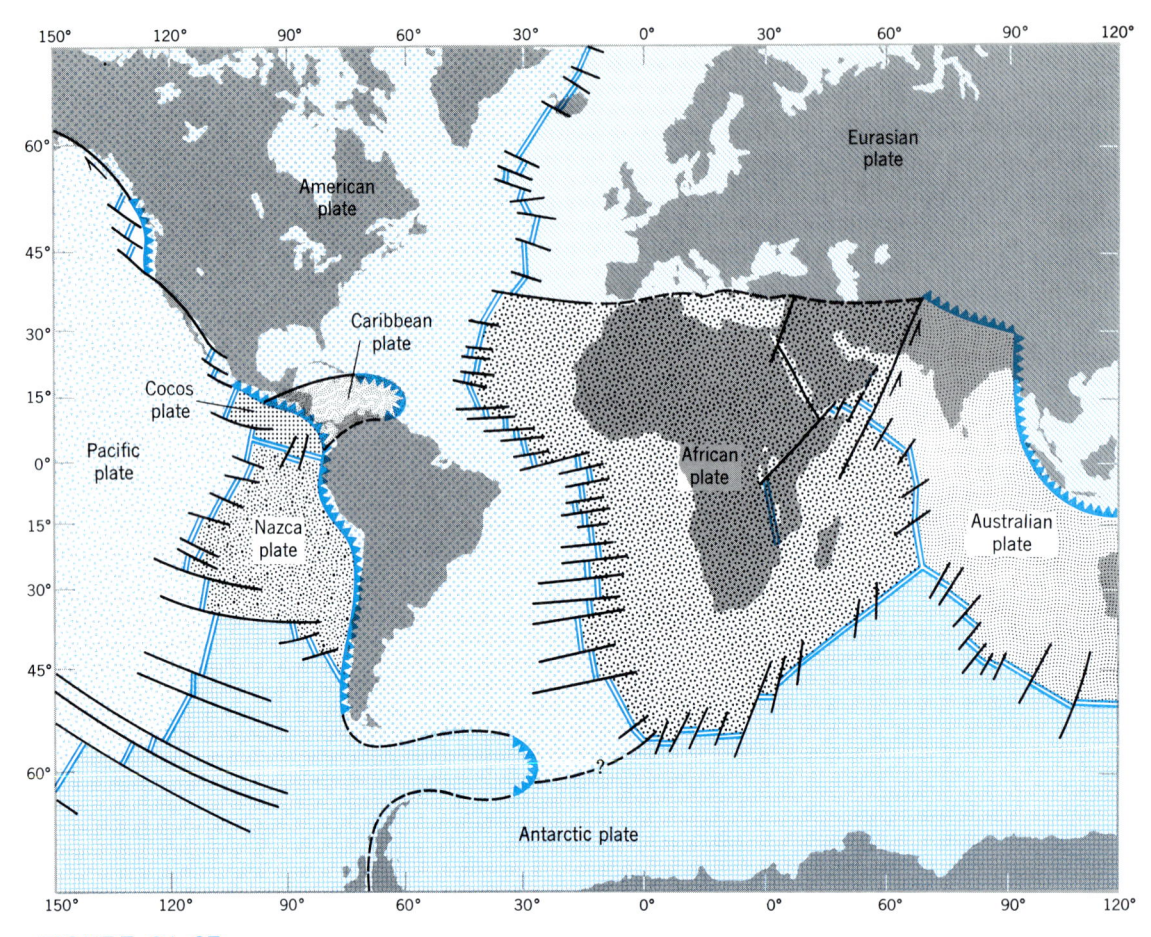

FIGURE 24–27
World map of lithospheric plates.

distance offshore. As shown in Figure 24–29, subduction of the plate beneath the ocean basin is the cause of the trench, while melting of the downbent plate produces magma pockets beneath the edge of the adjacent plate. Being less dense than the surrounding rock, the magma bodies slowly rise through the continental crust and emerge as extrusive igneous rocks to form the volcanic mountain range. Typically, the extrusive rock is of felsic composition and produces lava consisting of rhyolite and andesite rock.

Figure 24–30 shows the case of an island arc, typical of arcs of the western Pacific Ocean. In this case, oceanic lithosphere lies on both sides of the subduction zone. Volcanic islands rise by the upbuilding of volcanoes from the ocean floor, adjacent to the trench. Between the island arc and the continent lies a broad marginal ocean basin. Sediment, derived by erosion of the volcanic islands, accumulates on the floor of the marginal ocean basin.

24.24
CONTINENTAL COLLISION

Plate tectonics explains alpine structure through the process of *continental collision*, illustrated in Figure 24–31. In stage *A*, subduction is in progress. The lithospheric plate on the right is moving toward the left, bringing the two continents closer together, while the ocean basin is being reduced in width. Sediment is accumulating both on the deep ocean floor and at the continental margin, under the continental shelf. In stage *B*, narrowing of the ocean basin continues. Sediment is being scraped off the descending plate margin and is becoming crumpled and faulted in the trench. In stage *C* the two continents have collided, squeezing the sediment mass strongly, and throwing it into complicated folds with thrust faults. Now the ocean basin has disappeared entirely and a high alpine mountain range has come into existence. In stage *D*

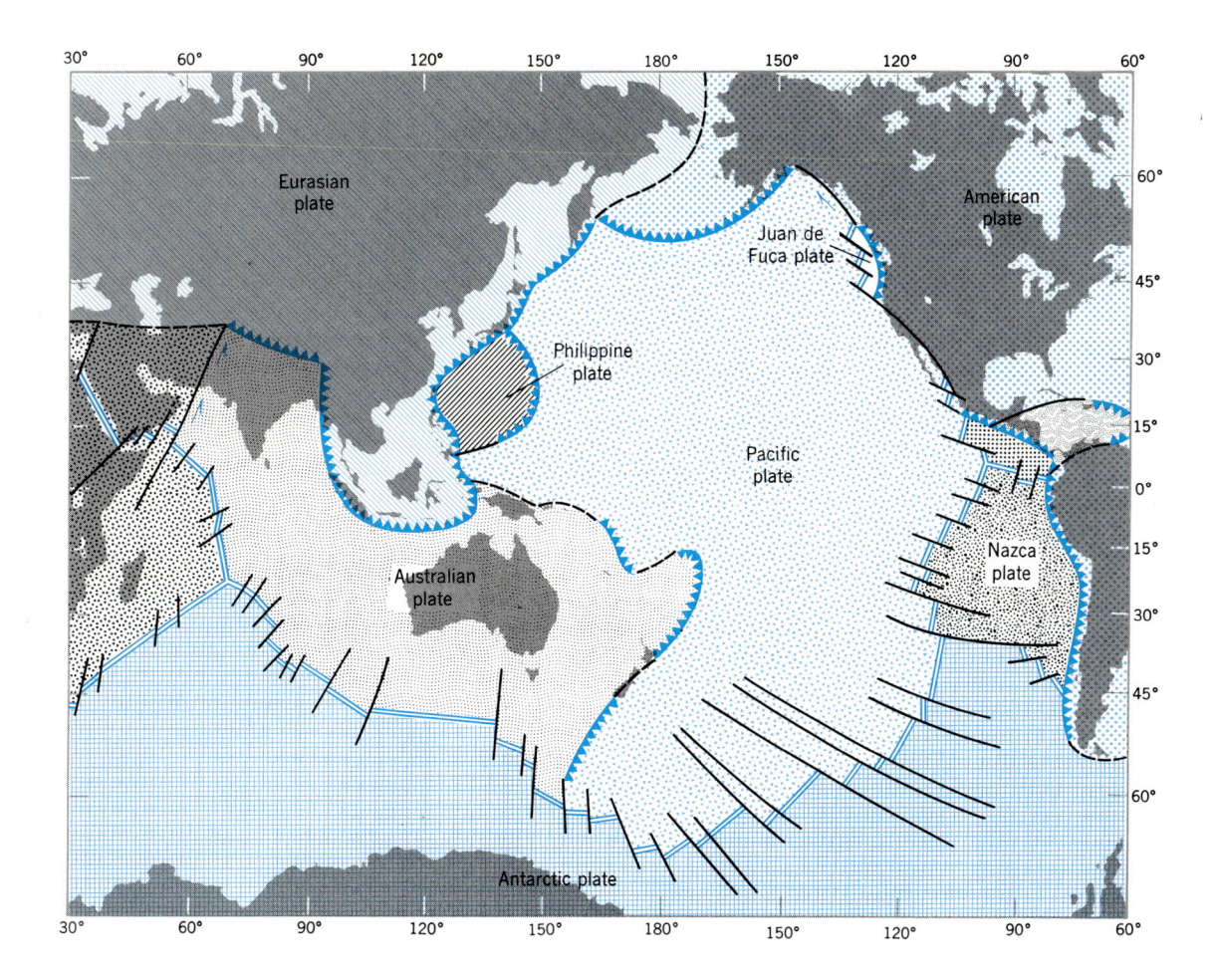

24.24 CONTINENTAL COLLISION

the oceanic crust has been eliminated entirely from the crust. The alpine zone remaining is called a *suture;* the process by which it is produced is called *continental suturing.*

Examples of alpine mountain ranges that have been quite recently formed by continental suturing are the European Alps and Himalaya Range of southern Asia.

FIGURE 24-28
Simplified model of a convection system in the mantle in relation to overlying lithospheric plates.

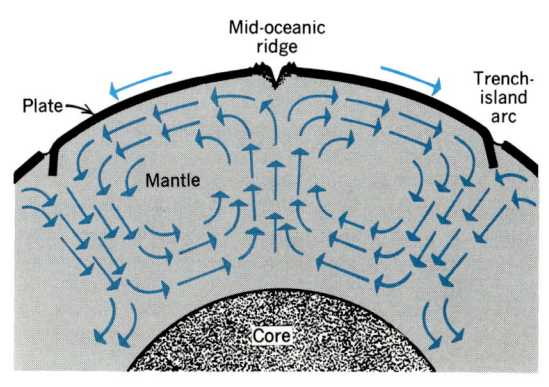

FIGURE 24-29
Subduction of oceanic lithosphere beneath the margin of a continental lithospheric plate, producing a volcanic mountain chain. Landforms are greatly exaggerated in vertical scale. This case illustrates the origin of the Andes Mountains of South America. (From PRINCIPLES OF PHYSICAL GEOLOGY by Arthur N. Strahler. Copyright © 1977 by Arthur N. Strahler. By permission of Harper & Row, Publishers, Inc.)

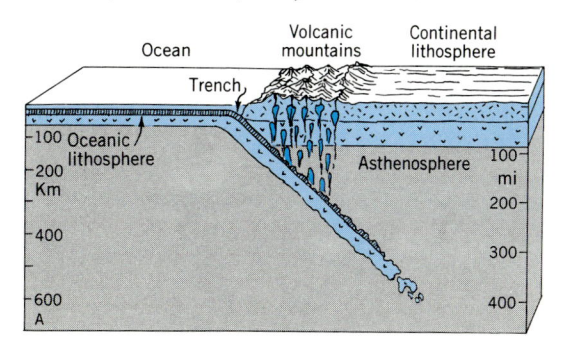

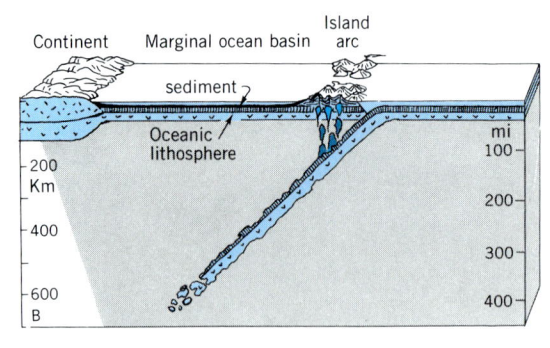

FIGURE 24-30
Subduction of oceanic lithosphere of one plate beneath oceanic lithosphere of another plate, leading to the formation of a volcanic island arc. (From PRINCIPLES OF PHYSICAL GEOLOGY by Arthur N. Strahler. Copyright © 1977 by Arthur N. Strahler. By permission of Harper & Row, Publishers, Inc.)

FIGURE 24-31
Stages in collision of two continental lithospheric plates. (After J. F. Dewey and J. M. Bird, 1970.)

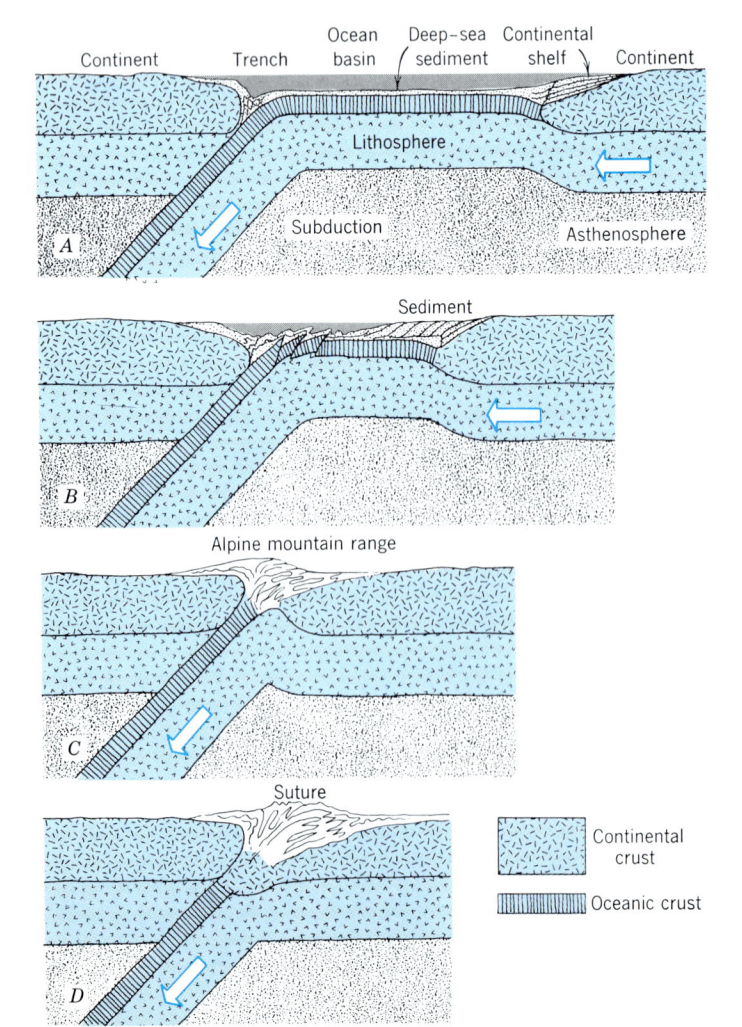

24.25 EVOLUTION OF THE CONTINENTAL CRUST

Origin of the continents has long been a major problem of geology. Most modern thought on this problem favors the supposition that the continental crust was not present when the growth of the earth as a planet was completed, some $4\frac{1}{2}$ billion years ago. Instead, the continents were formed later of rock of felsic mineral composition gradually segregated from an original crustal rock of mafic composition. Perhaps this original rock was similar to basalt of the present oceanic crust.

Our scenario of continental evolution requires that, initially, small elevated masses of granitic rock were added to the crust each time that sediments were crumpled by impact of a moving lithospheric plate during continental collision. Thus, the continents grew in size throughout the early part of geologic time. The continental interiors are composed largely of metamorphic and intrusive igneous rocks of great geologic age. Such rocks represent the roots of mountain ranges produced in a succession of continental collisions.

These interior regions are known as *continental shields*. (Figure 24-20 shows the distribution of shields.) Most of the shield rock is older than one billion years. Younger strata form thin covers over large areas of the shields, but these are superficial. Younger tectonic belts typically surround the shields along the continental margins.

24.26 CONTINENTAL RUPTURE

Not only did lithospheric plates collide repeatedly through geologic time, they also broke apart and separated many times. *Continental rupture* is illustrated in Figure 24-32. Diagram *A* shows the start of rupture. From deep beneath the continental plate, a column of heated mantle rock begins to rise. This column is called a *mantle plume*. The plume lifts the plate above it, causing the plate to fracture. The fracture system rapidly spreads across the entire plate, dividing it into two parts, which begin to separate. An axial *rift valley* is now formed. Basalt magma rises to pour out upon the floor of the rift valley. As spreading continues, the rift widens and deepens, admitting ocean water and becoming a narrow ocean, as shown in Diagram *B*. New oceanic lithosphere is being formed in the widening ocean basin. As shown in Diagram *C*, a broad ocean basin has been produced, with a characteristic mid-oceanic ridge.

The Red Sea, a narrow straight-sided trough lying between Africa and the Arabian Peninsula,

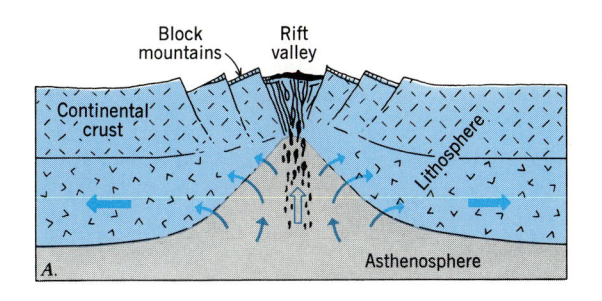

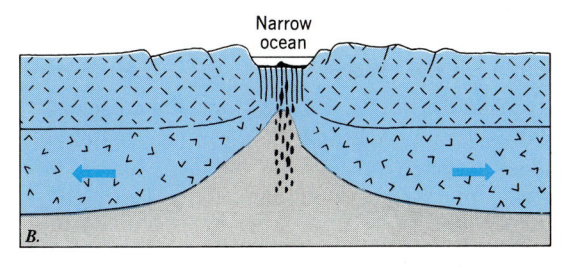

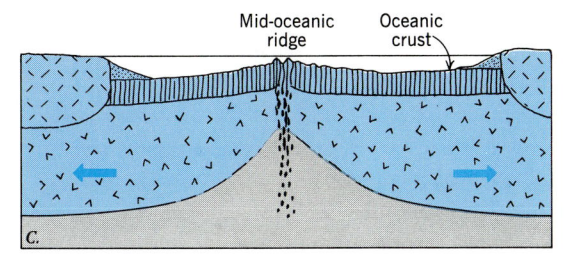

TERMS AND CONCEPTS

Earthquakes. Release of stored elastic strain in rocks adjacent to faults generates earthquakes, in which seismic waves are sent out from the focus. Analysis of primary, secondary, and surface waves as recorded by the seismograph yields information about the earth's interior and allows the earthquake intensity to be measured on the Richter scale.

Sea-floor spreading. A great mid-oceanic ridge, centrally located throughout the ocean basins, is a line of sea-floor spreading. Here lithospheric plates are pulling apart, while rising basalt magma solidifies to form new crust. Evidence of spreading rates lies in symmetrical magnetic stripes, formed during epochs of polarity alternations.

Plate tectonics. A global system of major lithospheric plates in relative motion explains tectonic activity and volcanism. Convection currents within the mantle are regarded as a probable driving mechanism for plate motions. Volcanic arcs lie above subduction zones, where oceanic lithospheric plates descend beneath the margins of continental lithospheric plates. Continental collision results in suturing and the formation of alpine mountain ranges, while continental rupture produces new ocean basins.

FIGURE 24–32
Stages in continental rupture and the formation of a new ocean basin. (From PRINCIPLES OF EARTH SCIENCE by Arthur N. Strahler. Copyright © 1976 by Arthur N. Strahler. By permission of Harper & Row, Publishers, Inc.)

is interpreted by geologists as a young ocean being formed by continental rupturing.

In our next chapter, we shall review a history of continental rupturing and continental separation on a vast scale, occurring throughout the last 150 million years of earth history.

SUMMARY

Earth's interior. A mantle of silicate rock of olivine composition surrounds the earth's core of iron. The outer core is liquid, the inner core solid. Temperatures in the core reach over 2700 K. Internal heat is sustained by radioactivity.

Earth's crust. The crust, a layer lying above the Moho, is thickest beneath the continents, where a granitic upper layer overlies a basaltic lower layer. Beneath the oceans is a thinner, basaltic crust.

Lithosphere and asthenosphere. A strong, rigid lithosphere about 60 km thick overlies a highly heated, soft asthenosphere within the upper mantle. Lithospheric plates move by slow shearing within the soft layer.

TERMS AND CONCEPTS

crust
mantle
core
continental crust
oceanic crust
granitic rock
basaltic rock
Moho
crustal roots
plastic layer (soft layer) of mantle
asthenosphere
lithosphere
shearing
lithospheric plates
tectonic activity
tectonics
primary arcs
circum-Pacific belt
island arcs
Eurasian-Melanesian belt
alpine system
oceanic trenches
earthquake
faulting
fault plane
normal fault
fault scarp
transcurrent fault

rift
strain of rock
elastic limit
focus
epicenter
seismic waves
primary waves (P-waves)
longitudinal wave
secondary waves (S-waves)
transverse wave
surface waves
seismograph
seismogram
Richter scale
ocean basins
continental margins
ocean-basin floors
mid-oceanic ridge
continental slope
continental rise
abyssal plain
seamounts
axial rift
sea-floor spreading
geomagnetic axis
north, south magnetic pole
dynamo theory
paleomagnetism
magnetic reversals
normal, reversed epoch
magnetic anomalies
plate tectonics
subduction
oceanic lithosphere
continental lithosphere
convection currents
continental collision
suture
continental suturing
continental shields
continental rupture
mantle plume
rift valley

QUESTIONS

1. Describe the earth's interior structure. Is the core entirely liquid or entirely solid? What evidence suggests the existence of an iron core? How high are temperatures in the mantle and core? What is the source of this heat?

2. How does the structure and composition of continental crust differ from that of oceanic crust? What is the significance of the Moho? Where are crustal roots deepest?

3. How deep is the soft layer in the mantle? Why is mantle rock weak and capable of slow flowage in this layer? Explain how the lithosphere is capable of moving over the asthenosphere. What significance has this concept?

4. What evidence can one find of tectonic activity? Where is tectonic activity now concentrated on a global scale? Name and describe the two great primary arcs. What relationships do trenches bear to island arcs?

5. How are earthquakes related to faults? Explain fully. What two common types of faults are the sources of earthquakes? What directions of motion are associated with each fault type?

6. Name and describe three basic types of seismic waves. What particle motion characterizes each? Which type of wave travels most rapidly? Which most slowly?

7. What principle is used in the design of a seismograph? How is the seismogram interpreted to yield the distance to the epicenter? How do seismic waves reveal the existence of a liquid outer core?

8. Describe the Richter scale of earthquake magnitudes. Has this scale a fixed upper limit? Describe the world pattern of distribution of earthquakes. Which belts account for the greatest energy release?

9. Describe the major divisions of the ocean basins. What feature marks the outer limit of the continental shelves? Why is the abyssal plain so flat? What feature marks the axis of the mid-oceanic ridge? What crustal activity does this axial feature suggest?

10. Describe the earth's magnetic field, including axis and poles. What is the probable mechanism of earth magnetism? How is paleomagnetism developed in rocks? What is the significance of magnetic polarity reversals in terms of sea-floor spreading? How are magnetic anomalies interpreted in terms of lithospheric plate motions?

11. Give a general description of the plate theory of global tectonics. What plate motions are possible? What changes in the state of rock materials are associated with sea-floor spreading and with subduction? Name several major lithospheric plates. How can convection currents be called upon to explain plate motions? Describe continental collision and continental rupture.

CHAPTER 25
EARTH HISTORY IN REVIEW

25.1 HISTORICAL GEOLOGY

In more than one sense this chapter is a step backward into time. It is, to be sure, an incursion into vast spans of geologic time. Less obviously, it summarizes progress through many decades in the history of geology as a science. *Historical geology* is a general term for the reconstruction of an orderly sequence of physical and biological events throughout the past $3\frac{1}{2}$ billion years or more of our planet's history. These events include tectonic and igneous activity, sediment deposition, and the evolution of life.

Starting early in the 1800s, historical geology was dominated by interpretation of sedimentary strata. This is the phase of historical geology with which we begin our chapter. A full century was to elapse before advances in physics opened the way to establish the ages of rocks in actual years of time before the present. Then another half century was to follow before the theory of plate tectonics took its place as the unifying concept for all earth history.

25.2 STRATIGRAPHY

Stratigraphy is a branch of historical geology dealing with the sequence of events in the earth's history as interpreted from evidence found in sedimentary rocks. Included in stratigraphy are the records of deposition of sedimentary strata, of the past geographic distributions of land and sea, and of the past conditions of climate and terrain.

The pursuit of stratigraphy depends on *paleontology*, the study of ancient life based on fossil remains of animals and plants. The stratigrapher uses fossils both to correlate strata in age of deposition and to establish the physical and chemical environments of deposition.

Our first stratigraphic principle is so simple as to seem self-evident: within a series of sedimentary strata whose attitude is approximately horizontal, each bed is younger than the bed beneath, but older than the bed above it (Figure 25–1). This age relationship could not be otherwise in the case of sediment layers deposited from suspension in water or air.

The age-layering concept is called the *principle of superposition*. Simple as it seems, there is cause for concern. Someone might object that the strata have been bodily overturned during tectonic activity, as may happen in close folding of strata. In an overturned sequence the uppermost beds are actually the oldest. The geologist routinely checks against this possibility of error by examining closely certain critical details of the sedimentary rock.

A means of ascertaining whether strata in two localities are of the same or different age is to travel the ground from one locality to the other, observing the strata continuously along the line of march. If one can actually walk upon the same rock layer throughout the entire distance, the similarity of age is proved by the *principle of continuity*. A simple case is shown in Figure 25–2, where the same layer can be followed for miles in the rim of a series of canyons and cliffs.

FIGURE 25-1
Sedimentary strata exposed in the walls of Grand Canyon, Arizona. This view spans a total thickness of about 1000 m of sandstones, shales, and limestones. (A. N. Strahler.)

The principles of superposition and continuity are attributed to a Danish physician, Nicolaus Steno, who worked out the evidence while serving as a physician to the Duke of Florence in the seventeenth century. Steno studied sedimentary strata exposed along the walls of the Arno Valley, not far southeast of Florence. His findings were published in book form in 1669, comprising the first fruitful effort to work out the geologic history of a region by interpretation of strata.

FIGURE 25-2
Principle of correlation of strata by direct continuity. The bed at *A* can be traced without interruption to a distant location, *B*. (From A. N. Strahler, 1971, *The Earth Sciences*, 2nd ed., Harper & Row, New York.)

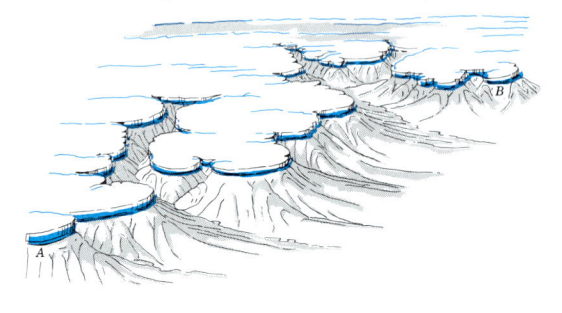

25.3 FOSSILS

Of all sources of information, perhaps the most helpful to the stratigrapher are *fossils*, those ancient plant and animal remains or impressions preserved by burial in sedimentary strata. About the year 1800 an English civil engineer and geologist, William Smith, collected fossils in strata exposed in canal excavations. He found fossils of like species to be present in all parts of a single formation that could be proved to be one and the same bed by direct continuity. Fossil species in strata above or below were found to be distinctively different, but to occur consistently in the same order in widely separate localities.

Once the order of fossils was established by direct observation, the fossils themselves became the evidence for matching like ages of strata elsewhere in the world. For example, the fossils in certain strata in Wales were studied early in the nineteenth century, and these rocks became established as the original standard for the Cambrian Period of geologic time (Cambria is the Latin name for Wales). One distinctive fossil animal, the trilobite, was abundant in the Cambrian seas, and consequently some of its various species serve as guide fossils for the Cambrian Period throughout the world (Figure 25-3).

FIGURE 25-3
This fossil trilobite from the lower walls of Grand Canyon establishes the formation as being of Cambrian age. The head is to the left. About $1\frac{1}{2}$ times natural size. (Department of the Interior, Grand Canyon National Park.)

The value of fossils in telling us the age of rock strata arises from the fact that all forms of plant and animal life have continually and systematically undergone changes with passage of time, a process termed *organic evolution*. If we have before us a complete, or nearly complete, description of past life forms as determined from fossils, and if we know the geologic age to which each fossil form belongs, it is often a simple matter to give the age of any sedimentary layer merely by extracting a few fossils from the rock and comparing them with the reference forms.

Worldwide studies by stratigraphers over the last 150 years have yielded an extremely detailed and nearly complete reference table of the divisions and subdivisions of geologic time, together with distinctive fossils for all ages from very ancient to recent. (The various time divisions are shown in Table 25.1.)

25.4 CATASTROPHISM AND UNIFORMITARIANISM

Early in the nineteenth century, before modern geologic science was fully established along today's lines, one group of naturalists explained folding and faulting as the result of one or more sudden catastrophes. The leader of this school was an able French student of fossil life forms, Baron Cuvier, who wrote: "The dislocation and overturning of older strata show without any doubt that the causes which brought them into the position which they now occupy, were sudden and violent . . . the evidences of those great and terrible events are everywhere to be clearly seen by anyone who knows how to read the record of the rocks." Those who supported this view were known as *catastrophists*. They held that not only were our present mountains, cliffs, and canyons formed by violent catastrophe, but also all the animals whose shells and bones we now find as fossils in the strata were suddenly killed in the cataclysm. This theory, *catastrophism*, was tenable only as long as evidence consisted of a few observations in only one region, but it gradually became clear that no single worldwide catastrophe can explain all known relations among strata and their fossil content. Furthermore, the length of time needed for the various events of geologic history was grossly underestimated by the catastrophists.

A sound basis for reconstructing geologic events is provided by the principle that processes acting in the past have been essentially the same as those seen in action over the face of the earth today. This principle, which was strongly maintained by a Scottish geologist, James Hutton (1726–1797), was termed *uniformitarianism*. It is often summarized in the statement that "the present holds the key to the past." Geologists believe that if they can watch a volcano in eruption and see the molten lava pour down the side of the mountain into the sea, they will learn the explanation for similar kinds of lava forms now found enclosed in ancient rocks. If they study the manner in which sediments are being laid down around the mouth of the Mississippi River today, they can learn how to interpret similar layers of shale and siltstone, exposed in the walls of canyons far inland from present-day shores. Uniformitarianism signifies the belief in similar processes acting to cause similar resultant products throughout all of geologic time.

25.5 PRECAMBRIAN TIME

All of geologic time older than 570 million years (-570 m.y.), and extending back to include the oldest known rocks—about $3\frac{1}{2}$ billion years old ($-3\frac{1}{2}$ b.y.)—is known as *Precambrian time*. In this vast and obscure block of time the oceans came into existence and most of the continental crust was formed. Lithospheric plates of many configurations pulled apart, creating new basaltic oceanic crust, while subduction zones consumed large expanses of plates bearing continental crust. Many ancient arcs of mountain building were pushed up and later obliterated. The record of Precambrian time lies in the metamorphic and igneous rocks remaining in the continental shields, but is extremely difficult to unravel.

TABLE 25.1
TABLE OF GEOLOGICAL HISTORY

Era	Period	Epoch	Absolute age in years before present	Major geologic events in United States given in order of increasing age	Distinctive features of plant and animal life	
Cenozoic		Recent (Holocene)	10,000	Minor changes in landforms by work of streams, waves, wind	Rise of civilization	Age of Humans
		Pleistocene	1,000,000	Four stages of spread of continental ice sheets and mountain glaciers	Development of humans; extinction of large mammals	
				Cascadian orogeny: Cascade and Sierra Nevada ranges uplifted; volcanoes built		
		Pliocene	13,000,000	Marine sediments deposited on Atlantic and Gulf coastal plain; stream deposits spread over Great Plains and Rocky Mountain basins; thick marine sediments deposited in Pacific coastal region	Early evolution of humans; dominance of elephants, horses, and large carnivores	Age of Mammals
		Miocene	25,000,000		Development of whales, bats, monkeys	
		Oligocene	36,000,000		Rise of anthropoids	
		Eocene	58,000,000		Development of primitive mammals; rise of grasses, cereals, fruits	
		Paleocene	65,000,000	Laramide orogeny; Rocky Mountains formed	Earliest horses	
Mesozoic	Cretaceous		136,000,000	Marine sediment deposition over Atlantic and Gulf coastal plain and in geosyncline of Rocky Mountain region	Extinction of dinosaurs; development of flowering plants	Age of Reptiles
				Nevadian orogeny: intrusion of batholith of Sierra Nevada region		
	Jurassic		190,000,000	Marine sediment deposition in seas of western United States; desert sands deposited in Colorado Plateau	Culmination of dinosaurs; first birds appear	
				Palisadian disturbance: block faulting in eastern United States		
	Triassic		225,000,000	Deposition of red beds in fault basins of eastern United States and in shallow basins of western United States	First dinosaurs; first primitive mammals; spread of cycads and conifers	
				Appalachian orogeny: folding of Paleozoic strata of Appalachian geosyncline		
Paleozoic	Permian		280,000,000	Deposition of red shales and limestones in southwestern United States; much salt and gypsum (glaciation of southern hemisphere continents)	Conifers abundant; reptiles developed; spread of insects and amphibians; trilobites become extinct.	Age of Amphibians
	Carboniferous — Pennsylvanian		310,000,000	Deposition of coal-bearing strata in eastern and central United States	Widespread forests of coal-forming spore-bearing plants; first reptiles; abundant insects	
	Carboniferous — Mississippian		345,000,000	Deposition of limy, shaly sediments in widespread, shallow seas of central and eastern United States	Spread of sharks; culmination of crinoids	
				Acadian orogeny: Folding and igneous rock intrusion in New England		
	Devonian		395,000,000	Deposition of thick marine strata in geosynclines of eastern and western United States	First amphibians; many corals; earliest forests spread over lands	Age of Fishes
	Silurian		430,000,000		First land plants and air-breathing animals; development of fishes	
				Taconian orogeny: folding of rocks in eastern United States, Nevada, and Utah		
	Ordovician		500,000,000	Deposition of thick marine strata in geosynclines of eastern and western United States	Life only in seas; spread of mollusks; culmination of trilobites	Age of Marine Invertebrates
	Cambrian		570,000,000		Trilobites predominant; many marine invertebrates	
	Precambrian time; age goes back to over four billion years			Many periods of sediment deposition alternating with orogeny	Earliest known forms of life; few fossils known	

It is speculated that the first life forms may have appeared as early as $3\frac{1}{2}$ billion years ago ($-3\frac{1}{2}$ b.y.). The oldest materials that can be interpreted as fossil remains of primitive life forms (possibly primitive algae) are in rocks dated -3.2 b.y., but major deposits produced by algae do not appear until about -2 b.y. From that time until near the close of Precambrian time (-570 m.y.), complex higher life forms seem not to have existed; in any case, they have not been found preserved in strata.

The reason it evidently took so long for even very primitive forms of life to emerge is that a living system, both in structure and operation, is extremely complicated, with all parts having to function together or none can long survive. No one believes that even a simple living cell just happened by an accidental coming together of all the requisite atoms in just the right arrangement driven by just the right energy. Far more likely would be the appearance on the sidewalk of a deck of cards all arranged in order by suits following someone's tossing the deck, card by card, into the wind from a tall building. Instead, the chemicals needed to make a cell almost certainly were made from simple materials in the early ocean long before anything like a cell appeared. A long period of chemical evolution preceded the emergence of living cells and the long succeeding period of their biological evolution. Each stage in both forms of evolution represented a very small change and therefore had a relatively high probability of happening in terms of the laws of chemistry and physics as we know them, laws we have every reason to believe have operated throughout all time.

Very likely the primitive atmosphere contained water, carbon dioxide, sulfur, nitrogen (as N_2), methane, ammonia, hydrogen, and carbon monoxide. Very little free oxygen was present. In fact, the earliest stages of organic evolution probably could not have occurred in our present atmosphere, because oxygen attacks so many things. Salts, including phosphates essential to living systems, were leached from the lithosphere, and the ocean's salinity developed.

The earth was bathed in solar radiation and unprotected by a stratospheric ozone layer. The ozone shield probably formed after the atmosphere became richer in oxygen as the result of photosynthesis after primitive photosynthesizing algae emerged. High-energy ultraviolet rays reached to the earth's surface. These can crack molecules of nitrogen.

$$N_2 + UV \rightarrow 2N$$
Nitrogen molecule → Nitrogen atoms

Nitrogen atoms can attack methane.

$$2N + 2CH_4 \rightarrow 2HCN + 3H_2$$
Methane → Hydrogen cyanide + Hydrogen

There are other ways hydrogen cyanide can form from the simpler chemicals. Both it and ammonia are very soluble in water, and the oceans absorbed them in vast quantities as solar energy continued to stream in.

A landmark experiment in 1953 by Stanley Miller, then a student of Harold Urey, showed that when a mixture of hot water vapor, methane, ammonia, and hydrogen is circulated past an electric spark for a period of a week, 5 percent of the carbon (as methane) is converted to amino acids (Section 19.12) and other compounds that are biochemically necessary for life. Carbon monoxide can be used instead of methane and nitrogen in place of ammonia with very similar results. Amino acids form, from which proteins can be generated; some of these might have been primitive catalysts (enzymes). In other experiments, it is known that hydrogen cyanide can be changed into compounds that lead to the side-chain amines of DNA and RNA, the chemicals of heredity. Compare, for example, the relation of adenine, one of the amines needed for DNA (Section 19.19) and a seemingly very complex molecule, to hydrogen cyanide.

Hydrogen cyanide (five molecules) → several steps → Adenine

No step involves the coming together of more than two molecules at a time.

Carbon monoxide is relatively easily converted to formaldehyde.

$$CO + H_2 \rightarrow \begin{array}{c} H \\ \diagdown \\ C=O \\ \diagup \\ H \end{array}$$
Carbon monoxide + Hydrogen → Formaldehyde

Formaldehyde can eventually be changed to the basic system found in molecules of simple sugars.

$$nCH_2O \xrightarrow{\text{several steps}} (CH_2O)_n$$
Formaldehyde → Basic unit of sugars
(if $n = 5$, ribose and isomers needed for nucleic acids
$n = 6$, glucose and isomers used for chemical energy and making other compounds)

From this we can visualize the formation in a hydrogen-rich environment of several substances having carbon chains, including lipids. It is thus altogether plausible in terms of the laws of physics and chemistry and the great time period available that chemicals essential to cells could have formed in the primordial soup of simple organic and inorganic molecules and ions in a world ocean bathed by high-energy ultraviolet rays.

Experiments by S. W. Fox, A. I. Oparin, L. Orgel, and others have demonstrated that cell-like structures will form when hot solutions containing certain kinds of simple proteins and other biochemicals are allowed to cool. No doubt these primitive cellular bodies formed and broke up innumerable times in the prebiotic ocean. It is not contrary to known scientific laws, however, to imagine that some survived long enough to get a very elementary kind of internal chemical machinery—metabolism—going. We are in an area now, however, of pure speculation. Understandably, the fossil record has left no clues. Only a few fossils are known from Precambrian times, but even these represent extraordinarily complex cells. Only with the march of succeeding eras of geologic time do forms of life appear that could withstand the ravages of fossilization.

25.6 ERAS OF GEOLOGIC TIME

Geologic time younger than Precambrian time is made up of three *geologic eras*—the *Paleozoic Era*, the *Mesozoic Era*, and the *Cenozoic Era*, listed in order from earliest to latest. Translating from the Greek roots of these three titles, they can be paraphrased as the eras of ancient (*paleos*), middle (*mesos*), and recent (*kainos*) life, respectively.

The table of geologic time gives ages and durations of the three eras, together with their subdivisions into *periods* (Table 25.1). Notice that the Paleozoic Era with six periods had a duration of 346 m.y., the Mesozoic Era with three periods lasted only 160 m.y., while the Cenozoic Era has been too short (65 m.y.) to warrant subdivision by periods.

In general, each geologic period represented a time of more or less continuous accumulation of strata in thick sequences referred to by geologists as geosynclines, and each period was brought to an end by an episode of tectonic activity known as an *orogeny*. Table 25.1 lists a number of orogenies important in the history of North America. We can interpret certain of these orogenies as collisions between lithospheric plates, whereas others may represent crustal block-faulting in rift zones where plates were spreading apart.

Keep in mind, of course, that the eras and periods of geologic time were unraveled throughout the 1800s through the methods of stratigraphy and paleontology, and that absolute ages of strata and orogenies were entirely unknown. Speculations as to the age in years of the eras and periods were indulged in by many geologists and several physicists, but were grossly in error; all fell far short of the figures we have given. Even so distinguished a physicist as Lord Kelvin was far wrong in his estimates of geologic time; he concluded all earth history from the beginning of Precambrian time was accomplished in a timespan of from 20 to 40 m.y. How could he have been so grossly in error? The answer is that the phenomenon of radioactivity was not known to Kelvin's generation. The great breakthrough in dating of geologic events came as a result of the discovery of radioactivity by Henri Bequerel in 1896, followed by the isolation of radium by Marie and Pierre Curie in 1898.

Dramatically, the search for a reliable age of the earth and the duration of geologic eras and periods was solved by the application of principles of radioactivity. Using these principles, the first reliable age determinations of rocks were made in 1907 by B. B. Boltwood, a chemist. His figures have required only minor adjustments to the present day. The oldest rock age found by Boltwood was about $1\frac{1}{2}$ billion years.

25.7 RADIOMETRIC AGE DETERMINATION

Determining the age in years of a mineral or rock is a procedure of science known as *geochronometry*. Ages thus determined are referred to as *radiometric ages*. At the time of crystallization of an igneous rock from its liquid state, minute amounts of minerals containing radioactive isotopes are entrapped within the crystal lattices of the common rock-forming minerals, in some cases forming distinctive radioactive minerals. At this initial point, none of the stable daughter products that constitute the end of the decay series are present. However, as time passes, the stable end member of each series is produced at a constant rate and accumulates in place.

Knowing the half-life of the decay system, we can estimate closely the time elapsed since mineral crystallization occurred. An accurate chemical determination of the ratio between the radioactive isotope and its stable daughter product must be made.

Take, for example, the uranium-lead series, uranium-238 to lead-206, which has a half-life of $4\frac{1}{2}$ b.y. Quantities of both uranium-238 and lead-206 are measured from a sample of uranium-bearing minerals or from a common mineral enclosing the radioactive isotopes. The ratio of lead to uranium is entered into an equation and easily solved for age in years.

Similar age determinations can be made using the series uranium-235 to lead-207. Because both series of uranium-lead isotopes are normally present in the same mineral sample, age analysis of one series can serve as a crosscheck on the other. It is possible to determine the absolute age of a sample of uranium-bearing mineral to within about 2 percent of the true value, and in some cases to within 1 percent. But this level of accuracy also assumes that none of the components in the decay series have been lost from the sample. Use of the uranium-lead systems for age determination can be applied to the oldest rocks known, as well as to meteorites. Age of meteorites is close to $-4\frac{1}{2}$ b.y., about 1 b.y. older than the oldest rocks of the earth's crust that have thus far been dated.

Also of great importance in age determination is the potassium-argon series, potassium-40 to argon-40, with a half-life of $1\frac{1}{3}$ b.y. The potassium-argon series gives reliable minimum ages for fine-grained volcanic rocks (lavas), which cannot be dated by other methods.

The rubidium-strontium decay series, rubidium-87 to strontium-87, has an extremely long half-life of 47 b.y. This series is of great value in dating both individual minerals and whole rock samples. It has proved successful in dating metamorphic rocks and thus in dating the tectonic events that produced the metamorphism.

The radiometric ages given for various events in the timetable of the earth's history are now accepted by geologists as valid within small percentages of error. Success of the radiometric age determinations of rocks stands as a striking scientific achievement based on the application of principles of chemistry and physics to geology.

25.8
THE PALEOZOIC ERA

Let us now begin a review of the major highlights of earth history, starting with the Paleozoic Era, which began at -570 m.y. The configuration of the continents at that point in time is not at all well known, but they were probably clustered fairly closely. Individual lithospheric plates were in sporadic motion with respect to one another. Here and there, continental crust ruptured along spreading zones, causing widening rifts, floored by oceanic crust. These widening seas were geosynclines, in which sediment was deposited in enormously thick layers as the basins were deepening. However, the seas remained shallow and were favorable to the evolution of a great variety of organisms. The animal forms were all invertebrates, and there were many primitive marine plants as well. No terrestrial life existed at the opening of the Paleozoic Era. Perhaps there did not then exist a sufficiently dense ozone layer in the stratosphere to shield land organisms from solar ultraviolet radiation. It is suspected that the level of atmospheric free oxygen was very small at that time, perhaps not more than 1 percent of the present value.

As the Cambrian Period began, initiating the Paleozoic Era, all of the major classes of invertebrate animals were present. They must have evolved rapidly in the closing stages of Precambrian time, but there is no fossil record of this event and it still remains enshrouded in mystery. Figure 25-4 shows a reconstruction of organisms of the Cambrian sea floor, well documented by abundant fossil remains. Among the groups present were sponges, coral, jellyfish, segmented worms, and primitive crustaceans. Two other invertebrate forms—common then, but either

FIGURE 25-4
Restoration of the middle Cambrian sea floor. Sponges are shown at far left and right, trilobites and arthropods in the center and at left, segmented worms on the bottom, and a jellyfish at upper left. (The Smithsonian Institution.)

unknown or very few in number today—were the trilobites and the brachiopods. Now extinct, the trilobites were exceptionally abundant in the Cambrian Period. They were bottom-feeding animals having a segmented outer skeleton of chitin; they outwardly resembled the "horseshoe crab" of today. The brachiopods were shellfish, outwardly resembling modern clams.

The Cambrian Period was followed by the Ordovician Period, which witnessed evolutionary development of many of the invertebrate forms, although trilobites were declining. In the ensuing Silurian Period, corals had become increasingly abundant and were producing important reef deposits of limestone.

Late in the Silurian Period the level of atmospheric oxygen had perhaps risen to proportions comparable to that of the present. Life forms emerged from the seas and spread to the lands. Plants first made the advance to the land environment; their primitive forms can be found as fossils in strata of upper Silurian age. By Devonian time, land plants had evolved to large, complex forms, including club mosses, horsetails, and ferns. These plants grew to the size of tall trees, and by the middle of the Devonian Period had spread over the lands in rich forests. In the meantime, vertebrate animals were continually evolving in the seas, first as fishes (abundant in Silurian time) then as amphibians, which took to the lands as air-breathers in late Devonian time.

The Carboniferous Period began at -345 m.y. In North America, this period is usually divided into two periods, the Mississippian and the Pennsylvanian (Table 25.1). Perhaps the most remarkable feature of the Carboniferous Period was the luxuriant growth of forests in vast swampy lowlands in many parts of the continents (at that time joined into a single supercontinent). Figure 25–5 is a reconstruction of a Carboniferous forest. The stucture and succulent foliage of these forest plants suggest that they grew in a warm, moist climate such as that found in low latitudes today. Widespread coal seams occur throughout the Carboniferous strata and attest to the enormous rate of production of organic matter during that period. Among the animals, the expansion in species of insects and arachnids (spiders and scorpions) during the Carboniferous Period was particularly remarkable. The first primitive reptiles evolved from the amphibians during this period, and in so doing became completely independent of the marine environment. These same evolutionary patterns persisted throughout the Permian Period, last of the periods of the Paleozoic Era.

FIGURE 25–5
Restoration of a Carboniferous forest. Seedless trees and seed-ferns dominate. *F*, Seed-fern. *L*, Lycopsid. *S*, Sphenopsid. (Illinois State Museum.)

25.9 PANGAEA IN LATE PALEOZOIC TIME

From at least the Carboniferous Period through to the middle of the Mesozoic Era, a single great continent existed on the globe. It was named *Pangaea* by those geologists who, in the 1920s, first presented strong evidence for the existence of a supercontinent. Pangaea consisted of the nested continental shields. Figure 25-6 shows a modern reconstruction of these nested shields. Shaded areas are the continental nuclei.

The arrangement of nuclei into two groups suggests that there were originally two centers of earlier continental crust accumulation: *Laurasia* in the northern hemisphere and *Gondwana* in the southern hemisphere (Asia is not shown in this reconstruction). Peninsular India, western Australia, Madagascar, and Antarctica are closely clustered beside the African continent to form Gondwana.

Evidence for the former unity of the continental shields tkes a variety of forms. Matching of similar rock types and rock ages from the margin of one continent to another provides one line of evidence. The case of South America and Africa is particularly interesting. As Figure 25-6 shows, small fragments of continental nuclei in South America seem to fit with larger nuclei in Africa. Moreover, the trends of Precambrian tectonic structures in the area between these nuclei are continuous from one continent to another. Today, in contrast, these linear structures project directly out toward the ocean basin and appear to have been abruptly truncated. In the North Atlantic, tectonic structures of the Appalachian belt, passing through Nova Scotia and Newfoundland, appear to line up with corresponding structures of the same geologic age in the British Isles and Norway.

Matching of fragments of the Gondwana nuclei in a single continental mass has been based in part on similarities of sedimentary rocks and their contained fossils of late Paleozoic and Mesozoic age. Distinctive and specialized plants of the same genus are found as fossils in sedimentary strata of Carboniferous age. According to those who first supported the hypothesis of continental drift, the simultaneous development of these plants on widely separated continents would have been an impossibility.

A similar argument for Gondwana has been based on distribution of animals thought to be incapable of migrating from one continent to another over deep ocean water (Figure 25-7). Key evidence has recently come from a mammallike reptile of the genus *Lystrosaurus*. This small animal somewhat resembled a hippopotamus, with massive wideset legs. Fossil remains of *Lystrosaurus* are abundant in Triassic strata of southern Africa and are also found in India, Russia, and China. Search for *Lystrosaurus* fossils in Triassic rocks of Antarctica met with success in 1969 when remains of *Lystrosaurus* were found in the Transantarctic Mountains, about 640 km from the south pole. The fossil find was hailed as one of the most significant in modern times, for it threw paleontologic evidence strongly in favor of the existence of a unified single landmass of Gondwana.

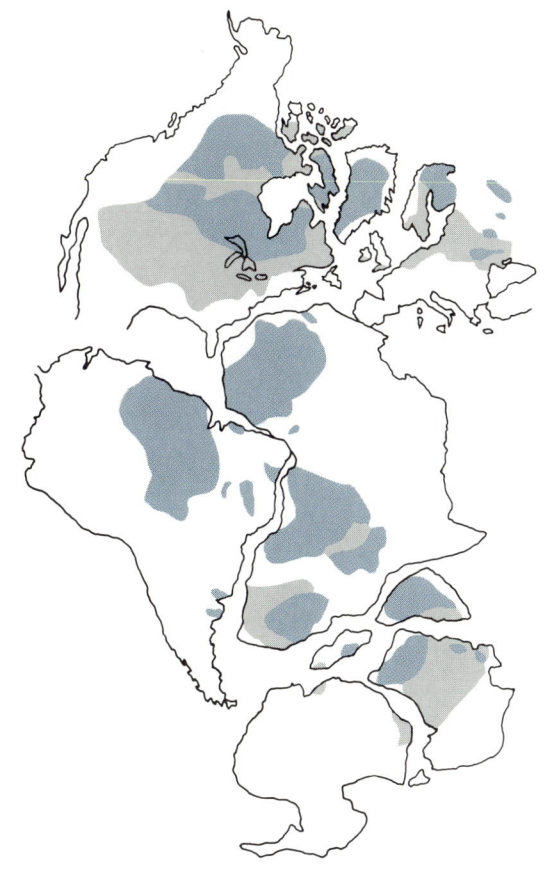

FIGURE 25-6
Continents reassembled as they may have been prior to the start of continental drift. Oldest shield rocks (older than -1.7 b.y.) shown by dark pattern; rocks ranging from -0.8 to -1.7 b.y., by light pattern. (Redrawn from a map by P. M. Hurley and J. R. Rand, 1969, *Science*, Vol. 164, p. 1237, Figure 8. Copyright 1969 by the American Association for the Advancement of Science.)

25.10 A GREAT GLACIATION

Supporters of a united supercontinent put forth a most interesting line of evidence that the continents were once united. This evidence consists

FIGURE 25-7

Reconstruction of a mammallike reptile, of late Permian time. These animals roamed widely over Gondwana. (Painting by John C. Germann, American Museum of Natural History.)

of finding the markings and deposits of an ancient ice sheet on all parts of what are now the fragments of Gondwana. Rock surfaces show scratches and grooves unmistakably made by ice abrasion (Figure 25-8). Lithified glacial materials, known as tillites, are also found overlying the abraded rock surfaces.

Geologists agree that these evidences point to a major glaciation and that it occurred during the late Carboniferous Period, some 300 million years ago. The glacial period may also have extended into the Permian Period, which followed. The question they argued was this: did the glaciation consist of individual ice sheets, each on its own isolated continent, or was there a single great ice sheet on Gondwana?

Support of a united continent with a single great ice sheet has come through the application of paleomagnetism, explained in Chapter 24. The locked magnetism in rocks of Carboniferous and Permian ages allows the position of the magnetic poles to be calculated for those periods. The magnetic poles are believed to have always coincided rather closely with the geographic poles of earth rotation. When the position of one pole is plotted for the major fragments of Gondwana, joined into a single continent, they converge on a single global point for a given point in geologic time. Moreover, the plotted pole point moves through geologic time across Gondwana, as shown in Figure 25-8. This map of Gondwana also shows the distribution of features of glacial action with respect to the pole path. The inference is strong that a single continent was centered over a polar region in late Paleozoic time.

25.11
THE MESOZOIC ERA

The Mesozoic Era began with the Triassic Period, a time when the reptiles began their rise to dominance as land animals. Among the true reptiles to appear at this time were turtles, lizards, snakes, and two groups of marine reptiles now extinct: the plesiosaurs and the ichthyosaurs (Figure 25-9). However, the reptiles destined for dominance on the lands were the ruling reptiles: dinosaurs, soaring reptiles, and crocodiles.

By the Jurassic Period, and enduring through the Cretaceous Period, dinosaurs were the largest land animals. Some were carnivores, such as *Tyrannosaurus* (Figure 25-10). Others were herbivores, such as *Brontosaurus* (Figure 25-11). There were also remarkable gliding and soaring reptiles, the pterosaurs. The earliest ancestors of the birds appeared in the Jurassic Period. Mammals first appeared in Jurassic time, but they were small creatures and did not evolve into dominance until the Cenozoic Era;

During the Mesozoic Era plants were also undergoing a major evolutionary development. Cycads and ginkgoes dominated the forests early in the Mesozoic Era, but by the Cretaceous Period there was a sudden emergence of the angiosperms, or flowering plants with covered seeds, that were to become the dominant land plants of the succeeding era.

Perhaps the most striking event affecting the living world at the close of the Mesozoic Era was the wholesale extinction of the ruling reptiles.

FIGURE 25-8

In this hypothetical restoration of Gondwana, the pole path crosses several continents, passing across the central region of Carboniferous glaciation. (Data of Du Toit and others. From A. N. Strahler, 1971, *The Earth Sciences,* 2nd ed., Harper & Row, New York.)

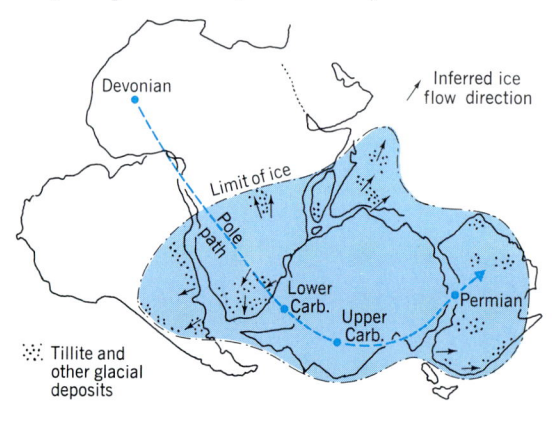

25.11 THE MESOZOIC ERA

FIGURE 25-9
Reconstruction of plesiosaurs and ichthyosaurs. (From a painting by C. R. Knight, American Museum of Natural History.)

FIGURE 25-10
Tyrannosaurus (*right*) confronting *Triceratops* (*left*). (Restoration of a late Cretaceous landscape by C. R. Knight, Field Museum of Natural History, Chicago.)

FIGURE 25-11
Reconstruction of *Brontosaurus,* a herbivorous dinosaur of Jurassic age. Its overall length reached about 20 m) (From a painting by C. R. Knight, Field Museum of Natural History, Chicago.)

All of the dinosaurs vanished, as did the great marine reptiles and soaring reptiles. Yet the primitive mammals made a successful transition into the Cenozoic Era, as did the flowering plants.

25.12
CONTINENTAL DRIFT— THE BREAKUP OF PANGAEA

Leader among the proponents of a united supercontinent was Alfred Wegener, a German scientist. In the 1920s Wegener published his hypothesis of Pangaea and its breakup into fragments, which drifted apart to become the continents as we know them today. Among English-speaking scientists Wegener's hypothesis became known as *continental drift*. Although there were a few strong supporters of his hypothesis on both sides of the Atlantic, most geologists strongly opposed the concept. Perhaps the major reason for disbelief was that Wegener visualized the continents as "floating" rafts of felsic crust drifting through a "sea" of denser mafic (basaltic) rock. For very good reasons, this mechanism was judged physically unsound. However, when the new geologic revolution later demonstrated that sea-floor spreading is actually taking place today, the separation of continents became an inescapable consequence. Of course, we know now that both the crust and upper mantle move as a solid unit within a given lithospheric plate; the continents simply ride along on the top of the plates.

Late in the Mesozoic Era, Pangaea began to break apart through the separation of several lithospheric plates. Figure 25–12 reconstructs the breakup of Pangaea in terms of plate tectonics. Continental separation may have begun along the western northern margin of Africa in Midtriassic time, about −200 m.y. South America was finally separated from southern Africa in the Cretaceous Period, about −130 m.y. As the Americas drew away from Africa and Europe, new oceanic crust was formed by rise of mantle rock in the midoceanic-ridge axis. Thus the entire Atlantic Ocean crust has formed since about early Cretaceous time and cannot be much older than about −130–140 m.y. Similarly, separation of Antarctica, Australia, and peninsular India from eastern Africa is depicted as having taken place to the accompaniment of crustal spreading along the midoceanic ridge in the Indian Ocean.

25.13
THE CENOZOIC ERA

The Cenozoic Era, or "Era of Recent Life," consists of only 65 m.y. of geologic time—less than the 70 m.y. span of the entire Cretaceous Period that preceded it. As shown in Table 25.1, the Cenozoic Era is subdivided into seven *epochs* of geologic time.

Because of the phenomenal rise of mammals to ascendency as the principal land animals, the Cenozoic Era is often called the "Age of Mammals" (Figure 25–13). A great variety of grazing mammals evolved to occupy extensive grassland plains, and with them a large variety of carnivores.

As the lithospheric plates of North America and South America moved westward relative to the Pacific plate, subduction off their western margins generated severe mountain-making activity in the belt of the Cordilleran Ranges and the Andes. At the same time, the African plate and the shield area of India moved northward against southern Eurasia, causing subduction and severe tectonic activity along the Eurasian-Melanesian belt. The great mountain ranges of southern Asia, including the Himalayas, originated in this way. These events are shown in Figures 25–12*D* and *E*.

Humans, as a genus of mammals, arrived on the scene late in the Cenozoic Era. Primitive, apelike animals called hominoids seem to have developed late in the Oligocene Epoch (−25–20 m.y.). Sometime in the ensuing Miocene Epoch, the hominoids (including the gibbons and apes) gave rise to hominids, considered human in distinction to the apes. The oldest known of fossil humans is the genus *Ramapithecus,* who flourished in late Miocene time; this genus was followed by *Australopithecus,* identified in Africa fossils as having lived as far back as −4–5 m.y., in late Pliocene time. *Homo,* the third and final human genus, appeared in the early part of the Pleistocene Epoch, roughly −700,000 years.

25.14
THE PLEISTOCENE EPOCH

Although the Pleistocene Epoch takes in only about the last one million years of geologic time—a trivially small part of the record of life on earth—it looms large in the development and environment of *Homo sapiens*. One single great geologic event dominated the environment of evolving modern humans in Europe: the great *glaciation* or *Ice Age*. On at least four separate occasions within the latter half of the Pleistocene Epoch, vast accumulations of glacial ice formed over North America and northern Europe, spreading south to marginal positions well within what is now the central United States and central Europe (Figures 25–14 and 25–15).

Not only did these ice advances have a pro-

25.14 THE PLEISTOCENE EPOCH

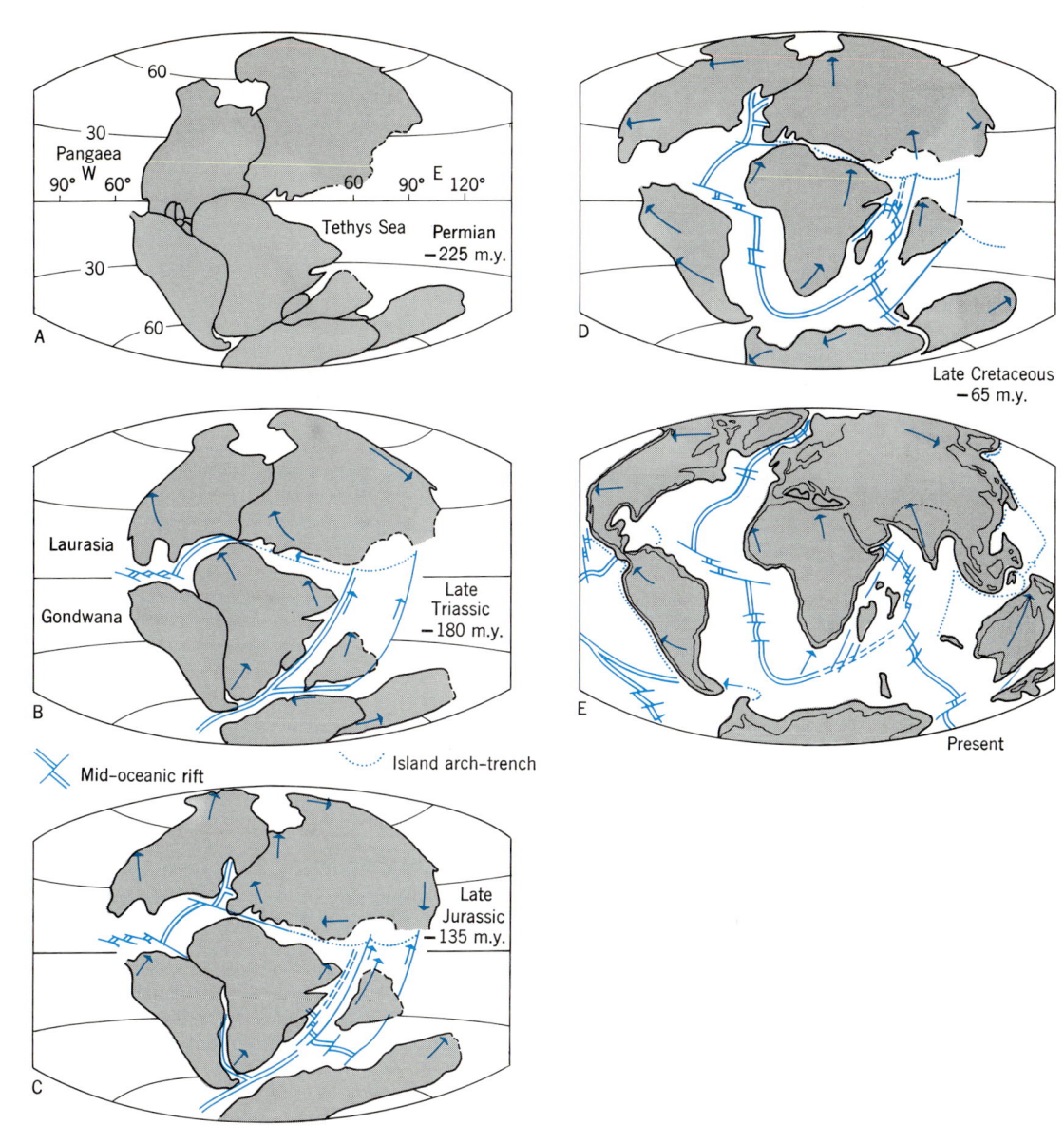

FIGURE 25-12

The breakup of Pangaea is shown in five stages. Inferred motion of lithospheric plates indicated by arrows. (Redrawn and simplified from maps by R. S. Dietz and J. C. Holden, 1970, *Jour. Geophys. Research*, Vol. 75, pp. 4943–4951, Figures 2–6. Copyrighted by American Geophysical Union.)

found effect on the terrain over which they spread; also, the withdrawals of seawater to form the ice sheets caused worldwide lowerings of sea level on the order of 100 m, exposing broad areas of continental shelves. Severe arctic climates near the ice border placed strong environmental stresses on all animals of Pleistocene time, including humans (Figure 25-16). Climate changes were also global in scope, shifting the boundaries between dry and wet climates in the tropical zones. Another major consequence of the growth of great ice sheets was to depress the crust hundreds of meters under the heavy load of ice. Following the final disappearance of the ice, these depressed crustal areas, located in the Hudson Bay and Baltic Sea regions, have been steadily rising.

The final epoch of the Cenozoic Era is called the Holocene Epoch. Its duration has been only some 10,000–20,000 years, which is the time elapsed since disappearance of the last ice sheets. During this time global climate has undergone fluctuations from periods warmer than today to periods substantially colder than today.

FIGURE 25-13
Restoration of a landscape in Nebraska during late Miocene time. *Left:* Short-legged rhinoceros. *Right:* Four-tusked mastodon. (Painting by C. R. Knight, Field Museum of Natural History, Chicago.)

25.15 CAUSES OF CONTINENTAL GLACIATION

The causes of glaciation remain uncertain despite all efforts of modern science to find a satisfactory explanation. Many hypotheses have been put forward and debated. The subject is extremely complex, and we can only touch the topic lightly.

We know that glaciation has occurred a number of times in the geologic past, as in the case of the Carboniferous-Permian glaciation described earlier in this chapter. Any theory of glaciation must therefore account for occasional and seemingly sporadic repetitions of glaciation throughout all recorded geologic time.

A general requirement of glaciation is a lowering of the earth's average atmospheric temperature along with sustained or increased levels of precipitation. It is well established by worldwide evidence that during the Pleistocene Epoch the snow line (elevation above which snowbanks remain throughout the year) was lowered in elevation by 900–1200 m in middle and high latitudes. This worldwide phenomenon clearly indicates a generally colder climate for the earth as a whole at times of glaciation. This conclusion is strongly reinforced by the data of deep-sea sediment cores. A reduced average temperature would, in general, reduce rates of wastage of snow at those places where snow could accumulate in large quantities.

Another requirement of ice-sheet growth is that there be present an elevated landmass—a plateau or mountain range—favorably situated to receive snowfall. Low-lying continental plains would not be likely to accumulate enough snowfall to initiate ice-sheet growth, even if the cli-

FIGURE 25-14
Pleistocene ice sheets of North America at their maximum spread reached as far south as the present Ohio and Missouri Rivers. (After R. F. Flint.)

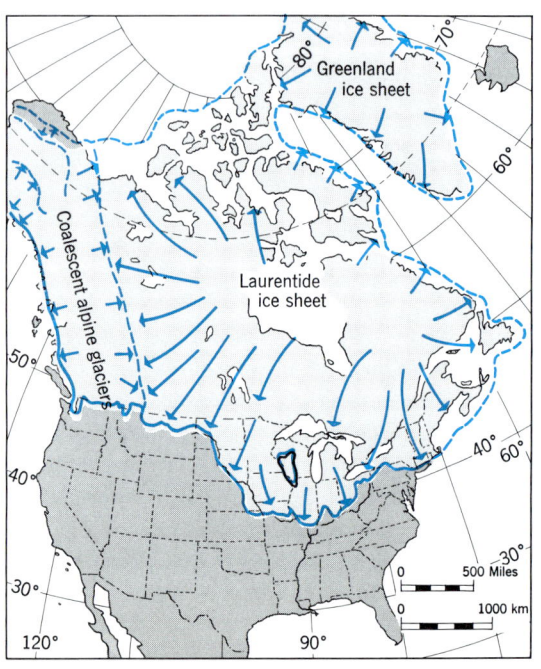

25.15 CAUSES OF CONTINENTAL GLACIATION

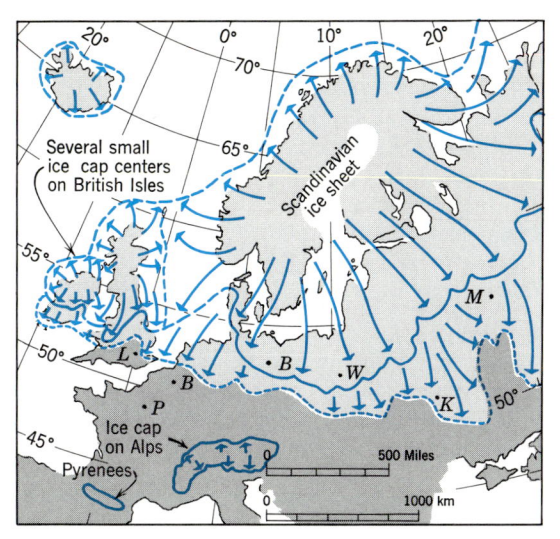

FIGURE 25-15

The Scandinavian ice sheet dominated northern Europe during the Pleistocene glaciations. Solid line shows limits of ice in the last glacial stage; dotted line on land shows maximum extent at any time. (After R. F. Flint.)

mate were sufficiently cold. For this reason some large arctic areas of Siberia were never glaciated. Now, we know that Pliocene and early Pleistocene times saw great tectonic and volcanic activity, resulting in growth of mountain ranges and creating topographic conditions favorable to the growth of Pleistocene ice sheets.

A change in rate of output of solar energy has been suggested as the basis for several hypotheses of glaciation. Although minor fluctuations are observed in incoming radiation, we do not as yet have evidence of any long-range trend of change in its value. There is no reason to doubt the constancy of the sun's energy output for the span of geologic time in which glaciations have left a record. But we can nevertheless speculate that reductions of the sun's energy output have occurred and that the planetary temperature was correspondingly lowered, bringing on the growth of ice sheets.

The hypothesis of reduced solar-energy output, combined with the favorable topographic effect of uplift of mountain and plateau areas, represents the *solar-topographic hypothesis*.

A different approach has been to invoke systematic changes in tilt of the earth's axis and form of the earth's orbit. These cyclic changes are known through astronomical observations. Astronomical cycles have been calculated to show that the solar energy received during summer in middle and high northerly latitudes has varied substantially. Peak values occurred every 20,000 years, but with unusually high peaks coming about 80,000 years apart. The *astronomical hypothesis* of glaciation holds that glaciations were set off by cyclic occurrence of lower-than-average absorption of solar radiation at the earth's surface in high latitudes.

A widely supported hypothesis, relatively simple in concept, attributes worldwide temperature drop and ice-sheet growth to a decrease in the carbon dioxide content of the atmosphere. The average content of carbon dioxide is very small, about 0.033 percent by volume. We noted in Chapter 21 that carbon dioxide is an important gas in causing the greenhouse effect in which longwave terrestrial radiation is absorbed

FIGURE 25-16

Restorations of the woolly mammoth (*left* and *center*) and woolly rhinoceros (*far right*) in a European landscape of Pleistocene time. (From a painting by C. R. Knight, Field Museum of Natural History, Chicago.)

in the lower atmosphere. In this way the average air temperature is considerably increased over what it would be without this gas. Estimates have been made to show that if the carbon dioxide content of the atmosphere were reduced to half of the existing quantity, the earth's average surface temperature would drop by about 4 C°. This drop is thought to be enough to bring on the growth of ice sheets under favorable topographic conditions.

Also involving change in atmospheric composition is the hypothesis of glaciation brought about by increase in quantity of volcanic dust in the upper atmosphere. Should there be an episode of unusually great volcanic activity, the greatly increased atmospheric dust would reflect back into space a greater part of the incoming solar radiation. The effect would be to reduce the quantity of solar energy received at the earth's surface and to lower the average atmospheric temperature. However, periods of exceptional volcanic activity have not been shown to be correlated with periods of glaciation.

Another prominent hypothesis requiring atmospheric changes is based on the suggestion that floating sea ice of the Arctic Ocean has at one time or another disappeared, leaving the water surface exposed to rapid evaporation. Moisture thus provided to the atmosphere was available to nourish ice sheets and set off glaciations. When the sea ice cover was restored, the ice sheets wasted away.

SUMMARY

Stratigraphy. Earth history was first unraveled through application of stratigraphy, using fossils and the principle of superposition to place strata in relative order of age. Fossils show slow changes through the evolutionary process from older to younger strata.

Radiometric ages. Dating of rocks in terms of absolute age in years was made possible by geochronometry, in which ratios of daughter-to-parent radioactive isotopes are measured. Uranium-to-lead ratios are particularly important in dating the most ancient rocks.

Precambrian time. Precambrian time, older than about −600 m.y., witnessed the origin of life in shallow seas and the slow evolution of primitive life forms.

Paleozoic Era. Complex invertebrate life forms were present at the opening of the Paleozoic Era, about −570 m.y. Throughout the Cambrian and Ordovician Periods, marine life evolved rapidly. The first terrestrial plants appeared in the Silurian Period; air-breathing animals evolved in the Carboniferous Period, along with great coal-producing forests. Worldwide glaciation occurred in the Permian Period.

Mesozoic Era. Great reptiles, dinosaurs, and many other groups rose to dominance in the Triassic and Jurassic Periods of the Mesozoic Era, only to become extinct at the close of the era. A single great continent, Pangaea, existed well into the Mesozoic Era, but began to break up into fragments in late Jurassic time. Continental drift through Cretaceous time opened up the Atlantic Ocean basin, separating the Americas from Europe and Africa.

Cenozoic Era. The Cenozoic Era, starting at −65 m.y., saw the rise to dominance of mammals. Continental drift continued, causing wide separation of the continents. Humans evolved along with other hominoids late in the Cenozoic Era, with modern humans appearing early in the Pleistocene Epoch. A great glaciation occurred in the latter half of the Pleistocene Epoch, covering much of Europe and North America under ice sheets.

TERMS AND CONCEPTS

historical geology
stratigraphy
paleontology
principle of superposition
principle of continuity
fossils
organic evolution
uniformitarianism
catastrophism
Precambrian time
geologic eras
Paleozoic Era
Mesozoic Era
Cenozoic Era
orogeny
geochronometry
radiometric ages
Pangaea
Laurasia
Gondwana
continental drift
epoch of geologic time
glaciation
Ice Age
solar-topographic hypothesis
astronomical hypothesis

QUESTIONS

1. Explain how the principles of superposition and continuity can be applied to determining relative ages of strata. How are fossils used in establishing the ages of strata?

2. What is the geologic doctrine of uniformitarianism? Give an example. What primitive concept did it replace?

3. How old is Precambrian time? What life forms are known to have existed in Precambrian time?

4. Name the three eras younger in age than Precambrian time. About how long did each era endure? How many periods has each of the eras? What is the significance of an orogeny in the geologic time scale?

5. Describe the method of determining the age of a rock by means of geochronometry. How is the principle of spontaneous decay of radioisotopes used in age calculation? Which decay series are most commonly used to date very ancient rocks?

6. Give a brief sketch of evolution of life forms in the Paleozoic Era. Why was life of the Cambrian and Ordovician Periods limited to the oceans? When did life make a transition to the terrestrial environment? What was significant about life of the Carboniferous Period? What was the status of the continents as a unit in late Paleozoic time?

7. Describe Pangaea and its two major divisions. What forms of geologic evidence support the concept of a united subcontinent of Gondwana? What evidence favors a single great ice sheet during the Carboniferous-Permian glaciation? What role does paleomagnetism play in this interpretation?

8. Give a brief sketch of changing life forms throughout the Mesozoic Era. Which animal forms dominated the lands? What happened to these animals at the close of the Mesozoic Era?

9. Describe the breakup of Pangaea. When did it begin? Which continents first began to separate? What ocean basin was formed by this separation? How did the northward movement of the Indian subcontinent result in tectonic activity in southern Asia?

10. How long did the Cenozoic Era last? Into what units of time is it subdivided? What life forms dominated during the Cenozoic Era? When did humans evolve as an animal species, and from what contemporary animals? When did the genus *Homo* first appear on the scene?

11. Describe glaciations of the Pleistocene Epoch. How far south did the Pleistocene ice sheets spread over North America and Europe? Review the various current hypotheses of the cause of continental glaciation. Which explanation do you find most convincing?

CHAPTER 26
GEOMORPHIC PROCESSES AND LANDSCAPE EVOLUTION

26.1 GEOMORPHIC PROCESSES

The study of the origin and evolution of the relief features of the landscape, or *landforms,* is a branch of geology known as *geomorphology.* In the study of landforms the various landscape features are sorted out according to processes of origin—the *geomorphic processes.*

In the broadest sense, all landforms fall into two great classes, the products of two great classes of geomorphic processes. The *internal processes* are tectonic activity, which dislocates the crust, and volcanism, which constructs new landforms by extrusions of magma. New relief features created by tectonic and volcanic activity are designated as *initial landforms* (Figure 26–1). The initial landforms are acted upon by the *external processes,* powered by solar energy acting through the earth's atmosphere and hydrosphere. The external processes create the *sequential landforms,* which are carved from initial landforms or built of sediment derived from the initial landforms.

In summarizing this introduction to geomorphology, it might be said that all landscapes of the continents reflect the existing stage in an unending conflict between internal and external processes. Where internal processes have been recently active, through tectonic activity and volcanism, there exist rugged alpine mountain and volcanic chains, or high plateaus. Where external processes have been given the opportunity to operate with little disturbance for vast spans of time, the continental surfaces have been reduced to low plains, such as we find in the shield areas.

The external geomorphic processes include rock weathering, spontaneous mass wasting under gravity, and the activity of the fluid agents—running water, waves and currents, glaciers, and winds. We shall touch on each of these activities briefly, describing some of the typical landforms produced by each.

26.2 BEDROCK REGOLITH AND SOIL

Examination of a freshly cut cliff, such as that in a new highway excavation or quarry wall, may

FIGURE 26–1
Initial and sequential landforms.

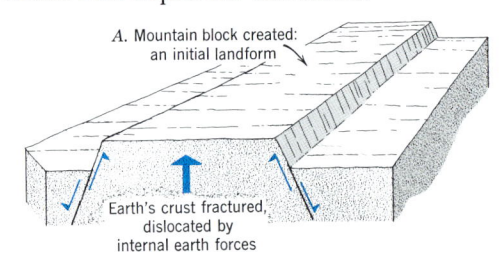

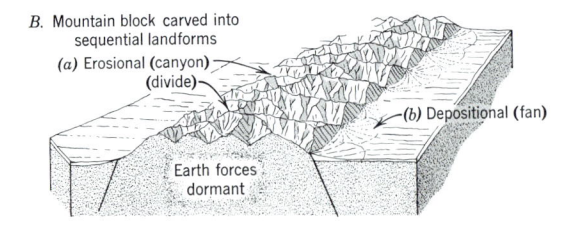

reveal several kinds of earth materials (Figure 26–2). Solid hard rock that is still in place and relatively unchanged is called *bedrock*. It grades upward into a zone where the rock is partly decayed and has disintegrated into clay and sand particles. This material is called the *regolith*. At the top is a layer of true *soil,* often called "topsoil" by farmers and gardeners. It is usually less than 1 m thick and may be relatively dark in color in comparison to the regolith below. A distinctive feature of the soil is the presence of a layered structure formed of two or three horizons. Over the soil may be a protective layer of grass, trees, or other vegetation.

26.3 NATURE OF WEATHERING

The weathering processes may be thought of as leading to the preparation of parent matter of the soil and, from the geologic standpoint, as the preparation of sediment for transportation and eventual accumulation as sedimentary rock. One aspect of weathering is the breaking up of hard bedrock, occurring in large blocks, into particles ranging down through the various size grades to clay colloids and ions. As breakdown occurs, the total surface area of the particles in a given bulk volume is enormously increased, facilitating complex chemical changes.

Another aspect of weathering, explained in Chapter 23, is the change in chemical composition of the rock-forming minerals, through reaction with acids and water, to yield new minerals that will remain stable indefinitely without further change under the conditions of temperature, pressure, and moisture prevailing at the earth's surface.

A third aspect of weathering is the continual agitation of the soil and weathered overburden as soil-moisture content increases and decreases seasonally and as soil temperatures rise and fall daily and seasonally. Drying and wetting, freezing and thawing, growth and decay of plant roots, and the burrowing and trampling of the soil by animals continually agitate the soil. Such disturbances affect the soil rhythmically long after the mineral matter has been reduced to minute particles and the principal chemical changes have largely occurred.

26.4 PHYSICAL WEATHERING

Consider first *physical weathering,* in which mechanical stresses act on rock, causing disintegration. These processes constitute the primary breakdown of bedrock into fragments whose mineral surfaces are in turn exposed to chemical weathering.

Most bedrock is so fractured by systems of joints that it is rare to find flawless bodies of rock (monoliths) more than a few meters across. Most joints occur in parallel sets, and there are often two or more sets intersecting at large angle. Consequently, most bedrock is already broken into blocks from a few centimeters to a few meters across. When stresses are exerted upon jointed rock, the rock comes apart readily along the joint planes.

In climates of the middle and high latitudes and at high altitudes, alternate freezing and melting of water in the soil and rock provide a powerful mechanism of rock breakup. Soil water and water that has penetrated the joint planes and other openings of the rock are transformed into ice crystals of needlelike form. Growing masses of crystals exert great pressures on the confining rock walls, causing joint blocks and layers to be heaved up and pried free of the parent mass.

The results of disintegration of bedrock by freezing water, a process commonly referred to simply as *frost action,* are most conspicuous above the timberline in high mountains and at lower levels in arctic latitudes.

In dry climates an important agent of rock disintegration is *salt-crystal growth,* a process quite similar physically to ice-crystal growth. Such climates have long drought periods in which evaporation can occur. Water films are drawn surfaceward by capillary-film tension. Moisture is steadily evaporated, permitting dissolved salts

FIGURE 26–2
Regolith and soil overlie the bedrock. Weathering progresses downward, aided by natural rock fractures, called joints.

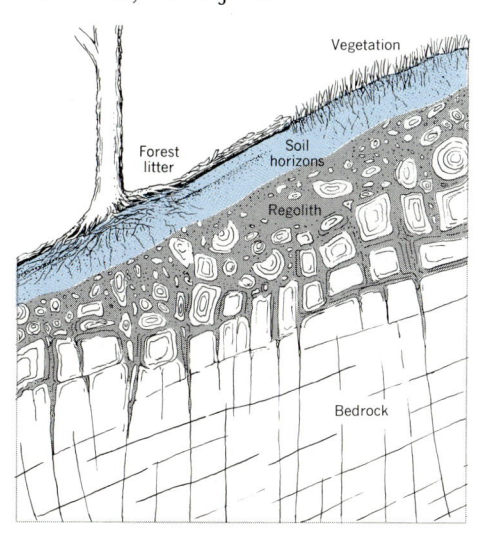

to be deposited in openings in the rock and soil. The growing salt crystals are capable of exerting powerful stresses. Even the hardest rocks (also concrete, mortar, and brick) can be reduced to a sand by continued action of the process. Niches and shallow caves near the base of a sandstone cliff are produced by this process.

Rock disintegration may occur through temperature changes alone, because most crystalline solids expand when heated and contract when cooled. Heating of rock causes expansion of the minerals and the rock may be broken. Sudden and intense heating by forest and brush fires causes severe flaking and scaling of exposed rocks. Also, we know that primitive mining methods included the building of fires on a quarry floor to cause slabs to break free. It is doubtful that the daily temperature cycle under solar heating and nightly cooling produces sufficiently great stresses to cause fresh hard rock to break apart.

Still another physical-weathering process is that of the action of growing plant roots, exerting pressure on the confining walls of soil or rock. This process is of importance in the breakup of rock already affected by other physical and chemical processes.

26.5
CHEMICAL WEATHERING

Chemical weathering was described in Chapter 23. Recall that silicate minerals in contact with atmospheric oxygen and water undergo oxidation and hydrolysis, processes that lead to the formation of the stable clay minerals and oxides of iron. Decay of igneous rock by hydrolysis and oxidation commonly results in *spheroidal weathering;* in which thin concentric shells of softened rock form within a single joint block (Figure 26-3).

Another important chemical weathering process is *carbonic acid action*. Atmospheric carbon dioxide in solution in rainwater and soil water forms a weak solution of carbonic acid, which reacts readily with calcium carbonate of limestone. Limestone surfaces commonly show the effects of this process in the form of small cups, ridges, and points (Figure 26–4). The products of carbonic acid action on limestone are calcium ions and carbonate ions; these are carried away in streams and are eventually precipitated as new carbonate sediment in lakes or in the ocean.

26.6
WEATHERING AND SOILS

All terrestrial plant and animal life depends for survival on the soil, a very shallow mineral layer over much of the continental surface. Prolonged action of weathering processes combined with organic activity of plants and animals brings the soil layer into physical and chemical equilibrium with the prevailing climatic factors of heat and moisture. Place-to-place variations in character of the soil are apparent through differences both in chemical composition and physical texture of the soil itself and in the characteristic forms of natural vegetation it bears. At the risk of oversimplification, we shall attempt only to distinguish three important soil-forming processes. Figure 26–5 is a schematic representation of soil profiles produced by these processes.

Consider first the soil-forming process of *laterization*, which operates in an environment of prevailingly warm temperatures and abundant precipitation occurring all year or in a long rainy season. This is the environment of equatorial lands and of those parts of tropical lands having a monsoon climate. High temperatures favor intense bacterial activity, so that fallen leaves and branches are rapidly oxidized and there is little organic matter in these soils.

The percolation of large volumes of rainfall through the soil causes silica (SiO_2) to be removed, a process termed *desilication* (Figure 26–5A). Soluble salts, including cations of such bases as calcium, sodium, and potassium, are largely removed. What finally remains in these tropical soils is a group of highly stable oxides and hydroxides of iron, manganese, and aluminum. These form such minerals as limonite (hydrous iron oxide) and bauxite (hydrous aluminum

FIGURE 26–3
Spheroidal weathering has produced many thin concentric shells in a basaltic igneous rock. (U.S. Geological Survey.)

26.6 WEATHERING AND SOILS 441

FIGURE 26-4
Carbonic acid action has developed deep pits in this exposed limestone. (Douglas Johnson.)

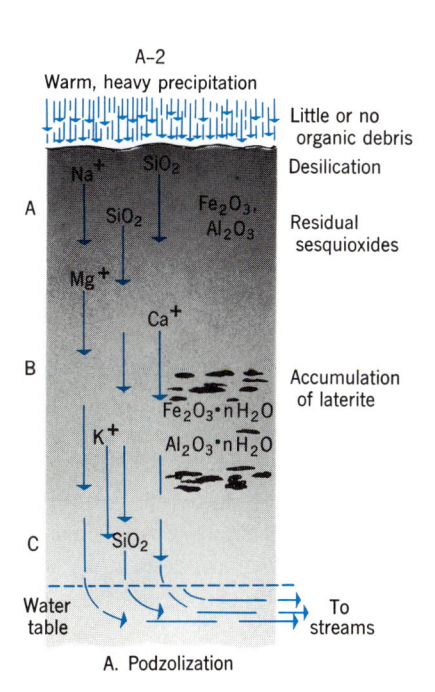

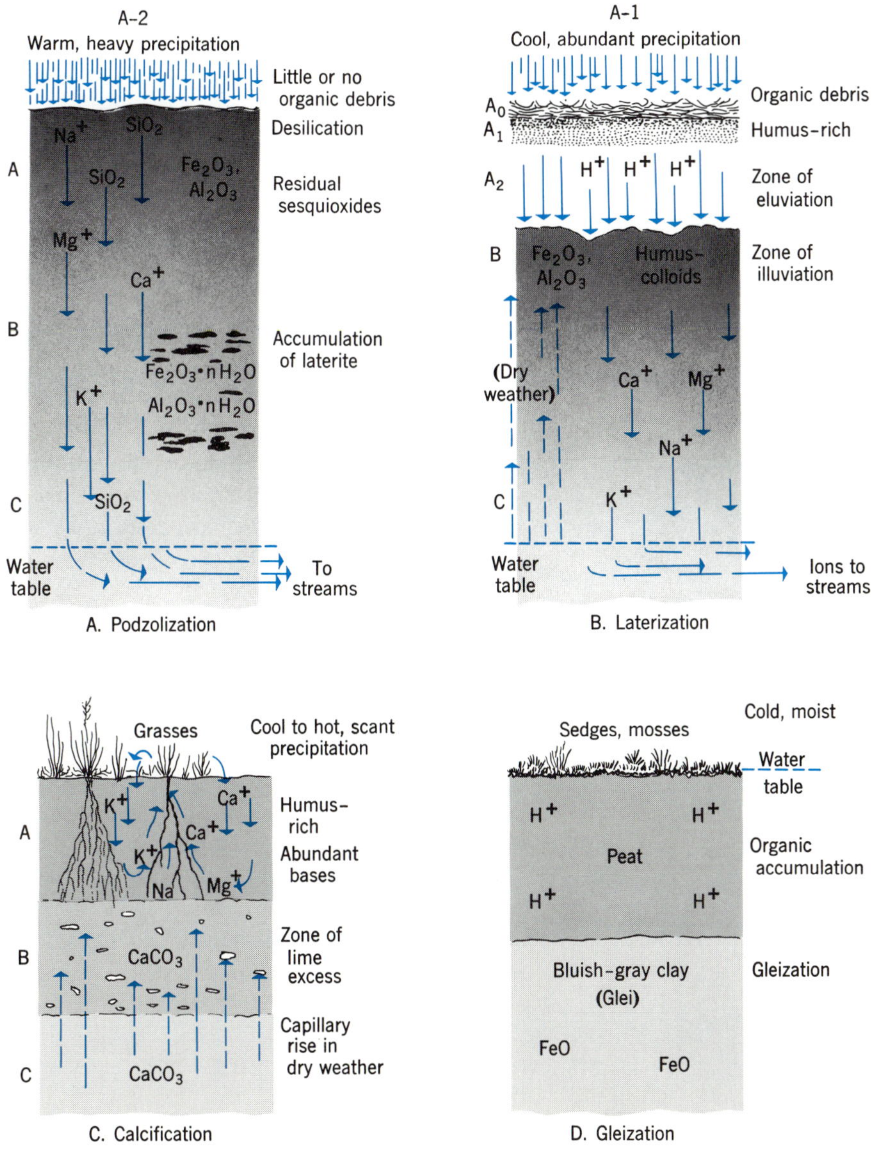

FIGURE 26-5
Schematic diagrams of three major soil-forming processes.

A-2 Warm, heavy precipitation — Little or no organic debris; Desilication; Residual sesquioxides; Accumulation of laterite; Water table; To streams.
A. Podzolization

A-1 Cool, abundant precipitation — Organic debris; Humus-rich; Zone of eluviation; Zone of illuviation; Water table; Ions to streams.
B. Laterization

Grasses — Cool to hot, scant precipitation; Humus-rich; Abundant bases; Zone of lime excess; Capillary rise in dry weather.
C. Calcification

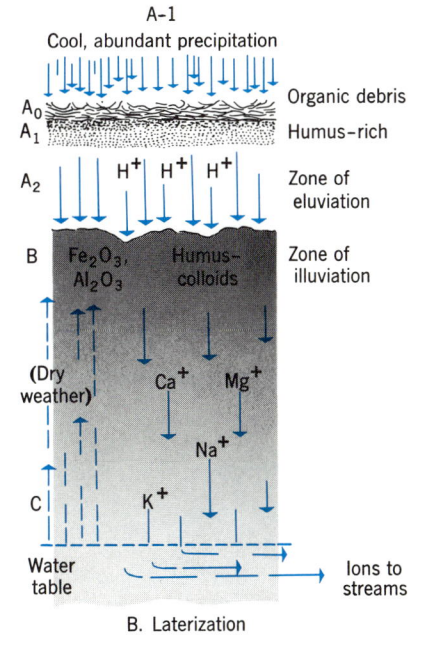

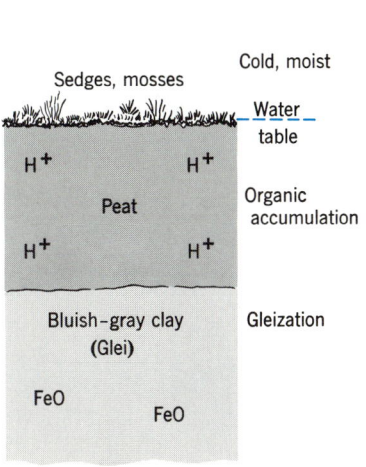

Sedges, mosses — Cold, moist; Water table; Peat; Organic accumulation; Bluish-gray clay (Glei); Gleization; FeO.
D. Gleization

oxide); these are not soluble in the soil water of the warm humid climates. Soils containing hydroxide minerals in abundance are known generally as *latosols.*

The color of latosols is typically reddish-brown or chocolate-brown; horizons are not apparent. Small, irregularly shaped nodules of hydroxides are distributed throughout the soil. The soil is favorable for the growth of a native vegetation consisting of rainforest but has little of the nutrients needed for cultivation of food crops.

In areas of alternately wet and dry (monsoon) climate, latosols exhibit a remarkable property of becoming hardened to bricklike consistency after the soil has been cut into blocks and exposed to the atmosphere. Such material, called *laterite,* has been widely used as a building material, particularly in Southeast Asia.

Latosols contain valuable mineral deposits where conditions have been favorable to the concentration of layers or lenses of bauxite, limonite, and manganite (manganese hydroxide). These minerals belong to a group termed *residual ores,* which accumulate near the surface because they are not readily dissolved by soil water.

FIGURE 26–6

A. Podzol soil profile developed on a sandy glacial deposit in Maine. *B.* Dark brown soil profile (chernozem) developed on grasslands in North Dakota. Scales are in feet. (Division of Soil Survey, U.S. Dept. of Agriculture.)

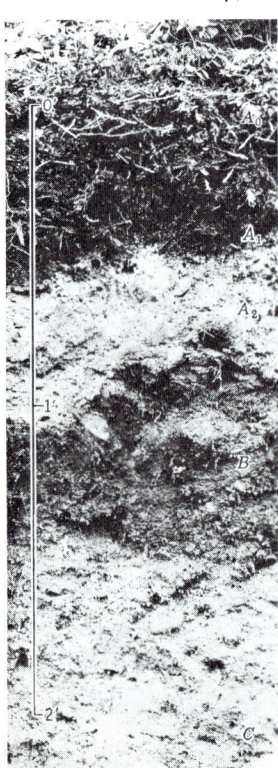

A second soil-forming process is *podzolization,* characteristic of moist climates with long cold winters. Such climates are widespread in latitudes of 45–65 degrees north in North America and Eurasia. Because of low temperatures, the production of organic matter by forest trees exceeds the rate of its destruction by soil bacteria. As a result, finely divided, dark-colored organic matter, called *humus,* accumulates in substantial quantities, forming a thick top layer. Organic acids produced in the decomposition of plant matter pass downward through the soil. Hydrogen ions of the acid solution replace the base cations, which are leached from the soil and are exported from the region as runoff in streams. Colloidal mineral and humus particles are also carried down from a thin upper layer of the soil, resulting in a characteristic ash-gray horizon, labeled A_2 in Figure 26–5B). These materials accumulate in the underlying zone forming a dense horizon designated as the B horizon. Figure 26–6A is a photograph of a typical *podzol* soil exhibiting the distinctive horizons produced under the regime of podzolization.

Podzols are low in fertility for agricultural purposes because strong leaching has removed the nutrient bases and has resulted in an acid soil. The addition of lime and fertilizers is necessary for successful crop production.

A third process, *calcification,* is characteristic of soils in regions of semiarid climate where annual evaporation on the average exceeds annual precipitation. The effects of this moisture imbalance are illustrated in Figure 26–5C. During dry periods or dry seasons, soil water rises toward the surface and is evaporated, leaving behind calcium carbonate, which forms nodules or lenses in the soil. This zone of carbonate accumulation is designated the B horizon. Figure 26–6B shows a profile of one of the major varieties of soils produced under a regime of calcification.

The most widespread type of natural vegetation in areas of calcification is grasslands. The grasses are deep-rooted and bring up to the surface the base cations they require for growth. Thus, the nutrients are recycled. Partial decay of grass roots adds substantial amounts of humus to the uppermost or A horizon, which is typically brown to black. These black soils are called *chernozems,* a word of Russian origin applied to soils of the Ukraine region. Chernozems are extraordinarily rich in nutrients needed for the cultivation of grains and are the soils of major wheat-producing regions of the world.

26.7
MASS WASTING

The *mass-wasting* processes include all forms of downslope movement of soil, regolith, or bedrock under the direct influence of gravity, but without the action of a moving fluid such as water, ice, or wind. Mass wasting represents the spontaneous yielding of earth materials when gravitational force exceeds the internal strength of the material. Therefore, mass wasting involves the sliding, rolling, and flowage of masses of soil, regolith, and bedrock to lower positions. Wherever the ground surface has a measurable inclination, or slope, with respect to the horizontal, a proportion of the acceleration of gravity is directed downslope parallel with the surface. Every particle has at least some tendency to roll or slide downhill and will do so whenever the downslope force exceeds the resisting forces of friction and cohesion that tend to bind the particle to the rest of the mass.

The forms of mass wasting include great catastrophic slides in alpine mountains, involving millions of cubic meters of rock and capable of wiping out a whole town (Figure 26–7). Small flows of saturated soil are seen commonly along the highways in early spring. But extremely slow movement of soil, imperceptible from one year to the next, also acts on almost every hillside. Another form of mass wasting, seen in high mountains, is the fall of countless blocks of bedrock broken free from cliffs by frost action. The blocks accumulate in conical piles, forming steep *talus slopes* (Figure 26–8).

FIGURE 26–7
Seen from the air, the Madison Slide forms a great dam of rubble across the Madison River Canyon. The rock mass which slid away from the mountain summit at the left was 600 m long and 300 m high. Momentum carried the tumbled rock debris over 120 m up the opposite side of the canyon. (U.S. Geological Survey.)

FIGURE 26–8
Talus cones at the base of a frost-shattered cirque headwall. Moraine Lake in the Canadian Rockies. (Ray Atkeson.)

26.8 SLOPE EROSION BY OVERLAND FLOW

We turn next to geomorphic processes described as *fluvial*, meaning that the process involves water flowing over the surfaces of the ground (overland flow) and in stream channels (channel flow). The flow of a sheet of water over the soil surface exerts a *shearing stress*, or *drag*, on the mineral grains. If this stress is sufficient to overcome the cohesive forces binding a grain to the parent mass, the grain is entrained into the flowing layer and is rolled, dragged, or carried downslope. The progressive removal of grains in this manner is described as *slope erosion*, or often *soil erosion*.

Although slope erosion by overland flow is a nearly universal process, occurring wherever rain falls on the land, it is most effective in eroding soil, regolith, or weak bedrock in areas of dry climate, where plant cover is sparse or absent. Clay and shale bedrock in arid lands is easily eroded by overland flow and gives rise to spectacular *badlands* (Figure 26–9). Cultivation of hillslopes in humid climates also leads to severe soil erosion, resembling that seen in arid lands.

FIGURE 26-9
Rapid erosion of weak clay formations has resulted in the Big Badlands of South Dakota. (Douglas Johnson.)

26.9
STREAM FLOW

A *stream* is a flow of water to lower levels within a trenchlike *channel,* confined between banks. (The word *river* is interchangeable with *stream*). Every particle of water in a stream is drawn vertically downward under the force of gravity, so that the water exerts a pressure on the channel walls proportional to the water depth. A part of the force of gravity acts in the downstream direction, parallel with the streambed, tending to cause flow of one water layer over the next lower layer in a type of motion known as *shear.* The fluid may be thought of as having almost infinitely thin layers of water molecules, each layer slipping over the layer below, much as playing cards slip over one another when the deck of cards is pushed along a table top. The layer immediately in contact with the solid bed does not slip, but each higher layer slips over the one below, so that the forward speed, or velocity V, increases from the bed upward into the stream.

Curved lines in Figure 26-10 show the successive positions that would be occupied by water particles starting out together on a vertical straight line. We see that velocity increases very rapidly from the bed upward, then increases less rapidly, so that the maximum velocity is found at a point close to the stream surface. Similarly, on the stream surface, velocity increases from zero at the banks to a maximum near the centerline.

These statements imply that each particle of water moves downstream in a direct simple path. This would be the case in true *laminar flow,* or *streamline flow,* which occurs in fluids when their motion is very slow. In most forms of runoff, including most overland flow and nearly all stream-channel flow, the water particles describe highly irregular paths of travel, resembling a tortuous corkscrew motion including sideways and vertical movements. Such motion, described as *turbulent flow,* consists of innumerable eddies of various sizes and intensities continually forming and dissolving. The velocity V, referred to above, and the simple paths of flow shown by the arrows in Figure 26-10 are merely the average velocities and average paths of the particles at given levels in the stream.

Turbulent flow in fluids is of great importance in the processes of erosion by running water, waves and tidal currents, and wind, because the transportation of fine particles held in *suspension* in the fluid depends on the upward currents in turbulence to support the particles. Without turbulence, particles could only be rolled or dragged on the bed or lifted a short distance above it.

Because of the differences in average flow velocity from point to point in a stream, a single statement of velocity is needed to apply to the stream as a whole. This is the *mean velocity* and is approximately equivalent to six-tenths of the maximum velocity.

The quantity of water that flows through a stream channel in a given period of time, the

FIGURE 26-10
Velocity of flow within a stream.

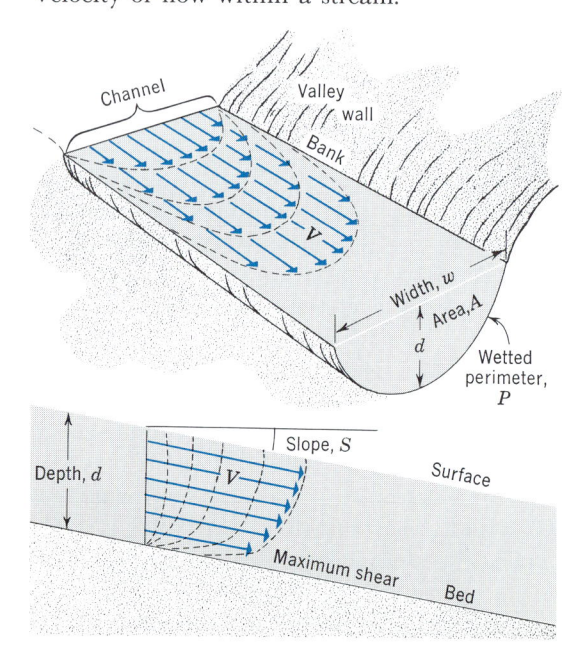

discharge, is a most important characteristic of the flow from the standpoint of describing the magnitude of the stream. Discharge, Q, is defined as the volume of water passing through a cross section in a short unit of time. Units are cubic meters per second. Discharge is computed by multiplying the mean velocity V times the cross-sectional area A in the formula $Q = AV$.

If a long stream channel is to conduct a given discharge through its entire course, the discharge must be constant at all cross sections, otherwise water would accumulate by ponding. It follows that the product of cross-sectional area and mean velocity must be constant in all cross sections along the stream (Figure 26–11). Where the stream becomes narrower, with reduced cross section, it must have a proportional increase of velocity. If the velocity should increase because of a steepened gradient, the cross-sectional area of the stream will become smaller. The same river that flows slowly in a broad channel on a low gradient will flow swiftly in a narrow stream when it enters a gorge of steep gradient. The equation $Q = AV$ is known as the *equation of continuity* of flow, because a stream that is neither gaining nor losing water at any point on its course must keep the discharge constant by appropriate combinations of cross-sectional area and velocity.

26.10
DRAINAGE SYSTEMS

Stream channels are organized into *drainage systems* evolved over long spans of time to drain both water and mineral matter from a given area of land. From the single stream mouth, where all discharge exits, the system perimeter is located by following out the natural *drainage divide*, a continuous line with respect to which

FIGURE 26–11
Schematic diagram of relationships among cross-sectional area, mean velocity, and gradient. (From THE EARTH SCIENCES, 2nd Edition by Arthur N. Strahler. Copyright © 1963, 1971 by Arthur N. Strahler. By permission of Harper & Row, Publishers, Inc.)

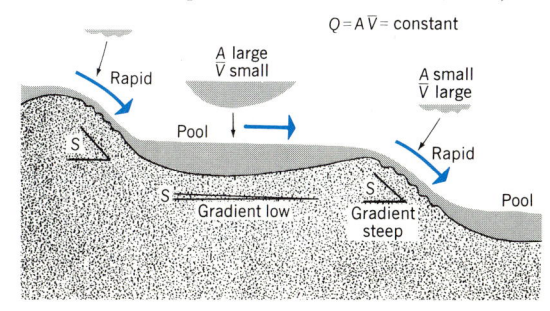

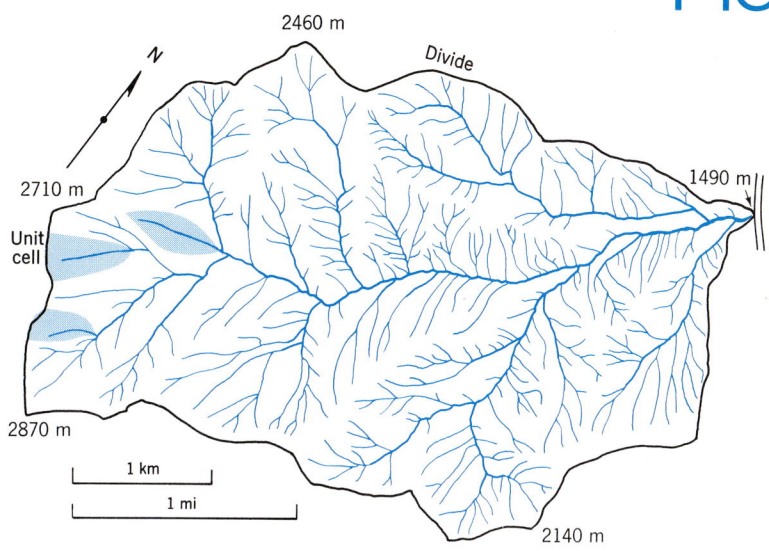

FIGURE 26–12
Channel network of a drainage basin in Utah. Elevations are given in meters. (Data of U.S. Geological Survey and Mark A. Melton.)

overland flow is directed either toward the system or away from it (Figure 26–12). The drainage divide outlines a natural *drainage basin*, which is commonly of elliptical or pear-shaped outline.

Dominating the drainage basin is a treelike, branched system of stream channels. Overland flow, originating at the divide of the drainage basin and along the many subsidiary divides between channel branches, makes its way by the most direct downslope trajectory to the nearest channel. A given element of channel receives its sustenance both from the channel lying upstream and from the adjacent land surfaces.

26.11
GROUND WATER

Water that sinks into the land surface during times of heavy precipitation and snowmelt percolates through the soil and underlying bedrock, moving downward under the force of gravity. As Figure 26–13 shows, the percolating water eventually reaches the *saturated zone*, in which all pore spaces in the rock are filled with water. This water body is called the *ground water*. The upper surface of the ground water body is called the *water table*. In a well, water will stand at the level of the water table. Above the water table is the *unsaturated zone* in which water is held in the form of capillary films, but does not occupy all of the pore spaces.

Ground water moves very slowly in deep paths, as shown in Figure 26–14. The water emerges in springs in stream channels, or along the shores of lakes or the ocean. In this way

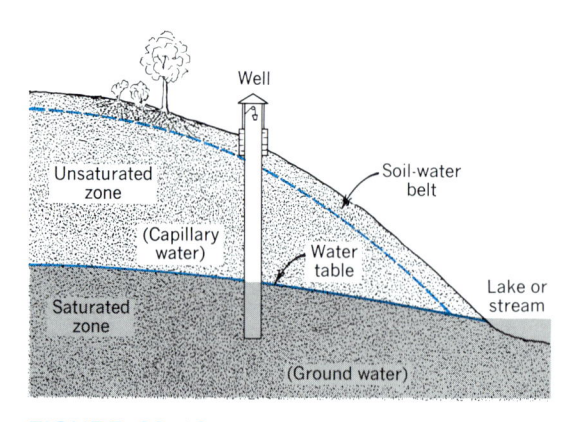

446   26 GEOMORPHIC PROCESSES AND LANDSCAPE EVOLUTION

FIGURE 26-13
Zones of subsurface water.

ground water eventually becomes surface water and is an important source of stream flow. In regions of humid climate, the outflow of ground water into stream channels sustains stream flow in the summer season in dry periods between rains.

26.12
GEOLOGIC WORK OF STREAMS

Streams perform three closely interrelated forms of geologic work: erosion, transportation, and deposition. *Stream erosion* is the progressive removal of mineral particles from the surfaces of a stream channel. *Stream transportation* is the movement of eroded particles in chemical solution, in turbulent suspension, or by rolling and dragging along the bed. Particles in stream transport constitute the *stream load*. *Stream deposition* consists of the accumulation of any transported particles on the streambed, on the adjoining floodplain, or on the floor of a body of standing water into which the stream empties. These phases of geologic work cannot be separated from each other, because where erosion occurs there must be at least some transportation, and eventually the transported particles must come to rest.

FIGURE 26-14
Theoretical paths of ground water movement under divides and valleys. (After M. K. Hubbert.)

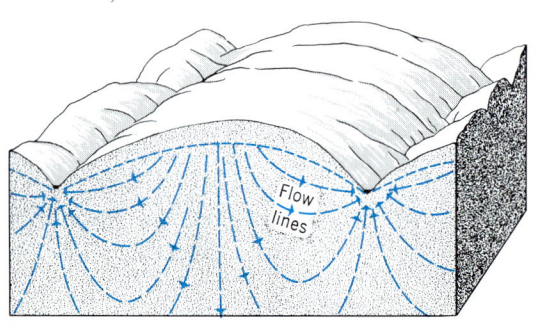

The nature of stream erosion depends on the materials of which the channel is composed and the means of erosion available to the stream. One simple form of erosion is by *hydraulic action*, the effect of pressure and shearing force of flowing water exerted upon grains projecting from the bed and banks. In flood stage the swift, highly turbulent flow on the outside of stream bends undermines the channel wall, causing masses of sand, gravel, silt, or clay to slump and slide into the channel, an activity described as *bank caving*. Huge volumes of sediment are thus incorporated into the stream flow in times of high stage, and the channel may shift laterally by many yards in a single flood (Figure 26-15).

Mechanical wear, termed *abrasion*, occurs through the impact of rock particles carried in the current striking against the exposed bedrock of the channel surfaces. Small particles are further reduced by crushing and grinding when caught between larger cobbles and boulders. Chemical reactions between ions, carried in solution in stream water, and the exposed mineral surfaces result in another form of erosion, *corrosion*, which is essentially the same process as chemical rock weathering.

Three forms of stream transportation can be distinguished. First, chemical reactions yield ions that may travel downstream indefinitely. Such matter, constituting the *dissolved solids*, does not

FIGURE 26-15
A river in flood eroded this huge trench at Cavendish, Vermont, in November 1927. An area 1.6 km wide and 4.8 km long, once occupied by eight farms, was cut away by the flood waters. Bank caving was rapid because the material consisted of sand and gravel which offered little resistance. (Wide World Photos.)

appreciably affect the mechanical behavior of the stream.

Second, particles of clay, silt, and sometimes fine sand are carried in suspension, a form of transport in which the upward currents in eddies of turbulent flow are capable of holding the particles indefinitely in the body of the stream. Material carried in suspension is referred to as the *suspended load* and constitutes a large share of the total load of most streams.

The third mode of transportation is rolling or sliding of grains along the streambed, a motion that can be conveniently included in the term *traction*. Particles thus in motion are referred to collectively as the *bed load* of the stream.

26.13
STREAM GRADATION

Streams and stream systems that have been in operation for millions of years have made a remarkable adjustment in form to perform the work of transporting both water and mineral matter. To understand this adjustment, it is helpful to consider the case of a stream that is at first poorly adjusted and gradually improves its operation. Suppose that crustal movements have recently and rapidly brought above sea level a new mass of land. We can assume an irregular, steplike land surface with various downsagged portions (Figure 26–16). The surface does, despite these irregularities, descend to the sea from a crudely delimited divide far inland. Soon a continuous chain of flow is established leading to the sea, and a trunk stream is formed. Falls and rapids are quickly eroded away. The stream rapidly develops a smoothly descending channel and is said to be *graded*. In this condition the stream is able to transport all of the load supplied to it from the drainage basin, but it is no longer engaged in rapid downcutting.

26.14
ALLUVIAL RIVERS

Graded streams are free to cut sidewise into the adjacent valley walls. As this *lateral cutting* proceeds, a flat valley floor comes into existence called the *floodplain*. The shifting stream leaves behind deposits of silt, sand, and gravel, collectively called *alluvium*.

Large, graded streams occupying broad floodplains are called *alluvial rivers*. Some common features of such rivers are shown in Figure 26–17. The broad, flat floodplain confined between steep *bluffs* is subject to inundation by floods usually occurring annually. The sinuous river channel is characterized by large loops, called *meanders*. Typically, a meander loop develops a

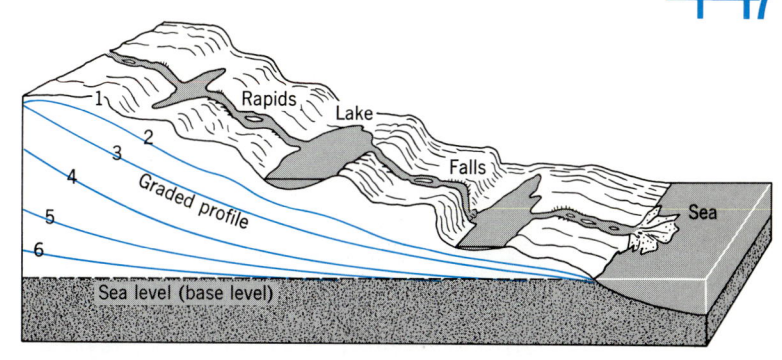

FIGURE 26–16
Schematic diagram of gradation of a stream. Originally the channel consists of a succession of lakes, falls, and rapids. (From PLANET EARTH: ITS PHYSICAL SYSTEMS THROUGH GEOLOGIC TIME by Arthur N. Strahler. Copyright © 1972 by Arthur N. Strahler. By permission of Harper & Row, Publishers, Inc.)

constriction, called a meander *neck*, and this is severed to result in a meander *cutoff*. The abandoned meander loop now becomes an *oxbow lake*. Numerous meanders and oxbows are shown in the air photograph, Figure 26–18. During times of flood, when the entire floodplain is inundated, silt and sand settle from suspension in floodwater adjacent to the channel, building low embankments called *natural levees*. Large *back-swamp* areas of floodplain are poorly drained and marshy much of the time.

26.15
BRAIDED STREAMS
AND ALLUVIAL FANS

In regions of arid climate, stream channels carry large loads of coarse particles—sand, gravel, and boulders—supplied from barren mountain

FIGURE 26–17
Landforms of an alluvial river floodplain.

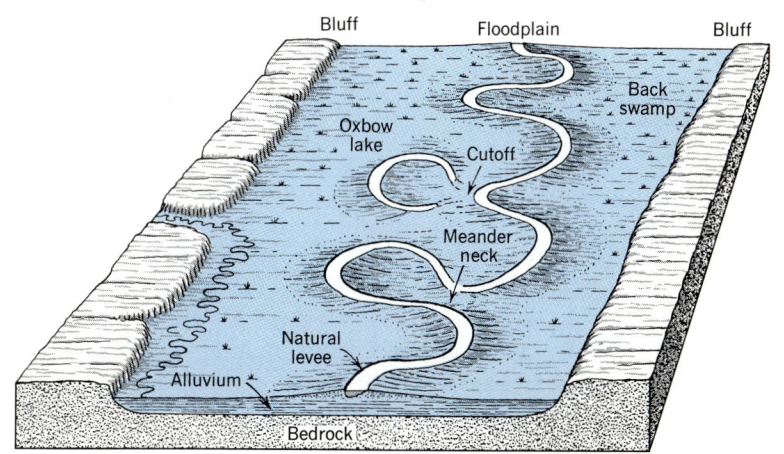

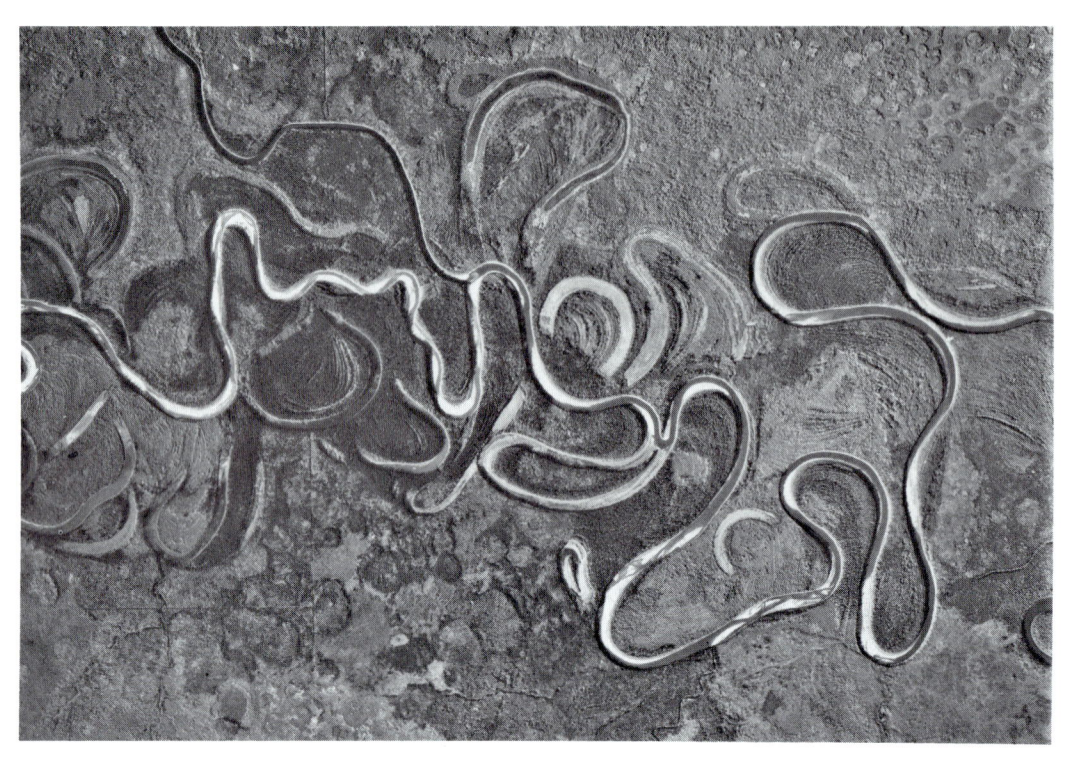

FIGURE 26-18
Vertical air photograph, taken from an altitude of about 6 km showing meanders, cutoffs, oxbow lakes and swamps, and floodplain of the Hay River, Alberta. (National Air Photo Library, Surveys and Mapping Branch, Canada Department of Energy, Mines and Resources.)

slopes in times of torrential rains. Channels choked with such debris develop a *braided pattern* in which the stream flow repeatedly divides and joins, like a braided cord. Where a braided stream emerges from a canyon at the foot of a mountain range to flow out upon a valley floor, the bedload accumulates in the shape of a low, conical deposit, called an *alluvial fan* (Figure 26-19). Much of the water of the stream percolates into the porous fan gravels, so that the stream may disappear entirely at the outer limit of the fan.

26.16
DELTAS

A stream reaching a body of standing water, whether a lake or the ocean, builds a deposit, the *delta*, composed of the stream's load. The growth of a simple delta can be followed in stages, shown in Figure 26-20. The stream enters the standing water body as a jet whose velocity is rapidly checked. Sediment is deposited in lateral embankments in zones of less turbulence on either side of the jet, thus extending the stream channel into the open water. The stream repeatedly breaks through the embankments to occupy different radii and in time produces a deposit of semicircular form, closely analogous to the alluvial fan (which is in a sense a terrestrial delta).

In cross section the simple delta consists largely of steeply sloping layers of sands that grade outward into thin layers of silt and clay (Figure 26-21).

26.17
STAGES OF LANDMASS EROSION

Using the concept of the graded stream, we can trace the development of a large land area, or *landmass*, as millions of years of weathering, mass wasting, and water erosion make significant changes. Three stages in this process are illustrated in Figure 26-22. The first stage shown is one of high, rugged mountains with steep crests and peaks. Erosion is rapid, and large amounts of sediment are exported from the area by streams with steep gradients. However, the land surface is gradually lowered and the slopes become more gentle, as shown in the second block. Erosion rates are now greatly reduced; weather-

FIGURE 26-19
A great alluvial fan in Death Valley, built of debris swept out of a large canyon. Notice the braided stream channels. (Spence Air Photos.)

ing and mass wasting subdue the divides into low, rounded forms.

Ultimately, as the third block shows, the landmass has been lowered to an undulating plain, called a *peneplain*. Streams of low gradient occupy broad floodplains. Because the land slopes are very faint, the erosion rate is now extremely slow and only fine clays and dissolved solids are carried out by the streams. This entire scenario of landmass erosion probably takes 40–70 million years and requires exceptional conditions of crustal stability with respect to sea level. Nevertheless, many peneplain surfaces can be identified over the continents today, and many more are preserved beneath sedimentary strata dating back as far as Precambrian time.

FIGURE 26-20
Stages in the formation of a simple delta. (After G. K. Gilbert.)

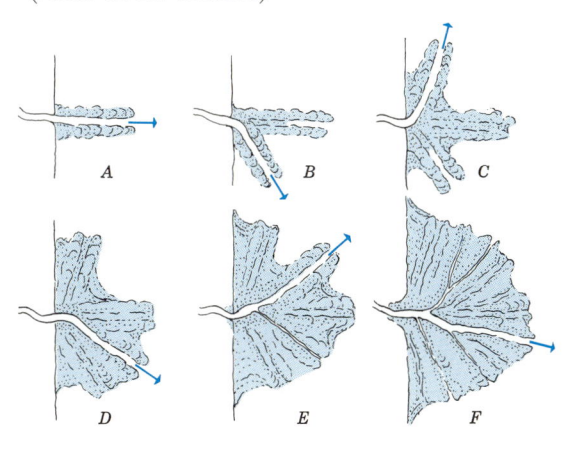

26.18 GLACIERS

A *glacier* is any large accumulation of land ice capable of slow movement from higher to lower levels. Although ice is a brittle crystalline substance when observed in small masses, it is capable of plastic flowage when under heavy load of a thick layer of overlying ice.

Glaciers come in two basic forms (Figure 26-23). The *alpine glacier* resembles a water stream, for it has branches. Actually, the alpine glacier pattern is imposed by a previous stream valley system, which the glacier has occupied (Figure 26-24). The second basic form is seen in the *ice cap* and its former extension into a vast *continental ice sheet*. We can think of the icecap as an essen-

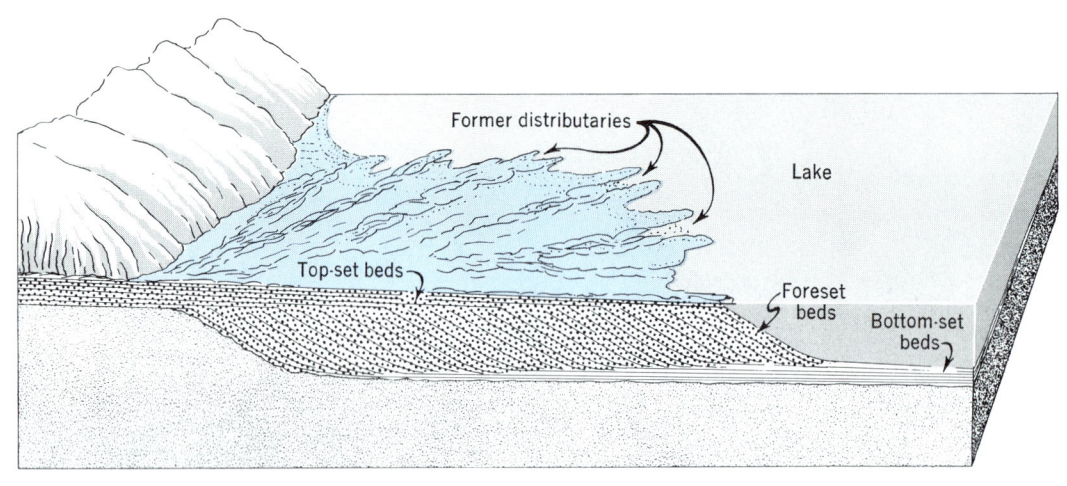

26 GEOMORPHIC PROCESSES AND LANDSCAPE EVOLUTION

FIGURE 26-21
Structure of a simple delta shown in a vertical section. (After G. K. Gilbert.)

FIGURE 26-22
Reduction of a landmass by fluvial erosion. (A) In early stages, relief is great, slopes are steep, and the rate of erosion is rapid. (B) In an advanced stage, relief is greatly reduced, slopes are gentle, and rate of erosion is slow. Soils are thick over the broadly rounded hill summits. (C) After many millions of years of fluvial denudation, a peneplain is formed. Slopes are very gentle and the landscape is an undulating plain. Floodplains are broad, and stream gradients are extremely low. All of the land surface lies close to sea level.

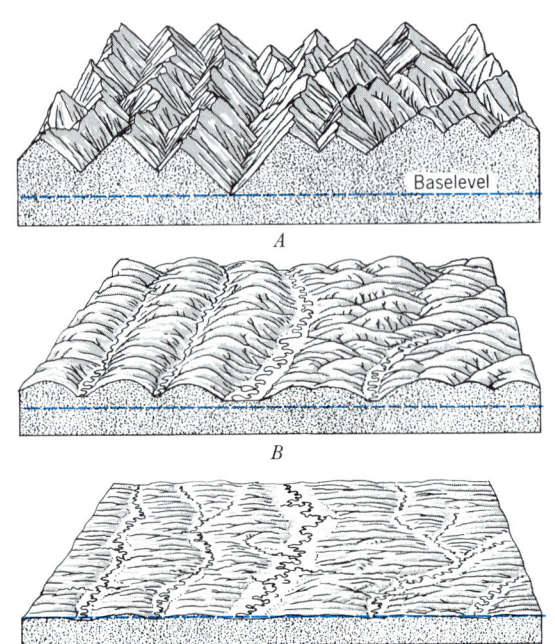

FIGURE 26-23
Schematic maps comparing the form of an alpine glacier with that of an icecap. (From PLANET EARTH: ITS PHYSICAL SYSTEMS THROUGH GEOLOGIC TIME by Arthur N. Strahler. Copyright © 1972 by Arthur N. Strahler. By permission of Harper & Row, Publishers, Inc.)

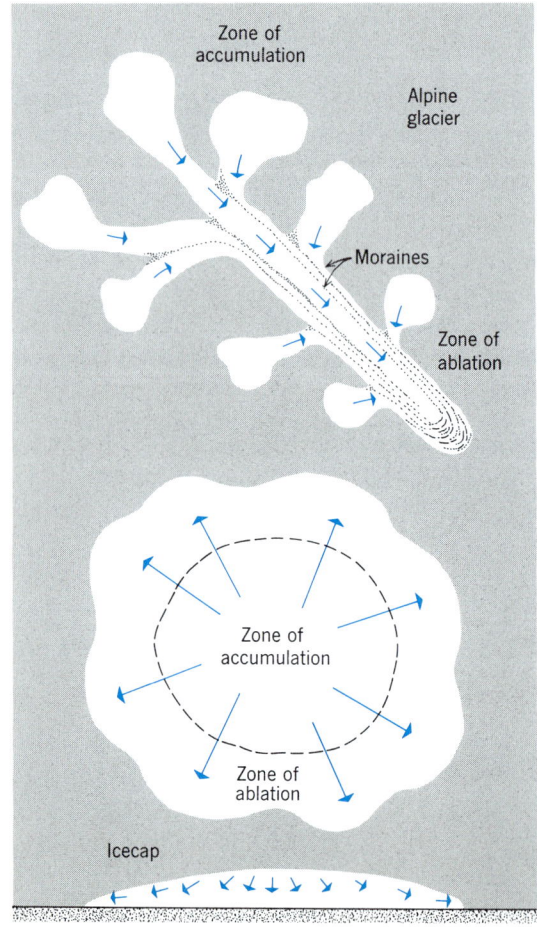

26.19 ICE SHEETS OF THE PRESENT

FIGURE 26-24
A large trunk alpine glacier with numerous branches; it heads in the Juneau Ice Field in Alaska. The dark bands are formed of rock debris resting on the glacier surface. (U.S. Army Air Force.)

FIGURE 26-25
A schematic longitudinal cross section of an alpine glacier, showing the components of input, output, and flowage. (From PLANET EARTH: ITS PHYSICAL SYSTEMS THROUGH GEOLOGIC TIME by Arthur N. Strahler. Copyright © 1972 by Arthur N. Strahler. By permission of Harper & Row, Publishers, Inc.)

tially circular plate of ice that spreads radially outward. Unlike the alpine glacier, which requires a steep down-valley gradient to cause flow, the icecap and ice sheet flow because the ice surface has a gradient from a higher central area to a lower periphery. A simple model for such flow would be a spoonful of pancake batter poured onto a skillet. As you add more batter at the center, the pancake increases in diameter.

Glacial ice is formed from layers of snow, gradually compacted into a dense state, and finally recrystallized into ice. Snow is received on the upper surface of the glacier in the zone of highest altitude. Here the rate of loss of snow by evaporation and melting in summer is, on the average, less than the rate at which the snow is received. This region is the *zone of accumulation* (Figure 26-25). By flowage the basal ice moves down-valley, the speed of motion of the glacier surface being on the order of a few centimeters per day. At the lower end or periphery of the glacier, loss by combined melting and evaporation exceeds the rate at which snow accumulates. This imbalance is greater as the glacier is followed to lower, warmer altitudes. Consequently, the ice disappears at the lower end or border (terminus) of the glacier. This region of net loss is known as the *zone of ablation*.

As a flow system, the motion of ice from the zone of nourishment to the zone of ablation depends on the ice thickness, which in turn depends on rates of nourishment and ablation.

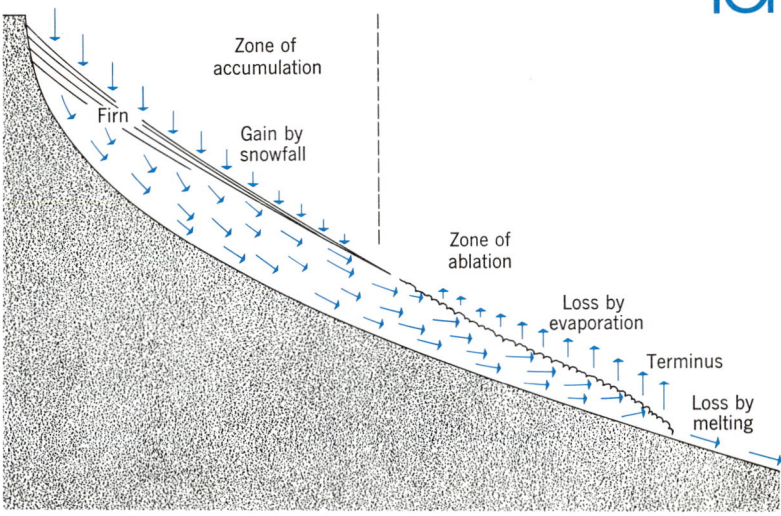

Consequently, a glacier easily adjusts its form and dimensions to reach equilibrium, in which the terminus remains essentially fixed in position for long periods of time.

Alpine glaciers are found today in many parts of the world, wherever high mountains exist. Most are in the high alpine chains described in Chapter 24. A few small, platelike icecaps also exist today—for example, on Baffin Island and on Iceland. However, there are only two great ice sheets in existence today—one on Greenland and the other on Antarctica.

26.19 ICE SHEETS OF THE PRESENT

The Greenland Ice Sheet occupies some 1,740,000 km² — which is 80 percent of the entire area of the island of Greenland — covering all but narrow land fringes (Figure 26-26). Altogether the ice sheet comprises some 2,800,000 km³ of ice. In a general way the ice forms a single, broadly arched, doubly convex ice lens, smoothly surfaced on the upper side and considerably rougher and less strongly curved on the underside. The mountainous terrain of the coast passes inland beneath the ice with steadily descending summit elevations, giving a central lowland area close to sea level in elevation. The ice thickness measures close to 3 km at its greatest.

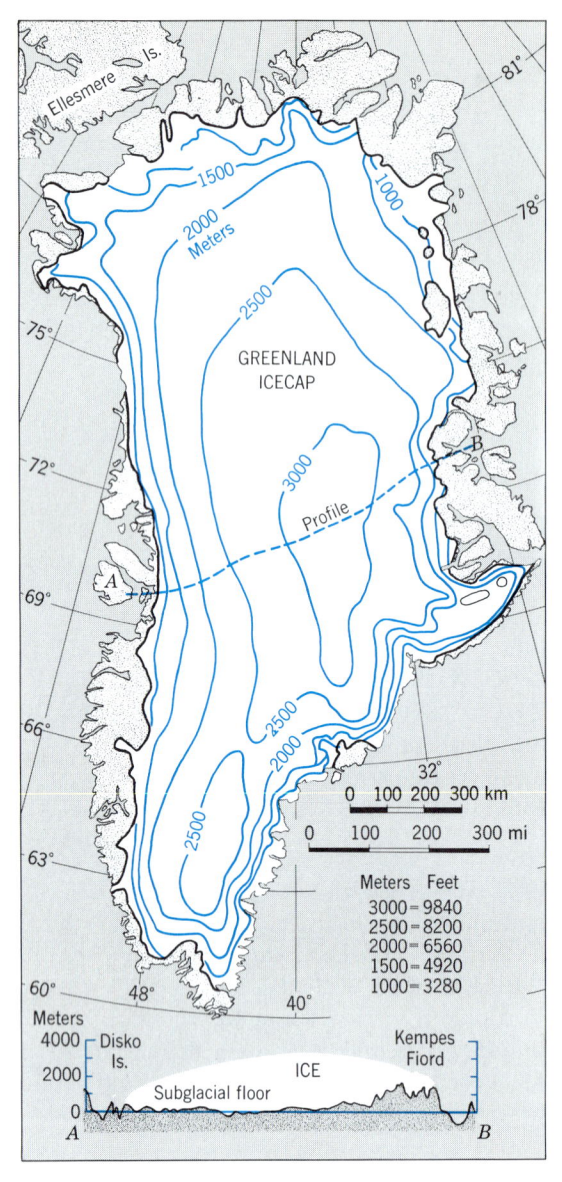

FIGURE 26-26
Generalized map of the Greenland Ice Sheet. (After R. F. Flint, *Glacial and Pleistocene Geology.*)

great plates of floating ice attached to the land (Figure 26-27). Largest of these is the Ross Ice Shelf, about 520,000 km² in area with its surface at an average altitude of about 70 m. (Ice sheets of the Pleistocene Epoch were described in Chapter 25.)

26.20
LANDFORMS MADE BY GLACIERS

Two block diagrams (Figure 26-28) show many of the distinctive landforms created by alpine glacier erosion. These include bowl-shaped depressions, called *cirques,* holding the accumulating snow at the heads of the glaciers. Sharp-crested divides culminate in pointed summits, called *horns.* Bands of rock debris, called *moraines,* are carried along in the moving ice. After the ice has disappeared, deep, steep-walled *glacial troughs* of U-shaped cross section are exposed.

Ice sheets create many surface features of erosion as the ice in motion grinds exposed rock hills and plucks joint blocks from the bedrock (Figure 26-29). Debris thus carried along by the ice is eventually deposited at the ice terminus, where it accumulates as a *terminal moraine.* The moraine consists of a heterogeneous mixture of boulders and clay, called *till.* Streams of meltwater issue from the wasting ice front, building up layers of sand and gravel in front of the terminus. These deposits form an *outwash plain.*

26.21
WAVE ACTION AS A GEOLOGIC AGENT

The shores of all continents and all inland lakes are shaped by the unceasing work of waves. Energy derived from winds is carried forward by deep-water waves. As waves reach the shallow waters of a coastline, their energy is transformed into currents and surges possessing great ability to erode rock and to transport sediment.

As an ocean wave nearing shore passes across a progressively shallower bottom, the wave becomes steeper until it collapses as a *breaker* (Figure 26-30). The orbital motion present in deep water is thus transformed into the forward surge of a water mass, which is carried landward up the sloping beach or rock platform as the *swash.* Forward motion of the swash is quickly brought to a halt by frictional resistance. The water then flows back down the beach slope under the force of gravity, becoming the *backwash.* The effect is that of an alternating water current, capable of dragging particles of rock in alternate landward and seaward motions.

Like Greenland, the antarctic continent is almost entirely buried beneath glacial ice. This is an ice area of just over 13 million km², or about 1½ times the total area of the contiguous 48 United States (Figure 26-27). Ice volume is about 25 million km³, which is over 90 percent of the total volume of the earth's glacial ice. (In comparison, the Greenland Ice Sheet has about 8 percent.)

The Antarctic Ice Sheet reaches its highest elevation, almost 4 km, in a broadly rounded summit. Ice thickness is from 2 to 3 km thick over much of the continent; in a few places, it extends far below sea level.

A characteristic feature of the antarctic coast is the presence of numerous *ice shelves,* which are

26.21 WAVE ACTION AS A GEOLOGIC AGENT

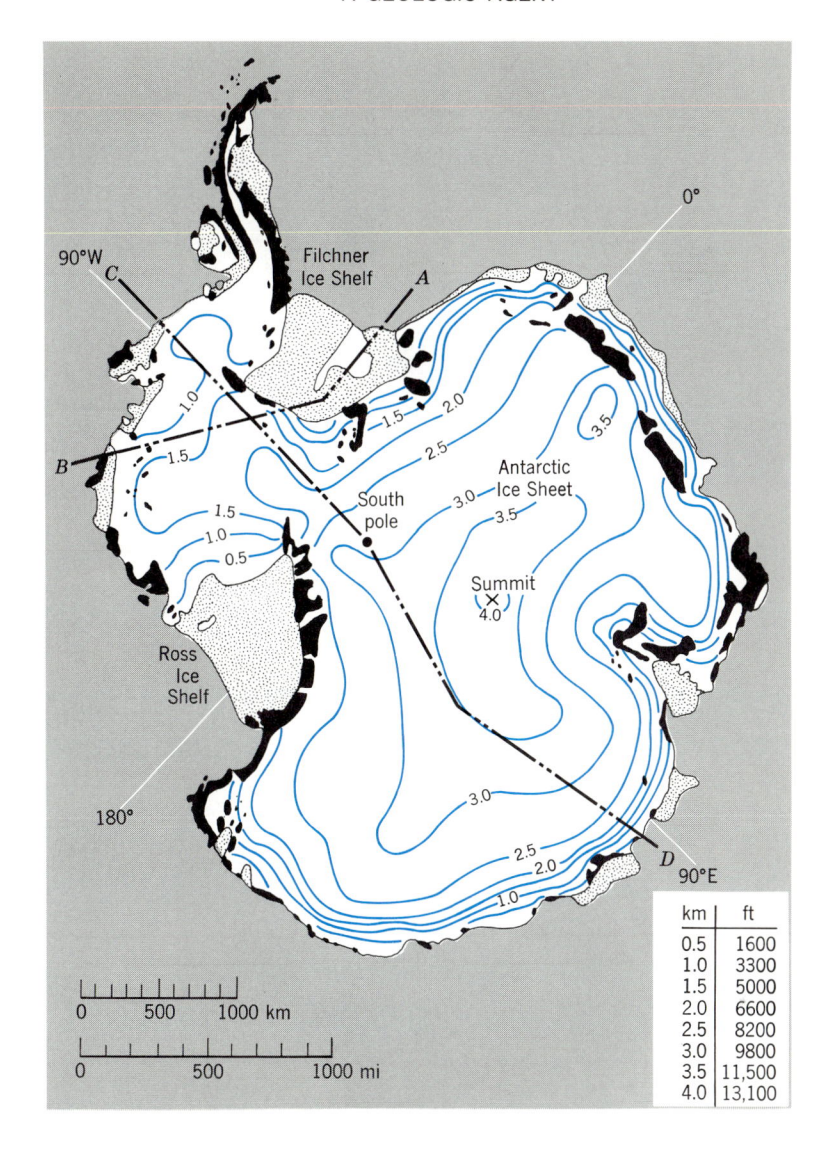

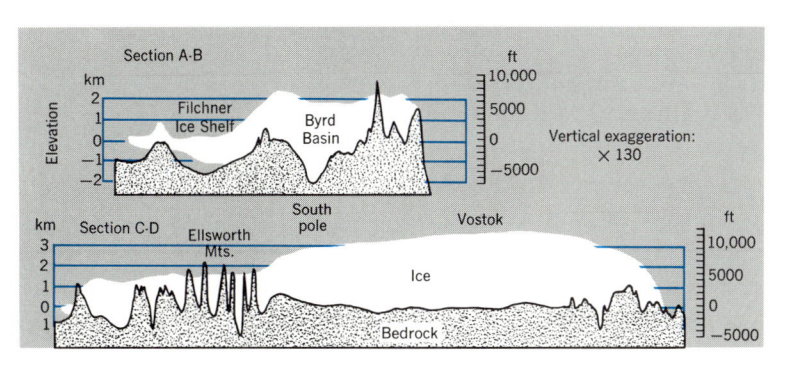

FIGURE 26-27
The Antarctic Ice Sheet and its ice shelves. (Based on data of American Geophysical Union.)

Where great storm waves are breaking on a shore, the swash can be a powerful agent of erosion. The landward thrust may be spent against a cliff, causing undermining and cliff recession. This form of marine erosion is not only an agent of continental erosion, but is also the source of sediment that is carried seaward to form marine sediments of the continental

454   26 GEOMORPHIC PROCESSES AND LANDSCAPE EVOLUTION

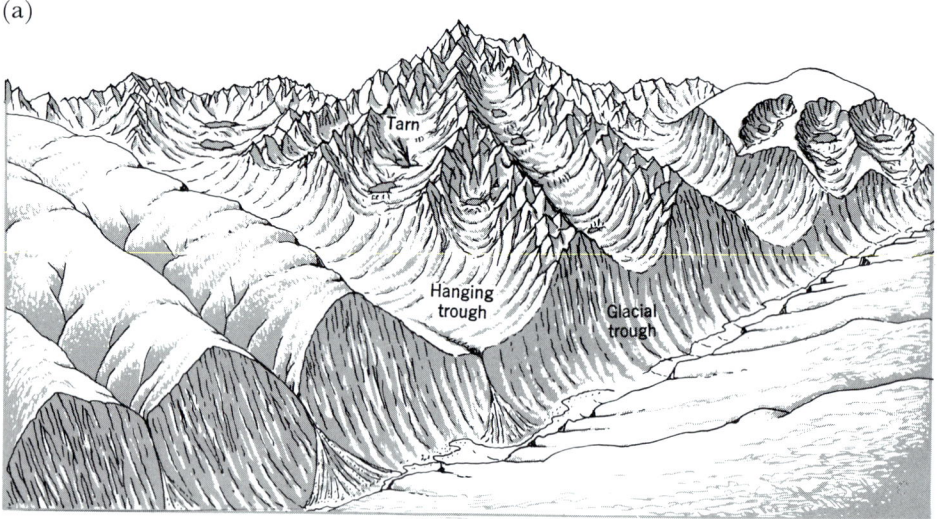

(a)

(b)

FIGURE 26-28
Landforms of alpine glaciation. The upper diagram shows a glacier system in full development. The lower diagram shows troughs exposed after the glaciers have disappeared.

FIGURE 26-29
Glacial erosion and transportation in the marginal zone of an ice sheet. (From A. N. Strahler, *A Geologist's View of Cape Cod*. © 1966 by A. N. Strahler. Reproduced by permission of Doubleday & Co., Inc.)

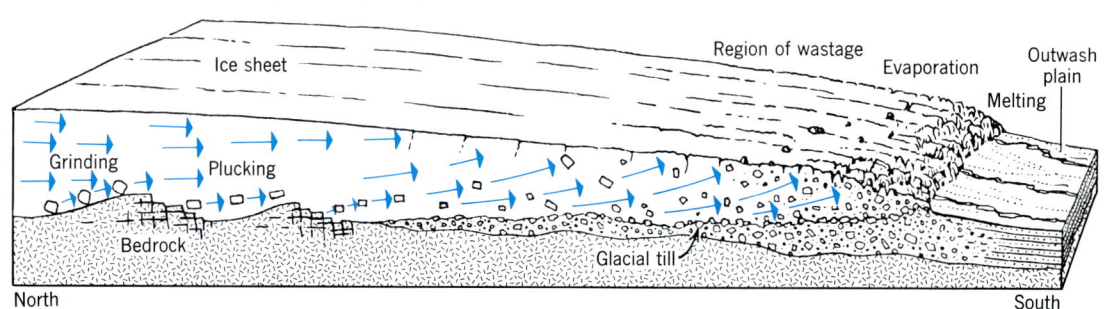

shelves. Many shorelines of the continents today are characterized by a *marine cliff* of wave erosion and a broad erosion platform extending seaward from the base of the cliff (Figure 26–31).

Where sediment of sand and gravel sizes has accumulated in substantial quantities in the zone of breaking waves, a *beach* is formed.

26.22 BEACH DRIFT AND SEDIMENT TRANSPORT

Along most beaches waves approach the shore obliquely at almost all times, causing the swash to be directed obliquely up the beach, as shown in Figure 26–32. As a result, particles carried in the swash ride obliquely up the beach face, but tend to be brought back in the direct downslope direction by the backwash. With each cycle of this movement, the particles are moved along the beach by an increment of distance that may amount to several meters. Multiplied by countless repetitions, this lateral movement, termed *beach drifting*, accounts for transport of vast quantities of sediment and is of primary importance in development of various kinds of beach deposits along a coast.

Along a straight coast, beach drifting will carry sediment continuously along the shore, often for many tens of kilometers (Figure 26–33). However, where a bay is encountered, sediment is carried out into open water to form a *sandspit*, a fingerlike extension of the beach.

The case of an embayed coast is also shown in Figure 26–33. Strong action of breakers against the headlands results in erosion of an abrasion platform and a marine cliff. Detritus thus produced moves by beach drifting along the sides of the bays, where wave approach is oblique. Sediment movement is directed along both sides of the bay toward the bayhead and accumulates there, producing a crescentic pocket beach.

26.23 WIND ACTION AS A GEOLOGIC AGENT

Wind is an agent of erosion and transportation, capable of producing distinctive landforms. These are found largely in desert areas and along coastlines.

One form of wind erosion is *sandblast action*, in which hard mineral grains of sand sizes are driven against exposed rock surfaces projecting above a plain. Because sand grains travel close to the ground, their erosive action is limited to surfaces lying within two meters of the flat ground over which the sand is being driven. Sandblast action is responsible for minor features such as

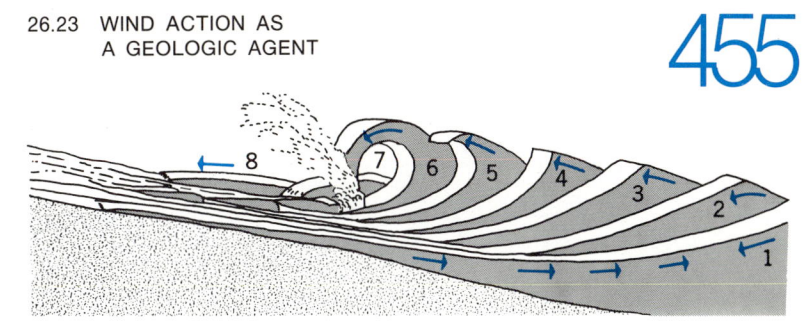

FIGURE 26–30
A breaking wave.

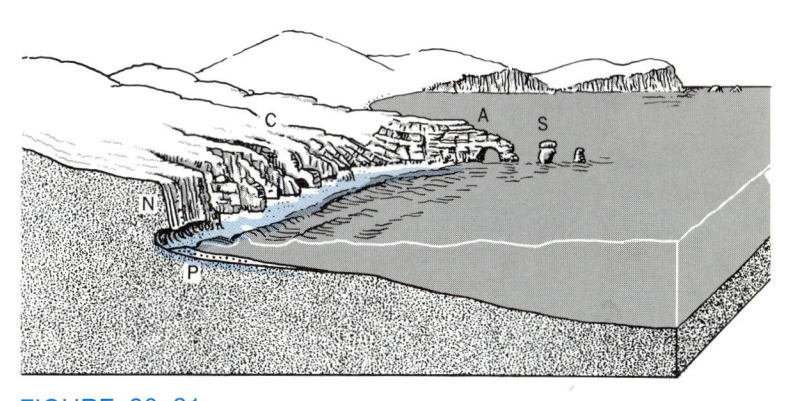

FIGURE 26–31
Details of a sea cliff. A = arch, S = stack, C = cave, N = notch, P = abrasion platform. (After E. Raisz.)

notches and hollows at the base of a cliff or a boulder.

A second form of wind erosion is *deflation*, the lifting and transport of loose particles of clay and silt sizes, collectively referred to as *dust*. Winds of high intensity and turbulence, blowing over plains and plateaus at times when the soil is dry, lift great quantites of dust into suspension in the atmosphere, giving rise to a *dust storm* (Figure 26–34). The smaller particles may quickly dif-

FIGURE 26–32
Oblique approach of the swash causes beach drifting, a mass motion of sand particles parallel with the shoreline.

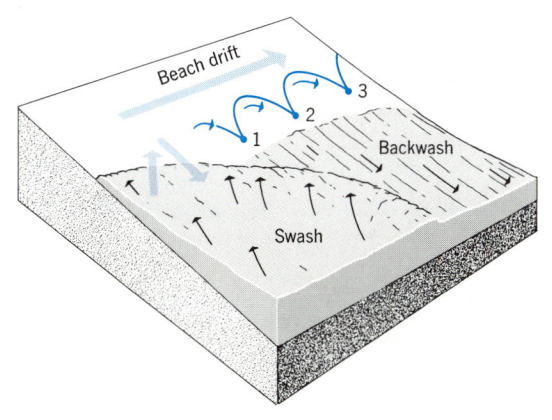

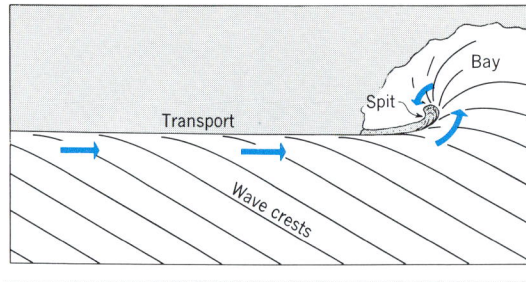

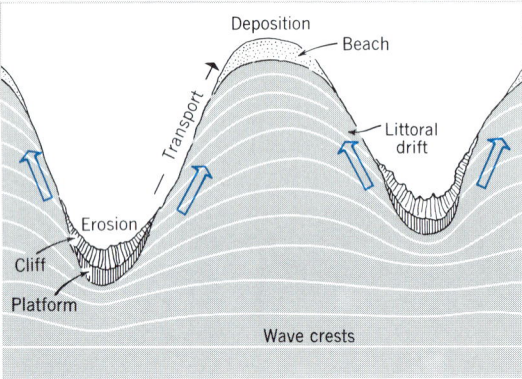

FIGURE 26-33
Beach drifting along a straight section of coastline, ending in a bay (*above*); along an embayed coast (*below*). (From THE EARTH SCIENCES, 2nd Edition by Arthur N. Strahler. Copyright © 1963, 1971 by Arthur N. Strahler. By permission of Harper & Row, Publishers, Inc.)

FIGURE 26-34
Front of an approaching dust storm, Coconino Plateau, Arizona. (D. L. Babenroth.)

fuse to heights of several kilometers and will travel for hundreds of kilometers before settling to earth.

Thick deposits of wind-transported dust can accumulate under favorable conditions to form the parent material for soil. Widespread in the middle latitudes are surface layers of *loess*, a porous, friable, yellowish silt. Loess is windblown dust of Pleistocene age carried from floodplains and outwash plains lying south of the limits of the ice sheets and from glacial deposits uncovered by glacial retreat.

26.24
SAND DUNES

Sand grains travel over a loose sand surface by a process known as *saltation*, in which individual grains make forward leaps, rebounding repeatedly off other grains. Most grains in saltation travel in a layer only a few centimeters in depth. There is an accompanying slow forward motion of surface sand grains under impact of leaping grains.

Where wind is free to act on large supplies of loose sand, distinctive hill-like landforms result—*sand dunes*, capable of downwind movement. Many varieties of sand dunes exist, each associated with a particular set of climatic and topographic conditions. Figure 26-35 illustrates a large area of vegetation-free dunes in a desert environment. Where the sand is in the form of a thick continuous cover, dunes take the form of transverse waves. Where the sand cover is discontinuous, individual dunes of crescent shape, known as *barchans*, are formed.

SUMMARY

Geomorphic processes. Internal and external geomorphic processes create initial landforms and sequential landforms, respectively. The external processes include weathering, mass wasting, and activity of the fluid agents: running water, waves, glaciers, and winds.

Weathering and soils. Physical weathering causes disintegration of bedrock; chemical weathering alters the surficial materials to form new, more stable compounds. Organic and physical soil-forming processes cause laterization in warm-moist climates, podzolization in cold-moist climates, and calcification in dry climates.

Mass wasting. Spontaneous downhill movements of soil, regolith, and bedrock result in many forms of mass wasting, ranging from very slow soil movement to large, rapid landslides of bedrock.

Fluvial erosion. Runoff of water in overland flow causes rapid soil erosion on susceptible slopes. Stream channel flow performs geologic work of erosion, transportation, and deposition. Channels are

FIGURE 26-35
Air photograph of a sand-dune field between Yuma, Arizona, and Calexico, California, shows a sand sea of transverse dunes in the background and a field of crescentic barchan dunes in the foreground. (Spence Air Photos.)

organized into drainage systems. Stream gradation works to lower the landmass to a low, undulating surface, a peneplain.

Ground water. Excess precipitation, entering the ground, percolates down by gravity through the unsaturated zone to the water table, becoming part of the ground water body of the saturated zone, and moving slowly to exit points in streams and shorelines.

Fluvial landforms. Alluvial rivers of low gradient, flowing on broad floodplains, have sinuous meanders with cutoffs and natural levees. In arid climates, stream channels are braided and build alluvial fans. At its terminus in a lake or the ocean, a stream builds a delta by deposition of its load.

Glaciers. Alpine glaciers are ice streams occupying former stream valleys; continental ice sheets are vast plates of ice spreading over entire continents during periods of glaciation. Both forms of ice leave distinctive erosional and depositional landforms.

Wave and wind action. Breaking waves attack the coastlines, producing marine cliffs and building beaches and sandspits. Wind acts as an erosional agent through deflation and sandblast action. Sand driven in saltation by winds is built into dunes, while silt carried high in the atmosphere settles out to form deposits of loess.

TERMS AND CONCEPTS

landforms
geomorphology
geomorphic processes
internal processes
initial landforms
external processes
sequential landforms
bedrock
regolith
soil
physical weathering
frost action
salt-crystal growth
chemical weathering
spheroidal weathering

26 GEOMORPHIC PROCESSES AND LANDSCAPE EVOLUTION

carbonic acid action
laterization
desilication
latosols
laterite
residual ores
podzolization
humus
podzol
calcification
chernozems
mass wasting
talus slopes
fluvial processes
shearing stress (drag)
slope erosion (soil erosion)
badlands
stream
channel
shear
laminar flow (streamline flow)
turbulent flow
suspension
mean velocity
discharge
equation of continuity
drainage systems
drainage divide
drainage basin
saturated zone
ground water
water table
unsaturated zone
stream erosion
stream transportation
stream load
stream deposition
hydraulic action
bank caving
abrasion
corrosion
dissolved solids
suspended load
traction
bed load
graded stream
lateral cutting
floodplain
alluvium
alluvial rivers
bluffs
meanders
meander neck
cutoff
oxbow lake

natural levees
back-swamp areas
braided pattern
alluvial fan
delta
landmass
peneplain
glacier
alpine glacier
icecap
continental ice sheet
zone of accumulation
zone of ablation
ice shelves
cirques
horns
moraines
glacial troughs
terminal moraine
till
outwash plain
breaker
swash
backwash
marine cliff
beach
beach drifting
sandspit
sandblast action
deflation
dust
dust storm
loess
saltation
sand dunes
barchans

QUESTIONS

1. What is the distinction between initial and sequential landforms? What processes are responsible for each of these classes? Name the external geomorphic processes.

2. What is the relationship of regolith to soil and bedrock? What is the nature of rock weathering? What is the importance of weathering? Describe the physical weathering processes and some of the forms they produce. Which are the dominant chemical weathering processes? What forms do they produce?

3. How are organic processes meshed with inorganic processes in the formation of the soil layer? Describe and explain the processes of laterization, podzolization, and calcification. With what climate factors is each associated? With what natural plant cover is each associated?

QUESTIONS

4. How do the mass wasting processes operate? What forms does mass wasting take? Give examples.

5. Explain how fluvial processes operate to produce slope erosion (soil erosion). What forms result from rapid slope erosion?

6. Describe the nature of water flow in a stream channel. What is the significance of stream turbulence? How is stream discharge calculated? How are changes in cross-sectional area and mean velocity related to discharge in a stream passing over rapids? Describe the organization of a drainage system.

7. What is the nature of ground water movement in the saturated zone? Describe the flow paths of ground water. Where does the ground water emerge?

8. What are the three interrelated forms of geologic activity carried on by streams? How does a stream erode its banks and bed? How does a stream carry its load? Explain the concept of stream gradation.

9. Describe the landforms found on the floodplain of an alluvial river. What is the explanation for natural levees? Why do streams in dry climates develop braided channels? How are alluvial fans formed? How are deltas constructed?

10. Describe the stages in landmass erosion leading to the formation of a peneplain. How long does this process take?

11. How does an alpine glacier differ from an icecap? Describe the flow pattern in each. What causes the ice to move? Describe a glacier as a flow system adjusted to rates of accumulation and ablation.

12. Describe the major features of the ice sheets of Greenland and Antarctica. How thick is the ice? How are ice shelves maintained?

13. Describe the typical landforms made by alpine glaciers. What depositional landforms are associated with ice sheets?

14. Explain how breaking waves shape a coastline. What forms result from wave erosion and deposition? How does beach drifting operate and what forms does it produce?

15. Evaluate wind as an agent of both erosion and sediment deposition. How important is deflation? What material is formed by accumulation of wind-transported dust? How are sand grains moved by wind? Describe some common dune forms.

CHAPTER 27
THE SOLAR SYSTEM

27.1 EARTH OR SUN AT THE CENTER?

Ancient astronomers were concerned largely with the solar system. The reason was obvious—without a telescope only the sun, moon, and planets could be seen to have relative motions among themselves. Only these objects showed measurable diameters, whereas the stars seemed to be fixed pinpoints of light. Moreover, the stars showed no motion with respect to one another. The question of which of two bodies, Earth or Sun,[1] was the fixed object about which the others revolved was disputed for perhaps 2000 years, to be settled only in the dawning years of modern astronomy.

The followers of Pythagoras, a Greek philosopher and mathematician who lived in the sixth century B.C., can be credited with first stating the *heliocentric theory* (from the Greek word *helios,* "sun") of the solar system. Under this theory, the Sun is at the center of planetary orbital motion.

The *geocentric theory* (from the Greek word *geos,* "earth"), places Earth at the center of the universe. This theory was put forward about the middle of the second century A.D. by Claudius Ptolemy, an astronomer and mathematician of Alexandria. Called also the *Ptolemaic system* of astronomy, the geocentric theory describes the individual apparent paths of Sun, Moon, planets, and stars as complex systems of cyclic motions relative to a stationary (nonrotating) Earth. It is possible in this way to represent the apparent motions of these bodies geometrically (Figure 27–1). But a good deal of ingenuity is required to cope with each new motion discovered by more precise measurements.

Heliocentric theory was revived by the Polish astronomer Nicolaus Copernicus (1473–1543). He developed what is now generally called the

FIGURE 27–1
The Ptolemaic system required epicycles superimposed on deferents to explain the retrograde motions of the planets. Notice that the epicycles of Mercury and Venus remain fixed on a line connecting Earth and Sun. (From THE EARTH SCIENCES, 2nd Edition by Arthur N. Strahler. Copyright © 1963, 1971 by Arthur N. Strahler. By permission of Harper & Row, Publishers, Inc.)

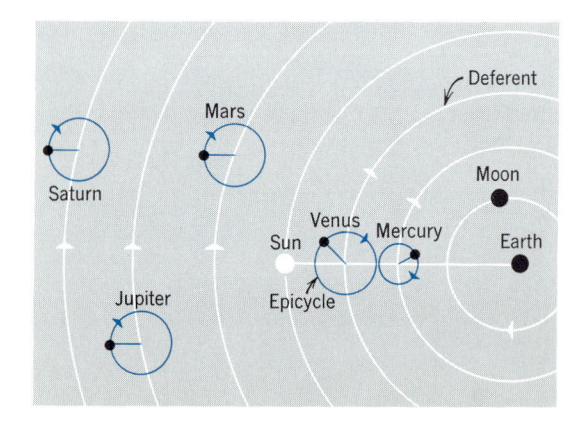

[1] In this and later chapters we capitalize the first letter of Earth, Moon, and Sun to make them proper names, consistent with the other planets.

Copernican system of planetary motions. Based largely on the argument that it was a much simpler way to explain the known facts, Copernican theory placed the Sun at the center of the solar system, with the planets revolving about the Sun in a set of circular orbits. Bear in mind that the telescope was not yet invented, nor were the laws of gravitation and motion known. There was no way Copernicus could marshal absolute evidence of the Earth's motions. Copernicus's contemporaries and successors were strongly divided on the merits of a heliocentric system.

Great strides in establishing the heliocentric theory came through the use of the first astronomical telescope by its inventor, the Italian scientist Galileo Galilei (1564-1642). In 1610 Galileo discovered that there are moons revolving around the planet Jupiter. You can see these moons for yourself with a pair of binoculars. The similarity in observed orbital motion of these moons to that required of the planets under Copernican theory could be regarded as favorable evidence by analogy.

Then Galileo discovered that Venus shows phases similar to those of the Moon and, moreover, changes apparent size with phase (Figure 27-2). Because the changes of phase are clearly the different proportions we see of a spherical surface illuminated by the Sun's rays, the only logical conclusion was that Venus revolves about the Sun.

Galileo was not free to support his own beliefs publicly, because Copernican theory was then regarded by church authorities as a religious heresy. He was compelled to recant his views and to retire into seclusion for his remaining years.

During this same period Johannes Kepler (1571-1630) carried the baton passed to him by Galileo. Using a long series of observations made by the Danish astronomer Tycho Brahe, Kepler discovered the three fundamental laws of planetary motion that describe the Copernican theory. Their physical explanation was not understood until Isaac Newton (1642-1727) discovered the underlying laws of gravitation and motion. Newton had formulated these laws perhaps as early as 1665, and he published them in 1687 in a volume entitled *Mathematical Principles of Natural Philosophy*. Despite such powerful supporting logic, opposition to Copernican theory persisted long beyond this time.

27.2
MEMBERS OF THE SOLAR FAMILY

The *solar system* consists of the planets and their satellites, the asteroids, meteoroids, and comets.

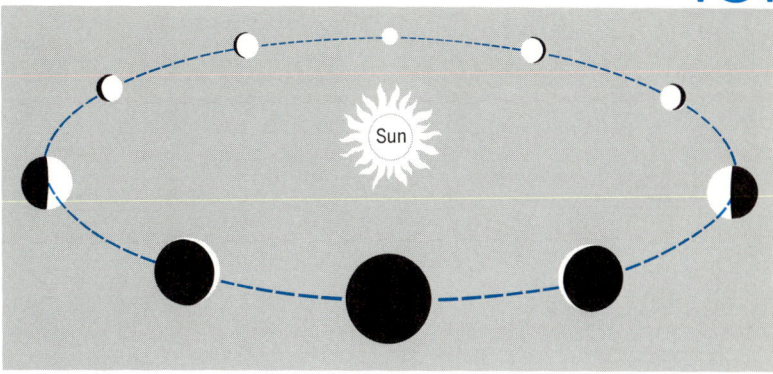

FIGURE 27-2
The orbit of Venus is shown in this perspective drawing as if viewed from a point in space above the Earth's north pole. See Figure 27-10 for photographs of these phases. (From PRINCIPLES OF EARTH SCIENCE by Arthur N. Strahler. Copyright © 1976 by Arthur N. Strahler. By permission of Harper & Row, Publishers, Inc.)

All of these objects move in the gravitational field of the Sun, the preeminent body and center of the entire system.

The term *planet* is limited in common usage to the nine largest bodies revolving about the Sun; they are also often called the *major planets*. In order of distance from the Sun they are: Mercury, Venus, Earth, Mars, Jupiter, Saturn, Uranus, Neptune, and Pluto. The asteroids, sometimes referred to as the minor planets, number in the thousands. All have diameters less than 800 km, and most are less than a few kilometers in diameter. In general, they are found between the orbits of Mars and Jupiter. The meteoroids are extremely minute solid particles traveling in swarms in orbits around the sun. A comet is a rather large, diffuse body of very small mass.

27.3
INNER AND OUTER PLANETS

Astronomers classify the nine major planets into two groups. The *inner planets,* also referred to as the *terrestrial planets,* are those four lying closest to the Sun—Mercury, Venus, Earth, and Mars (Figure 27-3). The four are grouped together because they are comparatively small and at the same time have orbits relatively close to one another and to the Sun.

The five *outer planets*—Jupiter, Saturn, Uranus, Neptune, and Pluto—move in orbits of vastly greater diameter than those of the four inner planets and, except for Pluto, are vastly greater in size. All nine planetary orbits cannot

FIGURE 27-3

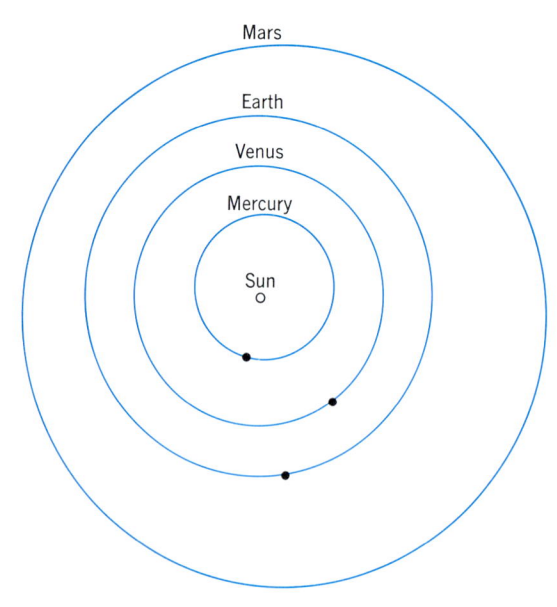

Orbits of the four inner planets. The black dots represent perihelion points. The orbits are drawn as circles, neglecting the true eliptical form. (From PRINCIPLES OF EARTH SCIENCE by Arthur N. Strahler. Copyright © 1976 by Arthur N. Strahler. By permission of Harper & Row, Publishers, Inc.)

FIGURE 27-4

Orbits of the outer planets. The innermost circle represents Mars's orbit, and the dashed circle represents the zone of asteroid orbits. Pluto will not collide with Neptune because Pluto's orbit is inclined more than 17 degrees with respect to the ecliptic plane. (From THE EARTH SCIENCES, 2nd Edition by Arthur N. Strahler. Copyright © 1963, 1971 by Arthur N. Strahler. By permission of Harper & Row, Publishers, Inc.)

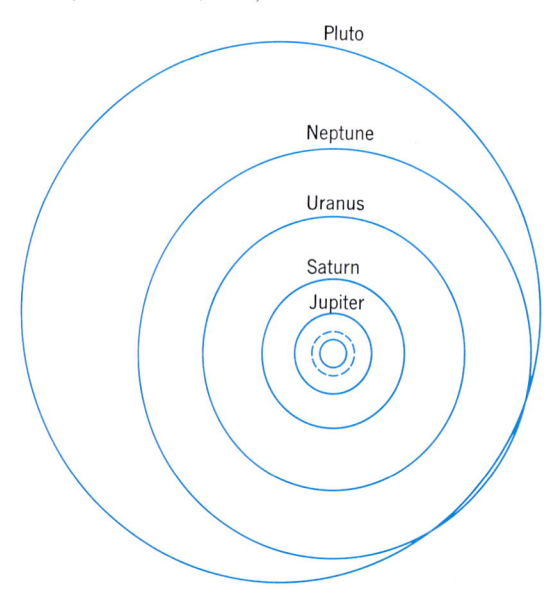

be shown to the same scale on one drawing; the scale of Figure 27-4, in which the orbits of the outer planets are shown, is about one-twentieth that used in Figure 27-3.

In order to convey a stronger impression of the size differences among the planets and Sun, the diameters of these bodies are drawn to a common scale in Figure 27-5, although only a small part of the Sun's disk can be shown.

Table 27.1 gives information in several categories for the nine major planets. It is obvious from the figures giving distance from Sun, diameter, and mass that the four inner, or terrestrial, planets form a group of quite similar bodies. Also, the first four outer planets—Jupiter, Saturn, Uranus, and Neptune—can appropriately be grouped as the *great planets*. In contrast, Pluto has a very much smaller size and may well be placed in a class by itself.

27.4
ORBITS OF THE PLANETS— KEPLER'S LAWS

The *orbit* of a planet is the path it follows around the Sun. The orbit of every planet is an *ellipse* (Figure 27-6). The Sun occupies one *focus* of the ellipse. An ellipse has a longest diameter, the *major axis*, and a shortest diameter, the *minor axis*. These two axes are at right angles to each other and intersect at the center of the ellipse. Along the major axis are two *foci* (plural of focus). Given any point on the ellipse, a straight line, known as a *radius vector*, can be drawn, one to each of the two foci.

A law of the geometry of the ellipse is that the sum of the two radius vectors remains a constant for all points on the ellipse. This law can be demonstrated by a device for drawing ellipses, using a loop of thread and two pins or thumbtacks on a drawing board (Figure 27-7). You can vary the degree of flattening of the ellipse by adjusting the spacing of the two foci in relation to the length of loop. As the two foci are brought closer to the center, the ellipse approaches a circle in form.

We owe the discovery of three laws of planetary motion to Johannes Kepler. Kepler's first law states simply that the orbit of each planet is an ellipse, with the Sun located at one focus of the ellipse.

Kepler's second law states that a planet moves in its orbit about the Sun at a varying velocity, such that the radius vector of the elliptical orbit sweeps over equal areas in equal times. This concept is illustrated in Figure 27-8, a greatly exaggerated ellipse. The circumference has been divided into 12 segments representing 12

27.4 ORBITS OF THE PLANETS—KEPLER'S LAWS

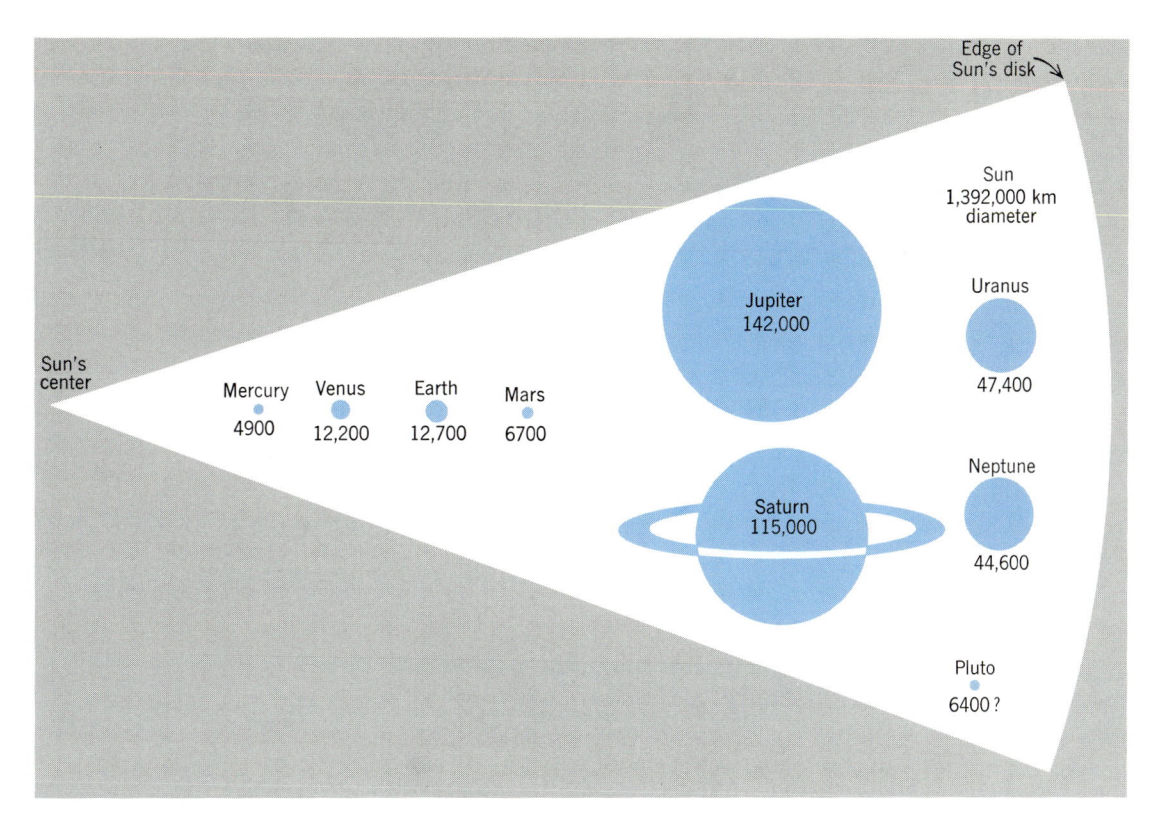

FIGURE 27-5
Relative sizes of the Sun and major planets. Figures give diameters in thousands of kilometers. (From PRINCIPLES OF EARTH SCIENCE by Arthur N. Strahler. Copyright © 1976 by Arthur N. Strahler. By permission of Harper & Row, Publishers, Inc.)

months, assumed exactly equal in time. The radius vectors enclose equal areas. Area *A* is equal to area *B*. In sweeping across area *A*, the radius vector must travel from *M* to *N*, a longer distance than from *P* to *Q*; consequently, a point on the orbit must travel at a greater average speed between *M* and *N* than between *P* and *Q*.

It is obvious from Figure 28-8 that the planet in its orbit is nearest the Sun when located at one end of the major axis of the ellipse. This closest point is termed *perihelion* (from the Greek words *peri*, "about or near," and *helios*, "sun"); the most distant point is at the opposite end of the major axis, a position termed *aphelion* (from the Greek *ap*, "away from," and *helios*). For example, Earth is at perihelion about January 3, and at aphelion about July 4 each year. At perihelion the radius vector is about 147 million km, at aphelion about 152 million km, giving a mean value of about 150 million km for the whole orbit.

It follows from Kepler's second law that the planet's velocity in its orbit must be continuously changing. From a maximum at perihelion, the velocity diminishes to a minimum at aphelion, then increases again to the next perihelion.

FIGURE 27-6
The orbit of every planet is an ellipse in which the sun occupies one focus. (From THE EARTH SCIENCES, 2nd Edition by Arthur N. Strahler. Copyright © 1963, 1971 by Arthur N. Strahler. By permission of Harper & Row, Publishers, Inc.)

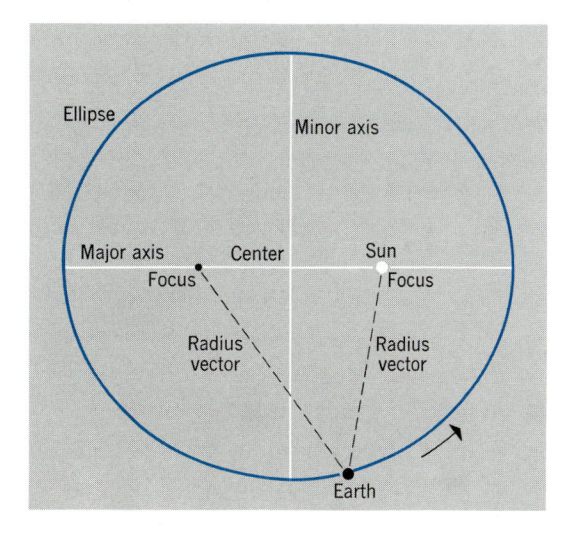

FIGURE 27-7
An ellipse can be easily constructed.

The more flattened a planet's elliptical orbit, the greater the *eccentricity* of its orbit. Eccentricity can be visualized as the extent to which the focus, represented by the Sun, lies to one side of the center point of the orbit ellipse. The eccentricities of the orbits of Earth and Venus are very small; their orbits are nearly perfect circles. The orbits of both Mercury and Mars have only moderate eccentricity. Eccentricities of the great planets' orbits are also small, but that of Pluto is abnormally great. In fact, the orbit of Pluto cuts inside the orbit of Neptune, as Figure 27-4 shows.

Kepler's third law, which he formulated nine years after the first two, is sometimes designated as the *harmonic law*. It states that for any two planets, the squares of the periods of revolution are proportional to the cubes of their mean distances from the Sun. In the following equation, the subscripts a and b denote any two planets, P is the period of revolution, and R the distance

$$\frac{P_a^2}{P_b^2} = \frac{R_a^3}{R_b^3}$$

In testing the law, calculations can be greatly simplified by using Earth as one planet, with its period of revolution taken as unity (one Earth year) and its distance from Sun as unity (one astronomical unit). The equation then simplifies to

$$P = \sqrt{R^3}$$

where P is the period of the planet in Earth years and R is the planet's distance from Sun in astronomical units. (One astronomical unit is about 150 million km.)

Referring to Table 27.1, and using the data for Neptune, we find that distance R is 30.1 astronomical units. Substituting in the simplified formula above,

$$P = \sqrt{(30.1)^3} = \sqrt{27,270.9} = 165 \text{ approx.}$$

The calculated period agrees closely with the observed value given in Table 27.1.

Strictly, Kepler's laws are correct only for the case of a single planet in relation to the Sun. Actually, the gravitational influence of other planets deflects a given planet slightly from its ideal elliptical orbit.

As a practical consequence of the third law, we see that the periods of revolution of the planets range from 88 days for Mercury, which lies closest to the Sun, to 248 years for Pluto, the most distant.

Kepler based his laws of planetary motion solely on observations of the planets themselves. A valid physical explanation for elliptical orbits and varying speeds was yet to come. Galileo, a contemporary of Kepler, was investigating by experiment the principles of the pendulum and the acceleration of falling bodies. Upon the foundations Galileo had laid, Sir Isaac Newton developed his laws of gravitation and motion, published in 1687. These were fully explained in Chapters 1 through 3.

27.5
REVOLUTION AND ROTATION

The motion of a planet in its orbit about the Sun is called *revolution*, whereas the spinning of the planet on an axis is called *rotation*. For Earth, one revolution in the orbit defines the *year*, whereas one complete turn on its axis defines the *day*. The other planets perform the same motions, but with differing periods. Table 27.1 gives the period of revolution of each planet in earth-days

FIGURE 27-8
An ellipse divided into twelve equal areas. (From THE EARTH SCIENCES, 2nd Edition by Arthur N. Strahler. Copyright © 1963, 1971 by Arthur N. Strahler. By permission of Harper & Row, Publishers, Inc.)

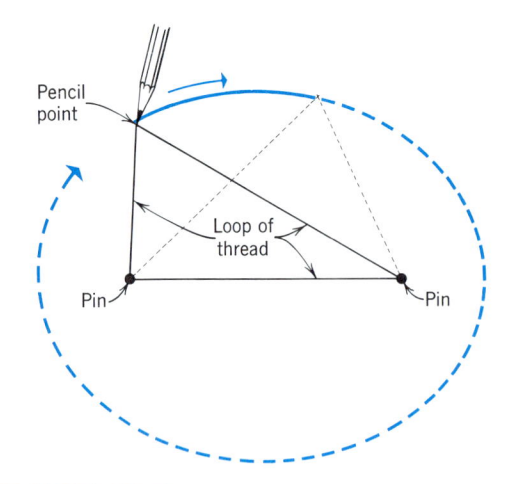

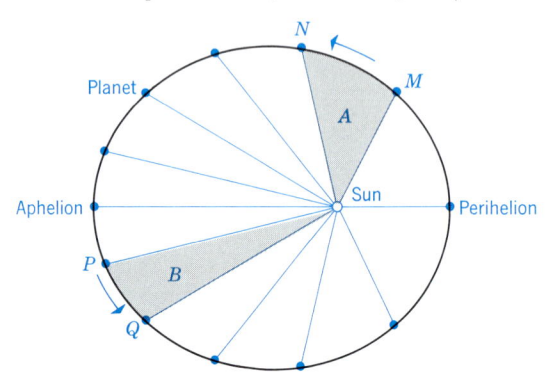

TABLE 27.1
THE PRINCIPAL PLANETS

Name	Distance from Sun, Millions of km (10^6 km)	Period of Revolution, Sidereal	Diameter, Thousands of km (10^3 km)	Mass, Relative to Earth	Mean Density g/cm^3	Period of Rotation	Number of Moons
Inner planets (terrestrial planets)		Days					
Mercury	58	88	4.9	0.06	5.4	58^{d}16^h	0
Venus	108	225	12.2	0.80	5.2	243^d	0
Earth	150	365¼	12.7	1.00	5.5	23^{h}56^m	1
Mars	228	687	6.7	0.11	4.0	24^{h}37^m	2
Outer planets (great planets)		Years					
Jupiter	779	12	142	315	1.3	9^{h}50^m	13
Saturn	1430	29½	115	94	0.7	10^{h}14^m	10
Uranus	2870	84	47.4	15	1.6	10^{h}42^m	5+
Neptune	4500	165	44.6	17	1.6	15^{h}48^m	2
Pluto	5900	248	6.4?	0.11	5.0?	6^d	0

and earth-years; it also gives the period of rotation of each planet in hours and minutes of earth-time.

As you would expect, the larger the planet's orbit, the longer is its period of revolution. Little Mercury orbits the Sun in only one-quarter the time required by Earth, while pluto requires 248 earth-years to make the circuit. When it comes to the periods of rotation, no such relationship to distance applies. Venus' day lasts for 243 earth-days, while Jupiter's day is less than half an earth-day. Actually, the rotation periods of the four great planets are quite similar—all between about 10 and 16 hours.

Direction of rotation of Earth is described as "eastward." This simply means that a point on the equator is traveling in an eastward direction. If we imagine ourselves to be out in space, situated at a point directly over the Earth's north pole, the direction of rotation would seem to be counterclockwise, as shown in Figure 27-9. From this same point of view, the direction of revolution is also counterclockwise about the Sun. Moreover, our Moon orbits the Earth in the same direction. Astronomers refer to this counterclockwise turning as *direct motion*. The opposite direction of turning is called *retrograde motion*.

All of the planets and most of their satellites move in direct motion. The Sun also rotates with direct motion. These facts strongly suggest that all members of the solar system received their motions at the time when the solar system came into existence, through condensation of a single rotating nebula (Chapter 28). This conclusion is strengthened by the fact that all planetary orbits lie approximately in the same plane, which is about the same as the plane of the Sun's equator. We would expect this uniformity if the solar nebula were a thin, flat disk, spinning with the Sun at its hub.

27.6
MERCURY

Mercury, fourth brightest of the planets, is the closest planet to the Sun. Its orbit has a radius only two-fifths that of Earth. In looking over the facts about Mercury in Table 27.1 we note that this planet is a mere pygmy compared to Earth.

Mercury's atmosphere is so rarefied as to be almost absent. The planet's surface gravity is only about three-tenths that of Earth, and any gases readily escape into space. Mercury's rate of rotation is very slow, and a day there lasts for 59 earth-days. Consequently, on the side of Mer-

FIGURE 27-9
Direction of Earth rotation and revolution.

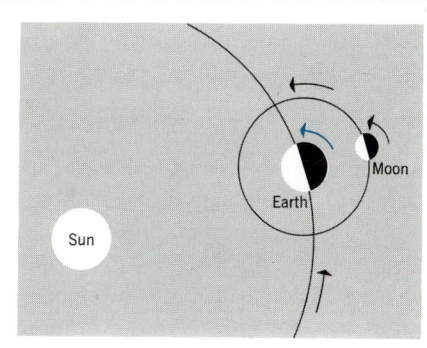

cury that happens to be facing the Sun, intense and prolonged heating bakes the surface. It is estimated that, in perihelion, surface temperatures on Mercury rise to perhaps 420°C, a value exceeding the melting points of tin and lead. In contrast, temperatures on the shadowed side of Mercury may fall nearly to absolute zero. No other planet has so vast a temperature range on its surface. Mercury's cratered surface can be assumed to be completely devoid of any forms of life.

Because Mercury is so close to the Sun, we can see it only on special occasions. Most of the time, the Sun's blinding rays conceal the planet. It can be seen as either a morning star or an evening star for the short period of time that the Sun is below the horizon and Mercury is above the horizon, and then only when the planet is not too nearly in line with the Sun.

27.7 VENUS

Venus is the most brilliant object in the sky except for the Sun and Moon. Venus approaches closer to Earth than any other planet; a distance of some 42 million km separates the two bodies at the minimum separation. The maximum distance is about 260 million km. As a result of this sixfold difference in separating distances, Venus seems to change greatly in diameter throughout its orbit (Figure 27–10). Moreover, the changing positions of Venus relative to Earth and Sun result in a series of phases of illumination ranging from a full disk to a thin crescent (Figure 27–2).

From the standpoint of diameter, mass, density, and length of year, Venus more closely resembles Earth than any other planet. Moreover, Venus has a dense atmosphere, held by a gravitational force almost as strong as that of Earth. Atmospheric pressure at the surface of Venus is about 100 times as great as that on Earth. One visible proof of the presence of an atmosphere on Venus is that at crescent phase a band of light extends entirely around the full circle, as you can see in Figure 27–10. This glow of light shows that sunlight is refracted around the sphere by a thick layer of gases.

Carbon dioxide constitutes 90–95 percent of the atmosphere of Venus. Oxygen has been found in substantial quantities, but water vapor and nitrogen have not been detected. Water was probably present on Venus at an early stage in the planet's history, but hydrogen atoms could not be held to the planet and most escaped into space.

Because of the presence of some kind of fine suspended particles in its outer atmosphere,

FIGURE 27–10
The planet Venus photographed at five different phases, showing its true relative sizes at various distances from Earth. Compare with Figure 27–2 to determine position in orbit. Note that in the largest view the atmospheric ring is complete. (Lowell Observatory photograph.)

Venus reflects sunlight brilliantly, and nothing can be seen of its surface. The suspended particles have been identified as an aerosol of sulfuric acid.

Temperatures on the surface of Venus have been obtained from temperature analysis of radio waves. Because of the dense atmosphere, acting as a blanket to hold heat, surface temperatures are about 480°C. The complete absence of water excludes the possibility of life forms such as those found on Earth. Venus has a hostile environment, but which human beings, if located on the dark side and suitably protected from the very high atmospheric pressure and possible strong winds, might be able to survive in it for limited periods.

Until radar was developed, it was practically impossible to determine the rotation period of Venus. Reflected radar signals have yielded a period of 243 days, which is somewhat longer than the planet's period of revolution of 225 days. Furthermore, the rotation of Venus is slowly clockwise, or retrograde, in contrast to the counterclockwise, or direct, rotation of the other planets. One effect of the very slow rotation is to give Venus a very simple system of atmospheric circulation in which the gas rises over the intensely heated equatorial area exposed to the sun. The center of rising air is seen in a dark spot called the Venusian Eye (Figure 27–11). At

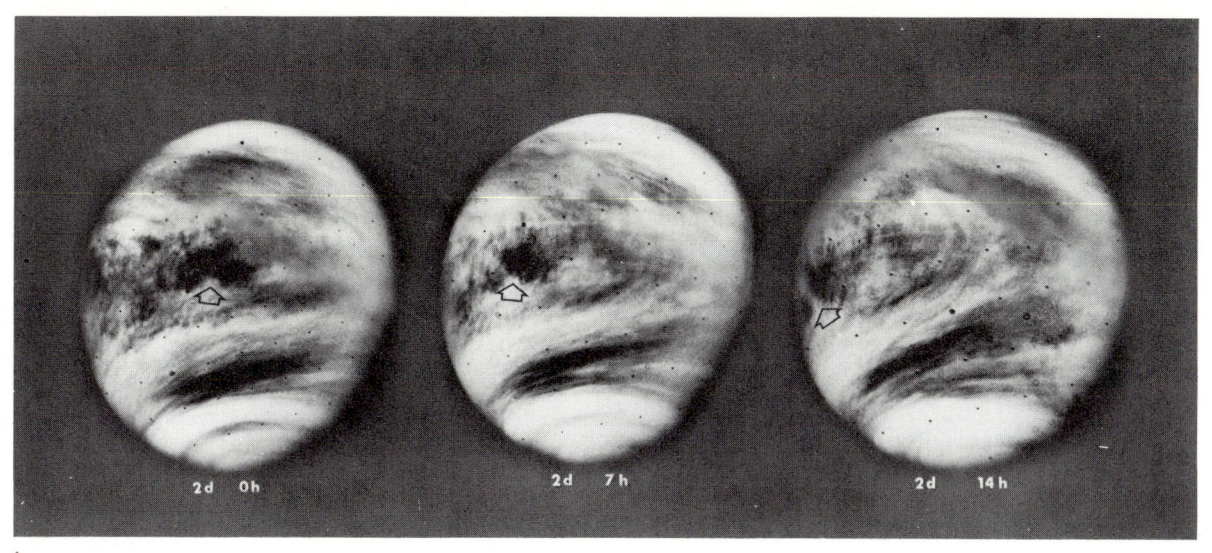

FIGURE 27-11
Venus shows spiral cloud bands, which change rapidly in form because of vigorous atmospheric circulation. The arrow points to the Venusian Eye, a huge turbulent convection cell in which heated gas is rising. (NASA.)

upper levels the gas spreads north and south, but sinks to the surface over the cold polar zones. Photographs of Mariner 10 show this atmospheric circulation through the changing shapes of spiral cloud bands seen in Figure 27-11. Rotation is strong enough, however, to produce a Coriolis effect, turning the flow into a west-to-east pattern with speeds up to 360 km/h. In this way the atmosphere spirals toward the poles.

27.8 MARS

Mars is the first planet on our list to be found in an orbit larger than that of Earth (Figure 27-12). Little more than half as large in diameter as Earth, the mass of Mars is only one-tenth that of Earth and the surface gravity only about one-third (Table 27.1).

Reddish in hue, Mars has definite surface features; there was recognition and speculation about these as early as the mid-seventeenth century. Study of these features enabled astronomers to measure with precision the period of rotation of Mars, which is about $24\frac{1}{2}$ hours, only a little longer than that of Earth. Also, the plane of Mars's equator is inclined about 25 degrees with respect to its orbital plane, a value very close to Earth's inclination of $23\frac{1}{2}$ degrees.

As seen from Earth, markings on the surface of Mars consist of dark and light areas in a permanent pattern. Much speculation arose over the significance of recurrent dark markings, which seemed to form an intersecting network of narrow bands. These were early interpreted as canals; their apparent straightness suggested to some astronomers that they were artificially produced by Martians, possibly to serve as irrigation canals.

FIGURE 27-12
Far-encounter photograph of Mars taken from Mariner 7 spacecraft at a distance of about 450,000 km. The dark spot at upper left is Olympia Mons, a great volcano. The white area at the bottom is the polar cap. (NASA.)

Now that Mariner and Viking spacecraft have circled Mars at close range, we know a great deal about the varied Martian surface, with its craters, volcanoes, troughs, dunes, and channels. These features are described in Chapter 28. Of great scientific interest is the seasonal growth and disappearance of the white polar caps on Mars. These caps are described in Chapter 28 and illustrated in Figure 28-15. Like Earth, Mars has a winter season in one hemisphere while it is summer in the opposite hemisphere. The polar cap grows during the autumn season of that hemisphere, spreading equatorward to a maximum in midwinter, then receding with the approach of spring. (Note that the Martian year lasts 687 earth-days, or nearly twice as long as an earth-year.) Carbon dioxide in solid form (dry ice) is now regarded as the most probable surface substance of the polar caps.

As seen through telescopes, the colors and color changes on Mars are remarkable. The dark areas are green, blue-green, or gray and show a seasonal browning to earth-red colors (brick-red to ocher). Color photographs taken at two Viking landing sites, occupied in 1976, reveal a uniformly brick-red color of the surface, attributed to the presence of coatings of hematite on the mineral surfaces.

The atmosphere of Mars is very thin as compared to that of Earth. Viking landers measured a surface barometric pressure of only about 7.6 mb—less than $\frac{1}{100}$ that at the Earth's surface. Carbon dioxide comprises 95 percent of the Martian atmosphere. Other gases measured by Viking landers are molecular nitrogen (2–3 percent), argon (1–2 percent), and oxygen (0.3 percent). Water vapor content of the atmosphere is extremely small—barely a trace as compared to the Earth's atmosphere. Perhaps there was originally much more free oxygen in the Martian atmosphere; if so, most of it may have been taken into storage in the form of iron oxides during rock weathering. In Chapter 28, we shall show evidence that free water may have been abundant at certain times in the past. Today, a great quantity of water is apparently held as ice beneath the polar caps.

According to records sent back by Viking landers, the surface temperature on Mars at low latitudes ranges daily from a low of $-85\,°C$ to a high of $-30\,°C$. Thus, under present climatic conditions, any water at the surface would be continuously frozen. It is possible, however, that surface temperatures above the freezing point are reached during midday in the equatorial zone.

It has long been realized that of all the planets, Mars offers the greatest possibility of harboring life, either at the present time or at some time in the geologic past. That life would of necessity be adapted to a very scanty supply of oxygen and water as well as to an extreme cold and to very low atmospheric density. Viking spacecraft that landed upon the Martian surface in 1976 undertook highly sophisticated experiments to determine the possible presence of life—living or extinct—through the presence of chemical substances of organic origin. The results of the Viking experiments have given little or no encouragement to scientists hoping to find positive signs of such life.

27.9 THE ASTEROIDS

As early as the seventeenth century, astronomers had recognized the possibility that there might be a small planet between Mars and Jupiter, but it was not until 1801 that a small planet was observed in this region and named Ceres. Shortly thereafter a second object, Pallas, was found, followed by Juno and Vesta. Forty years later a fifth object was found, then many more smaller ones. Now called *asteroids*, these bodies have also been referred to as *minor planets*. All follow the planetary laws and are true planets of the Sun in the mechanical sense, if not in size.

Diameters of the four largest asteroids, named above, follow almost in the order of discovery:

Ceres	770 km
Pallas	480 km
Vesta	385 km
Juno	190 km

Most are very much smaller—a few kilometers in diameter or less—and show only as points of light rather than as disks. The total number of asteroids runs into the tens of thousands, some 40,000 of which can be detected on photographs. The great majority of asteroids follow orbits between Mars and Jupiter, but some cut inside the orbit of Venus; one is known to sweep outward almost to Saturn's orbit. Their combined mass is perhaps $\frac{1}{1000}$ to $\frac{1}{500}$ that of Earth.

Of particular interest is Eros, an irregularly shaped asteroid about 25 km long. Its orbit is highly eccentric, at times bringing it as close as 22 million km to Earth. The asteroid Icarus came within about 6.5 million km of Earth on June 14, 1968. Observations showed that Icarus is irregular in shape, less than 2 km in width, and may be composed of iron. The smallest asteroids we can observe are about 1 km in diameter, but many are probably much smaller than this. It is entirely reasonable that many of the meteorites (Chapter 28) that strike Earth should be regarded as simply very small asteroids.

27.10 THE GREAT PLANETS

Strikingly unlike the terrestrial planets are the four great planets—Jupiter, Saturn, Uranus, and Neptune. Even the smallest of these, Neptune, has almost 4 times the diameter and 15 times the mass of Earth; whereas the giant of the group, Jupiter, has 11 times the diameter and 300 times the mass of Earth. Apart from their size, a second striking difference in these two groups of planets is that of density (Table 27.1). The least dense is Saturn (0.7 g/cm^3), about one-eighth the density of Earth and less than three-fourths that of liquid water. The other three have densities of 1.3 to 1.7 g/cm^3, values only one-fourth that of Earth.

A third striking difference in the two groups of planets is the extremely low prevailing temperature on the surfaces of the great planets, ranging from $-138°$C on Jupiter to $-201°$C on Neptune. A fourth striking difference is in composition. The four terrestrial planets are probably all composed of a rock mantle surrounding an iron core and have either no atmosphere or atmospheres of almost insignificant mass. In contrast, the four great planets have massive atmospheres of hydrogen and helium, with some ammonia, methane, and water. These substances in the solid state also make up most of the mass of each planet.

Jupiter appears as a somewhat flattened disk with dark and light bands extending across the surface in rough parallelism with the planet's equator (Figure 27-13). The bands are made irregular by cloudlike patches that, if observed over a period of days or weeks, show changing patterns. Apparently, the bands are produced by systems of flow in Jupiter's atmosphere analogous to Earth's planetary wind systems.

By analysis of ultraviolet light reflected from Jupiter's atmosphere, it has been determined that the outer layer consists of 84 percent hydrogen and 15 percent helium. The remainder is largely methane and ammonia. The abundance of hydrogen in the atmosphere suggests that the planet as a whole has hydrogen as the predominant constituent. Jupiter may have a core of hydrogen in an extremely dense, metallic state.

Saturn is well known to all through its distinctive rings, which are seen as concentric bands of light and dark color lying in a very thin zone in the plane of the planet's equator (Figure 27-14). The rings consist largely of individual solid fragments, each revolving about the planet in an orbit as if it were an independent satellite of the planet. The particles may be on the order of the size of gravel and some much larger, because they are capable of reflecting radar signals. Altogether the rings constitute not more than one-millionth of Saturn's mass.

Although believed to be generally similar to Jupiter in composition and structure, Saturn's proportion of hydrogen may be larger to yield the low average density of only 0.7.

Uranus and Neptune are nearly twins so far as diameter and mass are concerned. Because of their great distances from Earth, these planets are difficult to observe and show little or no surface marking. Under spectroscopic analysis both planets show methane to be the dominant atmospheric constituent, whereas ammonia appears only in a trace. In addition to hydrogen compounds and helium in the solid state, both Uranus and Neptune may have rock cores, since their densities are somewhat greater than those of Jupiter and Saturn.

FIGURE 27-13
Photographed from Pioneer 10 spacecraft, Jupiter shows a strongly banded upper atmosphere. The Great Red Spot lies just below the center of the photograph. (NASA.)

FIGURE 27-14
The planet Saturn and its rings, photographed with the 250-cm Hooker telescope on Mount Wilson. (The Hale Observatories.)

27.11 PLUTO

Pluto is in a class by itself, because it is on the same order of size as the terrestrial planets but is located in a highly eccentric orbit beyond the great planets. Although its existence was long suspected because of irregularities in the orbits of Uranus and Neptune, Pluto was discovered only in 1930. It appeared as a very faint object, found to have changed position among the stars on successive photographs taken six days apart. The mass of Pluto has been calculated, from its distortion of Neptune's orbit, to be 11 percent of Earth's mass. The planet is too small to permit its diameter to be measured, but an estimate places the figure at not over 6400 km. Pluto's density has been estimated at about 5 g/cm^3. Its surface temperature is judged to be not far above absolute zero.

27.12 METEOROIDS AND METEORS

Meteoroids are tiny particles of matter traveling at high velocities in space and entering the Earth's outer atmosphere in vast numbers. Meteoroids are of interest not only because they have a place in the solar family, but also because when they enter Earth's atmosphere they become luminous trails, called *meteors*. The trails can be studied to give valuable information about the physical properties and motions of the upper atmosphere (Figure 27–15).

It is estimated that most meteoroids have a mass less than 1 g and range downward in size to perhaps one-thousandth of that mass. Thousands of millions of such specks of solid matter strike Earth's atmosphere daily.

Meteoroids are observed to occur in large numbers at certain dates of the year; these events are called *meteor showers*. For each meteor shower, the meteors appear to diverge from a common point in the sky, showing that they are traveling in parallel paths. Recurrence of the showers on an annual basis indicates that the Earth passes through the line of space travel of each of several meteoroid groups; each belongs to a *meteoroid swarm* moving in its own orbit about the Sun.

Individual meteoroid swarms have been identified as groups of particles following highly eccentric orbits about the Sun (Figure 27–16). An example is the Leonid swarm, which seems to come from a point in the constellation of Leo. Each year Earth crosses the orbit of the Leonid meteoroid swarm in the period November 14–15. Most of the Leonid meteoroids move in a group that makes a complete circuit about every 33 years (next appearance in 1999). At least eight other important swarms are known. Some meteoroids will be found scattered around the entire orbit of a swarm, so that at least a few will be seen each time Earth passes through the orbit.

27.13 COMETS

Finally, in this brief review of the members of the solar family, we come to the comets—to many persons the most bizarre of the astronomical objects. The *comet* consists of a brightly luminous head, called the *coma*, from which a luminous *tail* streams off in a direction away from the Sun (Figure 27–17).

The apparent size of a comet varies greatly. The larger ones have tails as long as 80 million km, covering a wide arc in the sky. Within the head of a large comet there is often a bright, starlike center, known as the *nucleus*. The material constituting the coma and tail is so diffuse that stars shine through it with undiminished intensity.

Most comets, like meteoroid swarms, follow highly eccentric elliptical orbits around the Sun (Figure 27–18). Because of their high velocity near perihelion, at which time they are visible from Earth, comets are seen for only a very short time in comparison with the total period of their revolution. At aphelion the velocities are greatly diminished, so that a comet spends most of its time among or beyond the outer planets. Most

FIGURE 27–15
This meteor trail showed a sudden increase in brightness as it traveled toward the lower right. (Yerkes Observatory.)

comets have orbital periods of tens of thousands of years. However some, among them Halley's comet, reappear at regular intervals ranging from three years to a few centuries. About 100 such periodic comets are known.

A particular comet group, known as the *Jupiter family,* about 40 in all, have orbits whose most distant points are close to Jupiter's orbit. It is believed that the gravitational attraction of Jupiter entrapped these comets into their relatively small orbits. Their periods range from 3.3 years (Encke's comet) to 8.6 years.

Comets have an extremely low density; the coma and tail consist entirely of fine dust particles or of gaseous matter driven off from the nucleus. The nucleus itself, however, is believed to be an aggregation of small particles of solid matter and frozen gases. Spectroscopic analysis of its composition has determined this matter to be largely methane, ammonia, carbon dioxide, and water.

During close approach to the Sun, some of the gas is vaporized and ionized under the intense heat of the Sun's rays, diffusing outward to form the comet tail, and under pressure of the solar wind the tail is pushed away from the Sun. As a result of passage close to the Sun, a comet may lose part of its mass or be completely disrupted.

It is thought that the diffuse dust particles remaining from disintegrated comets may constitute the meteoroid swarms. This possibility is strengthened by the fact that the highly eccentric orbits of both meteoroid swarms and comets are much alike. Several cases are known in which the orbit of a meteoroid swarm is identical with the orbit of a previously known comet.

The source of matter for comets and meteoroids is a topic of speculation. The Dutch astronomer Jan H. Oort suggested that the comets originate in a vast cloud, or "reservoir," containing the substance of millions of comets and located far beyond the orbits of Neptune and Pluto. The matter in such a belt would be at a temperature close to absolute zero and would have been formed early in the history of the solar system.

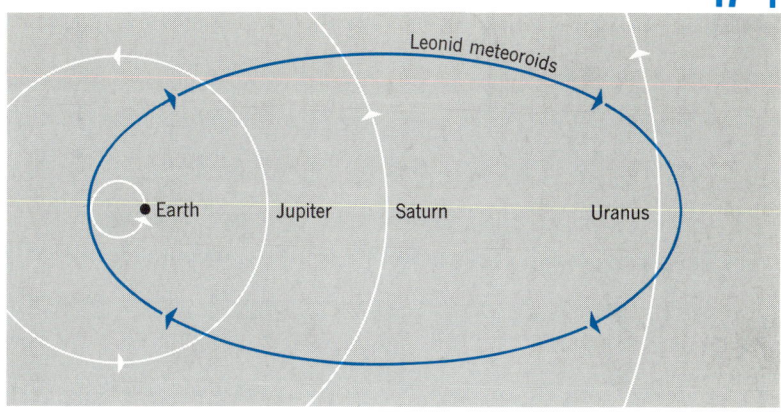

27.14 SATELLITES OF THE PLANETS

A *planetary satellite* is a solid object orbiting a planet. It is held in the planet's gravitational field. Altogether 33 satellites, or "moons," have been identified in orbits around the nine planets, but the distribution is highly varied in terms of numbers per planet (Table 27.1). The two innermost planets, Mercury and Venus, have no moons. Mars has two: Deimos and Phobos. Both are smaller than Earth's single moon and both orbit much closer to the parent body.

Jupiter's thirteen moons consist of four large ones and at least nine small ones. Galileo, using the first astronomical telescope, discovered the four large moons in 1610, and they have since been designated the *Galilean satellites.* Their

FIGURE 27-16
Orbit of the Leonid meteoroid swarm. The plane of the meteoroid orbit is inclined 17 degrees to the plane of the ecliptic. (From *Elements of Astronomy* by E. A. Fath. Copyright 1934 by the McGraw-Hill Book Company. Used by permission of the publisher.)

FIGURE 27-17
Halley's comet photographed on May 12 and 15, 1910, at Honolulu, Hawaii. The shorter tail (*right*) covers 30 degrees of arc, the longer tail (*left*) 40 degrees. (The Hale Observatories.)

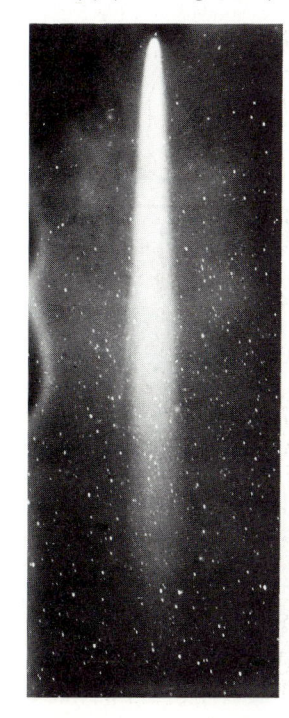

27 THE SOLAR SYSTEM

FIGURE 27-18

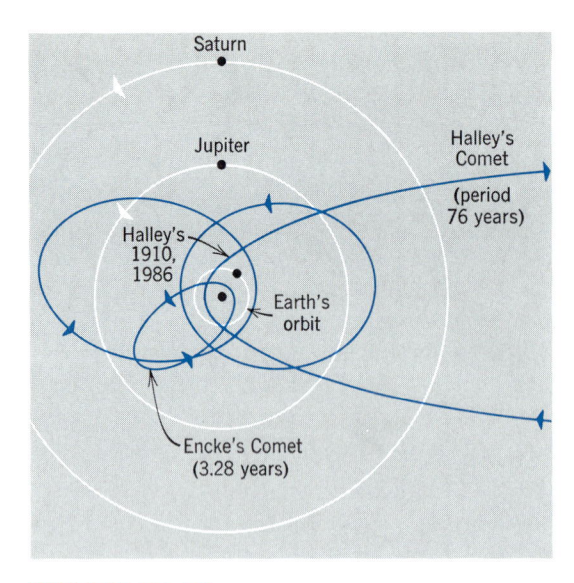

The orbits of Halley's comet and three comets of the Jupiter family. (From PRINCIPLES OF EARTH SCIENCE by Arthur N. Strahler. Copyright © 1976 by Arthur N. Strahler. By permission of Harper & Row, Publishers, Inc.)

names are Io, Europa, Ganymede, and Callisto, stated in order outward. Inside the orbit of Io is another very small satellite. The remaining seven lie far beyond the orbit of Callisto and are very small objects.

The Galilean satellites are easily seen through a low-powered telescope or good pair of binoculars (Figure 27-19). The apparent positions of these moons change greatly from night to night, but all revolve in about the same plane and in the same direction. Galileo's observation of these satellites was a powerful point in favor of the

FIGURE 27-19

The four Galilean satellites of Jupiter as they might appear through binoculars or a small telescope. Distances from Jupiter are shown to correct scale, as if all four satellites were in a line at right angles to the observer. Diameters are not to scale. (From PLANET EARTH: ITS PHYSICAL SYSTEMS THROUGH GEOLOGIC TIME by Arthur N. Strahler. Copyright © 1972 by Arthur N. Strahler. By permission of Harper & Row, Publishers, Inc.)

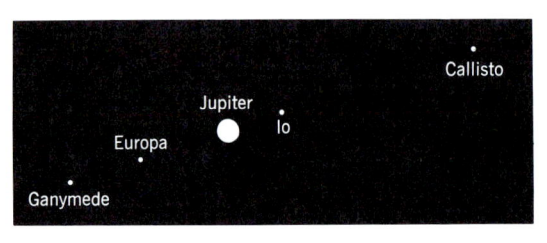

Copernican theory of the solar system, because it provided a small-scale model of planets orbiting a Sun. Saturn has ten large moons; the most distant is Phoebe, traveling in a retrograde orbit of 13 million km with a period of over 100 days. Little is known of the five large moons of Uranus and the two of Neptune.

27.15
THE EARTH'S MOON

Of all the satellites, Earth's Moon is unique in that it is a very large body in ratio with the planet which it orbits. With a diameter of 3476 km, compared with an Earth diameter of about 13,000 km, the ratio of diameters is about 1 to 4. Our Moon has a mass of about one-eighty-first that of Earth. Because of the Moon's relatively large size, compared to the planet it orbits, astronomers have commented that the Earth-Moon system can be considered a "binary planet."

Figure 27-20 shows the Moon's orbit and gives distances from Earth. The average radius of the Moon's orbit is about 384,000 km. The Moon's orbit is quite strongly eccentric. When closest to Earth, the Moon is said to be in *perigee*; when most distant, in *apogee*. These terms apply to corresponding positions of any satellite, including the artificial orbiting satellites. Notice that the Moon revolves about the Earth in a direction that can be described as counterclockwise (direct motion), when we imagine ourselves to be viewing the system from a point above the Earth's north pole. Rotation of both Earth and Moon is also direct in this sense.

The period of the Moon's revolution, calculated in terms of 360 degrees of angle with reference to the fixed stars, is 27.32 days. This period is the *sidereal month*. However, when we measure the revolution in terms of reference of the Sun's position in the sky (from one new moon to the next) the period averages $29\frac{1}{2}$ days. This period is the *synodic month*. We shall see that a monthly rhythm of the tides follows the synodic month.

Of interest in the history of both Earth and Moon is the fact that the Moon always shows the same face to observers on Earth. Until orbited by manned Apollo spacecraft, no human being had directly viewed the opposite side of the Moon, although photographs returned by unmanned satellites had previously revealed many details of that surface.

The fact that we can see only one side of the Moon requires that the Moon's period of rotation upon its axis be exactly the same as its period of revolution with respect to the stars. So, the Moon rotates once in 27.32 days. Such a coincidence could scarcely be ascribed to mere

27.17 LUNAR ECLIPSE

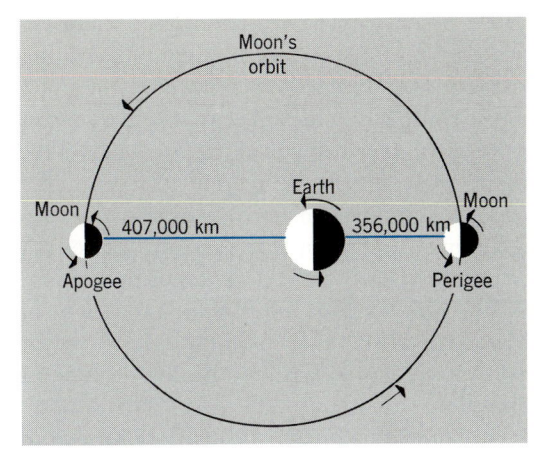

FIGURE 27-20
The Moon's orbit is an ellipse. Distances shown are from center of Earth to center of Moon.

chance. The Moon is slightly lopsided and has a concentration of denser rock bodies on the side facing the Earth. Gravitational attraction for this excess mass has locked the Moon's rotation into a period equal to its revolution.

27.16 PHASES OF THE MOON

We must know the simple geometric relations of the Earth, Moon, and Sun in order to understand the phases of the Moon, the eclipses, and the tide (Figure 27–21). Assume for the moment that the Moon's orbit lies in the plane of the

FIGURE 27-21
Relative position of Moon, Earth, and Sun determine the Moon's phases. (From PRINCIPLES OF EARTH SCIENCE by Arthur N. Strahler. Copyright © 1976 by Arthur N. Strahler. By permission of Harper & Row, Publishers, Inc.)

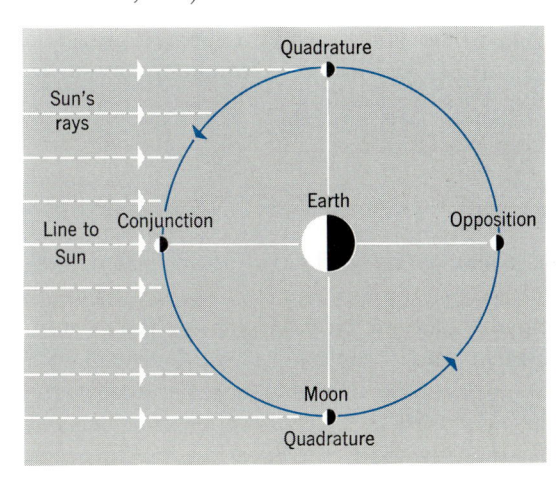

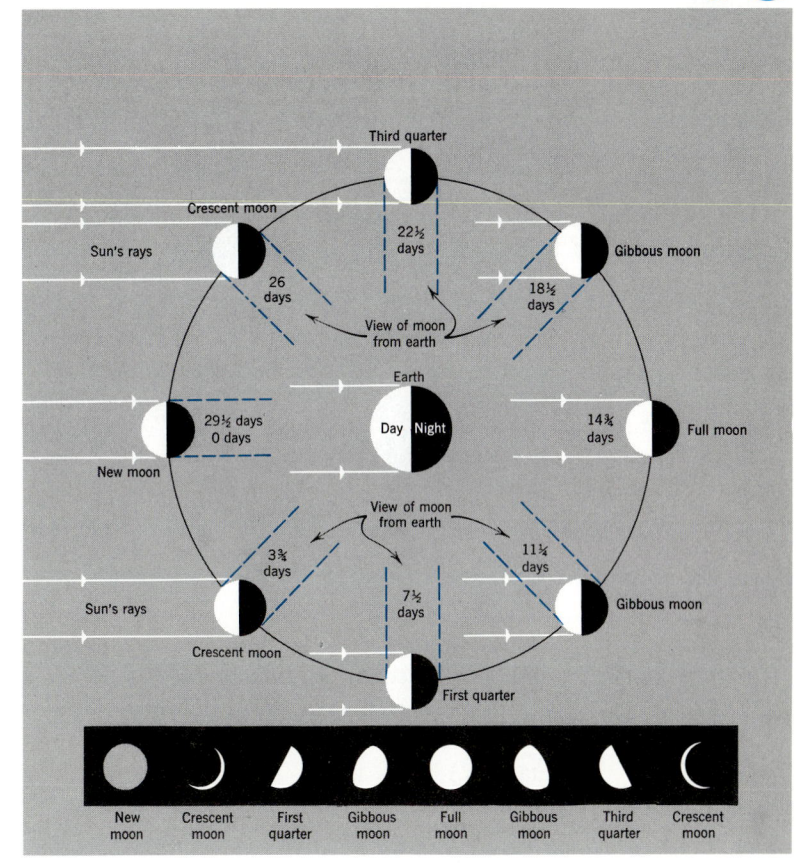

FIGURE 27-22
Phases of the Moon. Diagrams below show outline of Moon as seen in the southern half of the sky.

Earth's orbit, and consider only the alignment of the three bodies in a single plane.

When all three bodies lie along a single straight line, with Moon and Sun on the same side of Earth, the Moon and Sun are said to be in *conjunction;* when aligned with the Earth located between them, Moon and Sun are in *opposition.* A contrasting relation, called *quadrature,* exists when the line from Earth to Moon makes a right angle with respect to the line from Earth to Sun. In each synodic month the Moon will be found twice in quadrature.

The series of changes in appearance of the Moon throughout the synodic month are known as the *phases of the Moon* (Figure 27–22). Moonlight consists of sunlight falling on the Moon's surface and reflecting to Earth. The Moon is at all times divided into a sunlit hemisphere and a shadowed hemisphere. The phases of the Moon are simply the varying proportions of the sunlit and shadowed halves that we see from the Earth.

27.17 LUNAR ECLIPSE

When Sun and Moon are in opposition, the Moon may cross Earth's shadow, which is rela-

27 THE SOLAR SYSTEM

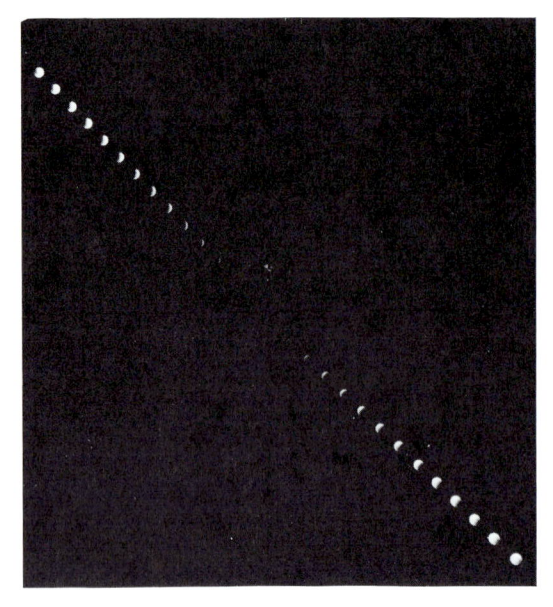

FIGURE 27-23
An eclipse of the Moon, recorded by multiple exposures at equal time intervals. The photograph has been condensed in the region of totality. (The American Museum–Hayden Planetarium.)

tively large in proportion to the Moon's diameter. The result is an eclipse of the Moon, or *lunar eclipse*. If the Earth's shadow merely passes across one edge of the Moon's disk without entirely enveloping it, a *partial eclipse* is said to occur; whereas if Earth's shadow completely covers the Moon, as shown in Figure 27-23, a *total eclipse* results.

FIGURE 27-24
Diagram of lunar eclipse. Scale of diameters of Earth, Moon, and umbra is about ten times the scale of separating distance. Point of the umbra cone lies 1,380,000 km from the Earth's center. (From PRINCIPLES OF EARTH SCIENCE by Arthur N. Strahler. Copyright © 1976 by Arthur N. Strahler. By permission of Harper & Row, Publishers, Inc.)

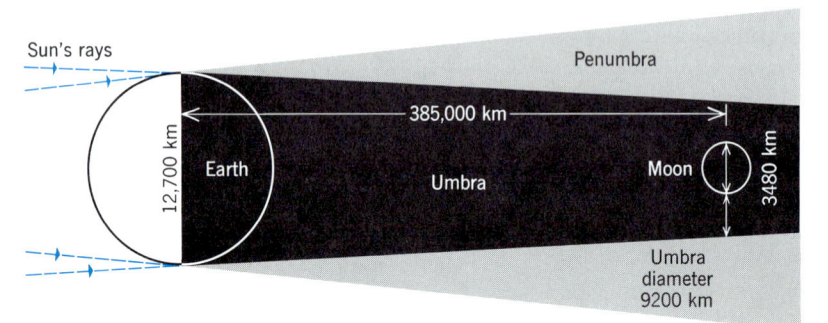

In a lunar eclipse two shadow zones are encountered (Figure 27-24). Because of the Sun's relatively large size, there is formed an inner cone of complete shadow, the *umbra*, surrounded by a zone of partial shadow, the *penumbra*. As the Moon enters the penumbra, it goes gradually from full illumination into a region of increasingly dim light; then it crosses an abrupt boundary into the nearly total darkness of the umbra.

If the Moon's orbit lay exactly in the plane of the earth's orbit, a total lunar eclipse would result once each synodic month at the time of opposition (full moon). However, the Moon's orbital plane is inclined by about 5 degrees from the plane of the Earth's orbit, so that a large share of the times that the Moon is in opposition it will pass above or below the Earth's shadow. As a result, there are only two periods per year within which an eclipse can occur.

27.18
SOLAR ECLIPSE

An eclipse of the Sun, or *solar eclipse*, occurs when the Moon is in conjunction with the Sun and casts its shadow on the Earth. For persons in the shadow zone on Earth, the Moon's disk seems to pass across the Sun's disk, partially or totally obscuring the Sun.

The Moon's core of total shadow forms only a very narrow track across the Earth—up to 270 km wide (Figure 27-25). For this reason, you will rarely see a total eclipse unless you make a special effort to travel to the predicted track. The shadow of totality travels across the Earth's surface at some 1600–6400 km/hr. The total phase lasts, at most, only $7\frac{1}{2}$ minutes. The region of partial eclipse, in contrast, is a very broad zone, up to several thousand kilometers wide.

On some occasions, when the separating distance between Sun and Earth is greater than average, the umbra fails to reach the Earth. This situation is shown in the lower diagram of Figure 27-25. Then only the penumbra reaches the Earth and only a partial eclipse occurs. As in the case of the lunar eclipse, there are only two periods in each years when conditions are favorable for a solar eclipse.

27.19
THE OCEAN TIDE

The periodic rise and fall of ocean level, or *ocean tide*, was known for centuries to be related to the Moon's path in the sky, but it was not until Newton published his law of gravitation in 1686 that the physical explanation became understood. Figure 27-26 shows the principle. The broad arrow represents the attraction the Moon

27.19 THE OCEAN TIDE

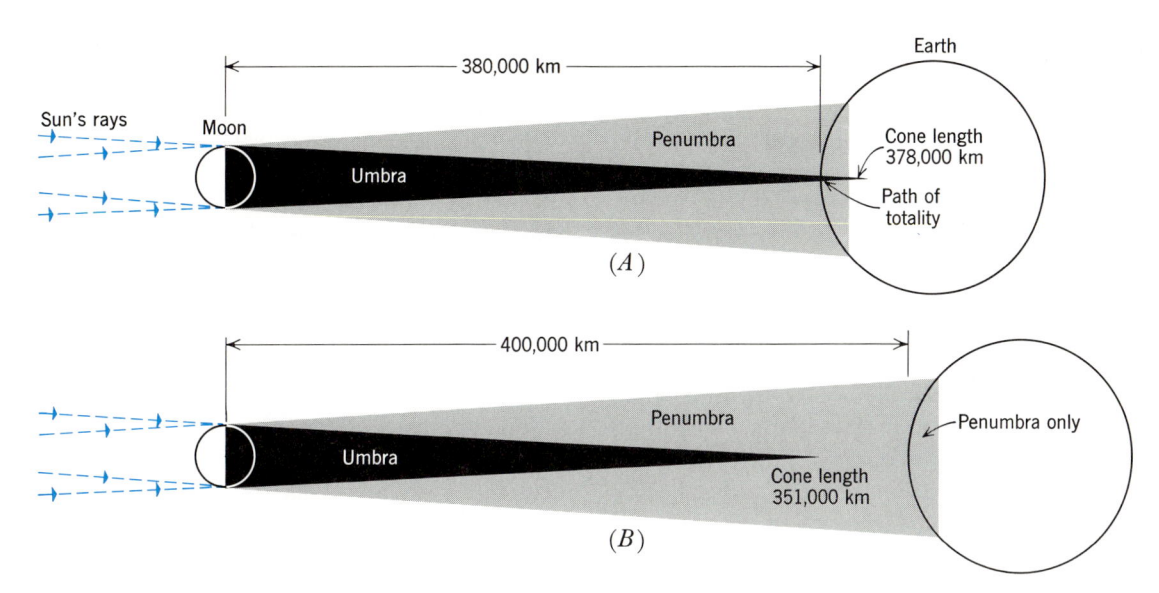

FIGURE 27-25
Diagrams of solar eclipses. (A) Long umbra cone and minimum separating distance allow a total eclipse to occur. (B) Short umbra cone and maximum separating distance allow partial eclipse only. (From PRINCIPLES OF EARTH SCIENCE by Arthur N. Strahler. Copyright © 1976 by Arthur N. Strahler. By permission of Harper & Row, Publishers, Inc.)

exerts on the Earth. On the side nearest the Moon, point T, the attraction is stronger than at the Earth's center, point C, because gravitational force decreases with an increase in the separating distance between two masses. For the same reason, the attractive force is even less at point A. The effect of these force differences is to distort the spherical shape of the Earth into a *prolate ellipsoid*, shaped somewhat like a United States football.

Although the solid lithosphere makes only a very small response to the earth-stretching tidal force, the oceans respond freely. As Figure 27-27 shows, the ocean water tends to move along the lines indicated by the surface arrows. The water

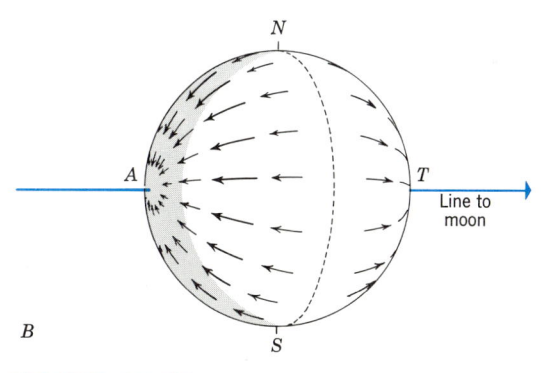

FIGURE 27-27
Ocean tides are caused by tractive forces directed along the Earth's surface toward two centers.

FIGURE 27-26
Gravitation is the basic tide-producing force.

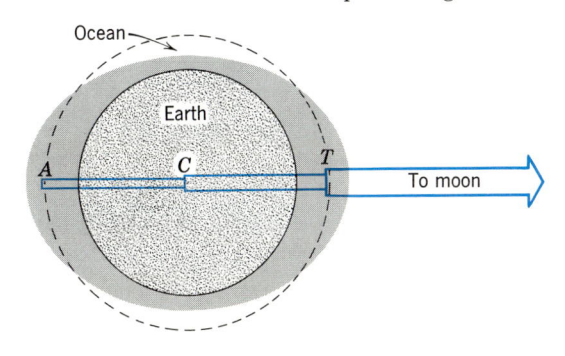

moves toward two centers, one at A and one at T, but moves away from a belt girdling the globe on a line passing about through the poles.

Next, we realize that the Earth is rotating on its axis. This means that the two centers of tidal accumulation, or "tidal bulges," will be traveling continually around the Earth. For a fixed point on the globe, the tidal bulges will pass by twice daily. Each bulge produces a *high water*. The actual interval is close to $12\frac{1}{2}$ hours between high waters. The belt of depressed surface will also pass by twice daily, so that we will have a *low water* $6\frac{1}{4}$ hours following a high water. This is the common tidal cycle found along most coastlines of the world.

Figure 27-28 shows a typical *tide curve* recorded throughout a 24-hour period. The difference in level between high water and low water, or *tide range,* was about 3 m in this particular case.

The Sun also produces a tide-raising force on the Earth, but it is not as strong as the Moon's force, because of the much greater distance between Earth and Sun. However, the Sun's tide-raising force is added to the Moon's force during both conjunction and opposition, when all three bodies are in a single line. At such times, the tide range is increased, an event called the *spring tide.* When Sun and Moon are in quadrature, pulling at right angles, the forces are diminished and tides of low range occur; these are called *neap tides.* Many astronomical variables influence the tide, and we cannot go into them in this very brief explanation.

27.20
THE SOLAR SYSTEM IN REVIEW

All members of the solar system, whether they be giant planets or specks of dust in meteoroid swarms, orbit the Sun in obedience to Newton's laws of gravitation and motion. The satellites orbit their master planets under these same laws. Outer space is so nearly devoid of matter that these motions are frictionless along the orbital pathways. Within the span of human history, the motions continue without measurable change. Nothing within human experience is so completely reliable, so exact in its schedule, and so beyond our power to alter as the apparent motions of the Sun, Moon, and planets.

Yet the motions of the solar system do undergo very slow changes. These arise from the phenomenon of tides, which affects all pairs of objects in the solar system. The invisible force of gravitational attraction deforms the planets and satellites as they revolve and rotate. This flexing by tidal force is a form of friction, and it robs the system of energy. So the orbits must change with time. Ultimately, the original energy of motion imparted to the solar system at the time of its formation will be converted into heat and dissipated into interstellar space. Perhaps, then, the ultimate demise of our solar system will take place as its members, one by one, fall into the Sun.

SUMMARY

Heliocentric theory—the Copernican system, in which planets revolve about the sun—replaced the Ptolemaic geocentric system through the efforts of Copernicus, Galileo, and Kepler, and with the ultimate support of Newton's laws of motion and gravitation.

Kepler's laws: I. The orbit of every planet is an ellipse with the sun at one focus. II. A planet moves at a varying velocity, such that the radius vector sweeps over equal areas in equal times. III. The squares of the periods of revolution of any two planets are proportional to the cubes of their mean distances from the Sun.

Inner planets. In order from Sun, the four inner, or terrestrial planets are Mercury, Venus, Earth, and Mars; all are comparitively small, with silicate rock mantles and iron cores. Mercury and Mars have little atmosphere; Venus has a dense, reflective atmosphere.

Asteroids. Numerous small planetary bodies, the minor planets, or asteroids, orbit largely between Mars and Jupiter. Possibly they represent a disrupted planet.

Great planets. In order outward, Jupiter, Saturn, Uranus, and Neptune comprise the great planets. All have masses several times greater than Earth but low densities compared with the inner planets. Surface temperatures are very low. Hydrogen and helium comprise their bulk; gaseous atmospheres are dense. Pluto, the outermost planet, is small and in a class by itself.

Meteoroids and comets. Meteoroids, tiny solid particles, travel in swarms in orbits about the Sun. They are intercepted by Earth and appear as meteors. Comets also follow highly eccentric orbits about the Sun. Comets consist of cold diffuse matter—gases and dust.

Planetary satellites. A total of 33 satellites, or moons, orbit the planets, most in direct motion. Jupiter's

FIGURE 27-28
Graph showing the height of water at Boston Harbor measured every half hour for a 24-hour period. (After H. A. Marmer.)

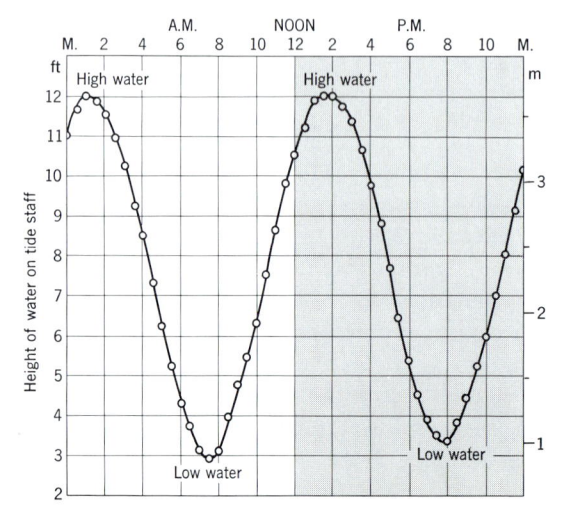

four Galilean moons comprise a small-scale solar system.

Earth's Moon. Unusually large as a satellite, Earth's Moon follows an eccentric orbit. Rotation period exactly equals the sidereal month, so that one side of the Moon always faces the Earth. Lunar phases and tides follow the synodic month with reference to the Sun. Eclipses occur at opposition or conjunction when Earth's shadow falls on Moon, or vice versa.

TERMS AND CONCEPTS

heliocentric theory
geocentric theory
Ptolemaic system
Copernican system
solar system
planet
major planets
inner planets
terrestrial planets
outer planets
great planets
orbit
ellipse
focus (foci) of ellipse
major axis
minor axis
radius vector
perihelion
aphelion
eccentricity
harmonic law
revolution
rotation
year
day
direct motion
retrograde motion
asteroids
minor planets
meteoroids
meteors
meteor shower
meteoroid swarm
comet
coma
tail of comet
nucleus of comet
Jupiter family
planetary satellite
Galilean satellites
perigee
apogee
sidereal month
synodic month
conjunction
opposition
quadrature
phases of Moon
lunar eclipse—partial, total
umbra
penumbra
solar eclipse
ocean tide
prolate ellipsoid
high water
low water
tide curve
tide range
spring tides
neap tides

QUESTIONS

1. Review the steps in development of the heliocentric and geocentric theories of planetary motion. What evidence favored the Copernican system over the Ptolemaic system?

2. What kinds of objects constitute the solar system? To what extent are the paths and directions of motion of these objects uniform? What is the significance of such uniformity?

3. Name the major planets in order outward from the Sun. Group the planets into two classes. In what respects are planets similar within each group? What are the major differences between the two groups?

4. Describe the orbit of a planet in terms of the properties of an ellipse. What is eccentricity? How can an ellipse be constructed?

5. State Kepler's first and second laws of planetary motion. How does the second law explain the difference in speeds of revolution at aphelion and perihelion?

6. State Kepler's third law of planetary motion. How can the astronomical unit be used to simplify statement of the law? Does a planet exert any influence on the motions of other planets?

7. Compare periods of planetary rotation in broad terms, giving approximate values. Is direction of rotation significant? Explain.

8. Give a brief résumé of each of the inner planets (excluding Earth), including such items as relative size, density, gravity, presence or absence of atmosphere, atmospheric composition, surface temperatures, and distinctive surface features. Then compare or contrast the surface environment of each planet with that of Earth. Which of the three is the most likely to sustain life forms?

9. Describe the asteroids in terms of orbits, size, numbers, and composition. How close to Earth do

they pass? What is known of the shapes of larger asteroids? What is their origin?

10. Describe Jupiter in some detail and then compare it with Saturn, Uranus, and Neptune in more general terms. What may be the internal structure and composition of these great planets? How do their surface temperatures compare with one another and with those of the inner planets? What is known about Pluto?

11. Distinguish between meteoroids, meteors, and meteorites. Describe meteors. What is a meteoroid swarm? A meteor shower?

12. Describe a typical comet. What is the coma? The tail? What is known of the composition of comets? What is the nature of their orbits? What are the Jupiter family of comets, and how did they gain their orbits? What effect has the solar wind on a comet? Where does the substance of comets originate?

13. What is the total number of satellites identified for the nine planets? Which planets have none? Which planet has the most? How does our Moon compare with other planetary satellites? What are the Galilean satellites? What influence did their discovery by Galileo have on conflicting views of his time concerning solar-system motions?

14. Compare the Moon with Earth in terms of diameter, mass, and surface gravity. Describe the Moon's orbit with regard to eccentricity. What are perigee and apogee? What is the Moon's period of revolution with respect to the stars? With respect to the Sun? How does the Moon's period of rotation compare with its period of revolution?

15. How are phases of the Moon related to its orbital position with respect to the Sun? What causes the phases? Explain the geometrical configurations required to produce a lunar eclipse and a solar eclipse.

16. What causes the ocean tide? Explain in terms of tide-raising forces. Describe a typical tide curve. How does the Sun's gravitational force modify the ocean tides?

CHAPTER 28
ASTROGEOLOGY— THE GEOLOGY OF OUTER SPACE

28.1 ASTROGEOLOGY

It may seem absurd at first thought to extend the word "geology" to include a study of the Moon, planets, and other solid objects of the solar system. The prefix "geo," after all, comes from the Greek word meaning "earth." Still, the Moon and closer planets are composed of minerals and rocks, and they can be expected to have some of the geologic features of our Earth, such as volcanoes and lava flows, faults and earthquakes, mantles and cores, and perhaps even tectonic belts. There may even exist sediments and sedimentary strata of some sorts on these bodies. Perhaps, then, it is not at all unreasonable to extend the science of geology to outer space, applying it wherever we can.

Astrogeology is the name established by the United States Geological Survey for the application of principles and methods of geology to all condensed matter and gases of the solar system outside the Earth.

Moon and Mercury, and to some extent Mars as well, have one feature in common that makes geologic investigations especially fruitful: they have practically no atmosphere. None of the three have running streams or oceans of free water. On Earth, processes of rock weathering, erosion by streams, waves and wind, and sedimentation quickly remove or bury features produced by volcanism, tectonic activity, and the impacts of large bodies from outer space. Where a planet has little or no atmosphere or hydrosphere, these processes either are absent or act very, very slowly. As a result, surface forms and structures produced by volcanic and tectonic activity, and by impact from outer space, can remain little changed through eons of geologic time.

Moon, Mars, and Mercury preserve on their faces the events of very early segments of geologic time. Study of these planetary features can shed much light on the early history of our own planet. For example, what was the composition of the earliest crust of the Earth? If none is preserved on our planet, perhaps it still remains on the Moon or on Mercury.

It is not surprising that earth scientists mounted a tremendous effort to explore the Moon's surface and to bring back samples for analysis. Lunar geology held great promise for new knowledge, not only of Moon itself, but for early Earth history as well. At least one leading scientist has likened the Moon to the Rosetta stone, which gave scholars the key to translation of Egyptian hieroglyphics, and so opened up the field of Egyptian history. In this chapter we apply geologic principles to understanding the Moon, the inner planets, and those fragments of a disrupted planet—the meteorites.

28.2 ORIGIN OF THE SOLAR SYSTEM

An overview of astrogeology would not be complete without at least a short sketch of the origin of the solar system; it makes a good background for understanding the geology of outer space.

Modern thinking runs along the lines that our solar system is a by-product of the formation of

28 ASTROGEOLOGY—
THE GEOLOGY OF OUTER SPACE

FIGURE 28-1
Artist's conception of a disklike solar nebula surrounding a central solar mass. (NASA.)

our Sun, an average-sized star with a history quite typical of many other stars of medium mass and temperature. (That history is given in Chapter 29.) A very diffuse body of hot interstellar dust and gas—a *nebula*—is taken as the starting point. The *solar nebula* began to contract under the mutual gravitational attraction of all its particles on one another. Gradually the nebula began to cool and assumed a wheellike shape—very thin and somewhat larger in diameter than our present solar system (Figure 28-1). A dense, luminous mass of gas that was to become the Sun formed an enlarged central hub.

Then, with further cooling of the nebular disk, condensation of gases began to occur rapidly. Minute particles of iron and silicates became solid first, then later the highly volatile compounds such as water, methane, and ammonia. As grains collided they began to stick together, a process called *accretion*. The resulting aggregations of matter are called *planetesimals*. They grew rapidly, but many were unstable and disrupted into countless small solid fragments; these orbited the sun along with the remaining large planetesimals. By a series of collisions, the larger planetesimals swept up most of the smaller fragments and quickly grew to their final dimensions. The impacts were not so high in energy release as to heat the growing planets to the point of complete melting.

A possible tenth planet may have formed between Mars and Jupiter and then later fragmented into solid bodies we know as the asteroids. Recall from Chapter 27 that most asteroids now orbit between the orbits of Mars and Jupiter. Possibly our Moon was another of the early planets, formed by accretion at the same time as Earth; at least, this is one hypothesis of the origin of the Moon.

A timetable for this phase of Earth history has been suggested as follows: the solar nebula was in process of rapid contraction about 5 billion years ago (-5.0 b.y.); accretion of the planets was largely completed at about -4.7 to -4.5 b.y.

Earth and the other three inner planets were probably not in a completely molten state at any time during accretion. However, there must have been local melting where exceptionally large objects impacted the surface. The Earth at this stage was a mixture of silicates and iron, not as yet differentiated into a core and mantle. That change was to follow as radiogenic heat accumulated and caused major episodes of melting and overturn. In any event, the Earth had acquired its layered structure and was probably internally stable by about -3.6 b.y., when the first crustal rocks of which we have any record were formed.

This brief overview of planetary origin may

raise more questions than it answers. What happened to the volatiles, such as methane and ammonia, probably abundantly present in early stages of accretion? These may have been driven off the growing planetesimals closest to the sun through heating and the pressure of the solar wind. Some water remained, however, locked up in the silicate minerals of the growing planet Earth. This water was later to emerge slowly by *outgassing* from volcanoes, ultimately to form the world ocean. Many other volatiles emerged with water, including nitrogen, carbon (as CO_2), chlorine, sulfur, argon, and fluorine. In this way the atmosphere developed into its modern state.

28.3 TESTIMONY OF THE METEORITES

Let us begin our geologic investigation of outer space with the meteorites. These fragments of rock and metal have been studied for many decades by geologists. Consequently, much was known about meteorites long before manned space vehicles went to the Moon.

We cannot obtain samples of the matter making up the Earth's mantle and core. However, for what it is worth in allowing controlled inferences, the evidence supplied by meteorites has been exploited with interesting results, as to both the composition of the Earth's interior and the age of the members of the solar system.

We described meteoroids and meteoroid swarms in Chapter 27. Most are, of course, extremely tiny bits of matter. Occasionally, a very large meteoroid enters the Earth's atmosphere. This object may be considered a small asteroid with a highly eccentric orbit. Though partially vaporized, it may reach the Earth's surface. These exotic solid objects are then called *meteorites*.

The fall of a meteorite is accompanied by a brilliant flash of light, called a *fireball*. There may also be loud sounds. Frictional resistance with the atmosphere causes the outer surface of the object to be intensely heated. However, this heat does not penetrate to the interior of a meteorite, which reaches the Earth with its original composition and structure unchanged. The single mass may explode before the impact, showering fragments over a wide area. The observed arrival of a meteorite and subsequent collection of the fragments is designated as a *fall*. Collection of a meteorite whose fall was not observed is designated as a *find*. Examples of very large meteorites are shown in Figure 28–2.

FIGURE 28–2
Two varieties of meteorites. (*A*) This stony meteorite, weighing 338 kg, is the largest single stony meteorite observed to fall. It struck the ground at Paragould, Arkansas, on February 17, 1930, forming a huge fireball visible over thousands of square kilometers. Height of the meteorite is about 0.6 m. (Yerkes Observatory.) (*B*) The Willamette meteorite, an iron meteorite, weighs 14 metric tons and is over 3 m long. The huge cavities were produced by rapid oxidation of the iron during fall through the atmosphere. (American Museum of Natural History–Hayden Planetarium.)

Meteorites have been intensively studied, not only as to chemical composition and structure, but also as to age. They fall into three classes.

1. The *irons* are composed almost entirely of a nickel-iron alloy, in which the nickel content ranges from 4–20 percent.

2. At the other end of the series are the *stones*, consisting largely of silicate minerals, mostly olivine and pyroxene, and with only 20 percent or less nickel-iron. Plagioclase feldspar may also be present.

3. An intermediate class of meteorites consists of the *stony irons*. In these, silicate minerals

and nickel-iron may form a continuous medium in which spherical bodies of silicate minerals are enclosed.

One group of stony meteorites possesses a coarse-grained structure resembling the structure of plutonic igneous rocks. We can infer that meteorites with this structure solidified from a magma.

When the meteorites of observed falls are cataloged, their relative abundance turns out to be about as follows: stones, 94 percent; irons, 4.5 percent; stony irons, 1.5 percent. The stony meteorites are thus preponderant in bulk. The nickel-iron meteorites can be interpreted as cores of those planetary bodies in which the process of differentiation had taken place. Stony meteorites having textures resembling terrestrial igneous rocks point to the possibility that melting and recrystallization had taken place to some degree in these original planetary bodies.

Age determinations of meteorites have been made using the uranium-lead, potassium-argon, and rubidium-strontium methods described in Chapter 25. There is a high degree of agreement in the results pointing to the time of formation of all types of meteorites as -4.5 b.y. before present. Note that this age is about 1 b.y. greater than that of the oldest known crustal rocks on earth. The composition and great age of meteorites has lent strong support to the accepted hypothesis that the Earth's core is composed of nickel-iron and the mantle of iron-magnesium silicates. This conclusion is greatly strengthened by independent evidence of the density and related physical properties of the Earth's interior derived from the study of earthquake waves, as we found in Chapter 24.

28.4 METEORITE IMPACT FEATURES ON EARTH

Have meteoroids of great size struck the Earth's surface to produce recognizable impact features? Perhaps the finest example of a large circular-rimmed crater of almost certain meteoritic impact origin is the Barringer Crater (formerly known as Meteor Crater) in Arizona (Figure 28-3). The diameter of this crater is 1200 m and its depth almost 180 m. The rim rises about 45 m above the surrounding plateau surface, which consists of almost horizontal limestone strata. Rock fragments have been found scattered over a radius of 10 km from the crater center, while meteoritic iron fragments numbering in the thousands have been collected from the immediate area.

Only a handful of known meteorite craters are found on the earth's surface today and we can safely draw the conclusion that impacts by large meteorites have been extremely rare events in recent geologic time. Perhaps the number has been approximately half a dozen occurrences per million years in the latest (Cenozoic) era.

We may also safely conclude that meteorites are not being produced in the modern solar system. If they represent space debris that continues to be swept up by planets and satellites,

FIGURE 28-3
Air view of the Barringer Crater in northern Arizona. (Yerkes Observatory.)

the frequency of impacts should be small in later geologic eras compared with a high frequency in the early stages of planetary formation.

So we must look to a celestial body on which impact features would have been preserved with little or no erosion for the entire 4.5 b.y. since Earth and other planets formed. Three such bodies are available for study—the Moon, Mars, and Mercury. Venus is so completely concealed by a dense cloudy atmosphere as to be an unlikely prospect.

28.5 THE LUNAR SURFACE ENVIRONMENT

To interpret correctly the physical features of the Moon's surface, we must first consider the surface environment of the satellite. Environmental factors include the Moon's gravity field, a lack of both atmosphere and hydrosphere, intensity of incoming and outgoing solar radiation, and extremes of surface temperatures. All of these factors show striking differences when compared with the environment of Earth.

Gravity on the Moon's surface is about one-sixth as great as on Earth. Therefore, an object that weighs 6 N on Earth will weigh only 1 N on the Moon. This relatively small gravity is of human interest when we watch films of our astronauts cavorting on the lunar surface. The small gravity is of great geologic importance in interpreting the Moon's surface and history. For example, rock of the same strength as rock on Earth could stand without collapse in much higher cliffs and peaks on the Moon than on Earth. Objects thrown upward at an angle from the Moon (as when the Moon is struck by a large meteoroid) will travel much higher and farther than under the same impetus on Earth.

Lack of sufficient gravity has cost the Moon the loss of any atmosphere it may have once possessed, since gas molecules of any earlier atmosphere could have escaped readily into space.

Lacking an atmosphere, the Moon's surface intercepts the Sun's radiation on a perpendicular surface with the full value of the Sun's output. Moreover, the Moon absorbs most of this incoming energy. The effect is to cause intense surface heating during the long lunar day of about two weeks duration. With the Sun's rays striking at a high angle for several days continuously, surface temperatures at lunar noon reach about 100°C. Correspondingly, conditions on the dark side of the Moon reach opposite extremes of cold during the long lunar night. With no atmosphere to block the escape of heat, surface temperatures drop to values estimated at below −170°C.

From the scientific standpoint, such vast temperature ranges are of interest because of the possible effect on minerals exposed at the Moon's surface. The expansion and contraction that crystalline minerals undergo when heated and cooled can bring about the disintegration of solid rock into small particles. The same volume changes can also cause loose particles to creep gradually to lower levels on a sloping ground surface. These effects may be unusually important on the Moon, because without an atmosphere and running water, ordinary terrestrial processes of weathering, erosion, and transportation cannot act on the Moon.

The Moon's surface contains no bodies of standing water, so there are no features of wave and current erosion. Without streams and water bodies, the Moon has no mechanisms or receiving areas for accumulation of water-laid sedimentary strata. Obviously, in comparison with Earth, surface changes must be extremely slow on the Moon.

28.6 A DECADE OF LUNAR EXPLORATION

The final splashdown of Apollo 17 in December 1972 brought to a close the first human exploration of the Moon's surface. The vast store of rock and soil samples, photographs, and other observational data accumulated during the *Apollo* program will take many years to analyze fully. Although certain basic geologic facts about the Moon were established beyond a doubt, many more deep and complex scientific puzzles have emerged than could have at first been imagined.

The decade of firsthand lunar exploration began in 1963 when United States Ranger spacecraft impacted the lunar surface, sending back thousands of pictures from a wide range of altitudes. There followed landings by Surveyor spacecraft, making direct physical and chemical tests of lunar surface materials as well as photographs of the ground immediately surrounding the vehicle. Then came a series of lunar orbiter vehicles on missions of photographing the lunar surface from heights as low as 56 km. Vertical photographs, at least 10 times sharper than the best taken by telescopes from Earth, were obtained from the nearside of the Moon, as well as almost complete coverage of the previously unknown farside. Thus, the detailed mapping of the Moon's entire surface was completed.

In 1969 and 1970 manned space vehicles of Apollo missions circled the Moon at low levels and descended to the lunar surface, permitting photographic negatives in color to be brought back to Earth. Extremely high-resolution photo-

28 ASTROGEOLOGY—
THE GEOLOGY OF OUTER SPACE

graphs were thus obtained, along with samples of lunar materials.

28.7
THE LUNAR MARIA

As anyone can easily see, using a small telescope or binoculars, the first major subdivision of the Moon's surface is into light-colored areas and dark-colored areas (Figure 28-4). The former constitute the relatively higher surfaces, or *highlands*. For many years the light-colored areas were called collectively the *terrae*. This Latin word reflects the earliest interpretations of Galileo—that these areas were dry lands. The dark-colored areas, the low-lying smooth lunar plains, are named the *maria*, plural of the Latin word *mare* for "sea"; Galileo applied this term to what he believed to be true seas of liquid. Of the nearside of the Moon, about 60 percent is terra and about 40 percent mare.

FIGURE 28-4

(*Left*) Sketch map of the major relief features of the Moon. This diagram may be used as an aid in identifying areas and subjects in the lunar photographs of this chapter. (From THE EARTH SCIENCES, 2nd Edition by Arthur N. Strahler. Copyright © 1963, 1971 by Arthur N. Strahler. By permission of Harper & Row, Publishers, Inc.) (*Right*) The Moon as it would appear if its whole disk could be simultaneously illuminated from a source at the same altitude. This is a composite of many photographs. (NASA.)

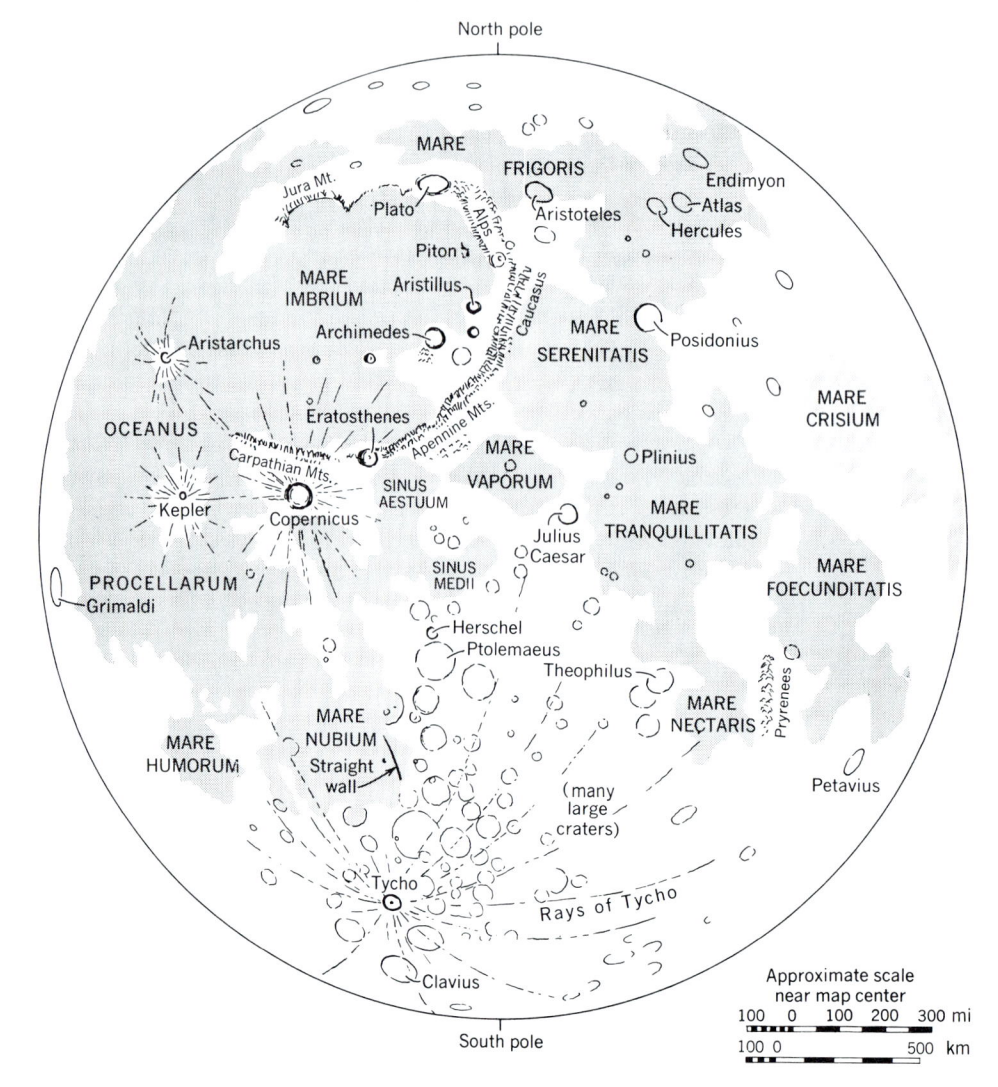

Lunar highland areas exhibit a wide range of relief features. Most outstanding are the great mountain ranges, of which there are 20 major groups on the nearside. Perhaps the most spectacular of these are the Appenines rimming the Mare Imbrium on the southwest side (Figure 28-5). Several peaks within the Appenines rise to heights of 4 to 5 km above the nearby mare surface. The highest and most massive mountains are those of the Leibnitz range, near the lunar south pole, which have peaks rising to heights of 11 km. Elsewhere, the highland terrain consists of gently rolling surfaces with low slopes and of rough areas with steep slopes.

Lunar maria of the nearside are divided into 10 major named areas, in addition to a single vast area, Oceanus Procellarum (Figure 28-4). Although mare outlines are in places highly irregular, with many bays, a circular outline is persistent for several, and particularly striking for Mare Imbrium.

28.8 THE LUNAR CRATERS

Most spectacular of the Moon's surface features are the *craters*. Even under the low magnification of a small telescope, the large craters form an awe-inspiring sight when seen in a partial phase of the Moon. Craters are abundant over both the highland and mare surfaces. Using telescopes alone, some 30,000 craters were counted on the nearside with diameters down to 3 km. An estimated 200,000 have been identified with space-vehicle photography. Included are recognizable craters as small as 0.6 m across. On the lunar nearside there are 150 craters of diameter larger than 80 km. Largest of these is *Clavius,* 235 km in diameter and surrounded by a rim rising 6 km above its floor (see south polar area in Figure 28-4). *Tycho* and *Copernicus* are among the most striking of the large craters (Figure 28-6).

The forms of large craters fall into several types. In some the floor is flat and smooth, and

28 ASTROGEOLOGY— THE GEOLOGY OF OUTER SPACE

FIGURE 28-5
Mare Imbrium, with its bordering mountains, the Jura, Alps, Caucasus, Apennines, and Carpathians; and the great craters Plato, Aristillus, Archimedes, and Eratosthenes. Note the spinelike peak, Piton. Identify these features with the aid of Figure 28-4. (The Hale Observatories.)

FIGURE 28-6
Copernicus, the great lunar crater lying south of Mare Imbrium (see Figure 28-4). Note the rays—radial streaks of lighter-colored material. The conspicuous crater lying east-northeast of Copernicus is Eratosthenes. (The Hale Observatories.)

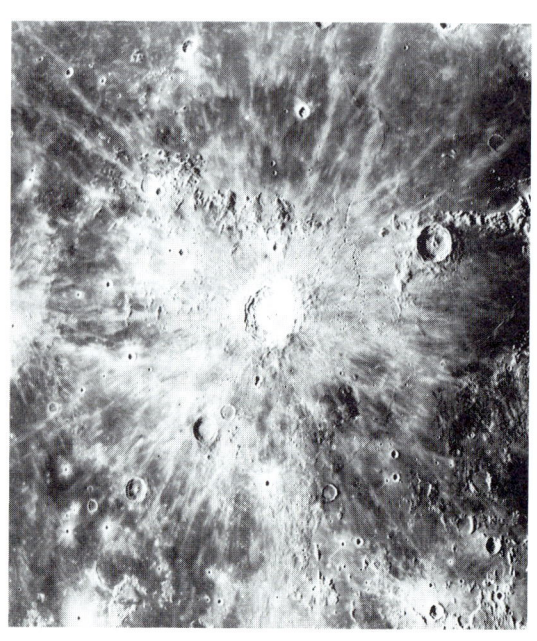

the rim is sharply defined and abrupt. In others, such as Copernicus, the floor is saucer-shaped, while the rim consists of multiple concentric ridges. Of particular interest are systems of *rays* of lighter-colored surface radiating from certain of the larger craters—for example, Copernicus (Figure 28-6). In several of the large craters there is a sharply defined *central peak*, which must be taken into account in interpreting the origin of craters. That of Eratosthenes shows up particularly well in Figure 28-6.

Small lunar craters are almost perfectly circular and have a cup-shaped interior and a prominent rim (Figure 28-7). However, these well-defined forms grade into less distinct craters with low slopes and into shallow depressions. The impact origin for almost all of these smaller craters is generally accepted. A few are fresh in appearance, with a litter of boulders on the rim and within the crater itself.

The farside of the Moon was photographed in detail by lunar orbiter spacecraft and by astronauts of Apollo spacecraft, the first human beings to see that side of the Moon. The lunar farside is heavily cratered but generally lacking in extensive maria.

28.9 RILLES AND WALLS

Another class of distinctive lunar-surface features are narrow, canyonlike features called *rilles*. Some are remarkably straight, taking the form

FIGURE 28-7
Censorinus (*arrow*), one of the freshest craters on the Moon's nearside, shows a sharp rim and surrounding zone of light-colored soil surface. Crater rim is about 7 km in diameter. (Apollo 10, NASA.)

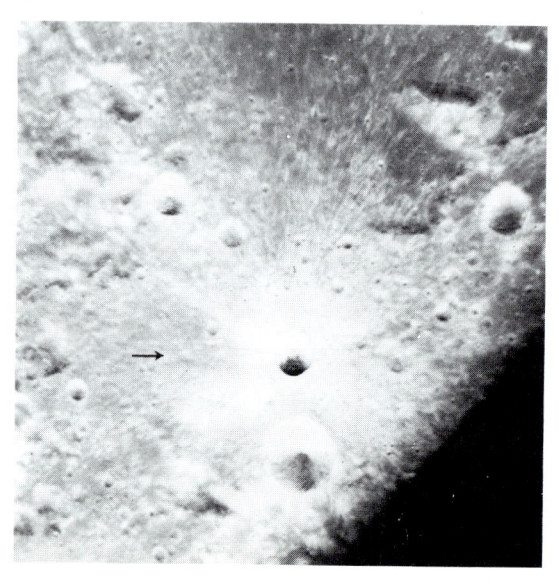

of a trench up to 240 km long. Others are irregular in plan, and a few are sinuous, suggestive of terrestrial meandering rivers (Figure 28-8). Astronauts of Apollo 15 inspected the walls of Hadley's Rille at close range. Layers of rock are exposed, suggesting that an erosion agent carved the rille into older deposits. It has also been suggested that the sinuous rilles were formed over lines of vents from which volcanic gases were emitted under high pressure.

Most of the straight rilles are interpreted as fracture features in brittle rock of the lunar crust. Related features are the straight cliffs, or *walls*, which may be fault escarpments. A particularly striking example is the Straight Wall in Mare Nubium (Figure 28-4), which is 240 m high and may represent a fault that broke the mare surface.

28.10 ORIGIN OF THE GREAT CRATERS AND MARIA

Although some evidence of volcanism can be found on the moon, it is fully agreed by astrogeologists that nearly all lunar craters are of impact origin, representing the infall of meteoroids of a vast range of sizes, including large asteroids.

Lunar craters are much larger than those on Earth and require explanation of the characteristic central peak. It has been suggested that the central peak lay directly beneath the center of impact-explosion. Because the shock wave was directed downward, the underlying rock remained intact while that surrounding it was blown outward (Figure 28-9). The rays that emanate from several large craters are explained as debris deposits thrown out over long distances from the explosion centers. The great mountain-rimmed maria, of which there are five on the lunar nearside, are considered to be the old impact scars of enormous masses, probably asteroids. The case of Mare Imbrium is particularly striking (Figure 28-5).

Lunar orbiter satellites have recorded the Moon's gravity field in great detail. The resulting map shows definite concentrations of mass, known as *mascons,* under the centers of the ringed maria of the lunar nearside. These concentrations are particularly strong under Mare Imbrium and Mare Serenitatis. They are interpreted as caused by saucer-shaped masses of basaltic lava lying below the surfaces of the maria. The dense basalt masses exert an abnormally great gravitational attraction.

28.11 LUNAR SURFACE MATERIALS

The first samples of lunar rock and soil were obtained in 1969 by astronauts of the Apollo 11 mission. There, on Mare Tranquillitatis, as at subsequent Apollo landing sites, surface materi-

FIGURE 28-8
Hadley's Rille, a sinuous canyonlike feature, crosses a cratered plain and ends in a highland area. North is at the bottom of this lunar orbiter photograph, spanning an area about 50 km wide. (NASA.)

FIGURE 28-9
This diagram suggests a possible origin for the central peak in a large lunar crater, under the impact hypothesis. (From PRINCIPLES OF EARTH SCIENCE by Arthur N. Strahler. Copyright © 1976 by Arthur N. Strahler. By permission of Harper & Row, Publishers, Inc.)

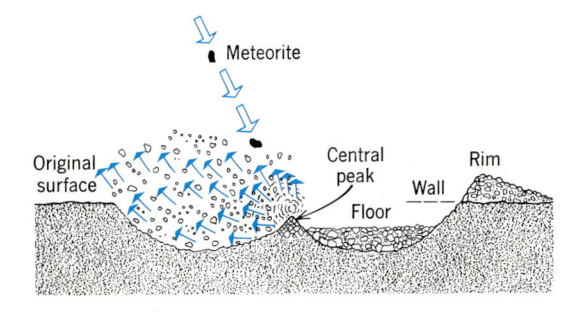

28 ASTROGEOLOGY—
THE GEOLOGY OF OUTER SPACE

als consist of unsorted fragmental debris. The fragments range in size from dust to blocks many meters in diameter. The layer of loose, dustlike mineral matter was called *regolith*. The uppermost layer—a few centimeters—is a brownish to grayish cohesive powdery substance; it consists of grains in the size range from silt to fine sand. The material is easily penetrated. Upon compaction it becomes stronger, easily supporting the weight of the astronauts and their equipment. Figure 28–10 shows the regolith compacted into a clear footprint.

All observed lunar material sufficiently hard to be called rock has proved to consist only of fragments or clusters of fragments. Nowhere did the astronauts find a massive outcropping of a large body of solid igneous rock in place. In other words, exposed igneous bedrock like that on Earth does not exist on the Moon, so far as is known.

All of the regolith and rock fragments have proved to be of igneous origin, with pyroxene, plagioclase feldspar, and olivine being the most abundant minerals. There are many other accessory minerals; these are of secondary importance so far as abundance is concerned. Some examples familiar from your study of igneous rocks in Chapter 23 are quartz, potash feldspar, and amphibole. Iron and titanium are constituent elements in quite a few of the accessory minerals. Some of the minerals contain uranium and thorium, which are radioactive.

28.12
LUNAR ROCKS

Lunar rock materials can be divided into two groups. First are fragments of igneous rocks, perhaps broken from some unknown parent mass of solidified magma. These come in both fine-grained and coarse-grained textures. Second are breccias. A *breccia* is a rock put together of many small angular fragments. Individual pieces of rock comprising the original rubble of the breccia are themselves fragments of igneous rock. Consequently, a single chunk of lunar breccia may contain representatives of igneous rock from many different sources.

The great majority of lunar rocks collected by the astronauts are breccias. It is believed that the lunar breccias originated from intense shock, binding together particles that had previously been fragmented by shock. This process has been called *impact metamorphism*. Meteoroid impacts are considered to be the shock mechanism, both for fragmenting rock into regolith and for bringing it together into breccia.

Of the igneous rock fragments, for the sake of simplicity, we identify only two major groups: basalt and anorthosite. Basalt is already familiar to you as a common terrestrial rock composed largely of calcic plagioclase feldspar, pyroxene, and often olivine as well. Lunar basalt, shown in Figure 28–11, is quite similar, except that the basalts underlying the maria are richer in iron and poorer in silica than Earth basalts. Maria basalt magma was highly fluid, more so than Earth basalt, and probably flowed very freely as

FIGURE 28–10
Astronaut's footprint in the lunar soil. Notice the many miniature craters in the surrounding surface. (Apollo 11, NASA.)

FIGURE 28–11
Named the Goodwill Rock by astronauts of Apollo 17 mission, this basalt specimen shows the original gas-bubble cavities formed as the magma cooled. The specimen is about 10 cm high. (NASA.)

it filled the maria basins. The second rock mentioned, *anorthosite,* is an igneous rock we did not include in the granite-gabbro series described in Chapter 23. Anorthosite occurs on Earth as a plutonic igneous rock. It consists largely of plagioclase feldspar. Consequently, anorthosite is a felsic rock, of lower density than basalt, which is classed as mafic.

If the Moon is literally smothered under a blanket of impact debris, how can we even guess at what kind of solid bedrock lies below, constituting the lunar crust? The answer lies in sampling the bouldery rim debris of craters (Figure 28–12). The scenario is that a large meteoroid, upon penetrating the regolith and exploding, tore loose and threw out boulders of the bedrock below. Acting on this supposition, astrogeologists have inferred that the uppermost part of the lunar crust consists in some places of basalt and in other places of anorthosite.

28.13 A GEOLOGIC HISTORY OF THE MOON

Radiometric age determination has been a mainstay in attempts to work out a geological history of the Moon. One surprising finding from lunar samples is that, so far, no igneous rock has turned up with an age of crystallization from its magma younger than -3 b.y. Impact fragmentation and formation of breccia has certainly been going on since that time, but it seems to have been only a mechanical reworking of the ancient igneous rocks. Practically all samples of breccia fragments have shown an age of crystallization in the time span from -4.2 to -3.1 b.y.

Geologists kept hoping that some lunar rock specimen would prove to be much older than -4.2 b.y. They were disappointed up to the very last minute. Finally, on the last *Apollo* mission, astronauts Eugene Cernan and Harrison Schmidt, the latter a geologist, brought home a boulder fragment of blue-gray breccia. From within the breccia several fragments of a greenish mineral were extracted. These proved to be fragments of the ultramafic rock dunite, composed mostly of the mineral olivine. Their age? No less than -4.6 b.y.! Here at last was evidence of the original igneous material with which the Moon was put together. On Earth, no rock of this great age has so far been discovered. In fact, no Earth rock is even as old as the most abundant lunar breccia fragments, in the age span -4.2 to -3.8 b.y. Clearly, the Moon's geologic record falls into a very different age bracket from that of Earth. Perhaps the Moon can tell us what may have happened on Earth before

FIGURE 28–12
Astronaut Harrison H. Schmitt stands beside a large boulder at Station 6, Taurus Littrow. (Apollo 17, NASA.)

-3.6 b.y., where the Earth record starts. The ancient record on Earth has probably vanished for good, consumed by crustal melting in subduction zones.

Astrogeologists are far from reaching any consensus on the Moon's geologic history. The following sequence of events has met with some measure of agreement, but is speculative to a large degree.

Accretion of the Moon as a small planet occurred at -4.7 to -4.6 b.y., about the same time as for the Earth. A major thermal event occurred at -4.5 to -4.3 b.y. At that time the outer zone of the Moon became highly heated and melted down to a depth of about 200 km. Between -4.2 and -4.0 b.y. the lunar crust solidified from molten magma. Possibly this crust consisted of an outer layer of anorthosite, with basalt beneath. Then, about -3.9 b.y., came a great bombardment by asteroids, producing the maria impact basins. Later, between -3.8 and -3.1 b.y. the maria basins were filled by outpourings of highly fluid basalt. This magma came from deep within the Moon. From -3.0 to the present no further significant igneous activity occurred. However, countless meteoroid impacts formed breccias by repeated fracturing, lifting, and dropping of older rock.

28.14
THE LUNAR INTERIOR

The Moon's interior structure and composition are still subject to much speculation, with some quite different interpretations being proposed simultaneously. One point of fact is established. The moon as a whole is deficient in iron, and almost certainly lacks an iron core like that present in Earth. The Moon's low average density, 3.34 g/cm^3, is about the same as for the Earth's mantle rock. If an iron core existed in the Moon, the average density would be considerably higher.

It is also agreed that the Moon has a low-density crust, its density averaging perhaps 2.9 g/cm^3. The average crustal composition is described by some researchers as "anorthositic-gabbro," by others as "gabbroic-anorthosite." Recall from Chapter 23 that gabbro is a plutonic rock of the same composition as basalt. Thus, the description of the crust implies a rock of composition intermediate between basalt and anorthosite.

Below the lunar crust is an interior region most probably consisting of olivene and pyroxene. In other words, the interior is made up of ultramafic rock, perhaps much like that of the Earth's mantle. The density of the lunar interior probably ranges from 3.2 g/cm^3 near the crust, to over 3.5 g/cm^3 near the center.

A major point of debate concerns the temperature of the Moon's interior. The hypothesis of a "cold" Moon, its temperature everywhere well above the melting point, had been widely held for many years. Even at a central temperature as high as 1400°C, the rock would remain solid, and this value was considered by some to be much higher than the actual value. However, some newer lines of evidence suggest that interior temperatures may reach a maximum value of 1600°C at a depth of about 1100 km. Perhaps at that depth there is a layer of partial melting, similar in some respects to the soft layer of the Earth's mantle. Most lunar earthquakes, called *moonquakes*, originate in this zone. Thus, there is some possibility for motion of the Moon's interior region with respect to an outer rigid layer. However, no crustal activity resembling plate tectonics on Earth is called for, and there is no surface evidence of the presence of plates or of plate motions.

28.15
ORIGIN OF THE MOON

One problem that seems to have remained entirely unsettled by the Apollo program is the Moon's origin. One hypothesis states that the Moon and Earth grew by accretion of nebular material as a binary planet, linked by gravitational attraction. However, the Moon's low density and lack of iron are hard to explain if accretion occurred at the same point in time as Earth under similar physical and chemical conditions.

A second possibility, widely debated, is known as the *capture hypothesis*. It is supposed that the Moon formed by accretion in another part of the solar nebula, where conditions were quite different from those prevailing where the inner planets were forming. Later the Earth captured the Moon, when by chance the Moon passed close to the Earth and was trapped by Earth's gravitational field. It has been suggested that capture occurred about −3.7 to −3.6 b.y. and corresponded with the rise of the maria flood basalts. However, the mechanism of capture of so large a planetary body by Earth faces strong objections.

A third hypothesis requires that the Moon was formed of material broken away from a fast-spinning Earth. This *fission hypothesis* was first put forward in the 1890s by Sir George Darwin, son of Charles Darwin and a noted authority on tides. He calculated that at some point in time the Earth was spinning much faster than today, and that the centrifugal force of rotation produced a large bulge in the Earth. The bulge then separated from the Earth and moved away to become the Moon. Under the fission hypothesis, the material thrown off from the Earth came from the mantle. This would explain the Moon's low density and lack of an iron core. However, even a modernized fission hypothesis faces many serious objections based on chemical composition differences between Earth rocks and Moon rocks. A satisfactory explanation of the Moon's origin seems almost as elusive as before the astronauts brought back their rock samples.

28.16
THE SURFACE OF MARS

Although Mariner 9, an orbiting spacecraft, circled Mars in 1971 and 1972, sending back thousands of excellent photographs, it was not until the later Viking missions that landing vehicles sent back data from the Martian surface. In 1976 two simultaneous Viking spacecraft orbited the planet continuously while their respective landing craft sent back data from the surface. The Chryse landing site of Viking I, located at about 25 degrees north latitude, was the first to be reached. Viking II landed farther poleward, at about 46 degrees north latitude. Thousands of

28.16 THE SURFACE OF MARS

high-quality photographs taken by two orbiters added to knowledge gained by the earlier Mariner 9 mission.

Many important Martian features are strikingly different from those found on either Earth or Moon, although some other features are very much like those found on Moon or Earth. One half of Mars, roughly equivalent to its southern hemisphere, is heavily cratered, much as is the Moon's surface, whereas the other (northern hemisphere) is only lightly cratered. It is thought that the history of cratering of Mars parallels that of the Moon in age and frequency of impacts.

Among the major geological features of Mars are volcanoes. One huge shield volcano, Olympia Mons, is 500 km wide at the base with a summit rising 25 km above the base (Figure 28–13). A caldera occupies the summit. Three other volcanoes are of comparable dimensions. These volcanoes indicate local accumulations of intense heat within the planet—possibly heat of radiogenic origin—leading to magma formation and extrusion comparatively recently in Mars' history. Many lava flows are clearly seen on the Martian surface.

Mars also has tectonic features in the form of numerous faults and a great rift valley system. The largest rift, Coprates, appears as a straight-walled trough 100 km wide and up to 6 km deep in places (Figure 28–14). The entire rift zone is some 5000 km long. Although the Martian lithosphere seems to have been pulled apart along the rift zone, there is no sign of subduction zones

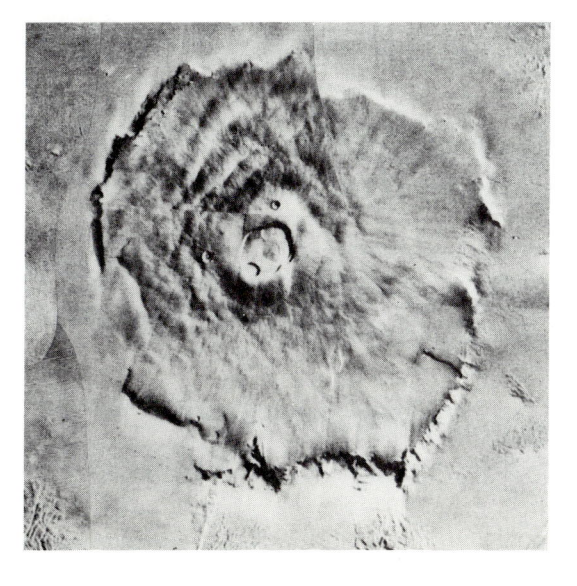

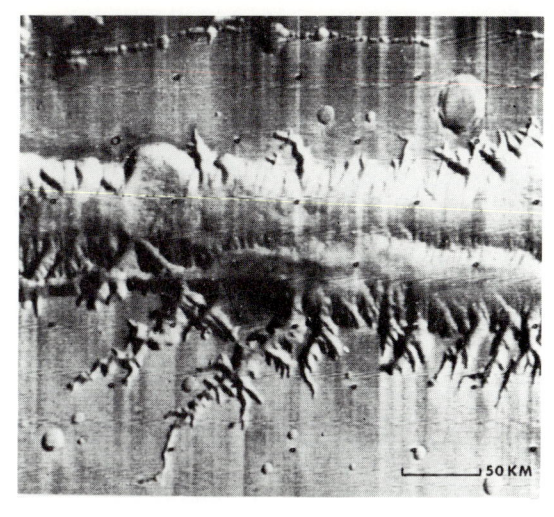

FIGURE 28–14

A deep, steep-sided trench near the equator of Mars. Width of the trench is about 100 km, its depth is as great as 6 km. (NASA.)

FIGURE 28–13

Olympia Mons is an enormous Martian shield volcano with a caldera at the summit. The basal diameter of the volcano is about 500 km. (NASA.)

and compressional mountain belts such as those on Earth. Narrow fracture lines, suggesting crustal tension, are found in many places.

Among the most puzzling surface features of Mars are what appear to be channels formed by fluid flow. These are largely concentrated in the area north of the rift zone. Some of the channels show a braided pattern in which flow lines repeatedly divide and join, like the shallow channels of Earth streams in dry regions. Mars is now devoid of surface water in liquid form, but the channels suggest the action of great water floods, acting over short periods. One suggestion is that water ice, held frozen in the ground, underwent an episode of sudden melting, releasing great floods. These floods eroded the channels and left the channel floors strewn with rock debris (Figure 28–15). Typically, the major channeled belts originate in steep-walled valley heads. Here, chaotic terrain suggests collapse as subterranean water from melting ground ice (permafrost) flowed rapidly out from beneath the surface during an episode of catastrophic floods.

Mars has an extremely thin atmosphere; the surface barometric pressure is only 7 to 8 mb, or less than $\frac{1}{100}$ that on Earth. Nevertheless, winds on Mars are capable of raising great dust storms. One such storm obscured the Martian surface for weeks following the arrival of Mariner 9. Signs of erosion of exposed rocks by impact of windblown sand and silt were observed at both Viking landing sites, while numerous small dunes and drifts were clearly shown. Enormous expanses of dunes, somewhat like sand seas of the Earth's deserts, also testify to the importance of wind as a geologic agent on Mars.

28 ASTROGEOLOGY— THE GEOLOGY OF OUTER SPACE

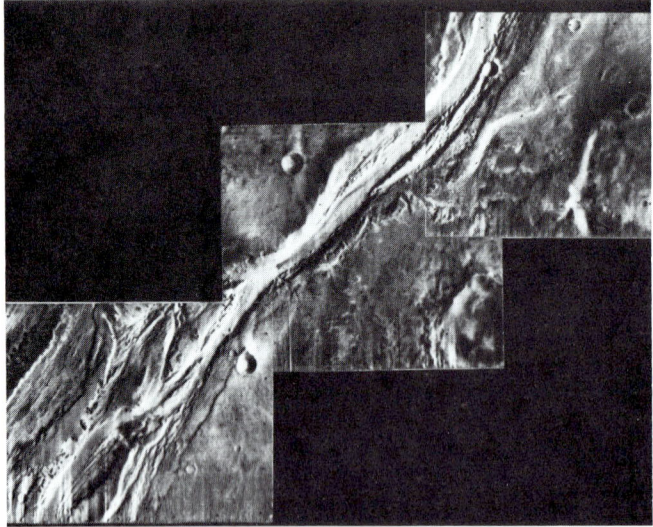

FIGURE 28-15
Eroded channels such as these on the Martian surface suggest the possibility of great water floods in the remote past. Large impact craters formed obstructions to the flood waters, which flowed from lower left to upper right. Many small craters have since pocked the channel floor. This Viking I orbiter photo was taken from a height of about 1600 km on June 23, 1976. (NASA.)

FIGURE 28-16
The Martian polar cap. Platelike terrain features near the pole (*center*) appear to overlap one another. (NASA.)

Perhaps the strangest of the many strange Martian features are the polar caps (Figure 28-16). The white polar caps consist of a thin surface coating of carbon-dioxide ice. Each cap undergoes a large seasonal change in size, expanding greatly during the winter of the respective pole and shrinking to a small area during the summer. The small residual polar cap, left at the end of summer, is thought to consist of an accumulation of water ice. The residual cap areas have strange, platelike terrain features, looking something like a collapsed stack of poker chips arranged in a spiral form. The plates consist of light and dark laminations, which are probably alternating layers of ice and dust. A vast belt of wavelike sand dunes surrounds the north polar cap.

At both Viking landing sites, details of the surface are quite similar (Figure 28-17). Rock fragments of a wide range of sizes from pebbles to boulders are strewn over a surface of fine-grained loose regolith. Many of the rocks are angular; many show the wind-eroded form of ventifacts common on the Earth's deserts. Many of the rock fragments show wind-tails of fine material formed as miniature drifts to the lee of each fragment. Some of the larger rocks, particularly at the Utopia site of Viking II, are highly porous and have what appears to be the vesicular structure of a scoriaceous lava. Other rock surfaces appear to be fine-grained. The color of all rock surfaces and much of the regolith is a uniform brick-red. This coloration has been attributed to thin coatings of limonite (a mixture of iron-oxide minerals).

Although the Viking landers were not equipped to sample rock fragments, small scoopfuls of regolith close to the vehicles were subjected to analysis, using an x-ray fluorescence spectrometer. The most abundant elements are iron (14 percent), silicon (15–30 percent), calcium (3–8 percent), and aluminum (2–7 percent). Also present are important amounts of potassium, sulfur, chlorine, and titanium. These data suggest that the rock may be of ultramafic composition, rich in olivine and pyroxene. Data for the two landing sites were almost identical. There seems to be general agreement that most Martian surface rock is probably mafic or ultramafic extrusive rock, a conclusion in agreement with the presence of volcanoes and lava flows.

The Martian surface features and materials lead to the conclusion that volcanic, tectonic, and erosional activities have all been active on Mars, but to a much lesser degree than on Earth. Thus, in many respects, Mars is geologically intermediate in geologic activity between Earth and Moon.

28.17 THE SURFACES OF MERCURY AND VENUS

Mariner 10 passed close to the planet Mercury on March 1974, sending back the first photographs ever to reveal the planetary surface. Strikingly like the Moon, the heavily cratered surface of Mercury showed countless lunar-like craters of many sizes and ages (Figure 28–18). Some craters are comparatively recent in age—light in color and having well-defined ray systems. Others are obviously very old large craters with flat floors. Also seen were dark plainlike areas suggestive of the lunar maria. Infrared radiation analyzed by instruments on the space vehicle suggest that Mercury has a regolith of low density, perhaps much like that on the Moon. Many long, straight scarps were identified, along with long, narrow ridges suggestive of crustal compression. However, no sinuous rilles, like those on the Moon, were observed.

As expected in the absence of water and all but an extremely tenuous atmosphere, the surface of Mercury revealed no signs of the eroding action of fluids, such as the braided channels of Mars.

The magnetic field of Mercury is extremely weak. Although Mercury may have an iron core, the planet's extremely slow rotation would probably not give rise to a dynamo effect such as that generated in the Earth's liquid core.

Because of its extremely dense atmosphere and high surface temperatures, Venus seemed for years a most unlikely candidate for clear and detailed surface photography. Then, in 1975, the Soviet spacecraft Venera 9 landed on Venus, sending back the first photograph of the solid planetary surface. What appeared on the image was a blocky ground surface that can be interpreted as composed of hard rocks fractured into angular blocks. Individual blocks seem to be on the order of 30–40 cm across; they cast sharp shadows.

Later in 1975 a second Soviet space vehicle, Venera 10, reached the surface of Venus and sent back another photographic image. The scene was described by Soviet news releases as an old mountain formation with smooth, rounded rocks. The two Soviet images offer, at best, very little information as to the full nature of the solid surface of Venus and the processes that may act there.

FIGURE 28–17
The surface of Mars, photographed from the Viking I lander in July 1976. Many of the larger rock fragments show effects of wind abrasion. (NASA.)

FIGURE 28–18
The heavily cratered surface of Mercury shows irregular fractures and ridges, forming a blocklike pattern. Width of the area shown is about 500 km. (Mariner 10, NASA.)

28.18 YES, THERE IS A GEOLOGY OF OUTER SPACE

Within only a decade of space exploration, the extension of geology as a science to the Moon and inner planets has passed from infancy to vigorous youth. Perhaps the greatest gain in our knowledge of the Earth's history comes from the very ancient lunar rocks, almost all of which crystallized from magma in a one-billion year period predating any known terrestrial rock. Earth's record of the first billion years of geologic time is totally missing because Earth has had an active lithosphere and vigorous atmospheric processes.

While Earth rocks were continually remelted and recycled, destroying all of the ancient crust, the primitive crustal features of the Moon, Mars, and Mercury went into cold storage, so to speak. Only meteoroid impacts have disturbed the surface of the Moon and Mercury since about -3.0 b.y., and even on Mars half the planet retains the early impacted surface. We thus have not one, but three, Rosetta stones from which to read the early history of our planet.

Of the inner planets, only Venus remains completely obscured from geologic examination by satellite cameras. Beneath the dense, clouded atmosphere of Venus may lie a varied surface showing intense tectonic and volcanic activity, for Venus is a close match to Earth in size and density and may well prove to have a "plate tectonics" of its own.

SUMMARY

Origin of solar system. Contraction of a solar nebula composed of hot dust and gases formed a thin, flat disk with a large central solar mass. Upon cooling, accretion of matter formed planetesimals, which joined by collisions to form the planets, about $-4\frac{1}{2}$ b.y.

Meteorites. Large meteoroids, possibly small asteroids, reach the Earth's surface to form meteorites. Some are of iron, others of silicate rock; all have an age of about $-4\frac{1}{2}$ b.y. Only a few large meteorite impact craters, such as Barringer Crater, can be found on Earth.

The lunar environment is characterized by a weak gravity, lack of atmosphere and hydrosphere, intense solar radiation, and extremes of surface temperature. Dark-colored maria alternate with light-colored highlands on the lunar nearside.

Lunar craters. Impact craters in vast numbers and a wide range of sizes cover the entire lunar surface; meteorite impact explains most. Radial rays of debris and central peaks are commonly associated with the larger craters. The great rimmed maria may represent impacts of large asteroids. Other surface features include rilles and walls.

Lunar rocks. All lunar rock materials are fragments or breccias affected by repeated impacts, causing impact metamorphism. Lunar regolith is a dustlike soil cover. Rock fragments are typically of basalt composition; some are anorthosite.

Moon's history. Moon's accretion occurred at about -4.6 b.y.; its crust was melted and resolidified at -4.5 to -4.0 b.y. Maria impacts occurred at -3.9 b.y., followed by maria basalt extrusion. Intense impacting continued thereafter to present. Hypotheses of Moon's origin include independent accretion, capture by Earth, and fission from Earth's mantle.

Mars and Mercury. Mars shows volcanic activity as well as intense cratering, with rift zones of faulting and braided channels suggestive of fluid flow. Mercury's surface is heavily cratered, much like Moon, and lacks volcanic or tectonic features.

TERMS AND CONCEPTS

astrogeology
solar nebula
accretion
planetesimals
outgassing
meteorite
fireball
fall
find
irons
stones
stony irons
highlands (terrae)
maria (mare)
craters
rays
central peak
rilles
walls
mascons
regolith, lunar
breccia
impact metamorphism
anorthosite
moonquakes
capture hypothesis
fission hypothesis

QUESTIONS

1. Describe the steps in origin of the planets and solar system according to the modern condensation hypothesis. How does this hypothesis account for

fundamental differences between the inner planets and outer planets?

2. Describe the fall of a large meteorite. Distinguish between a fall and a find. What are the three classes of meteorites? Describe the composition and relative abundance of each. What is the age of meteorites? What is the origin of meteorites? Relate meteorite composition to composition of the Earth's core.

3. Describe the Barringer Crater and its formation. Why are impact craters rare on the Earth's surface?

4. Describe the lunar-surface environment in terms of gravity, atmosphere, solar radiation, surface temperatures, and temperature ranges. Does the Moon's surface have a supply of free water? What processes operate on the Moon's surface? What geological processes are not active?

5. Review the history of lunar exploration by spacecraft. Into what two major terrain types is the Moon's surface subdivided? How did Galileo interpret each of these types? Describe other relief features on the Moon's surface, including mountains, craters, rays, rilles, and walls.

6. What evidence supports the hypothesis of meteorite impact for the origin of craters? Interpret the great rimmed maria. How does the presence of mascons relate to origin of the maria?

7. Describe the lunar regolith. How was it formed? What form do lunar rock materials take? Of what minerals are they composed? What is the significance of lunar breccia? Describe the process of impact metamorphism. What two major classes of igneous rocks have been identified on the Moon?

8. Recapitulate the major events in the geological history of the Moon. How old are the oldest lunar rocks? Compare this age with rocks of continental shields of our Earth. What is known or inferred about the chemical and physical nature of the Moon's interior?

9. What hypotheses of origin of the Moon as a planetary body are currently being considered? What objections does each hypothesis face?

10. Describe the surface of Mars, as revealed by Mariner and Viking spacecraft. Compare the intensity of volcanic and tectonic activity of Mars with that of Earth and Moon. What features have been recognized on the surface of Mercury?

CHAPTER 29
STARS, GALAXIES, AND THE UNIVERSE

29.1 THE SUN

Our Sun is an enormous sphere of incandescent gas, more than 100 times the diameter of Earth, with a mass more than 330,000 times that of Earth and a volume 1,300,000 times that of Earth. The Sun's surface gravity is 34 times as great as that of Earth.

The Sun's diameter is about 1,400,000 km, and it lies at an average distance from Earth of about 150 million km. Traveling at the speed of light, which is roughly 300,000 km/sec, solar radiation takes about $8\frac{1}{3}$ minutes to reach the Earth.

Like our Earth, the Sun rotates on an axis, but with an important difference: Earth is solid and has a uniform rate of rotation at its surface, whereas the Sun is a gaseous body and does not have the same rate of rotation from one part of its surface to another. From a study of the movements of sunspots we know that the equatorial region of the Sun rotates with a period of about 27 days; whereas at progressively higher latitudes, the rotation is slower.

The visible surface layer of the Sun is called the *photosphere*. The outer limit of the photosphere constitutes the edge of the Sun's disk as seen in white light. Gases in the photosphere are at a density less than that of Earth's atmosphere at sea level.

Temperature at the base of the photosphere is about 6000 K but decreases to about 4300 K at the outer photosphere boundary. Light production is extremely intense within the photosphere. Beneath the photosphere, temperatures and pressures increase to enormously high values in the Sun's interior, or *nucleus*. Here temperatures are between 13 to 18 million K.

Above the photosphere lies a low solar atmosphere, the *chromosphere*. The region includes rosy, spikelike clouds of hydrogen gas called *solar prominences*. Still farther above the Sun's surface is the *corona*, a region of pearly-gray streamers of light. The corona constitutes the Sun's outer atmosphere (Figure 29–1). At times the solar prominences reach far out into the corona as luminous archlike bodies (Figures 29–2 and 29–3) rising to heights of over one million km. Temperatures increase outward through the chromosphere and the corona until values as high as 2 million K are reached. Surprisingly, the photosphere, or Sun's surface, is its coolest layer.

The corona extends far out through the solar system and envelops the planets. It is known as the solar wind in the region surrounding the inner planets. This wind consists of charged particles—electrons and protons—derived from the breakup of solar hydrogen atoms. As we found in Chapter 21, the solar wind brings energetic particles to the magnetosphere, where they are entrapped by the lines of force of the Earth's magnetic field.

Although almost all the known elements can be detected by analysis of the Sun's rays, hydrogen is the predominant constituent of the Sun, with helium also abundant. It is estimated that hydrogen constitutes at least 90 percent of the Sun, and hydrogen and helium together total about 98 percent.

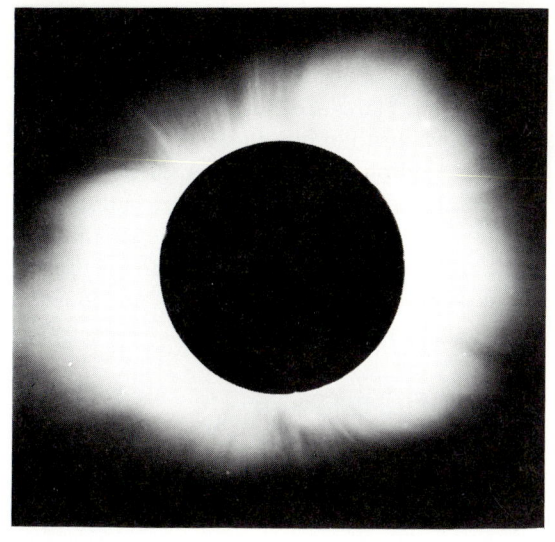

29.3 SUNSPOTS AND SOLAR FLARES

FIGURE 29–1
Photograph of the Sun's outer corona taken during a total eclipse. The moon's disk completely covers the Sun, permitting this pearly-white tenuous outer layer of gases to be seen. (The Hale Observatories.)

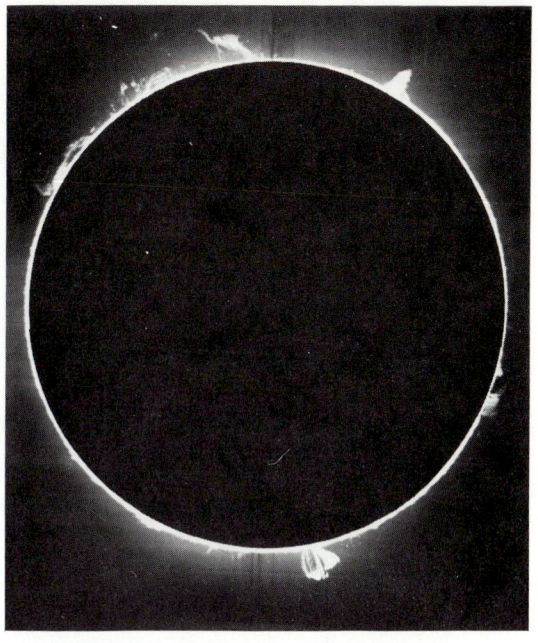

FIGURE 29–2
The entire edge of the Sun shows several prominences. (The Hale Observatories.)

29.2 THE SUN'S INTERIOR

The source of the Sun's energy is the conversion of hydrogen into helium within the Sun's interior. The process of production of energy within the Sun is that of nuclear fusion, in which hydrogen is transformed into helium. In the fusion process, described in Chapter 9, mass is converted into energy. At temperatures over 4 million K within the interior of a star, several forms of reactions occur in which helium is produced. The quantity of energy produced by conversion of matter into energy is enormous. At its present rate of energy production, the mass of the Sun will diminish only one-millionth part of its mass in 15 million years.

Heat produced in the Sun's innermost core region moves outward by a process of radiation through the extremely dense gas of the interior. In a zone nearer the Sun's exterior, a process of convection (mixing) transports the heat to the surface.

29.3 SUNSPOTS AND SOLAR FLARES

A *sunspot* is a dark spot on the Sun's photosphere. The spot normally consists of a darker central region, the *umbra*, surrounded by a somewhat lighter border, the *penumbra* (Figure 29–4). A single sunspot may be 800–80,000 km across and represents a strong disturbance extending far down into the Sun's interior. The spot has a somewhat lower temperature than the surrounding photosphere. Sunspots form and disappear over a timespan of several days to several weeks, during which time they can be seen to

FIGURE 29–3
A great solar prominence rising to a height of 225,000 km. (The Hale Observatories.)

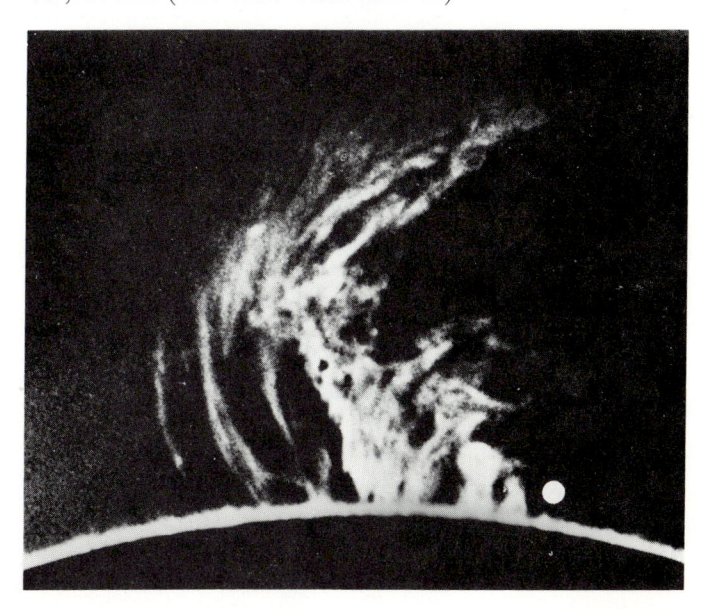

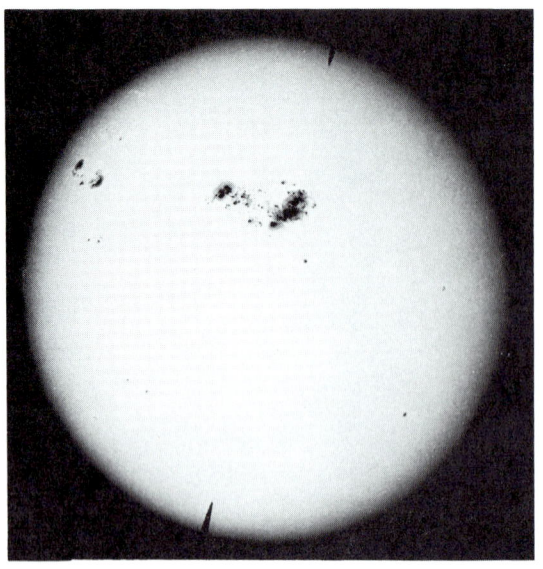

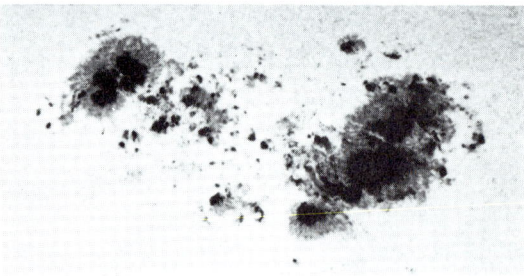

FIGURE 29-4

(*Above*) The whole disk of the Sun shows a large sunspot group. (*Below*) An enlargement of the group of spots. (The Hale Observatories.)

move with the Sun's rotation. The frequency of sunspots follows a cycle with an average period of about 11 years.

It has been found that the sunspots have powerful magnetic fields associated with them—several thousand times as great in intensity as the magnetic field at the Earth's surface. This magnetism takes the form of strong poles associated with the sunspots. Adjacent spots of a pair in the same hemisphere have opposite polarity.

The same intense magnetic fields that are associated with sunspots also produce *solar flares*, which are emissions of ionized hydrogen gas from the vicinity of the sunspots. It is from such flares that x-rays are sent out, followed by streams of charged particles. These particles reach the Earth about a day later. The emissions received from solar flares cause the Van Allen radiation belt to become intensely radioactive. At such times the magnetic field at the Earth's surface is severely disrupted. This phenomenon is a magnetic storm; it interferes seriously with radio communication (Chapter 21).

Solar flares occur in much greater numbers than sunspots. As many as 2000 to 4000 flares occur per year during times of maximum sunspot activity. Flares are thus about 20 times more frequent events than sunspots, but their duration is correspondingly much shorter. A single sunspot group in the course of its duration will produce as many as 40 flares. So we see that, in addition to the steady solar wind, the Earth intercepts intense bursts of x-rays and ionized particles at irregular intervals.

29.4 UNITS OF INTERSTELLAR DISTANCE

The vastness of interstellar space requires us to use units of length quite different from those applicable to the solar system. Consider the fact that the nearest star to our Sun, Alpha Centauri, is about 300,000 times more distant from the Sun than the Sun is from Earth. A convenient unit of interstellar distance is the *light-year;* it is the distance traveled by light in one year's time. Multiplying the speed of light, approximately 300,000 km/sec, by the number of seconds in the year gives a value of approximately 9 million million km as the distance equal to one light-year (exact value: 8.898×10^{12} km). Alpha Centauri is about 4.3 light-years distant from the Sun.

Astronomers also make use of another measure of distance. This measure is based on the principle of *stellar parallax*. As the Earth moves across its great orbital distance each year, the nearer stars should seem to change their apparent positions in relation to the more distant ones. *Parallax* is a word used in optics; it means a difference in the apparent relative positions of objects when viewed from different points. For example, as you walk along a road, a tree close by may appear in line with a distant tree. As you move farther along, the two trees will seem to become separated by a widening distance.

The principle of parallax is illustrated in Figure 29-5 (the angles are greatly exaggerated). A near star (*A*) may appear very close to a distant star (*B*) when viewed in the spring. But as the Earth moves in its orbit, star *A* will seem to shift its location in the sky, so that in the autumn star *A* may be separated from star *B* by a very small angle. The closer the star is to us, the greater the parallax effect.

A star having a parallax of exactly one second of arc would lie at a distance of about 32 million million km (3.2×10^{13} km) from the Sun. This distance is one *parsec*, a term coined from the words "parallax" and "second." Alpha Centauri

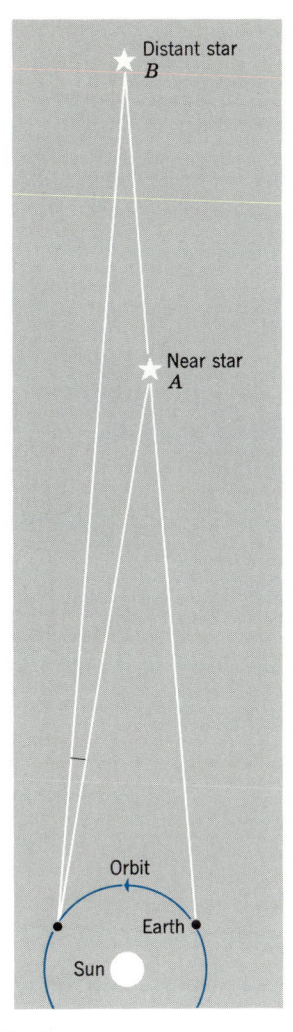

29.6 PROPERTIES OF STARS

FIGURE 29-5
The principle of stellar parallax. (From PRINCIPLES OF EARTH SCIENCE by Arthur N. Strahler. Copyright © 1976 by Arthur N. Strahler. By permission of Harper & Row, Publishers, Inc.)

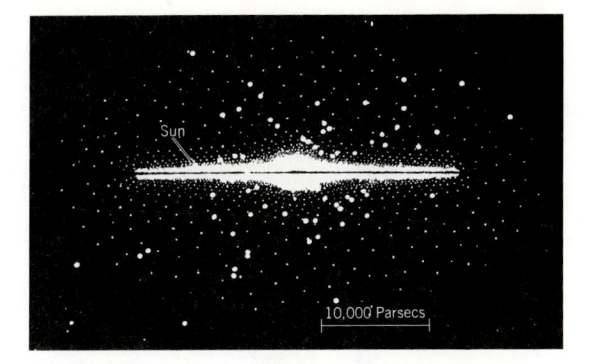

FIGURE 29-6
Schematic diagram of the Milky Way galaxy as viewed from a point in the plane of the spiral. Large spots represent star clusters; small spots represent stars. (Reprinted from O. Struve, *The Universe*, by permission of the M.I.T. Press, Cambridge, Massachusetts. © 1962 by the Massachusetts Institute of Technology.)

lies at a distance of 1.3 parsecs from the Sun. One parsec is equal to 3.26 light-years.

29.5 THE SUN IN OUR GALAXY

In its larger setting, our Sun is but one star among some billions of stars grouped into an assemblage called a *galaxy*. Our galaxy, in turn, is but one of a vast number of widely separated galaxies constituting the *universe*, which is the sum total of all matter and energy that exists.

Our galaxy has the form of a great disk, or wheel, with a marked central thickening at the hub (Figure 29-6). If it could be seen from an outside vantage point, our galaxy would probably be quite similar to the Whirlpool nebula and to the Great Spiral galaxy located in the constellation of Andromeda (Figure 29-7).

Our Sun occupies a position more than halfway out from the center toward the rim of the galaxy (Figure 29-6). As we look out into the plane of the disk, we see the stars of the galaxy massed in a great band, the Milky Way, which completely encircles the sky. For this reason our galaxy is named the Milky Way galaxy.

The Milky Way galaxy rotates about its hub, the center part turning more rapidly than the more distant outer regions. At the position occupied by our Sun, a full cycle of rotation requires about 200 million years. The velocity of the solar system in this circuit is about 800,000 km/h.

Thickness of the Milky Way galaxy is from 5,000 to 15,000 light-years, its diameter about 100,000 light-years. The galaxy has a system of *spiral arms*, comparable to those in the Andromeda spiral (Figure 29-7). Each arm consists of individual aggregations of stars, known as *star clouds*. Each cloud has a dimension of 5,000 to 20,000 light-years. Altogether, about 100 billion stars are contained in the galaxy.

The Milky Way galaxy also contains gas clouds and clouds of cosmic dust. Concentration of these clouds is particularly heavy in the plane of the galactic disk (Figure 29-6). Surrounding the disk is a vast *halo* of widely scattered stars and *globular star clusters* (Figure 29-8).

29.6 PROPERTIES OF STARS

To understand our Sun, we must compare it with other stars. We use the word *star* to refer to

FIGURE 29-7
Three spiral galaxies photographed with the 500-cm telescope. (*Left*) The Great Spiral galaxy, M 31, in the constellation Andromeda. (*Center*) Whirlpool nebula, spiral galaxy M 51. (*Right*) Spiral galaxy NGC 4565, seen edge on. (The Hale Observatories.)

discrete concentrations of matter in our galaxy, bound into single units by gravitation. In this way a star stands distinct from highly dispersed matter in the form of gas clouds and dust clouds.

Measurable properties that distinguish one star from another and enable classification to be made are mass, size, (volume, radius, or surface area), density, luminosity, and temperature. Temperature in turn determines the type of radiation emitted by the star.

FIGURE 29-8
Globular star cluster, M 13, in the constellation of Hercules. (The Hale Observatories.)

Mass of a star, which refers to the quantity of matter present, varies over a wide range. Taking the mass of our Sun as unity (1.0), the masses of stars range from as small as about one-tenth that of the Sun to about 20 times greater than the Sun. Stars also have a great range in diameter. For example, a small companion star to Sirius has a diameter only one-thirtieth that of the Sun, whereas the diameter of Antares is almost 500 times greater than that of the Sun.

Density of a star refers to the degree of concentration of mass within a given volume of space. Taking as a standard the density of water to be unity (1 g/cm^3), the average density of the Sun is about 1.4 g/cm^3 or only slightly more than the value for water at the Earth's surface. Stars show a truly enormous range in density, from less than one-millionth that of the Sun to more than 100 million times as great. The companion star to Sirius, referred to above as a small star, has a mass almost equal to that of the Sun and, consequently, a density 35,000 times that of water.

Luminosity of a star is the measure of its total radiant energy output as if measured at the star itself. Luminosity can be stated in reference to the luminosity of the Sun taken as unity (1.0). The range of luminosity among stars is from as low as one-millionth that of the Sun to as high as half a million times as great. However, for most stars the luminosity ranges between one ten-thousandth and 10,000 times that of the Sun.

Star temperature, always given in degrees Kelvin, refers to the surface temperature. Temperatures range from below 3500 K to 80,000 K. A star's color is closely related to its surface temperature: The hottest stars are blue; those only a little cooler are white; at progressively lower temperatures star color ranges from yellow through orange to red. (See Table 29.2 for star color in relation to temperature.)

29.7 STELLAR DISTANCES AND BRIGHTNESS

To the observer on Earth, the great range in brightness of the stars has long been recognized by designations of *star magnitude*. These designations are used for purposes of navigation and general descriptive astronomy. Many of you are familiar with a system used on star charts in which the brightest stars are classed as of the "first magnitude," those of lesser brightness as "second magnitude," and so on, down to the sixth magnitude.

When placed on an exact basis, the *apparent visual magnitude* of celestial objects resolves itself into a scale of numbers. In this magnitude scale, each integer value represents an increase in light intensity by a factor of 2.5 over the next larger integer. Thus, a star of magnitude 1 is 2.5 times as bright as one of magnitude 2, but 6.25 (2.5 × 2.5) times as bright as one of magnitude 3. The magnitude scale, which is a logarithmic (constant ratio) scale, extends through zero into negative numbers. According to this scale, the Sun's apparent visual magnitude is −26.7, the Moon when full, −12.7, and Venus in brightest phase, −4.5.

Table 29.1 gives the apparent visual magnitudes of the 15 brightest stars, together with information on luminosity and distance. Apparent visual magnitude is measured by sensitive photoelectric meters attached to telescopes. Magnitudes as faint as +24 can be measured. Consider next this concept. Apparent visual magnitude depends on two factors, luminosity of the star and its distance from Earth. Light emitted from a point source diminishes very rapidly with increasing distance. For two stars of equal distance from Earth, the one with the greater luminosity will appear to be the brighter. On the other hand, a near star of low luminosity might appear just as bright as a distant star of high luminosity. Check this point out by use of Table 29.1. For example, Betelgeuse and Achernar have nearly the same apparent visual magnitude, but Betelgeuse is vastly more luminous and is much farther away.

TABLE 29.1
THE FIFTEEN BRIGHTEST STARS

Name	Constellation	Apparent Visual Magnitude	Luminosity (Sun = 1)	Distance (light-years)
Sirius	Canis Major	−1.44	23	8.7
Canopus	Carina	−0.72	1,500	180
Alpha Centauri	Centaur	−0.27	1.5	4.3
Arcturus	Boötes	−0.05	110	36
Vega	Lyra	0.03	55	26.5
Capella	Auriga	0.09	170	47
Rigel	Orion	0.11	40,000	800
Procyon	Canis Minor	0.36	7.3	11.3
Betelguese	Orion	0.40	17,000	500
Achernar	Eridanus	0.49	200	65
Beta Centauri	Centaur	0.63	5,000	300
Altair	Aquila	0.77	11	16.5
Aldebaran	Taurus	0.80	100	53
Alpha Crucis	Southern Cross	0.83	4,000	400
Antares	Scorpius	0.94	5,000	400

To reduce the actual stellar luminosities to a scale that correlates with the scale of magnitudes, a system of absolute magnitudes is used. The *absolute magnitude* of a star is the apparent visual magnitude it would have if it were located at a distance of 10 parsecs from the Sun. In Figure 29–9, absolute magnitude is scaled on the left-hand side of the graph in numbers ranging from under −4 to over +16. By reading across to the right-hand side of the graph, you can read the corresponding value of luminosity.

29.8 STAR MASS AND LUMINOSITY

You might well reason that the larger a star, the greater will be its luminosity, since the area of radiating spherical surface increases greatly as the diameter increases. Your reasoning might be faulty if you ignored one important fact: a hot star may be radiating 10,000 times more strongly per square centimeter of its surface than a cool star.

Actually, there is a sound scientific reason to associate increased mass with increased luminosity. The more massive the star, the greater will be the gravitational pressure tending to cause contraction; consequently, the higher will be the star's internal temperature. As internal temperature increases, the rate of production of energy by the nuclear fusion processes also increases. For this reason, as a general rule, the larger the mass of a star, the greater will be its output of radiant energy.

Figure 29–9 is a graph in which luminosity (also absolute magnitude) is plotted against mass

29 STARS, GALAXIES, AND THE UNIVERSE

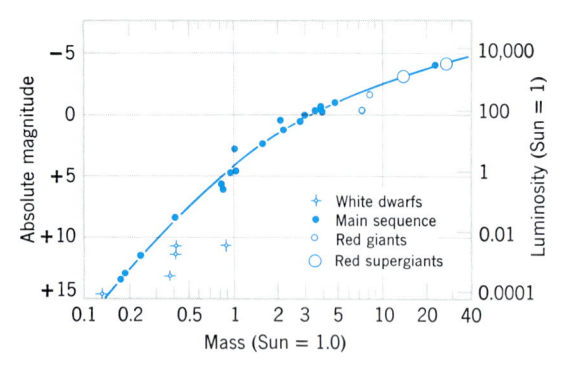

FIGURE 29-9

Mass-luminosity diagram. (From T. G. Mehlin, *Astronomy*. Copyright © 1959 by John Wiley & Sons, New York. Reprinted by permission of John Wiley & Sons, Inc.)

for a number of stars whose mass and luminosity have both been independently measured. For the most part, the stars fall on or close to a broadly curved line. At the upper right are enormous stars known as *red supergiants;* below them and to the left are *red giants*. In the middle of the graph are stars of the *main sequence*, ranging from 100 times to about one-five-hundredth of the Sun's mass. However, you see on the graph a group of stars known as the *white dwarfs*, whose plotted positions lie far off the typical curve. These are very small stars of extremely high density. They produce far less heat from thermonuclear processes than do stars of the main sequence having equivalent masses. Apparently, the white dwarfs have largely exhausted their supplies of hydrogen and have contracted into an abnormally dense state.

The mass-luminosity curve is useful to the astronomer because it enables an estimate of the mass of a star when its luminosity is known, or an estimate of the luminosity if only the mass is known.

29.9
SPECTRAL CLASSES OF STARS

The radiation spectrum produced on the photosphere of a star consists of the full sequence of wavelengths appropriate to the temperature of the radiating surface. We discussed these concepts in Chapter 21 under the subject of the Sun's radiation spectrum. As this radiation passes through the star's atmosphere (chromosphere), the various elements that make up the atmospheric gas absorb certain wavelengths. Where absorption occurs, black lines show on the color spectrum.

Each element has its particular set of absorption lines on the spectrum and can be identified with certainty. Moreover, it is possible to determine the physical state of the absorbing element, whether it exists as neutral atoms or in the ionized state. From these observations the temperature of the star's atmosphere can be quite accurately determined. The proportions in which each element is present can also be determined.

An instrument called the *slit spectroscope* is used to break up a light beam into its component colors (Figure 29-10). The essential item in this spectroscope is a prism that bends different wavelengths in different amounts. Attached to a telescope, the spectroscope is focused on a star and its spectrum photographed and analyzed. This procedure has been carried out on a large number of stars of our galaxy, with the result that they can be classified according to the *spectral class* to which each belongs. Arranged according to temperature, from hottest to coolest, the six major classes are designated *B, A, F, G, K,* and *M*. Table 29.2 summarizes the characteristics of the six major spectral classes. Figure 29-11 reproduces the actual spectra of six representative stars as photographed by telescope.

FIGURE 29-10

A slit spectrograph focuses the image of a narrow spectrum on a photographic plate. (From T. G. Mehlin, *Astronomy*. Copyright © 1959 by John Wiley & Sons, New York. Reprinted by permission of John Wiley & Sons, Inc.)

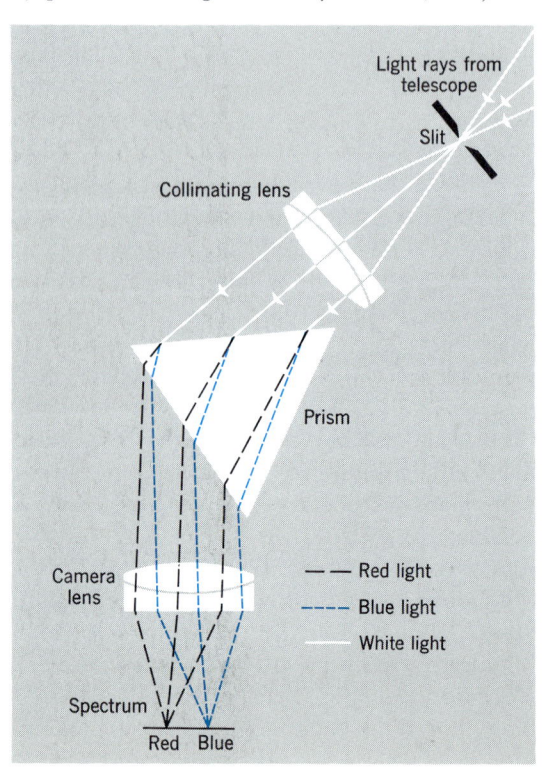

TABLE 29.2
CHARACTERISTICS OF THE SPECTRAL CLASSES

Spectral Class	Typical Stars	Color	Temperature (K)	Characteristics of Lines in Spectrum
B	Rigel, Spica	Blue-white	11,000–25,000	Helium and hydrogen strong
A	Sirius, Vega	White	7,500–11,000	Lines of hydrogen reach greatest intensity
F	Canopus, Procyon	Yellow-white	6,000–7,500	Hydrogen weakening, metals strengthening
G	Capella, The Sun	Yellow	5,000–6,000	Metals, particularly calcium, very strong
K	Arcturus, Aldebaran	Reddish	3,500–5,000	Maximum metallic lines, molecular bands appear
M	Betelgeuse, Antares	Red	2,000–3,500	Many molecular bands, violet spectrum weak

SOURCE: T. G. Mehlin, 1959, *Astronomy,* New York, Wiley, p. 50.

In addition to the six main classes, five spectral classes are added to accommodate a few stars that do not fit into the main temperature sequence.

29.10
SPECTRUM-LUMINOSITY RELATIONSHIPS

In about 1910, two astronomers, Hertzsprung and Russell, working independently, plotted star luminosity against position in the main spectral-temperature sequence. They found that a distinct and meaningful relationship exists. Figure 29–12 is the *Hertzsprung-Russell diagram* (or simply *H-R diagram*), in which each point represents a star. Luminosity is scaled on the vertical axis and a corresponding scale in terms of absolute magnitude is given as well. On the horizontal axis, spectral classes are arranged in sequence from highest temperature, on the left, to lowest temperature, on the right.

It is obvious that most of the stars plotted on the *H-R* diagram lie in a diagonal band commencing with high temperature and great luminosity at the upper left and ending with low temperature and small luminosity at the lower right. This band is the main sequence. Our Sun lies about two-thirds of the way down this main sequence. A large isolated cluster of points above and to the right of the main sequence consists of the red supergiants and red giants. These are stars of enormous size that have great luminosity despite their cool temperatures. They fall into the spectral classes *K* and *M*. In the lower part of the diagram are a very few stars—the white dwarfs. We have already noted that these stars are very small but of extremely great density. They are relatively hot stars.

29.11
THE LIFE HISTORY OF A STAR

Information about stars we have reviewed thus far can be organized into a time-sequence pattern describing the life history of a star. Deferring for the moment a consideration of how the universe itself may have originated, we start with a galaxy already in existence.

Within our Milky Way galaxy there are clouds of cold gas and dust whose temperature is close to absolute zero. Certain of these clouds appear as dark globules on astronomical photo-

FIGURE 29–11
Representative spectra of stars of the major spectral classes. Symbols at left designate star and constellation. (Yerkes Observatory photograph, University of Chicago.)

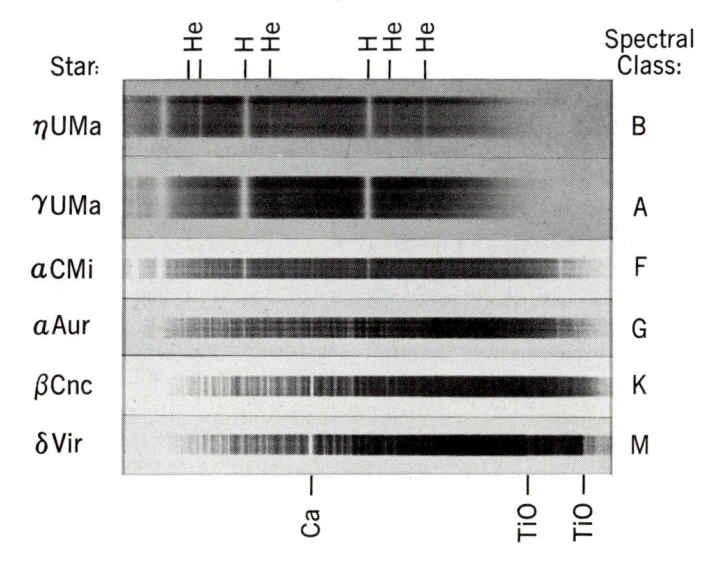

29 STARS, GALAXIES, AND THE UNIVERSE

FIGURE 29-12
The Hertzsprung-Russell spectrum-luminosity diagram. Each dot represents a star. Altogether a sample of 6700 stars is recorded on the diagram. (Yerkes Observatory photograph, University of Chicago.)

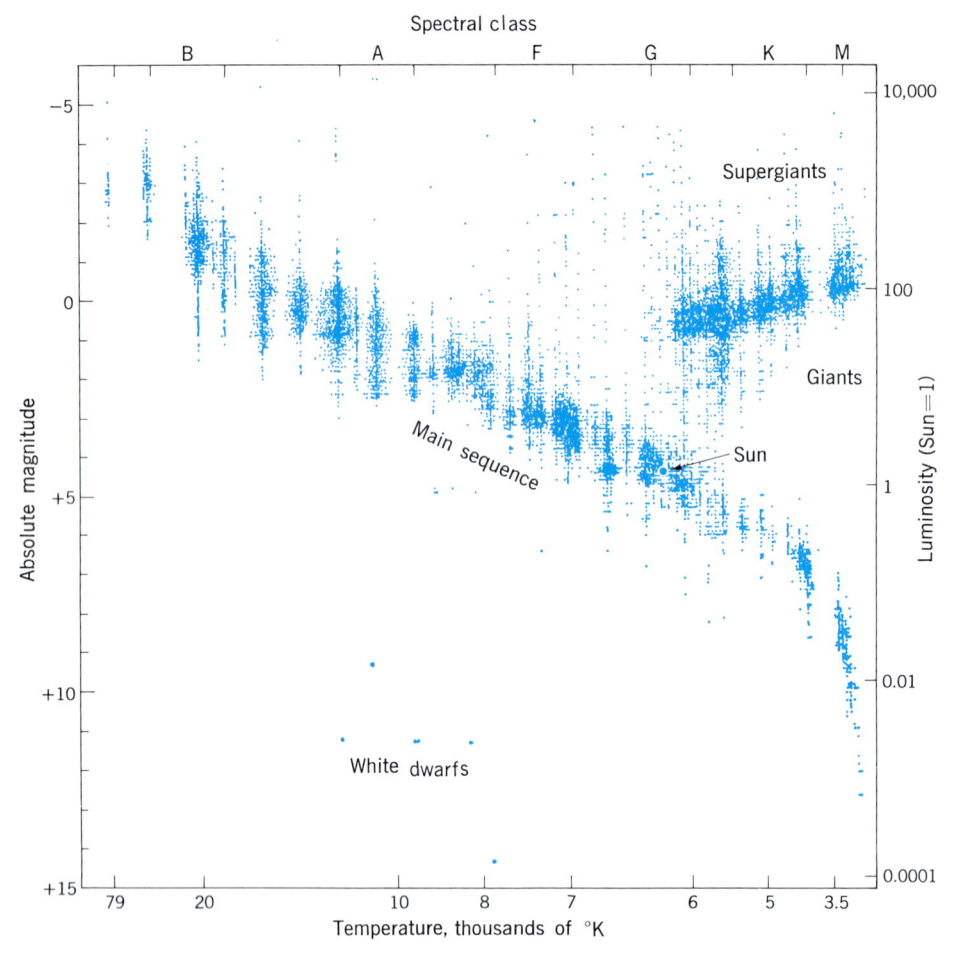

graphs because the gas effectively absorbs most or all of the starlight that would otherwise pass through from distant stars on the far side. Diameters of the dark globules are on the order of 100–1000 times the diameter of the solar system.

As a working hypothesis, we assume that the cloud of cold gas making up a dark globule represents the initial stage in the life history of a star. Through the gravitational attraction that all particles of the gas cloud exert on all other particles, the cloud begins to contract, occupying a smaller volume. The temperature of the contracting body of gas increases—particularly so near the center of the mass, where pressures are greatest.

Eventually, contraction forms a star with an interior temperature exceeding 1 million K. At this point, the first of a series of nuclear reactions begins to take place, converting matter into energy and causing the star to begin emitting large amounts of electromagnetic radiation. As contraction continues and interior temperatures rise, other forms of nuclear reactions develop and sustain a high level of energy production. A fully developed star such as our Sun has now come into existence.

As the Hertzsprung-Russell diagram shows, stars of the main sequence span a very great range in both temperature and mass. Those of small mass can attain only comparatively low temperatures and pressures. These small bodies therefore produce energy at a relatively slow rate, resulting in stars of faint luminosity. Such small stars will have an extremely long life because the utilization of the hydrogen supply takes place so very slowly.

On the other hand, stars at the high-temperature and large-mass end of the sequence are converting their hydrogen supply into energy at an extremely fast rate. Their life expectancies will be short. For example, a star of mass 10 times that of the Sun will radiate energy about

10,000 times as rapidly as the Sun. The life of such a large star must therefore be on the order of 1 percent of the life of our Sun, or as short as 100 million years. The small stars will correspondingly have lives vastly longer than the Sun, life spans as long as thousands of billions of years.

Figure 29-13 is a graph with essentially the same field as the *H-R* diagram, but it does not show the plots of the individual stars. The diagonal band shows the position of the main sequence. The chain of arrows represents the evolution of a single star of about the size of our Sun. The timepath enters from the right and moves horizontally toward the line of the main sequence. This horizontal timepath is covered comparatively rapidly and represents the stage of contraction of the gas cloud and its rise in temperature. When the star begins to consume its hydrogen by thermonuclear processes, it is located on the line of the main sequence. As appreciable amounts of the star's hydrogen are transformed into helium, the star may brighten slightly, moving slowly to a position slightly above its original main sequence location.

The next stage in the life history of a star comes when its hydrogen supply is seriously depleted. Nuclear activity ceases first in the central region of the star, which then contracts. Nuclear activity continues in a surrounding zone that gradually moves outward from the center toward the surface. As this happens, the star may expand greatly. Although the luminosity remains high, the surface temperature falls and the star spectrum changes toward the red region.

On the *H-R* diagram (Figure 29-12), this change requires that the plotted position of the star depart from the main sequence and move toward the upper right, occupying a position among the red giants and perhaps reaching the position of the supergiants. This timepath is shown on Figure 29-13.

The final hypothetical stage in the life of the star is suggested by a timepath that moves downward and to the left, then sharply downward to the region of the white dwarfs. These changes may be quite rapid. The star is now "burned out" and has only a faint luminosity despite its high temperature.

29.12
PULSARS

Most recently discovered of those stars whose brightness varies are the *pulsars*. These stars flash "on" and "off" rapidly, emitting both light waves and radio waves in the same rhythm. Light pulses range in frequency from about one pulse per four seconds in the slowest rhythm to as high as thirty pulses per second. In the case of the high rate of pulsation, the star appears to the eye and on photographs to be continuously bright, but special techniques can reveal the flashing on and off (Figure 29-14).

To explain the periodic emission of pulsars, it has been suggested that they are extremely small, dense, dwarf stars, of a kind called a *neutron star*. The star is rotating rapidly on an axis. The emitting source is situated at one spot on the star and thus gives forth a single turning

FIGURE 29-13
Simplified H-R diagram showing inferred evolution of an average star. (Reprinted from O. Struve, *The Universe*, by permission of the M.I.T. Press, Cambridge, Massachusetts. © 1962 by the Massachusetts Institute of Technology.)

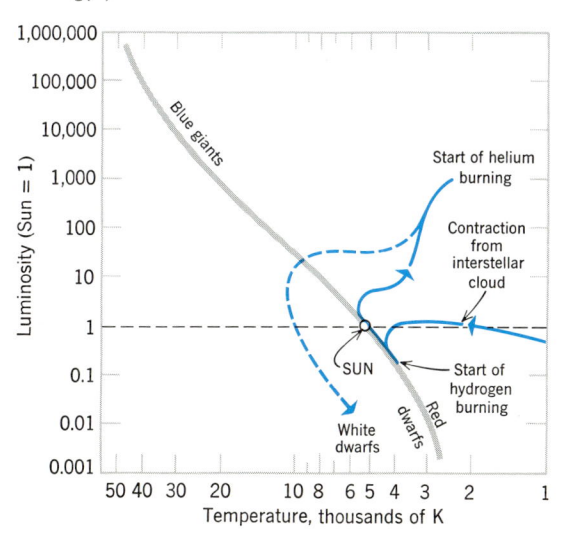

FIGURE 29-14
Comparison photographs of pulsar NP 0532 in "on" (*left*) and "off" (*right*) phases. (Lick Observatory photograph, University of California.)

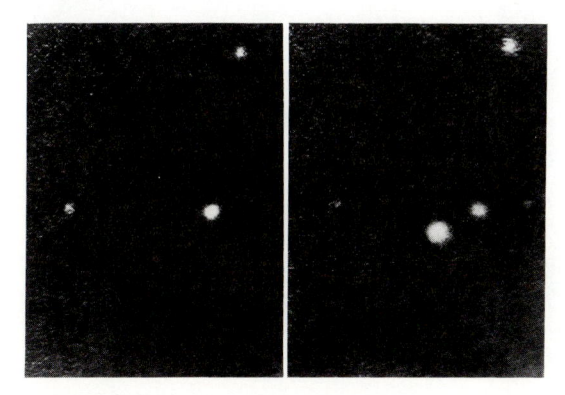

29.13
NOVAE, SUPERNOVAE, AND BLACK HOLES

On occasion, an extremely faint star bursts into intense brightness, then fades back to its original level. Such stars are known as *novae*, meaning "new stars," because they had not been observed to exist prior to the episode of brightness. A typical nova increases in brightness by 10 to 12 magnitudes in a timespan of a few hours to a few days. Immediately after the outburst the brightness falls off rapidly for a time and then tends to level off and to diminish gradually over many weeks. Within years the brightness has returned to the original level.

Increase in brightness of the nova is associated with an explosive increase in its size, which may be a diameter increase of from 100 to 200 times. This expansion takes place in the photosphere of the star and is not an explosive enlargement of the entire star interior. The expanded layer of gases gradually dissipates and is lost into space, revealing the main body of the star intact. Novae are interpreted as being stars in the white-dwarf stage, near the end of the stellar life cycle. The explosion represents a short period of instability during the final stages of contraction into an extremely dense small star.

A very rare type of nova is one that attains sudden brightness in the range between $\frac{1}{4}$ billion and $1\frac{1}{2}$ billion times the luminosity of the Sun. If this kind of star is in our own galaxy, its apparent magnitude may exceed that of the brightest planets. Such phenomena are known as *supernovae*. They occur within our galaxy with an average frequency of about two per hundred years. Following the outburst, an expanding cloud of gas and dust, or *nebula*, has been observed surrounding the site of the supernova. The Ring Nebula is an example of this feature (Figure 29-15).

The supernovae probably originate from extremely massive stars that have transformed so much of their hydrogen to heavier elements that they become explosively unstable. Unlike the typical nova, the outburst of a supernova blows off the major part of the material of the star, changing the nature of the star drastically.

In 1969 the central star of the Crab Nebula was shown to be a pulsar, or rapidly rotating neutron star, only a few kilometers in diameter, but with a density on the order of one million billion times the density of water. It emits extremely regular radio pulses at the rate of 30 per second. Thus, evidence is accumulating to show that a neutron star is the dense mass remaining after the outburst of a supernova.

Collapse of the dense matter remaining after a supernova explosion may, according to current theory, result in an object exerting such intense gravitational force that it pulls back into itself all light and other forms of electromagnetic radiation it produces. Under such conditions, the presence of the object could not be detected—it would be a *black hole* in space. Whether a particular supernova yields a neutron star or a black hole depends on the initial mass of the exploding star and the quantity of mass thrown off during the explosion. For a black hole to form, the collapsing matter would need to have a mass at least three times greater than that of our sun.

FIGURE 29-15
The Ring Nebula is the gaseous remains of a supernova, spreading rapidly outward. A compact dwarf star, seen at the center, is all that remains of the body which exploded to produce the nebula. (The Hale Observatories.)

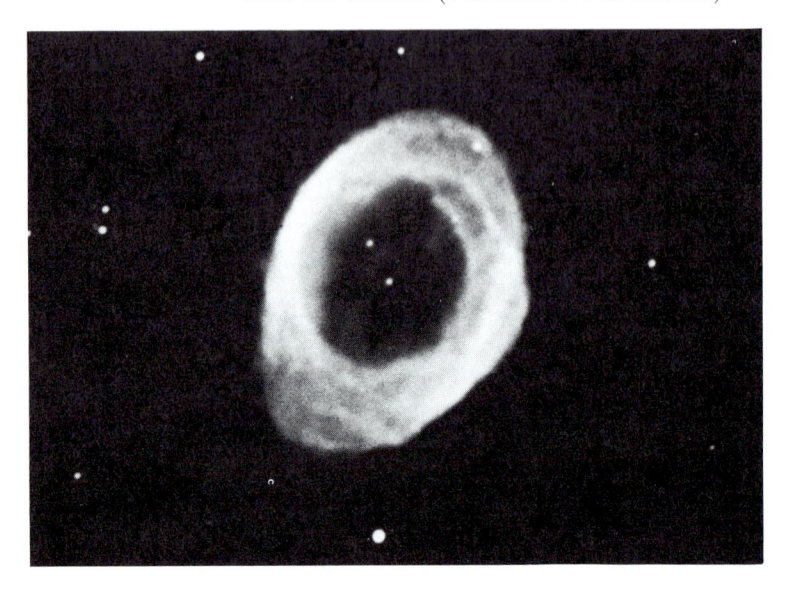

29.14
GALAXIES AND THE UNIVERSE

To the most distant limits of telescopic penetration, our universe consists of widely spaced galaxies, of which an estimated 10 billion can now be observed, but no outer limit to the universe can be recognized. The total extent of possible

29.14 GALAXIES AND THE UNIVERSE

FIGURE 29–16
A barred spiral galaxy in the constellation of Pegasus. (The Hale Observatories.)

observation of light from distant galaxies is estimated to be 10 billion light-years. Within this theoretical maximum radius of observation there may be as many as 100 billion galaxies.

Galaxies fall into several classes, according to their shapes. *Spiral galaxies,* such as our Milky Way, are illustrated by the Andromeda spiral, which is the closest galaxy to our own (Figure 29–7). Its distance is about two billion light-years, and its diameter is a bit larger than our Milky Way galaxy. Another class of galaxies is the *barred spiral,* in which the two arms uncoil from a central bar (Figure 29–16). Equally important are galaxies of the *elliptical* group (Figure 29–17). These are ellipsoidal or spherical masses having a high degree of symmetry. Their form suggests that they, like the spirals, are rotating. In addition, there are galaxies of highly irregular shape, but these are relatively few.

Within the nearer galaxies, individual stars and star clusters can be recognized. Clouds of dust and gas that are typical of the spiral galaxies seem to be absent from the elliptical types. Attempts have been made to arrange the several forms of galaxies into an evolutionary series. Edwin P. Hubble, the astronomer who did much of the pioneering work in galactic investigation, suggested a classification beginning with the almost spherical elliptical galaxies and then progressing to the more flattened systems. The series then branches into two parallel arms, one for the spirals and the other for barred spirals, and perhaps ends with the irregular galaxies (Figure 29–18).

When more was known about the galaxies and the ages of the stars in them, Harlow Shapley suggested a reversed evolutionary sequence. It might well begin with the irregular galaxies, developing into spiral systems in which the nucleus would move tightly into the arms with increasing age. Then, as the stars aged and the interstellar clouds of gas and dust were elimi-

FIGURE 29–17
An elliptical galaxy in the constellation of Andromeda. (The Hale Observatories.)

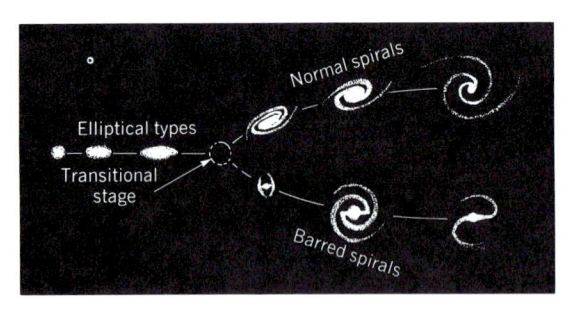

29 STARS, GALAXIES, AND THE UNIVERSE

FIGURE 29–18
Diagram of sequence of nebular types as arranged by E. Hubble. No nebulae have been recognized in the transitional stage, which is hypothetical. (From E. Hubble, 1936, *The Realm of the Nebulae*, New Haven, Yale University Press, p. 45.)

nated, the spirals might evolve into elliptical systems of varying degrees of flattening. It is still not understood why some spirals take the normal form and some become barred spirals.

29.15
THE DOPPLER EFFECT

One of the most remarkable findings about galaxies is evidence that they are rapidly moving away from Earth, and that the farther away they are, the faster they are receding. To understand the evidence for such a statement, we must investigate a principle of science applying to the light spectrum and other forms of wave motion, such as sound waves.

The principle involved is familiar in the *Doppler effect* on sound waves. You are familiar with the way in which the pitch of a sound of fixed vibration period seems higher as the emitting source is brought rapidly toward us and lower as it recedes from us. The sudden drop in pitch of a locomotive horn as it passes by at close range is a good example.

A very simple analogy may help to illustrate the Doppler principle. Suppose that we stand beside a long horizontal conveyer belt and place pebbles on the belt at uniform intervals of time. If the belt speed is constant, the pebbles will be uniformly spaced. Now, if as we place the pebbles we also walk slowly in the direction in which the belt is moving, the pebbles will be spaced closer together; whereas if we walk in a direction opposite to the belt motion, again placing pebbles at the same intervals of time, they will be spaced farther apart on the belt. In a similar way the frequency of light rays is changed as the emitting source moves.

Take now the case of a star emitting a given light spectrum. When the star is moving earthward, it appears to us that the frequencies of vibration constituting the light rays are all increased slightly. This increase in frequency results in a slight change in the color, since the color we observe is determined by the frequencies of light and these have been increased by the motion. When the star is moving away from the Earth a reverse effect occurs—the frequencies of vibration are reduced.

The Doppler effect has been an essential tool of astronomy. It has enabled astronomers to measure the speed of a star in the line of sight. This speed is called the *radial velocity*. Using this method, the speed of rotation of the Milky Way galaxy was determined.

29.16
THE RED SHIFT

In the case of a galaxy, its color spectrum is shifted toward the red end of the spectrum when the galaxy is moving away from us. For short, this effect is called the *red shift*. A particular line in the spectrum identified with a particular element, such as calcium, is found to be displaced toward the right in a spectrum such as that shown in Figure 29–11. The amount of the red shift can then be interpreted in terms of radial velocity.

Now, let us return to the observation that the red shift of the spectra of galaxies increases in direct proportion to their radial distance from our point of observation in the solar system. Assuming that the red shift is a true Doppler effect, all galaxies must be in radial motion, receding from our galaxy. Moreover, the speed of recession increases proportionately with increasing distance. This principle was discovered by Edwin Hubble and is known as *Hubble's law*.

29.17
THE ORIGIN OF THE UNIVERSE

The geometry of apparent radial outward motion of the galaxies can be visualized in terms of a universe that is expanding uniformly in volume. From any single vantage point in this system, all other objects will appear to be moving radially outward. Hubble's discovery of a law of increase of radial velocity proportionate with distance quickly led to a new theory of origin of the universe.

Cosmology as a science concerns itself with the nature and origin of the universe. Among the first to propose a cosmological theory based on Hubble's discovery was Canon Lemaître, a Belgian. He referred to the concept as a "fireworks theory." The theory requires an initial point in time at which all matter was concentrated into a small space. From this center it expanded explosively outward in all directions. The elements were created during this explosion and were later formed into the galaxies. Although now commonly referred to as the "big-bang" theory of the universe, the title of *evolutionary theory* is perhaps more fitting.

Using Hubble's first derived estimates of the rate of velocity increase with distance of separation, it could be calculated that all matter of the universe was concentrated into a small space about two billion years ago. This point in time was designated the *age of the universe*. Age of the universe in years would be equal to distance in light-years of the most rapidly traveling galaxy. In 1952 new data required that the age of the universe be placed at 5 billion years. The figures were repeatedly revised to increase the age. By 1977 the age of the universe had been confirmed as 20 billion years.

The evolutionary theory conforms to the principle that the distribution of galaxies is uniform in all directions throughout space. Under this concept, to an observer from any galaxy the average composition of the universe would appear the same. It is interesting to consider that under Hubble's principle the radial velocity of extremely distant galaxies, with respect to our observation point in the solar system, must reach and finally equal the speed of light. This distance would constitute the observable limit of the universe, beyond which we could receive no light or radio waves from the emitting sources.

The hypothesis of a *pulsating universe* has also been suggested as a modification of the "big-bang" hypothesis. Immediately after the initial explosion, all of the matter would be moving outward with high velocities, but the mutual gravitational attraction between all of the parts would tend to slow the outward motion, perhaps finally stopping it and causing the entire system to contract. All of the material would eventually come back to a central point in an implosion that would annihilate all forms of matter—stars, galaxies, and even individual atoms. The result would be another "cosmic bomb," which would explode and start the whole process over again. The interval for one complete cycle has been estimated to be something less than one hundred billion years.

A major rival theory of the universe holds that there was no single point in time at which matter was concentrated in one place. Instead, the production of matter has gone on throughout intergalactic space at a constant rate during all time. Rate of production of matter in the form of hydrogen atoms has been equaled by the rate at which matter is dispersed by the expansion of the universe. This *steady-state theory* of cosmology, proposed in 1948 by the astronomers H. Bondi, T. Gold, and F. Hoyle, has attracted great interest but is currently enjoying fewer supporters than it had two decades ago.

The most recent evaluations of information concerning galaxies and other distant objects seem to place the evolutionary theory of the universe in a stronger scientific position than the steady-state theory. However, we can anticipate modified and new cosmological theories of the universe to be brought forward from time to time as new information is gained from the development and use of newer tools of astronomy.

29.18 RADIO ASTRONOMY AND QUASARS

Part of the electromagnetic radiation spectrum—that in the longwave region—consists of radio waves. In the range of wavelengths between about 1 cm and about 20 m, radio waves can pass through our atmosphere and be received by *radio telescopes*. These instruments use a huge concave bowl-shaped (parabolic) antenna that can be aimed at a distant emitting source (Figure 29-19).

Thousands of radio-emitting sources have been discovered and their positions plotted, but only a few can be identified with stellar objects that appear on photographs. Some sources of radio emission lie within our Milky Way galaxy; others are in distant galaxies, referred to as *radio galaxies*. These radio galaxies, of which about 150 have been identified, are the most powerful of all radio-emission sources.

Hydrogen gas clouds within our galaxy are also radio-wave emitting sources. Our Sun shows strong radio-wave emission at those times when a solar flare is in progress. A number of stars are known to have flares of similar nature; at such times their brilliance is greatly increased. For this reason radio emissions received from these stars are believed to be associated with flares.

Among the most important of astronomical discoveries in recent years has been the finding of extremely small sources of intensely powerful radio emission not related to any surrounding galaxy. They were named *quasistellar radio sources,* but the term has since been reduced to *quasars*. These emission sources appear only as pinpoints of light. The distribution of the 100 or so quasars identified is quite uniform with respect to direction from the Earth.

FIGURE 29-19
This parabolic dish, 300 m in diameter, is the world's largest single radio telescope antenna unit. It lies nestled in a natural bowl-shaped valley in Puerto Rico, and belongs to the Arecibo Observatory. (Courtesy of the National Astronomy and Ionospheric Center, operated by Cornell University under contract with the National Science Foundation.)

A particularly striking feature of quasars is that the lines in their spectra show a very great shift toward the red. Although its use here may be questioned, if the same red shift-distance relationship developed for galaxies is applied to quasars, the extremely large red shift would lead to the conclusion that they are on the order of 1 to 10 billion light-years away. If so, they are the most distant known objects in the universe. The luminosities and energy outputs of the quasars must be truly enormous. One hypothesis explains the quasars as formed from gas clouds sent outward from the center of an exploding universe at a speed up to 80 percent that of light.

29.19
X-RAY ASTRONOMY

X-ray astronomy is one of the youngest fields of research, because it requires observation from orbiting space vehicles. The Earth's atmosphere effectively blocks nearly all x-ray, gamma ray, and ultraviolet radiation from terrestrial telescopes. Much of the infrared radiation is also blocked. Until orbiting satellites were available, astronomers had to depend largely on the visible light spectrum and on radio waves for their knowledge of stars, galaxies, and other energy-emitting objects of the universe. Now the situation has changed. Telescopes mounted on spacecraft orbit high above the atmosphere. They are highly specialized in function to receive specified bands of the electromagnetic spectrum over its entire range.

Launched in 1970, a satellite named Uhuru began to pick up sources of x-ray emission from many celestial sources. In 1972 another orbiting telescope package, named Copernicus, began its observations in the x-ray and ultraviolet bands. Uhuru identified more than 100 x-ray emission sources. Many of these x-ray sources lie in our galaxy but some are in other galaxies. Intense sources have been picked up from quasars and from pulsars in the remnants of supernovae. Other sources are suspected of being black holes, into which x-ray-emitting gases are being drawn from nearby stars.

29.20
COSMIC PARTICLES

The Earth's atmosphere is continually bombarded with elementary particles traveling at speeds approaching the speed of light and having enormous energy and penetrating power. This form of radiation from outer space is the *cosmic particle*, often called the "cosmic ray." It is an entirely independent phenomenon from the electromagnetic radiation spectrum of a star.

Cosmic particles are protons—that is, parts of the atomic nucleus. Approximately 90 percent are hydrogen nuclei, 9 percent are helium nuclei, and 1 percent are heavier nuclei. The energy of cosmic particles is enormous.

Cosmic particles approach the Earth from all directions. Their space paths seem to be quite at random, and they can be visualized as constituting a kind of cosmic "gas" in which particles undergo random collisions and can take an infinite variety of paths and a wide range of speeds. Sources of cosmic particles are considered to be varied. They are produced in solar flares, but most come from other sources, believed to be the explosions of supernova and other forms of explosive activity in the central parts of our own and other galaxies. It has been suggested that galaxies emitting radio waves are also sources of important amounts of cosmic radiation.

Cosmic particles are important in the environment of life on the Earth's surface. The extremely high energy of cosmic particles enables them to penetrate deep into the lower atmosphere and to reach the Earth's surface. This penetration is accompanied by an elaborate series of secondary nuclear reactions making up a shower of particles and secondary forms of radiation. When a high-speed cosmic particle impacts the nucleus of an atom within the atmosphere, there are produced neutrons and protons, mesons, and gamma radiations. The effect of such radiation on life forms is to induce genetic changes (mutations in genes), which are important in the process of organic evolution.

29.21 THE HUMAN PLACE IN THE UNIVERSE

Seen in its relative position among the other stars of the Milky Way galaxy, our Sun is a fairly typical star in most respects. It lies somewhat below the midpoint of the main sequence of stars, belonging to the spectral class *G*, which has moderate surface temperatures in terms of the total temperature range. Luminosity and mass are about midway on the scale of those values. Extreme constancy of energy output over vast spans of geologic time characterizes the Sun—a behavior in strong contrast to the changing energy emissions of the variable stars and novae.

Our Sun represents one of the basic forms of energy systems, that of conversion of matter to energy in nuclear reactions occurring within a gaseous medium under enormously high pressures and temperatures. The lifespan of our Sun is neither very short nor very long in comparison with the range found among stars, but it is long enough to assure that our terrestrial environment can continue with little change for a span of time vastly longer than that which has already elapsed as geologic time.

In reference to the total size of the Milky Way galaxy, our Sun is no more than an insignificant particle of matter, while in the context of the universe of galaxies, it comes infinitesimally close to being nothing at all. Within the universe there must be a very large number of stars quite similar to our Sun, and many of these must have planets resembling our own.

Reason leads us to suppose that spontaneous development of organic life and its evolution to highly complex states must have been replicated a great number of times on unknown planets. But we also realize that the vastnss of interstellar and intergalactic space reduces almost to zero the possibilities of identifying and communicating with even the closest of such organic complexes. Despite such odds, the possibility of a discovery that humans on planet Earth are not alone in the universe continues to fire the popular imagination.

SUMMARY

The Sun. An average star in mass and temperature, our Sun consists largely of hydrogen and helium with internal temperature over 4 million K. Energy of hydrogen fusion in the nucleus maintains the Sun's enormous output of energy from the surface layer, or photosphere. Beyond the solar atmosphere, or chromosphere, lies the corona, which gives rise to the solar wind. Sunspots and solar flares form frequently on the solar surface.

Intrstellar distance. Light-years measure interstellar distance. Also used is the parsec, based on stellar parallax.

Galaxies. Stars are organized into galaxies; that of our Sun is the Milky Way galaxy—of flattened, spiral form. Other common forms are barred spirals and elliptical galaxies.

Star properties. Star properties include mass, density, luminosity, temperature, and spectral class. Star magnitude may be stated on an apparent visual scale, or as absolute magnitude. Mass-luminosity relationships yield a main sequence of stars with red giants and white dwarfs occupying positions far off the main sequence curve. Spectral classes are arranged by temperature. The *R-H* diagram relates spectral class to luminosity.

Life history of a star. A new star enters the *R-H* field from the right, arrives at the main sequence, then moves up to the red giant region, and finally moves down to the white dwarf region, where fusion ceases and the star contracts to a high-density neutron star. Pulsars are identified as neutron stars and may achieve the state of black holes. Supernovae are exploding white-dwarf stars, passing into the neutron star condition.

29 STARS, GALAXIES, AND THE UNIVERSE

Galaxies and universe. Galaxies occupy the universe to an observed limit of about 10 billion light years. Galaxies may evolve through a series of forms ending in the elliptical form. Based on the Doppler-effect, red-shift of spectra, galaxies are interpreted as moving radially outward through the universe. The more distant the galaxy, the greater the red shift, or outward radial velocity.

Cosmology. Origin of the universe is postulated in the form of three major theories: evolutionary, or "big-bang" theory, pulsating universe, and steady state universe.

Quasars. Quasars are extremely distant, extremely powerful radio-emitting sources with extreme red shifts, possibly the most distant objects of a rapidly expanding universe.

Cosmic particles. Protons, mostly hydrogen nuclei, traveling at high speeds through interstellar space, impinge upon the earth as cosmic particles.

TERMS AND CONCEPTS

photosphere
nucleus
chromosphere
solar prominences
corona
sunspot
umbra
penumbra
solar flares
light-year
stellar parallax
parallax
parsec
galaxy
universe
spiral arms
star clouds
halo
globular star cluster
star
luminosity
star magnitude
apparent visual magnitude
absolute magnitude
red supergiants
red giants
main sequence
white dwarfs
slit spectroscope
spectral class
Hertzsprung-Russell (*H-R*) diagram
pulsar
neutron star
black hole
novae
supernovae
nebula
spiral galaxy
barred spiral
elliptical galaxy
Doppler effect
radial velocity
red shift
Hubble's law
cosmology
evolutionary theory
age of universe
pulsating universe
steady-state theory
radio telescope
radio galaxy
quasistellar radio source (quasar)
cosmic particle

QUESTIONS

1. Compare the Sun with Earth in respect to diameter and mass. In what respect is the Sun's rotation different from that of Earth? Describe the various layers, or shells, comprising the Sun and its atmosphere. Where are temperatures highest?

2. What process operates in the Sun's interior to produce energy from matter? Of what elements is the Sun largely composed? Describe sunspots. How are they related to magnetic fields? How are they related to solar flares?

3. What units are used in the measurement of interstellar distances? How far away is the nearest star? Describe the phenomenon of stellar parallax and show how it is used to measure distance to a star.

4. Describe the Milky Way galaxy and its motions. Where are spiral arms, star clouds, the halo, the globular star clusters located in this galaxy? Where in our galaxy is the Sun located?

5. What important physical properties of stars are subject to measurement? Compare our Sun with other stars in terms of mass, density, luminosity, and surface temperature.

6. What relation does luminosity bear to apparent visual magnitude? How is the absolute magnitude of a star determined? Why is it used?

7. Relate star mass to luminosity. Describe the mass-luminosity curve. Where on this curve are the red supergiants? The red giants? The main sequence? Where are the white dwarfs located with respect to the mass-luminosity curve? Explain their position.

8. What use is made of the spectrum of a star? What is the significance of the main spectral classes? What spectrum-luminosity relationships

emerge from the Hertzsprung-Russell diagram? Describe this diagram. How do the major groups of stars fit into the diagram?

9. Trace the evolution of a typical star, beginning with a cloud of dust and gas. What causes interior heating of a star? At what point do nuclear reactons begin to occur? Track the stages of development of a star on the *H-R* diagram. In what way is the life expectancy of a star related to its mass? What is the end stage in the life of a star?

10. Describe the emission of pulsars. What hypothesis explains the pulsation? To what class of stars do the pulsars belong? What is a black hole? How might a black hole be formed from a neutron star?

11. Describe the occurrence of a nova and explain what happends to the star. What event does a supernova represent in the life history of a star? What kind of star remains after a supernova has dispersed?

12. About how many galaxies can be observed with existing telescopes? How far away are the most distant of the galaxies from which light might be received? Within that maximum range, how many galaxies are estimated to exist?

13. Classify galaxies according to their forms. What sequence of development of galaxies was proposed by Hubble? by Shapley?

14. Explain the Doppler effect on the spectrum of a star. How is the Doppler effect used to measure radial velocity of a star?

15. Describe the red shift of spectra of the galaxies and relate it to their radial velocities. What is Hubble's law? Describe the geometry of a universe conforming to Hubble's law.

16. Describe the evolutionary ("big-bang") theory of the universe. How, according to this theory of cosmology, can the age of the universe be estimated? Compare earlier and recent estimates of age of the universe.

17. Describe the hypothesis of a pulsating universe. In what way does the steady-state theory differ from the evolutionary theory?

18. Describe the radio-emitting sources in space beyond the Earth. How are radio emissions received? Does the Sun emit radio waves? What are radio galaxies? What is remarkable about the emission spectra of quasars? How are quasars interpreted?

19. Describe the methods of x-ray astronomy. What x-ray sources have been discovered? What are cosmic particles? Of what forms of matter are they composed? What happens when a cosmic particle enters the Earth's atmosphere?

20. Summarize the characteristics of our Sun as compared with other stars, and comment on the importance of those characteristics in determining our planetary environment through geologic time.

21. What do you estimate to be the probability of existence of planets with life systems similar to those of planet Earth in other solar systems in our own and other galaxies? Would communication with advanced life forms on such distant planets be possible?

GLOSSARY

Abrasion is erosion of bedrock of a stream channel by impact of particles carried in a stream and by rolling of larger rock fragments over the streambed. Abrasion is also an activity of glacial ice, waves, and wind.

Abyssal plain is a large expanse of very smooth, flat ocean floor found at depths of 4600–5500 m.

Air mass is an extensive body of air within which upward gradients of temperature and moisture are fairly uniform over a large area.

Alluvial fan is a low, gently sloping, conical accumulation of coarse alluvium deposited by a braided stream undergoing aggradation below the point of emergence of the channel from a narrow canyon.

Alluvium is any stream-laid sediment deposit found in a stream channel and in low parts of a stream valley subject to flooding.

Alpine glacier is a long, narrow mountain glacier on a steep downgrade, occupying the floor of a troughlike valley.

Aluminosilicates are silicate minerals containing aluminum as an essential element.

Anticyclone is the center of high atmospheric pressure.

Aphelion is the point on Earth's elliptical orbit at which Earth is farthest from the sun.

Apogee is the point in Moon's orbit farthest from Earth.

Asteroids are solid bodies, numbering in the tens of thousands, orbiting the Sun between the orbits of Mars and Jupiter; also called minor planets.

Asthenosphere is the soft layer of the mantle, beneath the rigid lithosphere.

Astrogeology is a branch of geology applying principles and methods of geology to all condensed matter and gases of the solar system outside Earth.

Atmosphere is the envelope of gases surrounding Earth, held by gravity.

Atmospheric pressure is pressure exerted by the atmosphere because of the force of gravity acting on the overlying column of air.

Barchan dune is a sand dune of crescentic base outline with sharp crest and steep lee slip face, with crescent points (horns) pointing downwind.

Barometric pressure (see atmospheric pressure).

Basalt is extrusive igneous rock of gabbro composition; it occurs as lava.

Batholith is a large, deep pluton (body of intrusive igneous rock), usually with area of surface exposure greater than 100 km^2.

Beach is a thick, wedge-shaped accumulation of sand, gravel, or cobbles in the zone of breaking waves.

Beach drifting is transport of sand on a beach parallel with a shoreline by a succession of landward and seaward water movements at times when swash approaches obliquely.

Bed load is that fraction of the total load of a stream being moved in traction.

Bedrock is the solid rock in place wih respect to the surrounding and underlying rock and relatively unchanged by weathering processes.

GLOSSARY

Black hole is a neutron star that has attained a sufficiently high density to draw all electromagnetic radiation into itself through its enormously strong gravitation, and thus to lack any direct means of detection.

Breccia is a general term for sediment consisting of angular rock fragments in a matrix of finer sediment particles.

Catastrophism is the view held by naturalists of the late eighteenth century who explained all disruption of strata and extinction of organisms as occurring in a single great catastrophe.

Chemical weathering is chemical change in rock-forming minerals through exposure to atmospheric conditions in the presence of water, mainly chemical reactions of oxidation, hydrolysis, and carbonation, or direct solution.

Chromosphere is the low layer of the Sun's atmosphere, immediately above the photosphere; it contains solar prominences.

Cirque is a bowl-shaped depression holding the collecting ground and firn of an alpine glacier.

Clay minerals belong to a class of minerals, produced by alteration of silicate minerals, having plastic properties when moist.

Cleavage is the property of a mineral to split readily in a set of parallel planes, or along two or three sets of parallel planes.

Clouds are dense concentrations of suspended water or ice particles in diameter range 20–50 microns.

Cold front is a moving weather front along which a cold air mass is forcing itself beneath a warm air mass, causing the latter to be lifted.

Comet is a member of the solar system consisting of highly diffuse matter in the form of a brightly luminous coma and a tail, seen when passing close to the Sun.

Conjunction is alignment of Sun, Earth, and Moon, with Sun and Moon on same side of earth.

Continental drift is a hypothesis introduced by Alfred Wegener and others early in 1900s of the breakup of a parent continent, Pangaea, starting near the close of the Mesozoic era, and resulting in the present arrangement of continental shields and intervening ocean basin floors.

Continental shields are ancient crustal rock masses of the continents, largely igneous rock and metamorphic rock, and mostly of Precambrian age.

Convectional precipitation is formed within a spontaneously rising air column or air bubble, usually within a cumulonimbus cloud.

Copernican system is the heliocentric theory of the solar system devised by Nicolaus Copernicus in the early 1500s, replacing the Ptolemaic system.

Core of Earth is the spherical central mass of Earth composed largely of iron and consisting of an outer liquid zone and an interior solid zone.

Coriolis effect is a fictitious force that tends to deflect any object in motion toward the right of its direction of motion in the northern hemisphere and toward the left in the southern hemisphere.

Corona is the Sun's outer atmosphere above the chromosphere; a region of pearly white streamers of light reaching far out into the solar system.

Cosmic particle is radiation arriving from outer space in the form of a high-energy particle, which is a proton traveling at extremely high velocity; most are nuclei of hydrogen atoms.

Cosmology is the science dealing with the nature and origin of the universe.

Counterradiation is longwave radiation of atmosphere directed downward to Earth's surface.

Crust of earth is the outermost solid shell or layer of earth, composed largely of silicate minerals.

Cumulonimbus cloud is a large, dense cumuliform cloud yielding precipitation.

Cumulus clouds are clouds of globular shape, often with extended vertical development.

Cyclone is a center of low atmospheric pressure.

Deflation is lifting and transport in suspension by wind of loose particles of soil or regolith from dry ground surfaces.

Delta is a sediment deposit built by a stream entering a body of standing water and formed of the load carried by the stream.

Dew point temperature is the temperature of saturated air.

Discharge is the volume of flow moving through a given cross section of a stream in a given unit of time; commonly given in cubic meters per second.

Doppler effect is the change in pitch of sound waves as a sound-emitting source approaches and recedes from an observer.

Dry adiabatic lapse rate is the rate at which rising air is cooled by expansion when no condensation is occurring; $1.0°C/100$ m.

Earthquake is a trembling or shaking of the ground produced by the passage of seismic waves.

Ellipse is a geometric figure formed by a line lying in a plane and connecting all points so located that the sum of the two radius vectors is a constant.

Environmental temperature lapse rate is the rate of temperature decrease upward through the troposphere; standard value is 6.4°C/km.

Epicenter is the ground surface point directly above the focus of an earthquake.

Fault is a sharp break in rock with displacement (slippage) of block on one side with respect to an adjacent block.

Felsic minerals are quartz and feldspars treated as a mineral group of light color and relatively low density.

Floodplain is a belt of low, flat ground, present on one or both sides of a stream channel, subject to inundation by flood about once annually and underlain by alluvium.

Fog is a cloud layer in contact with a land or sea surface, or very close to that surface.

Fossils are ancient plant and animal remains or their traces or impressions preserved in sedimentary rocks.

Fossil fuel is a collective term for coal, petroleum, and natural gas capable of being utilized by combustion as energy sources.

Front is a surface of contact between two unlike air masses.

Gabbro is intrusive igneous rock consisting largely of pyroxene and calcic plagioclase feldspar, with variable amounts of olivine; a mafic igneous rock, occurs as a pluton.

Galaxy is an assemblage of stars, numbering several billions, widely separated from other galaxies and distributed throughout space of the universe.

Geocentric theory is the theory of solar system astronomy holding that the planets, Moon, and Sun revolve about the Earth.

Geochronometry is measurement of absolute ages of rocks, usually by radiometric age determinations.

Geomorphology is the science of landforms, including their history and processes of origin.

Geostrophic wind is wind at high levels above earth's surface blowing parallel with a system of straight, parallel isobars.

Geosyncline is a thick, lens-shaped accumulation of sedimentary strata deposited in a long, narrow belt of crustal subsidence.

Geothermal energy is heat energy of igneous origin drawn from steam or hot water beneath the Earth's surface.

Glaciation is (1) a general term for the total process of growth and landform modification of glaciers, and (2) a single episode or time period in which ice sheets formed, spread, and disappeared, as contrasted with interglaciation.

Glacier is a large natural accumulation of land ice affected by present or past flowage.

Granite is intrusive igneous rock consisting largely of quartz, potash feldspar, and sodic plagioclase feldspar, with minor amounts of biotite and hornblende; a felsic igneous rock, it occurs as a pluton.

Greenhouse effect is the accumulation of heat in the lower atmosphere through absorption of longwave radiation from Earth's surface.

Ground water is subsurface water occupying the saturated zone and moving under the force of gravity.

Hadley cell is an atmospheric circulation cell in low latitudes involving rising air over the equatorial trough and sinking air over subtropical high-pressure belts.

Heliocentric theory is the theory of solar system astronomy holding that the planets revolve about the Sun.

Hertzsprung-Russell diagram (*H-R* diagram) is a graphic representation of stars plotted on a field in which the vertical axis represents luminosity, the horizontal axis spectral class and star temperature.

Humidity is the general term for amount of water vapor present in the air.

Hurricane is a tropical cyclone of the western North Atlantic and Caribbean Sea.

Hydrolysis is chemical union of water molecules with minerals to form different, more stable mineral compounds.

Hydrosphere is the total water realm of the earth's surface zone, including the oceans, surface waters of the lands, ground water, and water held in the atmosphere.

Igneous rock is rock solidified from a high-temperature molten state; rock formed by cooling of magma.

Impact metamorphism is formation of lunar breccia by fragmentation and lithification of lunar rock during impacts that formed lunar craters.

Ionosphere is the atmospheric layer in the altitude range 80 to 400 km in which a large number of gas molecules undergo ionization by action of shortwave solar radiation.

Jet stream is the high-speed airflow in narrow bands within the upper-air westerlies and along certain other global latitude zones at high levels.

Kennelly-Heaviside layer is the layer in the lower part of the ionosphere capable of reflecting radio waves.

Landforms are configurations of the land surface taking distinctive forms and produced by natural processes.

Laterite is a rocklike ore deposit, formed as a layer beneath the soil, consisting of minerals rich in iron, aluminum, or manganese, such as limonite and bauxite.

Lava is magma emerging upon the earth's solid surface, exposed to air or water.

Light-year is the unit of interstellar distance equal to the distance traveled by light in one year's time.

Lithosphere is the general term for the entire solid earth realm; in plate tectonics, it refers to the strong, brittle outermost rock layer lying above the asthenosphere.

Lithospheric plate is a segment of the lithosphere moving as a unit, in contact with adjacent lithospheric plates along subduction zones, zones of crustal spreading, or transform faults.

Loess is the accumulation of yellowish to buff-colored, fine-grained sediment, largely of silt grade, on upland surfaces after transport in suspension (i.e., carried in a dust storm).

Longwave radiation is electromagnetic radiation emitted by earth, largely in the range from 3 to 50 microns.

Luminosity is the measure of the total radiant energy output of a star.

Lunar eclipse is the apparent darkening of the full Moon, occuring at opposition when Moon crosses Earth's shadow.

Lunar maria are dark-colored plains of the Moon's surface underlain by basalt (Latin *mare*, "sea").

Lunar regolith consists of particles of finely divided rock in a loose state forming a surface layer over most of the moon.

Mafic igneous rock is igneous rock dominantly composed of mafic minerals.

Mafic minerals (mafic mineral group) are minerals, largely silicates, rich in magnesium and iron, dark in color, and of relatively great density.

Magma is the mobile, high-temperature molten state of rock, usually of silicate mineral composition and with dissolved gases (volatiles).

Magnetic anomaly is any departure of magnetic intensity or inclination from a constant, normal value, as observed by use of the magnetometer.

Magnetopause is the outer boundary surface of the magnetosphere.

Magnetosphere is the external portion of earth's magnetic field, shaped by pressure of the solar wind and contained within the magnetopause.

Mantle is the rock layer or shell of Earth beneath the crust and surrounding the core; composed of ultramafic rock of silicate mineral composition.

Maria (see lunar maria).

Mass wasting is the spontaneous downward movement of soil, regolith, and bedrock under the influence of gravity; does not include the action of active agents.

Mercurial barometer is a barometer using the Torricelli principle in which atmospheric pressure counterbalances a column of mercury in a tube.

Mesopause is the upper limit of the mesosphere.

Mesosphere is the atmospheric layer of upwardly diminishing temperature, situated above the stratopause and below the mesopause.

Metamorphic rock is rock altered in physical structure and/or chemical (mineral) composition in the solid state by action of heat, pressure, shearing stress, or infusion of elements, all taking place at substantial depth beneath the surface.

Meteor is a light trail in the upper atmosphere produced by infall of a meteoroid.

Meteorite is a meteor that has landed on earth, becoming a variety of rock.

Meteoroid is a fragment of solid matter entering earth's atmosphere from outer space; most have a mass of less than one gram.

Meteorology is the science of the atmosphere; particularly the physics of the lower or inner atmosphere.

Micron is a length unit; one micron equals 0.0001 cm.

Midoceanic ridge is one of three major divisions of the ocean basins—the central belt of submarine mountain topography with a characteristic axial rift.

Millibar is a unit of atmospheric pressure; one-thousandth of a bar. Bar is a force of one million dynes per square centimeter.

Mineral is a naturally occurring inorganic substance, usually having a definite chemical composition and a characteristic atomic structure.

Moho is a contact surface between earth's crust and mantle; it is a contraction of Mohorovičić, the name of the seismologist who discovered this feature.

Moraine is an accumulation of rock debris carried by an alpine glacier or an ice sheet and deposited by the ice to become a depositional landform.

Neap tide is unusually a small tide range occurring when Sun and Moon are in quadrature.

Nebula is a diffuse interstellar cloud of gas and dust; it can be produced by explosion of a supernova.

Neutron star is an extremely small, extremely dense, dwarf star produced by gravitation collapse following exhaustion of nuclear fuel.

Nova (novae) is an event in the life of a star characterized by a sudden great increase in brightness occurring within a span of hours or days, followed by gradual decrease in brightness; associated with an explosive increase in diameter.

Nuclear fuels are radioactive isotopes, principally of uranium, used to furnish heat by controlled nuclear fission.

Nucleus of Sun is the Sun's interior region, lying beneath the photosphere.

Oil shale is shale containing dispersed hydrocarbon compounds, capable of yielding petroleum or natural gas by distillation when heated.

Opposition is the alignment of Sun, Earth, and Moon, with Sun and Moon on opposite sides of Earth.

Orogeny is a term in stratigraphy for a major episode of tectonic activity resulting in strata being deformed by folding and faulting.

Orographic precipitation is precipitation induced by the forced rise of moist air over a mountain barrier.

Outgassing is the process of exudation of water and other gases from earth's crust through volcanoes, to become a permanent part of earth's hydrosphere and atmosphere.

Oxbow lake is a crescent-shaped lake or swamp representing the abandoned channel left by cutoff of an alluvial meander.

Ozone layer is the layer in the stratosphere, mostly in the altitude range 20 to 35 km, in which a concentration of ozone is produced by the action of solar ultraviolet rays.

Paleomagnetism is relict magnetism within magnetic minerals in rocks, representing the state of earth's magnetic field at the time the rock was formed.

Paleontology is the study of ancient life based on fossil remains of plants and animals.

Pangaea is the hypothetical parent continent, enduring until near the close of the Mesozoic era, consisting of the continental shields of Laurasia and Gondwana joined into a single unit.

Parallax is an optical phenomenon of apparent moving apart or coming together of two distant objects because of changing alignment as viewed by a moving observer; applies to stars as seen from opposite points in Earth's orbit.

Parsec is a unit of interstellar distance equal to the distance from solar system to a star having a stellar parallax of one second of arc.

Peneplain is a land surface of low elevation and slight relief produced in the late stages of denudation of a landmass.

Penumbra is the zone of partial shadow from Sun's rays, surrounding the umbra of Earth and of Moon.

Perigee is the point in Moon's orbit closest to Earth.

Perihelion is the point on Earth's elliptical orbit at which Earth is nearest to Sun.

Photosphere is the visible outer surface of the Sun.

Physical weathering is the breakup of massive rock (bedrock) into small particles through the action of physical forces acting at or near the earth's surface.

Planetesimals are aggregations of matter formed into discrete solid bodies during accretion of the solar nebula.

Planets are solid objects, mostly of large mass, revolving about the Sun in an elliptical orbit.

Plate tectonics is a branch of tectonics or theory of tectonic activity dealing with lithospheric plates and their activity.

Pluton is a large body of coarse-grained intrusive igneous rock, typically of felsic minerals (example: batholith of granite).

Polar front is a front lying between cold polar and warm tropical air masses, often situated along a jet stream within the upper-air westerlies.

Precambrian time is all geologic time older than the beginning of the Cambrian Period—i.e., older than 600 million years.

Precipitation is particles of liquid water or ice that fall from the atmosphere and may reach the ground.

Pressure gradient is a change of atmospheric pressure measured along a line at right angles to the isobars.

Pressure-gradient force is a force acting horizontally, tending to move air in the direction of lower atmospheric pressure.

GLOSSARY

Ptolemaic system is the geocentric theory of the solar system put forth by Claudius Ptolemy in the second century A.D.

Pulsars are small neutron stars emitting regular pulses of light and radio waves in the same rhythm.

Quasistellar radio source (quasar) is an extremely distant, pinpoint source of extremely powerful radio wave emission, showing a very strong red shift in its light spectrum.

Radio galaxy is a galaxy that emits strong radio waves.

Radiometric age is the age of rock determined by measurement of ratios of radioactive isotopes to their daughter products.

Rain is a form of precipitation consisting of falling water drops, usually 0.5 mm or larger in diameter.

Rainshadow desert is a belt of arid climate to the lee of a mountain barrier, produced as a result of adiabatic warming of descending air.

Red giants are large red stars of mass intermediate between the red supergiants and most stars of the main sequence and of comparatively great luminosity and low temperature.

Red shift is a shift in lines of the spectrum of a star, galaxy, or quasar toward the red end of the spectrum as a result of its radial velocity away from the observer.

Regolith is the layer of mineral particles overlying the bedrock; it may be derived by weathering of underlying bedrock or transported from other locations by active agents.

Relative humidity is the ratio of water vapor present in the air to the maximum quantity possible for saturated air at the same temperature.

Revolution is the motion of a planet in its orbit about the Sun, or of a planetary satellite about a planet.

Richter scale is the scale of magnitude numbers describing the quantity of energy released by an earthquake.

Rilles are narrow, canyonlike features indenting into the Moon's surface; some are straight, others are sinuous.

Rock is a natural aggregate of minerals in the solid state, usually hard and consisting of one, two, or more mineral varieties.

Rossby waves (see upper-air waves).

Saltation is leaping, impacting, and rebounding of spherical sand grains transported over a sand or pebble surface by wind.

Saturated zone is the zone beneath the land surface in which all pores of the bedrock or regolith are filled with ground water.

Scoria are lava or tephra fragments containing numerous cavities produced by expanding gases during cooling.

Seafloor spreading is the pulling apart of the crust along a rift, typically the axial rift of the midoceanic ridge, and representing the separation of two lithospheric plates bearing oceanic crust.

Sediment is finely divided mineral matter and organic matter derived directly or indirectly from preexisting rock and from life processes.

Sedimentary rock is rock formed from accumulations of sediment.

Seismic waves are waves sent out during an earthquake by faulting or other crustal disturbance from an earthquake focus and propagated through the solid earth.

Seismograph is an instrument for detecting and recording seismic waves.

Shortwave radiation is electromagnetic radiation in the range from 0.2 to 3 microns, including most of the energy spectrum of soar radiation.

Solar constant is the intensity of solar radiation falling on a unit area of surface held at right angles to Sun's rays at a point outside Earth's atmosphere; equal to about 2 gram-calories per square centimeter per minute (2 cal/cm^2/min).

Solar eclipse is the apparent darkening of all or part of the Sun's disk, occurring at conjunction, when the Earth passes across the Moon's shadow.

Solar flare is the emission of ionized hydrogen gas from the vicinity of a sunspot, traveling outward through the solar system as a body of intense x-rays and ionized particles.

Solar nebula is a primordial body of diffuse gas and dust that, through condensation, gave rise to the solar system.

Solar wind is the flow of electrons and protons emanating from the Sun and traveling outward in all directions through the solar system.

Specific humidity is the mass of water vapor contained in a unit mass of air.

Spring tide is an unusually large tide range produced by combined tidal forces of Sun and Moon when in conjunction or opposition.

Star is a large, discrete concentration of matter within a galaxy, bound into a single unit by gravitation; contrasted with highly dispersed matter in form of gas clouds and dust clouds.

Star magnitude is the brightness (light intensity) of a star as compared with that of other stars and other celestial objects.

Stellar parallax is the changing arc of separation between two stars, one near and one distant, as observed at opposite seasons, because of parallax.

Strata are layers of sediment or sedimentary rock in which individual beds are separated from one another along stratification planes.

Stratigraphy is the branch of historical geology dealing with the sequence of events in earth's history as interpreted from the evidence found in sedimentary rocks.

Stratopause is the upper limit of the stratosphere, transitional upward into the mesosphere.

Stratosphere is the layer of atmosphere lying directly above the troposphere.

Stratus clouds are clouds of layered, blanketlike form.

Stream is a long, narrow body of flowing water occupying a stream channel and moving to lower levels under the force of gravity.

Subduction is the descent of the downbent edge of a lithospheric plate into the asthenosphere so as to pass beneath the edge of the adjoining plate.

Sunspot is a dark spot on the Sun's surface, forming and disappearing in a timespan of several days to several weeks, associated with a powerful magnetic field.

Supernova is a variety of nova in which brightness increase is extremely great, representing the explosive demolition of a small star and leaving only a neutron star.

Suspension is a form of stream transportation in which particles of clay, silt, or fine sand are held in upward currents in turbulent flow

Talus is accumulation of loose rock fragments derived by rockfall from a cliff.

Tectonics is a branch of geology specializing in the study of structural features of the earth's crust and their origin.

Thermocline is a water layer in which temperature changes rapidly in the vertical direction.

Thermosphere is the atmospheric layer of upwardly rising temperature, lying above the mesopause.

Thunderstorm is an intense, local convectional storm associated with a cumulonimbus cloud and yielding heavy precipitation, along with lightning and thunder, and sometimes the fall of hail.

Till is the heterogeneous mixture of rock fragments ranging in size from clay to boulders, deposited beneath moving glacial ice or directly from the melting in place of stagnant glacial ice.

Tornado is a small, very intense wind vortex with extremely low air pressure in the center, formed beneath a dense cumulonimbus cloud in proximity to a cold front.

Traction is the dragging action in which particles of sand, gravel, or cobbles are moved along the floor of a stream channel by force of the flowing water.

Transcurrent fault is a variety of fault on which the motion is dominantly horizontal along a near-vertical fault plane.

Tropical cyclone is an intense traveling cyclone of tropical latitudes, accompanied by high winds and heavy rainfall.

Tropopause is the boundary between the troposphere and stratosphere.

Troposphere is the lowermost layer of the atmosphere in which air temperature falls steadily with increasing altitude.

Ultramafic igneous rock is igneous rock composed almost entirely of mafic minerals, usually olivine or pyroxene.

Umbra is the long, conical zone of complete shadow from the Sun's rays, extending from the Earth and Moon.

Uniformitarianism is the concept, introduced by James Hutton in late 1700s, that geologic processes acting in the past are essentially the same as those seen in action today; opposed to the view of catastrophism.

Universe is the sum total of all matter and energy that exists.

Upper-air waves are large-scale horizontal undulations in the flow path of upper-air westerlies; also called Rossby waves.

Van Allen radiation belt is a doughnut-shaped belt of intense ionizing radiation surrounding the earth, within the inner magnetosphere.

Volcano is a conical, circular structure built by accumulation of lava flows and tephra, including volcanic ash.

Warm front is a moving weather front along which a warm air mass is sliding up over a cold air mass, leading to production of stratiform clouds and precipitation.

Water table is the upper boundary surface of the saturated zone; the upper limit of the ground water body.

Water vapor is the gaseous state of water.

Wave cyclone is a traveling, vortexlike cyclone involving interaction of cold and warm air masses along sharply defined fronts.

Wet adiabatic lapse rate is the reduced adiabatic lapse rate when condensation is taking place in rising air; value ranges from 0.3 to 0.6°C/100 m.

White dwarfs are very small stars of extremely high density and very low luminosity positioned far below the main sequence on the Hertsprung-Russell diagram.

Wind is air motion, dominantly horizontal relative to earth's surface.

APPENDIX

PREFIXES FOR MULTIPLES AND SUBMULTIPLES OF UNITS

Multiples

Prefix	Abbreviation	Multiplication Factor			
deka	da	ten	=	10	$= 10^1$
hecto	h	one hundred	=	100	$= 10^2$
kilo	k	one thousand	=	1,000	$= 10^3$
mega	M	one million	=	1,000,000	$= 10^6$
giga	G	one billion	=	1,000,000,000	$= 10^9$
tera	T	one trillion	=	1,000,000,000,000	$= 10^{12}$
peta	P	one million billion	=	1,000,000,000,000,000	$= 10^{15}$
exa	E	one billion billion	=	1,000,000,000,000,000,000	$= 10^{18}$

Submultiples

Prefix	Abbreviation	Multiplication Factor			
deci	d	one tenth	=	$\frac{1}{10}$	$= 10^{-1}$
centi	c	one hundredth	=	$\frac{1}{100}$	$= 10^{-2}$
milli	m	one thousandth	=	$\frac{1}{1,000}$	$= 10^{-3}$
micro	μ	one millionth	=	$\frac{1}{1,000,000}$	$= 10^{-6}$
nano	n	one billionth	=	$\frac{1}{1,000,000,000}$	$= 10^{-9}$
pico	p	one trillionth	=	$\frac{1}{1,000,000,000,000}$	$= 10^{-12}$
femto	f	one millionth of one billionth	=	$\frac{1}{1,000,000,000,000,000}$	$= 10^{-15}$
atto	a	one billionth of one billionth	=	$\frac{1}{1,000,000,000,000,000,000}$	$= 10^{-18}$

ACCURATE CONVERSION FACTORS

Length	1 inch = 2.540 cm
	1 foot = 0.3048 m
	1 yard = 0.9144 m
	1 mile = 1.609 km
	1 kilometer = 0.6214 mile
	1 parsec = 3.086×10^{16} m = 3.262 light years
Volume	1 liter = 1000 cm^3 = 10^{-3} m^3
	= 1.057 U.S. fluid quart
Speed	1 mile per hour = 0.4470 m/s
Mass and force	1 metric ton = 1000 kg
	1 pound = 4.448 N
	= weight of 0.4536 kg
Power	1 horsepower = 745.7 W
Heat	1 calorie = 4.186 J
Pressure	1 standard atmosphere = 1.013×10^5 N/m^2
	1 bar = 10^5 N/m^2
	1 pound per square inch = 6.895×10^3 N/m^2
	1 inch of mercury = 3.386×10^3 N/m^2
	1 centimeter of mercury = 1.333×10^3 N/m^2
	1 foot of water = 2.989×10^3 N/m^2
Temperature	T K = t°C + 273.15
	$= \frac{5}{9}$ (t°F + 459.67)
	t°C $= \frac{5}{9}$ (t°F − 32)
	t°F $= \frac{9}{5}$ t°C + 32

Broad Classification	Name	Particles Charge +e	Particles Charge 0	Particles Charge −e	Antiparticles Charge +e	Antiparticles Charge 0	Antiparticles Charge −e	Rest Mass (Units Such That Rest Mass of Electron = 1)	Spin. Units of $\frac{h}{2\pi}$
Photon	photon		γ			γ		0	1
Leptons	neutrinos		ν			$\bar{\nu}$		0	1/2
Leptons	μ-neutrinos		ν_μ			$\bar{\nu}_\mu$		0	1/2
Leptons	electrons			e^-	e^+			1	1/2
Leptons	muons			μ^-	μ^+			207	1/2
Hadrons — Mesons	π-mesons (pions)	π^+	π^0			π^0	π^-	264, 273	0
Hadrons — Mesons	K-mesons (Kaons)	K^+	K^0			$\bar{K}^0$	K^-	966, 974	0
Hadrons — Mesons	η-meson (eta)		η^0			η^0		1074	0
Hadrons — Baryons	nucleons	p	n			$\bar{n}$	$\bar{p}$	1836.1, 1838.6	1/2
Hadrons — Baryons	Lambda		Λ^0			$\bar{\Lambda}^0$		2183	1/2
Hadrons — Baryons	Sigma	Σ^+	Σ^0	Σ^-	$\bar{\Sigma}^+$	$\bar{\Sigma}^0$	$\bar{\Sigma}^-$	2328, 2334, 2343	1/2
Hadrons — Baryons	Xi		Ξ^0	Ξ^-	$\bar{\Xi}^+$	$\bar{\Xi}^0$		2573, 2586	1/2
Hadrons — Baryons	Omega			Ω^-	$\bar{\Omega}^+$			3276	3/2

THE GREEK ALPHABET

Alpha	A	α	Nu	N	ν
Beta	B	β	Xi	Ξ	ξ
Gamma	Γ	γ	Omicron	O	o
Delta	Δ	δ	Pi	Π	π
Epsilon	E	ε	Rho	P	ρ
Zeta	Z	ζ	Sigma	Σ	σ
Eta	H	η	Tau	T	τ
Theta	Θ	θ, ϑ	Upsilon	Υ	υ
Iota	I	ι	Phi	Φ	ϕ, φ
Kappa	K	κ	Chi	X	χ
Lambda	Λ	λ	Psi	Ψ	ψ
Mu	M	μ	Omega	Ω	ω

VECTOR CODE

This is not a standard convention. It is introduced with the hope that it might improve the clarity of the diagrams.

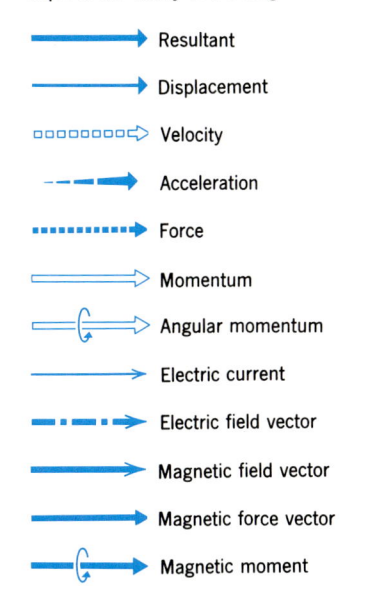

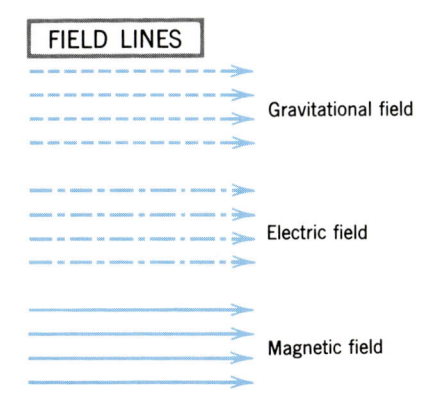

- Resultant
- Displacement
- Velocity
- Acceleration
- Force
- Momentum
- Angular momentum
- Electric current
- Electric field vector
- Magnetic field vector
- Magnetic force vector
- Magnetic moment

FIELD LINES

- Gravitational field
- Electric field
- Magnetic field

Ablation, 449
Abrasion, 444
Absolute rest, 100, 102
Absolute temperature, 46, 48
Absolute zero, of temperature, 46, 53
Abyssal plain, 408
Acceleration, 9
  of body moving in circle, 14
  due to gravity, 10, 21
Accretion, in solar nebula, 478
Acetic acid, 229, 285
Acetone, 286
Acetylsalicylic acid, 319
Acid, Arrhenius theory of, 229
  indicators for, 232, 233
  neutralization of, 232
  organic, 285
  properties of, 230
  reactions of, with bases, 230, 231
    with bicarbonates, 230
    with carbonates, 230
    with metals, 230
  table of, 229
Acid rain, 247
Actinides, 187
  electronic configurations of, 195
Action-at-a-distance, 59, 79
Activation, energy of, 216
Addition polymers, 317
Addition reaction, 283
Adeabatic cooling, atmospheric, 164
  heating, atmospheric, 164
  lapse rate, dry, saturation, and wet, 363
  process, 362

Adenine, 302, 423
Adrenaline, 320
Advection, fog, 365
Aerosols, Freon, 243
Age of universe, 507
Air, pure dry, 340
  saturated, 361
Air masses, 355, 368
  classification, 368
  continental, 368
  maritime, 369
  polar, 354, 369
  properties, 368
  tropical, 356, 369
  unstable, 369
Air pollution, air-temperature inversion and, 244
  carbon monoxide in, 352
  effect on materials of, 247
  nitrogen oxides in, 249
  ozone in, 244
  pollutants in, 245
  sulfuric acid in, 231
  sulfur oxides in, 169, 246
Air pressure, *see* Atmospheric pressure
Air temperature, *see* Atmospheric temperature
Albeds of earth, 349
Alcohol, 284
Aldehydes, 286
Algae, primitive, 423
Aliphatic compounds, 281
Alkaloids, 319
  coca family, 322
  ergot family, 322

Alkaloids (*continued*)
   opium family, 321
Alkanes, 281, 282
   chemical reactions of, 283
   table of straight-chain, 281
Alkenes, 282
   addition of water to, 283
   chemical reactions of, 283
   hydrogenation of, 283
   polymerization of, 283
Alkylation, gasoline from, 310
Allotropy, 169
Alloys, 166
Alluvial fan, 446
Alluvial rivers, 445
Alluvium, 445
Alpha helix, 299
Alpha-particle, 139
Alpine glacier, 447
Alpine system, 400
Alternating current, 65
Aluminosilicates, 379
Aluminum, 262
   alloys of, 263
   crustal abundance of, 262
   production of, 263
   recycling of, 264
   refining of, 262
Aluminum oxide, refining of, 262
Amides, 286
Amines, 285
   nucleic acid, 302
Amino acids, 295
   synthesis, 423
   table of, 296
para-Aminobenzoic acid, 318
Ammonia, 248
   amines and, 285
   as fertilizer, 248
   Haber synthesis of, 248
Ammonium carbonate, 234
Ammonium hydroxide, 230
Ammonium ion, structure of, 210
Ammonium nitrate fertilizer, 248
Amorphous solids, 214
Ampère, unit of electric current, 63
Amphetamines, 320, 321
Amphibale group, 380
Amplitude, of simple harmonic motion, 84
Amylopectin, 292
Amylose, 292
Analgesics, 319
Andesite, 385
Angular momentum, 30
   orbital and spin, 128
Aniline dyes, 285
Annihilation of matter, 141
Anode, 260

Anorthosite, lunar, 487
Antarctic ice sheet, 450
Antacids, 231
Anthracite, 307
Antibiotics, 318
Anticyclones, 353
   on weather map, 372
Antidepressants, 321
Antimatter, 143
Antimetabolites, 318
Antimony, 170, 248
Antineutrino, 140, 142
Antineutron, 142
Antiparticles, 142
Antiproton, 142
Antipyretics, 319
Anvil top, 367
Aphelion, 461
Apogee, 470
Apollo missions to moon, 481
Archimedes' principle, 161
Argon, 169
Aromatic compounds, 281
Arrhenius, Svante, 229
Arrhenius theory of acids and bases, 229
Arsenic, 170, 248
Arsenic poisoning, 300
Arterenol, 320
Ash, volcanic, 385
Asparagine, chirality of, 278
Aspirin, 319
Asteroids, 466
Asthenosphene, 399
Astrogeology, 477
Astronomical hypothesis of glaciation, 433
Astronomical unit, 462
Atmosphere, 239
   carbon dioxide level in, 253
   components, 340
   composition, 340
   of earth, 340
      and oceans compared, 345
   of mars, 466
   primitive, 423
   temperature conversion in, 244
Atmospheric circulation, global, 354
Atmospheric pressure, 47, 341
   decrease in, with altitude, 341
   standard sea-level, 342
   upward decrease in, 341
   and winds, 351
Atmospheric temperature, variation with altitude, 342
   vertical structure, 342
Atom, Dalton's theory of, 175
Atomic bomb, 137
Atomic clock, 6
Atomic number, 132

INDEX

Atomic orbitals, 187
   molecular orbitals from, 208
   order of filling of, 193
Atomic theory, 43
Atomic weight, 180
Atto, 7
Aufbau rules, 192
Augite, 380
Automobiles, nitrogen oxides from, 249
Avogadro, Amedeo, 181
Avogadro's law, 49
Avogadro's number, 181
Axial rift, 408

Back-swamp areas, 445
Backwash, 450
Badlands, 441
Baking soda, 234
Banding in rock, 389
Bank caving, 444
Barchan, 454
Barium sulfate, 234
Bar magnet, 69, 73
Barometer, mercurial, 342
Barometric pressure, see Atmospheric pressure
Barringer Crater, 480
Baryons, 145
Basalt, 385
   lunar, 486
Basaltic rock, 398
Bases, 232
   Arrhenius theory of, 229
   indicators for, 232, 233
   neutralization of, 231
   properties of, 231
   table of, 230
Basic oxygen process (BOP), 257
Batholith, 384
Battery, 62
Bauxite, 262, 438
Beach, 453
Beach drifting, 453
Bed load of stream, 445
Bedrock, 437
Benzedrine, 320, 321
Benzene, 283
   manufacture of, 310
   structure of, 284
   substitution reactions of, 283
Bessemer converter, 257
Beta-particle, 140
Bevatron, 143
Big-bang theory of universe, 507
Biosphere, 239
   organic compounds of, 289
Biotite, 380
Bismuth, 248
Bituminous coal, 307

Black body temperature, of earth and sun, 347
Black holes, 504
Blast furnace, 257
Blood, pH of, 232
Bluffs of floodplain, 445
Boiling, 165
Boltzmann's constant, K, 46
Bond, chemical, 202
   coordinate covalent, 210
   covalent, 207
   double, 208
   hydrogen, 225
   ionic, 203
   metallic, 213
   polar, 211
   triple, 208
Bond angle, 212
BOP (basic oxygen process), 257
Boric acid, 229
Boron, 169
Botulism, 300
Boyle's law, 48
Brahe, Tycho, 459
Braided pattern of stream, 446
Brass, 167
Breaker, 450
Breccia, lunar, 486
Breeding of nuclear fuel, 393
Brine, 345
Broglie waves, de, 120
Bromine, 169
Bromthymol blue, 233
Bronze, 166
Bronze Age, 167
Brownian movement, 222
Brunhes, B., 410
Buffer, acid-base, 232
Buoyancy, 161
Buoyant force, 161
Butane, isomer of, 273

Calcification, 440
Calcite, 388
Calcium carbonate, 388
   carbon cycle and, 253
Calcium hydroxide, 230
Calcium sulfate, 234, 235
Calomel, 234
Calorie, 50
Cambrian Period, 420, 425
Cancer, ozone depletion and, 244
Capture hypothesis of moon, 488
Carbohydrates, 290
   photosynthesis of, 289
Carbon, 169
Carbonates, carbon cycle and, 253
   sediments, 388

INDEX

Carbon chain, 271
Carbon dioxide, 252
   in carbon cycle, 253
   and glaciations, 433
   in oxygen cycle, 240
   in photosynthesis, 253
Carbon family, 252
Carbonic acid, 229
   action, 438
Carboniferous Period, 426
Carbon monoxide, air pollution by, 252
   iron refining and, 257
   poisoning by, 236
Carbon ring, 271
Carboxylic acids, 285
Catalyst, 217
Catalytic muffler, 217
Catastrophism, 421
Catastrophists, 421
Cathode, 260
Cellulose, 292
   hydrolysis, 290
Celsius temperature scale, 46
Cenozoic Era, 424, 430
Centi, 7
Central peak of crater, 484
Cesium, 169
CGS system of units, 8
Chain reaction, 137
Channel on Mars, 489
Channel of stream, 442
Charge, electric, 56
   on electron, e, 58
Charles' law, 49
Chemical bond, 202
Chemical charge, 160, 203
   laws of, 173
Chemical energy, 37, 39
Chemical equations, 178
   balanced, 179
Chemical formulas, 178
Chemical property, 160
   electronic configurations and, 199
Chemical reactions, 203
   energy features of, 215
Chemical weathering, 438
Chemistry, definition of, 157
Chernozems, 440
Chinook winds, heat in, 164
Chirality, 278
Chloramphenical, 320, 321
Chlorine, 169
   reaction of sodium with, 203
   in seawater, 345
   stratospheric, 244
Chlorophyll, 236
Chromium steel, 167
Chromosphere, 494
Circle of illumination, 347

Circular motion, 13, 24
   of charged particle in magnetic field, 73
Circum-Pacific belt, 400
Circum-Pacific ring, 385
Cirque, 450
Cirrus, 364
Citric acid, 285
Clay minerals, 387
Cleavage of minerals, 380
Cleopatra's needle, 247
Clocks, moving, 104
   in gravitational field, 113
Cloud, cumuliform, 363
   families, 364
   how formed, 363
   forms, 363
   of probability, 126
   seeding, 367
   stratiform, 363
Coal, 307, 388
   carbon dioxide from use of, 253
   coal oil from, 311
   coal tar from, 311
   coke from, 311
   destructive distillation of, 311
   formation of, 307
   gas, 311
   sulfur in, 248
   types of, 307
Coal measures, 388
Coal oil, 311
Coal reserves of world, 393
Coal seams, 388
Coal tar, 311
Cobalt steel, 167
Coca alkaloids, 322
Cocaine, 319, 321, 322
Codeine, 321
Coefficient, chemical, 179
Cofactor, enzyme, 299
Coke, 257
Cold front, 369
Colloidal dispersions, 222
   kinds of (table), 223
Colloidal state, 222
Color, 4, 89
Coma of comet, 468
Combustion, oxygen cycle and, 240
Comets, 468
Compass, 75
Components of vector, 13
Compound, 176
   coordination, 236
   formulas of, 178
   inorganic, 220
   ionic, 204
   molecular, 206
   nucleus, 136
   organic, 271

Compton effect, 119
Concentrated solution, 226
Concentrations, expressions for, 226
Condensation, and dew point, 362
   and precipitation, 362
   of water vapor, 362
Condensation polymers, 317
Conduction, of electricity, 62
   of heat, 162
Conductor, 63
Conglomerate, 387
Conjunction, 471
Conservation, of angular momentum, 32
   of charge, 56
   of energy, 33
   of linear momentum, 29
   of mass and energy, 107
Continent, origin, 416
Continental collision, 414
Continental crust, 398
Continental drift, 430
Continental evolution, 416
Continental ice sheet, 447
Continental lithosphere, 412
Continental margins, 406
Continental rise, 407
Continental rupture, 416
Continental separation, 430
Continental shelves, 406
Continental shields, 416
Continental slope, 407
Continental suturing, 415
Contraction, of moving object, 105
Convection, atmospheric, 366
   heat, 162
Convection currents in mantle, 412
Convergence in cyclones, 354
Coordinate covalent bond, 210
Coordination compounds, 236
Copernican system, 459
Copernicus, Nicolaus, 458
Copolymers, 317
Copper, 263
   alloys of, 166
   electroplating of, 260
   gold and silver from refining of, 265
   native, 166
   refining of ores of, 265, 266
   uses of, 265
Core of earth, 397
   and seismic waves, 405
Coriolis, G. G., 352
Coriolis effect, 351
   and ocean currents, 357
   on Venus, 465
   and winds, 351
Corona, 494
Corrosion, 444
Cosmic particles (rays), 508

Cosmology, 506
Cotton, cellulose in, 292
Coulomb, unit of electric charge, 58
Coulomb's law, 58
Counterradiation, 350
Country rock, 384
Couple, 31
Covalence, 207
Covalent bond, 207
Cracking, petroleum, 308
   catalytic, 310
   thermal, 308
Craters, impact, 480
   lunar, 483, 485
   on Mars, 489
   on Mercury, 489
   on meteorite, 480
   on moon, 483
   volcanic, 385
Cretaceous Period, 428
Crick, F. H. C., 302
Crick-Watson theory, 302
Crude oil, composition of, 307, 309
   gasoline from, 308
   refining of, 307, 309
Crustal roots, 398
Crust of earth, 397, 398, 416
   continental, 398, 416
   oceanic, 398
   origin, 416
   structure, 398
Crust of moon, 488
Cryolite, 262
Crystallization of magma, 382
Crystal structure, 214
   of minerals, 380
Crystal systems, 215
Cumulonimbus, 365
   and thunderstorms, 366
Cumulus, 365
Current, electric, 62
Cuvier, Baron, 421
Cyanide poisoning, 236
Cycle of rock transformation, 390
Cyclic compounds, 271
Cyclone, 353
   tropical, 373
   on weather map, 371
Cyclone families, 373
Cyclone tracks, 372
Cyclopropane, 272
Cytosine, 302

Dacron, 317
Dalton, John, 43, 175
Dalton's theory, 175
Darwin, George, 488
Davisson, C. J., 120
Day, 462

INDEX

Decay, carbon cycle, 253
    oxygen cycle, 240
Deci, 7
Deep environment of rocks, 390
Deflation, 453
Deka, 7
Delta, 446
Democritus, 43
Denaturation, 299
Denitrification, 249
Density, 8
    of earth, 397
    effect of temperature on, 161
Deoxyribonucleic acid (DNA), 301
Deoxyribose, 302
Deserts, rain-shadow and tropical, 368
Desilication, 438
Detergents, phosphate in, 236
Determinism, 16, 124
Deuterium, 133
Devonian Period, 426
Dew point, 362
Dew point lapse rate, 363
Dexedrine, 320, 321
Dextrose, 291
Diabetes, 301
Diamond, 169
    structure of, 215
Diesel oil, 309
Diethyl ether, 284
Diffraction, 94, 120
Digestion, carbohydrate, 290
    fat, 285
    liquid, 294
    protein, 286
Dilute solution, 226
Dimethyl ether, 273
Dinosaurs, 428
Diorite, 385
Dipeptide, 295
Diphosphate ion, 236
Dipole, electrical, 111
Direct motion, 463
Disacchorides, 290
Discharge of stream, 443
Disorder, 52
Displacement, 12
Dissolved solids in streams, 444
Divergence in anticyclones, 354
DNA, 301
    enzyme synthesis via, 303
    replication of, 304
Dolomite, 388
Doppler effect, and red shift, 506
    and star velocities, 506
Double helix, 302
Drainage basin, 443
Drainage divide, 443
Drainage systems, 443

Drug, 318
Dry adiabatic lapse rate, 363
Dry ice, 253
Dry stage, in adiabatic process, 363
Dual nature of light and matter, 115, 120, 122
Ductility, 165
Dunite, 385
Dust, atmospheric, 340, 341, 453
Dust storms, 453
    on Mars, 490
Dynamo theory, 409

Ear, 87
Earth, early history, 479
    as magnet, 409
    origin, 478
Earth-moon system, 470
Earthquake, 401, 402
    causes, 402
    energy, 405
    lunar, 488
    waves, 402
    world distrubution, 406
Earth resources, 391
Eccentricity of orbit, 462
Eclipse, lunar and solar, 472
    partial and total, 472
Einstein, Albert, 22, 78, 100, 102, 110, 118
Eka-silicon, production of, 186
Elastic limit, 402
Elastomers, 317
Electric charge, 56
Electric current, 62
Electric field, 60
Electricity consumption, by aluminum refining, 263
Electric potential, 60
Electric potential energy, 58
Electrochemistry, 257
Electrode, 260
Electrolysis, 260
Electromagnetic induction, 76
Electromagnetic waves, 80, 87
Electron, 5, 56, 58, 73, 128, 145
Electron-dot symbols, 207
Electronegativity, table of, 212
Electron-gas model, 213
Electronic configurations, 190, 193
    and chemical families, 199
    table of, 196–197
Electron-sea model, 213
Electron–transfer, 242
Electroplating, 260
Elements, chemical, 165
    combining weights of, 179
    electronic configurations of (table), 196–197
    periodic properties of, 183
    symbols of, 178

Elements of earth's crust, 378
   metallic, 391
Ellipse, 460
Ellipsoid, prolate, 473
Empedocles, 43
Emulsion, 223
Endocrine glands, 301
Energy, 32
   of activation, 216
   of electromagnetic radiation, 90
   levels, atomic, 187
     splitting of, 191
   sources, solar, 392
     sustained yield, 392
Entropy, 52
Environmental temperature lapse rate, 342
Enzymes, 299
   DNA directed synthesis of, 303
   lock and key theory of, 300
   metal ions in, 236
   poisons and, 300
Ephedrine, 320
Epicenter, 402
Epinephrine, 320, 321
Epochs, of geologic time, 430
   of magnetic polarity, 410
Epsom salt, 234
Equation, chemical, 178
   of continuity, 443
   ionic, 230
Equatorial current, 356
Equilibrium, dynamic, 228
Equinoxes, 348
Eras, of geologic time, 424
Ergot alkaloids, 372
Eros, 466
Erosion, of landmass, 446
   of slopes, 441
   of soil, 441
   by streams, 444
   by wind, 453
Escape velocity, 36
Escaping tendency, 164
Esters, hydrolysis of, 285
Ether, 59, 101, 284
Ethyl alcohol, 272, 273, 284
   esters from, 285
   manufacture of, 312
   oxidation of, 285
   reaction with hydriodic acid of, 273
   reaction with sodium of, 274
Ethylamine, 272
Ethylene, manufacture and uses of, 312
   polyethylene from, 314
Ethyl gasoline, 311
Ethyl mercaptan, 272
Euclidean geometry, 111
Eurasian-Melanesian belt, 400
Evaporation, global heat management by, 164

Evaporites, 388
Evolutionary theory of universe, 507
Exa, 7
Exclusion principle, 129
Exothermic charge, 216
External magnetic field, 344

Fall of meteorite, 479
Family, chemical, 185
   outside shell configurations in, 199
Faraday, Michael, 78
Ferenheit temperature scale, 46
Fats, animal, 293
   fatty acids from, 294
Fatty acids, 293–294
   table of, 294
Fault, normal and transcurrent, 401
Faulting, 401
Fault plane, 401
Fault scarp, 401
Feldspars, plagiodase and potash, 379
Felsic group, 380
Felsic igneous rock, 385
Femto, 7
Fermi, Enrico, 140
Field, 59
Field ion microscope, 44
Field lines, 60, 71
Find of meteorite, 479
Fireball, 479
Fission, 137
   hypothesis of moon, 488
Flint, composition of, 170
Floodplain, 445
Flotation process, 265, 267
Fluorides, air pollution by, 263
Fluorine, 169
Fluvial processes, 441
Focus, of earthquake, 402
   of ellipse, 460
Fog, 365
Foliation in rock, 389
Folic acid, 318
Force, 17, 28, 105, 146
   between moving charges, 69
Formaldehyde, 286, 423
Formula, chemical, 178, 205
   structural, 207
Formula unit, 181
Formula weight, 180
Fossil, 420
   fuel, 307, 391, 393
Fractionation, crude oil, 307, 309
Free rotation, 275
Freons, effect on ozone cycle of, 243
Frequency, 83
Fresnel, Augustin, 92, 94
Fronts, atmospheric, 356, 369
   cold, 369

Fronts (*continued*)
  occluded, 370
  polar, 356
  warm, 370
Frost action, 437
Froth flotation, 265, 267
Fructose, 292
Fuel oil, 308
Functional groups, organic, 280
Fundamental particles, 5, 56, 128, 145
Fundamental process, 140, 146
Fusion, 138
  heat of, 163

Gabbro, 385
Galaxies, 461, 504
  barred, 505
  elliptical, 505
  evolutionary sequence, 505
  Milky Way, 497
  radial velocities, 506
  radio, 507
  red shift, 506
  spiral, 505
Galilean satellites, 469
Galilei, Galileo, 459
Galileo, and lunar maria, 482
Gallium, 169
Galvanic action, 268
Gamma-rays, 141, 342
Gas, atmospheric, 340
  equation, 48
  ideal monatomic, 44
Gasoline, composition of, 308
  lead additives for, 311
  octane rating of, 311
  production of, 308
Gauss, unit of magnetic field, 72
Gel, 223
General relativity, 108, 110
Genes, DNA and, 301
  duplication of, 304
Genetic code, 302, 303
Geocentric theory, 458
Geochronometry, 424
Geologic eras, 424
Geologic time, divisions, 422
  eras, 424
  table, 422
Geology, historical, 419
  lunar, 477
  of planets, 477
Geomagnetic axis, 409
Germanium, predicted properties of, 186
Geomorphic processes, external and internal, 436
Geomorphology, 436
Geostrophic wind, 352

Geothermal energy, 393
Germer, L. H., 120
Giga, 7
Glacial trough, 450
Glaciation, Carboniferous-Permian, 427
  causes, 432
  Pleistocene, 430
Glaciers, alpine and continental, 447
Globular star clusters, 497
Glucose, 291
  alpha and beta forms of, 291
  from other carbohydrates, 290
Glycerol, esters in fats and oils of, 293
Glycogen, 292
  hydrolysis of, 290
Gneiss, 390
Gondwana, 427
Gram, 6
Gram–formula weight, 180
Granite, 384
Granite-gabbro rock series, 384
Granitic rock, 398
Graphite, 169
  structure of, 215
Gravitation, 10, 19, 20, 108, 110
  and ocean tides, 472
Gravitational constant, G, 20
Gravitational potential energy, 33, 39
Gravity, on moon, 481
Greenhouse effect, 350
Greenland ice sheet, 449
Grimaldi, Francesco Maria, 92
Ground water, 443
Group, chemical, 187
Guanine, 302
Gulf Stream, heat transfer by, 162
Gypsum, 388
Gyres, 356

Haber process, 248
Hadley, George, 355
Hadley cell, 355
Hadron, 145
Hail, 365
Hailstones, 367
Half-life time, 139
Halite, 388
Hall, Charles Martin, 262
Halley's comet, 469
Hallucinogens, 321
  LSD as, 322
Halo of stars, 497
Hard water, 236
Harmonic law, 462
Heat, 49, 52
  atmospheric management of, 164
  budget of earth, 358
  capacity, 162

INDEX

Heat (*continued*)
  human body, 163
  radiation, 40, 89, 116, 117
  within earth, 398
Hecto, 7
Heisenberg's uncertainty principle, 124
Heliocentric theory, 458
Helium, 169
Hematite, 256
Hemin, 236
Hemoglobin, 297
Henry, Joseph, 78
Heredity, nucleic acids and, 301
Heroin, 319, 321
Héroult, P. L. T., 262
Hertz, Heinrich, 88
Hertz, unit of frequency, 84
Hertzsprung-Russell (H-R) diagram, 501
Hexane, 272
Highlands of moon, 482
High water of tide, 473
Historical geology, 419
Holocene Epoch, 431
Homo sapiens, 430
Hormones, 300
Horn, 450
Hornblende, 380
Hubble's law, 506
Human evolution, 430
Human growth hormone, 301
Humidity, relative, 361
  specific, 362
Humus, 440
Hund's rule, 194
Hurricanes, 373
Hutton, James, 421
Huygens wavelets, 92
Hydrates, 235
Hydration, 225
Hydraulic action, 444
Hydriodic acid, 229
Hydrobromic acid, 229
Hydrocarbons, 281
  air pollution by, 245, 249
  aliphatic, 281
  aromatic, 281, 284
  chemical properties of, 283
  families of (table), 282
  and ozone formation in smog, 251
  petroleum source of, 306
  physical properties of, 282
  unsaturated (table), 283
Hydrochloric acid, 229
Hydrogen, 168
  addition to alkenes of, 283
  as future fuel, 169, 262
  in oleomargarine manufacture, 283
  synthesis of, 261

Hydrogen atom, 128
Hydrogen bomb, 140
Hydrogen bond, 225
  DNA and, 303
Hydrogen cyanide, 423
Hydrogen fluoride, air pollution by, 263
  aluminum refinery production of, 263
Hydrogen sulfide, structure of, 212
Hydrolysis, 388
  amide, 286
  ester, 285
Hydronium ion, 210
  as catalyst, 232
  effect on litmus, 232
  indicators for, 232
  neutralization of, 232
  pH and, 232
  in water, 226
Hydropower, 392
Hydrosphere, 239, 340
Hydrostatic pressure, 161
  in oceans, 346
Hydroxide ion, 210
  effect on litmus of, 232
  indicators for, 232
  neutralization of, 232
  pH and, 232
  in water, 226
Hygroscopic salt, 363

Icarus, 466
Ice, density of, 161
  of glaciers, 449
  point, 46
Ice Age, 430
Icecap, 447
Ice sheets, 430, 447
  causes, 432
  growth, 432
  of Pleistocene, 430
  of present, 449
Ice shelves, 450
Ideal gas equation, 48
Ideal monatomic gas, 44, 45, 47
Igneous rock, 379, 384
  extrusive, 384
  felsic, 385
  intrusive, 384
  mafic, 385
  texture, 383
  ultramafic, 385
Illite, 388
Impact metamorphism, 486
Indeterminism, 124
Indicators, acid-base, 232, 233
Infrared, 40, 89, 116, 117
  imagery, 359
  radiation, 347

INDEX

Initial landforms, 436
Inorganic compounds, 220
Insulator, 63
Insulin, 298
   action of, 301
Interatomic forces, 23
Interference, 92, 93
Internal energy, 45
International system of units, 8
Interstellar space, distance units, 496
Inverse square law, 20, 57
Inversion, air-temperature, 244
Iodine, 169
Ion, 203
   common reactions of, 236
   formation of, 203
   hydration of, 225
   important examples of, 203
   names of, 203, 208
   octet rule and, 205
   polyatomic, 208
   symbols of, 203
Ionic bond, 203
Ionic compound, 204
   symbols for, 205
Ionic equation, 230
Ionization, in upper atmosphere, 342
Ionizing radiation, 344
Ionosphere, 342
   and radio-waves, 343
Iron, 256
   alloys of, 167
   minerals of, 256
   refining of, 256
   tools from, 167
Iron (meteorite), 479
Iron Age, 167
Iron sulfide, 178
Island arcs, 400, 414
   origin, 414
Isobar, 351
Isobaric surface, 351
Isobutane, 273
Isomer, 271
Isomerism, 271
   geometric, 275
   optical, 275
   structural, 271
Isooctane, 272
   manufacture of, 310
   octane ratings and, 311
Isoprene, 316
Isotherm, 371
Isotope, 132

Jet stream, 355
Joints in rock, 437
Joule, unit of energy, 33
Jupiter, 467

Jupiter family of comets, 469
Jurassic Period, 428

Kaolinite, 388
Kaon, 145
Kelvin, unit of absolute temperature, 46
Kennelly-Heaviside layer, 343
Kepler, Johannes, 459, 460
Kepler's laws of planetary motion, first and
      second, 460
   third, 462
Kerogen, 393
Kerosene, 309
Ketones, 286
Kilo, 7
Kilogram, 6
Kinetic energy, 32
K-meson, 145
Krypton, 169

Lactose, 291
   hydrolysis of, 290
Lakes, fall overturn of, 162
Lambda-particle, 145
Laminar flow, 442
Landforms, 436
   of glacial action, 450
   initial, 436
   sequential, 436
Landmass, 446
Landmass erosion, stages, 446
Landslide, 441
Langmuir, Irving, 207
Lanthanide series, 187
   electronic configurations of, 195
Latent heat, 163
Latent heat of fusion, table of, 163
Latent heat of vaporization, table of, 163
Lateral cutting by stream, 445
Laterite, 440
Larterization, 438
Latosols, 440
Laurasia, 427
Lava, 384
Lavoisier, Antoine, 174
Law of conservation of mass, 174
Law of definite proportions, 174
   ionic compounds and, 206
Law of multiple proportions, 175
Lead, ores of, 265
   poisoning by, 267
   refining of, 265
   uses of, 267
   zinc, from refining of, 266
Le Châtelier, Henry Louis, 228
Le Châtelier's principle, 228
Lemaître, Canon, 506
Leonid meteoroids, 468
Lepton, 145

Level of condensation, 363
Lewis, G. N., 207
Lewis theory, 207
Life, primitive origins, 423
Ligands, 236
Light, 4, 80, 87, 89, 90, 91, 110, 115, 122, 123
Lightning, 367
Light-year, 496
Lignite, 307
Limestone, 388
  effect of acid on, 231
  in iron refining, 257
Limewater, 230
Limonite, 256, 438
Lipids, 293
  digestion of, 294
  fatty acids from, 294
  hydrogenation of, 293
Liquid, hydrostatic pressure in, 161
  kinetic view of, 161
  mechanical properties of, 161
  thermal properties of, 162
Listes, Joseph, 284
Lithification, 387
Lithosphere, 239, 378, 399
  continental, 412
  in motion, 399
  oceanic, 412
Lithospheric plates, 399
  global system, 412
  movement, 412
Litmus test, 232
Lock and key theory, 300
Loess, 454
London smog, 246
Longitudinal waves, 86, 402
Longwave radiation, 347, 349
Low water of tide, 474
LSD, 322
Lucite, 316
Luminosity of star, 498, 499
  and star mass, 499
Lunar craters, 483
  origin, 485
Lunar eclipse, total, 472
Lunar highlands, 482
Lunar maria, 482
Lunar regolith, 486
Lunar rock, 486
Lunar soil, 485
Lysergic acid diethylamide (LSD), 322
Lystrosaurus, 427

Mach, Ernst, 108
Machian force, 110
Mach's principle, 108
Mafic group, 380
Mafic igneous rocks, 385

Magma, 379
  crystallization, 382
  silicate, 382
Magnesium, hydroxide, 230
  sulfate, 234
Magnetic anomalies, 411
Magnetic atmosphere, 244
Magnetic compass, 75
Magnetic field, 70, 74, 75
  of earth, 344, 409
  external, 344
Magnetic force, 73, 76
Magnetic poles, 409
  migration, 428
Magnetic reversals, 410
Magnetic storm, 496
Magnetic stripes, 411
Magnetism of earth, 409
Magnetite, 75, 256
  recovery from taconite of, 256
Magnetomer surveys, 411
Magnetopause, 344
Magnetosphere, 344
Magnitude of star, absolute and apparent visual, 499
Main sequence of stars, 500
Major axis of ellipse, 460
Malleability, 165
Maltose, 291
  hydrolysis of, 290
Manganese, nodules, 392
  steel, 167
Manganite, 440
Mantle of earth, 397, 399
  soft layer, 399
Mantle plume, 416
Marble, 390
  effect of acid on, 231
Marginal ocean basin, 414
Maria, lunar, 482
  origin, 485
Marijuana, 322
Marine cliff, 453
Mariner spacecraft, 466, 488
Mars, 465
  atmosphere, 489
  polar caps, 490
  rocks, 490
  regolith, 490
  surface features, 488
Mass, 6, 19, 110
  and energy, 107
  variation with velocity, 105
Mass defect, 135
Mass number, 132
Mass-wasting processes, 440
Materialization of light, 142
Matter, classes of, 221
Maxwell, James Clerk, 79, 89

Maxwell's equations, 79, 87, 88, 89
Meander, 445
Meander cutoff, 445
Meander neck, 445
Mean velocity, 442
Mega, 7
Mendeliev, Dimitri, 183
Mercurial barometer, 342
Mercury, 169, 463, 491
　surface features, 491
Mercury (I) chloride, 234
Mercury (II) chloride, 234
Mescaline, 320, 321
Meson, 143, 145
Mesopause, 342
Mesosphere, 342
Mesozoic Era, 424, 428
Metal, 165
　in earth's crust, 391
　ions from, 205
　periodic chart locations of, 187
　refining of, 256
　uses of, 256
Metallic bond, 213
Metalliferous deposits, 391
Metalloids, 168, 170
Metamorphic rocks, 388
Metamorphism of rock, 388
　impact and lunar, 486
Meteor, 468
Meteroite, 397, 479
　age, 480
Meteorite craters, 480
Meteorite impact features, on earth, 480
Meteoroid, 468
Meteoroid swarm, 468
Meteorology, 340
Meteor showers, 468
Meter, 6
Methadone, 321
Methane, 272
　natural gas and, 283
Methedrine, 320, 321
Methyl alcohol, 284
Methyl orange, 233
Methyl salicylate, 319
Metric system of units, 6, 8
Meyer, Lothar, 183
Mica group, 380
Michelson–Morley experiment, 102
Micro, 7
Micron, 347
Microscope, 96
Microwaves, 90
Mid-Atlantic Ridge, 412
Mid-oceanic ridge, 406
　and sea-floor spreading, 411
Milk of magnesia, 230
Milk sugar, 291

Milky Way galaxy, 497
Miller, Stanley, 423
Milli, 7
Millibar, 342
Mineral, 256, 378
　felsic, 380
　human dietary, 236
　mafic, 380
　metamorphic, 389
　silicate, 379
Mineral cleavage, 380
Mineral oil, 309
Mineral resources, 391
　from sea, 392
Minor axis of ellipse, 460
Minor planets, 466
Mixtures, compounds versus, 176
　homogeneous, 220
　indefinite proportions in, 177
MKS system of units, 8
Mössbauer effect, 113
Moho, 398
Molar concentration, 226
Mole, 180
Molecular orbital, 208
Momentum, 28, 34, 90, 105, 122
Monatomic gas, 44
Monomer, 283
Monosaccharides, 290
Month, sidereal and synodic, 470
Moon, 470
　atmosphere, 481
　composition, 488
　exploration, 481
　geologic history, 487
　gravity, 481
　interior structure, 488
　orbit, 470
　origin, 478, 488
　phases, 471
　rock, 485
　rock ages, 487
　soil, 485
　surface environment, 481
　surface materials, 485
　surface temperatures, 481
Moonquakes, 488
Moons of planets, 469
Moraine, terminal, 450
Morphine, 319, 321
　preparation of, 321
Moseley, H. G. J., 186
Motion, 9, 13
Mountain arcs, 400
Mountain-making belts, 400
Müller, Erwin, W., 44
Mu-neutrino, 145
Muon, 145
Mylar, 318

Nano, 7
Natural gas, composition of, 283, 312
　formation of, 307
Natural levees, 445
Neap tide, 474
Nebula, solar, 478
　of supernova, 504
Neon, 169
Neo-Synephrine, 320, 321
Neptune, 467
Net radiation, and latitude, 350
Neutralization, acid-base, 232
Neutral solution, 228
Neutrino, 140, 142, 145
Neutron, 5, 56, 121, 128, 132, 140, 142, 145
　star, 503
Newlands, J. A. R., 183
Newton, Sir Issac, 3, 16, 19, 20, 91, 100, 108, 109, 459
Nickel steel, 167
Nimbostratus, 364
Nitrate, nitrogen cycle and, 249
Nitric acid, 229
　fertilizer from, 248
Nitric oxide, 249
　in ozone formation in smog, 251
　and stratospheric ozone level, 251
　from stratospheric SST flights, 251
Nitrification, 249
Nitrite ion, nitrates from, 249
　nitrogen cycle and, 249
Nitrogen, ammonia from, 248
　compounds of (table), 249
　oxides of, 249
　uses of, 248
Nitrogen cycle, 248
Nitrogen dioxide, and ozone in smog, 250
Nitrogen dioxide ozone cycle, 250
Nitrogen family, 248
Nitrogen fixation, 249
Nitrogen oxides, air pollution and, 249
Noble gas elements, 169
　stability of, 203
Non-Euclidean geometry, 111
Nonfunctional groups, organic, 280
Nonmetal, 168
　diatomic molecules of, 207
　ions from, 206
　periodic chart locations of, 187
Nonmetallic deposits, 391
Norepinephrine, 320
Normal epoch, 410
Normal fault, 401
North magnetic pole, 409
Novae, 504
Nuclear energy, as resource, 393
Nuclear fission, 137
Nuclear force, 133
Nuclear fuels, 391

Nuclear fusion in sun, 495
Nuclear reaction, 136
Nuclear reactor, 137, 139
Nucleic acids, 301
Nucleons, 132, 145
Nucleus, 5, 126, 132
　of comet, 468
　of sun, 494
Nylon, 317

Obsidian, 384
Occluded front, 388
Ocean, 340
　carbon dioxide in, 253
　physical behavior, 345
　pressure, 346
　temperature structure, 345
　volume, 345
Ocean basin, 406
　origin, 416
Ocean basin floors, 406
Ocean currents, 356
Oceanic crust, 398
Oceanic lithosphere, 412
Oceanic trenches, 400
Ocean tide, 472
Octane number, 311
Octet rule, 203
　ions and, 205
Oil, shale, 393
　vegetable, 293
　　fatty acids from, 294
　of wintergreen, 286
Olefin, 314
Olivine, 380
Omega-particle, 145
Oort, Jan H., 469
Opan, composition of, 170
Opium alkaloids, 321
Opposition, 471
Optical isomerism, 275
Orbit, 24
　of planets, Kepler's laws, 460
Orbital, angular momentum, 127
　atomic, 187
　molecular, 208
　quantum number, 127
d-Orbital, 191
f-Orbital, 191
p-Orbital, 190
s-Orbital, 190
Order, 52
Ordovician Period, 426
Ore, 168
　minerals and, 256
　residual, 440
Organic compounds, 220, 271
　aliphatic, 281
　aromatic, 281

INDEX

Organic compounds (*continued*)
  evolution in, 421
  examples of (table), 272–273
  families of, 279
  functional groups of, 280
  geometry of, 275
  nonfunctional groups of, 280
Orogeny, 424
Orographic precipitation, 366, 367
Outer octet, 203
Outgassing, 479
Outwash plain, 450
Oxbow lake, 445
Oxidation, 241
Oxidation-reduction, 242
  electrochemistry and, 257
Oxidizing agent, 241
Oxygen, 239
  crustal abundance of, 239
  in earth's crust, 378
  electrolytic synthesis of, 261
  family, 239
  global abundance of, 240
  as oxidizing agent, 241
  and ozone cycle, 242
  photosynthesis of, 289
  in steel making, 257
Oxygen cycle, 240
Ozone, 242
  air pollution and, 244
  effect on health of, 244
  effect on plants of, 244
  layer, 343
    and early life, 423
  in sewage treatment, 244
  smog and, 250
  stratospheric, 242, 251
Ozone cycle, Freon aerosols and, 243
  stratospheric, 242

Pair production, 142
Paleomagnetism, and sea floor spreading, 410
Paleontology, 419
Paleozoic Era, 424, 425
PAN (peroxacylnitrates), 246
Pangaea, 427
  breakup, 430
Paraffin wax, 309
Parallax, stellar, 496
Parsec, 496
Particle theory of light, 4, 91, 115, 118, 119, 121, 123, 145
Particulates, 247
Pascal, unit of pressure, 46
Pauli exclusion principle, 129, 193
Peat, 307
Pendulum, 83
Peneplain, 447

Penicillin, 318
Penumbra, of eclipse, 472
  of sunspot, 495
Pep pills, 320, 321
Peptide bond, 295
Percent concentration, 226
Perchloric acid, 229
Peridotite, 385
Perigee, 470
Perihelion, 461
Period, chemical, 187
  of geologic time, 424
  of simple harmonic motion, 83
Periodic chart, 185, 188–189
Periodic law, 186
Periodic property, 183
Peta, 7
Petrochemicals, 312
Petroleum, carbon dioxide from use of, 253
  composition of, 306, 309
  deposits, offshore, 392
  formation of, 306
  reserves of world, 393
  sulfur in, 248
PH, and pH meter, 232
Phases of moon, 471
Phenol, antiseptic properties of, 284
Phenolphthalein, 233
β-Phenylethylamines, 320
Phosphates, detergents and, 236
Phosphoric acid, 229
Phosphorus, 169, 248
Photoelectric effect, 120
Photolysis, 242
Photon, 4, 118, 119, 122, 123, 128, 145
Photosphere, 494
Photosynthesis, carbon cycle and, 253, 289
  chlorophyll in, 289
  oxygen cycle and, 240, 241
  petroleum from, 306
  plankton and, 289
  solar energy used by, 289
Physical change, 157
Physical oceanography, 340
Physical property, 157
Physical weathering, 437
Pico, 7
Pig iron, 257
Pion, 143, 145, 146
Pitch, 86
Plagioclase feldspars, calcic, intermediate, and sodic, 379
Planck, Max, 115, 117, 118
Planck's constant, h, 117
Planet, 459
  formation, 478
  great, 460, 467
  inner, 459
  major, 459

Planet (*continued*)
  minor, 466
  origin, 478
  outer, 459
  terrestrial, 459
Planetesimals, 478
Plaster of paris, 234, 235
Plastic, 313
Plasticizer, 313
Plate tectonics, 412
Pleistocene Epoch, 430
Plexiglas, 316
Plume in mantle, 416
Pluto, 467
Pluton, 384
Podzol, 440
Podzolization, 440
Poisons, enzymes and, 300
  legands as, 236
Polyamides, 317
Polatomic ions, 208
  table of, 210
Polyesters, 317
Polyethylene, 283
  uses of, 314
Polymers, 283, 313
  addition, 315
  alternate, 317
  block, 317
  condensation, 317
  drene, 316
  elastomer, 317
  graft, 317
  random, 317
  vinyl, 316
  vulcanized, 317
Polyolefins, 314
Polypeptides, 297
Polypropylene, 314
Polysaccharides, 290
Polystyrene, 316
Polyunsaturated vegetable oils, 283
Polyvinyl acetate, 316
Polyvinyl chloride (PVC), 316
Positron, 141, 145
Potash feldspar, 377
Potassium-argon series, 425
Potassium hydroxide, 230
Potassium permanganate, 234
Potential, 60
Potential energy, 33, 39, 58
Precambrian time, 421
Precipitation, 362
  cause of, 366
  convectional, 366
  forms, 365
  frontal, 366
  latent heat from, 164
  orographic, 367

Pressure, 46, 47
  cell, 356
  hydrostatic, 161
  of light, 90
Pressure gradient, and winds, 351
Pressure gradient force, 351
Prevailing westerlies, 354
Primary arcs, 400
Primary waves (P-waves), 402
Principle of continuity, 419
Principle of quantum number, 126
Principle of superposition, 419
Prism, 3
Probability, 122, 124, 126
Products, chemical, 173
Prolate ellipsoid, 473
Property, definition of, 157
Propylene, manufacture of, 312
  polypropylene from, 314
  uses of, 312, 313
Proteins, 295
  alpha-helix structure of, 299
  amino acids from (table), 296
  denaturation of, 299
Proton, 5, 56, 120, 126, 128, 132, 141, 145
  as cosmic particle, 509
Proust, Joseph, 175
Ptolemaic system, 458
Ptolemy, Claudius, 458
Pulsars, 503
Pyroxene group, 380
Pythagoras, 458

Quadrature, 471
Quantum, 117
  numbers, 126
  theory, 115
Quartz, 379
  composition of, 170
Quasars, 507
Quasistellar radio sources, *see* Quasars
Quinine, 319

Radar, 89, 90, 112
Radial velocity of star, 506
Radiation, ionizing, 344
  longwave and shortwave, 347
Radiation balance, 347
  of earth, 347, 350
Radiation deficit, annual, 350
Radiation fog, 365
Radiation spectrum, of stars, 500
Radiation surplus, annual, 350
Radioactive decay, 424
Radioactive isotopes, 425
Radioactivity, 139, 424
  and earth's internal heat, 398
Radio astronomy, 507
Radio galaxies, 507

Radiometric age, determination, 424
Radio telescopes, 507
Radio waves, 88, 90
  transmission, 343
Radius vector, 460
Radon, 169
Rain, 365
Rain-shadow desert, 368
Rain stage, 363
Ranger spacecraft, 481
Rare earth elements, 187
Ray, from lunar craters, 484
Rayleigh scattering low, 223
Rayon, 292
Reactants, 173
Reactor, nuclear, 137, 139
Red giants, 500
Red shift, 506
Red supergiants, 500
Reduction, 242
Reforming, crude oil, 310
Regolith, 437
  lunar, 486
  on Mars, 490
Relative humidity, and air temperature, 361
Relativity, 22, 78, 100
Resid, 308
Residual ores, 440
Resin, synthetic, 313
Respiration, carbon cycle and, 253
  oxygen cycle and, 240
Rest mass, 106, 122, 140, 145
Resultant vector, 13
Retrograde motion, 463
Reversed epoch, 410
Revolution, planetary, periods, 462
Rhyolite, 385
Ribonucleic acid (RNA), 301
Ribose, 302
Richter, C. F., 406
Richter scale, 406
Rift, 402
Rift valleys, 416
  on Mars, 489
Rilles, lunar, 484
Ripples on water, 84, 85
River, 442
RNA, 301
Rock, 378
  basaltic, 398
  granite-gabbro series, 384
  granitic, 398
  igneous, 379
  lunar, 486
  metamorphic, 388
  plutonic, 384
  sedimentary, 386
Rock ages, determination, 424
Rock cycle, 390

Rockets, 34
Rock metamorphism, 388
  by impact, 486
Rock salt, 388
Rock texture, 383
Rock weathering, 386
Rossby, C.-G., 355
Rossby waves, 355
Rotation, 31
  of planet, 462
Rubber, natural, 316
  vulcanized, 317
Rubidium-strontium decay series, 425

Salicylates, 318
Salicylic acid, drugs from, 318
Salinity of seawater, 345
Salol, 319
Salt, 234
  atmospheric, 363
  hygroscopic, 363
  in seawater, 345
  solubility rules for, 235
  table of, 234
Saltation, 454
Salt-crystal growth, 437
Sand, silica in, 170
Sandblast action, 453
Sand dunes, 454
  on Mars, 490
Sandspit, 453
Sandstone, 387
Saran, 316
Satellite, 24
  of planets, 469
Saturated compound, 281
Saturated solution, 226
Saturated zone, 443
Saturn, 467
Scalar, 12
Scattering of solar radiation, 349
Schist, 389
Scoria, 384
Sea floor spreading, 409
  and paleomagnetism, 410
Seamounts, 408
Seawater, composition, density, and salinity, 345
  as mineral resource, 392
Second, 6
Secondary waves (S-waves), 403
Sediment, 386
  carbonate, 388
  chemically precipitated, 386, 388
  elastic, 386
  organic, 387, 388
Sedimentary rocks, 387
  chemically precipitated, 388
  elastic, 387

Sedimentary rocks (*continued*)
  stratified, 387
Seismic waves, 402
  and earth's core, 405
Seismogram, 403
Seismograph, 403
Selenium, 170
Semimetals, 168, 170
Sequential landforms, 436
Shale, 387
Shapley, Harlow, 505
Shear, in stream flow, 442
Shearing in mantle, 399
Shearing stress, 441
Shield, of continent, 416
Shield volcanoes, 385
  on Mars, 489
Shock front, 344
Shortwave radiation, 347
Sickle cell anemia, 297
Sidereal month, 470
Siderite, 256
Sigma-particle, 145
Silicate, 379
Silicate magma, 382
Silicate minerals, 379
  atomic structure, 380
Silicon, 170
Silicon dioxide, 170
Silicon-oxygen tetrahedron, 380
Silicon steel, 167
Siltstone, 387
Silurian Period, 426
Silver nitrate, 234
Simple harmonic motion, 83
Sinusoidal, 84
SI units, 8
Sky, colors of, 222
Slag, 257
Slate, 389
Sleet, 365
Slit spectroscope, 500
Slope erosion, 441
Smith, William, 420
Smog, air-temperature inversion in, 244
  carbon monoxide in, 245, 252
  coal-type, 246
  hydrocarbons in, 245
  nitrogen oxides in, 245, 249
  ozone in, 244, 250
  PAN in, 246
  particulates in, 245, 247
  photochemical, 245, 246
  sulfur oxides in, 245, 246
Snow, 365
Soda, 234
Sodium, reaction of chlorine with, 203
Sodium bicarbonate, 234
Sodium carbonate, 234

Sodium hydroxide, 230
  properties of, 234
Soft layer in mantle, 399
Soil, 437, 438
Soil erosion, 441
Soil-forming processes, 438
Sol, 223
Solar constant, 347
Solar eclipse, 472
Solar energy, 392, 347, 395
  hydrogen fuel from use of, 169
  quantity received by earth of, 289
  source, 495
  spectrum, 342
Solar flares, 496
Solar nebula, 478
Solar prominences, 494
Solar radiation, 347
  absorption of, 342, 349
  depletion, 349
  and latitude, 347
  and seasons, 347
  over spherical earth, 347
Solar system, 5, 459, 477
  origin, 477
Solar-topographic hypothesis, 433
Solar wind, 344, 494
Solids, kinetic view of, 161
Solstices, 348
Solubilities, table of, 225
Solute, 221
Solutions, 221
  kinds of (table), 222
Solvent, 221
Sound, 86
Source region of air mass, 368
South magnetic pole, 409
Specific heat, 162
  table of, 163
Specific humidity, of air masses, 362
Spectral class of star, 500
Spectroscope, 3, 500
Spectrum, 4, 125
  of stars, 500
Spectrum-luminosity diagram, 501
Speech, 86
Speed, 9, 12
  of light, c, 72, 80, 88, 100, 102, 106, 122
Spheroidal weathering, 438
Spin of fundamental particles, 128, 145
Spiral arms of galaxy, 497
Spiral galaxies, barred, 505
Spring tide, 474
Standard sea-level pressure, 342
Stannous fluoride, 234
Star, 497
  density, 498
  exploding, 504
  life history, 501

Star (*continued*)
   luminosity, 498, 499
   main sequence, 500
   mass, 498, 499
   radial velocity, 506
   spectral classes, 500
   temperature, 498
Starch, digestion of, 292
   hydrolysis of, 290
Star clouds, 497
Star magnitude, absolute, and apparent visual, 499
Steady-state theory of universe, 507
Steam point, 46
Steel, 167
   manufacture of, 257
Stellar parallax, 496
Steno, Nicolaus, 420
Stones (meteorites), 479
Stony irons, 479
Storage battery, 62
Storms, tropical, 373
Storm track, 372
Strain in rock, 402
Strata, 387
   relative ages, 419
Stratification planes, 387
Stratigraphy, 419
Stratopause, 342
Stratosphere, 342
   chlorine in, 244
   Freon airosols in, 243
   ozone cycle in, 242
Stratus, 364
Stream, 442
   geologic work, 444
   graded, 445
Stream deposition, 444
Stream erosion, 444
Stream flow, 442
Stream gradation, 445
Streamline flow, 442
Stream load, 444
Stream transportation, 444
Stream velocity, 442
Structural formulas, 207
   rules for condensing, 272–273
Structure, chemical, 207
Subduction, 412
Sublimation, 170
Subsolar point, 347
Substitution reaction, aromatic, 284
Subtropical high-pressure belts, 354
Sucrose, 292
   hydrolysis of, 290
Sugar, *see* Sucrose
Sugar beets, 290
Sugar cane, 292

Sulfa drugs, 318
Sulfur, 169, 246
   important compounds of (table), 247
   occurrence, 246
   oxides of, 246
Sulfur dioxide, 246
   air pollution by, 169, 246
   effect on materials of, 247
   effect on vegetation of, 247
   sulfur trioxide from, 247
Sulfuric acid, 229
   air pollution by, 231
   importance of, 169, 229
Sulfur oxides, 246
Sulfur trioxide, 246
   air pollution by, 169, 246
   effect of particulates on, 247
   origin of, 246
Sun, composition, structure, and temperature of, 494
   dimensions, 494
   interior, 494
   mass, 494
   rotation, 494
Sun-earth-space radiation system, 346
Sunspot, 495
Sunspot cycle, 496
Sun's radiation spectrum, 347
Superimposibility, 275
Supernovae, 504
Supersonic transport (SST), stratospheric ozone level and, 251
Surface environment of rocks, 390
Surface waves, 403
Surface winds, 354
Surveyor spacecraft, 481
Suspended load, 445
Suspension, 442
Suture, 415
Swash, 450
Synodic month, 470
Synthetics, 306
Systeme International d'Unités, 8

Taconite, 256
   steel from, 257, 261
Tail of comet, 468
Talus slopes, 441
Tedlar, 316
Teflon, 316
Temperature, 46
Terminal moraine, 450
Tesla, unit of magnetic field, 72
Tetrahedra, of silicates, 380
Tetrahedral carbon, 275, 277
Tetrahydrocannabinol, 322
THC, 322
Thermal contact, 162

Thermocline, 345
Thermodynamics, first law of, 50
  second law of, 52
  third law of, 53
Thermometer, 48
Thermonuclear reaction, 138
Thermosphere, 342
Thomson, G. P., 120
Thunder, 367
Thunderstorms, 366
Thymine, 302
Tidal power, 393
Tide curve, 474
Tide range, 474
Till, 450
Time, 6, 104
Tin, bronze from, 166
Tincture of iodine, 171
Tinplate, 257
  by electroplating, 260
Tornadoes, 372
Torque, 31
Torricelli experiment, 342
Traction in streams, 445
Trades, 354
Trade winds, 354
Transcurrent fault, 401
Transition elements, 187
  electronic configurations of, 195, 198
Transuranium elements, 187
Transverse waves, 84, 88, 403
Triangle of vectors, 12
Triassic Period, 428
Triglycerides, digestion of, 294
  hydrogenation of, 293
Triphosphate ion, 236
Tritium, 133
Tropical air mass, 356
Tropical cyclones, 373
Tropical easterlies, 354
Tropopause, 342
Troposphere, 342
Tungsten steel, 167
Turbulent flow, 442
Tycho Brahe, 459
Tyndall effect, 222
Typhoon, 373

Uhuru satellite, 508
Ultramafic rocks, 385
Ultraviolet light, 90
  stratospheric ozone cycle and, 242
Ultraviolet radiation, and origin of life, 423
Ultraviolet rays, 342
Umbra, of eclipse, 472
  of sunspot, 495
Uncertainty principle, 124
Uniformitarianism, 421

Units, 6, 8
Universe, 497, 506
  age, 507
  origin, 506, 507
  pulsating, 507
  in steady-state, 507
Unsaturated compound, 281
Unsaturated solution, 226
Unsaturated zone, 443
Upper-air waves, 355
Upper-air westerlies, 355
Uracil, 302
Uranium-lead decay series, 425
Uranium ore, 393
Uranium reserves, 393
Uranus, 467

Valence, 185
Vanadium steel, 167
Van Allen radiation belt, 344, 496
Vanillin, 286
Vaporization, heat of, 163
Vapor pressure, 164
Vectors, 12
Velocity, 9, 12
  of light, c, 72, 80, 88, 100, 102, 106, 122
  of wave, 85
Venice, air pollution in, 247
Venus, 464
  phases, 459
  surface features, 491
Venutian Eye, 464
Viking spacecraft, 466, 488
Vinegar, 285
Vinyl polymers, 314
  table of, 316
Visible light, 3, 89
Vitamins, enzymes and, 300
Vocal cords, 86
Volatiles, outgassing, 479
Volcanic arcs, origin, 413
Volcanic cone, 385
Volcanic glass, 384
Volcanism, 385
Volcanoes, composite, 385
  on Mars, 489
Volt, 62
Vulcanization, 317

Walls, lunar, 485
Warm front, 370
Warped space, 111, 112
Water, 212
  addition to alkenes of, 283
  cracking of, 160
  density versus temperature of, 161
  digestive reactions of, 285, 286
  evaporation from body of, 164

Water (*continued*)
   hardness ions in, 236
   heat of fusion of, 163
   heat of vaporization of, 163
   Hydrogen bonds in, 225
   ionization of, 226
   ionization constant of, 228
   molecular polarity of, 210
   reaction with amides of, 286
   reaction with di- and polysaccharides of, 290
   reaction with esters of, 285
   solvent properties of, 223
   specific heat of, 163
   structure of, 212
   supercooled, 363
   vapor pressure of (table), 164
Water table, 443
Water vapor, 340
   atmospheric, 361
Water waves, 84
Watson, J. D., 302
Wave, 83, 120
Wave action, 450
Wave cyclones, evolution stages, 370
   on weather map, 371
Wave function, 121, 122, 123, 126
Wave nature of matter, 120
Wave theory of light, 4, 91, 115
Wave theory of particles, 115, 120
Wave velocity, 85
Wavefront, 92
Wavelength, 4, 85, 122
Wavelets, 92
Weather fronts, *see* Fronts, atmospheric
Weathering of rock, 386, 437
   chemical, 438
   physical, 437
   and soils, 438
   spheriodal, 438
Wegener, A., 430
Weight, 21, 38, 39

Weightlessness, 25
West-wind drift, 357
Wet adiabatic lapse rate, 363
White dwarfs, 500, 501
White light, 3
Wind, 351
   and Coriolis effect, 351
   at earth's surface, 354
   geotrophic, 352
   global, 354
   surface, 354
   upper-air, 354
   westerly, 354
Wind action, 453
   on Mars, 490
Wind power, 392
Winkler, C., 186
Wood alcohol, 284
World ocean, 345

Xenon, 169
Xerography, use of selenium in, 170
X_i-particle, 145
X-ray, 44, 90, 119, 120, 342
   photograph of molecule, 44
   solar, 496
X-ray astronomy, by satellite, 508
X-ray diffraction, crystal structure and, 214
X-ray diffractometer, 214
X-ray spectra, periodic law and, 186

Year, 462
Young, Thomas, 92

Ziegler catalyst, 312
Zinc, 268
   brass from, 167
   lead refining and, 266
   uses of, 268
Zone of ablation of glacier, 449
Zone of accumulation of glacier, 449